ELEMENTS OF MATHEMATICS

NICOLAS BOURBAKI

ELEMENTS OF MATHEMATICS

Lie Groups and Lie Algebras

Chapters 7–9

Springer

Originally published as
ÉLÉMENTS DE MATHÉMATIQUE,
GROUPES ET ALGÈBRES DE LIE 7–8 et 9

Translator
Andrew Pressley
Department of Mathematics
King's College London
Strand, London WC2R 2LS
United Kingdom
andrew.pressley@kcl.ac.uk

First softcover printing of the 1st English edition of 2005

ISBN 978-3-540-68851-8

Library of Congress Control Number: 2008931305

Mathematics Subject Classification (2000): 17B10, 17B15, 17-01, 22C05, 22E60, 22E46

Cover-design: WMXDesign GmbH, Heidelberg

Printed on acid-free paper

9 8 7 6 5 4 3 2 1

springer.com

CONTENTS

CHAPTER VIII SPLIT SEMI-SIMPLE LIE ALGEBRAS

CHAPTER IX COMPACT REAL LIE GROUPS

CHAPTER VII
Cartan Subalgebras and Regular Elements

In this chapter, k denotes a (commutative) field. By a "vector space", we mean a "vector space over k"; similarly for "Lie algebra", etc. All Lie algebras are assumed to be finite dimensional.

§ 1. PRIMARY DECOMPOSITION OF LINEAR REPRESENTATIONS

1. DECOMPOSITION OF A FAMILY OF ENDOMORPHISMS

Let V be a vector space, S a set, and r a map from S to End(V). Denote by P the set of maps from S to k. If $\lambda \in \mathrm{P}$, denote by $\mathrm{V}_\lambda(\mathrm{S})$ (resp. $\mathrm{V}^\lambda(\mathrm{S})$) the set of $v \in \mathrm{V}$ such that, for all $s \in \mathrm{S}$, $r(s)v = \lambda(s)v$ (resp. $(r(s) - \lambda(s))^n v = 0$ for n sufficiently large). The sets $\mathrm{V}_\lambda(\mathrm{S})$ and $\mathrm{V}^\lambda(\mathrm{S})$ are vector subspaces of V, and $\mathrm{V}_\lambda(\mathrm{S}) \subset \mathrm{V}^\lambda(\mathrm{S})$. We say that $\mathrm{V}_\lambda(\mathrm{S})$ is the *eigenspace* of V relative to λ (and to r), that $\mathrm{V}^\lambda(\mathrm{S})$ is the *primary subspace* of V relative to λ (and to r), and that $\mathrm{V}^0(\mathrm{S})$ is the *nilspace* of V (relative to r). We say that λ is a *weight* of S in V if $\mathrm{V}^\lambda(\mathrm{S}) \neq 0$.

In particular, if S reduces to a single element s, P can be identified with k; we use the notations $\mathrm{V}_{\lambda(s)}(s)$ and $\mathrm{V}^{\lambda(s)}(s)$, or $\mathrm{V}_{\lambda(s)}(r(s))$ and $\mathrm{V}^{\lambda(s)}(r(s))$, instead of $\mathrm{V}_\lambda(\{s\})$, $\mathrm{V}^\lambda(\{s\})$; we speak of eigenspaces, primary subspaces and the nilspace of $r(s)$; an element v of $\mathrm{V}_{\lambda(s)}(s)$ is called an *eigenvector* of $r(s)$, and, if $v \neq 0$, $\lambda(s)$ is called the corresponding *eigenvalue* (cf. *Algebra*, Chap. VII, §5).

For all $\lambda \in \mathrm{P}$, the following relations are immediate:

$$\mathrm{V}^\lambda(\mathrm{S}) = \bigcap_{s \in \mathrm{S}} V^{\lambda(s)}(s), \tag{1}$$

$$\mathrm{V}_\lambda(\mathrm{S}) = \bigcap_{s \in \mathrm{S}} V_{\lambda(s)}(s). \tag{2}$$

Let k' be an extension of k. The canonical map from End(V) to $\mathrm{End}(V \otimes_k k')$ gives, by composition with r, a map $r' : \mathrm{S} \to \mathrm{End}(V \otimes_k k')$. Similarly, every map λ from S to k defines canonically a map, also denoted by λ, from S to k'. With these notations, we have the following proposition:

PROPOSITION 1. *For all* $\lambda \in \mathrm{P}$,

$$(\mathrm{V} \otimes_k k')^{\lambda}(\mathrm{S}) = \mathrm{V}^{\lambda}(\mathrm{S}) \otimes_k k' \quad \text{and} \quad (\mathrm{V} \otimes_k k')_{\lambda}(\mathrm{S}) = \mathrm{V}_{\lambda}(\mathrm{S}) \otimes_k k'.$$

Let (a_i) be a basis of the k-vector space k'. If $v \in \mathrm{V} \otimes_k k'$, v can be expressed uniquely in the form $\sum v_i \otimes a_i$ where (v_i) is a finitely-supported family of elements of V. For all $s \in \mathrm{S}$,

$$(r'(s) - \lambda(s))^n(v) = \sum (r(s) - \lambda(s))^n v_i \otimes a_i.$$

It follows that

$$v \in (\mathrm{V} \otimes_k k')^{\lambda}(\mathrm{S}) \Longleftrightarrow v_i \in \mathrm{V}^{\lambda}(\mathrm{S}) \text{ for all } i,$$
$$v \in (\mathrm{V} \otimes_k k')_{\lambda}(\mathrm{S}) \Longleftrightarrow v_i \in \mathrm{V}_{\lambda}(\mathrm{S}) \text{ for all } i,$$

which implies the proposition.

PROPOSITION 2. *Let* $\mathrm{V}, \mathrm{V}', \mathrm{W}$ *be vector spaces. Let* $r : \mathrm{S} \to \mathrm{End}(\mathrm{V})$, $r' : \mathrm{S} \to \mathrm{End}(\mathrm{V}')$ *and* $q : \mathrm{S} \to \mathrm{End}(\mathrm{W})$ *be maps.*

(i) *Let* $f : \mathrm{V} \to \mathrm{W}$ *be a linear map such that* $q(s)f(v) = f(r(s)v)$ *for* $s \in \mathrm{S}$ *and* $v \in \mathrm{V}$. *Then, for all* $\lambda \in \mathrm{P}$, f *maps* $\mathrm{V}^{\lambda}(\mathrm{S})$ (resp. $\mathrm{V}_{\lambda}(\mathrm{S})$) *into* $W^{\lambda}(\mathrm{S})$ (resp. $W_{\lambda}(\mathrm{S})$).

(ii) *Let* $\mathrm{B} : \mathrm{V} \times \mathrm{V}' \to \mathrm{W}$ *be a bilinear map such that*

$$q(s)\mathrm{B}(v, v') = \mathrm{B}(r(s)v, v') + \mathrm{B}(v, r'(s)v')$$

for $s \in \mathrm{S}$, $v \in \mathrm{V}$, $v' \in \mathrm{V}'$. *Then, for all* $\lambda, \mu \in \mathrm{P}$, B *maps* $V^{\lambda}(\mathrm{S}) \times \mathrm{V}'^{\mu}(\mathrm{S})$ (resp. $\mathrm{V}_{\lambda}(\mathrm{S}) \times \mathrm{V}'_{\mu}(\mathrm{S})$) *into* $W^{\lambda+\mu}(\mathrm{S})$ (resp. $W_{\lambda+\mu}(\mathrm{S})$).

(iii) *Let* $\mathrm{B} : \mathrm{V} \times \mathrm{V}' \to \mathrm{W}$ *be a bilinear map such that*

$$q(s)\mathrm{B}(v, v') = \mathrm{B}(r(s)v, r'(s)v')$$

for $s \in \mathrm{S}$, $v \in \mathrm{V}$, $v' \in \mathrm{V}'$. *Then, for all* $\lambda, \mu \in \mathrm{P}$, B *maps* $V^{\lambda}(\mathrm{S}) \times \mathrm{V}'^{\mu}(\mathrm{S})$ (resp. $\mathrm{V}_{\lambda}(\mathrm{S}) \times \mathrm{V}'_{\mu}(\mathrm{S})$) *into* $W^{\lambda\mu}(\mathrm{S})$ (resp. $W_{\lambda\mu}(\mathrm{S})$).

In case (i), $(q(s) - \lambda(s))^n f(v) = f((r(s) - \lambda(s))^n v)$ for $s \in \mathrm{S}$ and $v \in \mathrm{V}$, hence the conclusion. In case (ii),

$$(q(s) - \lambda(s) - \mu(s))\mathrm{B}(v, v') = \mathrm{B}((r(s) - \lambda(s))v, v') + \mathrm{B}(v, (r'(s) - \mu(s))v')$$

for $s \in \mathrm{S}$, $v \in \mathrm{V}$, $v' \in \mathrm{V}'$, hence by induction on n

$$(q(s) - \lambda(s) - \mu(s))^n \mathrm{B}(v, v') = \sum_{i+j=n} \binom{n}{i} \mathrm{B}((r(s) - \lambda(s))^i v, (r'(s) - \mu(s))^j v').$$

The assertions in (ii) follow immediately. In case (iii),

$$(q(s)-\lambda(s)\mu(s))\mathrm{B}(v,v')=\mathrm{B}((r(s)-\lambda(s))v,r'(s)v')+\mathrm{B}(\lambda(s)v,(r'(s)-\mu(s))v')$$

for $s \in \mathrm{S}$, $v \in \mathrm{V}$, $v' \in \mathrm{V}'$, hence by induction on n

$$(q(s) - \lambda(s)\mu(s))^n\mathrm{B}(v,v') = \sum_{i+j=n} \binom{n}{i} \mathrm{B}(\lambda(s)^j(r(s) - \lambda(s))^i v, r'(s)^i(r'(s) - \mu(s))^j v').$$

The assertions in (iii) follow immediately.

PROPOSITION 3. *The sum* $\sum_{\lambda\in \mathrm{P}} \mathrm{V}^{\lambda}(\mathrm{S})$ *is direct. The sum* $\sum_{\lambda\in \mathrm{P}} \mathrm{V}_{\lambda}(\mathrm{S})$ *is direct.*

The second assertion is a consequence of the first; hence it suffices to prove that. We distinguish several cases.

a) S *is empty.* The assertion is trivial.

b) S *is reduced to a single element s.* Let $\lambda_0, \lambda_1, \dots, \lambda_n$ be distinct elements of k. For $i = 0, 1, \dots, n$, let $v_i \in \mathrm{V}^{\lambda_i}(s)$ and assume that $v_0 = v_1 + \cdots + v_n$. It suffices to prove that $v_0 = 0$. For $i = 0, \dots, n$, there exists an integer $q_i > 0$ such that $(r(s) - \lambda_i)^{q_i} v_i = 0$. Consider the polynomials $\mathrm{P}(\mathrm{X}) = \prod_{i\geq 1}(\mathrm{X} - \lambda_i)^{q_i}$ and $\mathrm{Q}(\mathrm{X}) = (\mathrm{X} - \lambda_0)^{q_0}$. We have $\mathrm{Q}(r(s))v_0 = 0$, and $\mathrm{P}(r(s))v_0 = \sum_{i=1}^{n} \mathrm{P}(r(s))v_i = 0$. Since P and Q are relatively prime, the Bezout identity proves that $v_0 = 0$.

c) S *is finite and non-empty.* We argue by induction on the cardinal of S. Let $s \in \mathrm{S}$ and $\mathrm{S}' = \mathrm{S} - \{s\}$. Let $(v_\lambda)_{\lambda\in\mathrm{P}}$ be a finitely-supported family of elements of V such that $\sum_{\lambda\in\mathrm{P}} v_\lambda = 0$ and $v_\lambda \in \mathrm{V}^{\lambda}(\mathrm{S})$. Let $\lambda_0 \in \mathrm{P}$. Let P$'$ be the set of $\lambda \in \mathrm{P}$ such that $\lambda|\mathrm{S}' = \lambda_0|\mathrm{S}'$. By the induction hypothesis applied to S$'$, we have $\sum_{\lambda\in\mathrm{P}'} v_\lambda = 0$. If λ, μ are distinct elements of P$'$, $\lambda(s) \neq \mu(s)$. Since the sum $\sum_{\alpha\in k} \mathrm{V}^{\alpha}(s)$ is direct by *b*), and since $v_\lambda \in \mathrm{V}^{\lambda(s)}(s)$, $v_\lambda = 0$ for all $\lambda \in \mathrm{P}'$, and in particular $v_{\lambda_0} = 0$, which we had to prove.

d) *General case.* Let $(v_\lambda)_{\lambda\in\mathrm{P}}$ be a finitely-supported family of elements of V such that $\sum_{\lambda\in\mathrm{P}} v_\lambda = 0$ and $v_\lambda \in \mathrm{V}^{\lambda}(\mathrm{S})$. Let P$'$ be the finite set of $\lambda \in \mathrm{P}$ such that $v_\lambda \neq 0$, and let S$'$ be a finite subset of S such that the conditions $\lambda \in \mathrm{P}'$, $\mu \in \mathrm{P}'$, $\lambda|\mathrm{S}' = \mu|\mathrm{S}'$ imply that $\lambda = \mu$. We have $v_\lambda \in \mathrm{V}^{\lambda|\mathrm{S}'}(\mathrm{S}')$; applying *c*), we see that $v_\lambda = 0$ for $\lambda \in \mathrm{P}'$, which completes the proof.

Recall that, if $x \in \mathrm{End}(\mathrm{V})$, we denote by $\mathrm{ad}\, x$ the map $y \mapsto xy - yx = [x, y]$ from End(V) to itself.

Lemma 1. Let $x, y \in \mathrm{End}(\mathrm{V})$.

(i) *Assume that* V *is finite dimensional. Then* x *is triangularizable if and only if* $\mathrm{V} = \sum_{a\in k} \mathrm{V}^{a}(x)$.

(ii) *If there exists an integer* n *such that* $(\operatorname{ad} x)^n y = 0$, *each* $V^a(x)$ *is stable under* y.

(iii) *Assume that* V *is finite dimensional. If* $V = \sum_{a \in k} V^a(x)$ *and if each* $V^a(x)$ *is stable under* y, *there exists an integer* n *such that* $(\operatorname{ad} x)^n y = 0$.

Part (i) follows from *Algebra*, Chap. VII, §5, no. 2, Prop. 3.

Let $E = \operatorname{End}(V)$. Let B be the bilinear map $(u, v) \mapsto u(v)$ from $E \times V$ to V. By the definition of $\operatorname{ad} x$,

$$x(B(u, v)) = B(u, x(v)) + B((\operatorname{ad} x)(u), v)$$

for $x \in E$, $u \in E$, $v \in V$. Let x operate on E via $\operatorname{ad} x$. By Prop. 2 (ii), $B(E^0(x), V^a(x)) \subset V^a(x)$ for all $a \in k$. If $(\operatorname{ad} x)^n y = 0$, then $y \in E^0(x)$, so $y(V^a(x)) \subset V^a(x)$, which proves (ii).

To prove (iii), we can replace V by $V^a(x)$ and x (resp. y) by its restriction to $V^a(x)$. Replacing x by $x - a$, we can assume that x is nilpotent. Then, $(\operatorname{ad} x)^{2 \dim V - 1} = 0$ (Chap. I, §4, no. 2), which proves (iii).

Remark. The argument proves that, if V is finite dimensional and if there exists an integer n such that $(\operatorname{ad} x)^n y = 0$, then $(\operatorname{ad} x)^{2 \dim V - 1} y = 0$.

In the sequel, we shall say that the map $r : S \to \operatorname{End}(V)$ satisfies condition (AC) ("almost commutative") if:

(AC) *For every pair* (s, s') *of elements of* S, *there exists an integer* n *such that*

$$(\operatorname{ad} r(s))^n r(s') = 0.$$

THEOREM 1. *Assume that* V *is finite dimensional. The following conditions are equivalent:*

(i) *Condition* (AC) *is satisfied and, for all* $s \in S$, $r(s)$ *is triangularizable.*

(ii) *For all* $\lambda \in P$, $V^\lambda(S)$ *is stable under* $r(S)$, *and* $V = \sum_{\lambda \in P} V^\lambda(S)$.

If $V = \sum_{\lambda \in P} V^\lambda(S)$, then $V = \sum_{a \in k} V^a(s)$ for all $s \in S$, and it follows from Lemma 1 that (ii) implies (i). Assume that condition (i) is satisfied. Lemma 1 and formula (1) imply that each $V^\lambda(S)$ is stable under $r(S)$. It remains to prove that $V = \sum_{\lambda \in P} V^\lambda(S)$. We argue by induction on $\dim V$. We distinguish two cases.

a) For all $s \in S$, $r(s)$ has a single eigenvalue $\lambda(s)$. Then $V = V^\lambda(S)$.

b) There exists $s \in S$ such that $r(s)$ has at least two distinct eigenvalues. Then V is the direct sum of the $V^a(s)$ for $a \in k$, and $\dim V^a(s) < \dim V$ for all a. Each $V^a(s)$ is stable under $r(S)$, and it suffices to apply the induction hypothesis.

COROLLARY 1. *Assume that* V *is finite dimensional and that condition* (AC) *is satisfied. Let* k' *be an extension of* k. *Assume that, for all* $s \in S$, *the endomorphism* $r(s) \otimes 1$ *of* $\mathrm{V} \otimes_k k'$ *is triangularizable. Let* P' *be the set of maps from* S *to* k'. *Then* $\mathrm{V} \otimes_k k' = \sum_{\lambda' \in \mathrm{P}'} (\mathrm{V} \otimes_k k')^{\lambda'}(\mathrm{S})$.

Let $r' : \mathrm{S} \to \mathrm{End}(\mathrm{V} \otimes_k k')$ be the map defined by r. If $s_1, s_2 \in \mathrm{S}$, there exists an integer n such that $(\operatorname{ad} r(s_1))^n r(s_2) = 0$, hence $(\operatorname{ad} r'(s_1))^n r'(s_2) = 0$. It now suffices to apply Th. 1.

COROLLARY 2. *Assume that* V *is finite dimensional and that condition* (AC) *is satisfied. Denote by* $\mathrm{V}^+(\mathrm{S})$ *the vector subspace* $\sum_{s \in \mathrm{S}} \left(\bigcap_{i \geq 1} r(s)^i \mathrm{V} \right)$.
Then:

(i) $\mathrm{V}^0(\mathrm{S})$ *and* $\mathrm{V}^+(\mathrm{S})$ *are stable under* $r(\mathrm{S})$;

(ii) $\mathrm{V} = \mathrm{V}^0(\mathrm{S}) \oplus \mathrm{V}^+(\mathrm{S})$;

(iii) *every vector subspace* W *of* V, *stable under* $r(\mathrm{S})$ *and such that* $\mathrm{W}^0(\mathrm{S}) = 0$, *is contained in* $\mathrm{V}^+(\mathrm{S})$;

(iv) $\sum_{s \in \mathrm{S}} r(s)\mathrm{V}^+(\mathrm{S}) = \mathrm{V}^+(\mathrm{S})$.

Moreover, $\mathrm{V}^+(\mathrm{S})$ *is the only vector subspace of* V *with properties* (i) *and* (ii). *For any extension* k' *of* k, $(\mathrm{V} \otimes_k k')^+(\mathrm{S}) = \mathrm{V}^+(\mathrm{S}) \otimes_k k'$.

The last assertion is immediate. Thus, taking Prop. 1 into account, in proving the others we can assume that k is algebraically closed. By Th. 1, $\mathrm{V} = \sum_{\lambda \in \mathrm{P}} \mathrm{V}^\lambda(\mathrm{S})$, and the $\mathrm{V}^\lambda(\mathrm{S})$ are stable under $r(\mathrm{S})$. If $s \in \mathrm{S}$, the characteristic polynomial of $r(s)|\mathrm{V}^\lambda(\mathrm{S})$ is $(\mathrm{X} - \lambda(s))^{\dim \mathrm{V}^\lambda(\mathrm{S})}$; it follows that $\bigcap_{i \geq 1} r(s)^i \mathrm{V}^\lambda(s)$ is zero if $\lambda(s) = 0$ and is equal to $\mathrm{V}^\lambda(\mathrm{S})$ if $\lambda(s) \neq 0$; hence,

$$\mathrm{V}^+(\mathrm{S}) = \sum_{\lambda \in \mathrm{P}, \lambda \neq 0} \mathrm{V}^\lambda(\mathrm{S}), \tag{3}$$

which proves (i), (ii) and (iv). If W is a vector subspace of V stable under $r(\mathrm{S})$, then $\mathrm{W} = \sum_{\lambda \in \mathrm{P}} \mathrm{W}^\lambda(\mathrm{S})$ and $\mathrm{W}^\lambda(\mathrm{S}) = \mathrm{W} \cap \mathrm{V}^\lambda(\mathrm{S})$. If $\mathrm{W}^0(\mathrm{S}) = 0$, we see that $\mathrm{W} \subset \mathrm{V}^+(\mathrm{S})$, which proves (iii).

Let V' be a vector subspace of V stable under $r(\mathrm{S})$ and such that $\mathrm{V}' \cap \mathrm{V}^0(\mathrm{S}) = 0$. Then $\mathrm{V}'^0(\mathrm{S}) = 0$, so $\mathrm{V}' \subset \mathrm{V}^+(\mathrm{S})$ by (iii). If, in addition, $\mathrm{V} = \mathrm{V}^0(\mathrm{S}) + \mathrm{V}'$, we see that $\mathrm{V}' = \mathrm{V}^+(\mathrm{S})$. Q.E.D.

We sometimes call $(\mathrm{V}^0(\mathrm{S}), \mathrm{V}^+(\mathrm{S}))$ the *Fitting decomposition* of V, or of the map $r : \mathrm{S} \to \mathrm{End}(\mathrm{V})$. If S reduces to a single element s, we write $\mathrm{V}^+(s)$ or $\mathrm{V}^+(r(s))$ instead of $\mathrm{V}^+(\{s\})$. We have that $\mathrm{V} = \mathrm{V}^0(s) \oplus \mathrm{V}^+(s)$, $\mathrm{V}^0(s)$ and $\mathrm{V}^+(s)$ are stable under $r(s)$, $r(s)|\mathrm{V}^0(s)$ is nilpotent and $r(s)|\mathrm{V}^+(s)$ is bijective.

COROLLARY 3. *Let* V *and* V' *be finite dimensional vector spaces, and let* $r : \mathrm{S} \to \mathrm{End}(\mathrm{V})$ *and* $r' : \mathrm{S} \to \mathrm{End}(\mathrm{V}')$ *be maps satisfying condition* (AC). *Let* $f : \mathrm{V} \to \mathrm{V}'$ *be a surjective linear map such that* $f(r(s)v) = r'(s)f(v)$ *for* $s \in \mathrm{S}$ *and* $v \in \mathrm{V}$. *Then* $f(\mathrm{V}^\lambda(\mathrm{S})) = \mathrm{V}'^\lambda(\mathrm{S})$ *for all* $\lambda \in \mathrm{P}$.

In view of Prop. 1, we are reduced to the case in which k is algebraically closed. We have $V = \bigoplus_{\lambda \in P} V^{\lambda}(S)$, $V' = \bigoplus_{\lambda \in P} V'^{\lambda}(S)$ by Th. 1, and $V' = f(V) = \sum_{\lambda \in P} f(V^{\lambda}(S))$. Finally, $f(V^{\lambda}(S)) \subset V'^{\lambda}(S)$ by Prop. 2 (i), hence the corollary.

PROPOSITION 4. *Assume that k is perfect. Let* V *be a finite dimensional vector space, u an element of* End(V), *u_s, u_n the semi-simple and nilpotent components of u* (*Algebra*, Chap. VII, §5, no. 8).

(i) *For all $\lambda \in k$, $V^{\lambda}(u) = V^{\lambda}(u_s) = V_{\lambda}(u_s)$.*

(ii) *If* V *has an algebra structure and if u is a derivation of* V, *u_s and u_n are derivations of* V.

(iii) *If* V *has an algebra structure and if u is an automorphism of* V, *then u_s and $1 + u_s^{-1}u_n$ are automorphisms of* V.

In view of Prop. 1, we can assume that k is algebraically closed, so

$$V = \sum_{\lambda \in k} V^{\lambda}(u).$$

The semi-simple component of $u|V^{\lambda}(u)$ is the homothety with ratio λ in $V^{\lambda}(u)$. This proves (i).

Assume from now on that V has an algebra structure. Let $x \in V^{\lambda}(u)$, $y \in V^{\mu}(u)$.

If u is a derivation of V, then $xy \in V^{\lambda+\mu}(u)$ (Prop. 2 (ii)), so

$$u_s(xy) = (\lambda + \mu)(xy) = (\lambda x)y + x(\mu y) = (u_s x)y + x(u_s y).$$

This proves that u_s is a derivation of V. Then $u_n = u - u_s$ is a derivation of V.

If u is an automorphism of V, $\mathrm{Ker}(u_s) = V^0(u) = 0$, so u_s is bijective. On the other hand, $xy \in V^{\lambda\mu}(u)$ (Prop. 2 (iii)), so

$$u_s(xy) = (\lambda\mu)(xy) = (\lambda x)(\mu y) = u_s(x).u_s(y).$$

This proves that u_s is an automorphism of V; but then so is

$$1 + u_s^{-1}u_n = u_s^{-1}u.$$

2. THE CASE OF A LINEAR FAMILY OF ENDOMORPHISMS

Assume now that S has a *vector space* structure, that the map $r : S \to \mathrm{End}(V)$ is *linear*, and that V and S are *finite dimensional.*

PROPOSITION 5. *Assume that condition* (AC) *is satisfied, and let $\lambda : S \to k$ be such that $V^{\lambda}(S) \neq 0$. If k is of characteristic 0, the map λ is linear. If k is of characteristic $p \neq 0$, there exists a power q of p dividing* $\dim V^{\lambda}(S)$, *and a homogeneous polynomial function $P : S \to k$ of degree q, such that $\lambda(s)^q = P(s)$ for all $s \in S$.*

Since $V^\lambda(S)$ is stable under $r(S)$ (Lemma 1 and formula (1) of no. 1), we can assume that $V = V^\lambda(S)$. Let $n = \dim V$. Thus, for $s \in S$,

$$\det(X - r(s)) = (X - \lambda(s))^n.$$

On the other hand, the expansion of the determinant shows that

$$\det(X - r(s)) = X^n + a_1(s)X^{n-1} + \cdots + a_i(s)X^{n-i} + \cdots$$

where $a_i : S \to k$ is a homogeneous polynomial function of degree i. Write $n = qm$ where q is a power of the characteristic exponent of k and $(q, m) = 1$. Then $(X - \lambda(s))^n = (X^q - \lambda(s)^q)^m$; hence $-m\lambda(s)^q = a_q(s)$, which implies the result.

PROPOSITION 6. *Assume that k is infinite and that condition* (AC) *is satisfied. Let k' be an extension of k. Put $V' = V \otimes_k k'$, $S' = S \otimes_k k'$. Let $r' : S' \to \mathrm{End}(V')$ be the map obtained from r by extension of scalars. Then*

$$V^0(S) \otimes_k k' = V'^0(S) = V'^0(S').$$

The first equality follows from Prop. 1. To prove the second, we can assume that $V = V^0(S)$ and so $V' = V'^0(S)$. Let $(s_1, \ldots, s_m)$ be a basis of S and $(e_1, \ldots, e_n)$ a basis of V. There exist polynomials $P_{ij}(X_1, \ldots, X_m)$ such that

$$r'(a_1 s_1 + \cdots + a_m s_m)^n e_j = \sum_{i=1}^{n} P_{ij}(a_1, \ldots, a_m) e_i$$

for $1 \leq j \leq n$ and $a_1, \ldots, a_m \in k'$. By hypothesis, $r'(s)^n = 0$ for all $s \in S$, in other words $P_{ij}(a_1, \ldots, a_m) = 0$ for $1 \leq i, j \leq n$ and $a_1, \ldots, a_m \in k$. Since k is infinite, $P_{ij} = 0$. Consequently, every element of $r'(S')$ is nilpotent and $V' = V'^0(S')$.

PROPOSITION 7. *Assume that k is infinite and that condition* (AC) *is satisfied. Let $\tilde{S}$ be the set of $s \in S$ such that $V^0(s) = V^0(S)$. If $s \in S$, let $P(s)$ be the determinant of the endomorphism of $V/V^0(S)$ defined by $r(s)$* (no. 1, Cor. 2 (i) of Th. 1).

(i) *The function $s \mapsto P(s)$ is polynomial on S. We have $\tilde{S} = \{s \in S | P(s) \neq 0\}$; this is an open subset of S in the Zariski topology* (App. 1).

(ii) *$\tilde{S}$ is non-empty, and $V^+(s) = V^+(S)$ for all $s \in \tilde{S}$.*

The fact that $s \mapsto P(s)$ is polynomial follows from the linearity of r. If $s \in S$, $V^0(s) \supset V^0(S)$, with equality if and only if $r(s)$ defines an automorphism of $V/V^0(S)$, hence (i).

Now let k' be an algebraic closure of k, and introduce V', S', r' as in Prop. 6. We remark that S' satisfies condition (AC) by continuation of the polynomial identity $(\mathrm{ad}\, r(s_1))^{2\dim V - 1} r(s_2) = 0$ valid for $s_1, s_2 \in S$ (no. 1, *Remark*). Applying Th.1, we deduce a decomposition

$$\mathrm{V}' = \mathrm{V}'^0(\mathrm{S}') \oplus \sum_{i=1}^{m} \mathrm{V}'^{\lambda_i}(\mathrm{S}')$$

with $\lambda_i \neq 0$ for $1 \leq i \leq m$. For $1 \leq i \leq m$, there exists a polynomial function P_i non-zero on S' and an integer q_i such that $\lambda_i^{q_i} = \mathrm{P}_i$ (Prop. 5). Since k is infinite, there exists $s \in \mathrm{S}$ such that $(\mathrm{P}_1 \ldots \mathrm{P}_m)(s) \neq 0$, cf. *Algebra*, Chap. IV, §2, no. 3, Cor. 2 of Prop. 9. Then $\lambda_i(s) \neq 0$ for all i, so $\mathrm{V}'^0(\mathrm{S}') = \mathrm{V}'^0(s)$ and consequently $\mathrm{V}^0(\mathrm{S}) = \mathrm{V}^0(s)$ (Prop. 6), which shows that $\tilde{\mathrm{S}} \neq \varnothing$. If $s \in \tilde{\mathrm{S}}$, the fact that $\mathrm{V}^+(\mathrm{S})$ is stable under $r(s)$ and is a complement of $\mathrm{V}^0(s)$ in V implies that $\mathrm{V}^+(\mathrm{S}) = \mathrm{V}^+(s)$ (Cor. 2 of Th. 1).

3. DECOMPOSITION OF REPRESENTATIONS OF A NILPOTENT LIE ALGEBRA

Let $\mathfrak{h}$ be a Lie algebra and M an $\mathfrak{h}$-module. For any map λ from $\mathfrak{h}$ to k, we have defined in no. 1 vector subspaces $\mathrm{M}^\lambda(\mathfrak{h})$ and $\mathrm{M}_\lambda(\mathfrak{h})$ of M. In particular, if $\mathfrak{g}$ is a Lie algebra containing $\mathfrak{h}$ as a subalgebra, and if $x \in \mathfrak{g}$, we shall often employ the notations $\mathfrak{g}^\lambda(\mathfrak{h})$ and $\mathfrak{g}_\lambda(\mathfrak{h})$; it will then be understood that $\mathfrak{h}$ operates on $\mathfrak{g}$ by the adjoint representation $\mathrm{ad}_{\mathfrak{g}}$.

PROPOSITION 8. *Let $\mathfrak{h}$ be a Lie algebra, and* L, M, N *$\mathfrak{h}$-modules. Denote by* P *the set of maps from $\mathfrak{h}$ to k.*

(i) *The sum* $\sum_{\lambda \in \mathrm{P}} \mathrm{L}^\lambda(\mathrm{P})$ *is direct.*

(ii) *If $f : \mathrm{L} \to \mathrm{M}$ is a homomorphism of $\mathfrak{h}$-modules,* $f(\mathrm{L}^\lambda(\mathfrak{h})) \subset \mathrm{M}^\lambda(\mathfrak{h})$ *for all $\lambda \in \mathrm{P}$.*

(iii) *If $f : \mathrm{L} \times \mathrm{M} \to \mathrm{N}$ is a bilinear $\mathfrak{h}$-invariant map,*

$$f(\mathrm{L}^\lambda(\mathfrak{h}) \times \mathrm{M}^\mu(\mathfrak{h})) \subset \mathrm{N}^{\lambda+\mu}(\mathfrak{h})$$

for all $\lambda, \mu \in \mathrm{P}$.

This follows from Props. 2 and 3.

PROPOSITION 9. *Let $\mathfrak{h}$ be a nilpotent Lie algebra and* M *a finite dimensional $\mathfrak{h}$-module. Denote by* P *the set of maps from $\mathfrak{h}$ to k.*

(i) *Each $\mathrm{M}^\lambda(\mathfrak{h})$ is an $\mathfrak{h}$-submodule of* M. *If x_M is triangularizable for all $x \in \mathfrak{h}$, then* $\mathrm{M} = \sum_{\lambda \in \mathrm{P}} \mathrm{M}^\lambda(\mathfrak{h})$.

(ii) *If k is infinite, there exists $x \in \mathfrak{h}$ such that* $\mathrm{M}^0(x) = \mathrm{M}^0(\mathfrak{h})$.

(iii) *If k is of characteristic* 0, *and if $\lambda \in \mathrm{P}$ is such that* $\mathrm{M}^\lambda(\mathfrak{h}) \neq 0$, *then λ is a linear form on $\mathfrak{h}$ vanishing on $[\mathfrak{h}, \mathfrak{h}]$, and* $\mathrm{M}_\lambda(\mathfrak{h}) \neq 0$.

(iv) *If $f : \mathrm{M} \to \mathrm{N}$ is a surjective homomorphism of finite dimensional $\mathfrak{h}$-modules, then* $f(\mathrm{M}^\lambda(\mathfrak{h})) = \mathrm{N}^\lambda(\mathfrak{h})$ *for all $\lambda \in \mathrm{P}$.*

(v) *If* N *is a finite dimensional $\mathfrak{h}$-module, and* B *a bilinear form on* $\mathrm{M} \times \mathrm{N}$ *invariant under $\mathfrak{h}$, then $\mathrm{M}^\lambda(\mathfrak{h})$ and $\mathrm{N}^\mu(\mathfrak{h})$ are orthogonal relative to*

B *if* $\lambda + \mu \neq 0$. *Moreover, if* B *is non-degenerate then so is its restriction to* $\mathrm{M}^{\lambda}(\mathfrak{h}) \times \mathrm{N}^{-\lambda}(\mathfrak{h})$ *for all* $\lambda \in \mathrm{P}$.

Part (i) follows from no. 1, Lemma 1 and Th. 1. Part (ii) follows from no. 2, Prop. 7. Part (iv) follows from no. 1, Cor. 3 of Th. 1. We prove (iii). We can assume that $\mathrm{M} = \mathrm{M}^{\lambda}(\mathfrak{h})$. Then, for all $x \in \mathfrak{h}$, $\lambda(x) = (\dim \mathrm{M})^{-1}\mathrm{Tr}(x_{\mathrm{M}})$; this proves that λ is linear (which also follows from Prop. 5) and that λ vanishes on $[\mathfrak{h}, \mathfrak{h}]$. Consider the map $\rho : \mathfrak{h} \to \mathrm{End}_k(\mathrm{M})$ defined by

$$\rho(x) = x_{\mathrm{M}} - \lambda(x)1_{\mathrm{M}};$$

from the above, this is a representation of $\mathfrak{h}$ on M, and $\rho(x)$ is nilpotent for all $x \in \mathfrak{h}$. By Engel's theorem (Chap. I, §4, no. 2, Th. 1), there exists $m \neq 0$ in M such that $\rho(x)m = 0$ for all $x \in \mathfrak{h}$, so $m \in \mathrm{M}_{\lambda}(\mathfrak{h})$.

The first assertion of (v) follows from no. 1, Prop. 2 (ii). To prove the second, we can assume that k is algebraically closed in view of Prop. 1 of no. 1; it then follows from the first and the fact that $\mathrm{M} = \sum_{\lambda} \mathrm{M}^{\lambda}(\mathfrak{h})$, $\mathrm{N} = \sum_{\lambda} \mathrm{N}^{\lambda}(\mathfrak{h})$, cf. (i).

Remark. Assume that k is perfect and of characteristic 2. Let $\mathfrak{h} = \mathfrak{sl}(2, k)$, and let M be the $\mathfrak{h}$-module k^2 (for the identity map of $\mathfrak{h}$). If $x = \begin{pmatrix} a & b \\ c & a \end{pmatrix}$ is an arbitrary element of $\mathfrak{h}$, denote by $\lambda(x)$ the unique $\lambda \in k$ such that $\lambda^2 = a^2 + bc$. A calculation shows immediately that $\mathrm{M} = \mathrm{M}^{\lambda}(\mathfrak{h})$; on the other hand, $\mathrm{M}_{\lambda}(\mathfrak{h}) = 0$ and λ is neither linear nor zero on $[\mathfrak{h}, \mathfrak{h}]$, even though $\mathfrak{h}$ is nilpotent.

COROLLARY. *Let* $\mathfrak{h}$ *be a nilpotent Lie algebra, and* M *a finite dimensional* $\mathfrak{h}$-*module such that* $\mathrm{M}^0(\mathfrak{h}) = 0$. *Let* $f : \mathfrak{h} \to \mathrm{M}$ *be a linear map such that*

$$f([x, y]) = x.f(y) - y.f(x) \quad \text{for } x, y \in \mathfrak{h}.$$

There exists $a \in \mathrm{M}$ *such that* $f(x) = x.a$ *for all* $x \in \mathfrak{h}$.

Let $\mathrm{N} = \mathrm{M} \times k$. Make $\mathfrak{h}$ operate on N by the formula

$$x.(m, \lambda) = (x.m - \lambda f(x), 0).$$

The identity satisfied by f implies that N is an $\mathfrak{h}$-module (Chap. I, §1, no. 8, Example 2). The map $(m, \lambda) \mapsto \lambda$ from N to k is a homomorphism from N to the trivial $\mathfrak{h}$-module k. By Prop. 9 (iv), it follows that $\mathrm{N}^0(\mathfrak{h})$ contains an element of the form $(a, 1)$ with $a \in \mathrm{M}$. In view of the hypothesis on M,

$$(\mathrm{M} \times 0) \cap \mathrm{N}^0(\mathfrak{h}) = 0,$$

so $\mathrm{N}^0(\mathfrak{h})$ is of dimension 1 and hence is annihilated by $\mathfrak{h}$. Thus, $x.a - f(x) = 0$ for all $x \in \mathfrak{h}$, which proves the corollary.

PROPOSITION 10. *Let* $\mathfrak{g}$ *be a Lie algebra,* $\mathfrak{h}$ *a nilpotent subalgebra of* $\mathfrak{g}$. *Denote by* P *the set of maps from* $\mathfrak{h}$ *to* k.

(i) *For* $\lambda, \mu \in \mathrm{P}$, $[\mathfrak{g}^{\lambda}(\mathfrak{h}), \mathfrak{g}^{\mu}(\mathfrak{h})] \subset \mathfrak{g}^{\lambda+\mu}(\mathfrak{h})$; *in particular,* $\mathfrak{g}^0(\mathfrak{h})$ *is a Lie subalgebra of* $\mathfrak{g}$ *containing* $\mathfrak{h}$, *and the* $\mathfrak{g}^{\lambda}(\mathfrak{h})$ *are stable under* $\mathrm{ad}\,\mathfrak{g}^0(\mathfrak{h})$. *Moreover,* $\mathfrak{g}^0(\mathfrak{h})$ *is its own normalizer in* $\mathfrak{g}$.

(ii) *If* M *is a* $\mathfrak{g}$*-module,* $\mathfrak{g}^{\lambda}(\mathfrak{h})\mathrm{M}^{\mu}(\mathfrak{h}) \subset \mathrm{M}^{\lambda+\mu}(\mathfrak{h})$ *for* $\lambda, \mu \in \mathrm{P}$; *in particular, each* $\mathrm{M}^{\lambda}(\mathfrak{h})$ *is a* $\mathfrak{g}^0(\mathfrak{h})$*-module.*

(iii) *If* B *is a bilinear form on* $\mathfrak{g}$ *invariant under* $\mathfrak{h}$, $\mathfrak{g}^{\lambda}(\mathfrak{h})$ *and* $\mathfrak{g}^{\mu}(\mathfrak{h})$ *are orthogonal relative to* B *for* $\lambda + \mu \neq 0$. *Assume that* B *is non-degenerate. Then, for all* $\lambda \in \mathrm{P}$, *the restriction of* B *to* $\mathfrak{g}^{\lambda}(\mathfrak{h}) \times \mathfrak{g}^{-\lambda}(\mathfrak{h})$ *is non-degenerate; in particular, the restriction of* B *to* $\mathfrak{g}^0(\mathfrak{h}) \times \mathfrak{g}^0(\mathfrak{h})$ *is non-degenerate.*

(iv) *Assume that* k *is of characteristic* 0. *Then, if* $x \in \mathfrak{g}^{\lambda}(\mathfrak{h})$ *with* $\lambda \neq 0$, $\mathrm{ad}\,x$ *is nilpotent.*

The map $(x, y) \mapsto [x, y]$ from $\mathfrak{g} \times \mathfrak{g}$ to $\mathfrak{g}$ is $\mathfrak{g}$-invariant by the Jacobi identity, hence $\mathfrak{h}$-invariant. The first part of (i) thus follows from Prop. 2 (ii). Part (ii) is proved similarly.

If x belongs to the normalizer of $\mathfrak{g}^0(\mathfrak{h})$ in $\mathfrak{g}$, $(\mathrm{ad}\,y).x = -[x, y] \in \mathfrak{g}^0(\mathfrak{h})$ for all $y \in \mathfrak{h}$, so $(\mathrm{ad}\,y)^n.x = 0$ for n sufficiently large. This proves that $x \in \mathfrak{g}^0(\mathfrak{h})$. Assertion (i) is now completely proved.

Assertion (iii) follows from Prop. 9 (v).

To prove (iv), we can assume that k is algebraically closed. Let $x \in \mathfrak{g}^{\lambda}(\mathfrak{h})$, with $\lambda \neq 0$. For all $\mu \in \mathrm{P}$ and any integer $n \geq 0$, $(\mathrm{ad}\,x)^n \mathfrak{g}^{\mu}(\mathfrak{h}) \subset \mathfrak{g}^{\mu+n\lambda}(\mathfrak{h})$; let P_1 be the finite set of $\mu \in \mathrm{P}$ such that $\mathfrak{g}^{\mu}(\mathfrak{h}) \neq 0$; if k is of characteristic 0 and $\lambda \neq 0$, $(\mathrm{P}_1 + n\lambda) \cap \mathrm{P}_1 = \varnothing$ for n sufficiently large, so $(\mathrm{ad}\,x)^n = 0$.

Lemma 2. Assume that k *is of characteristic* 0. *Let* $\mathfrak{g}$ *be a semi-simple Lie algebra over* k, B *the Killing form of* $\mathfrak{g}$, $\mathfrak{m}$ *a subalgebra of* $\mathfrak{g}$. *Assume that the following conditions are satisfied:*

1) *the restriction of* B *to* $\mathfrak{m}$ *is non-degenerate;*

2) *if* $x \in \mathfrak{m}$, *the semi-simple and nilpotent components*[1] *of* x *in* $\mathfrak{g}$ *belong to* $\mathfrak{m}$.

Then $\mathfrak{m}$ *is reductive in* $\mathfrak{g}$ (Chap. I, § 6, no. 6).

By Chap. I, §6, no. 4, Prop. 5 *d*), $\mathfrak{m}$ is reductive. Let $\mathfrak{c}$ be the centre of $\mathfrak{m}$. If $x \in \mathfrak{c}$ is nilpotent, then $x = 0$; indeed, for all $y \in \mathfrak{m}$, $\mathrm{ad}\,x$ and $\mathrm{ad}\,y$ commute, their composition $\mathrm{ad}\,x \circ \mathrm{ad}\,y$ is nilpotent, and $\mathrm{B}(x, y) = 0$, so $x = 0$. Now let x be an arbitrary element of $\mathfrak{c}$; let s and n be its semi-simple and nilpotent components. We have $n \in \mathfrak{m}$. Since $\mathrm{ad}\,n$ is of the form $\mathrm{P}(\mathrm{ad}\,x)$, where P is a polynomial with no constant term, $(\mathrm{ad}\,n).\mathfrak{m} = 0$ and so $n \in \mathfrak{c}$, and then $n = 0$ by the above. Thus $\mathrm{ad}\,x$ is semi-simple. Consequently, the restriction to $\mathfrak{m}$ of the adjoint representation of $\mathfrak{g}$ is semi-simple (Chap. I, §6, no. 5, Th. 4 *b*)).

PROPOSITION 11. *Assume that* k *is of characteristic* 0. *Let* $\mathfrak{g}$ *be a semi-simple Lie algebra,* $\mathfrak{h}$ *a nilpotent subalgebra of* $\mathfrak{g}$. *The algebra* $\mathfrak{g}^0(\mathfrak{h})$ *satisfies conditions* (1) *and* (2) *of Lemma 2; it is reductive in* $\mathfrak{g}$.

---

[1] By Chap. I, §6, no. 3, Th. 3, every $x \in \mathfrak{g}$ can be written uniquely as the sum of a semi-simple element s and a nilpotent element n that commute with each other; the element s (resp. n) is called the *semi-simple* (resp. *nilpotent*) *component* of x.

Let $x, x' \in \mathfrak{g}$, s and s' their semi-simple components, n and n' their nilpotent components. We have

$$\begin{aligned} x' \in \mathfrak{g}^0(x) &\iff (\operatorname{ad} s)(x') = 0 \quad \text{(Prop. 4)} \\ &\iff (\operatorname{ad} x')(s) = 0 \\ &\Longrightarrow (\operatorname{ad} s')(s) = 0 \\ &\iff (\operatorname{ad} s)(s') = 0 \\ &\iff s' \in \mathfrak{g}^0(x) \quad \text{(Prop. 4)} \end{aligned}$$

so $x' \in \mathfrak{g}^0(x) \Rightarrow n' \in \mathfrak{g}^0(x)$ and (2) is proved. The Killing form of $\mathfrak{g}$ is non-degenerate, so its restriction to $\mathfrak{g}^0(\mathfrak{h})$ is non-degenerate (Prop. 10 (iii)). The fact that $\mathfrak{g}^0(\mathfrak{h})$ is reductive in $\mathfrak{g}$ thus follows from Lemma 2.

4. DECOMPOSITION OF A LIE ALGEBRA RELATIVE TO AN AUTOMORPHISM

PROPOSITION 12. *Let $\mathfrak{g}$ be a Lie algebra, a an automorphism of $\mathfrak{g}$.*

(i) *For $\lambda, \mu \in k$, $[\mathfrak{g}^\lambda(a), \mathfrak{g}^\mu(a)] \subset \mathfrak{g}^{\lambda\mu}(a)$; in particular, $\mathfrak{g}^1(a)$ is a subalgebra of $\mathfrak{g}$.*

(ii) *If B is a symmetric bilinear form on $\mathfrak{g}$ invariant under a, $\mathfrak{g}^\lambda(a)$ and $\mathfrak{g}^\mu(a)$ are orthogonal relative to B for $\lambda\mu \neq 1$. Assume that B is non-degenerate. Then, if $\lambda \neq 0$, the restriction of B to $\mathfrak{g}^\lambda(a) \times \mathfrak{g}^{1/\lambda}(a)$ is non-degenerate.*

Assertion (i) and the first half of (ii) follow from Prop. 2 (iii) applied to the composition law $\mathfrak{g} \times \mathfrak{g} \to \mathfrak{g}$ and the bilinear form B. To prove the second half of (ii), we can assume that k is algebraically closed. Then $\mathfrak{g} = \bigoplus_{\nu \in k} \mathfrak{g}^\nu(a)$. In view of the above, $\mathfrak{g}^\lambda(a)$ is orthogonal to $\mathfrak{g}^\nu(a)$ if $\lambda\nu \neq 1$; since B is non-degenerate, it follows that its restriction to $\mathfrak{g}^\lambda(a) \times \mathfrak{g}^{1/\lambda}(a)$ is also.

COROLLARY. *Assume that k is of characteristic zero and that $\mathfrak{g}$ is semi-simple. Then the subalgebra $\mathfrak{g}^1(a)$ satisfies conditions* (1) *and* (2) *of Lemma 2; it is reductive in $\mathfrak{g}$.*

Condition (1) follows from part (ii) of Prop. 12; condition (2) follows from Prop. 4 of no. 1.

5. INVARIANTS OF A SEMI-SIMPLE LIE ALGEBRA RELATIVE TO A SEMI-SIMPLE ACTION

In this no., k is assumed to be of characteristic zero.

PROPOSITION 13. *Let $\mathfrak{g}$ be a semi-simple Lie algebra, $\mathfrak{a}$ a subalgebra of $\mathfrak{g}$ reductive in $\mathfrak{g}$, and $\mathfrak{m}$ the commutant of $\mathfrak{a}$ in $\mathfrak{g}$. The subalgebra $\mathfrak{m}$ satisfies conditions* (1) *and* (2) *of Lemma 2 of no. 3; it is reductive in $\mathfrak{g}$.*

By Prop. 6 of Chap. I, §3, no. 5, applied to the $\mathfrak{a}$-module $\mathfrak{g}$, we have $\mathfrak{g} = \mathfrak{m} \oplus [\mathfrak{a}, \mathfrak{g}]$. Let B be the Killing form of $\mathfrak{g}$, and let $x \in \mathfrak{a}, y \in \mathfrak{m}, z \in \mathfrak{g}$. Then,

$$\mathrm{B}([z,x],y) = \mathrm{B}(z,[x,y]) = 0 \quad \text{since } [x,y] = 0,$$

which shows that $\mathfrak{m}$ is orthogonal to $[\mathfrak{a}, \mathfrak{g}]$ relative to B. Since B is non-degenerate, and since $\mathfrak{g} = \mathfrak{m} \oplus [\mathfrak{a}, \mathfrak{g}]$, this implies that the restriction of B to $\mathfrak{m}$ is non-degenerate; condition (1) of Lemma 2 is thus satisfied.

Now let $x \in \mathfrak{m}$ and let s and n be its semi-simple and nilpotent components. The semi-simple component of $\operatorname{ad} x$ is $\operatorname{ad} s$, cf. Chap. I, §6, no. 3. Since $\operatorname{ad} x$ is zero on $\mathfrak{a}$, so is $\operatorname{ad} s$, by Prop. 4 (i). Thus $s \in \mathfrak{m}$, so $n = x - s \in \mathfrak{m}$, and condition (2) of Lemma 2 is satisfied.

Remark. The commutant of $\mathfrak{m}$ in $\mathfrak{g}$ does not necessarily reduce to $\mathfrak{a}$, cf. Exerc. 4.

PROPOSITION 14. *Let $\mathfrak{g}$ be a semi-simple Lie algebra,* A *a group and r a homomorphism from* A *to* $\operatorname{Aut}(\mathfrak{g})$. *Let $\mathfrak{m}$ be the subalgebra of $\mathfrak{g}$ consisting of the elements invariant under $r(\mathrm{A})$. Assume that the linear representation r is semi-simple. Then $\mathfrak{m}$ satisfies conditions* (1) *and* (2) *of Lemma 2 of no. 3; it is reductive in $\mathfrak{g}$.*

The proof is analogous to that of the preceding proposition:

Let $\mathfrak{g}^+$ be the vector subspace of $\mathfrak{g}$ generated by the $r(a)x - x$, $a \in \mathrm{A}, x \in \mathfrak{g}$. The vector space $\mathfrak{g}' = \mathfrak{m} + \mathfrak{g}^+$ is stable under $r(\mathrm{A})$. Let $\mathfrak{n}$ be a complement of $\mathfrak{g}'$ in $\mathfrak{g}$ stable under $r(\mathrm{A})$. If $x \in \mathfrak{n}, a \in \mathrm{A}$, $r(a)x - x \in \mathfrak{n} \cap \mathfrak{g}^+ = 0$, so $x \in \mathfrak{m}$ and then $x = 0$ since $\mathfrak{m} \cap \mathfrak{n} = 0$. Thus, $\mathfrak{g} = \mathfrak{g}' = \mathfrak{m} + \mathfrak{g}^+$. Let B be the Killing form of $\mathfrak{g}$ and let $y \in \mathfrak{m}, a \in \mathrm{A}, x \in \mathfrak{g}$. Then

$$\begin{aligned}\mathrm{B}(y, r(a)x - x) &= \mathrm{B}(y, r(a)x) - \mathrm{B}(y,x) \\ &= \mathrm{B}(r(a^{-1})y, x) - \mathrm{B}(y,x) \\ &= \mathrm{B}(y,x) - \mathrm{B}(y,x) = 0.\end{aligned}$$

Thus $\mathfrak{m}$ and $\mathfrak{g}^+$ are orthogonal relative to B. It follows that the restriction of B to $\mathfrak{m}$ is non-degenerate; hence condition (1) of Lemma 2. Condition (2) is immediate by transport of structure.

§2. CARTAN SUBALGEBRAS AND REGULAR ELEMENTS OF A LIE ALGEBRA

From no. 2 onwards, the field k is assumed to be infinite.

1. CARTAN SUBALGEBRAS

DEFINITION 1. *Let $\mathfrak{g}$ be a Lie algebra. A Cartan subalgebra of $\mathfrak{g}$ is a nilpotent subalgebra of $\mathfrak{g}$ equal to its own normalizer.*

Later we shall obtain the following results:

1) if k is infinite, $\mathfrak{g}$ has Cartan subalgebras (no. 3, Cor. 1 of Th. 1);

2) if k is of characteristic zero, all Cartan subalgebras of $\mathfrak{g}$ have the same dimension (§3, no. 3, Th. 2);

3) if k is algebraically closed and of characteristic 0, all Cartan subalgebras of $\mathfrak{g}$ are conjugate under the group of elementary automorphisms of $\mathfrak{g}$ (§3, no. 2, Th. 1).

Examples. 1) If $\mathfrak{g}$ is nilpotent, the only Cartan subalgebra of $\mathfrak{g}$ is $\mathfrak{g}$ itself (Chap. I, §4, no. 1, Prop. 3).

2) Let $\mathfrak{g} = \mathfrak{gl}(n, k)$, and let $\mathfrak{h}$ be the set of diagonal matrices belonging to $\mathfrak{g}$. We show that $\mathfrak{h}$ is a Cartan subalgebra of $\mathfrak{g}$. First, $\mathfrak{h}$ is commutative, hence nilpotent. Let (E_{ij}) be the canonical basis of $\mathfrak{gl}(n, k)$, and let $x = \sum \mu_{ij} E_{ij}$ be an element of the normalizer of $\mathfrak{h}$ in $\mathfrak{g}$. If $i \neq j$, formulas (5) of Chap. I, §1, no. 2 show that the coefficient of E_{ij} in $[E_{ii}, x]$ is μ_{ij}. Since $E_{ii} \in \mathfrak{h}$, $[E_{ii}, x] \in \mathfrak{h}$, and the coefficient in question is zero. Thus $\mu_{ij} = 0$ for $i \neq j$, so $x \in \mathfrak{h}$, which shows that $\mathfrak{h}$ is indeed a Cartan subalgebra of $\mathfrak{g}$.

3) Let $\mathfrak{h}$ be a Cartan subalgebra of $\mathfrak{g}$ and let $\mathfrak{g}_1$ be a subalgebra of $\mathfrak{g}$ containing $\mathfrak{h}$. Then $\mathfrak{h}$ is a Cartan subalgebra of $\mathfrak{g}_1$; this follows immediately from Def. 1.

PROPOSITION 1. *Let $\mathfrak{g}$ be a Lie algebra and let $\mathfrak{h}$ be a Cartan subalgebra of $\mathfrak{g}$. Then $\mathfrak{h}$ is a maximal nilpotent subalgebra of $\mathfrak{g}$.*

Let $\mathfrak{h}'$ be a nilpotent subalgebra of $\mathfrak{g}$ containing $\mathfrak{h}$. Then $\mathfrak{h}$ is a Cartan subalgebra of $\mathfrak{h}'$ (Example 3), so $\mathfrak{h} = \mathfrak{h}'$ (Example 1).

There exist maximal nilpotent subalgebras that are not Cartan subalgebras (Exerc. 2).

PROPOSITION 2. *Let $(\mathfrak{g}_i)_{i \in I}$ be a finite family of Lie algebras and $\mathfrak{g} = \prod_{i \in I} \mathfrak{g}_i$. The Cartan subalgebras of $\mathfrak{g}$ are the subalgebras of the form $\prod_{i \in I} \mathfrak{h}_i$, where $\mathfrak{h}_i$ is a Cartan subalgebra of $\mathfrak{g}_i$.*

If $\mathfrak{h}_i$ is a subalgebra of $\mathfrak{g}_i$ with normalizer $\mathfrak{n}_i$, then $\prod \mathfrak{h}_i$ is a subalgebra of $\mathfrak{g}$ with normalizer $\prod \mathfrak{n}_i$; if the $\mathfrak{h}_i$ are nilpotent, $\prod \mathfrak{h}_i$ is nilpotent; thus, if $\mathfrak{h}_i$ is a Cartan subalgebra of $\mathfrak{g}_i$ for all i, $\prod \mathfrak{h}_i$ is a Cartan subalgebra of $\mathfrak{g}$. Conversely, let $\mathfrak{h}$ be a Cartan subalgebra of $\mathfrak{g}$; the projection $\mathfrak{h}_i$ of $\mathfrak{h}$ onto $\mathfrak{g}_i$ is a nilpotent subalgebra of $\mathfrak{g}_i$, and $\prod \mathfrak{h}_i$ is a nilpotent subalgebra of $\mathfrak{g}$ containing $\mathfrak{h}$; hence $\mathfrak{h} = \prod \mathfrak{h}_i$ (Prop. 1); thus, for all i, $\mathfrak{h}_i$ is its own normalizer in $\mathfrak{g}_i$, and so is a Cartan subalgebra of $\mathfrak{g}_i$.

Example 4. If k is of characteristic 0, $\mathfrak{gl}(n,k)$ is the product of the ideals $\mathfrak{sl}(n,k)$ and $k.1$. It follows from Example 2 and Prop. 2 that the set of diagonal matrices of trace 0 in $\mathfrak{sl}(n,k)$ is a Cartan subalgebra of $\mathfrak{sl}(n,k)$.

PROPOSITION 3. *Let $\mathfrak{g}$ be a Lie algebra, $\mathfrak{h}$ a subalgebra of $\mathfrak{g}$, and k' an extension of k. Then $\mathfrak{h}$ is a Cartan subalgebra of $\mathfrak{g}$ if and only if $\mathfrak{h} \otimes_k k'$ is a Cartan subalgebra of $\mathfrak{g} \otimes_k k'$.*

Indeed, $\mathfrak{h}$ is nilpotent if and only if $\mathfrak{h} \otimes_k k'$ is (Chap. I, §4, no. 5). On the other hand, if $\mathfrak{n}$ is the normalizer of $\mathfrak{h}$ in $\mathfrak{g}$, the normalizer of $\mathfrak{h} \otimes_k k'$ in $\mathfrak{g} \otimes_k k'$ is $\mathfrak{n} \otimes_k k'$ (Chap. I, §3, no. 8).

PROPOSITION 4. *Let $\mathfrak{g}$ be a Lie algebra, $\mathfrak{h}$ a nilpotent subalgebra of $\mathfrak{g}$. Then $\mathfrak{h}$ is a Cartan subalgebra of $\mathfrak{g}$ if and only if $\mathfrak{g}^0(\mathfrak{h}) = \mathfrak{h}$.*

If $\mathfrak{g}^0(\mathfrak{h}) = \mathfrak{h}$, $\mathfrak{h}$ is its own normalizer (§1, Prop. 10 (i)), so $\mathfrak{h}$ is a Cartan subalgebra of $\mathfrak{g}$. Assume that $\mathfrak{g}^0(\mathfrak{h}) \neq \mathfrak{h}$. Consider the representation of $\mathfrak{h}$ on $\mathfrak{g}^0(\mathfrak{h})/\mathfrak{h}$ obtained from the adjoint representation by passage to the quotient. By applying Engel's theorem (Chap. I, §4, no. 2, Th. 1), we see that there exists $x \in \mathfrak{g}^0(\mathfrak{h})$ such that $x \notin \mathfrak{h}$ and $[\mathfrak{h}, x] \subset \mathfrak{h}$; then x belongs to the normalizer of $\mathfrak{h}$ in $\mathfrak{g}$, so $\mathfrak{h}$ is not a Cartan subalgebra of $\mathfrak{g}$.

COROLLARY 1. *Let $\mathfrak{g}$ be a Lie algebra, $\mathfrak{h}$ a Cartan subalgebra of $\mathfrak{g}$. If k is infinite, there exists $x \in \mathfrak{h}$ such that $\mathfrak{h} = \mathfrak{g}^0(x)$.*

Indeed, $\mathfrak{h} = \mathfrak{g}^0(\mathfrak{h})$ and we can apply Prop. 9 (ii) of §1.

COROLLARY 2. *Let $f : \mathfrak{g} \to \mathfrak{g}'$ be a surjective homomorphism of Lie algebras. If $\mathfrak{h}$ is a Cartan subalgebra of $\mathfrak{g}$, $f(\mathfrak{h})$ is a Cartan subalgebra of $\mathfrak{g}'$.*

Indeed, $f(\mathfrak{h})$ is a nilpotent subalgebra of $\mathfrak{g}'$. On the other hand, consider the representation $x \mapsto \operatorname{ad} f(x)$ of $\mathfrak{h}$ on $\mathfrak{g}'$. By Prop. 9 (iv) of §1, no. 3, $f(\mathfrak{g}^0(\mathfrak{h})) = \mathfrak{g}'^0(\mathfrak{h})$. Now $\mathfrak{g}^0(\mathfrak{h}) = \mathfrak{h}$, and on the other hand it is clear that $\mathfrak{g}'^0(\mathfrak{h}) = \mathfrak{g}'^0(f(\mathfrak{h}))$. Hence, $f(\mathfrak{h}) = \mathfrak{g}'^0(f(\mathfrak{h}))$ and it suffices to apply Prop. 4.

COROLLARY 3. *Let $\mathfrak{h}$ be a Cartan subalgebra of a Lie algebra $\mathfrak{g}$, and let $\mathscr{C}^n\mathfrak{g}$ $(n \geq 1)$ be a term of the descending central series of $\mathfrak{g}$* (Chap. I, §1, no. 5). *Then $\mathfrak{g} = \mathfrak{h} + \mathscr{C}^n\mathfrak{g}$.*

Indeed, Corollary 2 shows that the image of $\mathfrak{h}$ in $\mathfrak{g}/\mathscr{C}^n\mathfrak{g}$ is a Cartan subalgebra of $\mathfrak{g}/\mathscr{C}^n\mathfrak{g}$, hence is equal to $\mathfrak{g}/\mathscr{C}^n\mathfrak{g}$ since $\mathfrak{g}/\mathscr{C}^n\mathfrak{g}$ is nilpotent (Example 1).

COROLLARY 4. *Let $\mathfrak{g}$ be a Lie algebra, $\mathfrak{h}$ a Cartan subalgebra of $\mathfrak{g}$, and $\mathfrak{a}$ a subalgebra of $\mathfrak{g}$ containing $\mathfrak{h}$.*

(i) *$\mathfrak{a}$ is equal to its own normalizer in $\mathfrak{g}$.*

(ii) *Assume that $k = \mathbf{R}$ or $\mathbf{C}$; let G be a Lie group with Lie algebra $\mathfrak{g}$, A the integral subgroup of G with Lie algebra $\mathfrak{a}$. Then A is a Lie subgroup of G, and it is the identity component of the normalizer of A in G.*

Let $\mathfrak{n}$ be the normalizer of $\mathfrak{a}$ in $\mathfrak{g}$. Since $\mathfrak{h}$ is a Cartan subalgebra of $\mathfrak{n}$ (Example 3), $\{0\}$ is a Cartan subalgebra of $\mathfrak{n}/\mathfrak{a}$ (Cor. 2), hence is equal to its normalizer in $\mathfrak{n}/\mathfrak{a}$; in other words, $\mathfrak{n} = \mathfrak{a}$. Assertion (ii) follows from (i) and Chap. III, §9, no. 4, Cor. of Prop. 11.

COROLLARY 5. *Let $\mathfrak{g}$ be a Lie algebra, E a subset of $\mathfrak{g}$. Let E operate on $\mathfrak{g}$ by the adjoint representation. Then E is a Cartan subalgebra of $\mathfrak{g}$ if and only if* $\mathrm{E} = \mathfrak{g}^0(\mathrm{E})$.

The condition is necessary (Prop. 4). Assume now that $\mathrm{E} = \mathfrak{g}^0(\mathrm{E})$. By Prop. 2 (ii) of §1, no. 1, E is then a subalgebra of $\mathfrak{g}$. If $x \in \mathrm{E}$, $\mathrm{ad}_{\mathrm{E}}x$ is nilpotent since $\mathrm{E} \subset \mathfrak{g}^0(\mathrm{E})$; hence the algebra E is nilpotent. But then E is a Cartan subalgebra by Prop. 4.

COROLLARY 6. *Let $\mathfrak{g}$ be a Lie algebra, let k_0 be a subfield of k such that $[k : k_0] < +\infty$, and let $\mathfrak{g}_0$ be the Lie algebra obtained from $\mathfrak{g}$ by restricting the field of scalars to k_0. Let $\mathfrak{h}$ be a subset of $\mathfrak{g}$. Then $\mathfrak{h}$ is a Cartan subalgebra of $\mathfrak{g}$ if and only if $\mathfrak{h}$ is a Cartan subalgebra of $\mathfrak{g}_0$.*

This follows from Cor. 5, since the condition $\mathfrak{h} = \mathfrak{g}^0(\mathfrak{h})$ does not involve the base field.

PROPOSITION 5. *Let $\mathfrak{g}$ be a Lie algebra, $\mathfrak{c}$ its centre, $\mathfrak{h}$ a vector subspace of $\mathfrak{g}$. Then $\mathfrak{h}$ is a Cartan subalgebra of $\mathfrak{g}$ if and only if $\mathfrak{h}$ contains $\mathfrak{c}$ and $\mathfrak{h}/\mathfrak{c}$ is a Cartan subalgebra of $\mathfrak{g}/\mathfrak{c}$.*

Assume that $\mathfrak{h}$ is a Cartan subalgebra of $\mathfrak{g}$. Since $[\mathfrak{c}, \mathfrak{g}] \subset \mathfrak{h}$, we have $\mathfrak{c} \subset \mathfrak{h}$. On the other hand, $\mathfrak{h}/\mathfrak{c}$ is a Cartan subalgebra of $\mathfrak{g}/\mathfrak{c}$ by Cor. 2 of Prop. 4.

Assume that $\mathfrak{h} \supset \mathfrak{c}$ and that $\mathfrak{h}/\mathfrak{c}$ is a Cartan subalgebra of $\mathfrak{g}/\mathfrak{c}$. Let f be the canonical morphism from $\mathfrak{g}$ to $\mathfrak{g}/\mathfrak{c}$. The algebra $\mathfrak{h}$, which is a central extension of $\mathfrak{h}/\mathfrak{c}$, is nilpotent. Let $\mathfrak{n}$ be the normalizer of $\mathfrak{h}$ in $\mathfrak{g}$. If $x \in \mathfrak{n}$, $[f(x), \mathfrak{h}/\mathfrak{c}] \subset \mathfrak{h}/\mathfrak{c}$, hence $f(x) \in \mathfrak{h}/\mathfrak{c}$, and so $x \in \mathfrak{h}$. This proves that $\mathfrak{h}$ is a Cartan subalgebra of $\mathfrak{g}$.

COROLLARY. *Let $\mathscr{C}_\infty\mathfrak{g}$ be the union of the ascending central series of the Lie algebra $\mathfrak{g}$* (Chap. I, §1, no. 6). *The Cartan subalgebras of $\mathfrak{g}$ are the inverse images of the Cartan subalgebras of $\mathfrak{g}/\mathscr{C}_\infty\mathfrak{g}$.*

Indeed, the centre of $\mathfrak{g}/\mathscr{C}_i\mathfrak{g}$ is $\mathscr{C}_{i+1}\mathfrak{g}/\mathscr{C}_i\mathfrak{g}$, and the corollary follows immediately from Prop. 5 by induction.

Remark. $\mathscr{C}_\infty\mathfrak{g}$ is the smallest ideal $\mathfrak{n}$ of $\mathfrak{g}$ such that the centre of $\mathfrak{g}/\mathfrak{n}$ is zero; it is a characteristic and nilpotent ideal of $\mathfrak{g}$.

2. REGULAR ELEMENTS OF A LIE ALGEBRA

[Recall that k is assumed to be infinite from now on.]

Let $\mathfrak{g}$ be a Lie algebra of dimension n. If $x \in \mathfrak{g}$, write the characteristic polynomial of $\operatorname{ad} x$ in the form

$$\det(\mathrm{T} - \operatorname{ad} x) = \sum_{i=0}^{n} a_i(x)\mathrm{T}^i, \quad \text{with } a_i(x) \in k.$$

We have $a_i(x) = (-1)^{n-i}\mathrm{Tr}\left(\bigwedge^{n-i} \operatorname{ad} x\right)$, cf. *Algebra*, Chap. III, §8, no. 11. This shows that $x \mapsto a_i(x)$ is a homogeneous polynomial map of degree $n-i$ from $\mathfrak{g}$ to k (*Algebra*, Chap. IV, §5, no. 9).

Remarks. 1) If $\mathfrak{g} \neq \{0\}$, $a_0 = 0$ since $(\operatorname{ad} x)(x) = 0$ for all $x \in \mathfrak{g}$.

2) Let k' be an extension of k. Write $\det(\mathrm{T} - \operatorname{ad} x') = \sum_{i=0}^{n} a'_i(x')\mathrm{T}^i$ for $x' \in \mathfrak{g} \otimes_k k'$. Then $a'_i|\mathfrak{g} = a_i$ for all i.

DEFINITION 2. *The rank of $\mathfrak{g}$, denoted by $\mathrm{rk}(\mathfrak{g})$, is the smallest integer l such that $a_l \neq 0$. An element x of $\mathfrak{g}$ is called regular if $a_l(x) \neq 0$.*

For all $x \in \mathfrak{g}$, $\mathrm{rk}(\mathfrak{g}) \leq \dim \mathfrak{g}^0(x)$, and equality holds if and only if x is regular.

The set of regular elements is dense and open in $\mathfrak{g}$ for the Zariski topology (App. I).

Examples. 1) If $\mathfrak{g}$ is nilpotent, $\mathrm{rk}(\mathfrak{g}) = \dim \mathfrak{g}$ and all elements of $\mathfrak{g}$ are regular.

2) Let $\mathfrak{g} = \mathfrak{sl}(2, k)$. If $x = \begin{pmatrix} \gamma & \alpha \\ \beta & -\gamma \end{pmatrix} \in \mathfrak{g}$, an easy calculation gives

$$\det(\mathrm{T} - \operatorname{ad} x) = \mathrm{T}^3 - 4(\alpha\beta + \gamma^2)\mathrm{T}.$$

If the characteristic of k is $\neq 2$, then $\mathrm{rk}(\mathfrak{g}) = 1$ and the regular elements are those x such that $\alpha\beta + \gamma^2 \neq 0$.

3) Let V be a vector space of finite dimension n, and $\mathfrak{g} = \mathfrak{gl}(\mathrm{V})$. Let $\overline{k}$ be an algebraic closure of k. Let $x \in \mathfrak{g}$, and let $\lambda_1, \ldots, \lambda_n$ be the roots in $\overline{k}$ of the characteristic polynomial of x (each root being written a number of times equal to its multiplicity). The canonical isomorphism from $\mathrm{V}^* \otimes \mathrm{V}$ to $\mathfrak{g}$ is compatible with the $\mathfrak{g}$-module structures of these two spaces, in other words it takes $1 \otimes x - {}^t x \otimes 1$ to $\operatorname{ad} x$ (Chap. I, §3, no. 3, Prop. 4). In view of §1, Prop. 4 (i), it follows that the roots of the characteristic polynomial of $\operatorname{ad} x$ are the $\lambda_i - \lambda_j$ for $1 \leq i \leq n, 1 \leq j \leq n$ (each root being written a number of times equal to its multiplicity). Thus, the rank of $\mathfrak{g}$ is n, and x is regular if and only if each λ_i is a simple root of the characteristic polynomial of x.

PROPOSITION 6. *Let* $\mathfrak{g}$ *be a Lie algebra,* k' *an extension of* k, *and* $\mathfrak{g}' = \mathfrak{g} \otimes_k k'$.

(i) *An element* x *of* $\mathfrak{g}$ *is regular in* $\mathfrak{g}$ *if and only if* $x \otimes 1$ *is regular in* $\mathfrak{g}'$.

(ii) $\mathrm{rk}(\mathfrak{g}) = \mathrm{rk}(\mathfrak{g}')$.

This follows from Remark 2.

PROPOSITION 7. *Let* $(\mathfrak{g}_i)_{i\in \mathrm{I}}$ *be a finite family of Lie algebras, and let* $\mathfrak{g} = \prod_{i\in\mathrm{I}} \mathfrak{g}_i$.

(i) *An element* $(x_i)_{i\in\mathrm{I}}$ *of* $\mathfrak{g}$ *is regular in* $\mathfrak{g}$ *if and only if, for all* $i \in \mathrm{I}$, x_i *is regular in* $\mathfrak{g}_i$.

(ii) $\mathrm{rk}(\mathfrak{g}) = \sum_{i\in\mathrm{I}} \mathrm{rk}(\mathfrak{g}_i)$.

Indeed, for any $x = (x_i)_{i\in\mathrm{I}} \in \mathfrak{g}$, the characteristic polynomial of $\mathrm{ad}_{\mathfrak{g}} x$ is the product of the characteristic polynomials of the $\mathrm{ad}_{\mathfrak{g}_i} x_i$.

PROPOSITION 8. *Let* $f : \mathfrak{g} \to \mathfrak{g}'$ *be a surjective homomorphism of Lie algebras.*

(i) *If* x *is a regular element of* $\mathfrak{g}$, $f(x)$ *is regular in* $\mathfrak{g}'$. *The converse is true if* $\mathrm{Ker}\, f$ *is contained in the centre of* $\mathfrak{g}$.

(ii) $\mathrm{rk}(\mathfrak{g}) \geq \mathrm{rk}(\mathfrak{g}')$.

Put $\mathrm{rk}(\mathfrak{g}) = r, \mathrm{rk}(\mathfrak{g}') = r'$. Let $x \in \mathfrak{g}$. The characteristic polynomials of $\mathrm{ad}\, x$, $\mathrm{ad}\, f(x)$ and $\mathrm{ad}\, x|\mathrm{Ker}\, f$ are of the form

$$\mathrm{P(T)} = \mathrm{T}^n + a_{n-1}(x)\mathrm{T}^{n-1} + \cdots + a_r(x)\mathrm{T}^r,$$

$$\mathrm{Q(T)} = \mathrm{T}^{n'} + b_{n'-1}(x)\mathrm{T}^{n'-1} + \cdots + b_{r'}(x)\mathrm{T}^{r'},$$

$$\mathrm{R(T)} = \mathrm{T}^{n''} + c_{n''-1}(x)\mathrm{T}^{n''-1} + \cdots + c_{r''}(x)\mathrm{T}^{r''},$$

where the a_i, b_i, c_i are polynomial functions on $\mathfrak{g}$, with $a_r \neq 0, b_{r'} \neq 0, c_{r''} \neq 0$. We have $\mathrm{P} = \mathrm{QR}$, so $r = r' + r''$ and $a_r(x) = b_{r'}(x)c_{r''}(x)$, which proves (ii) and the first assertion of (i). If $\mathrm{Ker}\, f$ is contained in the centre of $\mathfrak{g}$, $\mathrm{R(T)} = \mathrm{T}^{n''}$ and so $a_r(x) = b_{r'}(x)$, hence the second assertion of (i).

COROLLARY. *Let* $\mathscr{C}_n\mathfrak{g}$ $(n \geq 0)$ *be a term of the ascending central series of* $\mathfrak{g}$ (Chap. I, §1, no. 6). *The regular elements of* $\mathfrak{g}$ *are those whose image in* $\mathfrak{g}/\mathscr{C}_n\mathfrak{g}$ *is regular.*

PROPOSITION 9. *Let* $\mathfrak{g}$ *be a Lie algebra,* $\mathfrak{g}'$ *a subalgebra of* $\mathfrak{g}$. *Every element of* $\mathfrak{g}'$ *regular in* $\mathfrak{g}$ *is regular in* $\mathfrak{g}'$.

For $x \in \mathfrak{g}'$, the restriction of $\mathrm{ad}_{\mathfrak{g}} x$ to $\mathfrak{g}'$ is $\mathrm{ad}_{\mathfrak{g}'} x$, and so defines an endomorphism $u(x)$ of the vector space $\mathfrak{g}/\mathfrak{g}'$ by passage to the quotient. Let $d_0(x)$ (resp. $d_1(x)$) be the dimension of the nilspace of $\mathrm{ad}_{\mathfrak{g}'}(x)$ (resp. of $u(x)$), and let c_0 (resp. c_1) be the minimum of $d_0(x)$ (resp. $d_1(x)$) when x belongs to $\mathfrak{g}'$. There exist non-zero polynomial maps p_0, p_1 from $\mathfrak{g}'$ to k such that

$$d_0(x) = c_0 \iff p_0(x) \neq 0, \quad d_1(x) = c_1 \iff p_1(x) \neq 0.$$

Since k is infinite, the set S of $x \in \mathfrak{g}'$ such that $d_0(x) = c_0$ and $d_1(x) = c_1$ is non-empty. Every element of S is regular in $\mathfrak{g}'$. On the other hand, S is the set of elements of $\mathfrak{g}'$ such that the nilspace of $\mathrm{ad}_{\mathfrak{g}} x$ has minimum dimension, and thus contains every element of $\mathfrak{g}'$ regular in $\mathfrak{g}$.

Remark. 3) Elements of $\mathfrak{g}'$ regular in $\mathfrak{g}$ do not necessarily exist. If at least one does exist, the set of these elements is precisely the set denoted by S in the above proof.

3. CARTAN SUBALGEBRAS AND REGULAR ELEMENTS

THEOREM 1. *Let $\mathfrak{g}$ be a Lie algebra.*

(i) *If x is a regular element of $\mathfrak{g}$, $\mathfrak{g}^0(x)$ is a Cartan subalgebra of $\mathfrak{g}$.*

(ii) *If $\mathfrak{h}$ is a maximal nilpotent subalgebra of $\mathfrak{g}$, and if $x \in \mathfrak{h}$ is regular in $\mathfrak{g}$, then $\mathfrak{h} = \mathfrak{g}^0(x)$.*

(iii) *If $\mathfrak{h}$ is a Cartan subalgebra of $\mathfrak{g}$, then* $\dim(\mathfrak{h}) \geq \mathrm{rk}(\mathfrak{g})$.

(iv) *The Cartan subalgebras of $\mathfrak{g}$ of dimension* $\mathrm{rk}(\mathfrak{g})$ *are the $\mathfrak{g}^0(x)$ where x is a regular element.*

Let x be a regular element of $\mathfrak{g}$ and let $\mathfrak{h} = \mathfrak{g}^0(x)$. Clearly $\mathfrak{h}^0(x) = \mathfrak{h}$. Since x is regular in $\mathfrak{h}$ (Prop. 9), $\mathrm{rk}(\mathfrak{h}) = \dim(\mathfrak{h})$, so $\mathfrak{h}$ is nilpotent. On the other hand, $\mathfrak{h} = \mathfrak{g}^0(x) \supset \mathfrak{g}^0(\mathfrak{h}) \supset \mathfrak{h}$, so $\mathfrak{h} = \mathfrak{g}^0(\mathfrak{h})$ is a Cartan subalgebra of $\mathfrak{g}$ (Prop. 4). This proves (i).

If $\mathfrak{h}$ is a maximal nilpotent subalgebra of $\mathfrak{g}$, and if $x \in \mathfrak{h}$ is regular in $\mathfrak{g}$, then $\mathfrak{h} \subset \mathfrak{g}^0(x)$ and $\mathfrak{g}^0(x)$ is nilpotent by (i), so $\mathfrak{h} = \mathfrak{g}^0(x)$, which proves (ii).

If $\mathfrak{h}$ is a Cartan subalgebra of $\mathfrak{g}$, there exists $x \in \mathfrak{h}$ such that $\mathfrak{h} = \mathfrak{g}^0(x)$ (Cor. 1 of Prop. 4), so $\dim(\mathfrak{h}) \geq \mathrm{rk}(\mathfrak{g})$, which proves (iii). If in addition $\dim(\mathfrak{h}) = \mathrm{rk}(\mathfrak{g})$, x is regular. Finally, if x' is regular in $\mathfrak{g}$, $\mathfrak{g}^0(x')$ is a Cartan subalgebra by (i), and is obviously of dimension $\mathrm{rk}(\mathfrak{g})$. This proves (iv).

We shall see in §3, Th. 2 that, when k is of characteristic zero, all the Cartan subalgebras of $\mathfrak{g}$ have dimension $\mathrm{rk}(\mathfrak{g})$.

COROLLARY 1. *Every Lie algebra $\mathfrak{g}$ has Cartan subalgebras, and the rank of $\mathfrak{g}$ is the minimum dimension of a Cartan subalgebra.*

COROLLARY 2. *Let $f : \mathfrak{g} \to \mathfrak{g}'$ be a surjective homomorphism of Lie algebras. If $\mathfrak{h}'$ is a Cartan subalgebra of $\mathfrak{g}'$, there exists a Cartan subalgebra $\mathfrak{h}$ of $\mathfrak{g}$ such that $\mathfrak{h}' = f(\mathfrak{h})$.*

Let $\mathfrak{a} = f^{-1}(\mathfrak{h}')$. By Cor. 1, $\mathfrak{a}$ has a Cartan subalgebra $\mathfrak{h}$. By Cor. 2 of Prop. 4, $f(\mathfrak{h}) = \mathfrak{h}'$. We show that $\mathfrak{h}$ is a Cartan subalgebra of $\mathfrak{g}$. Let $\mathfrak{n}$ be the normalizer of $\mathfrak{h}$ in $\mathfrak{g}$. It is enough to prove that $\mathfrak{h} = \mathfrak{n}$. If $x \in \mathfrak{n}$, $f(x)$ belongs to the normalizer of $\mathfrak{h}'$ in $\mathfrak{g}'$, so $f(x) \in \mathfrak{h}'$ and $x \in \mathfrak{a}$; but $\mathfrak{h}$ is its own normalizer in $\mathfrak{a}$, so $x \in \mathfrak{h}$.

COROLLARY 3. *Every Lie algebra $\mathfrak{g}$ is the sum of its Cartan subalgebras.*

The sum $\mathfrak{s}$ of the Cartan subalgebras of $\mathfrak{g}$ contains the set of regular elements of $\mathfrak{g}$ (Th. 1 (i)). Since this set is dense in $\mathfrak{g}$ for the Zariski topology, $\mathfrak{s} = \mathfrak{g}$.

PROPOSITION 10. *Let* $\mathfrak{g}$ *be a Lie algebra,* $\mathfrak{a}$ *a commutative subalgebra of* $\mathfrak{g}$ *and* $\mathfrak{c}$ *the commutant of* $\mathfrak{a}$ *in* $\mathfrak{g}$. *Assume that* $\mathrm{ad}_{\mathfrak{g}}x$ *is semi-simple for all* $x \in \mathfrak{a}$. *Then the Cartan subalgebras of* $\mathfrak{c}$ *are the Cartan subalgebras of* $\mathfrak{g}$ *containing* $\mathfrak{a}$.

Let $\mathfrak{h}$ be a Cartan subalgebra of $\mathfrak{c}$. Since $\mathfrak{a}$ is contained in the centre $\mathfrak{z}$ of $\mathfrak{c}$, $\mathfrak{a} \subset \mathfrak{z} \subset \mathfrak{h}$ (Prop. 5). Let $\mathfrak{n}$ be the normalizer of $\mathfrak{h}$ in $\mathfrak{g}$. Then

$$[\mathfrak{a}, \mathfrak{n}] \subset [\mathfrak{h}, \mathfrak{n}] \subset \mathfrak{h}.$$

Since the $\mathrm{ad}_{\mathfrak{g}}x$, $x \in \mathfrak{a}$, are semi-simple and commute with each other, it follows from *Algebra*, Chap. VIII, §5, no. 1, that there exists a vector subspace $\mathfrak{d}$ of $\mathfrak{n}$ stable under $\mathrm{ad}_{\mathfrak{g}}\mathfrak{a}$ and such that $\mathfrak{n} = \mathfrak{h} \oplus \mathfrak{d}$. Then $[\mathfrak{a}, \mathfrak{d}] \subset \mathfrak{h} \cap \mathfrak{d} = 0$, so $\mathfrak{d} \subset \mathfrak{c}$. Thus, $\mathfrak{n}$ is the normalizer of $\mathfrak{h}$ in $\mathfrak{c}$, and hence $\mathfrak{n} = \mathfrak{h}$, so $\mathfrak{h}$ is a Cartan subalgebra of $\mathfrak{g}$ containing $\mathfrak{a}$.

Conversely, let $\mathfrak{h}$ be a Cartan subalgebra of $\mathfrak{g}$ containing $\mathfrak{a}$. Then $\mathfrak{h} = \mathfrak{g}^0(\mathfrak{h}) \subset \mathfrak{g}^0(\mathfrak{a})$, and by hypothesis $\mathfrak{g}_0(\mathfrak{a}) = \mathfrak{g}^0(\mathfrak{a}) = \mathfrak{c}$. Hence $\mathfrak{a} \subset \mathfrak{h} \subset \mathfrak{c}$ and $\mathfrak{h}$ is a Cartan subalgebra of $\mathfrak{c}$ (for it is equal to its own normalizer in $\mathfrak{g}$, and so *a fortiori* in $\mathfrak{c}$).

PROPOSITION 11. *Let* $\mathfrak{n}$ *be a nilpotent subalgebra of a Lie algebra* $\mathfrak{g}$. *There exists a Cartan subalgebra of* $\mathfrak{g}$ *contained in* $\mathfrak{g}^0(\mathfrak{n})$.

Put $\mathfrak{a} = \mathfrak{g}^0(\mathfrak{n})$. Then $\mathfrak{n} \subset \mathfrak{a}$ since $\mathfrak{n}$ is nilpotent. If $x \in \mathfrak{a}$, let $\mathrm{P}(x)$ be the determinant of the endomorphism of $\mathfrak{g}/\mathfrak{a}$ defined by $\mathrm{ad}\,x$. Denote by $\mathfrak{a}'$ the set of $x \in \mathfrak{a}$ such that $\mathrm{P}(x) \neq 0$, which is an open subset of $\mathfrak{a}$ in the Zariski topology; the relations $x \in \mathfrak{a}'$ and $\mathfrak{g}^0(x) \subset \mathfrak{a}$ are equivalent. By Prop. 7 (ii) of §1, no. 2, there exists $y \in \mathfrak{n}$ such that $\mathfrak{g}^0(y) = \mathfrak{a}$, and $y \in \mathfrak{a}'$ so $\mathfrak{a}'$ is non-empty. Since $\mathfrak{a}'$ is open, its intersection with the set of regular elements of $\mathfrak{a}$ is non-empty. Let x be an element of this intersection. Then $\mathfrak{g}^0(x) \subset \mathfrak{a}$ and $\mathfrak{g}^0(x)$ is a Cartan subalgebra of $\mathfrak{a}$, hence is nilpotent. On the other hand, Prop. 10 (i) of §1, no. 3, shows that $\mathfrak{g}^0(x)$ is its own normalizer in $\mathfrak{g}$; it is therefore a Cartan subalgebra of $\mathfrak{g}$, which completes the proof.

4. CARTAN SUBALGEBRAS OF SEMI-SIMPLE LIE ALGEBRAS

THEOREM 2. *Assume that* k *is of characteristic* 0. *Let* $\mathfrak{g}$ *be a semi-simple Lie algebra,* $\mathfrak{h}$ *a Cartan subalgebra of* $\mathfrak{g}$. *Then* $\mathfrak{h}$ *is commutative, and all of its elements are semi-simple in* $\mathfrak{g}$ (Chap. I, §6, no. 3, Def. 3).

Since $\mathfrak{h} = \mathfrak{g}^0(\mathfrak{h})$, $\mathfrak{h}$ is reductive (§1, Prop. 11), hence commutative since it is nilpotent. On the other hand, the restriction of the adjoint representation of $\mathfrak{g}$ to $\mathfrak{h}$ is semi-simple (*loc. cit.*), so the elements of $\mathfrak{h}$ are semi-simple in $\mathfrak{g}$ (*Algebra*, Chap. VIII, §5, no. 1).

COROLLARY 1. *If $x \in \mathfrak{h}$ and $y \in \mathfrak{g}^\lambda(\mathfrak{h})$, we have $[x, y] = \lambda(x)y$.*

Indeed, $\mathfrak{g}^{\lambda(x)}(x) = \mathfrak{g}_{\lambda(x)}(x)$ since $\operatorname{ad} x$ is semi-simple.

COROLLARY 2. *Every regular element of $\mathfrak{g}$ is semi-simple.*

Indeed, such an element belongs to a Cartan subalgebra (no. 3, Th. 1 (i)).

COROLLARY 3. *Let $\mathfrak{h}$ be a Cartan subalgebra of a reductive Lie algebra $\mathfrak{g}$.*

a) $\mathfrak{h}$ is commutative.

b) If ρ is a finite dimensional semi-simple representation of $\mathfrak{g}$, the elements of $\rho(\mathfrak{h})$ are semi-simple.

Let $\mathfrak{c}$ be the centre of $\mathfrak{g}$, and $\mathfrak{s}$ its derived algebra. Then $\mathfrak{g} = \mathfrak{c} \times \mathfrak{s}$, so $\mathfrak{h} = \mathfrak{c} \times \mathfrak{h}'$, where $\mathfrak{h}'$ is a Cartan subalgebra of $\mathfrak{s}$ (Prop. 2). In view of Th. 2, $\mathfrak{h}'$ is commutative, hence so is $\mathfrak{h}$. Moreover, $\rho(\mathfrak{h}')$ consists of semi-simple elements and so does $\rho(\mathfrak{c})$ (Chap. I, §6, no. 5, Th. 4); assertion *b*) follows.

§3. CONJUGACY THEOREMS

In this paragraph, the base field k is of characteristic 0.

1. ELEMENTARY AUTOMORPHISMS

Let $\mathfrak{g}$ be a Lie algebra. Denote its group of automorphisms by $\operatorname{Aut}(\mathfrak{g})$. If $x \in \mathfrak{g}$ and if $\operatorname{ad} x$ is nilpotent, $e^{\operatorname{ad} x} \in \operatorname{Aut}(\mathfrak{g})$ (Chap. I, §6, no. 8).

DEFINITION 1. *A finite product of automorphisms of $\mathfrak{g}$ of the form $e^{\operatorname{ad} x}$ with $\operatorname{ad} x$ nilpotent is called an elementary automorphism of $\mathfrak{g}$. The group of elementary automorphisms of $\mathfrak{g}$ is denoted by $\operatorname{Aut}_e(\mathfrak{g})$.*

If $u \in \operatorname{Aut}(\mathfrak{g})$, $ue^{\operatorname{ad} x}u^{-1} = e^{\operatorname{ad} u(x)}$. It follows that $\operatorname{Aut}_e(\mathfrak{g})$ is a normal subgroup of $\operatorname{Aut}(\mathfrak{g})$. If $k = \mathbf{R}$ or $\mathbf{C}$, $\operatorname{Aut}_e(\mathfrak{g})$ is contained in the group $\operatorname{Int}(\mathfrak{g})$ of inner automorphisms of $\mathfrak{g}$ (Chap. III, §6, no. 2, Def. 2).

* In the general case, $\operatorname{Aut}_e(\mathfrak{g})$ is contained in the identity component of the algebraic group $\operatorname{Aut}(\mathfrak{g})$.*

Lemma 1. Let V *be a finite dimensional vector space, $\mathfrak{n}$ a Lie subalgebra of $\mathfrak{a} = \mathfrak{gl}(\mathrm{V})$ consisting of nilpotent elements.*

(i) *The map $x \mapsto \exp x$ is a bijection from $\mathfrak{n}$ to a subgroup* N *of* $\mathbf{GL}(\mathrm{V})$ *consisting of unipotent elements* (Chap. II, §6, no. 1, Remark 4). *We have $\mathfrak{n} = \log(\exp \mathfrak{n})$. The map $f \mapsto f \circ \log$ is an isomorphism from the algebra of* *polynomial functions on* $\mathfrak{n}$ *to the algebra of restrictions to* N *of polynomial functions on* End(V).

(ii) *If* $x \in \mathfrak{n}$ *and* $a \in \mathfrak{a}$,

$$(\exp \operatorname{ad}_{\mathfrak{a}} x).a = (\exp x)a(\exp(-x)).$$

(iii) *Let* V′ *be a finite dimensional vector space,* $\mathfrak{n}'$ *a Lie subalgebra of* $\mathfrak{gl}(V')$ *consisting of nilpotent elements,* ρ *a homomorphism from* $\mathfrak{n}$ *to* $\mathfrak{n}'$. *Let* π *be the map* $\exp x \mapsto \exp \rho(x)$ *from* $\exp \mathfrak{n}$ *to* $\exp \mathfrak{n}'$. *Then* π *is a group homomorphism.*

By Engel's theorem, we can identify V with k^n in such a way that $\mathfrak{n}$ is a subalgebra of $\mathfrak{n}(n,k)$ (the Lie subalgebra of $\mathbf{M}_n(k)$ consisting of the lower triangular matrices with zeros on the diagonal). For $s \geq 0$, let $\mathfrak{n}_s(n,k)$ be the set of $(x_{ij})_{1 \leq i,j \leq n} \in \mathbf{M}_n(k)$ such that $x_{ij} = 0$ for $i - j < s$. Then

$$[\mathfrak{n}_s(n,k), \mathfrak{n}_{s'}(n,k)] \subset \mathfrak{n}_{s+s'}(n,k)$$

(Chap. II, §4, no. 6, Remark), and the Hausdorff series defines a polynomial map $(a,b) \mapsto \mathrm{H}(a,b)$ from $\mathfrak{n}(n,k) \times \mathfrak{n}(n,k)$ to $\mathfrak{n}(n,k)$ (Chap. II, §6, no. 5, Remark 3); this map makes $\mathfrak{n}(n,k)$ into a group (Chap. II, §6, no. 5, Prop. 4). By Chap. II, §6, no. 1, Remark 4, the maps $x \mapsto \exp x$ from $\mathfrak{n}(n,k)$ to $1+\mathfrak{n}(n,k)$ and $y \mapsto \log y$ from $1+\mathfrak{n}(n,k)$ to $\mathfrak{n}(n,k)$ are inverse bijections and are polynomial; by Chap. II, §6, no. 5, Prop. 3, these maps are isomorphisms of groups if $\mathfrak{n}(n,k)$ is given the group law $(a,b) \mapsto \mathrm{H}(a,b)$ and if $1+\mathfrak{n}(n,k)$ is considered as a subgroup of $\mathbf{GL}_n(k)$. Assertions (i) and (iii) of the lemma now follow. Let $x \in \mathfrak{n}$. Denote by $\mathrm{L}_x, \mathrm{R}_x$ the maps $u \mapsto xu, u \mapsto ux$ from $\mathfrak{a}$ to $\mathfrak{a}$, which commute and are nilpotent. We have $\operatorname{ad}_{\mathfrak{a}} x = \mathrm{L}_x - \mathrm{R}_x$, so, for all $a \in \mathfrak{a}$,

$$\begin{aligned}(\exp \operatorname{ad}_{\mathfrak{a}} x)a &= (\exp(\mathrm{L}_x - \mathrm{R}_x))a = (\exp \mathrm{L}_x)(\exp \mathrm{R}_{-x})a \qquad (1)\\ &= \sum_{i,j \geq 0} \frac{\mathrm{L}_x^i}{i!}\frac{\mathrm{R}_{-x}^j}{j!}a = (\exp x)a(\exp(-x)).\end{aligned}$$

With the notation in Lemma 1, π is called the linear representation of $\exp \mathfrak{n}$ *compatible* with the given representation ρ of $\mathfrak{n}$ on V′. When k is $\mathbf{R}$, $\mathbf{C}$, or a non-discrete complete ultrametric field, $\rho = \mathrm{L}(\pi)$ by the properties of exponential maps (Chap. III, §4, no. 4, Cor. 2 of Prop. 8).

PROPOSITION 1. *Let* $\mathfrak{g}$ *be a Lie algebra,* $\mathfrak{n}$ *a subalgebra of* $\mathfrak{g}$ *such that* $\operatorname{ad}_{\mathfrak{g}} x$ *is nilpotent for all* $x \in \mathfrak{n}$. *Then* $e^{\operatorname{ad}_{\mathfrak{g}} \mathfrak{n}}$ *is a subgroup of* $\operatorname{Aut}_e(\mathfrak{g})$.

This follows immediately from Lemma 1 (i).

In particular, if $\mathfrak{n}$ is the nilpotent radical of $\mathfrak{g}$, $e^{\operatorname{ad}_{\mathfrak{g}} \mathfrak{n}}$ is the group of *special automorphisms* of $\mathfrak{g}$ (Chap. I, §6, no. 8, Def. 6).

Remarks. 1) Let V be a finite dimensional vector space, $\mathfrak{g}$ a Lie subalgebra of $\mathfrak{a} = \mathfrak{gl}(V)$, x an element of $\mathfrak{g}$ such that $\mathrm{ad}_{\mathfrak{g}}x$ is nilpotent. Then there exists a nilpotent element n of $\mathfrak{a}$ such that $\mathrm{ad}_{\mathfrak{a}}n$ extends $\mathrm{ad}_{\mathfrak{g}}x$. Indeed, let s, n be the semi-simple and nilpotent components of x; then $\mathrm{ad}_{\mathfrak{a}}s$ and $\mathrm{ad}_{\mathfrak{a}}n$ are the semi-simple and nilpotent components of $\mathrm{ad}_{\mathfrak{a}}x$ (Chap. I, §5, no. 4, Lemma 2), so $\mathrm{ad}_{\mathfrak{a}}s$ and $\mathrm{ad}_{\mathfrak{a}}n$ leave $\mathfrak{g}$ stable, and $\mathrm{ad}_{\mathfrak{a}}s|\mathfrak{g}$ and $\mathrm{ad}_{\mathfrak{a}}n|\mathfrak{g}$ are the semi-simple and nilpotent components of $\mathrm{ad}_{\mathfrak{g}}x$; consequently, $\mathrm{ad}_{\mathfrak{g}}x = \mathrm{ad}_{\mathfrak{a}}n|\mathfrak{g}$, which proves our assertion. In view of Lemma 1 (ii), every elementary automorphism of $\mathfrak{g}$ extends to an automorphism of $\mathfrak{a}$ of the form $u \mapsto mum^{-1}$ where $m \in \mathbf{GL}(V)$.

2) Let V be a finite dimensional vector space. For all $g \in \mathbf{SL}(V)$, let $\varphi(g)$ be the automorphism $x \mapsto gxg^{-1}$ of $\mathfrak{gl}(V)$. Then

$$\mathrm{Aut}_e(\mathfrak{gl}(V)) = \varphi(\mathbf{SL}(V)).$$

Indeed, by (1), $\mathrm{Aut}_e(\mathfrak{gl}(V))$ is contained in $\varphi(\mathbf{SL}(V))$, and the opposite inclusion follows from *Algebra*, Chap. III, §8, no. 9, Prop. 17 and (1). An analogous argument shows that $\mathrm{Aut}_e(\mathfrak{sl}(V))$ is the set of restrictions of elements of $\varphi(\mathbf{SL}(V))$ to $\mathfrak{sl}(V)$.

2. CONJUGACY OF CARTAN SUBALGEBRAS

Let $\mathfrak{g}$ be a Lie algebra, $\mathfrak{h}$ a nilpotent subalgebra of $\mathfrak{g}$ and R the set of non-zero weights of $\mathfrak{h}$ in $\mathfrak{g}$, in other words the set of linear forms $\lambda \neq 0$ on $\mathfrak{h}$ such that $\mathfrak{g}^{\lambda}(\mathfrak{h}) \neq 0$, cf. §1, no. 3, Prop. 9 (iii). Assume that

$$\mathfrak{g} = \mathfrak{g}^0(\mathfrak{h}) \oplus \sum_{\lambda \in \mathrm{R}} \mathfrak{g}^{\lambda}(\mathfrak{h}),$$

which is the case if k is algebraically closed (§1, no. 3, Prop. 9 (i)). For $\lambda \in \mathrm{R}$ and $x \in \mathfrak{g}^{\lambda}(\mathfrak{h})$, $\mathrm{ad}\, x$ is nilpotent (§1, no. 3, Prop. 10 (iv)). Denote by $\mathrm{E}(\mathfrak{h})$ the subgroup of $\mathrm{Aut}_e(\mathfrak{g})$ generated by the $e^{\mathrm{ad}\, x}$ where x is of the form above. If $u \in \mathrm{Aut}(\mathfrak{g})$, it is immediate that $u\mathrm{E}(\mathfrak{h})u^{-1} = \mathrm{E}(u(\mathfrak{h}))$.

Lemma 2. (i) *Let $\mathfrak{h}_r$ be the set of $x \in \mathfrak{h}$ such that $\mathfrak{g}^0(x) = \mathfrak{g}^0(\mathfrak{h})$; this is the set of $x \in \mathfrak{h}$ such that $\lambda(x) \neq 0$ for all $\lambda \in \mathrm{R}$, and $\mathfrak{h}_r$ is open and dense in $\mathfrak{h}$ in the Zariski topology.*

(ii) *Put $\mathrm{R} = \{\lambda_1, \lambda_2, \ldots, \lambda_p\}$ where the λ_i are mutually distinct. Let F be the map from $\mathfrak{g}^0(\mathfrak{h}) \times \mathfrak{g}^{\lambda_1}(\mathfrak{h}) \times \cdots \times \mathfrak{g}^{\lambda_p}(\mathfrak{h})$ to $\mathfrak{g}$ defined by the formula*

$$\mathrm{F}(h, x_1, \ldots, x_p) = e^{\mathrm{ad}\, x_1} \ldots e^{\mathrm{ad}\, x_p} h.$$

Then F *is a dominant polynomial map* (App. I).

Assertion (i) is clear. We prove (ii). Let $n = \dim \mathfrak{g}$. If $\lambda \in \mathrm{R}$ and $x \in \mathfrak{g}^{\lambda}(\mathfrak{h})$, we have $(\mathrm{ad}\, x)^n = 0$. It follows that $(y, x) \mapsto e^{\mathrm{ad}\, x}y$ is a polynomial map from $\mathfrak{g} \times \mathfrak{g}^{\lambda}(\mathfrak{h})$ to $\mathfrak{g}$; it follows by induction that F is polynomial. Let $h_0 \in \mathfrak{h}_r$ and let DF be the tangent linear map of F at $(h_0, 0, \ldots, 0)$; we show that DF is surjective. For $h \in \mathfrak{g}^0(\mathfrak{h})$, $\mathrm{F}(h_0+h,0,\dots,0) = h_0+h$, so $\mathrm{DF}(h,0,\dots,0) = h$ and $\mathrm{Im}(\mathrm{DF}) \supset \mathfrak{g}^0(\mathfrak{h})$. On the other hand, for $x \in \mathfrak{g}^{\lambda_1}(\mathfrak{h})$,

$$\mathrm{F}(h_0,x,0,\dots,0) = e^{\operatorname{ad} x}h_0 = h_0 + (\operatorname{ad} x).h_0 + \frac{(\operatorname{ad} x)^2}{2!}h_0 + \cdots$$

so $\mathrm{DF}(0,x,0,\dots,0) = (\operatorname{ad} x).h_0 = -(\operatorname{ad} h_0)x$; since $\operatorname{ad} h_0$ induces an automorphism of $\mathfrak{g}^{\lambda_1}(\mathfrak{h})$, $\mathrm{Im}(\mathrm{DF}) \supset \mathfrak{g}^{\lambda_1}(\mathfrak{h})$. Similarly,

$$\mathrm{Im}(\mathrm{DF}) \supset \mathfrak{g}^{\lambda_i}(\mathfrak{h})$$

for all i, hence the surjectivity of DF. Prop. 4 of App. I now shows that F is dominant.

PROPOSITION 2. *Assume that k is algebraically closed. Let $\mathfrak{g}$ be a Lie algebra, $\mathfrak{h}$ and $\mathfrak{h}'$ Cartan subalgebras of $\mathfrak{g}$. There exist $u \in \mathrm{E}(\mathfrak{h})$ and $u' \in \mathrm{E}(\mathfrak{h}')$ such that $u(\mathfrak{h}) = u'(\mathfrak{h}')$.*

We retain the notation of Lemma 2. From the fact that $\mathfrak{h}$ and $\mathfrak{h}'$ are Cartan subalgebras, it follows that $\mathfrak{g}^0(\mathfrak{h}) = \mathfrak{h}$ and $\mathfrak{g}^0(\mathfrak{h}') = \mathfrak{h}'$. By Lemma 2 and Prop. 3 of App. I, $\mathrm{E}(\mathfrak{h})\mathfrak{h}_r$ and $\mathrm{E}(\mathfrak{h}')\mathfrak{h}'_r$ contain open dense subsets of $\mathfrak{g}$ in the Zariski topology. Thus $\mathrm{E}(\mathfrak{h})\mathfrak{h}_r \cap \mathrm{E}(\mathfrak{h}')\mathfrak{h}'_r \neq \varnothing$. In other words, there exist $u \in \mathrm{E}(\mathfrak{h}), u' \in \mathrm{E}(\mathfrak{h}'), h \in \mathfrak{h}_r, h' \in \mathfrak{h}'_r$ such that $u(h) = u'(h')$; then

$$u(\mathfrak{h}) = u(\mathfrak{g}^0(\mathfrak{h})) = \mathfrak{g}^0(u(h)) = \mathfrak{g}^0(u'(h')) = u'(\mathfrak{h}').$$

COROLLARY. $\mathrm{E}(\mathfrak{h}) = \mathrm{E}(\mathfrak{h}')$.

Let u, u' be as in Prop. 2. Then

$$\mathrm{E}(\mathfrak{h}) = u\mathrm{E}(\mathfrak{h})u^{-1} = \mathrm{E}(u(\mathfrak{h})) = \mathrm{E}(u'(\mathfrak{h}')) = u'\mathrm{E}(\mathfrak{h}')u'^{-1} = \mathrm{E}(\mathfrak{h}'),$$

hence the corollary.

Because of this result, if k is algebraically closed we shall denote simply by E the group $\mathrm{E}(\mathfrak{h})$, where $\mathfrak{h}$ is a Cartan subalgebra of $\mathfrak{g}$.

In general, $\mathrm{Aut}_e(\mathfrak{g}) \neq \mathrm{E}$ (for example, if $\mathfrak{g}$ is nilpotent, E reduces to the identity element, even though non-trivial elementary automorphisms exist provided $\mathfrak{g}$ is non-commutative). However, it can be shown (Chap. VIII, §10, Exerc. 5) that $\mathrm{Aut}_e(\mathfrak{g}) = \mathrm{E}$ for $\mathfrak{g}$ semi-simple.

THEOREM 1. *Assume that k is algebraically closed. Let $\mathfrak{g}$ be a Lie algebra. The group E is normal in $\mathrm{Aut}(\mathfrak{g})$ and operates transitively on the set of Cartan subalgebras of $\mathfrak{g}$.*

Let $\mathfrak{h}$ be a Cartan subalgebra of $\mathfrak{g}$, and $v \in \mathrm{Aut}(\mathfrak{g})$. Then

$$v\mathrm{E}(\mathfrak{h})v^{-1} = \mathrm{E}(v(\mathfrak{h})) = \mathrm{E}(\mathfrak{h}),$$

so $\mathrm{E}(\mathfrak{h}) = \mathrm{E}$ is normal in $\mathrm{Aut}(\mathfrak{g})$. If $\mathfrak{h}'$ is another Cartan subalgebra of $\mathfrak{g}$, then, in the notation of Prop. 2, $u'^{-1}u(\mathfrak{h}) = \mathfrak{h}'$, and $u'^{-1}u \in \mathrm{E}$.

3. APPLICATIONS OF CONJUGACY

THEOREM 2. *Let* $\mathfrak{g}$ *be a Lie algebra.*

(i) *The Cartan subalgebras of* $\mathfrak{g}$ *are all of the same dimension, namely* $\mathrm{rk}(\mathfrak{g})$, *and the same nilpotency class.*

(ii) *An element* $x \in \mathfrak{g}$ *is regular if and only if* $\mathfrak{g}^0(x)$ *is a Cartan subalgebra of* $\mathfrak{g}$; *every Cartan subalgebra is obtained in this way.*

To prove (i), we can assume that k is algebraically closed (cf. §2, Prop. 3 and Prop. 6), in which case it follows from Th. 1 of no. 2. Assertion (ii) follows from (i) and §2, Th. 1 (i) and (iv).

PROPOSITION 3. *Let* $\mathfrak{g}$ *be a Lie algebra,* $\mathfrak{g}'$ *a subalgebra of* $\mathfrak{g}$. *The following conditions are equivalent:*

(i) $\mathfrak{g}'$ *contains a regular element of* $\mathfrak{g}$, *and* $\mathrm{rk}(\mathfrak{g}) = \mathrm{rk}(\mathfrak{g}')$;

(ii) $\mathfrak{g}'$ *contains a Cartan subalgebra of* $\mathfrak{g}$;

(iii) *every Cartan subalgebra of* $\mathfrak{g}'$ *is a Cartan subalgebra of* $\mathfrak{g}$.

(i) $\Longrightarrow$ (ii): Assume that $\mathrm{rk}(\mathfrak{g}) = \mathrm{rk}(\mathfrak{g}')$, and that there exists $x \in \mathfrak{g}'$ regular in $\mathfrak{g}$. Put $\mathfrak{h} = \mathfrak{g}^0(x), \mathfrak{h}' = \mathfrak{g}'^0(x) = \mathfrak{h} \cap \mathfrak{g}'$. Then

$$\mathrm{rk}(\mathfrak{g}') \leq \dim \mathfrak{h}' \leq \dim \mathfrak{h} = \mathrm{rk}(\mathfrak{g}) = \mathrm{rk}(\mathfrak{g}')$$

so $\mathfrak{h} = \mathfrak{h}' \subset \mathfrak{g}'$. This proves (ii).

(ii) $\Longrightarrow$ (iii): Assume that $\mathfrak{g}'$ contains a Cartan subalgebra $\mathfrak{h}$ of $\mathfrak{g}$, and let $\mathfrak{h}_1$ be a Cartan subalgebra of $\mathfrak{g}'$. To prove that $\mathfrak{h}_1$ is a Cartan subalgebra of $\mathfrak{g}$, we can assume that k is algebraically closed. Let $\mathrm{E}(\mathfrak{h})$ and $\mathrm{E}'(\mathfrak{h})$ be the groups of automorphisms of $\mathfrak{g}$ and $\mathfrak{g}'$ associated to $\mathfrak{h}$ (no. 2). By Th. 1, there exists $f \in \mathrm{E}'(\mathfrak{h})$ such that $f(\mathfrak{h}) = \mathfrak{h}_1$. Now every element of $\mathrm{E}'(\mathfrak{h})$ is induced by an element of $\mathrm{E}(\mathfrak{h})$; indeed, it suffices to verify this for $e^{\mathrm{ad}\, x}$, with $x \in \mathfrak{g}'^\lambda(\mathfrak{h})$, $\lambda \neq 0$, in which case it follows from the inclusion $\mathfrak{g}'^\lambda(\mathfrak{h}) \subset \mathfrak{g}^\lambda(\mathfrak{h})$. Thus $\mathfrak{h}_1$ is a Cartan subalgebra of $\mathfrak{g}$.

(iii) $\Longrightarrow$ (i): Assume that condition (iii) is satisfied. Let $\mathfrak{h}$ be a Cartan subalgebra of $\mathfrak{g}'$. Since this is a Cartan subalgebra of $\mathfrak{g}$, it contains a regular element of $\mathfrak{g}$ (Th. 2 (ii)), and on the other hand $\mathrm{rk}(\mathfrak{g}) = \dim(\mathfrak{h}) = \mathrm{rk}(\mathfrak{g}')$.

COROLLARY. *Let* $\mathfrak{h}$ *be a nilpotent subalgebra of* $\mathfrak{g}$. *The subalgebra* $\mathfrak{g}^0(\mathfrak{h})$ *has properties* (i), (ii), (iii) *in Prop. 3.*

Indeed, Prop. 11 of §2, no. 3, shows that $\mathfrak{g}^0(\mathfrak{h})$ has property (ii).

PROPOSITION 4. *Let* $\mathfrak{g}$ *be a Lie algebra,* l *the rank of* $\mathfrak{g}$, c *the nilpotency class of the Cartan subalgebras of* $\mathfrak{g}$, *and* $x \in \mathfrak{g}$. *There exists an* l*-dimensional subalgebra of* $\mathfrak{g}$ *whose nilpotency class is* $\leq c$ *and which contains* x.

Let T be an indeterminate. Let $k' = k(\mathrm{T})$ and $\mathfrak{g}' = \mathfrak{g} \otimes_k k'$. If $\mathfrak{h}$ is a Cartan subalgebra of $\mathfrak{g}$, $\mathfrak{h} \otimes_k k'$ is a Cartan subalgebra of $\mathfrak{g}'$, hence the rank of $\mathfrak{g}'$ is l and the nilpotency class of the Cartan subalgebras of $\mathfrak{g}'$ is c.

Choose a regular element y of $\mathfrak{g}$. With the notations of §2, no. 2, we have $a_l(y) \neq 0$. Denote also by a_l the polynomial function on $\mathfrak{g}'$ that extends a_l. Then the element $a_l(x + \mathrm{T}y)$ of $k[\mathrm{T}]$ has dominant coefficient $a_l(y)$. In particular, $x + \mathrm{T}y$ is regular in $\mathfrak{g}'$. Let $\mathfrak{h}'$ be the nilspace of $\mathrm{ad}(x + \mathrm{T}y)$ in $\mathfrak{g}'$. Then $\dim \mathfrak{h}' = l$ and the nilpotency class of $\mathfrak{h}'$ is c. Put $\mathfrak{k} = \mathfrak{h}' \cap (\mathfrak{g} \otimes_k k[\mathrm{T}])$; then $\mathfrak{k} \otimes_{k[\mathrm{T}]} k(\mathrm{T}) = \mathfrak{h}'$.

Let φ be the homomorphism from $k[\mathrm{T}]$ to k such that $\varphi(\mathrm{T}) = 0$, and let ψ be the homomorphism $1 \otimes \varphi$ from $\mathfrak{g} \otimes_k k[\mathrm{T}]$ to $\mathfrak{g}$. Then $\psi(\mathfrak{k})$ is a subalgebra of $\mathfrak{g}$ whose nilpotency class is $\leq c$ and which contains $\psi(x + \mathrm{T}y) = x$.

In the free $k[\mathrm{T}]$-module $\mathfrak{g} \otimes_k k[\mathrm{T}]$, $\mathfrak{k}$ is a submodule of rank l, and $(\mathfrak{g} \otimes_k k[\mathrm{T}])/\mathfrak{k}$ is torsion free, so the submodule $\mathfrak{k}$ is a direct summand of $\mathfrak{g} \otimes_k k[\mathrm{T}]$ (*Algebra*, Chap. VII, §4, no. 2, Th. 1). Hence $\dim_k \psi(\mathfrak{k}) = l$, which completes the proof.

4. CONJUGACY OF CARTAN SUBALGEBRAS OF SOLVABLE LIE ALGEBRAS

Let $\mathfrak{g}$ be a solvable Lie algebra. Denote by $\mathscr{C}^\infty(\mathfrak{g})$ the intersection of the terms of the descending central series of $\mathfrak{g}$ (Chap. I, §1, no. 5). This is a characteristic ideal of $\mathfrak{g}$, and is the smallest ideal $\mathfrak{m}$ of $\mathfrak{g}$ such that $\mathfrak{g}/\mathfrak{m}$ is nilpotent. Since $\mathscr{C}^\infty(\mathfrak{g}) \subset [\mathfrak{g}, \mathfrak{g}]$, $\mathscr{C}^\infty(\mathfrak{g})$ is a nilpotent ideal of $\mathfrak{g}$ (Chap. I, §5, no. 3, Cor. 5 of Th. 1). By Prop. 1 of no. 1, the set of $e^{\mathrm{ad}\, x}$, for $x \in \mathscr{C}^\infty(\mathfrak{g})$, is a subgroup of $\mathrm{Aut}(\mathfrak{g})$ contained in the group of special automorphisms (Chap. I, §6, no. 8, Def. 6).

THEOREM 3. *Let $\mathfrak{g}$ be a solvable Lie algebra, and let $\mathfrak{h}, \mathfrak{h}'$ be Cartan subalgebras of $\mathfrak{g}$. There exists $x \in \mathscr{C}^\infty(\mathfrak{g})$ such that $e^{\mathrm{ad}\, x}\mathfrak{h} = \mathfrak{h}'$.*

We argue by induction on $\dim \mathfrak{g}$, the case where $\mathfrak{g} = 0$ being trivial. Let $\mathfrak{n}$ be a minimal non-zero commutative ideal of $\mathfrak{g}$. Let $\varphi : \mathfrak{g} \to \mathfrak{g}/\mathfrak{n}$ be the canonical morphism. Then $\varphi(\mathscr{C}^\infty\mathfrak{g}) = \mathscr{C}^\infty(\mathfrak{g}/\mathfrak{n})$ (Chap. I, §1, no. 5, Prop. 4). Since $\varphi(\mathfrak{h})$ and $\varphi(\mathfrak{h}')$ are Cartan subalgebras of $\mathfrak{g}/\mathfrak{n}$ (§2, no. 1, Cor. 2 of Prop. 4), there exists, by the induction hypothesis, an $x \in \mathscr{C}^\infty(\mathfrak{g})$ such that $e^{\mathrm{ad}\, \varphi(x)}\varphi(\mathfrak{h}) = \varphi(\mathfrak{h}')$. Replacing $\mathfrak{h}$ by $e^{\mathrm{ad}\, x}\mathfrak{h}$, we can assume that $\varphi(\mathfrak{h}) = \varphi(\mathfrak{h}')$, in other words that

$$\mathfrak{h} + \mathfrak{n} = \mathfrak{h}' + \mathfrak{n}.$$

Then $\mathfrak{h}$ and $\mathfrak{h}'$ are Cartan subalgebras of $\mathfrak{h} + \mathfrak{n}$. If $\mathfrak{h} + \mathfrak{n} \neq \mathfrak{g}$, the assertion to be proved follows from the induction hypothesis. Assume from now on that $\mathfrak{h} + \mathfrak{n} = \mathfrak{h}' + \mathfrak{n} = \mathfrak{g}$.

By the minimality of $\mathfrak{n}$, $[\mathfrak{g}, \mathfrak{n}] = \{0\}$ or $[\mathfrak{g}, \mathfrak{n}] = \mathfrak{n}$. If $[\mathfrak{g}, \mathfrak{n}] = \{0\}$, then $\mathfrak{n} \subset \mathfrak{h}$ and $\mathfrak{n} \subset \mathfrak{h}'$ (§2, no. 1, Prop. 5), so $\mathfrak{h} = \mathfrak{h} + \mathfrak{n} = \mathfrak{h}' + \mathfrak{n} = \mathfrak{h}'$. It remains to consider the case where $[\mathfrak{g}, \mathfrak{n}] = \mathfrak{n}$, so $\mathfrak{n} \subset \mathscr{C}^\infty(\mathfrak{g})$. The ideal $\mathfrak{n}$ is a simple $\mathfrak{g}$-module; since $\mathfrak{g} = \mathfrak{h} + \mathfrak{n}$, and since $[\mathfrak{n}, \mathfrak{n}] = \{0\}$, it follows that $\mathfrak{n}$ is a simple $\mathfrak{h}$-module. If $\mathfrak{h} \cap \mathfrak{n} \neq \{0\}$, then $\mathfrak{n} \subset \mathfrak{h}$, so $\mathfrak{g} = \mathfrak{h}$ and $\mathfrak{h}' = \mathfrak{h}$. Assume now that

$\mathfrak{h} \cap \mathfrak{n} = \{0\}$. Then $\mathfrak{g} = \mathfrak{h} \oplus \mathfrak{n}$ and hence $\mathfrak{g} = \mathfrak{h}' \oplus \mathfrak{n}$, since $\mathfrak{h}$ and $\mathfrak{h}'$ have the same dimension.

For all $x \in \mathfrak{h}$, let $f(x)$ be the unique element of $\mathfrak{n}$ such that $x - f(x) \in \mathfrak{h}'$; if $x, y \in \mathfrak{h}$,

$$[x,y] - [x, f(y)] - [f(x), y] = [x - f(x), y - f(y)] \in \mathfrak{h}',$$

so $f([x,y]) = [x, f(y)] + [f(x), y]$. By §1, no. 3, Cor. of Prop. 9, there exists $a \in \mathfrak{n}$ such that $f(x) = [x, a]$ for all $x \in \mathfrak{h}$. We have $(\operatorname{ad} a)^2(\mathfrak{g}) \subset (\operatorname{ad} a)(\mathfrak{n}) = 0$, so, for all $x \in \mathfrak{h}$,

$$e^{\operatorname{ad} a} x = x + [a, x] = x - f(x).$$

Thus $e^{\operatorname{ad} a}(\mathfrak{h}) = \mathfrak{h}'$. Since $a \in \mathscr{C}^{\infty}(\mathfrak{g})$, this completes the proof.

Lemma 3. Let $\mathfrak{g}$ be a Lie algebra, $\mathfrak{r}$ its radical, φ the canonical homomorphism from $\mathfrak{g}$ to $\mathfrak{g}/\mathfrak{r}$, v an elementary automorphism of $\mathfrak{g}/\mathfrak{r}$. There exists an elementary automorphism u of $\mathfrak{g}$ such that $\varphi \circ u = v \circ \varphi$.

We can assume that v is of the form $e^{\operatorname{ad} b}$, where $b \in \mathfrak{g}/\mathfrak{r}$ and $\operatorname{ad} b$ is nilpotent. Let $\mathfrak{s}$ be a Levi subalgebra of $\mathfrak{g}$ (Chap. I, §6, no. 8, Def. 7) and let a be the element of $\mathfrak{s}$ such that $\varphi(a) = b$. Since $\operatorname{ad}_{\mathfrak{s}} a$ is nilpotent, $\operatorname{ad}_{\mathfrak{g}} a$ is nilpotent (Chap. I, §6, no. 3, Cor. of Prop. 3), and $u = e^{\operatorname{ad}_{\mathfrak{g}} a}$ is an elementary automorphism of $\mathfrak{g}$ such that $\varphi \circ u = v \circ \varphi$.

PROPOSITION 5. *Let $\mathfrak{g}$ be a Lie algebra, $\mathfrak{r}$ its radical, $\mathfrak{h}$ and $\mathfrak{h}'$ Cartan subalgebras of $\mathfrak{g}$, and φ the canonical homomorphism from $\mathfrak{g}$ to $\mathfrak{g}/\mathfrak{r}$. The following conditions are equivalent:*

(i) *$\mathfrak{h}$ and $\mathfrak{h}'$ are conjugate by an elementary automorphism of $\mathfrak{g}$;*

(ii) *$\varphi(\mathfrak{h})$ and $\varphi(\mathfrak{h}')$ are conjugate by an elementary automorphism of $\mathfrak{g}/\mathfrak{r}$.*

(i) $\Longrightarrow$ (ii): This is clear.

(ii) $\Longrightarrow$ (i): We assume that condition (ii) is satisfied and prove (i). By Lemma 3, we are reduced to the case where $\varphi(\mathfrak{h}) = \varphi(\mathfrak{h}')$. Put $\mathfrak{k} = \mathfrak{h} + \mathfrak{r} = \mathfrak{h}' + \mathfrak{r}$, which is a solvable subalgebra of $\mathfrak{g}$. Then $\mathfrak{h}$ and $\mathfrak{h}'$ are Cartan subalgebras of $\mathfrak{k}$, so there exists $x \in \mathscr{C}^{\infty}(\mathfrak{k})$ such that $e^{\operatorname{ad}_{\mathfrak{k}} x}\mathfrak{h} = \mathfrak{h}'$ (Th. 3). Since $\mathfrak{k}/\mathfrak{r}$ is nilpotent, $\mathscr{C}^{\infty}(\mathfrak{k}) \subset \mathfrak{r}$; on the other hand, $\mathscr{C}^{\infty}(\mathfrak{k}) \subset [\mathfrak{k}, \mathfrak{k}] \subset [\mathfrak{g}, \mathfrak{g}]$, so $x \in \mathfrak{r} \cap [\mathfrak{g}, \mathfrak{g}]$; by Chap. I, §5, no. 3, Th. 1, $\operatorname{ad}_{\mathfrak{g}} x$ is nilpotent, so $e^{\operatorname{ad}_{\mathfrak{g}} x}$ is an elementary automorphism of $\mathfrak{g}$ transforming $\mathfrak{h}$ to $\mathfrak{h}'$.

5. LIE GROUP CASE

PROPOSITION 6. *Assume that k is* **R**, **C** *or a non-discrete complete ultrametric field of characteristic 0. Let G be a finite dimensional Lie group over k, e its identity element, $\mathfrak{g}$ its Lie algebra, $\mathfrak{h}$ a Cartan subalgebra of $\mathfrak{g}$, $\mathfrak{h}_r$ the set of regular elements of $\mathfrak{g}$ belonging to $\mathfrak{h}$.*

(i) *Let* $\mathfrak{s}$ *be a vector space complement of* $\mathfrak{h}$ *in* $\mathfrak{g}$, $\mathfrak{s}_0$ *a neighbourhood of* 0 *in* $\mathfrak{s}$ *on which an exponential map is defined, and* $h_0 \in \mathfrak{h}_r$. *The map* $(s,h) \mapsto \mathrm{F}(s,h) = (\exp \mathrm{ad}\, s).h$ *from* $\mathfrak{s}_0 \times \mathfrak{h}$ *to* $\mathfrak{g}$ *is étale at* $(0, h_0)$.

(ii) *The map* $(g,h) \mapsto \mathrm{F}'(g,h) = (\mathrm{Ad}\, g).h$ *from* $\mathrm{G}\times\mathfrak{h}_r$ *to* $\mathfrak{g}$ *is a submersion. In particular, its image* Ω *is open. For all* $x \in \Omega$, $\mathfrak{g}^0(x)$ *is a Cartan subalgebra of* $\mathfrak{g}$ *conjugate to* $\mathfrak{h}$ *under* Ad(G).

(iii) *Let* $h_0 \in \mathfrak{h}_r$. *For any neighbourhood* U *of* e *in* G, *the set* $\bigcup_{a\in \mathrm{U}} (\mathrm{Ad}\, a)(\mathfrak{h}_r)$ *is a neighbourhood of* h_0 *in* $\mathfrak{g}$.

Let h_0 and $\mathfrak{s}$ be as in (i). Let T be the tangent linear map of F at $(0, h_0)$. Then $\mathrm{F}(0,h) = h$ for all $h \in \mathfrak{h}$, so $\mathrm{T}(0,h) = h$ for all $h \in \mathfrak{h}$. On the other hand, for $\mathfrak{s}_0$ sufficiently small, the tangent linear map at 0 of the map $s \mapsto \exp \mathrm{ad}\, s$ from $\mathfrak{s}_0$ to $\mathrm{End}(\mathfrak{g})$ is the map $s \mapsto \mathrm{ad}\, s$ from $\mathfrak{s}$ to $\mathrm{End}(\mathfrak{g})$. Thus $\mathrm{T}(s,0) = [s, h_0]$ for all $s \in \mathfrak{s}$. Now the map from $\mathfrak{g}/\mathfrak{h}$ to $\mathfrak{g}/\mathfrak{h}$ induced by $\mathrm{ad}\, h_0$ by passage to the quotient is bijective. It follows that T is bijective, hence (i). Since $\exp \mathrm{ad}\, s = \mathrm{Ad} \exp s$ for all $s \in \mathfrak{s}$ sufficiently close to 0, (iii) and the first assertion of (ii) follow. Every $x \in \Omega$ is of the form $(\mathrm{Ad}\, a)(h)$ with $a \in \mathrm{G}$ and $h \in \mathfrak{h}_r$, so $\mathfrak{g}^0(x) = (\mathrm{Ad}\, a)(\mathfrak{g}^0(h)) = (\mathrm{Ad}\, a)(\mathfrak{h})$ is a subalgebra of $\mathfrak{g}$ conjugate to $\mathfrak{h}$ under Ad(G).

§4. REGULAR ELEMENTS OF A LIE GROUP

In nos. 1, 2 and 3 of this paragraph, we assume that k *is* **R**, **C** *or a non-discrete complete ultrametric field of characteristic* 0. *We denote by* G *a finite dimensional Lie group over* k, *by* $\mathfrak{g}$ *its Lie algebra, and by* e *its identity element. If* $a \in \mathrm{G}$, *we denote by* $\mathfrak{g}^1(a)$ *the nilspace of* $\mathrm{Ad}\,(a) - 1$, *in other words the space* $\mathfrak{g}^1(\mathrm{Ad}(a))$ (cf. §1, no. 1).

1. REGULAR ELEMENTS FOR A LINEAR REPRESENTATION

Lemma 1. Let M *be an analytic manifold over* k *and* $a = (a_0, \dots, a_{n-1}, a_n = 1)$ *a sequence of analytic functions on* M. *For all* $x \in \mathrm{M}$, *let* $r_a(x)$ *be the upper bound of those* $i \in [0, n]$ *such that* $a_j(x) = 0$ *for* $j < i$ *and let* $r_a^0(x)$ *be the upper bound of those* $i \in [0, n]$ *such that* a_j *is zero on a neighbourhood of* x *for* $j < i$.

(i) *The function* r_a *is upper semi-continuous.*

(ii) *For all* $x \in \mathrm{M}$, $r_a^0(x) = \liminf_{y\to x} r_a(y)$.

(iii) *The function* r_a^0 *is locally constant.*

(iv) *The set of points* $x \in \mathrm{M}$ *such that* $r_a^0(x) = r_a(x)$ *is the set of points of* M *on a neighbourhood of which* r_a *is constant. This is a dense open subset of* M. *If* $k = \mathbf{C}$ *and* M *is finite dimensional and connected, it is open and connected.*

(i) If $r_a(x) = i$, then $a_i(x) \neq 0$ and, for all y in a neighbourhood of x, we have $a_i(y) \neq 0$, so $r_a(y) \leq i$.

(ii) If $r_a^0(x) = i$, the functions $a_0, \ldots, a_{i-1}$ are zero on a neighbourhood of x and, for any y in this neighbourhood, $r_a(y) \geq i$. Consequently, $\liminf_{y \to x} r_a(y) \geq i$. Every neighbourhood of x contains a point y such that $a_i(y) \neq 0$ and hence $r_a(y) \leq i$. Thus $\liminf_{y \to x} r_a(y) = i$.

(iii) Let $i = r_a^0(x)$ and let V be a neighbourhood of x such that $a_j(y) = 0$ for all $y \in \mathrm{V}$ and all $j < i$. Then $x \in \mathrm{M} - \mathrm{Z}$, where Z denotes the set of points of M in a neighbourhood of which the function a_i is zero. Since Z is closed in M (*Differentiable and Analytic Manifolds, Results*, 5.3.5), $\mathrm{V} \cap (\mathrm{M} - \mathrm{Z})$ is a neighbourhood of x. For every point y in this neighbourhood, $r_a^0(y) = i$.

(iv) The function $r_a - r_a^0$ is upper semi-continuous and its value at any point is ≥ 0. If $r_a(x) = r_a^0(x)$, $r_a - r_a^0$ is zero on a neighbourhood of x, which shows that r_a is constant on a neighbourhood of x by (iii). Conversely, if r_a is constant on a neighbourhood of x, then $r_a^0(x) = r_a(x)$ by (ii). The set of points $x \in \mathrm{M}$ such that $r_a^0(x) = r_a(x)$ is thus an open subset Ω of M. If $x \in \mathrm{M}$ and if $r_a^0(x) < r_a(x)$, every neighbourhood of x contains a point y such that $r_a(y) < r_a(x)$ and $r_a^0(y) = r_a^0(x)$. Every neighbourhood of x thus contains a point y such that

$$r_a(y) - r_a^0(y) < r_a(x) - r_a^0(x).$$

It follows that Ω is dense in M.

If M is connected and if p is the value of r_a^0 on M, the points of Ω are the points $x \in \mathrm{M}$ such that $a_p(x) \neq 0$. If $k = \mathbf{C}$, this implies that Ω is connected by Lemma 3 of Appendix II.

Let ρ be an analytic linear representation of G on a vector space V of finite dimension n over k. Put

$$\det(\mathrm{T} - \rho(g) + 1) = a_0(g) + a_1(g)\mathrm{T} + \cdots + a_{n-1}(g)\mathrm{T}^{n-1} + \mathrm{T}^n.$$

The functions r_a and r_a^0 associated to the sequence $(a_0, a_1, \ldots, a_{n-1}, 1)$ will be denoted by r_ρ and r_ρ^0, respectively. Then, for all $g \in \mathrm{G}$,

$$r_\rho(g) = \dim \mathrm{V}^1(\rho(g))$$
$$r_\rho^0(g) = \lim \inf_{g' \to g} \dim \mathrm{V}^1(\rho(g')).$$

Lemma 2. Let $0 \to \mathrm{V}' \to \mathrm{V} \to \mathrm{V}'' \to 0$ *be an exact sequence of* G*-modules defined by analytic linear representations* ρ', ρ, ρ'' *of* G*, respectively. Then:*

$$r_\rho = r_{\rho'} + r_{\rho''}, \quad \text{and} \quad r_\rho^0 = r_{\rho'}^0 + r_{\rho''}^0.$$

Indeed, for all $g \in \mathrm{G}$, there is (§1, no. 1, Cor. 3 of Th.1) an exact sequence

$$0 \to (\mathrm{V}')^1(\rho'(g)) \to \mathrm{V}^1(\rho(g)) \to (\mathrm{V}'')^1(\rho''(g)) \to 0,$$

which proves the first assertion. The second follows from it since, by Lemma 1 (iv), $r^0_\rho = r_\rho, r^0_{\rho'} = r_{\rho'}$ and $r^0_{\rho''} = r_{\rho''}$ on a dense open subset of G.

DEFINITION 1. *An element $g \in$ G is called* regular *for the linear representation ρ if $r_\rho(g) = r^0_\rho(g)$.*

PROPOSITION 1. *The regular points for an analytic linear representation ρ of* G *are the points of* G *in a neighbourhood of which r_ρ is constant. They constitute a dense open subset of* G. *If $k = \mathbf{C}$ and* G *is connected, the set of regular points for ρ is connected.*

This follows from Lemma 1 (iv).

Remark. Let G^* be an open subgroup of G. An element $a \in \mathrm{G}^*$ is a regular element of G for the linear representation ρ of G if and only if it is a regular element of G^* for the linear representation $\rho|\mathrm{G}^*$.

2. REGULAR ELEMENTS OF A LIE GROUP

DEFINITION 2. *An element of* G *is said to be regular if it is regular for the adjoint representation of* G.

In other words (Prop. 1), an element $g \in$ G is regular if, for all elements g' in a neighbourhood of g in G, the dimension of the nilspace of $\mathrm{Ad}(g') - 1$ is equal to the dimension of the nilspace of $\mathrm{Ad}(g) - 1$.

PROPOSITION 2. *Let* G′ *be a finite dimensional Lie group over k and f an open morphism from* G *to* G′. *The image under f of a regular element of* G *is a regular element of* G′. *If the kernel of f is contained in the centre of* G, *an element $g \in$ G is regular if and only if $f(g)$ is regular.*

Indeed, let $\mathfrak{g}'$ be the Lie algebra of G′ and $\mathfrak{h}$ the ideal in $\mathfrak{g}$ given by the kernel of $\mathrm{T}f|\mathfrak{g}$. Let ρ be the linear representation of G on $\mathfrak{h}$ defined by $\rho(g) = \mathrm{Ad}\, g|\mathfrak{h}$ for all $g \in$ G, and let $\mathrm{Ad} \circ f$ be the linear representation of G on $\mathfrak{g}'$ given by the composite of f with the adjoint representation of G′. These linear representations define an exact sequence of G-modules:

$$0 \to \mathfrak{h} \to \mathfrak{g} \to \mathfrak{g}' \to 0.$$

By Lemma 2, $r_{\mathrm{Ad}} = r_\rho + r_{\mathrm{Ad} \circ f}$. Since $r_{\mathrm{Ad} \circ f} = r_{\mathrm{Ad}} \circ f$ and since f is an open map, $r^0_{\mathrm{Ad} \circ f} = r^0_{\mathrm{Ad}} \circ f$. Consequently:

$$r_{\mathrm{Ad}} - r^0_{\mathrm{Ad}} = r_\rho - r^0_\rho + (r_{\mathrm{Ad}} - r^0_{\mathrm{Ad}}) \circ f.$$

Thus, if g is regular, $(r_{\mathrm{Ad}} - r^0_{\mathrm{Ad}})(f(g)) = 0$, which means that $f(g)$ is regular. If the kernel of f is contained in the centre of G,

$$r_\rho(g) = r^0_\rho(g) = \dim \mathfrak{h}$$

for all $g \in \mathrm{G}$. Consequently, if $f(g)$ is regular, $r_{\mathrm{Ad}}(g) = r^0_{\mathrm{Ad}}(g)$, in other words, g is regular.

PROPOSITION 3. *Let* G_1 *and* G_2 *be two finite dimensional Lie groups over* k. *An element* (g_1, g_2) *of* $\mathrm{G}_1 \times \mathrm{G}_2$ *is regular if and only if* g_1 *and* g_2 *are regular elements of* G_1 *and* G_2*, respectively.*

The condition is necessary by Prop. 2. We show that it is sufficient. For all $g = (g_1, g_2) \in \mathrm{G}_1 \times \mathrm{G}_2$, $r_{\mathrm{Ad}}(g) = r_{\mathrm{Ad}}(g_1) + r_{\mathrm{Ad}}(g_2)$. In view of Lemma 1 (ii), it follows that $r^0_{\mathrm{Ad}}(g) = r^0_{\mathrm{Ad}}(g_1) + r^0_{\mathrm{Ad}}(g_2)$. If g_1 and g_2 are regular, $r^0_{\mathrm{Ad}}(g_1) = r_{\mathrm{Ad}}(g_1)$ and $r^0_{\mathrm{Ad}}(g_2) = r_{\mathrm{Ad}}(g_2)$, so $r^0_{\mathrm{Ad}}(g) = r_{\mathrm{Ad}}(g)$, which means that g is regular.

Lemma 3. Let $a \in \mathrm{G}$ *and let* $\mathfrak{m}$ *be a complement of* $\mathfrak{g}^1(a)$ *in* $\mathfrak{g}$. *Let* U *be a neighbourhood of* 0 *in* $\mathfrak{g}$ *and* exp *an exponential map from* U *to* G. *The map*

$$f : (x, y) \mapsto (\exp y) a (\exp x)(\exp y)^{-1}$$

from $(\mathfrak{g}^1(a) \times \mathfrak{m}) \cap \mathrm{U}$ *to* G *is étale at* $(0,0)$.

The tangent linear maps at 0 of the maps $x \mapsto a(\exp x)$ and $y \mapsto (\exp y)a(\exp y)^{-1}$ are the maps $x \mapsto ax$ and $y \mapsto ya - ay = a(a^{-1}ya - y)$ from $\mathfrak{g}$ to $\mathrm{T}_a\mathrm{G} = a\mathfrak{g}$ (Chap. III, §3, no. 12, Prop. 46). Consequently, the tangent map of f at $(0,0)$ is the map $(x, y) \mapsto ax + a(a^{-1}ya - y) = a(x + a^{-1}ya - y)$ from $\mathfrak{g}^1(a) \times \mathfrak{m}$ to $a\mathfrak{g}$. This map is injective. Indeed, if $x \in \mathfrak{g}^1(a), y \in \mathfrak{m}$ and if $x + a^{-1}ya - y = 0$, then $(\mathrm{Ad}(a) - 1)y = \mathrm{Ad}(a)x \in \mathfrak{g}^1(a)$ since $\mathrm{Ad}(a)\mathfrak{g}^1(a) \subset \mathfrak{g}^1(a)$. This implies that $y \in \mathfrak{g}^1(a)$ and consequently that $y = 0$. Since $\mathrm{Ad}(a)$ is injective on $\mathfrak{g}^1(a)$, it follows that $x = 0$. Since $\dim \mathfrak{g} = \dim \mathfrak{g}^1(a) + \dim \mathfrak{m}$, this shows that f is étale at $(0,0)$.

PROPOSITION 4. *Let* $a \in \mathrm{G}$ *and* H *be a Lie subgroup germ of* G *with Lie algebra* $\mathfrak{g}^1(a)$. *The map* $(b, c) \mapsto cabc^{-1}$ *from* $\mathrm{H} \times \mathrm{G}$ *to* G *is a submersion at* (e, e).

Indeed, let $\mathfrak{m}$ be a complement of $\mathfrak{g}^1(a)$ in $\mathfrak{g}$ and exp an exponential map of G defined on an open neighbourhood U of 0 in $\mathfrak{g}$. We can choose U so that $\exp(\mathrm{U} \cap \mathfrak{g}^1(a)) \subset \mathrm{H}$. The map $f : (x, y) \mapsto (\exp x, \exp y)$ is an analytic map on a neighbourhood of $(0,0)$ in $\mathfrak{g}^1(a) \times \mathfrak{m}$ with values in $\mathrm{H} \times \mathrm{G}$. By Lemma 3, the composite of f with the map $\varphi : (b, c) \mapsto cabc^{-1}$ is étale at $(0,0)$. It follows that φ is a submersion at $f(0,0) = (e, e)$.

PROPOSITION 5. *Let* $a \in \mathrm{G}$ *and let* W *be a neighbourhood of* e *in* G. *There exists a neighbourhood* V *of* a *with the following property: for all* $a' \in \mathrm{V}$*, there exists an element* $g \in \mathrm{W}$ *such that* $\mathfrak{g}^1(a') \subset \mathrm{Ad}(g)\mathfrak{g}^1(a)$.

Put $\mathfrak{g}^1 = \mathfrak{g}^1(a)$ and let $\mathfrak{g} = \mathfrak{g}^1 + \mathfrak{g}^+$ be the Fitting decomposition of $\mathrm{Ad}(a) - 1$ (§1, no. 1). Let H be a Lie subgroup germ of G with Lie algebra $\mathfrak{g}^1$. For all $h \in \mathrm{H}$, $\mathrm{Ad}(h)\mathfrak{g}^1 \subset \mathfrak{g}^1$. Since $[\mathfrak{g}^1, \mathfrak{g}^+] \subset \mathfrak{g}^+$, there exists a neighbourhood U of e in H such that $\mathrm{Ad}(h)\mathfrak{g}^+ \subset \mathfrak{g}^+$ for all $h \in \mathrm{H}$. Since the restriction of $\mathrm{Ad}(a) - 1$ to $\mathfrak{g}^+$ is bijective, U can be chosen so that the restriction of $\mathrm{Ad}(ah) - 1$ to $\mathfrak{g}^+$ is bijective for all $h \in \mathrm{U}$. Then $\mathfrak{g}^1(ah) \subset \mathfrak{g}^1(a) = \mathfrak{g}^1$ for all $h \in \mathrm{U}$. By Proposition 4, $\mathrm{Int(W)}(a\mathrm{U})$ is a neighbourhood of a in G. If $a' \in \mathrm{Int(W)}(a\mathrm{U})$, then $a' = g(ah)g^{-1}$ with $g \in \mathrm{W}$ and $h \in \mathrm{U}$; it follows that $\mathfrak{g}^1(a') = \mathrm{Ad}(g)\mathfrak{g}^1(ah) \subset \mathrm{Ad}(g)\mathfrak{g}^1(a)$.

COROLLARY. *Let* G^* *be an open subgroup of* G. *If* $a \in \mathrm{G}$ *is regular, there exists a neighbourhood* V *of* a *such that, for all* $a' \in \mathrm{V}$, $\mathfrak{g}^1(a')$ *is conjugate to* $\mathfrak{g}^1(a)$ *under* $\mathrm{Ad}(\mathrm{G}^*)$.

3. RELATIONS WITH REGULAR ELEMENTS OF THE LIE ALGEBRA

PROPOSITION 6. *Let* V *be an open subgroup of* $\mathfrak{g}$ *and let* $\exp : \mathrm{V} \to \mathrm{G}$ *be an exponential map defined on* V.

(i) *There exists a neighbourhood* W *of* 0 *in* V *such that* $\mathfrak{g}^1(\exp x) = \mathfrak{g}^0(x)$ *for all* $x \in \mathrm{W}$.

(ii) *If* $k = \mathbf{R}$ *or* $\mathbf{C}$, $\mathfrak{g}^1(\exp x) \supset \mathfrak{g}^0(x)$ *for all* $x \in \mathfrak{g}$.

By Cor. 3 of Prop. 8 of Chap. III, §4, no. 4, there exists a neighbourhood V' of 0 in V such that, for all $x \in \mathrm{V}'$, $\exp(\mathrm{ad}(x)) = \sum_{n=0}^{\infty} \frac{1}{n!}\mathrm{ad}(x)^n$ is defined and $\mathrm{Ad}(\exp x) = \exp(\mathrm{ad}(x))$. If $\mathrm{P} \in k[\mathrm{X}]$ and $\alpha \in \mathrm{End}(\mathfrak{g})$, it is easy to check that $\mathfrak{g}^\lambda(\alpha) \subset \mathfrak{g}^{\mathrm{P}(\lambda)}(\mathrm{P}(\alpha))$ for all $\lambda \in k$. Consequently,

$$\mathfrak{g}^0(\mathrm{ad}(x)) \subset \mathfrak{g}^1(\exp(\mathrm{ad}(x))) = \mathfrak{g}^1(\mathrm{Ad}(\exp x)) = \mathfrak{g}^1(\exp x)$$

for all $x \in \mathrm{V}'$. If $k = \mathbf{R}$ or $\mathbf{C}$, $\mathrm{V} = \mathfrak{g}$ and we can take $\mathrm{V}' = \mathrm{V}$, which proves (ii). We prove (i). Let U be a neighbourhood of 0 in $\mathrm{End}(\mathfrak{g})$ such that $\mathrm{Log}(1+\alpha) = \sum_{n>0} (-1)^{n+1}\frac{1}{n}\alpha^n$ is defined for all $\alpha \in \mathrm{U}$. Then $\mathrm{Log} \circ \exp = 1$ on a neighbourhood of 0 and $\mathfrak{g}^1(1+\alpha) \subset \mathfrak{g}^0(\mathrm{Log}(1+\alpha))$ for all $\alpha \in \mathrm{U}$. Let W be the neighbourhood of 0 in $\mathfrak{g}$ consisting of those $x \in \mathrm{V}'$ such that $\exp \mathrm{ad}\, x \in 1 + \mathrm{U}$ and

$$\mathrm{Log}(\exp(\mathrm{ad}(x))) = \mathrm{ad}(x).$$

Then, for all $x \in \mathrm{W}$,

$$\begin{aligned}\mathfrak{g}^1(\exp x) = \mathfrak{g}^1(\mathrm{Ad}(\exp x)) &= \mathfrak{g}^1(\exp(\mathrm{ad}(x)))\\ &\subset \mathfrak{g}^0(\mathrm{Log}(\exp(\mathrm{ad}(x)))) = \mathfrak{g}^0(\mathrm{ad}(x)) = \mathfrak{g}^0(x).\end{aligned}$$

This shows that $\mathfrak{g}^1(\exp x) = \mathfrak{g}^0(x)$ for all $x \in \mathrm{W}$.

Lemma 4. Let U *be a neighbourhood of* 0 *in* $\mathfrak{g}$ *and* exp *an exponential map from* U *to* G, *étale at every point of* U *and such that* $\mathfrak{g}^1(\exp x) = \mathfrak{g}^0(x)$ *for all* $x \in \mathrm{U}$.

(i) *The function* r^0_{Ad} *is constant and equal to the rank of* $\mathfrak{g}$ *on* $\exp(\mathrm{U})$.

(ii) *If* $x \in \mathrm{U}$, $\exp x$ *is regular if and only if* x *is a regular element of* $\mathfrak{g}$.

(iii) *An element* $a \in \exp(\mathrm{U})$ *is regular if and only if* $\mathfrak{g}^1(a)$ *is a Cartan subalgebra of* $\mathfrak{g}$.

Let $l = \mathrm{rk}(\mathfrak{g})$. If $x \in \mathrm{U}$ is a regular element of $\mathfrak{g}$,

$$r_{\text{Ad}}(\exp x) = \dim \mathfrak{g}^1(\exp x) = \dim \mathfrak{g}^0(x) = l.$$

Since the regular elements of $\mathfrak{g}$ belonging to U constitute a neighbourhood of x and exp is étale at x, this shows that $\exp x$ is regular and that $r^0_{\text{Ad}}(\exp x) = l$. The regular elements of $\mathfrak{g}$ belonging to U being dense in U, we have $r^0_{\text{Ad}}(a) = l$ for all $a \in \exp(\mathrm{U})$. Let $a \in \exp(\mathrm{U})$ be a regular element of G and let $x \in \mathrm{U}$ be such that $a = \exp x$. Since $\mathfrak{g}^0(x) = \mathfrak{g}^1(a)$, $\dim \mathfrak{g}^0(x) = r^0_{\text{Ad}}(a) = l$. Consequently, x is a regular element of $\mathfrak{g}$ and $\mathfrak{g}^1(a)$ is a Cartan subalgebra of $\mathfrak{g}$. Finally, if $a \in \exp(\mathrm{U})$ and $\mathfrak{g}^1(a)$ is a Cartan subalgebra of $\mathfrak{g}$,

$$r_{\text{Ad}}(a) = \dim \mathfrak{g}^1(a) = l = r^0_{\text{Ad}}(a),$$

so a is regular.

PROPOSITION 7. *Let* V *be a neighbourhood of* e *in* G. *Every Cartan subalgebra of* $\mathfrak{g}$ *is of the form* $\mathfrak{g}^1(a)$ *where* a *is a regular element of* G *belonging to* V.

By Prop. 6, there exists an open neighbourhood U of 0 in $\mathfrak{g}$ and an exponential map $\exp : \mathrm{U} \to \mathrm{G}$ satisfying the conditions of Lemma 4. If $\mathfrak{h}$ is a Cartan subalgebra of $\mathfrak{g}$, there exists a regular element $x \in \mathfrak{h}$ such that $\mathfrak{h} = \mathfrak{g}^0(x)$ (§3, Th. 2). On the other hand, there exists an element $t \in k^*$ such that $tx \in \mathrm{U}$ and $\exp(tx) \in \mathrm{V}$. Then $\mathfrak{h} = \mathfrak{g}^0(x) = \mathfrak{g}^0(tx) = \mathfrak{g}^1(\exp(tx))$, and by Lemma 4 (ii), $\exp(tx)$ is a regular element of G.

PROPOSITION 8. *Let* l *be the rank of* $\mathfrak{g}$. *There exists an open subgroup* G^* *of* G *such that:*

(i) *the function* r^0_{Ad} *is constant on* G^* *and its value is* l;

(ii) *an element* $a \in \mathrm{G}^*$ *is regular if and only if* $\mathfrak{g}^1(a)$ *is a Cartan subalgebra of* $\mathfrak{g}$;

(iii) *if* $a \in \mathrm{G}^*$, *every Cartan subalgebra of* $\mathfrak{g}^1(a)$ *is a Cartan subalgebra of* $\mathfrak{g}$.

(i) By Prop. 6, there exists an open neighbourhood U of 0 in $\mathfrak{g}$ and an exponential map exp from U to G satisfying the conditions of Lemma 4. In what follows, G^* will denote the identity component of G if $k = \mathbf{R}$ or $\mathbf{C}$ and an open subgroup of G contained in $\exp(\mathrm{U})$ if k is ultrametric. Since r^0_{Ad} is locally constant and its value at any point of $\exp(\mathrm{U})$ is l (Lemma 4 (i)), it follows that r^0_{Ad} is constant and equal to l on G^*.

(ii) Let R^* (resp. S^*) be the set of regular elements of G^* (resp. the set of elements $a \in \mathrm{G}^*$ such that $\mathfrak{g}^1(a)$ is a Cartan subalgebra of $\mathfrak{g}$). Then $\mathrm{S}^* \subset \mathrm{R}^*$. Indeed, if $a \in \mathrm{S}^*$, then $r_{\text{Ad}}(a) = l = r^0_{\text{Ad}}(a)$. We show that $\mathrm{R}^* \subset \mathrm{S}^*$. If k is ultrametric, this follows from the inclusion $G^* \subset \exp(U)$ and Lemma 4 (iii). Assume that $k = \mathbf{C}$. By the Cor. of Prop. 5, if $a \in R^*$, then for every a' belonging to a neighbourhood of a, $\mathfrak{g}^1(a')$ is conjugate to $\mathfrak{g}^1(a)$ by an automorphism of $\mathfrak{g}$. This proves that S^* and $R^* - S^*$ are open subsets of G^*. We have seen that S^* contains all the regular elements in a neighbourhood of e (Lemma 4 (iii)); consequently, S^* is non-empty. Since G^* is connected, so is R^* (Prop. 1) and consequently $S^* = R^*$.

It remains to study the case $k = \mathbf{R}$. Assume first of all that G^* is an integral subgroup of $\mathbf{GL}(E)$ where E denotes a finite dimensional real vector space. Let G_c^* be the integral subgroup of $\mathbf{GL}(E \otimes_{\mathbf{R}} \mathbf{C})$ with Lie algebra $\mathfrak{g}_c = \mathfrak{g} \otimes \mathbf{C}$. There exists an analytic function on G_c^* whose set of zeros is the complement of the open set of regular elements of G_c^*. By *Differentiable and Analytic Manifolds, Results*, 3.2.5, this function cannot vanish at every point of G^*. Consequently, G^* contains a regular element of G_c^*. Let Ad_c be the adjoint representation of G_c^*. For any $a \in G^*$, $\mathfrak{g}_c^1(a) = \mathfrak{g}^1(a) \otimes \mathbf{C}$, so $r_{\mathrm{Ad}_c}(a) = r_{\mathrm{Ad}}(a)$. If $a \in G^*$ is a regular element of G_c^*, this is a regular element of G^* and $r^0_{\mathrm{Ad}_c}(a) = r^0_{\mathrm{Ad}}(a)$. The functions $r^0_{\mathrm{Ad}_c}$ and r^0_{Ad} being constant on G_c^* and on G^*, respectively, it follows that the regular elements of G^* are the regular elements of G_c^* belonging to G^*. From the above, if a is a regular element of G^*, $\mathfrak{g}_c^1(a) = \mathfrak{g}^1(a) \otimes \mathbf{C}$ is a Cartan subalgebra of $\mathfrak{g}_c$; this implies that $\mathfrak{g}^1(a)$ is a Cartan subalgebra of $\mathfrak{g}$ (§2, Prop. 3).

Assume now that G is simply connected. There exists a finite dimensional real vector space E and an étale morphism f from G to an integral subgroup G' of $\mathbf{GL}(E)$ (Chap. III, §6, no. 1, Cor. of Th. 1). By Prop. 2, if $a \in G$ is regular, $f(a)$ is regular. By the preceding, $\mathfrak{g}'^1(f(a))$ is a Cartan subalgebra of the Lie algebra $\mathfrak{g}'$ of G'. Since $\mathfrak{g}'^1(f(a)) = (\mathrm{T}f)\mathfrak{g}^1(a)$ and $\mathrm{T}f$ is an isomorphism from $\mathfrak{g}$ to $\mathfrak{g}'$, this proves that $\mathfrak{g}^1(a)$ is a Cartan subalgebra of $\mathfrak{g}$.

We turn finally to the general case ($k = \mathbf{R}$). Let $\tilde{G}$ be a universal covering of G^*, $\tilde{\mathfrak{g}} = L(\tilde{G})$, and q the canonical map from $\tilde{G}$ to G^*. Since the kernel of q is contained in the centre of $\tilde{G}$, if $a \in G^*$ is regular and if $a' \in q^{-1}(a)$, then a' is regular (Prop. 2). By the preceding, $\tilde{\mathfrak{g}}^1(a')$ is a Cartan subalgebra of $\tilde{\mathfrak{g}}$. Since $\mathfrak{g}^1(a) = (\mathrm{T}q)\tilde{\mathfrak{g}}^1(a')$ and since $\mathrm{T}q$ is an isomorphism from $\tilde{\mathfrak{g}}$ to $\mathfrak{g}$, this proves that $\mathfrak{g}^1(a)$ is a Cartan subalgebra of $\mathfrak{g}$.

(iii) By Prop. 5, there exists a neighbourhood V of a such that, for all $a' \in V$, $\mathfrak{g}^1(a')$ is conjugate to a subalgebra of $\mathfrak{g}^1(a)$ by an automorphism of $\mathfrak{g}$. Since every neighbourhood of a contains a regular element of G^*, it follows from (ii) that $\mathfrak{g}^1(a)$ contains a Cartan subalgebra of $\mathfrak{g}$. Thus, by Prop. 3 of §3, every Cartan subalgebra of $\mathfrak{g}^1(a)$ is a Cartan subalgebra of $\mathfrak{g}$.

Remark. If $k = \mathbf{C}$, the subalgebras $\mathfrak{g}^1(a)$, for a regular and belonging to a connected component M of G, are conjugate under $\mathrm{Int}(\mathfrak{g})$. Indeed, let R be the set of regular elements of G. For all $a \in R \cap M$, let M_a be the set of those $b \in R \cap M$ such that $\mathfrak{g}^1(a)$ is conjugate to $\mathfrak{g}^1(a)$ under $\mathrm{Int}(\mathfrak{g})$. We have $\mathrm{Int}(\mathfrak{g}) = \mathrm{Ad}(G^0)$, where G^0 is the identity component of G. By the Corollary to Prop. 5, M_a is open in R. It follows that M_a is open and closed in R. Since $k = \mathbf{C}$, $R \cap M$ is connected (Lemma 1), hence $M_a = R \cap M$.

4. APPLICATION TO ELEMENTARY AUTOMORPHISMS

PROPOSITION 9. *Let k be a field of characteristic 0 and $\mathfrak{g}$ a Lie algebra over k. If $a \in \mathrm{Aut}_e(\mathfrak{g})$, the dimension of the nilspace of $a - 1$ is greater than or equal to the rank of $\mathfrak{g}$.*

By the "Lefschetz principle" (*Algebra*, Chap. V, §14, no. 6, Cor. 2 of Th. 5), k is an ascending directed union of subfields $(k_i)_{i \in I}$ which admit $\mathbf{C}$ as extension field. Let (e_α) be a basis of $\mathfrak{g}$ over k and $x_1, \ldots, x_m$ elements of $\mathfrak{g}$ such that $\mathrm{ad}(x_1), \ldots, \mathrm{ad}(x_m)$ are nilpotent and $a = e^{\mathrm{ad}(x_1)} \ldots e^{\mathrm{ad}(x_m)}$. Let $c_{\alpha\beta}^{\gamma}$ be the structure constants of $\mathfrak{g}$ with respect to the basis (e_α) and (x_r^α) the components of x_r with respect to this basis ($1 \leq r \leq m$). There exists an index $j \in I$ such that the $c_{\alpha\beta}^{\gamma}$ and the x_r^α all belong to k_j. Let $\mathfrak{g}_j = \sum_\alpha k_j e_\alpha$; this is a Lie algebra over k_j containing $x_1, \ldots, x_m$, and the restriction a_j of a to $\mathfrak{g}_j$ is an elementary automorphism of $\mathfrak{g}_j$. The extension of a_j to $\mathfrak{g}_j \otimes_{k_j} \mathbf{C}$ is an elementary automorphism $a_j \otimes 1$ of $\mathfrak{g}_j \otimes \mathbf{C}$. So let G_j be a connected complex Lie group with Lie algebra $\mathfrak{g}_j \otimes \mathbf{C}$, and s an element of G_j such that $\mathrm{Ad}(s) = a_j \otimes 1$. Prop. 8, applied to the pair (G_j, s), shows that the nilspace of $a_j \otimes 1 - 1$ is of dimension n, so

$$n \geq \mathrm{rk}(\mathfrak{g}_j \otimes \mathbf{C}) = \mathrm{rk}(\mathfrak{g}_j) = \mathrm{rk}(\mathfrak{g}).$$

But this nilspace has the same dimension as that of $a_j - 1$ and that of $a - 1$. Hence the proposition.

§5. DECOMPOSABLE LINEAR LIE ALGEBRAS

In this paragraph, k is assumed to be of characteristic 0. We denote by V a finite dimensional vector space.

1. DECOMPOSABLE LINEAR LIE ALGEBRAS

DEFINITION 1. Let $\mathfrak{g}$ be a Lie subalgebra of $\mathfrak{gl}(V)$. Then $\mathfrak{g}$ is said to be decomposable if $\mathfrak{g}$ contains the semi-simple and nilpotent components of each of its elements (*Algebra*, Chap. VII, §5, no. 8).

Examples. 1) Let V' and V'' be vector subspaces of V such that $V'' \supset V'$. The set of $x \in \mathfrak{gl}(V)$ such that $x(V'') \subset V'$ is a decomposable Lie subalgebra of $\mathfrak{gl}(V)$; indeed, for all $x \in \mathfrak{gl}(V)$, the semi-simple and nilpotent components of x are of the form $P(x)$ and $Q(x)$, where P and Q are polynomials without constant term.

2) Assume that V has an algebra structure. The set of derivations of V is a decomposable Lie subalgebra of $\mathfrak{gl}(V)$ (§1, no. 1, Prop. 4 (ii)).

3) *More generally, it can be shown that the Lie algebra of any algebraic subgroup of $\mathbf{GL}(V)$ is decomposable.*

PROPOSITION 1. *Let $\mathfrak{g}$ be a decomposable Lie subalgebra of $\mathfrak{gl}(V)$, $x \in \mathfrak{g}$, s and n the semi-simple and nilpotent components of x.*

(i) *The semi-simple and nilpotent components of* $\mathrm{ad}_{\mathfrak{g}} x$ *are* $\mathrm{ad}_{\mathfrak{g}} s$ *and* $\mathrm{ad}_{\mathfrak{g}} n$*, respectively.*

(ii) *x is regular in $\mathfrak{g}$ if and only if s is.*

(iii) *If $\mathfrak{g}'$ is a subalgebra of $\mathfrak{gl}(V)$ containing $\mathfrak{g}$, every elementary automorphism of $\mathfrak{g}$ extends to an elementary automorphism of $\mathfrak{g}'$.*

Put $\mathfrak{a} = \mathfrak{gl}(V)$. By Chap. I, §5, no. 4, Lemma 2, the semi-simple and nilpotent components of $\mathrm{ad}_{\mathfrak{a}} x$ are $\mathrm{ad}_{\mathfrak{a}} s$ and $\mathrm{ad}_{\mathfrak{a}} n$; assertion (i) follows from this. We deduce that the characteristic polynomials of $\mathrm{ad}_{\mathfrak{g}} x$ and $\mathrm{ad}_{\mathfrak{g}} s$ are the same; hence (ii). If $\mathrm{ad}_{\mathfrak{g}} x$ is nilpotent, $\mathrm{ad}_{\mathfrak{g}} x = \mathrm{ad}_{\mathfrak{g}} n$, so $\mathrm{ad}_{\mathfrak{g}'} n$ extends $\mathrm{ad}_{\mathfrak{g}} x$, and n is a nilpotent element of $\mathfrak{g}'$, hence (iii).

Let $\mathfrak{g}$ be a Lie subalgebra of $\mathfrak{gl}(V)$. We know (Chap. I, §6, no. 5, Th. 4) that the following conditions are equivalent:

(i) the identity representation of $\mathfrak{g}$ is semi-simple;

(ii) $\mathfrak{g}$ is reductive and every element of the centre of $\mathfrak{g}$ is a semi-simple endomorphism.

These conditions are actually equivalent to the following:

(iii) $\mathfrak{g}$ is a reductive subalgebra in $\mathfrak{gl}(V)$.

Indeed, (i) $\Longrightarrow$ (iii) by Chap. I, §6, no. 5, Cor. 3 of Th. 4, and (iii) $\Longrightarrow$ (i) by Chap. I, §6, no. 6, Cor. 1 of Prop. 7. We are going to show that if $\mathfrak{g}$ satisfies these conditions, $\mathfrak{g}$ is decomposable. More generally:

PROPOSITION 2. *Let $\mathfrak{g}$ be a Lie subalgebra of $\mathfrak{gl}(V)$ reductive in $\mathfrak{gl}(V)$, E a finite dimensional vector space and $\pi : \mathfrak{g} \to \mathfrak{gl}(E)$ a semi-simple linear representation of $\mathfrak{g}$ on E. Then:*

(i) *$\mathfrak{g}$ and $\pi(\mathfrak{g})$ are decomposable.*

(ii) *The semi-simple (resp. nilpotent) elements of $\pi(\mathfrak{g})$ are the images under π of the semi-simple (resp. nilpotent) elements of $\mathfrak{g}$.*

(iii) *If $\mathfrak{h}$ is a decomposable subalgebra of $\mathfrak{gl}(V)$ contained in $\mathfrak{g}$, $\pi(\mathfrak{h})$ is a decomposable subalgebra of $\mathfrak{gl}(E)$.*

(iv) *If $\mathfrak{h}'$ is a decomposable subalgebra of $\mathfrak{gl}(E)$, $\pi^{-1}(\mathfrak{h}')$ is a decomposable subalgebra of $\mathfrak{gl}(V)$.*

Let $\mathfrak{s} = [\mathfrak{g}, \mathfrak{g}]$ and let $\mathfrak{c}$ be the centre of $\mathfrak{g}$. Then $\mathfrak{g} = \mathfrak{s} \times \mathfrak{c}$, and $\pi(\mathfrak{g}) = \pi(\mathfrak{s}) \times \pi(\mathfrak{c})$ by Chap. I, §6, no. 4, Cor. of Prop. 5. Let $y \in \mathfrak{s}, z \in \mathfrak{c}, y_s$ and y_n the semi-simple and nilpotent components of y. Then $y_s, y_n \in \mathfrak{s}$ (Chap. I, §6, no. 3, Prop. 3), $y_s + z$ is semi-simple (*Algebra*, Chap. VII, §5, no. 7, Cor. of Prop. 16), and y_n commutes with $y_s + z$. Hence, the semi-simple and nilpotent components of $y + z$ are $y_s + z$ and y_n. Thus, $\mathfrak{g}$ is decomposable. Since $\pi(\mathfrak{g})$ is reductive in $\mathfrak{gl}(\mathrm{E})$, the same argument applies to $\pi(\mathfrak{g})$ and shows that $\pi(\mathfrak{g})$ is decomposable. Moreover, the nilpotent elements of $\mathfrak{g}$ (resp. $\pi(\mathfrak{g})$) are the nilpotent elements of $\mathfrak{s}$ (resp. $\pi(\mathfrak{s})$). Hence the nilpotent elements of $\pi(\mathfrak{g})$ are the images under π of the nilpotent elements of $\mathfrak{g}$ (Chap. I, §6, no. 3, Prop. 4). The semi-simple elements of $\mathfrak{g}$ (resp. $\pi(\mathfrak{g})$) are the sums of the semi-simple elements of $\mathfrak{s}$ (resp. $\pi(\mathfrak{s})$) and the elements of $\mathfrak{c}$ (resp. $\pi(\mathfrak{c})$). Thus the semi-simple elements of $\pi(\mathfrak{g})$ are the images under π of the semi-simple elements of $\mathfrak{g}$ (Chap. I, *loc. cit.*). Hence (ii).

Assertions (iii) and (iv) follow immediately from (i) and (ii).

Remarks. 1) The semi-simplicity assumption on π is equivalent to saying that $\pi(x)$ is semi-simple for all $x \in \mathfrak{c}$. Note that this assumption is satisfied when π is obtained from the identity representation $\mathfrak{g} \to \mathfrak{gl}(\mathrm{V})$ by the successive application of the following operations: tensor product, passage to the dual, to a subrepresentation, to a quotient, to a direct sum.

2) Let $\mathfrak{g} \subset \mathfrak{gl}(\mathrm{V})$, $\mathfrak{g}' \subset \mathfrak{gl}(\mathrm{V}')$ be decomposable Lie algebras, φ an isomorphism from $\mathfrak{g}$ to $\mathfrak{g}'$. Note that φ does not necessarily transform semi-simple (resp. nilpotent) elements of $\mathfrak{g}$ to semi-simple (resp. nilpotent) elements of $\mathfrak{g}'$ (Exerc. 2). However, this is the case if $\mathfrak{g}$ is semi-simple (Chap. I, §6, no. 3, Th. 3).

PROPOSITION 3. *Let $\mathfrak{a}$ be a decomposable Lie subalgebra of $\mathfrak{gl}(\mathrm{V})$ and let $\mathfrak{b}$ and $\mathfrak{c}$ be vector subspaces of $\mathfrak{gl}(\mathrm{V})$ such that $\mathfrak{b} \subset \mathfrak{c}$. Let $\mathfrak{a}'$ be the set of $x \in \mathfrak{a}$ such that $[x, \mathfrak{c}] \subset \mathfrak{b}$. Then $\mathfrak{a}'$ is decomposable.*

Put $\mathfrak{g} = \mathfrak{gl}(\mathrm{V})$; the subalgebra $\mathfrak{h}'$ of $\mathfrak{gl}(\mathfrak{g})$ consisting of the $z \in \mathfrak{gl}(\mathfrak{g})$ such that $z(\mathfrak{c}) \subset \mathfrak{b}$ is decomposable (Example 1). Let $\pi : \mathfrak{g} \to \mathfrak{gl}(\mathfrak{g})$ be the adjoint representation of $\mathfrak{g}$. Prop. 2 (iv), applied to π, shows that $\pi^{-1}(\mathfrak{h}')$ is decomposable. Hence so is $\mathfrak{a}' = \mathfrak{a} \cap \pi^{-1}(\mathfrak{h}')$.

COROLLARY 1. *If $\mathfrak{a}$ is a decomposable Lie subalgebra of $\mathfrak{gl}(\mathrm{V})$, and $\mathfrak{n}$ a Lie subalgebra of $\mathfrak{a}$, the normalizer* (resp. *centralizer*) *of $\mathfrak{n}$ in $\mathfrak{a}$ is decomposable.*

This follows from Prop. 3 by taking $\mathfrak{c} = \mathfrak{n}, \mathfrak{b} = \mathfrak{n}$ (resp. $\mathfrak{c} = \mathfrak{n}, \mathfrak{b} = \{0\}$).

COROLLARY 2. *The Cartan subalgebras of a decomposable Lie subalgebra of $\mathfrak{gl}(\mathrm{V})$ are decomposable.*

This follows from Corollary 1.

Remark. We shall prove later (no. 5, Th. 2) a converse of Cor. 2.

2. DECOMPOSABLE ENVELOPE

The intersection of a family of decomposable Lie subalgebras of $\mathfrak{gl}(V)$ is clearly decomposable. Consequently, if $\mathfrak{g}$ is a Lie subalgebra of $\mathfrak{gl}(V)$, the set of decomposable Lie subalgebras of $\mathfrak{gl}(V)$ containing $\mathfrak{g}$ has a smallest element, called the *decomposable envelope* of $\mathfrak{g}$; in this paragraph, this envelope will be denoted by $e(\mathfrak{g})$.

PROPOSITION 4. *Let* $\mathfrak{g}$ *be a Lie subalgebra of* $\mathfrak{gl}(V)$ *and* $\mathfrak{n}$ *an ideal of* $\mathfrak{g}$. *Then* $\mathfrak{n}$ *and* $e(\mathfrak{n})$ *are ideals of* $e(\mathfrak{g})$, *and* $[e(\mathfrak{g}), e(\mathfrak{n})] = [\mathfrak{g}, \mathfrak{n}]$.

Let $\mathfrak{g}_1$ be the set of $x \in \mathfrak{gl}(V)$ such that $[x, \mathfrak{n}] \subset [\mathfrak{g}, \mathfrak{n}]$. This is a decomposable Lie subalgebra of $\mathfrak{gl}(V)$, containing $\mathfrak{g}$ and hence $e(\mathfrak{g})$, cf. no. 1, Prop. 3; in other words, $[e(\mathfrak{g}), \mathfrak{n}] \subset [\mathfrak{g}, \mathfrak{n}]$. Let $\mathfrak{n}_1$ be the set of $y \in \mathfrak{gl}(V)$ such that

$$[e(\mathfrak{g}), y] \subset [\mathfrak{g}, \mathfrak{n}].$$

This is a decomposable Lie subalgebra of $\mathfrak{gl}(V)$ containing $\mathfrak{n}$ by the preceding, and hence containing $e(\mathfrak{n})$; in other words $[e(\mathfrak{g}), e(\mathfrak{n})] \subset [\mathfrak{g}, \mathfrak{n}]$, so

$$[e(\mathfrak{g}), e(\mathfrak{n})] = [\mathfrak{g}, \mathfrak{n}].$$

It follows that $[e(\mathfrak{g}), \mathfrak{n}] \subset [e(\mathfrak{g}), e(\mathfrak{n})] \subset \mathfrak{n}$, so $\mathfrak{n}$ and $e(\mathfrak{n})$ are ideals of $e(\mathfrak{g})$.

COROLLARY 1. (i) $\mathscr{D}^i\mathfrak{g} = \mathscr{D}^i e(\mathfrak{g})$ *for* $i \geq 1$, *and* $\mathscr{C}^i\mathfrak{g} = \mathscr{C}^i e(\mathfrak{g})$ *for* $i \geq 2$.

(ii) *If* $\mathfrak{g}$ *is commutative* (resp. *nilpotent*, resp. *solvable*), *then* $e(\mathfrak{g})$ *is commutative* (resp. *nilpotent*, resp. *solvable*).

Assertion (i) follows from Prop. 4 by induction on i and (ii) follows from (i).

COROLLARY 2. *Let* $\mathfrak{r}$ *be the radical of* $\mathfrak{g}$. *If* $\mathfrak{g}$ *is decomposable,* $\mathfrak{r}$ *is decomposable.*

Indeed, $e(\mathfrak{r})$ is a solvable ideal of $\mathfrak{g}$ by Prop. 4 and Cor. 1, hence $e(\mathfrak{r}) = \mathfrak{r}$.

3. DECOMPOSITIONS OF DECOMPOSABLE ALGEBRAS

If $\mathfrak{g}$ is a Lie subalgebra of $\mathfrak{gl}(V)$ with radical $\mathfrak{r}$, the set of nilpotent elements of $\mathfrak{r}$ is a nilpotent ideal of $\mathfrak{g}$, the largest nilpotency ideal of the identity representation of $\mathfrak{g}$ (Chap. I, §5, no. 3, Cor. 6 of Th. 1). In this paragraph, we shall denote this ideal by $\mathfrak{n}_V(\mathfrak{g})$. It contains the nilpotent radical $[\mathfrak{g}, \mathfrak{g}] \cap \mathfrak{r}$ of $\mathfrak{g}$ (Chap. I, §5, no. 3, Th. 1).

PROPOSITION 5. *Let* $\mathfrak{g}$ *be a decomposable nilpotent Lie subalgebra of* $\mathfrak{gl}(V)$. *Let* $\mathfrak{t}$ *be the set of semi-simple elements of* $\mathfrak{g}$. *Then* $\mathfrak{t}$ *is a central subalgebra of* $\mathfrak{g}$, *and* $\mathfrak{g}$ *is the product of* $\mathfrak{t}$ *and* $\mathfrak{n}_V(\mathfrak{g})$ *as Lie algebras.*

If $x \in \mathfrak{t}$, $\mathrm{ad}_{\mathfrak{g}} x$ is semi-simple and nilpotent, hence zero, so that x is central in $\mathfrak{g}$. Consequently, $\mathfrak{t}$ is an ideal of $\mathfrak{g}$, and $\mathfrak{t} \cap \mathfrak{n}_V(\mathfrak{g}) = 0$. Since $\mathfrak{g}$ is decomposable, $\mathfrak{g} = \mathfrak{t} + \mathfrak{n}_V(\mathfrak{g})$, hence the proposition.

PROPOSITION 6. *Let $\mathfrak{g}$ be a decomposable Lie subalgebra of $\mathfrak{gl}(V)$. Let $\mathscr{T}$ be the set of commutative subalgebras of $\mathfrak{g}$ consisting of semi-simple elements, and $\mathscr{T}_1$ the set of maximal elements of $\mathscr{T}$. Let $\mathscr{H}$ be the set of Cartan subalgebras of $\mathfrak{g}$.*

(i) *For $\mathfrak{h} \in \mathscr{H}$, let $\varphi(\mathfrak{h})$ be the set of semi-simple elements of $\mathfrak{h}$. Then $\varphi(\mathfrak{h}) \in \mathscr{T}_1$.*

(ii) *For $\mathfrak{t} \in \mathscr{T}_1$, let $\psi(\mathfrak{t})$ be the commutant of $\mathfrak{t}$ in $\mathfrak{g}$. Then $\psi(\mathfrak{t}) \in \mathscr{H}$.*

(iii) *The maps φ and ψ are inverse bijections from $\mathscr{H}$ to $\mathscr{T}_1$ and from $\mathscr{T}_1$ to $\mathscr{H}$.*

(iv) *If k is algebraically closed, $\mathrm{Aut}_e(\mathfrak{g})$ operates transitively on $\mathscr{T}_1$.*

Let $\mathfrak{h} \in \mathscr{H}$, and put $\mathfrak{t} = \varphi(\mathfrak{h})$. By Prop. 5 and Cor. 2 of Prop. 3, $\mathfrak{t} \in \mathscr{T}$ and $\mathfrak{h} = \mathfrak{t} \times \mathfrak{n}_V(\mathfrak{h})$. For any subalgebra $\mathfrak{u}$ of $\mathfrak{g}$, we denote by $\psi(\mathfrak{u})$ the commutant of $\mathfrak{u}$ in $\mathfrak{g}$. Then $\mathfrak{h} \subset \psi(\mathfrak{t})$, and $\psi(\mathfrak{t}) \subset \mathfrak{g}^0(\mathfrak{h})$ since the elements of $\mathfrak{n}_V(\mathfrak{h})$ are nilpotent, so $\mathfrak{h} = \psi(\mathfrak{t})$. If $\mathfrak{t}' \in \mathscr{T}$ and $\mathfrak{t} \subset \mathfrak{t}'$, we have $\mathfrak{t}' \subset \psi(\mathfrak{t}) = \mathfrak{h}$ so $\mathfrak{t}' = \mathfrak{t}$, and hence $\mathfrak{t} \in \mathscr{T}_1$.

Let $\mathfrak{t} \in \mathscr{T}_1$, and put $\mathfrak{c} = \psi(\mathfrak{t})$. Let $\mathfrak{h}$ be a Cartan subalgebra of $\mathfrak{c}$. By §2, no. 3, Prop. 10, $\mathfrak{h} \in \mathscr{H}$ and $\mathfrak{t} \subset \mathfrak{h}$. Put $\mathfrak{t}_1 = \varphi(\mathfrak{h}) \in \mathscr{T}$. Then $\mathfrak{t} \subset \mathfrak{t}_1$ so $\mathfrak{t} = \mathfrak{t}_1$, and $\mathfrak{h} = \psi(\mathfrak{t}_1) = \psi(\mathfrak{t}) = \mathfrak{c}$ by the preceding. Thus, $\psi(\mathfrak{t}) \in \mathscr{H}$, and $\varphi(\psi(\mathfrak{t})) = \mathfrak{t}$.

We have thus proved (i), (ii) and (iii). Assume that k is algebraically closed. Since $\mathrm{Aut}_e(\mathfrak{g})$ operates transitively on $\mathscr{H}$ (§3, no. 2, Th. 1), $\mathrm{Aut}_e(\mathfrak{g})$ operates transitively on $\mathscr{T}_1$.

COROLLARY 1. *The Cartan subalgebras of $\mathfrak{g}$ are the centralizers of the regular semi-simple elements of $\mathfrak{g}$.*

If $x \in \mathfrak{g}$ is regular, $\mathfrak{g}^0(x)$ is a Cartan subalgebra of $\mathfrak{g}$ (§2, no. 3, Th. 1 (i)); moreover, if x is semi-simple $\mathfrak{g}^0(x)$ is the centralizer of x in $\mathfrak{g}$. Conversely, let $\mathfrak{h}$ be a Cartan subalgebra of $\mathfrak{g}$. There exists $\mathfrak{t} \in \mathscr{T}_1$ such that $\mathfrak{h} = \psi(\mathfrak{t})$. By §1, no. 2, Prop. 7, there exists $x \in \mathfrak{t}$ such that $\mathfrak{h} = \mathfrak{g}^0(x)$; since $x \in \mathfrak{t}$, $\mathfrak{g}^0(x) = \mathfrak{g}_0(x)$. By §3, no. 3, Th. 2 (ii), x is regular.

COROLLARY 2. *Assume in addition that $\mathfrak{g}$ is solvable. Then:*

(i) *The subgroup of $\mathrm{Aut}(\mathfrak{g})$ consisting of the $e^{\mathrm{ad}\, x}, x \in \mathscr{C}^\infty\mathfrak{g}$ (cf. §3, no. 4), operates transitively on $\mathscr{T}_1$.*

(ii) *If $\mathfrak{t} \in \mathscr{T}_1$, $\mathfrak{g}$ is the semi-direct product of $\mathfrak{t}$ and $\mathfrak{n}_V(\mathfrak{g})$.*

Assertion (i) follows from the fact that the group of the $e^{\mathrm{ad}\, x}$, $x \in \mathscr{C}^\infty\mathfrak{g}$, operates transitively on $\mathscr{H}$ (§3, no. 4, Th. 3).

We prove (ii). Let $\mathfrak{t} \in \mathscr{T}_1$, and let $\mathfrak{h} = \psi(\mathfrak{t})$ be the corresponding Cartan subalgebra of $\mathfrak{g}$. In view of Prop. 5, $\mathfrak{h} = \mathfrak{t} + \mathfrak{n}_V(\mathfrak{h}) \subset \mathfrak{t} + \mathfrak{n}_V(\mathfrak{g})$. On the other hand, $\mathfrak{g} = \mathfrak{h} + [\mathfrak{g}, \mathfrak{g}]$ (§2, no. 1, Cor. 3 of Prop. 4) and $[\mathfrak{g}, \mathfrak{g}] \subset \mathfrak{n}_V(\mathfrak{g})$, so $\mathfrak{g} = \mathfrak{t} + \mathfrak{n}_V(\mathfrak{g})$. But it is clear that $\mathfrak{t} \cap \mathfrak{n}_V(\mathfrak{g}) = \{0\}$. The algebra $\mathfrak{g}$ is thus the semi-direct product of $\mathfrak{t}$ and the ideal $\mathfrak{n}_V(\mathfrak{g})$.

PROPOSITION 7. *Let* $\mathfrak{g}$ *be a decomposable Lie subalgebra of* $\mathfrak{gl}(V)$.

(i) *There exists a Lie subalgebra* $\mathfrak{m}$ *of* $\mathfrak{g}$, *reductive in* $\mathfrak{gl}(V)$, *such that* $\mathfrak{g}$ *is the semi-direct product of* $\mathfrak{m}$ *and* $\mathfrak{n}_V(\mathfrak{g})$.

(ii) *Any two Lie subalgebras of* $\mathfrak{g}$ *with the properties in* (i) *are conjugate under* $\mathrm{Aut}_e(\mathfrak{g})$.

The radical $\mathfrak{r}$ of $\mathfrak{g}$ is decomposable (no. 2, Cor. 2 of Prop. 4). By Cor. 2 of Prop. 6, there exists a commutative subalgebra $\mathfrak{t}$ of $\mathfrak{r}$, consisting of semi-simple elements, such that $\mathfrak{r} = \mathfrak{t} \oplus \mathfrak{n}_V(\mathfrak{r})$. Since $\mathrm{ad}_{\mathfrak{g}}\mathfrak{t}$ consists of semi-simple elements, $\mathfrak{g}$ is the direct sum of $[\mathfrak{t}, \mathfrak{g}]$ and the centralizer $\mathfrak{z}$ of $\mathfrak{t}$ (Chap. I, §3, no. 5, Prop. 6). Since $[\mathfrak{t}, \mathfrak{g}] \subset \mathfrak{r}$, $\mathfrak{g} = \mathfrak{z} + \mathfrak{r}$. Consequently, if $\mathfrak{s}$ is a Levi subalgebra of $\mathfrak{z}$ (Chap. I, §6, no. 8), $\mathfrak{g} = \mathfrak{s} + \mathfrak{r}$, so $\mathfrak{s}$ is a Levi subalgebra of $\mathfrak{g}$. Put $\mathfrak{m} = \mathfrak{s} \oplus \mathfrak{t}$. Since $[\mathfrak{s}, \mathfrak{t}] = \{0\}$, $\mathfrak{m}$ is a Lie subalgebra of $\mathfrak{g}$, reductive in $\mathfrak{gl}(V)$ by Chap. I, §6, no. 5, Th. 4. Moreover,

$$\mathfrak{g} = \mathfrak{s} \oplus \mathfrak{r} = \mathfrak{s} \oplus \mathfrak{t} \oplus \mathfrak{n}_V(\mathfrak{r}) = \mathfrak{s} \oplus \mathfrak{t} \oplus \mathfrak{n}_V(\mathfrak{g}) = \mathfrak{m} \oplus \mathfrak{n}_V(\mathfrak{g})$$

since $\mathfrak{n}_V(\mathfrak{g}) = \mathfrak{n}_V(\mathfrak{r})$. Hence (i).

Now let $\mathfrak{m}'$ be a Lie subalgebra of $\mathfrak{g}$ complementary to $\mathfrak{n}_V(\mathfrak{g})$ and reductive in $\mathfrak{gl}(V)$. We show that $\mathfrak{m}'$ is conjugate to $\mathfrak{m}$ under $\mathrm{Aut}_e(\mathfrak{g})$. We have $\mathfrak{m}' = \mathfrak{s}' \oplus \mathfrak{t}'$, where $\mathfrak{s}' = [\mathfrak{m}', \mathfrak{m}']$ is semi-simple and the centre $\mathfrak{t}'$ of $\mathfrak{m}'$ consists of semi-simple elements. Then $\mathfrak{r} = \mathfrak{t} \oplus \mathfrak{n}_V(\mathfrak{g}) = \mathfrak{t}' \oplus \mathfrak{n}_V(\mathfrak{g})$. In view of Cor. 2 of Prop. 6, we are reduced to the case $\mathfrak{t} = \mathfrak{t}'$. Then $\mathfrak{s}' \subset \mathfrak{z}$; since $\dim \mathfrak{s}' = \dim \mathfrak{s}$, $\mathfrak{s}'$ is a Levi subalgebra of $\mathfrak{z}$. By Chap. I, §6, no. 8, Th. 5, there exists $x \in \mathfrak{n}_V(\mathfrak{z})$ such that $e^{\mathrm{ad}\, x}(\mathfrak{s}) = \mathfrak{s}'$; since x commutes with $\mathfrak{t}$, we also have $e^{\mathrm{ad}\, x}(\mathfrak{t}) = \mathfrak{t}$.

4. LINEAR LIE ALGEBRAS OF NILPOTENT ENDOMORPHISMS

Lemma 1. Let $\mathfrak{n}$ *be a Lie subalgebra of* $\mathfrak{gl}(V)$ *consisting of nilpotent endomorphisms, and* N *the subgroup* $\exp \mathfrak{n}$ *of* $\mathbf{GL}(V)$ (§3, no. 1, Lemma 1).

(i) *Let* ρ *be a finite dimensional linear representation of* $\mathfrak{n}$ *on* W, *such that the elements of* $\rho(\mathfrak{n})$ *are nilpotent,* W′ *a vector subspace of* W *stable under* ρ, ρ_1 *and* ρ_2 *the subrepresentation and quotient representation of* ρ *defined by* W′, π, π_1, π_2 *the representations of* N *compatible with* ρ, ρ_1, ρ_2 (§3, no. 1). *Then* π_1, π_2 *are the subrepresentation and quotient representation of* π *defined by* W′.

(ii) *Let* ρ_1, ρ_2 *be finite dimensional linear representations of* $\mathfrak{n}$ *such that the elements of* $\rho_1(\mathfrak{n})$ *and* $\rho_2(\mathfrak{n})$ *are nilpotent, and* π_1, π_2 *the representations of* N *compatible with* ρ_1, ρ_2. *Then* $\pi_1 \otimes \pi_2$ *is the representation of* N *compatible with* $\rho_1 \otimes \rho_2$.

(iii) *Let* ρ_1, ρ_2 *be finite dimensional linear representations of* $\mathfrak{n}$ *on vector spaces* V_1, V_2, *such that the elements of* $\rho_1(\mathfrak{n})$ *and* $\rho_2(\mathfrak{n})$ *are nilpotent,* ρ *the representation of* $\mathfrak{n}$ *on* $\mathrm{Hom}(V_1, V_2)$ *determined by* ρ_1, ρ_2. *Let* π_1, π_2 *be the representations of* N *compatible with* ρ_1, ρ_2, *and* π *the representation of* N *on* $\mathrm{Hom}(\mathrm{V}_1, \mathrm{V}_2)$ *determined by* π_1, π_2. *Then* π *is the representation of* N *compatible with* ρ.

Assertion (i) is clear. Let $\rho_1, \rho_2, \pi_1, \pi_2$ be as in (ii). If $x \in \mathfrak{n}$, we have, since $\rho_1(x) \otimes 1$ and $1 \otimes \rho_2(x)$ commute,

$$\begin{aligned}\exp(\rho_1(x) \otimes 1 + 1 \otimes \rho_2(x)) &= \exp(\rho_1(x) \otimes 1). \exp(1 \otimes \rho_2(x)) \\ &= (\exp \rho_1(x)) \otimes 1.1 \otimes (\exp \rho_2(x)) \\ &= (\exp \rho_1(x)) \otimes (\exp \rho_2(x)) \\ &= \pi_1(\exp x) \otimes \pi_2(\exp x) \\ &= (\pi_1 \otimes \pi_2)(\exp x),\end{aligned}$$

hence (ii). Let $\rho_1, \rho_2, \rho, \pi_1, \pi_2, \pi, \mathrm{V}_1, \mathrm{V}_2$ be as in (iii). If $v_1 \in \mathrm{End}\mathrm{V}_1$ and $v_2 \in \mathrm{End}\mathrm{V}_2$, denote by R_{v_1} and L_{v_2} the maps $u \mapsto uv_1$ and $u \mapsto v_2 u$ from $\mathrm{Hom}(\mathrm{V}_1, \mathrm{V}_2)$ to itself; these maps commute and $\rho(x)u = (\mathrm{L}_{\rho_2(x)} - \mathrm{R}_{\rho_1(x)})u$, so

$$\begin{aligned}\exp \rho(x).u &= \exp \mathrm{L}_{\rho_2(x)}. \exp \mathrm{R}_{-\rho_1(x)}.u \\ &= \mathrm{L}_{\exp \rho_2(x)}.\mathrm{R}_{\exp(-\rho_1(x))}.u \\ &= \mathrm{L}_{\pi_2(\exp x)}.\mathrm{R}_{\pi_1(\exp(-x))}.u \\ &= \pi(\exp x).u,\end{aligned}$$

hence (iii).

Lemma 2[2]. (i) *Let* W *be a vector subspace of* V *of dimension* d, D *the line* $\bigwedge^d \mathrm{W} \subset \bigwedge^d \mathrm{V}$, θ *the canonical representation of* $\mathfrak{gl}(\mathrm{V})$ *on* $\bigwedge \mathrm{V}$ (Chap. III, App.). *Let* $x \in \mathfrak{gl}(\mathrm{V})$. *Then* $x(\mathrm{W}) \subset \mathrm{W}$ *if and only if* $\theta(x)(\mathrm{D}) \subset \mathrm{D}$.

(ii) *Let* $(e_1, \ldots, e_n)$ *be the canonical basis of* k^n, θ *the canonical representation of* $\mathfrak{gl}(n, k)$ *on* $\bigwedge(k^n)$, *and* $x \in \mathfrak{gl}(n, k)$. *Then* $x \in \mathfrak{n}(n, k)$ *if and only if*

$$\theta(x)(e_{n-d+1} \wedge \cdots \wedge e_n) = 0$$

for $1 \leq d \leq n$.

(i) If $x(\mathrm{W}) \subset \mathrm{W}$, it is clear that $\theta(x)\mathrm{D} \subset \mathrm{D}$. Conversely, assume that $\theta(x)\mathrm{D} \subset \mathrm{D}$. Let u be a non-zero element of D and let $y \in \mathrm{W}$. Then $y \wedge u = 0$. Since $\theta(x)$ is a derivation of $\bigwedge \mathrm{V}$, this implies

$$\theta(x)y \wedge u + y \wedge \theta(x)u = 0.$$

Now $\theta(x)u \in ku$, so $y \wedge \theta(x)u = 0$ and consequently $\theta(x)y \wedge u = 0$. By *Algebra*, Chap. III, §7, no. 9, Prop. 13, this implies that $\theta(x)y \in \mathrm{W}$, i.e. $x(y) \in \mathrm{W}$, which proves that $x(\mathrm{W}) \subset \mathrm{W}$.

(ii) The condition stated in (ii) is clearly necessary for $x \in \mathfrak{n}(n, k)$. Assume that it is satisfied. By (i), x leaves

[2] In this lemma, k can be an arbitrary (commutative) field.

$$ke_{n-d+1} + \cdots + ke_n$$

stable, and since this holds for $d = 1, \ldots, n$, x is lower triangular. Put

$$x = (x_{ij})_{1 \leq i,j \leq n}.$$

We have $0 = x(e_n) = x_{nn}e_n$, so $x_{nn} = 0$. Let $i < n$, and assume that we have proved that $x_{jj} = 0$ for $j > i$. Then

$$0 = \theta(x)(e_i \wedge e_{i+1} \wedge \cdots \wedge e_n) = x_{ii}(e_i \wedge e_{i+1} \wedge \cdots \wedge e_n),$$

so $x_{ii} = 0$. Thus, $x \in \mathfrak{n}(n, k)$.

PROPOSITION 8. *Let* $\mathfrak{n}$ *be a Lie subalgebra of* $\mathfrak{gl}(\mathrm{V})$ *consisting of nilpotent elements,* $\mathfrak{q}$ *the normalizer of* $\mathfrak{n}$ *in* $\mathfrak{gl}(\mathrm{V})$. *There exists a finite dimensional vector space* E, *a representation* ρ *of* $\mathfrak{gl}(\mathrm{V})$ *on* E, *and a vector subspace* F *of* E, *satisfying the following conditions:*

(i) *the image under* ρ *of a homothety of* V *is diagonalizable;*

(ii) F *is stable under* $\rho(\mathfrak{q})$;

(iii) $\mathfrak{n}$ *is the set of* $x \in \mathfrak{gl}(\mathrm{V})$ *such that* $\rho(x)(\mathrm{F}) = 0$.

Let $n = \dim \mathrm{V}$. By Engel's theorem, V can be identified with k^n in such a way that $\mathfrak{n} \subset \mathfrak{n}(n, k)$. Let P be the algebra of polynomial functions on $\mathfrak{gl}(n, k)$. For $i = 0, 1, \ldots$, let P_i be the set of elements of P homogeneous of degree i. Let $\mathrm{N} = \exp \mathfrak{n}$, which is a subgroup of the strictly lower triangular group T. Let J be the set of elements of P that are zero on N; this is an ideal in P. Let $\mathrm{N_J}$ be the set of $x \in \mathfrak{gl}(n, k)$ such that $p(x) = 0$ for all $p \in \mathrm{J}$. Then $\mathrm{N} \subset \mathrm{N_J}$. Conversely, let $x \in \mathrm{N_J}$. Denote by p_{ij} the polynomial functions giving the entries of an element of $\mathfrak{gl}(n, k)$. The ideal J contains the p_{ij} (for $i < j$) and the $p_{ii} - 1$; hence $x \in \mathrm{T}$. On the other hand, if u is a linear form on $\mathfrak{gl}(n, k)$ which is zero on $\mathfrak{n}$, there exists $p_u \in \mathrm{P}$ such that $p_u(z) = u(\log z)$ for all $z \in \mathrm{T}$ (§3, no. 1, Lemma 1 (i)); we have $p_u \in \mathrm{J}$, so $u(\log x) = 0$. It follows that $\log x$ belongs to $\mathfrak{n}$, so $x \in \mathrm{N}$, proving that $\mathrm{N} = \mathrm{N_J}$.

For all $p \in \mathrm{P}$ and $g \in \mathbf{GL}_n(k)$, let $\lambda(g)p$ be the function $x \mapsto p(g^{-1}x)$ on $\mathfrak{gl}(n, k)$; then $\lambda(g)p \in \mathrm{P}$, $\lambda(g)$ is an automorphism of the algebra P, and λ is a representation of $\mathbf{GL}_n(k)$ on P which leaves each P_i stable. We show that

$$\mathrm{N} = \{x \in \mathbf{GL}_n(k) \mid \lambda(x)\mathrm{J} = \mathrm{J}\}. \tag{1}$$

If $x \in \mathrm{N}, p \in \mathrm{J}, y \in \mathrm{N}$, then $(\lambda(x)p)(y) = p(x^{-1}y) = 0$ since $x^{-1}y \in \mathrm{N}$; thus $\lambda(x)p \in \mathrm{J}$, so $\lambda(x)\mathrm{J} = \mathrm{J}$. Let $x \in \mathbf{GL}_n(k)$ be such that $\lambda(x)\mathrm{J} = \mathrm{J}$; let $p \in \mathrm{J}$; then $p(x^{-1}) = (\lambda(x)p)(e) = 0$, so $x^{-1} \in \mathrm{N_J} = \mathrm{N}$ and $x \in \mathrm{N}$. This proves (i).

The ideal J is of finite type (*Commutative Algebra*, Chap. III, §2, no. 10, Cor. 2 of Th. 2). Hence, there exists an integer q such that, if $\mathrm{W} = \mathrm{P}_0 + \mathrm{P}_1 + \cdots + \mathrm{P}_q$, then $\mathrm{J} \cap \mathrm{W}$ generates J as an ideal. Denote by λ_j (resp. λ') the subrepresentation of λ defined by $\mathrm{P_J}$ (resp. by W). By (1),

$$\mathrm{N} = \{x \in \mathbf{GL}_n(k) \mid \lambda'(x)(\mathrm{J} \cap \mathrm{W}) = \mathrm{J} \cap \mathrm{W}\}. \tag{2}$$

We show that, for all j, there exists a representation σ_j of the Lie algebra $\mathfrak{gl}(n,k)$ on P_j such that:

$$\sigma_j|\mathfrak{n}(n,k) \text{ is compatible (§3, no. 1) with } \lambda_j|\mathrm{T}. \tag{3}$$

$$\text{For all } x \in k.1_n,\ \sigma_j(x) \text{ is a homothety.} \tag{4}$$

Since λ_j is the jth symmetric power of λ_1, it suffices to prove the existence of σ_1, cf. Lemma 1. Now λ_1 is the contragredient representation of the representation γ of $\mathbf{GL}_n(k)$ on $\mathfrak{gl}(n,k)$ given by

$$\gamma(x)y = xy, \quad x \in \mathbf{GL}_n(k),\ y \in \mathfrak{gl}(n,k).$$

Let c be the representation of the Lie algebra $\mathfrak{gl}(n,k)$ on $\mathfrak{gl}(n,k)$ given by

$$c(x)y = xy, \quad x, y \in \mathfrak{gl}(n,k).$$

It is immediate that $c|\mathfrak{n}(n,k)$ and $\gamma|\mathrm{T}$ are compatible, and that $c(x)$ is a homothety for all $x \in k.1_n$. Thus, it suffices to take for σ_1 the dual representation of c (Chap. I, §3, no. 3).

Now let σ' be the representation of $\mathfrak{gl}(n,k)$ on W given by the direct sum of the σ_j, $0 \le j \le q$. In view of (2) and the relations

$$\lambda'(\exp(x)) = \exp(\sigma'(x)) \text{ and } \sigma'(\log(y)) = \log(\lambda'(y)), \quad x \in \mathfrak{n}(n,k),\ y \in \mathrm{T},$$

we have

$$\mathfrak{n} = \{x \in \mathfrak{n}(n,k) \mid \sigma'(x)(\mathrm{J} \cap \mathrm{W}) \subset \mathrm{J} \cap \mathrm{W}\}. \tag{5}$$

Let $d = \dim(\mathrm{J} \cap \mathrm{W})$, and let $\tau = \bigwedge^d \sigma'$. Let $\mathrm{D} = \bigwedge^d(\mathrm{J} \cap \mathrm{W})$. By (5) and Lemma 2 (i),

$$\mathfrak{n} = \{x \in \mathfrak{n}(n,k) \mid \tau(x)(\mathrm{D}) \subset \mathrm{D}\}. \tag{6}$$

But $\tau(\mathfrak{n}(n,k))$ consists of nilpotent endomorphisms, so (6) can also be written

$$\mathfrak{n} = \{x \in \mathfrak{n}(n,k) \mid \tau(x)(\mathrm{D}) = 0\}. \tag{7}$$

Now let $\mathrm{E} = \bigwedge^d \mathrm{W} \oplus \bigwedge^1 \mathrm{V} \oplus \bigwedge^2 \mathrm{V} \oplus \cdots \oplus \bigwedge^n \mathrm{V}$; let ρ be the direct sum of τ and the canonical representations of $\mathfrak{gl}(n,k)$ on $\bigwedge^1 \mathrm{V}, \ldots, \bigwedge^n \mathrm{V}$. Let $E_0 \subset \mathrm{E}$ be the sum of $\mathrm{D} = \bigwedge^d(\mathrm{J} \cap \mathrm{W})$ and the lines generated by $e_{n-j+1} \wedge \cdots \wedge e_n$ for $j = 1, \ldots, n$. By (7) and Lemma 2 (ii),

$$\mathfrak{n} = \{x \in \mathfrak{gl}(\mathrm{V}) \mid \rho(x)(\mathrm{E}_0) = 0\}. \tag{8}$$

It is immediate that, if $x \in k.1_n$, $\rho(x)$ is diagonalizable. Finally, if F is the set of elements of E annihilated by $\rho(\mathfrak{n})$, F is stable under $\rho(\mathfrak{q})$ (Chap. I, §3, no. 5, Prop. 5), and by (8),

$$\mathfrak{n} = \{x \in \mathfrak{gl}(\mathrm{V}) \mid \rho(x)(\mathrm{F}) = 0\}. \tag{9}$$

5. CHARACTERIZATIONS OF DECOMPOSABLE LIE ALGEBRAS

Every decomposable Lie algebra is generated as a vector space (and *a fortiori* as a Lie algebra) by the set of its elements that are either semi-simple or nilpotent. Conversely:

THEOREM 1. *Let* $\mathfrak{g}$ *be a Lie subalgebra of* $\mathfrak{gl}(\mathrm{V})$ *and let* X *be a subset of* $\mathfrak{g}$ *generating* $\mathfrak{g}$ *as a Lie algebra over* k. *If every element of* X *is either semi-simple or nilpotent,* $\mathfrak{g}$ *is decomposable.*

a) $\mathfrak{g}$ *is commutative.*

The semi-simple (resp. nilpotent) elements of $\mathfrak{g}$ form a vector subspace $\mathfrak{g}_s$ (resp. $\mathfrak{g}_n$). The assumption is equivalent to $\mathfrak{g} = \mathfrak{g}_s \oplus \mathfrak{g}_n$, hence the fact that $\mathfrak{g}$ is decomposable.

b) $\mathfrak{g}$ *is reductive.*

Then $\mathfrak{g} = \mathfrak{g}' \times \mathfrak{c}$ with $\mathfrak{g}'$ semi-simple and $\mathfrak{c}$ commutative. By Prop. 2, $\mathfrak{g}'$ is decomposable. Let $x = a + b \in \mathfrak{g}$ with $a \in \mathfrak{g}'$, $b \in \mathfrak{c}$. Let a_s, a_n, b_s, b_n be the semi-simple and nilpotent components of a, b. Since a_s, a_n, b_s, b_n mutually commute, the semi-simple and nilpotent components of x are $a_s + b_s, a_n + b_n$. Now $a_s, a_n \in \mathfrak{g}'$. If x is semi-simple, $x = a_s + b_s$; since $a_s \in \mathfrak{g}'$, we have $b_s \in \mathfrak{g}$, so $b_s \in \mathfrak{c}$ since b_s commutes with $\mathfrak{g}$; consequently, $a = a_s$ and $b = b_s$. Similarly, if x is nilpotent, $a = a_n$ and $b = b_n$. It follows that the projections on $\mathfrak{c}$ of the elements of X are either semi-simple or nilpotent; by *a*), this implies that $\mathfrak{c}$ is decomposable. Retaining the preceding notation, but without the assumption on x, we now have $b_s, b_n \in \mathfrak{c}$, so $a_s + b_s, a_n + b_n \in \mathfrak{g}$, which proves the theorem in this case.

c) General case.

We assume that the theorem is proved for Lie algebras of dimension $< \dim \mathfrak{g}$ and prove it for $\mathfrak{g}$.

Let $\mathfrak{n}$ be the largest ideal of nilpotency of the identity representation of $\mathfrak{g}$. If $\mathfrak{n} = 0$, $\mathfrak{g}$ has an injective semi-simple representation, and so is reductive. Assume that $\mathfrak{n} \neq 0$. Let $\mathfrak{p}$ be the normalizer of $\mathfrak{n}$ in $\mathfrak{gl}(\mathrm{V})$. There exist E, ρ, F satisfying the conditions of Prop. 8. Since $\mathfrak{g} \subset \mathfrak{p}$, $\rho(\mathfrak{g})$ leaves F stable; let ρ_0 be the representation $u \mapsto \rho(u)|\mathrm{F}$ of $\mathfrak{g}$ on F; we have $\mathfrak{n} = \mathrm{Ker}\rho_0$. The image under ρ of every semi-simple (resp. nilpotent) element of $\mathfrak{gl}(\mathrm{V})$ is semi-simple (resp. nilpotent) (Prop. 2). The algebra $\rho_0(\mathfrak{g})$ is thus generated by its semi-simple elements and its nilpotent elements. By the induction hypothesis, $\rho_0(\mathfrak{g})$ is decomposable.

Let $x \in \mathfrak{g}$, and let x_s, x_n be its semi-simple and nilpotent components. By Prop. 2, the semi-simple and nilpotent components of $\rho(x)$ are $\rho(x_s), \rho(x_n)$. Since $\rho_0(\mathfrak{g})$ is decomposable, there exist $y, z \in \mathfrak{g}$ such that

$$\rho_0(y) = \rho(x_s)|\mathrm{F}, \quad \rho_0(z) = \rho(x_n)|\mathrm{F}.$$

Then $x_s \in y + \mathfrak{n}, x_n \in z + \mathfrak{n}$, so $x_s, x_n \in \mathfrak{g}$. Q.E.D.

COROLLARY 1. *Every subalgebra of* $\mathfrak{gl}(V)$ *generated by its decomposable subalgebras is decomposable.*

This is clear.

COROLLARY 2. *Let* $\mathfrak{g}$ *be a Lie subalgebra of* $\mathfrak{gl}(V)$. *Then* $[\mathfrak{g}, \mathfrak{g}]$ *is decomposable.*

Let $\mathfrak{r}$ be the radical of $\mathfrak{g}$, $\mathfrak{s}$ a Levi subalgebra of $\mathfrak{g}$ (Chap. I, §6, no. 8). Then

$$[\mathfrak{g}, \mathfrak{g}] = [\mathfrak{s}, \mathfrak{s}] + [\mathfrak{s}, \mathfrak{r}] + [\mathfrak{r}, \mathfrak{r}] = \mathfrak{s} + [\mathfrak{g}, \mathfrak{r}].$$

The algebra $[\mathfrak{g}, \mathfrak{r}]$ is decomposable since all of its elements are nilpotent (Chap. I, §5, no. 3). On the other hand, $\mathfrak{s}$ is decomposable (Prop. 2). It follows that $[\mathfrak{g}, \mathfrak{g}]$ is decomposable (Cor. 1).

COROLLARY 3. *Let* $\mathfrak{g}$ *be a Lie subalgebra of* $\mathfrak{gl}(V)$, *and let* X *be a subset of* $\mathfrak{g}$ *generating* $\mathfrak{g}$ *(as a Lie algebra over* k*).*

(i) *The decomposable envelope* $e(\mathfrak{g})$ *of* $\mathfrak{g}$ *is generated by the semi-simple and nilpotent components of the elements of* X.

(ii) *If* k' *is an extension of* k, $e(\mathfrak{g} \otimes_k k') = e(\mathfrak{g}) \otimes_k k'$; *and* $\mathfrak{g}$ *is decomposable if and only if* $\mathfrak{g} \otimes_k k'$ *is decomposable.*

Let $\tilde{\mathfrak{g}}$ be the subalgebra of $\mathfrak{gl}(V)$ generated by the semi-simple and nilpotent components of the elements of X. Then $\mathfrak{g} \subset \tilde{\mathfrak{g}} \subset e(\mathfrak{g})$; by Th. 1, $\tilde{\mathfrak{g}}$ is decomposable, so $\tilde{\mathfrak{g}} = e(\mathfrak{g})$, which proves (i). Assertion (ii) follows, since X generates the k'-algebra $\mathfrak{g} \otimes_k k'$.

COROLLARY 4. *Let* $\mathfrak{g}$ *be a decomposable Lie subalgebra of* $\mathfrak{gl}(V)$. *Let* $\mathscr{T}$ *be the set of commutative subalgebras of* $\mathfrak{g}$ *consisting of semi-simple elements* (cf. Prop. 6). *The maximal elements of* $\mathscr{T}$ *all have the same dimension.*

Let k' be an algebraically closed extension of k and $V' = V \otimes_k k'$, $\mathfrak{g}' = \mathfrak{g} \otimes_k k'$. Let $\mathfrak{t}_1, \mathfrak{t}_2$ be maximal elements of $\mathscr{T}$, $\mathfrak{t}'_i = \mathfrak{t}_i \otimes_k k'$, $\mathfrak{h}_i$ the commutant of $\mathfrak{t}_i$ in $\mathfrak{g}$, $\mathfrak{h}'_i = \mathfrak{h}_i \otimes_k k'$. Then $\mathfrak{h}_i$ is a Cartan subalgebra of $\mathfrak{g}$ (Prop. 6) so $\mathfrak{h}'_i$ is a Cartan subalgebra of $\mathfrak{g}'$. Then $\mathfrak{h}_i = \mathfrak{t}_i \times \mathfrak{n}_V(\mathfrak{h}_i)$, hence $\mathfrak{h}'_i = \mathfrak{t}'_i \times \mathfrak{n}_{V'}(\mathfrak{h}'_i)$, so that $\mathfrak{t}'_i$ is the set of semi-simple elements of $\mathfrak{h}'_i$. Since $\mathfrak{g}'$ is decomposable (Cor. 3), $\mathfrak{t}'_1$ and $\mathfrak{t}'_2$ are conjugate under $\mathrm{Aut}_e(\mathfrak{g}')$ (Prop. 6), so $\dim \mathfrak{t}_1 = \dim \mathfrak{t}_2$.

THEOREM 2. *Let* $\mathfrak{g}$ *be a Lie subalgebra of* $\mathfrak{gl}(V)$. *The following conditions are equivalent:*

(i) $\mathfrak{g}$ *is decomposable;*

(ii) *every Cartan subalgebra of* $\mathfrak{g}$ *is decomposable;*

(iii) $\mathfrak{g}$ *has a decomposable Cartan subalgebra;*

(iv) *the radical of* $\mathfrak{g}$ *is decomposable.*

(i) $\Longrightarrow$ (ii): This follows from Cor. 2 of Prop. 3.

(ii) $\Longrightarrow$ (i): This follows from Cor. 1 of Th. 1, since $\mathfrak{g}$ is generated by its Cartan subalgebras (§2, no. 3, Cor. 3 of Th. 1).

(ii) $\Longrightarrow$ (iii): This is clear.

(iii) $\Longrightarrow$ (ii): By Cor. 3 of Th. 1, we can assume that k is algebraically closed. The Cartan subalgebras of $\mathfrak{g}$ are then conjugate under the elementary automorphisms of $\mathfrak{g}$ (§3, no. 2, Th. 1); in view of Remark 1 of §3, no. 1, it follows that, if one of these is decomposable, they all are.

(i) $\Longrightarrow$ (iv): This follows from Cor. 2 of Prop. 4.

(iv) $\Longrightarrow$ (i): Assume that the radical $\mathfrak{r}$ of $\mathfrak{g}$ is decomposable. Let $\mathfrak{s}$ be a Levi subalgebra of $\mathfrak{g}$; it is decomposable (Prop. 2). Hence $\mathfrak{g} = \mathfrak{s} + \mathfrak{r}$ is decomposable (Cor. 1 of Th. 1).

APPENDIX I
POLYNOMIAL MAPS AND ZARISKI TOPOLOGY

In this appendix, k is assumed to be infinite.

1. ZARISKI TOPOLOGY

Let V be a finite dimensional vector space. We denote by A_V the algebra of polynomial functions on V with values in k (*Algebra*, Chap. IV, §5, no. 10, Def. 4). This is a graded algebra; its component of degree 1 is the *dual* V^* of V, and the injection of V^* into A_V extends to an *isomorphism from the symmetric algebra* $\mathbf{S}(V^*)$ *to* A_V (*Algebra*, Chap. IV, §5, no. 11, Remark 2).

If $(e_1, \dots, e_n)$ is a basis of V, and $(X_1, \dots, X_n)$ a sequence of indeterminates, the map from $k[X_1, \dots, X_n]$ to A_V that takes any element f of $k[X_1, \dots, X_n]$ to the function

$$\sum_{i=1}^{n} \lambda_i e_i \mapsto f(\lambda_1, \dots, \lambda_n)$$

is an isomorphism of algebras (*Algebra*, Chap. IV, §5, no. 10, Cor. of Prop. 19).

PROPOSITION 1. *Let* H *be the set of algebra homomorphisms from* A_V *to* k. *For any* $x \in V$, *let* h_x *be the homomorphism* $f \mapsto f(x)$ *from* A_V *to* k. *Then, the map* $x \mapsto h_x$ *is a bijection from* V *to* H.

Indeed, let H′ be the set of algebra homomorphisms from $k[X_1, \dots, X_n]$ to k. The map $\chi \mapsto (\chi(X_1), \dots, \chi(X_n))$ is clearly a bijection from H′ to k^n.

COROLLARY. *For any* $x \in V$, *let* $\mathfrak{m}_x = \operatorname{Ker}(h_x)$. *Then the map* $x \mapsto \mathfrak{m}_x$ *is a bijection from* V *to the set of ideals* $\mathfrak{m}$ *of* A_V *such that* $A_V/\mathfrak{m} = k$.

A subset F of V is said to be *closed* if there exists a family $(f_i)_{i \in I}$ of elements of A_V such that

$x \in \mathrm{F} \Longleftrightarrow x \in \mathrm{V}$ and $f_i(x) = 0$ for all $i \in \mathrm{I}$.

It is clear that $\varnothing$ and V are closed, and that any intersection of closed sets is closed. If F is defined by the vanishing of the f_i and F′ by that of the f'_j, $\mathrm{F} \cup \mathrm{F}'$ is defined by the vanishing of the $f_i f'_j$, and hence is closed. Thus, there exists a topology on V such that the closed sets for this topology are exactly the closed sets in the above sense. This topology is called the *Zariski topology* on V. For any $f \in \mathrm{A_V}$, we denote by V_f the set of $x \in \mathrm{V}$ such that $f(x) \neq 0$; this is an open subset of V. It is clear that the V_f form a base of the Zariski topology. (If k is a topological field, the canonical topology of V is finer than the Zariski topology.)

The map $x \mapsto \mathfrak{m}_x$ of the Cor. of Prop. 1 can be considered as a map ε from V to the prime spectrum $\mathrm{Spec}(\mathrm{A_V})$ of $\mathrm{A_V}$ (*Commutative Algebra*, Chap. II, §4, no. 3, Def. 4). It is immediate that the Zariski topology is the inverse image under ε of the topology of $\mathrm{Spec}(\mathrm{A_V})$.

PROPOSITION 2. *The vector space* V, *equipped with the Zariski topology, is an irreducible noetherian space. In particular, every non-empty open subset of* V *is dense.*

Since $\mathrm{A_V}$ is noetherian, $\mathrm{Spec}(\mathrm{A_V})$ is noetherian (*Commutative Algebra*, Chap. II, §4, no. 3, Cor. 7 of Prop. 11), and every subspace of a noetherian space is noetherian (*loc. cit.*, no. 2, Prop. 8). With the notation of the Cor. of Prop. 1, the intersection of the $\mathfrak{m}_x$ is $\{0\}$, and $\{0\}$ is a prime ideal of $\mathrm{A_V}$; thus V is irreducible (*loc. cit.*, no. 3, Prop. 14).

2. DOMINANT POLYNOMIAL MAPS

Let V, W be finite dimensional vector spaces. Let f be a polynomial map from V to W (*Algebra*, Chap. IV, §5, no. 10, Def. 4). If $\psi \in \mathrm{A_W}$, $\psi \circ f \in \mathrm{A_V}$ (*loc. cit.*, Prop. 17). The map $\psi \mapsto \psi \circ f$ is a homomorphism from $\mathrm{A_W}$ to $\mathrm{A_V}$, said to be *associated* to f. Its kernel consists of the functions $\psi \in \mathrm{A_W}$ which vanish on $f(\mathrm{V})$ (and hence also on the *closure* of $f(\mathrm{V})$ in the Zariski topology).

DEFINITION 1. *A polynomial map* $f : \mathrm{V} \to \mathrm{W}$ *is said to be dominant if the homomorphism from* $\mathrm{A_W}$ *to* $\mathrm{A_V}$ *associated to* f *is injective.*

In view of the preceding, f is dominant if and only if $f(\mathrm{V})$ is *dense* in W in the Zariski topology.

PROPOSITION 3. *Assume that k is algebraically closed. Let* $f : \mathrm{V} \to \mathrm{W}$ *be a dominant polynomial map. The image under f of any dense open subset of* V *contains a dense open subset of* W.

It suffices to prove that, for every non-zero element φ of $\mathrm{A_V}$, $f(\mathrm{V}_\varphi)$ contains a dense open subset of W. Identify $\mathrm{A_W}$ with a subalgebra of $\mathrm{A_V}$ by means of the homomorphism associated to f. There exists a non-zero element ψ of $\mathrm{A_W}$ such that every homomorphism $w : \mathrm{A_W} \to k$ which does not annihilate ψ extends to a homomorphism $v : \mathrm{A_V} \to k$ which does not annihilate φ (*Commutative Algebra*, Chap. V, §3, no. 1, Cor. 3 of Th. 1). Now such a w (resp. v) can be identified with an element of W_ψ (resp. of V_φ) and to say that v extends w means that $f(v) = w$. Hence, $\mathrm{W}_\psi \subset f(\mathrm{V}_\varphi)$. Q.E.D.

Let $f : \mathrm{V} \to \mathrm{W}$ be a polynomial map, and $x_0 \in \mathrm{V}$. The map $h \mapsto f(x_0+h)$ from V to W is polynomial. Decompose it into a finite sum of homogeneous polynomial maps:

$$f(x_0 + h) = f(x_0) + \mathrm{D}_1(h) + \mathrm{D}_2(h) + \cdots$$

where $\mathrm{D}_i : \mathrm{V} \to \mathrm{W}$ is homogeneous of degree i (*Algebra*, Chap. IV, §5, no. 10, Prop. 19). The linear map D_1 is called the *tangent linear map of f at x_0*. We denote it by $\mathrm{D}f(x_0)$.

PROPOSITION 4. *Let $f : \mathrm{V} \to \mathrm{W}$ be a polynomial map. Assume that there exists $x_0 \in \mathrm{V}$ such that $(\mathrm{D}f)(x_0)$ is surjective. Then f is dominant.*

Applying a translation in V and one in W, we can assume that $x_0 = 0$ and $f(x_0) = 0$. The decomposition of f as a sum of homogeneous elements can then be written

$$f = f_1 + f_2 + \cdots \text{ with } \deg f_i = i,$$

and the linear map f_1 is surjective by hypothesis. Suppose that f is not dominant. Then there exists a non-zero element ψ of $\mathrm{A_W}$ such that $\psi \circ f = 0$. Let $\psi = \psi_m + \psi_{m+1} + \cdots$ be the decomposition of ψ into homogeneous elements, with $\deg \psi_i = i$ and $\psi_m \neq 0$. Then

$$\begin{aligned} 0 = \psi \circ f &= \psi_m \circ f + \psi_{m+1} \circ f + \cdots \\ &= \psi_m \circ f_1 + \rho, \end{aligned}$$

where ρ is a sum of homogeneous polynomial maps of degrees $> m$. It follows that $\psi_m \circ f_1 = 0$. Since f_1 is surjective, $\psi_m = 0$, a contradiction.

COROLLARY. *If k is algebraically closed and if f satisfies the assumptions of Prop. 4, the image under f of any dense open subset of* V *contains a dense open subset of* W.

This follows from Props. 3 and 4.

APPENDIX II
A CONNECTEDNESS PROPERTY

Lemma 1. Let X *be a connected topological space and* Ω *a dense open subset of* X*. If, for any* $x \in \mathrm{X}$*, there exists a neighbourhood* V *of* x *such that* $\mathrm{V} \cap \Omega$ *is connected, then* Ω *is connected.*

Indeed, let Ω_0 be a non-empty open and closed subset of Ω. Let $x \in \mathrm{X}$ and let V be a neighbourhood of x such that $\mathrm{V} \cap \Omega$ is connected. If $x \in \overline{\Omega}_0$,

$$(\mathrm{V} \cap \Omega) \cap \Omega_0 = \mathrm{V} \cap \Omega_0 \neq \varnothing,$$

so $\mathrm{V} \cap \Omega \subset \Omega_0$. Thus, since Ω is dense in X, $\overline{\Omega}_0$ is a neighbourhood of x. Consequently, $\overline{\Omega}_0$ is non-empty, open and closed, and since X is connected, $\overline{\Omega}_0 = \mathrm{X}$. Since Ω_0 is closed in Ω, this implies that $\Omega_0 = \Omega \cap \overline{\Omega}_0 = \Omega$, which proves that Ω is connected.

Lemma 2. Let U *be an open ball in* $\mathbf{C}^n$ *and* $f : \mathrm{U} \to \mathbf{C}$ *a holomorphic function, not identically zero. Let* A *be a subset of* U *such that* $f = 0$ *on* A*. Then* $\mathrm{U} - \mathrm{A}$ *is dense in* U *and connected.*

The density of $\mathrm{U} - \mathrm{A}$ follows from *Differentiable and Analytic Manifolds, Results*, 3.2.5. Assume first that $n = 1$. If $a \in \mathrm{A}$, the power series expansion of f about a (*Differentiable and Analytic Manifolds, Results*, 3.2.1) is not reduced to 0, and it follows that there exists a neighbourhood V_a of a in U such that f does not vanish on $\mathrm{V}_a - \{a\}$. Thus, a is isolated in A, which proves that A is a *discrete* subset of U, hence countable since U is countable at infinity. Let $x, y \in \mathrm{U} - \mathrm{A}$. The union of the real affine lines joining x (resp. y) to a point of A is meagre (*General Topology*, Chap. IX, §5, no. 2, Def. 2). Hence, there exists $z \in \mathrm{U} - \mathrm{A}$ such that neither of the segments $[x, z]$ and $[y, z]$ meets A. The points x, y, z thus belong to the same connected component of $\mathrm{U} - \mathrm{A}$, which proves the lemma in the case $n = 1$. We turn to the general case. We can assume that A is the set of zeros of f (*General Topology*, Chap. I, §11, no. 1, Prop. 1). Let $x, y \in \mathrm{U} - \mathrm{A}$ and let L be an affine line containing x and y. The restriction of f to $\mathrm{L} \cap \mathrm{U}$ is not identically zero since $x \in \mathrm{L} \cap \mathrm{U}$. By what has already been proved, x and y belong to the same connected component of $(\mathrm{L} \cap \mathrm{U}) - (\mathrm{L} \cap \mathrm{A})$ and hence to the same connected component of $\mathrm{U} - \mathrm{A}$.

Lemma 3. Let X *be a finite dimensional connected complex-analytic manifold and let* A *be a subset of* X *satisfying the following condition:*

> *For any* $x \in \mathrm{X}$*, there exists an analytic function germ* f_x*, not vanishing at* x*, such that the germ of* A *at* x *is contained in the germ at* x *of the set of zeros of* f_x*.*

Then $\mathrm{X} - \mathrm{A}$ *is dense in* X *and connected.*

The density of $\mathrm{X} - \mathrm{A}$ follows from *Differentiable and Analytic Manifolds, Results*, 3.2.5. We can assume that A is closed (*General Topology*, Chap. I,

§11, no. 1, Prop. 1). For any $x \in \mathrm{X}$, there exists an open neighbourhood V of x and an isomorphism c from V to an open ball in $\mathbf{C}^n$ such that $c(\mathrm{A} \cap \mathrm{V})$ is contained in the set of zeros of a holomorphic function not identically zero on $c(\mathrm{V})$. Then, by Lemma 2, $\mathrm{V} \cap (\mathrm{X} - \mathrm{A})$ is connected. In view of Lemma 1, this proves that $\mathrm{X} - \mathrm{A}$ is connected.

EXERCISES

All Lie algebras and modules over them are assumed to be finite dimensional over k; from §3 onwards, k is assumed to be of characteristic zero.

§1

1) Assume that k has characteristic $p > 0$. Let V be a vector space, S a finite set. A map $r : \mathrm{S} \to \mathrm{End(V)}$ satisfies condition (AC) if and only if there exists a power q of p such that $[s^q, s'^q] = 0$ for all $s, s' \in \mathrm{S}$. (Use Chap. I, §1, Exerc. 19, formula (1).)

2) Assume that k is perfect. Let V be a finite dimensional vector space, and $u, v \in \mathrm{End(V)}$. Let u_s, u_n, v_s, v_n be the semi-simple and nilpotent components of u, v. The following conditions are equivalent: (i) there exists an integer m such that $(\mathrm{ad}\, u)^m v = 0$; (ii) u_s and v commute. (To prove (i) $\Longrightarrow$ (ii), reduce to the case where k is algebraically closed and use Lemma 1 (ii).)

3) We make the assumptions in no. 2. Assume that k is infinite and that condition (AC) is satisfied. Let k' be a perfect extension of k. Let $\lambda : \mathrm{S} \to k$ be such that $\mathrm{V}^{\lambda}(\mathrm{S}) \neq 0$. Put

$$\mathrm{V}' = \mathrm{V} \otimes_k k', \quad \mathrm{S}' = \mathrm{S} \otimes_k k'.$$

Let $r' : \mathrm{S}' \to \mathrm{End(V')}$ be the linear map obtained from r by extension of scalars. There exists a unique map $\lambda' : \mathrm{S}' \to k'$ such that $\mathrm{V}^{\lambda}(\mathrm{S}) \otimes_k k' = \mathrm{V}'^{\lambda'}(\mathrm{S}')$. (Reduce to the case where $\mathrm{V} = \mathrm{V}^{\lambda}(\mathrm{S})$. Let P be a polynomial function on S and q a power of the characteristic exponent of k dividing dim V, such that $\lambda^q = \mathrm{P}$. Let P$'$ be the polynomial function on S$'$ which extends P. For each $s' \in \mathrm{S}'$, there exists a $\lambda'(s') \in k'$ such that $\lambda'(s')^q = \mathrm{P}'(s')$. Show that the characteristic polynomial of $r'(s')$ is $(\mathrm{X} - \lambda'(s'))^{\dim \mathrm{V}}$.)

4) Assume that k has characteristic zero. Let $\mathfrak{g} = \mathfrak{sl}(3, k)$ and let $\mathfrak{a}$ be the subalgebra of $\mathfrak{g}$ generated by a diagonal matrix with eigenvalues $1, -1, 0$. Show that $\mathfrak{a}$ is reductive in $\mathfrak{g}$, that the commutant $\mathfrak{m}$ of $\mathfrak{a}$ in $\mathfrak{g}$ consists of the diagonal matrices of trace zero, and that the commutant of $\mathfrak{m}$ in $\mathfrak{g}$ is equal to $\mathfrak{m}$, and hence is distinct from $\mathfrak{a}$ (cf. no. 5, *Remark*).

¶ 5) Assume that k is infinite. Let $\mathfrak{g}$ be a Lie algebra and V a $\mathfrak{g}$-module. If n is an integer ≥ 0, denote by V_n the set of $v \in V$ such that $x^n v = 0$ for all $x \in \mathfrak{g}$.

a) Show that, if $v \in V_n$, $x, y \in \mathfrak{g}$, then

$$\left(\sum_{i=1}^{n} x^{n-i} y x^{i-1}\right) v = 0.$$

(Use the fact that $(x + ty)^n v = 0$ for all $t \in k$.)

Replacing y by $[x, y]$ in this formula, deduce[3] that $(x^n y - y x^n)v = 0$, and hence that $x^n y v = 0$.

b) Show that V_n is a $\mathfrak{g}$-submodule of V (use *a*)). In particular $V^0(\mathfrak{g}) = \bigcup_n V_n$ is a $\mathfrak{g}$-submodule of V.

c) Assume that $k = \mathbf{R}$ or $\mathbf{C}$, and denote by G a simply-connected Lie group with Lie algebra $\mathfrak{g}$; the action of $\mathfrak{g}$ on V defines a law of operation of G on V (Chap. III, §6, no. 1). Show that an element $v \in V$ belongs to V_n if and only if $(s - 1)^n v = 0$ for all $s \in G$; in particular $V^0(\mathfrak{g}) = V^1(G)$.

6) The notations are those of Exerc. 12 of Chap. I, §3. In particular, $\mathfrak{g}$ is a Lie algebra, M a $\mathfrak{g}$-module, and $H^p(\mathfrak{g}, M) = Z^p(\mathfrak{g}, M)/B^p(\mathfrak{g}, M)$ is the *cohomology space* of degree p of $\mathfrak{g}$ with values in M.

a) Show that $B^p(\mathfrak{g}, M)$ and $Z^p(\mathfrak{g}, M)$ are stable under the natural representation θ of $\mathfrak{g}$ on the space of cochains $C^p(\mathfrak{g}, M)$. It follows that there is a representation of $\mathfrak{g}$ on $H^p(\mathfrak{g}, M)$. Show that this representation is *trivial* (use the formula $\theta = di + id$, *loc. cit.*).

b) Let $\overline{k}$ be an algebraic closure of k. Let $x \in \mathfrak{g}$ and let x_M be the corresponding endomorphism of M. Let $\lambda_1, \ldots, \lambda_n$ (resp. $\mu_1, \ldots, \mu_m$) be the eigenvalues (in $\overline{k}$) of $\mathrm{ad}_{\mathfrak{g}} x$ (resp. of x_M), repeated according to their multiplicity. Show that the eigenvalues of the endomorphism $\theta(x)$ of $C^p(\mathfrak{g}, M)$ are the $\mu_j - (\lambda_{i_1} + \cdots + \lambda_{i_p})$, where $1 \leq j \leq m$ and

$$1 \leq i_1 < i_2 < \cdots < i_p \leq n.$$

Deduce, using *a*), that $H^p(\mathfrak{g}, M) = 0$ if none of the $\mu_j - (\lambda_{i_1} + \cdots + \lambda_{i_p})$ is zero.

c) Assume that the representation $\mathfrak{g} \to \mathrm{End}(M)$ is *faithful*, and that x satisfies the condition:

$$\mu_{j_1} + \cdots + \mu_{j_p} \neq \mu_{k_1} + \cdots + \mu_{k_{p+1}} \qquad (S_p)$$

for all $j_1, \ldots, j_p, k_1, \ldots, k_{p+1} \in [1, m]$.

Show that we then have $H^p(\mathfrak{g}, M) = 0$ (remark that the eigenvalues λ_i of $\mathrm{ad}_{\mathfrak{g}} x$ are of the form $\mu_j - \mu_k$, and apply *b*)).

---

[3] This proof was communicated to us by G. SELIGMAN.

7) Let $\mathfrak{g}$ be a nilpotent Lie algebra and V a $\mathfrak{g}$-module such that $V^0(\mathfrak{g}) = 0$. Show that $H^p(\mathfrak{g}, V) = 0$ for all $p \geq 0$. (Reduce to the case where $V = V^\lambda(\mathfrak{g})$, with $\lambda \neq 0$ and choose an element $x \in \mathfrak{g}$ such that $\lambda(x) \neq 0$. Apply Exerc. 6 *b*), remarking that the λ_i are all zero and the μ_j are all equal to $\lambda(x)$.)

Recover the Cor. of Prop. 9 (take $p = 1$).[4]

¶ 8) Assume that k is of characteristic $p > 0$. Let $\mathfrak{g}$ be a Lie algebra over k with basis $(e_1, \ldots, e_n)$. Denote by U the enveloping algebra of $\mathfrak{g}$ and C the centre of U. For $i = 1, \ldots, n$, choose a non-zero *p-polynomial* f_i, of degree d_i, such that $f_i(\operatorname{ad} e_i) = 0$; then $f_i(e_i) \in C$, cf. Chap. I, §7, Exerc. 5. Put $z_i = f_i(e_i)$.

a) Show that $z_1, \ldots, z_n$ are algebraically independent. If $A = k[z_1, \ldots, z_n]$, show that U is a free A-module with basis the monomials $e_1^{\alpha_1} \ldots e_n^{\alpha_n}$, where $0 \leq \alpha_i \leq d_i$. (Use the Poincaré-Birkhoff-Witt theorem.) The rank $[U : A]$ of U over A is equal to $d_1 \ldots d_n$; it is a power of p. Deduce that C is an A-module of finite type, hence a k-algebra of finite type and of dimension n (*Commutative Algebra*, Chap. VIII).

b) Let K be the field of fractions of A, and let

$$U_{(K)} = U \otimes_A K, \quad C_{(K)} = C \otimes_A K.$$

Then $U_{(K)} \supset C_{(K)} \supset K$. Show that $U_{(K)}$ is a field with centre $C_{(K)}$, and that this is the quotient field (both left and right) of U, cf. Chap. I, §2, Exerc. 10. Deduce that $[U_{(K)} : C_{(K)}]$ is of the from q^2, where q is a power of p; we have $[C_{(K)} : K] = q_C$, where q_C is a power of p, and $[U : A] = q_C q^2$.

c) Let d be a non-zero element of A, and let Λ be a subring of $U_{(K)}$ such that $U \subset \Lambda \subset d^{-1}U$. Show that $\Lambda = U$. [If $x = b/a$, $a \in A - \{0\}$, is an element of Λ, show by induction on m that the relation $b \in Ua + U_m$ implies that $b \in Ua + U_{m-1}$, where $\{U_m\}$ is the canonical filtration of U. (For this, use the fact that $\operatorname{gr} U$ is integrally closed, and argue as in Prop. 15 of *Commutative Algebra*, Chap. V, §1, no. 4.) For $m = 0$, this gives $b \in Ua$, i.e. $x \in U$.]

Deduce that C is *integrally closed.*

d) Assume that k is *algebraically closed.* Let $\rho : \mathfrak{g} \to \mathfrak{gl}(V)$ be an irreducible linear representation of $\mathfrak{g}$ and ρ_U the corresponding representation of U. The restriction of ρ_U to C is a homomorphism γ_ρ from C to k (identified with the homotheties of V); let α_ρ be its restriction to A. Show that for any homomorphism α (resp. γ) from the k-algebra A (resp. C) to k, there exists at least one irreducible representation ρ of $\mathfrak{g}$ such that $\alpha_\rho = \alpha$ (resp. $\gamma_\rho = \gamma$) and that there are only finitely-many such representations (up to equivalence). Show that $\dim V \leq q$, with the notation of *b*).[5]

[4] For more details, cf. J. DIXMIER, Cohomologie des algèbres de Lie nilpotentes, *Acta Sci. Math. Szeged*, Vol. XVI (1955), pp. 246-250.

[5] For more details, cf. H. ZASSENHAUS, The representations of Lie algebras of prime characteristic, *Proc. Glasgow Math. Assoc.*, Vol. II (1954), pp. 1-36.

¶ 9) We retain the notations of the preceding exercise, and assume further that $\mathfrak{g}$ is *nilpotent*.

a) Show that the basis $(e_1, \dots, e_n)$ can be chosen so that, for any pair (i, j), $[e_i, e_j]$ is a linear combination of the e_h for $h > \sup(i, j)$. Assume from now on that the e_i satisfy this condition. For $i = 1, \dots, n$ choose a power $q(i)$ of p such that $\operatorname{ad}(e_i)^{q(i)} = 0$, and put $z_i = e_i^{q(i)}$, $\mathrm{A} = k[z_1, \dots, z_n]$, cf. Exerc. 8.

b) Let $\rho : \mathfrak{g} \to \mathfrak{gl}(\mathrm{V})$ be a linear representation of $\mathfrak{g}$. Assume that $\rho(e_i)$ is nilpotent for $i = 1, \dots, n$. Show that $\rho(x)$ is nilpotent for all $x \in \mathfrak{g}$. (Argue by induction on $n = \dim \mathfrak{g}$ and reduce to the case where ρ is irreducible. Show that, in this case, $\rho(e_n) = 0$ and apply the induction hypothesis.)

c) Let $\rho_1 : \mathfrak{g} \to \mathfrak{gl}(\mathrm{V}_1)$ and $\rho_2 : \mathfrak{g} \to \mathfrak{gl}(\mathrm{V}_2)$ be two linear representations of $\mathfrak{g}$. Assume that V_1 and V_2 are $\neq 0$, and that $\mathrm{V}_1 = \mathrm{V}^{\lambda_1}(\mathfrak{g})$, $\mathrm{V}_2 = \mathrm{V}^{\lambda_2}(\mathfrak{g})$, where λ_1 and λ_2 are two functions on $\mathfrak{g}$, cf. no. 3. Show that, if $\lambda_1(e_i) = \lambda_2(e_i)$ for $i = 1, \dots, n$, then $\lambda_1 = \lambda_2$ and there exists a non-zero $\mathfrak{g}$-homomorphism from V_1 to V_2 (apply *b*) to the $\mathfrak{g}$-module $\mathrm{V} = \mathscr{L}(\mathrm{V}_1, \mathrm{V}_2)$ and use Engel's theorem to show that V contains a non-zero $\mathfrak{g}$-invariant element). Deduce that if, in addition, V_1 and V_2 are simple, they are isomorphic.

d) Assume that k is algebraically closed. Let R be the set of equivalence classes of irreducible representations of $\mathfrak{g}$. If $\rho \in \mathrm{R}$, put

$$x_\rho = (x_\rho(1), \dots, x_\rho(n)) \in k^n,$$

where $x_\rho(i)$ is the unique eigenvalue of $\rho(e_i)$. Show that $\rho \mapsto x_\rho$ is a *bijection from* R *to* k^n. (Injectivity follows from *c*), and surjectivity from Exerc. 8 *d*).) Deduce the following consequences:

(i) For any maximal ideal $\mathfrak{m}$ of A, the quotient of $\mathrm{U}/\mathfrak{m}\mathrm{U}$ by its radical is a matrix algebra.

(ii) The degree of any irreducible representation of $\mathfrak{g}$ is a power of p (this follows from (i) and the fact that $[\mathrm{U}/\mathfrak{m}\mathrm{U} : k]$ is a power of p).

(iii) Every homomorphism from A to k extends uniquely to a homomorphism from C to k (use the fact that $\mathrm{C}/\mathfrak{m}\mathrm{C}$ is contained in the centre of $\mathrm{U}/\mathfrak{m}\mathrm{U}$, which is a local k-algebra with residue field k).

(iv) There exists an integer $\mathrm{N} \geq 0$ such that $x^{p^{\mathrm{N}}} \in \mathrm{A}$ for all $x \in \mathrm{C}$ (this follows from (iii)).[6]

¶ 10) Assume that k is of characteristic $p > 0$. Denote by $\mathfrak{g}$ a Lie algebra with basis $\{e_1, e_2, e_3\}$, with $[e_1, e_2] = e_3$, $[e_1, e_3] = [e_2, e_3] = 0$.

a) Show that the centre of $\mathrm{U}\mathfrak{g}$ is $k[e_1^p, e_2^p, e_3]$.

b) Assume that k is algebraically closed. Show that, for all $(\lambda_1, \lambda_2, \lambda_3) \in k^3$, there exists (up to equivalence) a unique irreducible representation ρ of $\mathfrak{g}$

[6] For more details, cf. H. ZASSENHAUS, Über Liesche Ringe mit Primzahlcharakteristik, *Hamb. Abh.*, Vol. XIII (1939), pp. 1-100, and Darstellungstheorie nilpotenter Lie-Ringe bei Charakteristik $p > 0$, *Crelle's J.*, Vol. CLXXXII (1940), pp. 150-155.

such that λ_i is the unique eigenvalue of $\rho(e_i)$ $(i = 1, 2, 3)$; the degree of ρ is p if $\lambda_3 \neq 0$, and is 1 if $\lambda_3 = 0$. (Apply Exerc. 8 and 9, or argue directly.)

11) Let $\mathfrak{h}$ be a nilpotent Lie algebra, V an $\mathfrak{h}$-module not reduced to 0, and λ a function on $\mathfrak{h}$ such that $V = V^{\lambda}(\mathfrak{h})$. Prove the equivalence of the following properties:
(i) λ is a linear form on $\mathfrak{h}$, zero on $[\mathfrak{h}, \mathfrak{h}]$.
(ii) There exists a basis of V with respect to which the endomorphisms defined by the elements of $\mathfrak{h}$ are triangular.

(To prove that (i) $\Longrightarrow$ (ii), apply Engel's theorem to the $\mathfrak{h}$-module $\mathscr{L}(W_\lambda, V)$, where W is the 1-dimensional $\mathfrak{h}$-module defined by λ.)

Properties (i) and (ii) are true if k is of characteristic 0 (Prop. 9).

§2

1) The diagonal matrices of trace 0 form a Cartan subalgebra of $\mathfrak{sl}(n, k)$, except when $n = 2$ and k is of characteristic 2.

2) Let e be the element $\begin{pmatrix} 0 & 1 \\ 0 & 0 \end{pmatrix}$ of $\mathfrak{sl}(2, \mathbf{C})$. Show that $\mathbf{C}e$ is a maximal nilpotent Lie subalgebra of $\mathfrak{sl}(2, \mathbf{C})$, but not a Cartan subalgebra of $\mathfrak{sl}(2, \mathbf{C})$.

3) Assume that k is of characteristic 0. Let $\mathfrak{g}$ be a semi-simple Lie algebra. Let E be the set of commutative subalgebras of $\mathfrak{g}$ all of whose elements are semi-simple in $\mathfrak{g}$. Then the Cartan subalgebras of $\mathfrak{g}$ are the maximal elements of E. (Use Th. 2 and Prop. 10.)

In particular, the union of the Cartan subalgebras of $\mathfrak{g}$ is equal to the set of semi-simple elements of $\mathfrak{g}$.

4) Let $\mathfrak{g}$ be a Lie algebra with a basis (x, y, z) such that $[x, y] = y$, $[x, z] = z$, $[y, z] = 0$. Let $\mathfrak{a}$ be the ideal $ky + kz$ of $\mathfrak{g}$. Then $\mathrm{rk}(\mathfrak{a}) = 2$ and $\mathrm{rk}(\mathfrak{g}) = 1$.

5) Assume that k is of characteristic 0. Let $\mathfrak{g}$ be a Lie algebra, $\mathfrak{r}$ its radical, $\mathfrak{h}$ a Cartan subalgebra of $\mathfrak{g}$. Show that

$$\mathfrak{r} = [\mathfrak{g}, \mathfrak{r}] + (\mathfrak{h} \cap \mathfrak{r}).$$

(Observe that the image of $\mathfrak{h}$ in $\mathfrak{g}/[\mathfrak{g}, \mathfrak{r}]$ contains the centre $\mathfrak{r}/[\mathfrak{g}, \mathfrak{r}]$ of $\mathfrak{g}/[\mathfrak{g}, \mathfrak{r}]$.)

6) Let $\mathfrak{g}$ be a Lie algebra, $\mathfrak{h}$ a nilpotent subalgebra of $\mathfrak{g}$. If $\mathfrak{g}^0(\mathfrak{h})$ is nilpotent, $\mathfrak{g}^0(\mathfrak{h})$ is a Cartan subalgebra of $\mathfrak{g}$.

7) Let $\mathfrak{s}$ be a Lie algebra, $\mathfrak{a}$ a Cartan subalgebra of $\mathfrak{s}$ and V an $\mathfrak{s}$-module. Let $\mathfrak{g} = \mathfrak{s} \times V$ be the semi-direct product of $\mathfrak{s}$ by V. Show that $\mathfrak{a} \times V^0(\mathfrak{a})$ is a Cartan subalgebra of $\mathfrak{g}$.

8) Assume that k is of characteristic $p > 0$. Denote by $\mathfrak{s}$ a Lie algebra with basis $\{x, y\}$ such that $[x, y] = y$. Let V be a k-vector space with basis $\{e_i\}_{i \in \mathbf{Z}/p\mathbf{Z}}$.

a) Show that V has a unique $\mathfrak{s}$-module structure such that $xe_i = ie_i$ and $ye_i = e_{i+1}$ for all i. This $\mathfrak{s}$-module is simple.

b) Let $\mathfrak{g} = \mathfrak{s} \times \mathrm{V}$ be the semi-direct product of $\mathfrak{s}$ by V. Show that $\mathfrak{g}$ is a solvable algebra of rank 1 whose derived algebra is not nilpotent.

c) An element of $\mathfrak{g}$ is regular if and only if its projection on $\mathfrak{s}$ is of the form $ax + by$, with $ab \neq 0$.

d) We have $\mathrm{V}^0(x+y) = 0$ and $\mathrm{V}^0(x) = ke_0$. Deduce (cf. Exerc. 6) that $\mathfrak{g}$ has Cartan subalgebras of dimension 1 (for example that generated by $x+y$) and Cartan subalgebras of dimension 2 (for example that generated by x and e_0).

9) Let $\mathfrak{g}$ be a Lie algebra with a basis (x, y) such that $[x, y] = y$. Let $\mathfrak{k} = ky$, and $\varphi : \mathfrak{g} \to \mathfrak{g}/\mathfrak{k}$ the canonical morphism. The element 0 of $\mathfrak{g}/\mathfrak{k}$ is regular in $\mathfrak{g}/\mathfrak{k}$ but is not the image under φ of a regular element of $\mathfrak{g}$.

10) Assume that k is infinite. Let $\mathfrak{g}$ be a Lie algebra. Prove the equivalence of the following properties:
(i) $\mathrm{rk}(\mathfrak{g}) = \dim(\mathfrak{g})$.
(ii) $\mathfrak{g}$ is nilpotent.
(iii) $\mathfrak{g}$ has only finitely-many Cartan subalgebras of dimension $\mathrm{rk}(\mathfrak{g})$.
(iv) $\mathfrak{g}$ has only one Cartan subalgebra.

11) Let $\mathfrak{h}$ be a commutative Lie algebra $\neq 0$, P a finite subset of $\mathfrak{h}^*$ containing 0. Show that there exists a Lie algebra $\mathfrak{g}$ containing $\mathfrak{h}$ as a Cartan subalgebra, and such that the set of weights of $\mathfrak{h}$ in $\mathfrak{g}$ is P. (Construct $\mathfrak{g}$ as the semi-direct product of $\mathfrak{h}$ by the $\mathfrak{h}$-module V which is the direct sum of the 1-dimensional modules corresponding to the elements of $\mathrm{P} - \{0\}$, cf. Exerc. 7.)

An element x of $\mathfrak{h}$ is such that $\mathfrak{h} = \mathfrak{g}^0(x)$ if and only if x is not orthogonal to any element of $\mathrm{P} - \{0\}$.

12) Assume that k is finite. Construct an example of a Lie algebra $\mathfrak{g}$ having a Cartan subalgebra $\mathfrak{h}$ in which there exists no element x such that $\mathfrak{h} = \mathfrak{g}^0(x)$. (Use the preceding exercise, and take $\mathrm{P} = \mathfrak{h}^*$.)

¶ 13) Assume that k is *finite*. Denote by k' an infinite extension of k. Let $\mathfrak{g}$ be a Lie algebra over k. The *rank of* $\mathfrak{g}$, denoted by $\mathrm{rk}(\mathfrak{g})$, is the rank of the k'-Lie algebra $\mathfrak{g}' = \mathfrak{g} \otimes_k k'$; an element of $\mathfrak{g}$ is said to be *regular* if it is regular in $\mathfrak{g}'$; these definitions do not depend on the choice of k'. Show that, if

$$\mathrm{Card}(k) \geq \dim \mathfrak{g} - \mathrm{rk}(\mathfrak{g}),$$

$\mathfrak{g}$ contains a regular element (hence also a Cartan subalgebra).

(Use the following result: if a is a non-zero homogeneous element of $k[\mathrm{X}_1, \dots, \mathrm{X}_n]$, and if $\mathrm{Card}(k) \geq \deg(a)$, there exists $x \in k^n$ such that $a(x) \neq 0$.)

14) Assume that k is of characteristic zero. Let V be a finite dimensional k-vector space, $\mathfrak{g}$ a Lie subalgebra of $\mathfrak{gl}(\mathrm{V})$, $\mathfrak{h}$ a Cartan subalgebra of $\mathfrak{g}$ and $\mathfrak{n}_\mathrm{V}$ the largest ideal of nilpotency of the $\mathfrak{g}$-module V (Chap. I, §4, no. 3, Def. 2).

Show that an element of $\mathfrak{h}$ is nilpotent if and only if it belongs to $\mathfrak{n}_V$. (Reduce to the case where the $\mathfrak{g}$-module V is semi-simple, and use Cor. 3 of Th. 2.)

¶ 15) Assume that k is infinite. Let $\mathfrak{g}$ be a Lie algebra, $\mathscr{C}_\infty\mathfrak{g}$ the union of the ascending central series of $\mathfrak{g}$, and x an element of $\mathfrak{g}$. Prove the equivalence of the following properties:
(i) x belongs to every Cartan subalgebra of $\mathfrak{g}$.
(ii) $x \in \mathfrak{g}^0(y)$ for all $y \in \mathfrak{g}$ (i.e. $x \in \mathfrak{g}^0(\mathfrak{g})$).
(iii) $x \in \mathscr{C}_\infty\mathfrak{g}$.
(The implications (iii) $\Longrightarrow$ (ii) $\Longrightarrow$ (i) are immediate. To prove that (i) $\Longrightarrow$ (ii), remark that (i) is equivalent to saying that $x \in \mathfrak{g}^0(y)$ for every regular element y in $\mathfrak{g}$, and use the fact that the regular elements are dense in $\mathfrak{g}$ in the Zariski topology. To prove that (ii) $\Longrightarrow$ (iii), observe that $\mathfrak{n} = \mathfrak{g}^0(\mathfrak{g})$ is stable under $\mathfrak{g}$ (§1, Exerc. 5) and apply Engel's theorem to the $\mathfrak{g}$-module $\mathfrak{n}$; deduce that $\mathfrak{n}$ is contained in $\mathscr{L}_\infty\mathfrak{g}$.)

¶ 16) Let $\mathfrak{g}$ be a solvable complex Lie algebra, $\mathfrak{h}$ a Cartan subalgebra of $\mathfrak{g}$, $\mathfrak{g} = \bigoplus \mathfrak{g}^\lambda(\mathfrak{h})$ the corresponding decomposition of $\mathfrak{g}$ into primary subspaces, with $\mathfrak{g}^0(\mathfrak{h}) = \mathfrak{h}$.

a) Show that the restrictions to $\mathfrak{h}$ of the linear forms called roots of $\mathfrak{g}$ in Chap. III, §9, Exerc. 17 *c*) are the weights of $\mathfrak{h}$ in $\mathfrak{g}$, i.e. the λ such that $\mathfrak{g}^\lambda(\mathfrak{h}) \neq 0$; deduce that such a λ vanishes on $\mathfrak{h} \cap \mathscr{D}\mathfrak{g}$.

b) Let $(x, y) \mapsto [x, y]'$ be the alternating bilinear map from $\mathfrak{g} \times \mathfrak{g}$ to $\mathfrak{g}$ with the following properties:
(i) If $x \in \mathfrak{g}^\lambda(\mathfrak{h}), y \in \mathfrak{g}^\mu(\mathfrak{h})$, with $\lambda \neq 0, \mu \neq 0$, then $[x, y]' = [x, y]$;
(ii) if $x \in \mathfrak{g}^0(\mathfrak{h}), y \in \mathfrak{g}^\mu(\mathfrak{h})$, then $[x, y]' = [x, y] - \mu(x)y$.

Show that this gives a new Lie algebra structure on $\mathfrak{g}$ (use *a*)). Denote it by $\mathfrak{g}'$.

c) Show that, if $x \in \mathfrak{g}^\lambda(\mathfrak{h})$, the map $\mathrm{ad}'x : y \mapsto [x, y]'$ is nilpotent. Deduce that $\mathfrak{g}'$ is *nilpotent* (apply Exerc. 11 of Chap. I, §4 to the set E of $\mathrm{ad}'x$, where x belongs to the union of the $\mathfrak{g}^\lambda(\mathfrak{h})$).

§3

1) Let $\mathfrak{g}$ be a Lie algebra, $\mathfrak{g}'$ a Cartan subalgebra of $\mathfrak{g}$. Then the conditions of Prop. 3 are satisfied. But an element of $\mathfrak{g}'$, even if it is regular in $\mathfrak{g}'$, is not necessarily regular in $\mathfrak{g}$.

2) Let $\mathfrak{g}$ be a real Lie algebra of dimension n, U (resp. H) the set of regular elements (resp. of Cartan subalgebras) of $\mathfrak{g}$, and $\mathrm{Int}(\mathfrak{g})$ the group of inner automorphisms of $\mathfrak{g}$ (Chap. III, §6, no. 2, Def. 2).

a) Show that, if x and y belong to the same connected component of U, $\mathfrak{g}^0(x)$ and $\mathfrak{g}^0(y)$ are conjugate under $\mathrm{Int}(\mathfrak{g})$.

b) Show that the number of connected components of U is finite, and that this number is bounded by a constant $c(n)$ depending only on n (apply Exerc. 2 of App. II).

c) Deduce that the number of orbits of $\mathrm{Int}(\mathfrak{g})$ on H is $\leq c(n)$.

3) Let $\mathfrak{g}$ be a real Lie algebra, $\mathfrak{r}$ the radical of $\mathfrak{g}$, $\mathfrak{h}$ and $\mathfrak{h}'$ Cartan subalgebras of $\mathfrak{g}$, φ the canonical homomorphism from $\mathfrak{g}$ to $\mathfrak{g}/\mathfrak{r}$. The following conditions are equivalent:
(i) $\mathfrak{h}$ and $\mathfrak{h}'$ are conjugate under $\mathrm{Int}(\mathfrak{g})$;
(ii) $\varphi(\mathfrak{h})$ and $\varphi(\mathfrak{h}')$ are conjugate under $\mathrm{Int}(\mathfrak{g}/\mathfrak{r})$. (Imitate the proof of Prop. 5.)

4) Let $\mathfrak{g} = \mathfrak{sl}(2, \mathbf{R})$, $x = \begin{pmatrix} 1 & 0 \\ 0 & -1 \end{pmatrix}, y = \begin{pmatrix} 0 & -1 \\ 1 & 0 \end{pmatrix}$. Show that $\mathbf{R}x$ and $\mathbf{R}y$ are Cartan subalgebras of $\mathfrak{g}$ not conjugate under $\mathrm{Aut}(\mathfrak{g})$.

5) *a*) Show that there exists a Lie algebra $\mathfrak{g}$ over k with basis (x, y, z, t) such that

$$[x, y] = z, \ [x, t] = t, \ [y, t] = 0, \ [\mathfrak{g}, z] = 0.$$

Show that $\mathfrak{g}$ is solvable and that $\mathfrak{k} = kx + ky + kz$ is a subalgebra of $\mathfrak{g}$.

b) Show that the elementary automorphisms of $\mathfrak{g}$ are the maps of the form $1 + \lambda\, \mathrm{ad}_{\mathfrak{g}} y + \mu\, \mathrm{ad}_{\mathfrak{g}} t$ where $\lambda, \mu \in k$.

c) Show that $1 + \mathrm{ad}_{\mathfrak{k}} x$ is an elementary automorphism of $\mathfrak{k}$ which does not extend to an elementary automorphism of $\mathfrak{g}$.

d) Let $\mathfrak{s}$ be a semi-simple subalgebra of a Lie algebra $\mathfrak{a}$. Show that every elementary automorphism of $\mathfrak{s}$ extends to an elementary automorphism of $\mathfrak{a}$.

6) Every element of a reductive Lie algebra $\mathfrak{g}$ is contained in a commutative subalgebra of dimension $\mathrm{rk}(\mathfrak{g})$.

7) Let $\mathfrak{g}$ be a Lie algebra and $\mathfrak{g}'$ a subalgebra of $\mathfrak{g}$ reductive in $\mathfrak{g}$. Let $\mathfrak{a}$ be a Cartan subalgebra of $\mathfrak{g}'$. Show that there exists a Cartan subalgebra of $\mathfrak{g}$ which contains $\mathfrak{a}$ (use Prop. 10 of §2). Deduce that $\mathrm{rk}(\mathfrak{g}') \leq \mathrm{rk}(\mathfrak{g})$ and that equality holds if and only if $\mathfrak{g}'$ has properties (i), (ii), (iii) of Prop. 3.

8) Let $\mathfrak{g}$ be a Lie algebra, $\mathfrak{a}$ an ideal of $\mathfrak{g}$, $\mathfrak{h}$ a Cartan subalgebra of $\mathfrak{g}$ and $\mathscr{C}_\infty\mathfrak{g}$ the union of the ascending central series of $\mathfrak{g}$. Show that $\mathfrak{a} \subset \mathfrak{h}$ implies $\mathfrak{a} \subset \mathscr{C}_\infty\mathfrak{g}$ (in other words $\mathscr{C}_\infty\mathfrak{g}$ is the largest ideal of $\mathfrak{g}$ contained in $\mathfrak{h}$). (Reduce to the case where k is algebraically closed and remark that $\mathfrak{a}$ is stable under every elementary automorphism of $\mathfrak{g}$; the relation $\mathfrak{a} \subset \mathfrak{h}$ then implies that $\mathfrak{a}$ is contained in every Cartan subalgebra of $\mathfrak{g}$; conclude by means of Exerc. 15 of §2.)

¶ 9) Let $\mathfrak{g}$ be a Lie algebra, $\mathfrak{h}$ a Cartan subalgebra of $\mathfrak{g}$ and x an element of $\mathfrak{h}$. Let $\mathfrak{g} = \mathfrak{h} \oplus \mathfrak{g}^+$ be the Fitting decomposition (§1, no. 1) of $\mathfrak{g}$ with respect to $\mathfrak{h}$.

a) Let $\mathfrak{n}$ be the largest semi-simple $\mathfrak{h}$-submodule contained in $\mathfrak{g}^0(x) \cap \mathfrak{g}^+$. Show that $\mathfrak{n} = 0$ if and only if $\mathfrak{g}^0(x) = \mathfrak{h}$, i.e. x is regular in $\mathfrak{g}$.

b) Show that $\mathfrak{h} \oplus \mathfrak{n}$ is a subalgebra of $\mathfrak{g}$. If $\mathfrak{h}'$ is the intersection of $\mathfrak{h}$ with the commutant of $\mathfrak{n}$, show that $\mathfrak{h}'$ is an ideal of $\mathfrak{h} \oplus \mathfrak{n}$ which contains $\mathscr{D}\mathfrak{h}$ and x. Conclude (Exerc. 8) that $\mathfrak{h}' \subset \mathscr{C}_\infty(\mathfrak{h} \oplus \mathfrak{n})$, and hence that $x \in \mathscr{C}_\infty(\mathfrak{h} \oplus \mathfrak{n})$ and that x belongs to every Cartan subalgebra of $\mathfrak{h} \oplus \mathfrak{n}$.

c) If $\mathfrak{n} \neq 0$, $\mathfrak{h} \oplus \mathfrak{n}$ is not nilpotent and has infinitely-many Cartan subalgebras (§2, Exerc. 10). Conclude that x belongs to infinitely-many Cartan subalgebras of $\mathfrak{g}$.

d) An element of $\mathfrak{g}$ is regular if and only if it belongs to a unique Cartan subalgebra.

¶ 10) Let $\mathfrak{g}$ be a Lie algebra, $\mathfrak{r}$ its radical, $\mathfrak{n}$ its largest nilpotent ideal and $\mathfrak{a}$ one of its Levi subalgebras.

a) Put $\mathfrak{g}' = \mathfrak{n} + \mathscr{D}\mathfrak{g}$. Show that $\mathfrak{g}' = \mathfrak{n} \oplus \mathfrak{a}$. (Use the fact that $[\mathfrak{g}, \mathfrak{r}]$ is contained in $\mathfrak{n}$.) If $\mathfrak{g} \neq 0$, then $\mathfrak{g}' \neq 0$.

b) Assume that k is algebraically closed. Let $(\mathrm{V}_i)_{i \in \mathrm{I}}$ be the quotients of a Jordan-Hölder sequence of the $\mathfrak{g}$-module $\mathfrak{g}$ (with the adjoint representation). If $x \in \mathfrak{r}$, show that x_{V_i} is a homothety and that $x_{\mathrm{V}_i} = 0$ for all i if and only if x belongs to $\mathfrak{n}$. Deduce that an element $y \in \mathfrak{g}$ belongs to $\mathfrak{g}'$ if and only if $\mathrm{Tr}(y_{\mathrm{V}_i}) = 0$ for all $i \in \mathrm{I}$.

c) Denote by N the vector subspace of $\mathfrak{g}$ generated by the elements x such that $\operatorname{ad} x$ is nilpotent. Show that N is a subalgebra of $\mathfrak{g}$ (use the fact that N is stable under $\mathrm{Aut}_e(\mathfrak{g})$). Show, by using *b*), that $\mathrm{N} \subset \mathfrak{g}'$.

d) Let $\mathfrak{h}$ be a Cartan subalgebra of $\mathfrak{g}$. Assume that there exists a subset R of $\mathfrak{h}^*$ such that

$$\mathfrak{g} = \mathfrak{h} \oplus \bigoplus_{\alpha \in \mathrm{R}} \mathfrak{g}^\alpha(\mathfrak{h}),$$

an assumption which is satisfied, in particular, if k is algebraically closed. Show that N then contains the $\mathfrak{g}^\alpha(\mathfrak{h})$, $\mathscr{D}\mathfrak{h}$ and $\mathfrak{n}$; deduce that N contains $\mathfrak{g}'$, so $\mathrm{N} = \mathfrak{g}'$.

e) If k is algebraically closed and $\mathfrak{g} \neq 0$, $\mathfrak{g}$ contains an element $x \neq 0$ such that $\operatorname{ad} x$ is nilpotent. (Indeed, we then have $\mathfrak{g}' \neq 0$.)

¶ 11) Let $\mathfrak{g}$ be a Lie algebra, $\mathfrak{r}$ its radical.

a) Let $\mathfrak{s}$ be a Levi subalgebra of $\mathfrak{g}$, $\mathfrak{k}$ a Cartan subalgebra of $\mathfrak{s}$. Show that $\mathfrak{k}$ is contained in a Cartan subalgebra $\mathfrak{h}$ of $\mathfrak{g}$ which is the sum of $\mathfrak{k}$ and a subalgebra of $\mathfrak{r}$. (Use §2, Th. 2, Prop. 10 and Cor. 2 of Th. 1.)

b) Let $\mathfrak{h}'$ be a Cartan subalgebra of $\mathfrak{g}$. Show that there exists a Levi subalgebra $\mathfrak{s}'$ of $\mathfrak{g}$ such that $\mathfrak{h}'$ is the sum of a Cartan subalgebra of $\mathfrak{s}'$ and a subalgebra of $\mathfrak{r}$. (The subalgebras $\mathfrak{s}, \mathfrak{k}, \mathfrak{h}$ in *a*) can be chosen so that $\mathfrak{h} + \mathfrak{r} = \mathfrak{h}' + \mathfrak{r}$. Put $\mathfrak{a} = \mathfrak{h} + \mathfrak{r}$, which is solvable. By Th. 3, there exists $x \in \mathscr{C}^\infty(\mathfrak{a})$ such that

$e^{\mathrm{ad}_\mathfrak{a} x}\mathfrak{h} = \mathfrak{h}'$. Then $e^{\mathrm{ad}_\mathfrak{g} x}$ is a special automorphism of $\mathfrak{g}$ which transforms $\mathfrak{s}$ into the required Levi subalgebra.)

c) Let $\mathfrak{s}$ be a Levi subalgebra of $\mathfrak{g}$. Let $\mathfrak{h}$ be a Cartan subalgebra of $\mathfrak{g}$ which is the sum of a Cartan subalgebra $\mathfrak{k}$ of $\mathfrak{s}$ and a subalgebra $\mathfrak{l}$ of $\mathfrak{r}$. Let $\mathfrak{c}$ be the commutant of $\mathfrak{k}$ in $\mathfrak{r}$. Show that $\mathfrak{l}$ is a Cartan subalgebra of $\mathfrak{c}$. (For $x \in \mathfrak{k}$, $\mathrm{ad}_\mathfrak{g} x$ is semi-simple, but $\mathrm{ad}_\mathfrak{h} x$ is nilpotent, so $[\mathfrak{k}, \mathfrak{h}] = 0$. If $y \in \mathfrak{c}$ is such that $[y, \mathfrak{l}] \subset \mathfrak{l}$, then $[y, \mathfrak{h}] \subset \mathfrak{h}$ so $y \in \mathfrak{h} \cap \mathfrak{r} = \mathfrak{l}$.)

d) Let $\mathfrak{s}$ be a Levi subalgebra of $\mathfrak{g}$, $\mathfrak{k}$ a Cartan subalgebra of $\mathfrak{s}$, $\mathfrak{c}$ the commutant of $\mathfrak{k}$ in $\mathfrak{r}$, $\mathfrak{l}$ a Cartan subalgebra of $\mathfrak{c}$. Then $\mathfrak{h} = \mathfrak{k} + \mathfrak{l}$ is a Cartan subalgebra of $\mathfrak{g}$. (Let $x = y + z$ ($y \in \mathfrak{s}, z \in \mathfrak{r}$) be an element of the normalizer $\mathfrak{u}$ of $\mathfrak{h}$ in $\mathfrak{g}$. Show that $[y, \mathfrak{k}] \subset \mathfrak{k}$, so $y \in \mathfrak{k}$ and $z \in \mathfrak{u}$. Then show that $[z, \mathfrak{k}] \subset \mathfrak{h} \cap \mathfrak{r} \subset \mathfrak{c}$, so $[\mathfrak{k}, [\mathfrak{k}, z]] = 0$, and hence that $[\mathfrak{k}, z] = 0$ and $z \in \mathfrak{c}$. Finally, $[z, \mathfrak{l}] \subset \mathfrak{l}$ hence $z \in \mathfrak{l}$ and $x \in \mathfrak{h}$.)

e) Let $\mathfrak{s}, \mathfrak{k}, \mathfrak{c}$ be as in *d*), and $\mathfrak{q} = [\mathfrak{g}, \mathfrak{r}]$ the nilpotent radical of $\mathfrak{g}$. Let $x \in \mathfrak{q}$ and u the special automorphism $e^{\mathrm{ad}\, x}$. If $u(\mathfrak{k}) \subset \mathfrak{k} + \mathfrak{c}$, then $x \in \mathfrak{c}$. (Consider the adjoint representation ρ of $\mathfrak{s}$ on $\mathfrak{q}$, and let $\mathfrak{q}_i$ be a complement of $\mathscr{C}^{i+1}\mathfrak{q}$ in $\mathscr{C}^i\mathfrak{q}$ stable under ρ; let ρ_i be the subrepresentation of ρ defined by $\mathfrak{q}_i$. Let $\sigma_i = \rho_i|\mathfrak{k}$. Let $\mathfrak{q}'_i$ be the commutant of $\mathfrak{k}$ in $\mathfrak{q}_i$ and $\mathfrak{q}''_i$ a complement of $\mathfrak{q}'_i$ in $\mathfrak{q}_i$ stable under σ_i. Let $x = x'_1 + x''_1 + \cdots + x'_n + x''_n$ with $x'_i \in \mathfrak{q}'_i, x''_i \in \mathfrak{q}''_i$. Arguing by contradiction, assume that the x''_i are not all zero and $x''_1 = \cdots = x''_{p-1} = 0$, $x''_p \neq 0$, for example. If $h \in \mathfrak{k}$, $u(h) = h + [x''_p, h] + y$ with $y \in \mathscr{C}^{p+1}\mathfrak{q}$. Since $u(h) \in \mathfrak{k} + \mathfrak{c}$, this gives $[x''_p, h] + y \in \mathfrak{q}'_p + \mathfrak{q}'_{p+1} + \cdots + \mathfrak{q}'_n$, hence $[h, x''_p] \in \mathfrak{q}'_p$, so $[h, x''_p] = 0$. Then $x''_p = 0$, a contradiction.)

f) Let $\mathfrak{h}$ be a Cartan subalgebra of $\mathfrak{g}$. Then $\mathfrak{h}$ can be expressed uniquely as the sum of $\mathfrak{h} \cap \mathfrak{r}$ and a Cartan subalgebra of a Levi subalgebra of $\mathfrak{g}$. (For the uniqueness, use *e*) and Th. 5 of Chap. I, §6, no. 8.) The Levi subalgebra in question is not unique in general.

g) Let $\mathfrak{h}$ be a Cartan subalgebra of $\mathfrak{g}$, and $\mathfrak{t} = \mathfrak{g}^0(\mathfrak{h} \cap \mathfrak{r})$. Then $\mathfrak{h}$ is a Cartan subalgebra of $\mathfrak{t}$. We have $\mathfrak{g} = \mathfrak{t} + \mathfrak{r}$ (use a Fitting decomposition for the adjoint representation of $\mathfrak{h} \cap \mathfrak{r}$ on $\mathfrak{g}$). The algebra $\mathfrak{t} \cap \mathfrak{r}$ is the radical of $\mathfrak{t}$ and is nilpotent (use Exerc. 5 of §2).

h) Assume that k is algebraically closed. Let $\mathfrak{h}$ be a Cartan subalgebra of $\mathfrak{g}$. There exists a Levi subalgebra $\mathfrak{s}$ of $\mathfrak{g}$ such that, for all $\lambda \in \mathfrak{h}^*$,

$$\mathfrak{g}^\lambda(\mathfrak{h}) = (\mathfrak{g}^\lambda(\mathfrak{h}) \cap \mathfrak{s}) + (\mathfrak{g}^\lambda(\mathfrak{h}) \cap \mathfrak{r}).$$

(With the notations of *g*), take for $\mathfrak{s}$ a Levi subalgebra of $\mathfrak{t}$ such that $\mathfrak{h} = (\mathfrak{h} \cap \mathfrak{s}) + (\mathfrak{h} \cap \mathfrak{r})$; this exists by *b*).)[7]

¶ 12) *a*) Let $\mathfrak{g}$ be a solvable Lie algebra, and G a finite subgroup (resp. compact subgroup if $k = \mathbf{R}$ or $\mathbf{C}$) of $\mathrm{Aut}(\mathfrak{g})$. Show that there exists a Cartan subalgebra of $\mathfrak{g}$ stable under G. (Argue by induction on $\dim \mathfrak{g}$, and reduce to

[7] For more details, cf. J. DIXMIER, Sous-algèbres de Cartan et décompositions de Levi dans les algèbres de Lie, *Trans. Royal Soc. Canada*, Vol. L (1956), pp. 17-21.

the case in which $\mathfrak{g}$ is an extension of a nilpotent algebra $\mathfrak{g}/\mathfrak{n}$ by a commutative ideal $\mathfrak{n}$ which is a non-trivial simple $\mathfrak{g}/\mathfrak{n}$-module (cf. proof of Th. 3). The Cartan subalgebras of $\mathfrak{g}$ then form an affine space attached to $\mathfrak{n}$, on which G operates. Conclude by an argument with the barycentre.)

b) Let $\mathfrak{g}$ be a solvable Lie algebra and $\mathfrak{s}$ a Lie subalgebra of $\mathrm{Der}(\mathfrak{g})$. Assume that the $\mathfrak{s}$-module $\mathfrak{g}$ is semi-simple. Show that there exists a Cartan subalgebra of $\mathfrak{g}$ stable under $\mathfrak{s}$. (Same method.)

¶ 13) Let $\mathfrak{g}$ be a Lie algebra and G a finite subgroup of $\mathrm{Aut}(\mathfrak{g})$. Assume that G is *hyper-solvable* (*Algebra*, Chap. I, §6, Exerc. 26). Show that there exists a Cartan subalgebra of $\mathfrak{g}$ stable under G.

(Argue by induction on $\dim \mathfrak{g}$. Using Exerc. 12, reduce to the case where $\mathfrak{g}$ is semi-simple. If $G \neq \{1\}$, choose a normal subgroup C of G that is cyclic of prime order (*Algebra, loc. cit.*). The subalgebra $\mathfrak{s}$ consisting of the elements invariant under C is reductive in $\mathfrak{g}$ (§1, no. 5), and distinct from $\mathfrak{g}$. By the induction hypothesis, $\mathfrak{s}$ has a Cartan subalgebra $\mathfrak{a}$ stable under G. We have $\mathfrak{s} \neq 0$ (Chap. I, §4, Exerc. 21 *c*)), so $\mathfrak{a} \neq 0$. The commutant $\mathfrak{z}$ of $\mathfrak{a}$ in $\mathfrak{g}$ is distinct from $\mathfrak{g}$ and stable under G; choose a Cartan subalgebra $\mathfrak{h}$ of $\mathfrak{z}$ stable under G, and show that $\mathfrak{h}$ is a Cartan subalgebra of $\mathfrak{g}$, cf. the Cor. to Prop. 3.)

Construct a finite group of automorphisms of $\mathfrak{sl}(2, \mathbf{C})$ which is isomorphic to $\mathfrak{A}_4$ (and hence solvable) and which does not leave any Cartan subalgebra stable.

14) *Show that every irreducible complex (resp. real) linear representation of a hyper-solvable finite group G is induced by a representation of degree 1 (resp. of degree 1 or 2) of a subgroup of G. (Apply Exerc. 13 to the Lie algebras $\mathfrak{gl}(n, \mathbf{C})$ and $\mathfrak{gl}(n, \mathbf{R})$.)*

15) Assume that k is algebraically closed. Let $\mathfrak{g}$ be a Lie algebra, $\mathfrak{h}$ a Cartan subalgebra of $\mathfrak{g}$, and A a subset of $\mathfrak{g}$. Assume that A is dense in $\mathfrak{g}$ (in the Zariski topology) and stable under $\mathrm{Aut}_e(\mathfrak{g})$. Show that $A \cap \mathfrak{h}$ is dense in $\mathfrak{h}$. (Let X be the closure of $A \cap \mathfrak{h}$, and $U = \mathfrak{h} - X$. Assume that $U \neq \varnothing$. With the notations in Lemma 2, the image under F of $U \times \mathfrak{g}^{\lambda_1}(\mathfrak{h}) \times \cdots \times \mathfrak{g}^{\lambda_p}(\mathfrak{h})$ contains a non-empty open subset of $\mathfrak{g}$. Since this image is contained in $\mathfrak{g} - A$, this contradicts the fact that A is dense in $\mathfrak{g}$.)

16) Let V be a finite dimensional k-vector space, and $\mathfrak{g}$ a Lie subalgebra of $\mathfrak{gl}(V)$. We are going to show that the following three properties are equivalent:

(i) The Cartan subalgebras of $\mathfrak{g}$ are commutative and consist of semi-simple elements.

(ii) Every regular element of $\mathfrak{g}$ is semi-simple.

(iii) The semi-simple elements of $\mathfrak{g}$ are dense in $\mathfrak{g}$ in the Zariski topology.

a) Show that (i) $\Longrightarrow$ (ii) $\Longrightarrow$ (iii).

b) Let A be the set of semi-simple elements of $\mathfrak{g}$. Show that A is stable under $\mathrm{Aut}_e(\mathfrak{g})$.

c) Show that (iii) $\Longrightarrow$ (i). (Reduce (App. I, Exerc. 1) to the case where k is algebraically closed. If $\mathfrak{h}$ is a Cartan subalgebra of $\mathfrak{g}$, show, by using Exerc. 15, that $\mathrm{A} \cap \mathfrak{h}$ is dense in $\mathfrak{h}$; since $[x, y] = 0$ if $x \in \mathrm{A} \cap \mathfrak{h}, y \in \mathfrak{h}$, deduce that $\mathfrak{h}$ is commutative, from which (i) is immediate.)

d) Assume that k is $\mathbf{R}, \mathbf{C}$ or a non-discrete complete ultrametric field of characteristic zero. Give $\mathfrak{g}$ the topology defined by that of k. Show that properties (i), (ii), (iii) are equivalent to the following:
(iv) The semi-simple elements of $\mathfrak{g}$ are dense in $\mathfrak{g}$.
(Show that (iv) $\Longrightarrow$ (iii) and (ii) $\Longrightarrow$ (iv), cf. App. I, Exerc. 4.)

17) Assume that k is algebraically closed. Let $\mathfrak{g}$ be a Lie algebra, $\mathfrak{h}$ a Cartan subalgebra, and A a subset of the centre of $\mathfrak{h}$. Denote by $\mathrm{E}_{\mathfrak{g}}$ the subgroup of $\mathrm{Aut}(\mathfrak{g})$ denoted by E in no. 2. Show that, if s is an element of $\mathrm{Aut}(\mathfrak{g})$ such that $s\mathrm{A} = \mathrm{A}$, there exists $t \in \mathrm{E}_{\mathfrak{g}}$ such that $t\mathfrak{h} = \mathfrak{h}$ and $t|\mathrm{A} = \mathrm{Id}_{\mathrm{A}}$; in particular, $ts|\mathrm{A} = s|\mathrm{A}$. (Let $\mathfrak{a}$ be the commutant of A in $\mathfrak{g}$; since $\mathfrak{h}$ and $s\mathfrak{h}$ are Cartan subalgebras of $\mathfrak{a}$, there exists $\theta \in \mathrm{E}_{\mathfrak{a}}$ such that $\theta(s\mathfrak{h}) = \mathfrak{h}$; choose t from the elements of $\mathrm{E}_{\mathfrak{g}}$ that extend θ.)

¶ 18) Let $\mathfrak{g}$ be a Lie algebra, $\mathfrak{h}$ a Cartan subalgebra of $\mathfrak{g}$ and $\mathrm{U}\mathfrak{g}$ (resp. $\mathrm{U}\mathfrak{h}$) the enveloping algebra of $\mathfrak{g}$ (resp. $\mathfrak{h}$). A linear form φ on $\mathrm{U}\mathfrak{g}$ is said to be *central* if it vanishes on $[\mathrm{U}\mathfrak{g}, \mathrm{U}\mathfrak{g}]$, i.e. if $\varphi(a.b) = \varphi(b.a)$ for all $a, b \in \mathrm{U}\mathfrak{g}$.

a) Let $x, y \in \mathfrak{g}$. Assume that there exists $s \in \mathrm{Aut}_e(\mathfrak{g})$ such that $s(x) = y$. Show that

$$\varphi(x^n) = \varphi(y^n)$$

for all $n \in \mathbf{N}$ and for every central linear form φ on $\mathrm{U}\mathfrak{g}$.

b) Let φ be a central linear form on $\mathrm{U}\mathfrak{g}$ whose restriction to $\mathrm{U}\mathfrak{h}$ vanishes. Show that $\varphi = 0$. (We can assume that k is algebraically closed. Deduce from *a*) that we then have $\varphi(x^n) = 0$ for all $n \in \mathbf{N}$ and all regular $x \in \mathfrak{g}$; use a density argument to remove the assumption of regularity.)

c) Show that $\mathrm{U}\mathfrak{g} = [\mathrm{U}\mathfrak{g}, \mathrm{U}\mathfrak{g}] + \mathrm{U}\mathfrak{h}$.

d) Let V be a semi-simple $\mathfrak{g}$-module. Show that V is semi-simple as an $\mathfrak{h}$-module. In particular, $\mathrm{V}^{\lambda}(\mathfrak{h}) = \mathrm{V}_{\lambda}(\mathfrak{h})$ for all $\lambda \in \mathfrak{h}^*$.

e) Let V′ be a semi-simple $\mathfrak{g}$-module. Assume that V and V′ are isomorphic as $\mathfrak{h}$-modules. Show that they are isomorphic as $\mathfrak{g}$-modules. (If $a \in \mathrm{U}\mathfrak{h}$, remark that $\mathrm{Tr}(a_{\mathrm{V}}) = \mathrm{Tr}(a_{\mathrm{V}'})$. Deduce, by using *b*) or *c*), that $\mathrm{Tr}(x_{\mathrm{V}}) = \mathrm{Tr}(x_{\mathrm{V}'})$ for all $x \in \mathrm{U}\mathfrak{g}$, and conclude by using *Algebra*, Chap. VIII.)

If k is algebraically closed, the assumption that "V and V′ are $\mathfrak{h}$-isomorphic" is equivalent to saying that $\dim \mathrm{V}_{\lambda}(\mathfrak{h}) = \dim \mathrm{V}'_{\lambda}(\mathfrak{h})$ for all $\lambda \in \mathfrak{h}^*$.

§ 4

The notations and assumptions are those of nos. 1, 2, 3 *of* § 4.

1) Take $G = \mathbf{GL}_n(k)$, $n \geq 0$.

a) Show that $r^0_{\mathrm{Ad}}(g) = n$ for all $g \in G$.

b) Show that an element $g \in G$ is regular if and only if its characteristic polynomial $P_g(T) = \det(T - g)$ is separable; this is equivalent to saying that the discriminant (*Algebra*, Chap. IV, §1, no. 10) of $P_g(T)$ is $\neq 0$.

2) Construct a Lie group G such that the function r^0_{Ad} is not constant. (Take $\mathfrak{g}$ abelian $\neq 0$ and Ad non-trivial.)

3) Let $(\rho_i)_{i \in I}$ be a *countable* family of analytic linear representations of G. Prove that the elements of G which are regular for all the ρ_i constitute a *dense* subset of G. Construct an example showing that the countability assumption cannot be omitted.

4) Assume that $k = \mathbf{C}$ and that G is connected. Prove the equivalence of the following properties:
(i) G is nilpotent.
(ii) Every element $\neq 1$ of G is regular.
(Show first that (ii) implies
(ii)$'$ Every element $\neq 0$ of $\mathfrak{g}$ is regular.
Remark next that, if $\mathfrak{g} \neq 0$, then $\mathfrak{g}$ contains elements $x \neq 0$ such that $\mathrm{ad}\, x$ is nilpotent, cf. §3, Exerc. 10. Deduce that $\mathfrak{g}$ is nilpotent, hence (i).)

§ 5

1) Show that the solvable Lie algebra considered in Chap. I, §5, Exerc. 6 is not isomorphic to any decomposable Lie algebra.

2) Let u (resp. v) be a non-zero semi-simple (resp. nilpotent) endomorphism of V. Then the map $\lambda u \mapsto \lambda v$ ($\lambda \in k$) is an isomorphism from $\mathfrak{g} = ku$ to $\mathfrak{g}' = kv$ which does not take semi-simple elements to semi-simple elements.

3) Let u be an endomorphism of V that is neither semi-simple nor nilpotent. Then $\mathfrak{g} = ku$ is non-decomposable, but $\mathrm{ad}_{\mathfrak{g}}\mathfrak{g}$ is decomposable.

¶ 4) Let $\mathfrak{g}$ be a decomposable Lie subalgebra of $\mathfrak{gl}(V)$. Let $\mathfrak{q}$ be a Lie subalgebra of $\mathfrak{g}$ whose identity representation is semi-simple. There exists $a \in \mathrm{Aut}_e(\mathfrak{g})$ such that $a(\mathfrak{q})$ is contained in the subalgebra $\mathfrak{m}$ of Prop. 7. (Imitate the proof of Prop. 7 (ii).)

¶ 5) Let $\mathfrak{g}$ be a Lie subalgebra of $\mathfrak{gl}(V)$. Then $\mathfrak{g}$ is said to be *algebraic* if, for all $x \in \mathfrak{g}$, the replicas of x (Chap. I, §5, Exerc. 14) belong to $\mathfrak{g}$. Such an algebra is decomposable.

a) Denote by $a(\mathfrak{g})$ the smallest algebraic subalgebra of $\mathfrak{gl}(V)$ containing $\mathfrak{g}$. Then

$$a(\mathfrak{g}) \supset e(\mathfrak{g}) \supset \mathfrak{g}.$$

Give an example where $a(\mathfrak{g})$ and $e(\mathfrak{g})$ are distinct (take V of dimension 2 and $\mathfrak{g}$ of dimension 1).

b) Show that, if $\mathfrak{n}$ is an ideal of $\mathfrak{g}$, $\mathfrak{n}$ and $a(\mathfrak{n})$ are ideals of $a(\mathfrak{g})$, and $[a(\mathfrak{g}), a(\mathfrak{n})] = [\mathfrak{g}, \mathfrak{n}]$ (imitate the proof of Prop. 4). Deduce that $\mathscr{D}^i a(\mathfrak{g}) = \mathscr{D}^i \mathfrak{g}$ for $i \geq 1$ and that $\mathscr{C}^i a(\mathfrak{g}) = \mathscr{C}^i \mathfrak{g}$ for $i \geq 2$.

c) Show that every Lie algebra consisting of nilpotent elements is algebraic.[8]

¶ 6) Let $\mathfrak{g}$ be a semi-simple subalgebra of $\mathfrak{gl}(V)$, $\mathbf{T}(V) = \bigoplus_{n=0}^{\infty} \mathbf{T}^n(V)$ the tensor algebra of V, $\mathbf{T}(V)^{\mathfrak{g}}$ the set of elements of $\mathbf{T}(V)$ invariant under $\mathfrak{g}$ (cf. Chap. III, App.) and $\overline{\mathfrak{g}}$ the set of $u \in \mathfrak{gl}(V)$ such that $u.x = 0$ for all $x \in \mathbf{T}(V)^{\mathfrak{g}}$. We are going to show that $\overline{\mathfrak{g}} = \mathfrak{g}$.

a) Show that the representation of $\mathfrak{g}$ on the dual V^* of V is isomorphic to its representation on $\bigwedge^{p-1} V$, where $p = \dim V$ (use the fact that $\mathfrak{g}$ is contained in $\mathfrak{sl}(V)$). Deduce that every element of $\mathbf{T}_{n,m} = \mathbf{T}^n(V) \otimes \mathbf{T}^m(V^*)$ invariant under $\mathfrak{g}$ is invariant under $\overline{\mathfrak{g}}$, and that $\overline{\mathfrak{g}}$ is algebraic (Exerc. 5).

b) Let W be a vector subspace of one of the $\mathbf{T}_{n,m}$. Assume that W is stable under $\mathfrak{g}$. Show that W is stable under $\overline{\mathfrak{g}}$ (if $e_1, \ldots, e_r$ is a basis of W, remark that $e_1 \wedge \cdots \wedge e_r$ is invariant under $\mathfrak{g}$, and hence under $\overline{\mathfrak{g}}$).

Deduce that $\mathfrak{g}$ is an ideal of $\overline{\mathfrak{g}}$, and that $\overline{\mathfrak{g}}/\mathfrak{g}$ is commutative (cf. proof of Prop. 4). We have $\overline{\mathfrak{g}} = \mathfrak{g} \times \mathfrak{c}$, where $\mathfrak{c}$ is the centre of $\overline{\mathfrak{g}}$.

c) Let R be the associative subalgebra of $\mathfrak{gl}(V)$ generated by 1 and $\mathfrak{g}$. Show that $\mathfrak{c}$ is contained in the centre of R (remark that $\overline{\mathfrak{g}}$ is contained in the bicommutant of R, which is equal to R). Deduce that the elements of $\mathfrak{c}$ are semi-simple.

d) Let $x \in \mathfrak{c}$. Show that the replicas of x belong to $\mathfrak{c}$ (Chap. I, §5, Exerc. 14). Show that $\operatorname{Tr}(sx) = 0$ for all $s \in \mathfrak{g}$; deduce that $\operatorname{Tr}(sx) = 0$ for all $s \in \overline{\mathfrak{g}}$, and hence that x is nilpotent (*loc. cit*).

e) Show that $\mathfrak{c} = 0$ and $\overline{\mathfrak{g}} = \mathfrak{g}$ by combining *c*) and *d*).

7) Let $\mathfrak{g}$ be a Lie subalgebra of $\mathfrak{gl}(V)$. Let m, n be two integers ≥ 0, W and W′ two vector subspaces of $\mathbf{T}^m(V) \otimes \mathbf{T}^n(V^*)$ where V^* is the dual of V. Assume that $W' \subset W$ and that W and W′ are stable under the natural representation of $\mathfrak{g}$ on $\mathbf{T}^m(V) \otimes \mathbf{T}^n(V^*)$. Show that W and W′ are then stable under $e(\mathfrak{g})$. If π denotes the representation of $e(\mathfrak{g})$ on W/W' thus obtained, show that $\pi e(\mathfrak{g})$ is the decomposable envelope of $\pi(\mathfrak{g})$ (use Th. 1).

Deduce that $\operatorname{ad} e(\mathfrak{g})$ is the decomposable envelope of $\operatorname{ad} \mathfrak{g}$ in $\mathfrak{gl}(\mathfrak{g})$.

---

[8] For more details, cf. C. CHEVALLEY, Théorie des groupes de Lie, II, Groupes algébriques, Chap. II, §14, *Paris, Hermann*, 1951.

8) Let $\mathfrak{g}$ be a Lie subalgebra of $\mathfrak{gl}(V)$ and $\mathfrak{h}$ a Cartan subalgebra of $\mathfrak{g}$.

a) Show that $e(\mathfrak{g}) = e(\mathfrak{h}) + \mathscr{D}\mathfrak{g} = e(\mathfrak{h}) + \mathfrak{g}$.
(Remark that $e(\mathfrak{h})+\mathscr{D}\mathfrak{g}$ is decomposable (Cor. 1 of Th. 1), contains $\mathfrak{g} = \mathfrak{h}+\mathscr{D}\mathfrak{g}$, and is contained in $e(\mathfrak{g})$; thus, it is $e(\mathfrak{g})$.)

b) We have $e(\mathfrak{h}) \cap \mathfrak{g} = \mathfrak{h}$ (remark that $e(\mathfrak{h}) \cap \mathfrak{g}$ is nilpotent).

c) Let x be an element of the normalizer of $e(\mathfrak{h})$ in $e(\mathfrak{g})$. Show that $x \in e(\mathfrak{h})$. (Write $x = y+z$ with $y \in e(\mathfrak{h}), z \in \mathfrak{g}$, cf. *a*); remark that $[z, \mathfrak{h}] \subset e(\mathfrak{h}) \cap \mathfrak{g} = \mathfrak{h}$, hence $z \in \mathfrak{h}$.)

d) Show that $e(\mathfrak{h})$ is a Cartan subalgebra of $e(\mathfrak{g})$.

9) Let $\mathfrak{g}$ be a Lie subalgebra of $\mathfrak{gl}(V)$. Show that conditions (i), (ii), (iii) of Exerc. 16 of §3 are equivalent to:

(v) $\mathfrak{g}$ is decomposable and has the same rank as $\mathfrak{g}/\mathfrak{n}_V(\mathfrak{g})$.

(If $\mathfrak{h}$ is a Cartan subalgebra of $\mathfrak{g}$, the condition "$\mathfrak{g}$ has the same rank as $\mathfrak{g}/\mathfrak{n}_V(\mathfrak{g})$" is equivalent to saying that $\mathfrak{h} \cap \mathfrak{n}_V(\mathfrak{g}) = 0$, i.e. that $\mathfrak{h}$ has no nilpotent element $\neq 0$ (cf. §2, Exerc. 14). Deduce the equivalence of (i) and (v).)

10) Let k' be an extension of k and $\mathfrak{g}'$ a k'-Lie subalgebra of

$$\mathfrak{gl}(V \otimes_k k') = \mathfrak{gl}(V) \otimes_k k'.$$

a) Show that there exists a smallest Lie subalgebra $\mathfrak{g}$ of $\mathfrak{gl}(V)$ such that $\mathfrak{g} \otimes_k k'$ contains $\mathfrak{g}'$.

b) Assume that k' is algebraically closed and denote by G the group of k-automorphisms of k'; this group operates in a natural way on $V \otimes_k k'$. Show that $\mathfrak{g} \otimes_k k'$ is the Lie subalgebra generated by the conjugates of $\mathfrak{g}'$ under G (use the fact that k is the field of G-invariants in k').

c) Show that $\mathfrak{g}$ is decomposable if $\mathfrak{g}'$ is decomposable. (Reduce to the case where k' is algebraically closed, and use *b*) as well as Cor. 1 and 3 of Th. 1.)

¶ 11) Exceptionally, we assume in this exercise that k is a *perfect field of characteristic* $p > 0$.

Let $\mathfrak{g}$ be a Lie p-algebra (Chap. I, §1, Exerc. 20). If $x \in \mathfrak{g}$, denote by $\langle x \rangle$ the smallest Lie p-subalgebra containing x. It is commutative and generated as a k-vector space by the x^{p^i} where $i = 0, 1, \ldots$. Then x is said to be *nilpotent* (resp. *semi-simple*) if the p-map of $\langle x \rangle$ is nilpotent (resp. bijective).

a) Show that x can be decomposed uniquely in the form $x = s + n$, with $s, n \in \langle x \rangle$, s semi-simple and n nilpotent (apply Exerc. 23 of Chap. I, §1). If f is a p-homomorphism from $\mathfrak{g}$ to $\mathfrak{gl}(V)$, $f(s)$ and $f(n)$ are the semi-simple and nilpotent components of the endomorphism $f(x)$; this applies in particular to $f = \mathrm{ad}$.

b) A subalgebra of $\mathfrak{g}$ is said to be *decomposable* if it contains the semi-simple and nilpotent components of its elements. Show that, if $\mathfrak{b}$ and $\mathfrak{c}$ are vector subspaces of $\mathfrak{g}$ such that $\mathfrak{b} \subset \mathfrak{c}$, the set of $x \in \mathfrak{g}$ such that $[x, \mathfrak{c}] \subset \mathfrak{b}$ is decomposable (same proof as Prop. 3); in particular, every Cartan subalgebra of $\mathfrak{g}$ is decomposable.

c) Let $\mathfrak{t}$ be a commutative subalgebra of $\mathfrak{g}$ consisting of semi-simple elements, and maximal with this property. Let $\mathfrak{h}$ be the commutant of $\mathfrak{t}$ in $\mathfrak{g}$. Let $x \in \mathfrak{h}$ and let $x = s + n$ be its canonical decomposition; since $\mathfrak{h}$ is decomposable (cf. *b*)), $s, n \in \mathfrak{h}$. Show that the subalgebra generated by $\mathfrak{t}$ and s is commutative and consists of semi-simple elements, hence coincides with $\mathfrak{t}$. Deduce that $\mathrm{ad}_{\mathfrak{h}} x = \mathrm{ad}_{\mathfrak{h}} n$ is nilpotent, and hence that $\mathfrak{h}$ is nilpotent. Since $\mathfrak{h} = \mathfrak{g}^0(\mathfrak{h})$, $\mathfrak{h}$ is a Cartan subalgebra of $\mathfrak{g}$ (§2, Prop. 4).

In particular, every Lie p-algebra over a finite field has a Cartan subalgebra.[9]

Appendix I

Denote by V *a finite dimensional vector space over* k.

1) Let k' be an extension of k, and let $\mathrm{V}_{(k')} = \mathrm{V} \otimes_k k'$. Show that the Zariski topology on $\mathrm{V}_{(k')}$ induces the Zariski topology on V, and that V is dense in $\mathrm{V}_{(k')}$.

2) Assume that V is the product of two vector spaces V_1 and V_2.

a) The Zariski topology on V is finer than the product topology of the Zariski topologies on V_1 and V_2; it is strictly finer if $\mathrm{V}_1 \neq 0$ and $\mathrm{V}_2 \neq 0$.

b) If A_1 (resp. A_2) is a subset of V_1 (resp. V_2), the closure of $\mathrm{A}_1 \times \mathrm{A}_2$ is the product of the closures of A_1 and A_2.

3) Assume that k is algebraically closed. Let A and B be two closed subsets of V, and $\mathfrak{a}$ (resp. $\mathfrak{b}$) the set of $f \in \mathrm{A}_\mathrm{V}$ which vanish on A (resp. B). Prove the equivalence of the following properties:

(i) $\mathrm{A} \cap \mathrm{B} = \varnothing$.

(ii) $\mathfrak{a} + \mathfrak{b} = \mathrm{A}_\mathrm{V}$.

(iii) There exists a polynomial function f on V which is equal to 1 on A and to 0 on B.

(Use Hilbert's theorem of zeros (*Commutative Algebra*, Chap. V, §3, no. 3) to prove that (i) $\Longrightarrow$ (ii).)

4) Assume that k is a non-discrete complete valued field. Denote by $\mathscr{T}$ (resp. $\mathscr{Z}$) the Banach space (resp. Zariski) topology on V.

a) Show that $\mathscr{T}$ is finer than $\mathscr{Z}$ (and strictly finer if $\mathrm{V} \neq 0$).

b) Show that every non-empty $\mathscr{Z}$-open subset of V is $\mathscr{T}$-dense.

[9] For more details, cf. G. B. SELIGMAN, Modular Lie Algebras, Chap. V, §7, *Springer-Verlag*, 1967.

Appendix II

¶ 1) Let X be a locally connected topological space, $\mathscr{C}(X)$ the space of continuous real-valued functions on X, and d an integer ≥ 0. Let $F \in \mathscr{C}(X)[T]$ be a monic polynomial of degree d with coefficients in $\mathscr{C}(X)$:

$$F = T^d + T^{d-1}f_1 + \cdots + f_d, \quad f_i \in \mathscr{C}(X).$$

Identify F with a function on $\mathbf{R} \times X$ by putting

$$F(t,x) = t^d + t^{d-1}f_1(x) + \cdots + f_d(x) \ \text{ if } \ t \in \mathbf{R},\ x \in X.$$

Let $\Delta \in \mathscr{C}(X)$ be the *discriminant* of the polynomial F (*Algebra*, Chap. IV, §1, no. 10).

If U is an open subset of X, denote by Z_U the set of (t,x), with $t \in \mathbf{R}$ and $x \in U$, such that $F(t,x) = 0$; this is a closed subset of $\mathbf{R} \times U$.

a) Show that the projection $\mathrm{pr}_2 : Z_U \to U$ is proper (*General Topology*, Chap. I, §10).

b) Assume that U is connected and that $\Delta(x) \neq 0$ for all $x \in U$. Show that $Z_U \to U$ is a *covering* of U (*General Topology*, Chap. XI) of degree $\leq d$, and that the number of connected components of $\mathbf{R} \times U - Z_U$ is $\leq d+1$.

c) Let X′ be the set of points of X at which Δ is $\neq 0$. Assume that X′ is dense in X. Denote by $\mathscr{A}$ (resp. $\mathscr{B}$) the set of connected components of X′ (resp. of $\mathbf{R} \times X - Z_X$). Show that

$$\mathrm{Card}(\mathscr{B}) \leq (d+1)\mathrm{Card}(\mathscr{A}) \quad (\text{use } b)).$$

d) Assume that X is connected, and that $d \geq 1$. Show that

$$\mathrm{Card}(\mathscr{B}) \leq 1 + d\,\mathrm{Card}(\mathscr{A}).$$

¶ 2) Let V be a real vector space of finite dimension n, and F a polynomial function on V of degree d. Let V′ be the set of points of V at which $F \neq 0$. Show that the number of connected components of V′ is finite and bounded by a constant depending only on n and d. (Argue by induction on n. Reduce to the case in which F has no multiple factors, and show that V can be decomposed as $\mathbf{R} \times X$ in such a way that the results of Exerc. 1 are applicable to F.)[10]

[10]For other results in the same direction, cf. J. MILNOR, On the Betti numbers of real varieties, *Proc. Amer. Math. Soc.*, Vol. XV (1964), pp. 275-280.

CHAPTER VIII
Split Semi-simple Lie Algebras

In this chapter, k denotes a (commutative) field of characteristic 0. *Unless otherwise stated, by a "vector space", we mean a "vector space over k"; similarly for "Lie algebra", etc.*

§ 1. THE LIE ALGEBRA $\mathfrak{sl}(2,k)$ AND ITS REPRESENTATIONS

1. CANONICAL BASIS OF $\mathfrak{sl}(2,k)$

Lemma 1. Let A *be an associative algebra over* k, H *and* X *elements of* A *such that* $[H,X]=2X$.

(i) $[H,X^n]=2nX^n$ *for any integer* $n\geq 0$.

(ii) *If* Z *is an element of* A *such that* $[Z,X]=H$, *then, for any integer* $n>0$,

$$[Z,X^n]=nX^{n-1}(H+n-1)=n(H-n+1)X^{n-1}.$$

The map $T\mapsto [H,T]$ from A to A is a derivation, which implies (i). With the assumptions in (ii),

$$\begin{aligned}[Z,X^n] &= \sum_{i+j=n-1} X^iHX^j\\ &= \sum_{i+j=n-1}(X^iX^jH+X^i2jX^j)\\ &= nX^{n-1}H+2X^{n-1}\frac{n(n-1)}{2}\\ &= nX^{n-1}(H+n-1).\end{aligned}$$

On the other hand, $X^{n-1}(H+n-1)=(H-n+1)X^{n-1}$ by (i). Q.E.D.

Recall that we denote by $\mathfrak{sl}(2,k)$ the Lie algebra consisting of the square matrices of order 2, trace zero, and with entries in k. This Lie algebra is simple of dimension 3 (Chap. I, §6, no. 7, Example). The canonical basis of $\mathfrak{sl}(2,k)$ is the basis (X_+, X_-, H), where

$$X_+ = \begin{pmatrix} 0 & 1 \\ 0 & 0 \end{pmatrix} \quad X_- = \begin{pmatrix} 0 & 0 \\ -1 & 0 \end{pmatrix} \quad H = \begin{pmatrix} 1 & 0 \\ 0 & -1 \end{pmatrix}.$$

We have

$$[H, X_+] = 2X_+ \quad [H, X_-] = -2X_- \quad [X_+, X_-] = -H. \tag{1}$$

Since the identity representation of $\mathfrak{sl}(2,k)$ is injective, H is a semi-simple element of $\mathfrak{sl}(2,k)$ and X_+, X_- are nilpotent elements of $\mathfrak{sl}(2,k)$ (Chap. I, §6, no. 3, Th. 3). By Chap. VII, §2, no. 1, Example 4, kH is a Cartan subalgebra of $\mathfrak{sl}(2,k)$. The map $U \mapsto -{}^tU$ is an involutive automorphism of the Lie algebra $\mathfrak{sl}(2,k)$, called the *canonical involution of* $\mathfrak{sl}(2,k)$; it transforms (X_+, X_-, H) into $(X_-, X_+, -H)$.

Lemma 2. In the enveloping algebra of $\mathfrak{sl}(2,k)$,

$$[H, X_+^n] = 2nX_+^n \quad [H, X_-^n] = -2nX_-^n$$

for any integer $n \geq 0$, *and*

$$[X_-, X_+^n] = nX_+^{n-1}(H + n - 1) = n(H - n + 1)X_+^{n-1}$$
$$[X_+, X_-^n] = nX_-^{n-1}(-H + n - 1) = n(-H - n + 1)X_-^{n-1}$$

if $n > 0$.

The first and third relations follow from Lemma 1. The others can be deduced from them by using the canonical involution of $\mathfrak{sl}(2,k)$.

2. PRIMITIVE ELEMENTS OF $\mathfrak{sl}(2,k)$-MODULES

Let E be an $\mathfrak{sl}(2,k)$-module. If $A \in \mathfrak{sl}(2,k)$ and $x \in \mathrm{E}$, we shall often write Ax instead of $A_{\mathrm{E}}x$. Let $\lambda \in k$. If $Hx = \lambda x$ we say, by abuse of language, that x is an element of E *of weight* λ, or that λ is *the weight* of x. If E is finite dimensional, H_{E} is semi-simple, so the set of elements of weight λ is the primary subspace of E relative to H_{E} and λ (cf. Chap. VII, §1, no. 1).

Lemma 3. If x is an element of weight λ, then X_+x is an element of weight $\lambda + 2$ and X_-x is an element of weight $\lambda - 2$.

Indeed, $HX_+x = [H, X_+]x + X_+Hx = 2X_+x + X_+\lambda x = (\lambda + 2)X_+x$, and similarly $HX_-x = (\lambda - 2)X_-x$ (cf. also Chap. VII, §1, no. 3, Prop. 10 (ii)).

DEFINITION 1. *Let* E *be an* $\mathfrak{sl}(2,k)$*-module. An element of* E *is said to be primitive if it is a non-zero eigenvector of* H_{E} *and belongs to the kernel of* $X_{+\mathrm{E}}$.

A non-zero element e of E is primitive if and only if ke is stable under the operation of $kH + kX_+$; this follows for example from Lemma 3.

Examples The element X_+ is primitive of weight 2 for the adjoint representation of $\mathfrak{sl}(2, k)$. The element $(1, 0)$ of k^2 is primitive of weight 1 for the identity representation of $\mathfrak{sl}(2, k)$ on k^2.

Lemma 4. Let E *be a non-zero finite dimensional* $\mathfrak{sl}(2, k)$*-module. Then* E *has primitive elements.*

Since X_+ is a nilpotent element of $\mathfrak{sl}(2, k)$, $X_{+\mathrm{E}}$ is nilpotent. Assume that $X_{+\mathrm{E}}^{m-1} \neq 0$ and $X_{+\mathrm{E}}^m = 0$. By Lemma 2,

$$m(H_{\mathrm{E}} - m + 1)X_{+\mathrm{E}}^{m-1} = [X_{-\mathrm{E}}, X_{+\mathrm{E}}^m] = 0,$$

and hence the elements of $X_+^{m-1}(\mathrm{E}) - \{0\}$ are primitive.

PROPOSITION 1. *Let* E *be an* $\mathfrak{sl}(2, k)$*-module, and* e *a primitive element of* E *of weight* λ. *Put* $e_n = \frac{(-1)^n}{n} X_-^n e$ *for* $n \geq 0$, *and* $e_{-1} = 0$. *Then*

$$\begin{cases} He_n &= (\lambda - 2n)e_n \\ X_-e_n &= -(n+1)e_{n+1} \\ X_+e_n &= (\lambda - n + 1)e_{n-1}. \end{cases} \tag{2}$$

The first formula follows from Lemma 3, and the second from the definition of the e_n. We prove the third by induction on n. It is satisfied for $n = 0$ since $e_{-1} = 0$. If $n > 0$,

$$\begin{aligned} nX_+e_n &= -X_+X_-e_{n-1} = -[X_+, X_-]e_{n-1} - X_-X_+e_{n-1} \\ &= He_{n-1} - X_-(\lambda - n + 2)e_{n-2} \\ &= (\lambda - 2n + 2 + (n-1)(\lambda - n + 2))e_{n-1} \\ &= n(\lambda - n + 1)e_{n-1}. \end{aligned}$$

COROLLARY. *The submodule of* E *generated by* e *is the vector subspace generated by the* e_n.

This follows from the formulas (2).

The integers $n \geq 0$ such that $e_n \neq 0$ constitute an interval in $\mathbf{N}$, and the corresponding elements e_n form a basis over k of the submodule generated by e (indeed, they are linearly independent because they are non-zero elements of distinct weights). This basis will be said to be *associated* to the primitive element e.

PROPOSITION 2. *If the submodule* V *of* E *generated by the primitive element* e *is finite dimensional, then:*

(i) *the weight* λ *of* e *is integral and equal to* $\dim \mathrm{V} - 1$;

(ii) $(e_0, e_1, \ldots, e_\lambda)$ *is a basis of* V, *and* $e_n = 0$ *for* $n > \lambda$;

(iii) *the eigenvalues of* H_{V} *are* $\lambda, \lambda - 2, \lambda - 4, \ldots, -\lambda$; *they are all of multiplicity* 1;

(iv) *every primitive element of* V *is proportional to* e;

(v) *the commutant of the module* V *is reduced to the scalars; in particular,* V *is absolutely simple.*

Let m be the largest integer such that $e_m \neq 0$. Then $0 = X_+e_{m+1} = (\lambda - m)e_m$, so $\lambda = m$; since $(e_0, e_1, \ldots, e_m)$ is a basis of V, this proves (i) and (ii). Assertion (iii) follows from the equality $He_n = (\lambda - 2n)e_n$. We have $X_+e_n \neq 0$ for $1 \leq n \leq m$, hence (iv). Let c be an element of the commutant of the module V. Then $Hc(e) = cH(e) = \lambda c(e)$, so there exists $\mu \in k$ such that $c(e) = \mu e$; then

$$cX_-^q e = X_-^q ce = \mu X_-^q e$$

for all $q \geq 0$, so $c = \mu.1$, proving (v).

COROLLARY. *Let* E *be a finite dimensional* $\mathfrak{sl}(2,k)$*-module.*

(i) *The endomorphism* H_{E} *is diagonalizable and its eigenvalues are rational integers.*

(ii) *For any* $p \in \mathbf{Z}$, *let* E_p *be the eigenspace of* H_{E} *corresponding to the eigenvalue* p. *Let* i *be an integer* ≥ 0. *The map* $X_{-\mathrm{E}}^i|\mathrm{E}_p : \mathrm{E}_p \to \mathrm{E}_{p-2i}$ *is injective for* $i \leq p$, *bijective for* $i = p$, *and surjective for* $i \geq p$. *The map* $X_{+\mathrm{E}}^i|\mathrm{E}_{-p} : \mathrm{E}_{-p} \to \mathrm{E}_{-p+2i}$ *is injective for* $i \leq p$, *bijective for* $i = p$, *and surjective for* $i \geq p$.

(iii) *The length of* E *is equal to* $\dim \mathrm{Ker} X_{+\mathrm{E}}$ *and to* $\dim \mathrm{Ker} X_{-\mathrm{E}}$.

(iv) *Let* E′ (resp. E″) *be the sum of the* E_p *for* p *even* (resp. *odd*). *Then* E′ (resp. E″) *is the sum of the simple submodules of* E *of odd* (resp. *even*) *dimension; and* $\mathrm{E} = \mathrm{E}' \oplus \mathrm{E}''$. *The length of* E′ *is* $\dim \mathrm{E}_0$, *and that of* E″ *is* $\dim \mathrm{E}_1$.

(v) $\mathrm{Ker} X_{+\mathrm{E}} \cap \mathrm{Im} X_{+\mathrm{E}} \subset \sum_{p>0} \mathrm{E}_p$ *and* $\mathrm{Ker} X_{-\mathrm{E}} \cap \mathrm{Im} X_{-\mathrm{E}} \subset \sum_{p<0} \mathrm{E}_p$.

If E is simple, E is generated by a primitive element (Lemma 4), and it suffices to apply Propositions 1 and 2. The general case follows since every finite dimensional $\mathfrak{sl}(2,k)$-module is semi-simple.

3. THE SIMPLE MODULES V(m)

Let (u, v) be the canonical basis of k^2. For the identity representation of $\mathfrak{sl}(2,k)$,

$$X_+u = 0 \quad Hu = u \quad X_-u = -v$$
$$X_+v = u \quad Hv = -v \quad X_-v = 0.$$

Consider the symmetric algebra $\mathbf{S}(k^2)$ of k^2 (*Algebra*, Chap. III, §6, no. 1, Def. 1). The elements of $\mathfrak{sl}(2,k)$ extend uniquely to derivations of $\mathbf{S}(k^2)$, giving $\mathbf{S}(k^2)$ the structure of an $\mathfrak{sl}(2,k)$-module (Chap. I, §3, no. 2). Let $\mathrm{V}(m)$ be the set of homogeneous elements of $\mathbf{S}(k^2)$ of degree m. Then $\mathrm{V}(m)$ is an $\mathfrak{sl}(2,k)$-submodule of $\mathbf{S}(k^2)$ of dimension $m+1$, the mth symmetric power of $\mathrm{V}(1) = k^2$ (Chap. III, Appendix). If m, n are integers such that $0 \leq n \leq m$, put

$$e_n^{(m)} = \binom{m}{n} u^{m-n} v^n \in \mathrm{V}(m).$$

PROPOSITION 3. *For any integer $m \geq 0$, $\mathrm{V}(m)$ is an absolutely simple $\mathfrak{sl}(2,k)$-module. In this module, $e_0^{(m)} = u^m$ is primitive of weight m.*

We have $X_+ u^m = 0$ and $Hu^m = mu^m$, so u^m is primitive of weight m. The submodule of $\mathrm{V}(m)$ generated by u^m is of dimension $m+1$ (Prop. 2 (i)) and so is equal to $\mathrm{V}(m)$. By Prop. 2 (v), $\mathrm{V}(m)$ is absolutely simple.

THEOREM 1. *Every simple $\mathfrak{sl}(2,k)$-module of finite dimension n is isomorphic to $\mathrm{V}(n-1)$. Every finite dimensional $\mathfrak{sl}(2,k)$-module is a direct sum of submodules isomorphic to the modules $\mathrm{V}(m)$.*

This follows from Lemma 4 and Prop. 1, 2 and 3.

Remarks. 1) The adjoint representation of $\mathfrak{sl}(2,k)$ defines on $\mathfrak{sl}(2,k)$ the structure of a simple $\mathfrak{sl}(2,k)$-module. This module is isomorphic to $\mathrm{V}(2)$ by an isomorphism that takes u^2 to X_+, $2uv$ to $-H$, and v^2 to X_-.

2) For $n \geq 0$ and $m > n$,

$$X_- e_n^{(m)} = -(m-n)\binom{m}{n} u^{m-n-1} v^{n+1} = -(n+1) e_{n+1}^{(m)}.$$

Hence, $(e_0^{(m)}, e_1^{(m)}, \ldots, e_m^{(m)})$ is the basis of $\mathrm{V}(m)$ associated to the primitive element $e_0^{(m)}$.

3) Let Φ be the bilinear form on $\mathrm{V}(m)$ such that

$$\Phi(e_n^{(m)}, e_{n'}^{(m)}) = 0 \quad \text{if } n + n' \neq m$$
$$\Phi(e_n^{(m)}, e_{m-n}^{(m)}) = (-1)^n \binom{m}{n}.$$

If $x = au + bv$ and $y = cu + dv$, then $\Phi(x^m, y^m) = (ad - bc)^m$. It is now easy to check that Φ is invariant, and that Φ is symmetric for m even, and alternating for m odd.

PROPOSITION 4. *Let E be a finite dimensional $\mathfrak{sl}(2,k)$-module, m an integer ≥ 0, P_m the set of primitive elements of weight m. Let L be the vector space of homomorphisms from the $\mathfrak{sl}(2,k)$-module $\mathrm{V}(m)$ to the $\mathfrak{sl}(2,k)$-module E. The map $f \mapsto f(u^m)$ from L to E is linear, injective, and its image is $\mathrm{P}_m \cup \{0\}$.*

This map is clearly linear, and it is injective because u^m generates the $\mathfrak{sl}(2,k)$-module $\mathrm{V}(m)$. If $f \in \mathrm{L}$,

$$X_+(f(u^m)) = f(X_+u^m) = 0, \qquad H(f(u^m)) = f(Hu^m) = mf(u^m)$$

so $f(u^m) \in \mathrm{P}_m \cup \{0\}$. Let $e \in \mathrm{P}_m$, and V the submodule of E generated by e. By Prop. 1, there exists an isomorphism from the module $\mathrm{V}(m)$ to the module V that takes u^m to e. Then $\mathrm{L}(u^m) = \mathrm{P}_m \cup \{0\}$.

COROLLARY. *The isotypical component of* E *of type* $\mathrm{V}(m)$ *has length*

$$\dim(\mathrm{P}_m \cup \{0\}.$$

4. LINEAR REPRESENTATIONS OF THE GROUP $\mathbf{SL}(2,k)$

Recall (*Algebra*, Chap. III, §8, no. 9) that we denote by $\mathbf{SL}(2,k)$ the group of square matrices of order 2 with coefficients in k whose determinant is equal to 1. If $x \in \mathfrak{sl}(2,k)$ is nilpotent, then $x^2 = 0$ (*Algebra*, Chap. VII, §5, Cor. 3 of Prop. 5) and $e^x = 1 + x \in \mathbf{SL}(2,k)$. If E is a finite dimensional vector space and ρ is a linear representation of $\mathfrak{sl}(2,k)$ on E, then $\rho(x)$ is nilpotent and so $e^{\rho(x)}$ is defined (Chap. I, §6, no. 3).

DEFINITION 2. *Let* E *be a finite dimensional vector space, and* ρ (resp. π) *a linear representation of* $\mathfrak{sl}(2,k)$ (resp. $\mathbf{SL}(2,k)$) *on* E. *Then* ρ *and* π *are said to be compatible if, for every nilpotent element* x *of* $\mathfrak{sl}(2,k)$, $\pi(e^x) = e^{\rho(x)}$.

In other words, ρ *and* π *are compatible if, for every nilpotent element* x *of* $\mathfrak{sl}(2,k)$, *the restriction of* ρ *to* kx *is compatible with the restriction of* π *to the group* $1 + kx$ (Chap. VII, §3, no. 1).

If ρ and π are compatible, so are the dual representations, the mth tensor powers, and the mth symmetric powers of ρ and π, respectively (Chap. VII, §5, no. 4, Lemma 1 (i) and (ii)). Similarly for the representations induced by ρ and π on a vector subspace stable under ρ and π (*loc. cit.*).

In particular, the representation ρ_m of $\mathfrak{sl}(2,k)$ on $\mathrm{V}(m)$ (no. 3) is compatible with the mth symmetric power π_m of the identity representation π_1 of $\mathbf{SL}(2,k)$. Putting $e_n^{(m)} = \binom{m}{n} u^{m-n}v^n$ as above, we have

$$\pi_m(s)e_n^{(m)} = \binom{m}{n}(su)^{m-n}(sv)^n \tag{3}$$

for $s \in \mathbf{SL}(2,k)$ and $0 \leq n \leq m$.

THEOREM 2. *Let* ρ *be a linear representation of* $\mathfrak{sl}(2,k)$ *on a finite dimensional vector space* E.

(i) *There exists a unique linear representation* π *of* $\mathbf{SL}(2, k)$ *on* E *that is compatible with* ρ.

(ii) *A vector subspace* F *of* E *is stable under* π *if and only if it is stable under* ρ.

(iii) *Let* $x \in$ E. *Then* $\pi(s)x = x$ *for all* $s \in \mathbf{SL}(2, k)$ *if and only if* x *is invariant under* ρ *(that is,* $\rho(a)x = 0$ *for all* $a \in \mathfrak{sl}(2, k)$*).*

The existence of π follows from the preceding and Th. 1. On the other hand, we know that the group $\mathbf{SL}(2, k)$ is generated by the elements of the form

$$e^{tX_+} = \begin{pmatrix} 1 & t \\ 0 & 1 \end{pmatrix} \quad e^{-tX_-} = \begin{pmatrix} 1 & 0 \\ t & 1 \end{pmatrix}$$

where $t \in k$ (*Algebra*, Chap. III, §8, no. 9, Prop. 17). This proves the uniqueness of π.

Assertions (ii) and (iii) follow from what we have said, together with Chap. VII, §3, no. 1, Lemma 1 (i). Q.E.D.

Every finite dimensional $\mathfrak{sl}(2, k)$-module therefore has a unique $\mathbf{SL}(2, k)$-module structure, which is said to be *associated* to its $\mathfrak{sl}(2, k)$-module structure.

Remark. When k is $\mathbf{R}$ or $\mathbf{C}$ or a complete non-discrete ultrametric field, $\mathfrak{sl}(2, k)$ is the Lie algebra of $\mathbf{SL}(2, k)$. Let ρ and π be as in Th. 2. The homomorphism π is a homomorphism *of Lie groups* from $\mathbf{SL}(2, k)$ to $\mathbf{GL}(\mathrm{E})$: this is clear when $\mathrm{E} = \mathrm{V}(m)$, and the general case follows, in view of Th. 1. By Chap. VII, §3, no. 1, $\rho(X_+) = \mathrm{L}(\pi)(X_+)$, $\rho(X_-) = \mathrm{L}(\pi)(X_-)$. Hence $\rho = \mathrm{L}(\pi)$ (for a converse, see Exerc. 18).

PROPOSITION 5. *Let* E, F *be finite dimensional* $\mathfrak{sl}(2, k)$*-modules, and let* $f \in \mathrm{Hom}_k(\mathrm{E}, \mathrm{F})$. *The following conditions are equivalent:*

(i) f *is a homomorphism of* $\mathfrak{sl}(2, k)$*-modules;*

(ii) f *is a homomorphism of* $\mathbf{SL}(2, k)$*-modules.*

Condition (i) means that f is an invariant element of the $\mathfrak{sl}(2, k)$-module $\mathrm{Hom}_k(\mathrm{E}, \mathrm{F})$, and condition (ii) means that f is an invariant element of the $\mathbf{SL}(2, k)$-module $\mathrm{Hom}_k(\mathrm{E}, \mathrm{F})$. Since these module structures are associated by Chap. VII, §5, no. 4, Lemma 1 (iii), the proposition follows from Th. 2 (iii).

DEFINITION 3. *The adjoint representation of the group* $\mathbf{SL}(2, k)$ *is the linear representation* Ad *of* $\mathbf{SL}(2, k)$ *on* $\mathfrak{sl}(2, k)$ *defined by*

$$\mathrm{Ad}(s).a = sas^{-1}$$

for all $a \in \mathfrak{sl}(2, k)$ *and all* $s \in \mathbf{SL}(2, k)$.

When k is $\mathbf{R}$ or $\mathbf{C}$ or a complete non-discrete ultrametric field, we recover Def. 7 of Chap. III, §3, no. 12 (cf. *loc. cit.*, Prop. 49).

By Chap. VII, §5, no. 4, Lemma 1 (i) and (ii), the adjoint representations of $\mathfrak{sl}(2,k)$ and $\mathbf{SL}(2,k)$ *are compatible.* By Chap. VII, §3, no. 1, Remark 2, $\mathrm{Ad}(\mathbf{SL}(2,k)) = \mathrm{Aut}_e(\mathfrak{sl}(2,k))$.

5. SOME ELEMENTS OF SL(2, k)

For any $t \in k^*$, put

$$\begin{aligned}\theta(t) &= e^{tX_+}e^{t^{-1}X_-}e^{tX_+}\\ &= \begin{pmatrix}1 & t\\ 0 & 1\end{pmatrix}\begin{pmatrix}1 & 0\\ -t^{-1} & 1\end{pmatrix}\begin{pmatrix}1 & t\\ 0 & 1\end{pmatrix}\\ &= \begin{pmatrix}0 & t\\ -t^{-1} & 0\end{pmatrix}\\ &= e^{t^{-1}X_-}e^{tX_+}e^{t^{-1}X_-}.\end{aligned}$$

With the notations of no. 3,

$$\theta(t)u = -t^{-1}v \qquad \theta(t)v = tu$$

so

$$\theta(t)e_n^{(m)} = (-1)^{m-n}t^{2n-m}e_{m-n}^{(m)}. \tag{4}$$

Hence, the element $\theta(t)^2 = \begin{pmatrix}-1 & 0\\ 0 & -1\end{pmatrix}$ operates by $(-1)^m$ on $\mathrm{V}(m)$. If E is an odd-dimensional simple $\mathfrak{sl}(2,k)$-module, $\theta(t)_{\mathrm{E}}$ is thus an involutive automorphism of the vector space E. In particular, taking E to be the adjoint representation:

$$\theta(t)_{\mathrm{E}}X_+ = t^{-2}X_- \quad \theta(t)_{\mathrm{E}}X_- = t^2X_+ \quad \theta(t)_{\mathrm{E}}H = -H \tag{5}$$

so that $\theta(1)_{\mathrm{E}} = \theta(-1)_{\mathrm{E}}$ is the canonical involution of $\mathfrak{sl}(2,k)$.

For any $t \in k^*$, put

$$h(t) = \begin{pmatrix}t & 0\\ 0 & t^{-1}\end{pmatrix} = \theta(t)\theta(-1).$$

Then $h(t)u = tu$, $h(t)v = t^{-1}v$, so

$$h(t)e_n^{(m)} = t^{m-2n}e_n^{(m)}. \tag{6}$$

PROPOSITION 6. *Let* E *be a finite dimensional* $\mathfrak{sl}(2,k)$*-module, and* $t \in k^*$. *Let* E_p *be the set of elements of* E *of weight* p.

(i) $\theta(t)_{\mathrm{E}}|\mathrm{E}_p$ *is a bijection from* E_p *to* E_{-p}.

(ii) *$h(t)_{\mathrm{E}}|\mathrm{E}_p$ is the homothety with ratio t^p on E_p.*

If $\mathrm{E} = \mathrm{V}(n)$, the proposition follows from formulas (4) and (6). The general case follows from Th. 1.

COROLLARY. *Let $\mathrm{E} = \mathrm{E}' \oplus \mathrm{E}''$ be the decomposition of* E *defined in the Cor. of Prop. 2. The element* $\begin{pmatrix} -1 & 0 \\ 0 & -1 \end{pmatrix}$ *of* $\mathbf{SL}(2, k)$ *operates by* $+1$ *on* E′ *and by* -1 *on* E″.

This follows from (ii), applied to $t = -1$.

§ 2. ROOT SYSTEM OF A SPLIT SEMI-SIMPLE LIE ALGEBRA

1. SPLIT SEMI-SIMPLE LIE ALGEBRAS

DEFINITION 1. *Let $\mathfrak{g}$ be a semi-simple Lie algebra. A Cartan subalgebra $\mathfrak{h}$ of $\mathfrak{g}$ is called splitting if, for all $x \in \mathfrak{h}$, $\mathrm{ad}_{\mathfrak{g}}x$ is triangularizable. A semi-simple Lie algebra is called splittable if it has a splitting Cartan subalgebra. A split semi-simple Lie algebra is a pair $(\mathfrak{g}, \mathfrak{h})$ where $\mathfrak{g}$ is a semi-simple Lie algebra and $\mathfrak{h}$ is a splitting Cartan subalgebra of $\mathfrak{g}$.*

Remarks. 1) Let $\mathfrak{g}$ be a semi-simple Lie algebra, $\mathfrak{h}$ a Cartan subalgebra of $\mathfrak{g}$. For all $x \in \mathfrak{h}$, $\mathrm{ad}_{\mathfrak{g}}x$ is semi-simple (Chap. VII, §2, no. 4, Th. 2). Thus, to say that $\mathfrak{h}$ is splitting means that $\mathrm{ad}_{\mathfrak{g}}x$ is diagonalizable for all $x \in \mathfrak{h}$.

2) If k is algebraically closed, every semi-simple Lie algebra $\mathfrak{g}$ is splittable, and every Cartan subalgebra of $\mathfrak{g}$ is splitting. When k is not algebraically closed, there exist non-splittable semi-simple Lie algebras (Exerc. 2 *a*)); moreover, if $\mathfrak{g}$ is splittable, there may exist Cartan subalgebras of $\mathfrak{g}$ that are not splitting (Exerc. 2 *b*)).

3) Let $\mathfrak{g}$ be a semi-simple Lie algebra, $\mathfrak{h}$ a Cartan subalgebra of $\mathfrak{g}$, and ρ a finite dimensional injective representation of $\mathfrak{g}$ such that $\rho(\mathfrak{h})$ is diagonalizable. Then $\mathrm{ad}_{\mathfrak{g}}x$ is diagonalizable for all $x \in \mathfrak{h}$ (Chap. VII, §2, no. 1, Example 2), so $\mathfrak{h}$ is splitting.

4) We shall see (§3, no. 3, Cor. of Prop. 10) that if $\mathfrak{h}$, $\mathfrak{h}'$ are splitting Cartan subalgebras of $\mathfrak{g}$, there exists an elementary automorphism of $\mathfrak{g}$ transforming $\mathfrak{h}$ into $\mathfrak{h}'$.

5) Let $\mathfrak{g}$ be a reductive Lie algebra. Then $\mathfrak{g} = \mathfrak{c} \times \mathfrak{s}$ where $\mathfrak{c}$ is the centre of $\mathfrak{g}$ and $\mathfrak{s} = \mathscr{D}\mathfrak{g}$ is semi-simple. The Cartan subalgebras of $\mathfrak{g}$ are the subalgebras of the form $\mathfrak{h} = \mathfrak{c} \times \mathfrak{h}'$ where $\mathfrak{h}'$ is a Cartan subalgebra of $\mathfrak{s}$ (Chap. VII, §2, no. 1, Prop. 2). Then $\mathfrak{h}$ is called splitting if $\mathfrak{h}'$ is splitting relative to $\mathfrak{s}$. This leads in an obvious way to the definition of splittable or split reductive algebras.

2. ROOTS OF A SPLIT SEMI-SIMPLE LIE ALGEBRA

In this number, $(\mathfrak{g}, \mathfrak{h})$ *denotes a split semi-simple Lie algebra.*

For any $\lambda \in \mathfrak{h}^*$, denote by $\mathfrak{g}^\lambda(\mathfrak{h})$, or simply by $\mathfrak{g}^\lambda$, the primary subspace of $\mathfrak{g}$ relative to λ (cf. Chap. VII, §1, no. 3). Recall that $\mathfrak{g}^0 = \mathfrak{h}$ (Chap. VII, §2, no. 1, Prop. 4), that $\mathfrak{g}$ is the direct sum of the $\mathfrak{g}^\lambda$ (Chap. VII, §1, no. 3, Prop. 8 and 9), that $\mathfrak{g}^\lambda$ is the set of $x \in \mathfrak{g}$ such that $[h, x] = \lambda(h)x$ for all $h \in \mathfrak{h}$ (Chap. VII, §2, no. 4, Cor. 1 of Th. 2), and that the weights of $\mathfrak{h}$ on $\mathfrak{g}$ are the linear forms λ on $\mathfrak{h}$ such that $\mathfrak{g}^\lambda \neq 0$ (Chap. VII, §1, no. 1).

DEFINITION 2. *A root of* $(\mathfrak{g}, \mathfrak{h})$ *is a non-zero weight of* $\mathfrak{h}$ *on* $\mathfrak{g}$.

Denote by $\mathrm{R}(\mathfrak{g}, \mathfrak{h})$, or simply by R, the set of roots of $(\mathfrak{g}, \mathfrak{h})$. We have

$$\mathfrak{g} = \mathfrak{h} \oplus \bigoplus_{\alpha \in \mathrm{R}} \mathfrak{g}^\alpha.$$

PROPOSITION 1. *Let* α, β *be roots of* $(\mathfrak{g}, \mathfrak{h})$ *and let* $\langle \cdot, \cdot \rangle$ *be a non-degenerate invariant symmetric bilinear form on* $\mathfrak{g}$ *(for example the Killing form of* $\mathfrak{g}$*).*

(i) *If* $\alpha + \beta \neq 0$, $\mathfrak{g}^\alpha$ *and* $\mathfrak{g}^\beta$ *are orthogonal. The restriction of* $\langle \cdot, \cdot \rangle$ *to* $\mathfrak{g}^\alpha \times \mathfrak{g}^{-\alpha}$ *is non-degenerate. The restriction of* $\langle \cdot, \cdot \rangle$ *to* $\mathfrak{h}$ *is non-degenerate.*

(ii) *Let* $x \in \mathfrak{g}^\alpha$, $y \in \mathfrak{g}^{-\alpha}$ *and* $h \in \mathfrak{h}$. *Then* $[x, y] \in \mathfrak{h}$ *and*

$$\langle h, [x, y] \rangle = \alpha(h)\langle x, y \rangle.$$

Assertion (i) is a particular case of Prop. 10 (iii) of Chap. VII, §1, no. 3. If $x \in \mathfrak{g}^\alpha$, $y \in \mathfrak{g}^{-\alpha}$ and $h \in \mathfrak{h}$, we have $[x, y] \in \mathfrak{g}^{\alpha-\alpha} = \mathfrak{h}$, and

$$\langle h, [x, y] \rangle = \langle [h, x], y \rangle = \langle \alpha(h)x, y \rangle = \alpha(h)\langle x, y \rangle.$$

THEOREM 1. *Let* α *be a root of* $(\mathfrak{g}, \mathfrak{h})$.

(i) *The vector space* $\mathfrak{g}^\alpha$ *is of dimension* 1.

(ii) *The vector subspace* $\mathfrak{h}_\alpha = [\mathfrak{g}^\alpha, \mathfrak{g}^{-\alpha}]$ *of* $\mathfrak{h}$ *is of dimension* 1. *It contains a unique element* H_α *such that* $\alpha(H_\alpha) = 2$.

(iii) *The vector subspace* $\mathfrak{s}_\alpha = \mathfrak{h}_\alpha + \mathfrak{g}^\alpha + \mathfrak{g}^{-\alpha}$ *is a Lie subalgebra of* $\mathfrak{g}$.

(iv) *If* X_α *is a non-zero element of* $\mathfrak{g}^\alpha$, *there exists a unique* $X_{-\alpha} \in \mathfrak{g}^{-\alpha}$ *such that* $[X_\alpha, X_{-\alpha}] = -H_\alpha$. *Let* φ *be the linear map from* $\mathfrak{sl}(2, k)$ *to* $\mathfrak{g}$ *that takes* X_+ *to* X_α, X_- *to* $X_{-\alpha}$, *and* H *to* H_α; *then* φ *is an isomorphism from the Lie algebra* $\mathfrak{sl}(2, k)$ *to the Lie algebra* $\mathfrak{s}_\alpha$.

a) Let h_α be the unique element of $\mathfrak{h}$ such that $\alpha(h) = \langle h_\alpha, h \rangle$ for all $h \in \mathfrak{h}$. By Prop. 1, $[x, y] = \langle x, y \rangle h_\alpha$ for all $x \in \mathfrak{g}^\alpha$, $y \in \mathfrak{g}^{-\alpha}$; on the other hand $\langle \mathfrak{g}^\alpha, \mathfrak{g}^{-\alpha} \rangle \neq 0$. Hence $\mathfrak{h}_\alpha = [\mathfrak{g}^\alpha, \mathfrak{g}^{-\alpha}] = kh_\alpha$.

b) Choose $x \in \mathfrak{g}^\alpha$, $y \in \mathfrak{g}^{-\alpha}$ such that $\langle x, y \rangle = 1$, so $[x, y] = h_\alpha$. Recall that $[h_\alpha, x] = \alpha(h_\alpha)x$, $[h_\alpha, y] = -\alpha(h_\alpha)y$. If $\alpha(h_\alpha) = 0$, it follows that

$kx + ky + kh_\alpha$ is a nilpotent subalgebra $\mathfrak{t}$ of $\mathfrak{g}$; since $h_\alpha \in [\mathfrak{t}, \mathfrak{t}]$, $\mathrm{ad}_\mathfrak{g} h_\alpha$ is nilpotent (Chap. I, §5, no. 3, Th. 1), which is absurd since $\mathrm{ad}_\mathfrak{g} h_\alpha$ is non-zero semi-simple. So $\alpha(h_\alpha) \neq 0$. Hence there exists a unique $H_\alpha \in \mathfrak{h}_\alpha$ such that $\alpha(H_\alpha) = 2$, which proves (ii).

c) Choose a non-zero element X_α of $\mathfrak{g}^\alpha$. There exists $X_{-\alpha} \in \mathfrak{g}^{-\alpha}$ such that $[X_\alpha, X_{-\alpha}] = -H_\alpha$ (since $[X_\alpha, \mathfrak{g}^{-\alpha}] = \mathfrak{h}_\alpha$ by *b*)). Then

$$[H_\alpha, X_\alpha] = \alpha(H_\alpha)X_\alpha = 2X_\alpha, \quad [H_\alpha, X_{-\alpha}] = -\alpha(H_\alpha)X_{-\alpha} = -2X_{-\alpha},$$
$$[X_\alpha, X_{-\alpha}] = -H_\alpha;$$

hence $kX_\alpha + kX_{-\alpha} + kH_\alpha$ is a subalgebra of $\mathfrak{g}$ and the linear map φ from $\mathfrak{sl}(2,k)$ to $kX_\alpha + kX_{-\alpha} + kH_\alpha$ such that $\varphi(X_+) = X_\alpha$, $\varphi(X_-) = X_{-\alpha}$, $\varphi(H) = H_\alpha$ is an isomorphism of Lie algebras.

d) Assume that $\dim \mathfrak{g}^\alpha > 1$. Let y be a non-zero element of $\mathfrak{g}^{-\alpha}$. There exists a non-zero element X_α of $\mathfrak{g}_\alpha$ such that $\langle y, X_\alpha\rangle = 0$. Choose $X_{-\alpha}$ as in *c*), and consider the representation $\rho : u \mapsto \mathrm{ad}_\mathfrak{g}\varphi(u)$ from $\mathfrak{sl}(2,k)$ to $\mathfrak{g}$. We have

$$\rho(H)y = [\varphi(H), y] = [H_\alpha, y] = -2y$$
$$\rho(X_+)y = [\varphi(X_+), y] = [X_\alpha, y] = \langle X_\alpha, y\rangle h_\alpha = 0.$$

Thus, y is primitive for ρ, of weight -2, which contradicts Prop. 2 of §1, no. 2. This proves (i).

e) Assertion (iii) is now a consequence of *c*). On the other hand, if X_α is a non-zero element of $\mathfrak{g}^\alpha$, the element $X_{-\alpha}$ constructed in *c*) is the unique element of $\mathfrak{g}^{-\alpha}$ such that $[X_\alpha, X_{-\alpha}] = -H_\alpha$ since $\dim \mathfrak{g}^{-\alpha} = 1$. The last assertion of (iv) is a consequence of *c*). Q.E.D.

The notations h_α, H_α, $\mathfrak{s}_\alpha$ will be retained in what follows. (To define h_α, we take $\langle\cdot,\cdot\rangle$ equal to the Killing form.) If X_α is a non-zero element of $\mathfrak{g}^\alpha$, the isomorphism φ of Th. 1 and the representation $u \mapsto \mathrm{ad}_\mathfrak{g}\varphi(u)$ of $\mathfrak{sl}(2,k)$ on $\mathfrak{g}$ will be said to be *associated* to X_α.

COROLLARY. *Let Φ be the Killing form of $\mathfrak{g}$. For all $a, b \in \mathfrak{h}$,*

$$\Phi(a,b) = \sum_{\gamma \in \mathrm{R}} \gamma(a)\gamma(b).$$

Indeed, $\mathrm{ad}\, a.\mathrm{ad}\, b$ leaves each $\mathfrak{g}^\gamma$ stable, and its restriction to $\mathfrak{g}^\gamma$ is the homothety with ratio $\gamma(a)\gamma(b)$; if $\gamma \neq 0$, $\dim \mathfrak{g}^\gamma = 1$.

PROPOSITION 2. *Let $\alpha, \beta \in \mathrm{R}$.*

(i) $\beta(H_\alpha) \in \mathbf{Z}$.

(ii) *If Φ denotes the Killing form of $\mathfrak{g}$, $\Phi(H_\alpha, H_\beta) \in \mathbf{Z}$.*

Let X_α be a non-zero element of $\mathfrak{g}^\alpha$, and let ρ be the representation of $\mathfrak{sl}(2,k)$ on $\mathfrak{g}$ associated to X_α. The eigenvalues of $\rho(\mathrm{H})$ are 0 and the $\beta(H_\alpha)$

for $\beta \in \mathrm{R}$. Hence (i) follows from §1, no. 2, Cor. of Prop. 2. Assertion (ii) follows from (i) and the Cor. of Th. 1. Q.E.D.

Let $\alpha \in \mathrm{R}$, X_α a non-zero element of $\mathfrak{g}^\alpha$, $X_{-\alpha}$ the element of $\mathfrak{g}^{-\alpha}$ such that $[X_\alpha, X_{-\alpha}] = -H_\alpha$, and ρ the representation of $\mathfrak{sl}(2,k)$ on $\mathfrak{g}$ associated to X_α. Let π be the representation of $\mathbf{SL}(2,k)$ on $\mathfrak{g}$ compatible with ρ (§1, no. 4, Th. 2). Since $\operatorname{ad} X_\alpha$ is nilpotent (Chap. VII, §1, no. 3, Prop. 10 (iv)), $\pi(e^{X_+}) = e^{\operatorname{ad} X_\alpha}$ is an elementary automorphism of $\mathfrak{g}$. Similarly, $\pi(e^{X_-}) = e^{\operatorname{ad} X_{-\alpha}}$ is an elementary automorphism of $\mathfrak{g}$. Hence $\pi(\mathbf{SL}(2,k)) \subset \operatorname{Aut}_e(\mathfrak{g})$. We make use of the notation $\theta(t)$ of §1, no. 5. For $t \in k^*$, put

$$\theta_\alpha(t) = \pi(\theta(t)) = e^{\operatorname{ad} tX_\alpha} e^{\operatorname{ad} t^{-1}X_{-\alpha}} e^{\operatorname{ad} tX_\alpha}. \tag{1}$$

Lemma 1. (i) *For all $h \in \mathfrak{h}$, $\theta_\alpha(t).h = h - \alpha(h)H_\alpha$.*

(ii) *For all $\beta \in \mathrm{R}$, $\theta_\alpha(t)(\mathfrak{g}^\beta) = \mathfrak{g}^{\beta - \beta(H_\alpha)\alpha}$.*

(iii) *If $\alpha, \beta \in \mathrm{R}$, $\beta - \beta(H_\alpha)\alpha \in \mathrm{R}$.*

Let $h \in \mathfrak{h}$. If $\alpha(h) = 0$, $[X_\alpha, h] = [X_{-\alpha}, h] = 0$, so $\theta_\alpha(t).h = h$. On the other hand, the formulas (5) of §1, no. 5 show that $\theta_\alpha(t).H_\alpha = -H_\alpha$. This proves assertion (i). It follows that $\theta_\alpha(t)^2|\mathfrak{h} = \mathrm{Id}$. If $x \in \mathfrak{g}^\beta$ and $h \in \mathfrak{h}$,

$$\begin{aligned}
[h, \theta_\alpha(t)x] &= \theta_\alpha(t).[\theta_\alpha(t)h, x] - \beta(\theta_\alpha(t)h).\theta_\alpha(t)x \\
&= (\beta(h) - \alpha(h)\beta(H_\alpha)).\theta_\alpha(t)x \\
&= (\beta - \beta(H_\alpha)\alpha)(h).\theta_\alpha(t)x
\end{aligned}$$

so $\theta_\alpha(t)x \in \mathfrak{g}^{\beta - \beta(H_\alpha)\alpha}$. This proves (ii). Assertion (iii) follows from (ii).

THEOREM 2. (i) *The set $\mathrm{R} = \mathrm{R}(\mathfrak{g}, \mathfrak{h})$ is a reduced root system in $\mathfrak{h}^*$.*

(ii) *Let $\alpha \in \mathrm{R}$. The map $s_{\alpha, H_\alpha} : \lambda \mapsto \lambda - \lambda(H_\alpha)\alpha$ from $\mathfrak{h}^*$ to $\mathfrak{h}^*$ is the unique reflection s of $\mathfrak{h}^*$ such that $s(\alpha) = -\alpha$ and $s(\mathrm{R}) = \mathrm{R}$. For all $t \in k^*$, s is the transpose of $\theta_\alpha(t)|\mathfrak{h}$.*

First, R generates $\mathfrak{h}^*$, for if $h \in \mathfrak{h}$ is such that $\alpha(h) = 0$ for all $\alpha \in \mathrm{R}$, then $\operatorname{ad} h = 0$ and hence $h = 0$ since the centre of $\mathfrak{g}$ is zero. By definition, $0 \notin \mathrm{R}$. Let $\alpha \in \mathrm{R}$. Since $\alpha(H_\alpha) = 2$, $s = s_{\alpha, H_\alpha}$ is a reflection such that $s(\alpha) = -\alpha$. Then $s(\mathrm{R}) = \mathrm{R}$ by Lemma 1 (iii), and $\beta(H_\alpha) \in \mathbf{Z}$ for all $\beta \in \mathrm{R}$ (Prop. 2 (i)). This shows that R is a root system in $\mathfrak{h}^*$. For all $h \in \mathfrak{h}$ and $\lambda \in \mathfrak{h}^*$,

$$\langle s(\lambda), h\rangle = \langle \lambda - \lambda(H_\alpha)\alpha, h\rangle = \langle \lambda, h - \alpha(h)H_\alpha\rangle = \langle \lambda, \theta_\alpha(t)h\rangle$$

so s is the transpose of $\theta_\alpha(t)|\mathfrak{h}$. Finally, we show that the root system R is reduced. Let $\alpha \in \mathrm{R}$ and $y \in \mathfrak{g}^{2\alpha}$. Since $3\alpha \notin \mathrm{R}$ (Chap. VI, §1, no. 3, Prop. 8), $[X_\alpha, y] = 0$; on the other hand, $[X_{-\alpha}, y] \in \mathfrak{g}^{-\alpha+2\alpha} = \mathfrak{g}^\alpha = kX_\alpha$, so $[X_\alpha, [X_{-\alpha}, y]] = 0$; thus

$$4y = 2\alpha(H_\alpha)y = [H_\alpha, y] = -[[X_\alpha, X_{-\alpha}], y] = 0$$

so $y = 0$ and $\mathfrak{g}^{2\alpha} = 0$. In other words, 2α is not a root. Q.E.D.

Identify $\mathfrak{h}$ canonically with $\mathfrak{h}^{**}$. With the notations of Chap. VI, §1, no. 1, we then have, by Th. 2 (ii),

$$H_\alpha = \alpha^\vee \quad \text{for all } \alpha \in \mathrm{R}. \tag{2}$$

The H_α thus form the root system $\mathrm{R}^\vee$ in $\mathfrak{h}$ inverse to R.

We shall call $\mathrm{R}(\mathfrak{g}, \mathfrak{h})$ the *root system of* $(\mathfrak{g}, \mathfrak{h})$. The reflections s_{α, H_α} will be denoted simply by s_α. The Weyl group, group of weights, Coxeter number ... of $\mathrm{R}(\mathfrak{g}, \mathfrak{h})$ are called *the Weyl group, group of weights, Coxeter number ... of* $(\mathfrak{g}, \mathfrak{h})$. As in Chap. VI, §1, no. 1, we consider the Weyl group as operating not only on $\mathfrak{h}^*$, but also on $\mathfrak{h}$ by transport of structure, so that $s_\alpha = \theta_\alpha(t)|\mathfrak{h}$. Since the $\theta_\alpha(t)$ are elementary automorphisms of $\mathfrak{g}$, we have:

COROLLARY. *Every element of the Weyl group of* $(\mathfrak{g}, \mathfrak{h})$, *operating on* $\mathfrak{h}$, *is the restriction to* $\mathfrak{h}$ *of an elementary automorphism of* $\mathfrak{g}$.

For a converse of this result, see §5, no. 2, Prop. 4.

Remark 1. If $\mathfrak{h}_{\mathbf{Q}}$ (resp. $\mathfrak{h}^*_{\mathbf{Q}}$) denotes the $\mathbf{Q}$-vector subspace of $\mathfrak{h}$ (resp. $\mathfrak{h}^*$) generated by the H_α (resp. the α), where $\alpha \in \mathrm{R}$, then $\mathfrak{h}$ (resp. $\mathfrak{h}^*$) can be identified canonically with $\mathfrak{h}_{\mathbf{Q}} \otimes_{\mathbf{Q}} k$ (resp. with $\mathfrak{h}^*_{\mathbf{Q}} \otimes_{\mathbf{Q}} k$) and $\mathfrak{h}^*_{\mathbf{Q}}$ can be identified with the dual of $\mathfrak{h}_{\mathbf{Q}}$ (Chap. VI, §1, no. 1, Prop. 1). We call $\mathfrak{h}_{\mathbf{Q}}$ and $\mathfrak{h}^*_{\mathbf{Q}}$ the canonical $\mathbf{Q}$-structures on $\mathfrak{h}$ and $\mathfrak{h}^*$ (*Algebra*, Chap. II, §8, no. 1, Def. 1). When we mention $\mathbf{Q}$-rationality for a vector subspace of $\mathfrak{h}$, for a linear form on $\mathfrak{h}$, etc., we shall mean these structures, unless we indicate otherwise. When we mention Weyl chambers, or facets, of $\mathrm{R}(\mathfrak{g}, \mathfrak{h})$, we shall work in $\mathfrak{h}_{\mathbf{Q}} \otimes_{\mathbf{Q}} \mathbf{R}$ or $\mathfrak{h}^*_{\mathbf{Q}} \otimes_{\mathbf{Q}} \mathbf{R}$, that we shall denote by $\mathfrak{h}_{\mathbf{R}}$ and $\mathfrak{h}^*_{\mathbf{R}}$.

Remark 2. The root system $\mathrm{R}^\vee$ in $\mathfrak{h}$ defines a non-degenerate symmetric bilinear form β on $\mathfrak{h}$ (Chap. VI, §1, no. 1, Prop. 3), namely the form $(a, b) \mapsto \sum_{\alpha \in \mathrm{R}} \langle \alpha, a \rangle \langle \alpha, b \rangle$. By the Cor. to Th. 1, this form is just the restriction of the Killing form to $\mathfrak{h}$. The extension of $\beta | \mathfrak{h}_{\mathbf{Q}} \times \mathfrak{h}_{\mathbf{Q}}$ to $\mathfrak{h}_{\mathbf{Q}} \otimes_{\mathbf{Q}} \mathbf{R}$ is positive non-degenerate (Chap. VI, §1, no. 1, Prop. 3). On the other hand, we see that the inverse form on $\mathfrak{h}^*$ of the restriction to $\mathfrak{h}$ of the Killing form on $\mathfrak{g}$ is *the canonical bilinear form* Φ_{R} *of* R (Chap. VI, §1, no. 12).

Let $(\mathfrak{g}_1, \mathfrak{h}_1)$, $(\mathfrak{g}_2, \mathfrak{h}_2)$ be split semi-simple Lie algebras, φ an isomorphism from $\mathfrak{g}_1$ to $\mathfrak{g}_2$ such that $\varphi(\mathfrak{h}_1) = \mathfrak{h}_2$. By transport of structure, the transpose of the map $\varphi|\mathfrak{h}_1$ takes $\mathrm{R}(\mathfrak{g}_2, \mathfrak{h}_2)$ to $\mathrm{R}(\mathfrak{g}_1, \mathfrak{h}_1)$.

PROPOSITION 3. *Let* $\mathfrak{g}$ *be a semi-simple Lie algebra,* $\mathfrak{h}_1$ *and* $\mathfrak{h}_2$ *splitting Cartan subalgebras of* $\mathfrak{g}$. *There exists an isomorphism from* $\mathfrak{h}_1^*$ *to* $\mathfrak{h}_2^*$ *that takes* $\mathrm{R}(\mathfrak{g}, \mathfrak{h}_1)$ *to* $\mathrm{R}(\mathfrak{g}, \mathfrak{h}_2)$.

(For more precise results, see §3, no. 3, Cor. of Prop. 10, and §5, no. 3, Prop. 5).

Let k' be an algebraic closure of k, $\mathfrak{g}' = \mathfrak{g}\otimes_k k'$, $\mathfrak{h}'_i = \mathfrak{h}_i \otimes_k k'$. Then $\mathrm{R}(\mathfrak{g}', \mathfrak{h}'_i)$ is the image of $\mathrm{R}(\mathfrak{g}, \mathfrak{h}_i)$ under the map $\lambda \mapsto \lambda \otimes 1$ from $\mathfrak{h}_i^*$ to $\mathfrak{h}_i^* \otimes_k k' = \mathfrak{h}_i'^*$. By Chap. VII, §3, no. 2, Th. 1, there exists an automorphism of $\mathfrak{g}'$ taking $\mathfrak{h}'_1$ to $\mathfrak{h}'_2$, hence an isomorphism φ from $\mathfrak{h}_1'^*$ to $\mathfrak{h}_2'^*$ that takes $\mathrm{R}(\mathfrak{g}', \mathfrak{h}'_1)$ to $\mathrm{R}(\mathfrak{g}', \mathfrak{h}'_2)$. Then $\varphi|\mathfrak{h}_1^*$ takes $\mathrm{R}(\mathfrak{g}, \mathfrak{h}_1)$ to $\mathrm{R}(\mathfrak{g}, \mathfrak{h}_2)$, and hence $\mathfrak{h}_1^*$ to $\mathfrak{h}_2^*$. Q.E.D.

In view of Prop. 3, the root system of $(\mathfrak{g}, \mathfrak{h})$ depends, up to isomorphism, only on $\mathfrak{g}$ and not on $\mathfrak{h}$. In the same way, the Weyl group, group of weights ... of $(\mathfrak{g}, \mathfrak{h})$ are simply called, by abuse of language, the Weyl group, group of weights ... of $\mathfrak{g}$ (cf. also §5, no. 3, Remark 2). If the Dynkin graph of $\mathfrak{g}$ is of type A_l, or B_l, ... (cf. Chap. VI, §4, no. 2, Th. 3), we say that $\mathfrak{g}$ is of type A_l, or B_l, ...

Recall that, if α and β are linearly independent roots, the set of $j \in \mathbf{Z}$ such that $\beta + j\alpha \in \mathrm{R}$ is an interval $[-q, p]$ of $\mathbf{Z}$ containing 0, with $p - q = -\langle \beta, \alpha^\vee \rangle = -\beta(H_\alpha)$ (Chap. VI, §1, no. 3, Prop. 9).

PROPOSITION 4. *Let α and β be linearly independent roots. Let p (resp. q) be the largest integer j such that $\beta + j\alpha$ (resp. $\beta - j\alpha$) is a root.*

(i) *The vector subspace $\sum_{-q \leq j \leq p} \mathfrak{g}^{\beta + j\alpha}$ of $\mathfrak{g}$ is a simple $\mathfrak{s}_\alpha$-module of dimension $p + q + 1$.*

(ii) *If $\alpha + \beta$ is a root, then $[\mathfrak{g}^\alpha, \mathfrak{g}^\beta] = \mathfrak{g}^{\alpha+\beta}$.*

Let X_α (resp. x) be a non-zero element of $\mathfrak{g}^\alpha$ (resp. $\mathfrak{g}^{\beta + p\alpha}$). Then

$$[X_\alpha, x] \in \mathfrak{g}^{\beta + (p+1)\alpha} = 0$$
$$[H_\alpha, x] = (\beta(H_\alpha) + p\alpha(H_\alpha))x = (-p + q + 2p)x = (p + q)x.$$

Thus, x is primitive of weight $p + q$ for the representation of $\mathfrak{sl}(2, k)$ on $\mathfrak{g}$ associated to X_α; but the $\mathfrak{sl}(2, k)$-module $\sum_{-q \leq j \leq p} \mathfrak{g}^{\beta + j\alpha}$ is of dimension $p + q + 1$; hence it is simple (§1, no. 2, Prop. 2). If $\alpha + \beta \in \mathrm{R}$, then $p \geq 1$, so the elements of $\mathfrak{g}^\beta$ are not primitive, and hence $[X_\alpha, \mathfrak{g}^\beta] \neq 0$. Since $[\mathfrak{g}^\alpha, \mathfrak{g}^\beta] \subset \mathfrak{g}^{\alpha+\beta}$, we see finally that $[\mathfrak{g}^\alpha, \mathfrak{g}^\beta] = \mathfrak{g}^{\alpha+\beta}$.

Remark 3. Recall that, by Chap. VI, §1, no. 3, Cor. of Prop. 9, the integer $p + q + 1$ can only take the values $1, 2, 3, 4$.

Remark 4. Let $(\mathfrak{g}, \mathfrak{h})$ be a split reductive Lie algebra, $\mathfrak{c}$ the centre of $\mathfrak{g}$, $\mathfrak{g}' = \mathscr{D}\mathfrak{g}$, $\mathfrak{h}' = \mathfrak{h} \cap \mathfrak{g}'$. Then $\mathfrak{h} = \mathfrak{c} \times \mathfrak{h}'$, and we identify $\mathfrak{h}'^*$ with a vector subspace of $\mathfrak{h}^*$. For any $\lambda \in \mathfrak{h}^*$ such that $\lambda \neq 0$, the primary subspace $\mathfrak{g}^\lambda$ relative to λ is equal to $\mathfrak{g}'^{\lambda|\mathfrak{h}'}$. A non-zero weight of $\mathfrak{h}$ on $\mathfrak{g}$ is called a *root* of $(\mathfrak{g}, \mathfrak{h})$; every root vanishes on $\mathfrak{c}$. Denote by $\mathrm{R}(\mathfrak{g}, \mathfrak{h})$ the set of roots of $(\mathfrak{g}, \mathfrak{h})$; it can be identified canonically with $\mathrm{R}(\mathfrak{g}', \mathfrak{h}')$. Let $\alpha \in \mathrm{R}(\mathfrak{g}, \mathfrak{h})$. We define h_α, H_α, $\mathfrak{s}_\alpha$, the isomorphisms $\mathfrak{sl}(2, k) \to \mathfrak{s}_\alpha$, and the representations of $\mathfrak{sl}(2, k)$ on $\mathfrak{g}$ associated to α, as in the semi-simple case.

3. INVARIANT BILINEAR FORMS

PROPOSITION 5. *Let* $(\mathfrak{g}, \mathfrak{h})$ *be a split semi-simple Lie algebra,* Φ *an invariant symmetric bilinear form on* $\mathfrak{g}$, *and* W *the Weyl group of* $(\mathfrak{g}, \mathfrak{h})$. *Then the restriction* Φ' *of* Φ *to* $\mathfrak{h}$ *is invariant under* W. *Moreover, if* Φ *is non-degenerate, so is* Φ'.

Let $\alpha \in \mathrm{R}$, let X_α be a non-zero element of $\mathfrak{g}^\alpha$, ρ the associated representation of $\mathfrak{sl}(2, k)$ on $\mathfrak{g}$, and π the representation of $\mathbf{SL}(2, k)$ on $\mathfrak{g}$ compatible with ρ. Then Φ is invariant under ρ, and hence under π (§1, no. 4). In particular, Φ' is invariant under $\theta_\alpha(t)|\mathfrak{h}$ (no. 2), and hence under W. The last assertion follows from Prop. 1 (i).

PROPOSITION 6. *Let* $(\mathfrak{g}, \mathfrak{h})$ *be a split semi-simple Lie algebra,* Φ *a non-degenerate invariant symmetric bilinear form on* $\mathfrak{g}$. *For all* $\alpha \in \mathrm{R}$, *let* X_α *be a non-zero element of* $\mathfrak{g}^\alpha$. *Let* $(H_i)_{i \in I}$ *be a basis of* $\mathfrak{h}$, *and* $(H'_i)_{i \in I}$ *the basis of* $\mathfrak{h}$ *such that* $\Phi(H_i, H'_j) = \delta_{ij}$. *The Casimir element associated to* Φ *in the enveloping algebra of* $\mathfrak{g}$ (Chap. I, §3, no. 7) *is then*

$$\sum_{\alpha \in \mathrm{R}} \frac{1}{\Phi(X_\alpha, X_{-\alpha})} X_\alpha X_{-\alpha} + \sum_{i \in I} H_i H'_i.$$

Indeed, by Prop. 1, $\Phi(H_i, X_\alpha) = \Phi(H'_i, X_\alpha) = 0$ for all $i \in I$, $\alpha \in \mathrm{R}$, and $\Phi\left(\frac{1}{\Phi(X_\alpha, X_{-\alpha})} X_\alpha \, , \, X_{-\beta}\right) = \delta_{\alpha\beta}$ for all $\alpha, \beta \in \mathrm{R}$.

4. THE COEFFICIENTS $\mathrm{N}_{\alpha\beta}$

In this number, we again denote by $(\mathfrak{g}, \mathfrak{h})$ *a split semi-simple Lie algebra.*

Lemma 2. There exists a family $(X_\alpha)_{\alpha \in \mathrm{R}}$ *such that, for all* $\alpha \in \mathrm{R}$,

$$X_\alpha \in \mathfrak{g}^\alpha \quad \text{and} \quad [X_\alpha, X_{-\alpha}] = -H_\alpha.$$

Let R_1 be a subset of R such that $\mathrm{R} = \mathrm{R}_1 \cup (-\mathrm{R}_1)$ and $\mathrm{R}_1 \cap (-\mathrm{R}_1) = \varnothing$. For $\alpha \in \mathrm{R}_1$, choose an arbitrary non-zero element X_α of $\mathfrak{g}^\alpha$. There exists a unique $X_{-\alpha} \in \mathfrak{g}^{-\alpha}$ such that $[X_\alpha, X_{-\alpha}] = -H_\alpha$ (Th. 1 (iv)). Then

$$[X_{-\alpha}, X_\alpha] = H_\alpha = -H_{-\alpha}. \qquad \text{Q.E.D.}$$

If $(X_\alpha)_{\alpha \in \mathrm{R}}$ is one family satisfying the conditions of Lemma 2, the most general family satisfying these conditions is $(t_\alpha X_\alpha)_{\alpha \in \mathrm{R}}$ where $t_\alpha \in k^*$ and $t_\alpha t_{-\alpha} = 1$ for all $\alpha \in \mathrm{R}$.

In the remainder of this number, we denote by $(X_\alpha)_{\alpha \in \mathrm{R}}$ *a family satisfying the conditions of Lemma 2. We denote by* $\langle \cdot, \cdot \rangle$ *a non-degenerate invariant symmetric bilinear form on* $\mathfrak{g}$.

Every $x \in \mathfrak{g}$ can be written uniquely in the form

$$x = h + \sum_{\alpha \in R} \mu_\alpha X_\alpha \quad (h \in \mathfrak{h},\ \mu_\alpha \in k).$$

The bracket of two such elements can be calculated by means of the following formulas:

$$[h, X_\alpha] = \alpha(h) X_\alpha$$

$$[X_\alpha, X_\beta] = \begin{cases} 0 & \text{if } \alpha + \beta \notin R \cup \{0\} \\ -H_\alpha & \text{if } \alpha + \beta = 0 \\ N_{\alpha\beta} X_{\alpha+\beta} & \text{if } \alpha + \beta \in R \end{cases}$$

the $N_{\alpha\beta}$ being non-zero elements of k.

Lemma 3. For all $\alpha \in R$,

$$\langle X_\alpha, X_{-\alpha} \rangle = -\frac{1}{2} \langle H_\alpha, H_\alpha \rangle.$$

Indeed,

$$\begin{aligned} 2\langle X_\alpha, X_{-\alpha} \rangle &= \langle \alpha(H_\alpha) X_\alpha, X_{-\alpha} \rangle = \langle [H_\alpha, X_\alpha], X_{-\alpha} \rangle \\ &= \langle H_\alpha, [X_\alpha, X_{-\alpha}] \rangle = -\langle H_\alpha, H_\alpha \rangle. \end{aligned}$$

Lemma 4. Let $\alpha, \beta \in R$ be such that $\alpha + \beta \in R$. Let p (resp. q) be the largest integer j such that $\beta + j\alpha \in R$ (resp. $\beta - j\alpha \in R$). Then,

$$N_{\alpha,\beta} N_{-\alpha,\alpha+\beta} = -p(q+1) \tag{3}$$

$$N_{-\alpha,\alpha+\beta} \langle H_\beta, H_\beta \rangle = -N_{-\alpha,-\beta} \langle H_{\alpha+\beta}, H_{\alpha+\beta} \rangle \tag{4}$$

$$N_{\alpha,\beta} N_{-\alpha,-\beta} = (q+1)^2. \tag{5}$$

Let ρ be the representation of $\mathfrak{sl}(2, k)$ on $\mathfrak{g}$ defined by X_α. The element $e = X_{\beta + p\alpha}$ is primitive of weight $p + q$ (Prop. 4 (i)). Put

$$e_n = \frac{(-1)^n}{n!} \rho(X_-)^n e \quad \text{for } n \geq 0.$$

By Prop. 1 of §1,

$$\begin{aligned} (\operatorname{ad} X_\alpha) e_p &= (q+1) e_{p-1} \\ (\operatorname{ad} X_{-\alpha})(\operatorname{ad} X_\alpha) e_p &= -p(q+1) e_p. \end{aligned}$$

This proves (3) since e_p is a non-zero element of $\mathfrak{g}^\beta$.

The form $\langle \cdot, \cdot \rangle$ being invariant, we have

$$\langle [X_{-\alpha}, X_{\alpha+\beta}], X_{-\beta} \rangle = -\langle X_{\alpha+\beta}, [X_{-\alpha}, X_{-\beta}] \rangle$$

so

$$\mathrm{N}_{-\alpha,\alpha+\beta}\langle X_\beta, X_{-\beta}\rangle = -\mathrm{N}_{-\alpha,-\beta}\langle X_{\alpha+\beta}, X_{-\alpha-\beta}\rangle$$

which, in view of Lemma 3, proves (4).

The restriction of $\langle\cdot,\cdot\rangle$ to $\mathfrak{h}$ is non-degenerate and invariant under the Weyl group (Prop. 5). Identify $\mathfrak{h}$ and $\mathfrak{h}^*$ by means of this restriction. If $\gamma \in \mathrm{R}$, H_γ is identified with $2\gamma/\langle\gamma,\gamma\rangle$ (Chap. VI, §1, no. 1, Lemma 2); hence, for all $\gamma, \delta \in \mathrm{R}$,

$$\frac{\langle\gamma,\gamma\rangle}{\langle\delta,\delta\rangle} = \frac{\langle H_\delta, H_\delta\rangle}{\langle H_\gamma, H_\gamma\rangle}. \tag{6}$$

Now, by Chap. VI, §1, no. 3, Prop. 10,

$$\frac{\langle\alpha+\beta,\alpha+\beta\rangle}{\langle\beta,\beta\rangle} = \frac{q+1}{p} \tag{7}$$

so, by (3), (4), (6), (7),

$$\begin{aligned}\mathrm{N}_{\alpha,\beta}\mathrm{N}_{-\alpha,-\beta} &= -\mathrm{N}_{\alpha,\beta}\mathrm{N}_{-\alpha,\alpha+\beta}\frac{\langle H_\beta, H_\beta\rangle}{\langle H_{\alpha+\beta}, H_{\alpha+\beta}\rangle}\\ &= -\mathrm{N}_{\alpha,\beta}\mathrm{N}_{-\alpha,\alpha+\beta}\frac{q+1}{p} = (q+1)^2.\end{aligned}$$

DEFINITION 3. *A Chevalley system for* $(\mathfrak{g},\mathfrak{h})$ *is a family* $(X_\alpha)_{\alpha\in\mathrm{R}}$ *such that*

(i) $X_\alpha \in \mathfrak{g}^\alpha$ *for all* $\alpha \in \mathrm{R}$;

(ii) $[X_\alpha, X_{-\alpha}] = -H_\alpha$ *for all* $\alpha \in \mathrm{R}$;

(iii) *the linear map from* $\mathfrak{g}$ *to* $\mathfrak{g}$ *which is equal to* -1 *on* $\mathfrak{h}$ *and which takes* X_α *to* $X_{-\alpha}$ *for all* $\alpha \in \mathrm{R}$ *is an automorphism of* $\mathfrak{g}$.

The extension of this definition to the case where $(\mathfrak{g},\mathfrak{h})$ is split reductive is immediate.

We shall show (§4, no. 4, Cor. of Prop. 5) that Chevalley systems for $(\mathfrak{g},\mathfrak{h})$ *exist*.

PROPOSITION 7. *Let* $(X_\alpha)_{\alpha\in\mathrm{R}}$ *be a Chevalley system for* $(\mathfrak{g},\mathfrak{h})$. *We retain the notation of Lemma 4. Then,* $\mathrm{N}_{-\alpha,-\beta} = \mathrm{N}_{\alpha,\beta}$ *and* $\mathrm{N}_{\alpha,\beta} = \pm(q+1)$ *for* $\alpha, \beta, \alpha+\beta \in \mathrm{R}$.

Let φ be the automorphism of $\mathfrak{g}$ considered in Def. 3 (iii). Then

$$\begin{aligned}\mathrm{N}_{-\alpha,-\beta}X_{-\alpha-\beta} &= [X_{-\alpha}, X_{-\beta}] = [\varphi(X_\alpha), \varphi(X_\beta)] = \varphi([X_\alpha, X_\beta])\\ &= \varphi(\mathrm{N}_{\alpha,\beta}X_{\alpha+\beta}) = \mathrm{N}_{\alpha,\beta}X_{-\alpha-\beta}\end{aligned}$$

so $\mathrm{N}_{-\alpha,-\beta} = \mathrm{N}_{\alpha,\beta}$. Now $\mathrm{N}_{\alpha,\beta} = \pm(q+1)$ by (5).

PROPOSITION 8. *Let* $(X_\alpha)_{\alpha\in\mathrm{R}}$ *be a Chevalley system for* $(\mathfrak{g},\mathfrak{h})$. *Let* M *be a* **Z***-submodule of* $\mathfrak{h}$ *containing the* H_α *and contained in the group of weights* *of* $R^\vee$. *Let* $\mathfrak{g}_{\mathbf{Z}}$ *be the* $\mathbf{Z}$*-submodule of* $\mathfrak{g}$ *generated by* M *and the* X_α. *Then* $\mathfrak{g}_{\mathbf{Z}}$ *is a* $\mathbf{Z}$*-Lie subalgebra of* $\mathfrak{g}$, *and the canonical map from* $\mathfrak{g}_{\mathbf{Z}} \otimes_{\mathbf{Z}} k$ *to* $\mathfrak{g}$ *is an isomorphism.*

If $\alpha, \beta \in R$ are such that $\alpha + \beta \in R$, then $N_{\alpha,\beta} \in \mathbf{Z}$ (Prop. 7). On the other hand, if $\alpha \in R$ and $h \in M$, then $\alpha(h) \in \mathbf{Z}$ (Chap. VI, §1, no. 9). This proves that $\mathfrak{g}_{\mathbf{Z}}$ is a $\mathbf{Z}$-Lie subalgebra of $\mathfrak{g}$. On the other hand, M is a free abelian group of rank $\dim \mathfrak{h}$ (*Algebra*, Chap. VII, §3, Th. 1), so $\mathfrak{g}_{\mathbf{Z}}$ is a free abelian group of rank $\dim \mathfrak{g}$; this implies the last assertion.

§3. SUBALGEBRAS OF SPLIT SEMI-SIMPLE LIE ALGEBRAS

In this paragraph, we denote by $(\mathfrak{g}, \mathfrak{h})$ *a split semi-simple Lie algebra, and by* R *its root system.*

1. SUBALGEBRAS STABLE UNDER ad $\mathfrak{h}$

Lemma 1. Let V *be a vector subspace of* $\mathfrak{g}$ *and* R(V) *the set of* $\alpha \in R$ *such that* $\mathfrak{g}^\alpha \subset V$. *Then,* $(V \cap \mathfrak{h}) + \sum_{\alpha \in R(V)} \mathfrak{g}^\alpha$ *is the largest vector subspace of* V *stable under* ad $\mathfrak{h}$.

A vector subspace W of V is stable under ad $\mathfrak{h}$ if and only if

$$W = (W \cap \mathfrak{h}) + \sum_{\alpha \in R} (W \cap \mathfrak{g}^\alpha)$$

(*Algebra*, Chap. VII, §2, no. 2, Cor. 1 of Th. 1). The largest vector subspace of V stable under ad $\mathfrak{h}$ is thus $(V \cap \mathfrak{h}) + \sum_{\alpha \in R} (V \cap \mathfrak{g}^\alpha)$. But $V \cap \mathfrak{g}^\alpha = \mathfrak{g}^\alpha$ for $\alpha \in R(V)$, and $V \cap \mathfrak{g}^\alpha = 0$ for $\alpha \notin R(V)$ since $\dim \mathfrak{g}^\alpha = 1$. Q.E.D.

For any subset P of R, put

$$\mathfrak{g}^P = \sum_{\alpha \in P} \mathfrak{g}^\alpha \qquad \mathfrak{h}_P = \sum_{\alpha \in P} \mathfrak{h}_\alpha.$$

If $P \subset R$ and $Q \subset R$, we clearly have

$$[\mathfrak{h}, \mathfrak{g}^P] \subset \mathfrak{g}^P \tag{1}$$

$$[\mathfrak{g}^P, \mathfrak{g}^Q] = \mathfrak{g}^{(P+Q) \cap R} + \mathfrak{h}_{P \cap (-Q)}. \tag{2}$$

Recall (Chap. VI, §1, no. 7, Def. 4) that a subset P of R is said to be *closed* if the conditions $\alpha \in P, \beta \in P, \alpha + \beta \in R$ imply $\alpha + \beta \in P$, in other words if $(P + P) \cap R \subset P$.

Lemma 2. Let $\mathfrak{h}'$ *be a vector subspace of* $\mathfrak{h}$ *and* P *a subset of* R. *Then* $\mathfrak{h}' + \mathfrak{g}^{\mathrm{P}}$ *is a subalgebra of* $\mathfrak{g}$ *if and only if* P *is a closed subset of* R *and*

$$\mathfrak{h}' \supset \mathfrak{h}_{\mathrm{P}\cap(-\mathrm{P})}.$$

Indeed,

$$[\mathfrak{h}' + \mathfrak{g}^{\mathrm{P}}, \mathfrak{h}' + \mathfrak{g}^{\mathrm{P}}] = [\mathfrak{h}', \mathfrak{g}^{\mathrm{P}}] + [\mathfrak{g}^{\mathrm{P}}, \mathfrak{g}^{\mathrm{P}}] = \mathfrak{h}_{\mathrm{P}\cap(-\mathrm{P})} + [\mathfrak{h}', \mathfrak{g}^{\mathrm{P}}] + \mathfrak{g}^{(\mathrm{P}+\mathrm{P})\cap\mathrm{R}}.$$

Hence $\mathfrak{h}' + \mathfrak{g}^{\mathrm{P}}$ is a subalgebra of $\mathfrak{g}$ if and only if

$$\mathfrak{h}_{\mathrm{P}\cap(-\mathrm{P})} \subset \mathfrak{h}' \quad \text{and} \quad \mathfrak{g}^{(\mathrm{P}+\mathrm{P})\cap\mathrm{R}} \subset \mathfrak{g}^{\mathrm{P}}$$

which proves the lemma.

PROPOSITION 1. (i) *The subalgebras of* $\mathfrak{g}$ *stable under* ad $\mathfrak{h}$ *are the vector subspaces of the form* $\mathfrak{h}' + \mathfrak{g}^{\mathrm{P}}$, *where* P *is a closed subset of* R *and* $\mathfrak{h}'$ *is a vector subspace of* $\mathfrak{h}$ *containing* $\mathfrak{h}_{\mathrm{P}\cap(-\mathrm{P})}$.

(ii) *Let* $\mathfrak{h}', \mathfrak{h}''$ *be vector subspaces of* $\mathfrak{h}$ *and* P, Q *closed subsets of* R, *with* $\mathfrak{h}' \supset \mathfrak{h}_{\mathrm{P}\cap(-\mathrm{P})}$, $\mathfrak{h}'' \subset \mathfrak{h}'$ *and* Q $\subset$ P. *Then* $\mathfrak{h}'' + \mathfrak{g}^{\mathrm{Q}}$ *is an ideal of* $\mathfrak{h}' + \mathfrak{g}^{\mathrm{P}}$ *if and only if*

$$(\mathrm{P}+\mathrm{Q}) \cap \mathrm{R} \subset \mathrm{Q} \quad \text{and} \quad \mathfrak{h}_{\mathrm{P}\cap(-\mathrm{Q})} \subset \mathfrak{h}'' \subset \bigcap_{\alpha\in\mathrm{P},\alpha\notin\mathrm{Q}} \operatorname{Ker}\alpha.$$

Assertion (i) follows immediately from Lemmas 1 and 2. Let $\mathfrak{h}', \mathfrak{h}''$, P, Q be as in (ii). Then

$$[\mathfrak{h}' + \mathfrak{g}^{\mathrm{P}}, \mathfrak{h}'' + \mathfrak{g}^{\mathrm{Q}}] = \mathfrak{h}_{\mathrm{P}\cap(-\mathrm{Q})} + [\mathfrak{h}', \mathfrak{g}^{\mathrm{Q}}] + [\mathfrak{h}'', \mathfrak{g}^{\mathrm{P}}] + \mathfrak{g}^{(\mathrm{P}+\mathrm{Q})\cap\mathrm{R}}.$$

Hence, $\mathfrak{h}'' + \mathfrak{g}^{\mathrm{Q}}$ is an ideal of $\mathfrak{h}' + \mathfrak{g}^{\mathrm{P}}$ if and only if

$$\mathfrak{h}_{\mathrm{P}\cap(-\mathrm{Q})} \subset \mathfrak{h}'', \quad [\mathfrak{h}'', \mathfrak{g}^{\mathrm{P}}] \subset \mathfrak{g}^{\mathrm{Q}}, \quad \mathfrak{g}^{(\mathrm{P}+\mathrm{Q})\cap\mathrm{R}} \subset \mathfrak{g}^{\mathrm{Q}}.$$

This implies (ii).

PROPOSITION 2. *Let* $\mathfrak{a}$ *be a subalgebra of* $\mathfrak{g}$ *stable under* ad $\mathfrak{h}$, *and let* $\mathfrak{h}' \subset \mathfrak{h}$, P $\subset$ R *be such that* $\mathfrak{a} = \mathfrak{h}' + \mathfrak{g}^{\mathrm{P}}$.

(i) *Let* $\mathfrak{k}$ *be the set of* $x \in \mathfrak{h}'$ *such that* $\alpha(x) = 0$ *for all* $\alpha \in \mathrm{P} \cap (-\mathrm{P})$. *The radical of* $\mathfrak{a}$ *is* $\mathfrak{k} + \mathfrak{g}^{\mathrm{Q}}$, *where* Q *is the set of* $\alpha \in$ P *such that* $-\alpha \notin$ P. *Moreover,* $\mathfrak{g}^{\mathrm{Q}}$ *is a nilpotent ideal of* $\mathfrak{a}$.

(ii) $\mathfrak{a}$ *is semi-simple if and only if* P = −P *and* $\mathfrak{h}' = \mathfrak{h}_{\mathrm{P}}$.

(iii) $\mathfrak{a}$ *is solvable if and only if* $\mathrm{P} \cap (-\mathrm{P}) = \varnothing$. *In that case* $[\mathfrak{a}, \mathfrak{a}] = \mathfrak{g}^{\mathrm{S}}$, *where*

$$\mathrm{S} = ((\mathrm{P}+\mathrm{P}) \cap \mathrm{R}) \cup \{\alpha \in \mathrm{P} | \alpha(\mathfrak{h}') \neq 0\}.$$

(iv) $\mathfrak{a}$ *is reductive in* $\mathfrak{g}$ *if and only if* P = −P.

(v) $\mathfrak{a}$ *consists of nilpotent elements if and only if* $\mathfrak{h}' = 0$. *Then* $\mathrm{P} \cap (-\mathrm{P}) = \varnothing$, *and* $\mathfrak{a}$ *is nilpotent.*

We prove (v). If $\mathfrak{a}$ consists of nilpotent elements, $\mathfrak{a}$ is clearly nilpotent, and $\mathfrak{h}' = 0$ since the elements of $\mathfrak{h}$ are semi-simple. Assume that $\mathfrak{h}' = 0$. By Prop. 1 (i), $\mathrm{P} \cap (-\mathrm{P}) = \varnothing$. By Chap. VI, §1, no. 7, Prop. 22, there exists a chamber C of R such that $\mathrm{P} \subset \mathrm{R}_+(\mathrm{C})$. Hence, there exists an integer $n > 0$ with the following properties: if $\alpha_1, \ldots, \alpha_n \in \mathrm{P}$ and $\beta \in \mathrm{R} \cup \{0\}$, then

$$\alpha_1 + \cdots + \alpha_n + \beta \notin \mathrm{R} \cup \{0\}.$$

This implies that every element of $\mathfrak{g}^{\mathrm{P}}$ is nilpotent, hence (v).

We prove (iii). If $\mathrm{P} \cap (-\mathrm{P}) = \varnothing$, $\mathfrak{g}^{\mathrm{P}}$ is a subalgebra of $\mathfrak{g}$ (Prop. 1 (i)), and is nilpotent by (v). Now

$$[\mathfrak{a}, \mathfrak{a}] = [\mathfrak{h}', \mathfrak{g}^{\mathrm{P}}] + [\mathfrak{g}^{\mathrm{P}}, \mathfrak{g}^{\mathrm{P}}] = [\mathfrak{h}', \mathfrak{g}^{\mathrm{P}}] + \mathfrak{g}^{(\mathrm{P}+\mathrm{P}) \cap \mathrm{R}} \subset \mathfrak{g}^{\mathrm{P}},$$

so $\mathfrak{a}$ is solvable and $[\mathfrak{a}, \mathfrak{a}]$ is given by the formula in the proposition. If $\mathrm{P} \cap (-\mathrm{P}) \neq \varnothing$, let $\alpha \in \mathrm{P}$ be such that $-\alpha \in \mathrm{P}$. Then $\mathfrak{h}_\alpha + \mathfrak{g}^\alpha + \mathfrak{g}^{-\alpha}$ is a simple subalgebra of $\mathfrak{a}$ so $\mathfrak{a}$ is not solvable.

We prove (i). Since P is closed, $(\mathrm{P} + \mathrm{Q}) \cap \mathrm{R} \subset \mathrm{P}$. If $\alpha \in \mathrm{P}, \beta \in \mathrm{Q}$ and $\alpha + \beta \in \mathrm{R}$, we cannot have $\alpha + \beta \in -\mathrm{P}$, for, P being closed, this would imply that $-\beta = -(\alpha + \beta) + \alpha \in \mathrm{P}$ whereas $\beta \in \mathrm{Q}$; thus, $(\mathrm{P} + \mathrm{Q}) \cap \mathrm{R} \subset \mathrm{Q}$. This proves that $\mathfrak{g}^{\mathrm{Q}}$ is an ideal of $\mathfrak{a}$, nilpotent by (v). We have $\mathrm{P} \cap (-\mathrm{Q}) = \varnothing$, and $\mathrm{P} \cap (-\mathrm{P}) = \mathrm{P} \cap \complement\mathrm{Q}$, so $\mathfrak{h}_{\mathrm{P} \cap (-\mathrm{Q})} \subset \mathfrak{k} \subset \bigcap_{\alpha \in \mathrm{P}, \alpha \notin \mathrm{Q}} \mathrm{Ker}\, \alpha$. By Prop. 1 (ii), $\mathfrak{k} + \mathfrak{g}^{\mathrm{Q}}$ is an ideal of $\mathfrak{a}$. Since $\mathrm{Q} \cap (-\mathrm{Q}) = \varnothing$, this ideal is solvable by (iii). It is therefore contained in the radical $\mathfrak{r}$ of $\mathfrak{a}$. Since $\mathfrak{r}$ is stable under every derivation of $\mathfrak{a}$, $\mathfrak{r}$ is stable under $\mathrm{ad}\, \mathfrak{h}$. Hence there exists a subset S of P such that $\mathfrak{r} = (\mathfrak{r} \cap \mathfrak{h}) + \mathfrak{g}^{\mathrm{S}}$. Suppose that $\alpha \in \mathrm{S}$ and that $-\alpha \in \mathrm{P}$. Then $\mathfrak{h}_\alpha = [\mathfrak{g}^\alpha, \mathfrak{g}^{-\alpha}] \subset \mathfrak{r}$, so $\mathfrak{g}^{-\alpha} = [\mathfrak{h}_\alpha, \mathfrak{g}^{-\alpha}] \subset \mathfrak{r} = 0$, so that $-\alpha \in \mathrm{S}$; by (iii), this contradicts the fact that $\mathfrak{r}$ is solvable. Consequently, $\mathrm{S} \subset \mathrm{Q}$. Finally, if $x \in \mathfrak{r} \cap \mathfrak{h}$ and if $\alpha \in \mathrm{P} \cap (-\mathrm{P})$, then $[x, \mathfrak{g}^\alpha] \subset \mathfrak{g}^\alpha \cap \mathfrak{r} = 0$, so $\alpha(x) = 0$; this shows that $x \in \mathfrak{k}$. Hence $\mathfrak{r} \subset \mathfrak{k} + \mathfrak{g}^{\mathrm{Q}}$ and the proof of (i) is complete.

We prove (iv). By (i), the adjoint representation of $\mathfrak{a}$ on $\mathfrak{g}$ is semi-simple if and only if $\mathrm{ad}_{\mathfrak{g}} x$ is semi-simple for all $x \in \mathfrak{k} + \mathfrak{g}^{\mathrm{Q}}$ (Chap. I, §6, no. 5, Th. 4); by (v), this is the case if and only if $\mathrm{Q} = \varnothing$, in other words $\mathrm{P} = -\mathrm{P}$.

We prove (ii). If $\mathfrak{a}$ is semi-simple, $\mathrm{P} = -\mathrm{P}$ by (i), so $\mathfrak{h}_{\mathrm{P}} \subset \mathfrak{h}'$; further, $\mathfrak{a} = [\mathfrak{a}, \mathfrak{a}] \subset \mathfrak{h}_{\mathrm{P}} + \mathfrak{g}^{\mathrm{P}}$ and consequently $\mathfrak{h}' = \mathfrak{h}_{\mathrm{P}}$. If $\mathrm{P} = -\mathrm{P}$ and $\mathfrak{h}' = \mathfrak{h}_{\mathrm{P}}$, $\mathfrak{a}$ is reductive by (iv), and $\mathfrak{a} = \sum_{\alpha \in \mathrm{P}} \mathfrak{s}_\alpha$, so $\mathfrak{a} = [\mathfrak{a}, \mathfrak{a}]$ and $\mathfrak{a}$ is semi-simple.

PROPOSITION 3. *Let* $\mathfrak{a}$ *be a semi-simple subalgebra of* $\mathfrak{g}$ *stable under* $\mathrm{ad}(\mathfrak{h})$ *and let* P *be the subset of* R *such that* $\mathfrak{a} = \mathfrak{h}_{\mathrm{P}} + \mathfrak{g}^{\mathrm{P}}$.

(i) $\mathfrak{h}_{\mathrm{P}}$ *is a splitting Cartan subalgebra of* $\mathfrak{a}$.

(ii) *The root system of* $(\mathfrak{a}, \mathfrak{h}_{\mathrm{P}})$ *is the set of restrictions to* $\mathfrak{h}_{\mathrm{P}}$ *of elements of* P.

Since $\mathfrak{h}_P$ is stable under ad $\mathfrak{h}$, its normalizer in $\mathfrak{a}$ is stable under ad $\mathfrak{h}$, and hence is of the form $\mathfrak{h}_P + \mathfrak{g}^Q$ where $Q \subset P$ (Lemma 1). If $\alpha \in Q$,

$$\mathfrak{g}^\alpha = [\mathfrak{h}_\alpha, \mathfrak{g}^\alpha] \subset [\mathfrak{h}_P, \mathfrak{g}^\alpha] \subset \mathfrak{h}_P,$$

which is absurd. Thus $Q = \varnothing$ and $\mathfrak{h}_P$ is its own normalizer in $\mathfrak{a}$. This proves that $\mathfrak{h}_P$ is a Cartan subalgebra of $\mathfrak{a}$. If $x \in \mathfrak{h}_P$, $\mathrm{ad}_{\mathfrak{g}} x$, and *a fortiori* $\mathrm{ad}_{\mathfrak{a}} x$, are triangularizable. Thus (i) is proved, and (ii) is clear.

By Prop. 1 (i), the subalgebras of $\mathfrak{g}$ containing $\mathfrak{h}$ are the sets $\mathfrak{h} + \mathfrak{g}^P$ where P is a closed subset of R. By Chap. VII, §3, Prop. 3, every Cartan subalgebra of $\mathfrak{h} + \mathfrak{g}^P$ is a Cartan subalgebra of $\mathfrak{g}$.

PROPOSITION 4. *Let $\mathfrak{a}$ be a subalgebra of $\mathfrak{g}$ containing $\mathfrak{h}$, x an element of $\mathfrak{a}$, s and n its semi-simple and nilpotent components. Then $s \in \mathfrak{a}$ and $n \in \mathfrak{a}$.*

We have $(\mathrm{ad}\, x)\mathfrak{a} \subset \mathfrak{a}$, so $(\mathrm{ad}\, s)\mathfrak{a} \subset \mathfrak{a}$ and $(\mathrm{ad}\, n)\mathfrak{a} \subset \mathfrak{a}$. Since $\mathfrak{a}$ is its own normalizer in $\mathfrak{g}$ (Chap. VII, §2, no. 1, Cor. 4 of Prop. 4), $s \in \mathfrak{a}$ and $n \in \mathfrak{a}$.

PROPOSITION 5. *Let* P *be a closed subset of* R.

(i) *$\mathfrak{h} + \mathfrak{g}^P$ is solvable if and only if $P \cap (-P) = \varnothing$. In that case, $[\mathfrak{h} + \mathfrak{g}^P, \mathfrak{h} + \mathfrak{g}^P] = \mathfrak{g}^P$.*

(ii) *$\mathfrak{h} + \mathfrak{g}^P$ is reductive if and only if $P = -P$.*

Assertion (i) follows from Prop. 2 (iii). If $P = -P$, $\mathfrak{h} + \mathfrak{g}^P$ is reductive (Prop. 2 (iv)). Assume that $\mathfrak{a} = \mathfrak{h} + \mathfrak{g}^P$ is reductive. Then

$$\mathfrak{g}^P = [\mathfrak{h}, \mathfrak{g}^P] \subset [\mathfrak{a}, \mathfrak{a}] \subset \mathfrak{h} + \mathfrak{g}^P,$$

so $[\mathfrak{a}, \mathfrak{a}]$ is of the form $\mathfrak{h}' + \mathfrak{g}^P$ with $\mathfrak{h}' \subset \mathfrak{h}$; since $[\mathfrak{a}, \mathfrak{a}]$ is semi-simple, $P = -P$ (Prop. 2 (ii)).

2. IDEALS

PROPOSITION 6. *Let $R_1, \ldots, R_p$ be the irreducible components of* R. *For $i = 1, \ldots, p$, put $\mathfrak{g}_i = \mathfrak{h}_{R_i} + \mathfrak{g}^{R_i}$. Then $\mathfrak{g}_1, \ldots, \mathfrak{g}_p$ are the simple components of $\mathfrak{g}$.*

The $\mathfrak{g}_i$ are ideals of $\mathfrak{g}$ (Prop. 1 (ii)). It is clear that $\mathfrak{g}$ is the direct sum of the $\mathfrak{g}_i$, hence the product of the $\mathfrak{g}_i$. Let $\mathfrak{a}$ and $\mathfrak{b}$ be complementary ideals of $\mathfrak{g}$. Then $\mathfrak{a}$ and $\mathfrak{b}$ are semi-simple and stable under ad $\mathfrak{h}$, so there exist subsets P, Q of R such that $\mathfrak{a} = \mathfrak{h}_P + \mathfrak{g}^P$, $\mathfrak{b} = \mathfrak{h}_Q + \mathfrak{g}^Q$. Then $\mathfrak{h}_P$, $\mathfrak{h}_Q$ are orthogonal complements of each other in $\mathfrak{h}$ for the Killing form, so P and Q are unions of irreducible components of R. This proves that the $\mathfrak{g}_i$ are minimal ideals of $\mathfrak{g}$.

COROLLARY 1. *$\mathfrak{g}$ is simple if and only if* R *is irreducible* (*in other words, its Dynkin graph is connected*).

This follows from Prop. 6.

A Lie algebra $\mathfrak{a}$ is said to be *absolutely simple* if, for every extension k' of k, the k'-Lie algebra $\mathfrak{a}_{(k')}$ is simple.

COROLLARY 2. *A splittable simple Lie algebra is absolutely simple.*

This follows from Cor. 1.

If $\mathfrak{g}$ is of type A_l $(l \geq 1)$ or B_l $(l \geq 1)$ or C_l $(l \geq 1)$ or D_l $(l \geq 3)$, $\mathfrak{g}$ is said to be a *classical* splittable simple Lie algebra. If $\mathfrak{g}$ is of type E_6, E_7, E_8, F_4, or G_2, $\mathfrak{g}$ is said to be an *exceptional* splittable simple Lie algebra.

3. BOREL SUBALGEBRAS

PROPOSITION 7. *Let $\mathfrak{b} = \mathfrak{h} + \mathfrak{g}^P$ be a subalgebra of $\mathfrak{g}$ containing $\mathfrak{h}$. The following conditions are equivalent:*

(i) *$\mathfrak{b}$ is a maximal solvable subalgebra of $\mathfrak{g}$;*

(ii) *there exists a chamber C of R such that $P = R_+(C)$;*

(iii) *$P \cap (-P) = \varnothing$ and $P \cup (-P) = R$.*

(i) $\Longrightarrow$ (ii): If $\mathfrak{b}$ is solvable, $P \cap (-P) = \varnothing$. Then there exists a chamber C of R such that $P \subset R_+(C)$ (Chap. VI, §1, no. 7, Prop. 22). Then $\mathfrak{h} + \mathfrak{g}^{R_+(C)}$ is a solvable subalgebra of $\mathfrak{g}$ containing $\mathfrak{b}$, hence equal to $\mathfrak{b}$ if $\mathfrak{b}$ is maximal.

(ii) $\Longrightarrow$ (iii): This is clear.

(iii) $\Longrightarrow$ (i): Assume that $P \cap (-P) = \varnothing$ and that $P \cup (-P) = R$. Then $\mathfrak{b}$ is solvable. Let $\mathfrak{b}'$ be a solvable subalgebra of $\mathfrak{g}$ containing $\mathfrak{b}$. There exists a subset Q of R such that $\mathfrak{b}' = \mathfrak{h} + \mathfrak{g}^Q$. Then $Q \cap (-Q) = \varnothing$ and $Q \supset P$, so $Q = P$ and $\mathfrak{b}' = \mathfrak{b}$.

DEFINITION 1. *A subalgebra of $\mathfrak{g}$ containing $\mathfrak{h}$ and satisfying the equivalent condition in Prop. 7 is called a Borel subalgebra of $(\mathfrak{g}, \mathfrak{h})$.*

A subalgebra $\mathfrak{b}$ of a splittable algebra $\mathfrak{g}$ is called a Borel subalgebra of $\mathfrak{g}$ if there exists a splitting Cartan subalgebra $\mathfrak{h}'$ of $\mathfrak{g}$ such that $\mathfrak{b}$ is a Borel subalgebra of $(\mathfrak{g}, \mathfrak{h}')$.

Let $(\mathfrak{g}, \mathfrak{h})$ be a split reductive Lie algebra. Let $\mathfrak{g} = \mathfrak{c} \times \mathfrak{s}$ with $\mathfrak{c}$ commutative and $\mathfrak{s}$ semi-simple. A subalgebra of $\mathfrak{g}$ of the form $\mathfrak{c} \times \mathfrak{b}$, where $\mathfrak{b}$ is a Borel subalgebra of $(\mathfrak{s}, \mathfrak{h} \cap \mathfrak{s})$, is called a Borel subalgebra of $(\mathfrak{g}, \mathfrak{h})$.

With the notations of Prop. 7, we also say that $\mathfrak{b}$ is the Borel subalgebra of $\mathfrak{g}$ defined by $\mathfrak{h}$ and C (or by $\mathfrak{h}$ and the basis of R associated to C).

Remark. The map which associates $R_+(C)$ to a chamber C of R is injective (Chap. VI, §1, no. 7, Cor. 1 of Prop. 20). Consequently, $C \mapsto \mathfrak{h} + \mathfrak{g}^{R_+(C)}$ is a bijection from the set of chambers of R to the set of Borel subalgebras of $(\mathfrak{g}, \mathfrak{h})$. Thus, the number of Borel subalgebras of $(\mathfrak{g}, \mathfrak{h})$ is equal to the order of the Weyl group of R (Chap. VI, §1, no. 5, Th. 2).

PROPOSITION 8. *Let* $\mathfrak{b}$ *be a subalgebra of* $\mathfrak{g}$, k' *an extension of* k. *Then* $\mathfrak{b} \otimes_k k'$ *is a Borel subalgebra of* $(\mathfrak{g} \otimes_k k', \mathfrak{h} \otimes_k k')$ *if and only if* $\mathfrak{b}$ *is a Borel subalgebra of* $(\mathfrak{g}, \mathfrak{h})$.

This is clear from condition (iii) of Prop. 7.

PROPOSITION 9. *Let* $\mathfrak{b}$ *be the Borel subalgebra of* $(\mathfrak{g}, \mathfrak{h})$ *defined by a chamber* C *of* R. *Let* $\mathfrak{n} = \mathfrak{g}^{\mathrm{R}_+(\mathrm{C})} = \sum_{\alpha \in \mathrm{R}, \alpha > 0} \mathfrak{g}^\alpha$. *Let* $l = \dim \mathfrak{h}$.

(i) *If* $h \in \mathfrak{h}$ *and* $x \in \mathfrak{n}$, *the characteristic polynomial of* $\mathrm{ad}_\mathfrak{g}(h + x)$ *is* $\mathrm{T}^l \prod_{\alpha \in \mathrm{R}} (\mathrm{T} - \alpha(h))$.

(ii) *The largest nilpotent ideal of* $\mathfrak{b}$ *is equal to* $\mathfrak{n}$ *and to* $[\mathfrak{b}, \mathfrak{b}]$. *This is also the set of elements of* $\mathfrak{b}$ *nilpotent in* $\mathfrak{g}$.

(iii) *Let* B *be the basis of* R *associated to* C. *For all* $\alpha \in \mathrm{B}$, *let* X_α *be a non-zero element of* $\mathfrak{g}^\alpha$. *Then* $(X_\alpha)_{\alpha \in \mathrm{B}}$ *generates the Lie algebra* $\mathfrak{n}$. *We have* $[\mathfrak{n}, \mathfrak{n}] = \sum_{\alpha \in \mathrm{R}, \alpha > 0, \alpha \notin \mathrm{B}} \mathfrak{g}^\alpha$.

There exists a total order on $\mathfrak{h}^*_\mathbf{Q}$ compatible with its vector space structure and such that the elements of $\mathrm{R}_+(\mathrm{C})$ are > 0 (Chap. VI, §1, no. 7). Let h, x be as in (i) and $y \in \mathfrak{g}^\alpha$. Then $[h + x, y] = \alpha(h)y + z$ where $z \in \sum_{\beta > \alpha} \mathfrak{g}^\beta$. Then, with respect to a suitable basis of $\mathfrak{g}$, the matrix of $\mathrm{ad}_\mathfrak{g}(h+x)$ has the following properties:

1) it is lower triangular;

2) the diagonal entries of the matrix are the number 0 (l times) and the $\alpha(h)$ for $\alpha \in \mathrm{R}$.

This proves (i). It also shows that the characteristic polynomial of $\mathrm{ad}_\mathfrak{b}(h + x)$ is $\mathrm{T}^l \prod_{\alpha \in \mathrm{R}_+(\mathrm{C})} (\mathrm{T} - \alpha(h))$. It follows from the preceding that the set of elements of $\mathfrak{b}$ nilpotent in $\mathfrak{g}$, as well as the largest nilpotent ideal of $\mathfrak{b}$, are equal to $\mathfrak{n}$. We have $\mathfrak{n} = [\mathfrak{b}, \mathfrak{b}]$ by Prop. 5 (i). Finally, assertion (iii) follows from §2, Prop. 4 (ii) and Chap. VI, §1, no. 6, Prop. 19.

COROLLARY. *Let* $\mathfrak{b}$ *be a Borel subalgebra of* $\mathfrak{g}$.

(i) *Every Cartan subalgebra of* $\mathfrak{b}$ *is a splitting Cartan subalgebra of* $\mathfrak{g}$.

(ii) *If* $\mathfrak{h}_1, \mathfrak{h}_2$ *are Cartan subalgebras of* $\mathfrak{b}$, *there exists* $x \in [\mathfrak{b}, \mathfrak{b}]$ *such that* $e^{\mathrm{ad}_\mathfrak{g} x} \mathfrak{h}_1 = \mathfrak{h}_2$.

Assertion (i) follows from Prop. 9 (i) and Chap. VII, §3, no. 3, Prop. 3. Assertion (ii) follows from Prop. 9 (ii) and Chap. VII, §3, no. 4, Th. 3.

PROPOSITION 10. *Let* $\mathfrak{b}, \mathfrak{b}'$ *be Borel subalgebras of* $\mathfrak{g}$. *There exists a splitting Cartan subalgebra of* $\mathfrak{g}$ *contained in* $\mathfrak{b} \cap \mathfrak{b}'$.

Let $\mathfrak{h}$ be a Cartan subalgebra of $\mathfrak{b}$, $\mathfrak{n} = [\mathfrak{b}, \mathfrak{b}]$, $\mathfrak{n}' = [\mathfrak{b}', \mathfrak{b}']$, $\mathfrak{p} = \mathfrak{b} \cap \mathfrak{b}'$, and $\mathfrak{s}$ a vector subspace of $\mathfrak{g}$ complementary to $\mathfrak{b} + \mathfrak{b}'$. Denote by $\mathfrak{s}^\perp, \mathfrak{b}^\perp, \mathfrak{b}'^\perp$ the orthogonal complements of $\mathfrak{s}, \mathfrak{b}, \mathfrak{b}'$ with respect to the Killing form of $\mathfrak{g}$. Put $l = \dim \mathfrak{h}, n = \dim \mathfrak{n}, p = \dim \mathfrak{p}$. Then $\dim \mathfrak{b} = \dim \mathfrak{b}' = l + n$,

$$\dim \mathfrak{s}^{\perp} = \dim(\mathfrak{b} + \mathfrak{b}') = 2(l+n) - p,$$

and so

$$\begin{aligned} \dim(\mathfrak{s}^{\perp} \cap \mathfrak{p}) &\geq \dim \mathfrak{s}^{\perp} + \dim \mathfrak{p} - \dim \mathfrak{g} \\ &= 2(l+n) - p + p - (l + 2n) = l. \end{aligned} \tag{3}$$

By Prop. 1 of §2, no. 2, $\mathfrak{n} \subset \mathfrak{b}^{\perp}, \mathfrak{n}' \subset \mathfrak{b}'^{\perp}$. The elements of $\mathfrak{p} \cap \mathfrak{n}$ are nilpotent in $\mathfrak{g}$ (Prop. 9 (ii)), and belong to $\mathfrak{b}'$, and hence to $\mathfrak{n}'$ (Prop. 9 (ii)). Consequently, $\mathfrak{p} \cap \mathfrak{n} \subset \mathfrak{n} \cap \mathfrak{n}' \subset \mathfrak{b}^{\perp} \cap \mathfrak{b}'^{\perp}$, so $\mathfrak{s}^{\perp} \cap \mathfrak{p} \cap \mathfrak{n} = 0$. In view of (3), we see that $\mathfrak{s}^{\perp} \cap \mathfrak{p}$ is a complement of $\mathfrak{n}$ in $\mathfrak{b}$. Let z be an element of $\mathfrak{h}$ regular in $\mathfrak{g}$; there exists $y \in \mathfrak{n}$ such that $y + z \in \mathfrak{s}^{\perp} \cap \mathfrak{p}$; by Prop. 9 (i), $\mathrm{ad}_{\mathfrak{g}}(y+z)$ has the same characteristic polynomial as $\mathrm{ad}_{\mathfrak{g}} z$, so $x = y + z$ is regular in $\mathfrak{g}$ and *a fortiori* in $\mathfrak{b}$ and $\mathfrak{b}'$ (Chap. VII, §2, no. 2, Prop. 9). Since $\mathfrak{g}, \mathfrak{b}, \mathfrak{b}'$ have the same rank, $\mathfrak{b}^0(x) = \mathfrak{g}^0(x) = \mathfrak{b}'^0(x)$ is simultaneously a Cartan subalgebra of $\mathfrak{b}$, of $\mathfrak{g}$ and of $\mathfrak{b}'$ (Chap. VII, §3, no. 3, Th. 2). Finally, this Cartan subalgebra of $\mathfrak{g}$ is splitting by the Cor. of Prop. 9.

COROLLARY. *The group* $\mathrm{Aut}_e(\mathfrak{g})$ *operates transitively on the set of pairs* $(\mathfrak{t}, \mathfrak{b})$ *where* $\mathfrak{t}$ *is a splitting Cartan subalgebra of* $\mathfrak{g}$ *and* $\mathfrak{b}$ *is a Borel subalgebra of* $(\mathfrak{g}, \mathfrak{t})$.

Let $(\mathfrak{t}_1, \mathfrak{b}_1)$ and $(\mathfrak{t}_2, \mathfrak{b}_2)$ be two such pairs. There exists a splitting Cartan subalgebra $\mathfrak{t}$ of $\mathfrak{g}$ contained in $\mathfrak{b}_1 \cap \mathfrak{b}_2$ (Prop. 10). By the Cor. of Prop. 9, we are reduced to the case in which $\mathfrak{t}_1 = \mathfrak{t}_2 = \mathfrak{t}$. Let S be the root system of $(\mathfrak{g}, \mathfrak{t})$. There exists bases B_1, B_2 of S such that $\mathfrak{b}_i$ is associated to B_i $(i = 1, 2)$, and there exists $s \in W(S)$ which transforms B_1 into B_2. Finally, there exists $a \in \mathrm{Aut}_e(\mathfrak{g})$ such that $a|\mathfrak{t} = s$ (§2, no. 2, Cor. of Th. 2). Then $a(\mathfrak{t}) = \mathfrak{t}$ and $a(\mathfrak{b}_1) = \mathfrak{b}_2$.

4. PARABOLIC SUBALGEBRAS

PROPOSITION 11. *Let* $\mathfrak{p} = \mathfrak{h} + \mathfrak{g}^{P}$ *be a subalgebra of* $\mathfrak{g}$ *containing* $\mathfrak{h}$. *The following conditions are equivalent:*

(i) $\mathfrak{p}$ *contains a Borel subalgebra of* $(\mathfrak{g}, \mathfrak{h})$;

(ii) *there exists a chamber* C *of* R *such that* $P \supset R_+(C)$;

(iii) P *is parabolic, in other words* (Chap. VI, §1, no. 7, Def. 4), $P \cup (-P) = R$.

Conditions (i) and (ii) are equivalent by Prop. 7. Conditions (ii) and (iii) are equivalent by Chap. VI, §1, no. 7, Prop. 20.

DEFINITION 2. *A subalgebra of* $\mathfrak{g}$ *containing* $\mathfrak{h}$ *and satisfying the equivalent conditions of Prop. 11 is called a parabolic subalgebra of* $(\mathfrak{g}, \mathfrak{h})$. *A parabolic subalgebra of* $\mathfrak{g}$ *is a parabolic subalgebra of* $(\mathfrak{g}, \mathfrak{h}')$ *where* $\mathfrak{h}'$ *is a splitting Cartan subalgebra of* $\mathfrak{g}$.

This definition extends immediately to the case in which $(\mathfrak{g}, \mathfrak{h})$ is a split reductive Lie algebra.

Remark. Let B be a basis of R, and $\mathfrak{b}$ the corresponding Borel subalgebra. If $\Sigma \subset \mathrm{B}$, denote by Q_Σ the set of roots that are linear combinations of elements of Σ with coefficients ≤ 0; put $\mathfrak{p}(\Sigma) = \mathrm{R}_+(\mathrm{B}) \cup \mathrm{Q}_\Sigma$ and $\mathfrak{p}_\Sigma = \mathfrak{h} \oplus \mathfrak{g}^{\mathrm{P}(\Sigma)}$. By Chap. VI, §1, no. 7, Lemma 3 and Prop. 20, $\mathfrak{p}_\Sigma$ is a parabolic subalgebra containing $\mathfrak{b}$ and every parabolic subalgebra of $\mathfrak{g}$ containing $\mathfrak{b}$ is obtained in this way.

Lemma 3. Let V *be a finite dimensional real vector space,* S *a root system in* V^*, $\mathscr{P}$ *the set of parabolic subsets of* S; *let* $\mathscr{H}$ *be the set of* $\operatorname{Ker}\alpha$ *for* $\alpha \in \mathrm{S}$, *and* $\mathscr{F}$ *the set of facets of* V *relative to* $\mathscr{H}$ (Chap. V, §1, no. 2, Def. 1).

If $\mathrm{P} \in \mathscr{P}$, *let* $\overline{\mathrm{F}}(\mathrm{P})$ *be the set of* $v \in \mathrm{V}$ *such that* $\alpha(v) \geq 0$ *for all* $\alpha \in \mathrm{P}$. *If* $\mathrm{F} \in \mathscr{F}$, *let* $\mathrm{P}(\mathrm{F})$ *be the set of* $\alpha \in \mathrm{R}$ *such that* $\alpha(v) \geq 0$ *for all* $v \in \mathrm{F}$.

Then $\mathrm{F} \mapsto \mathrm{P}(\mathrm{F})$ *is a bijection from* $\mathscr{F}$ *to* $\mathscr{P}$; *for all* $\mathrm{F} \in \mathscr{F}$, $\overline{\mathrm{F}}(\mathrm{P}(\mathrm{F}))$ *is the closure of* F.

a) Let $\mathrm{P} \in \mathscr{P}$. There exists a chamber C of S and a subset Σ of the basis B(C) such that $\mathrm{P} = \mathrm{S}_+(\mathrm{C}) \cup \mathrm{Q}$ where Q is the set of linear combinations of elements of Σ with non-positive integer coefficients (Chap. VI, §1, no. 7, Prop. 20). Put

$$\mathrm{B}(\mathrm{C}) = \{\alpha_1, \dots, \alpha_l\}, \ \Sigma = \{\alpha_1, \dots, \alpha_m\}.$$

If $v \in \mathrm{V}$, we have the following equivalences:

$$\begin{aligned} &\alpha(v) \geq 0 \text{ for all } \alpha \in \mathrm{P} \\ \iff &\alpha_1(v) \geq 0, \dots \alpha_l(v) \geq 0, \ \alpha_1(v) \leq 0, \dots, \alpha_m(v) \leq 0 \\ \iff &\alpha_1(v) = \cdots = \alpha_m(v) = 0, \ \alpha_{m+1}(v) \geq 0, \dots, \alpha_l(v) \geq 0, \end{aligned}$$

so $\overline{\mathrm{F}}(\mathrm{P})$ is the closure of the set

$$\{v \in \mathrm{V} \mid \alpha_1(v) = \cdots = \alpha_m(v) = 0, \ \alpha_{m+1}(v) > 0, \dots, \alpha_l(v) > 0\},$$

a set which is a facet F relative to $\mathscr{H}$ since every element of S is a linear combination of $\alpha_1, \dots, \alpha_l$ in which the coefficients are either all ≥ 0 or all ≤ 0. Moreover, if $\beta = u_1\alpha_1 + \cdots + u_l\alpha_l \in \mathrm{S}$,

$$\begin{aligned} &\beta \in \mathrm{P}(\mathrm{F}) \iff u_{m+1} \geq 0, \dots, u_l \geq 0 \\ \iff &\beta \in \mathrm{S}_+(\mathrm{C}) \text{ or } (-\beta \in \mathrm{S}_+(\mathrm{C}) \text{ and } u_{m+1} = \ldots = u_l = 0) \\ \iff &\beta \in \mathrm{S}_+(\mathrm{C}) \cup \mathrm{Q} = \mathrm{P}, \end{aligned}$$

so $\mathrm{P}(\mathrm{F}) = \mathrm{P}$.

b) Let $F \in \mathscr{F}$. It is clear that $P(F) \in \mathscr{P}$. On the other hand, F is contained in the closure of a chamber relative to $\mathscr{H}$ (Chap. V, §1, no. 3, formulas (6)), and so is a facet relative to the set of walls of this chamber (Chap. V, §1, no. 4, Prop. 9). Consequently, $\overline{F}$ is of the form $\{v \in V \mid \alpha(v) \geq 0 \text{ for all } \alpha \in T\}$, where T is a subset of S which we can clearly take to be equal to P(F). Thus, $\overline{F} = \overline{F}(P(F))$. Q.E.D.

If $P \in \mathscr{P}$, the facet F such that $P = P(F)$ is said to be *associated* to P; we denote it by F(P). We extend these conventions to the case in which $(\mathfrak{g}, \mathfrak{h})$ is split reductive.

PROPOSITION 12. *Let $\mathscr{H}$ be the set of hyperplanes of $\mathfrak{h}_{\mathbf{R}}$ consisting of the kernels of the roots in* R. *Let $\mathscr{F}$ be the set of facets of $\mathfrak{h}_{\mathbf{R}}$ relative to $\mathscr{H}$. Let $\mathscr{S}$ be the set of parabolic subalgebras of $(\mathfrak{g}, \mathfrak{h})$. For every $\mathfrak{p} = \mathfrak{h} + \mathfrak{g}^{P} \in \mathscr{S}$, let* $F(\mathfrak{p})$ *be the facet associated to* P. *Then $\mathfrak{p} \mapsto F(\mathfrak{p})$ is a bijection from $\mathscr{S}$ to $\mathscr{F}$. If $\mathfrak{p}_1, \mathfrak{p}_2 \in \mathscr{P}$,*

$$\mathfrak{p}_1 \supset \mathfrak{p}_2 \iff F(\mathfrak{p}_1) \subset \overline{F(\mathfrak{p}_2)}.$$

This follows immediately from Lemma 3.

Example. The facets corresponding to the parabolic subalgebras of $(\mathfrak{g}, \mathfrak{h})$ containing a Borel algebra $\mathfrak{b}$ are the facets contained in the closure of the chamber associated to $\mathfrak{b}$ (cf. the Remark above).

PROPOSITION 13. *Let $\mathfrak{p} = \mathfrak{h} + \mathfrak{g}^{P}$ be a parabolic subalgebra of $(\mathfrak{g}, \mathfrak{h})$, Q the set of $\alpha \in P$ such that $-\alpha \notin P$, and $\mathfrak{s} = \mathfrak{h} + \mathfrak{g}^{P \cap (-P)}$. Then $\mathfrak{p} = \mathfrak{s} \oplus \mathfrak{g}^{Q}$, $\mathfrak{s}$ is reductive in $\mathfrak{g}$, and $\mathfrak{g}^{Q}$ is the largest nilpotent ideal of $\mathfrak{p}$ and the nilpotent radical of $\mathfrak{p}$. The centre of $\mathfrak{p}$ is zero.*

By Prop. 2, $\mathfrak{s}$ is reductive in $\mathfrak{g}$ and $\mathfrak{g}^{Q}$ is a nilpotent ideal of $\mathfrak{p}$. If $\mathfrak{n}$ is the largest nilpotent ideal of $\mathfrak{p}$, $\mathfrak{g}^{Q} \subset \mathfrak{n} \subset \mathfrak{h} + \mathfrak{g}^{Q}$ (Prop. 2 (i)); if $x \in \mathfrak{n} \cap \mathfrak{h}$, $\mathrm{ad}_{\mathfrak{p}} x$ is nilpotent, so $\alpha(x) = 0$ for all $\alpha \in P$, and hence $x = 0$; this proves that $\mathfrak{n} = \mathfrak{g}^{Q}$. Since $[\mathfrak{h}, \mathfrak{g}^{Q}] = \mathfrak{g}^{Q}$, the nilpotent radical of $\mathfrak{p}$ contains $\mathfrak{g}^{Q}$ and consequently is equal to $\mathfrak{g}^{Q}$. Let $z = h + \sum_{\alpha \in P} u_\alpha$ (where $h \in \mathfrak{h}, u_\alpha \in \mathfrak{g}^{\alpha}$) be an element of the centre of $\mathfrak{p}$. For all $h' \in \mathfrak{h}$, $0 = [h', z] = \sum \alpha(h') u_\alpha$, so $u_\alpha = 0$ for all $\alpha \in P$; it follows that $[h, \mathfrak{g}^{\beta}] = 0$ for all $\beta \in P$, so $h = 0$.

5. NON-SPLIT CASE

PROPOSITION 14. *Let k' be an extension of k and $\mathfrak{g}' = \mathfrak{g} \otimes_k k'$. Let $\mathfrak{m}$ be a subalgebra of $\mathfrak{g}$ and $\mathfrak{m}' = \mathfrak{m} \otimes_k k'$. If $\mathfrak{m}'$ is a parabolic* (resp. *Borel*) *subalgebra of $\mathfrak{g}'$, $\mathfrak{m}$ is a parabolic* (resp. *Borel*) *subalgebra of $\mathfrak{g}$.*

By Prop. 8 and 11, it suffices to prove that $\mathfrak{m}$ contains a splitting Cartan subalgebra of $\mathfrak{g}$. Let $\mathfrak{b}$ be a Borel subalgebra of $\mathfrak{g}$. Then $\mathfrak{b}' = \mathfrak{b} \otimes_k k'$ is a Borel subalgebra of $\mathfrak{g}'$, so $\mathfrak{m}' \cap \mathfrak{b}'$ contains a Cartan subalgebra of $\mathfrak{g}'$ (Prop. 10). Let $\mathfrak{t}$ be a Cartan subalgebra of $\mathfrak{m} \cap \mathfrak{b}$. Then $\mathfrak{t} \otimes_k k'$ is a Cartan subalgebra of $\mathfrak{m}' \cap \mathfrak{b}'$, and hence of $\mathfrak{g}'$ (Chap. VII, §3, no. 3, Prop. 3). Consequently, $\mathfrak{t}$ is a Cartan subalgebra of $\mathfrak{g}$, and it is splitting since it is contained in $\mathfrak{b}$.

DEFINITION 3. *Let $\mathfrak{a}$ be a semi-simple (or more generally reductive) Lie algebra and $\bar{k}$ an algebraic closure of k. A subalgebra $\mathfrak{m}$ of $\mathfrak{a}$ is said to be parabolic* (resp. *Borel*) *if $\mathfrak{m} \otimes_k \bar{k}$ is a parabolic* (resp. *Borel*) *subalgebra of $\mathfrak{a} \otimes_k \bar{k}$.*

If $\mathfrak{a}$ is splittable, Prop. 14 shows that this definition is equivalent to Definition 2 (resp. to Definition 1).

PROPOSITION 15. *Let $\mathfrak{a}$ be a reductive Lie algebra, k' an extension of k, and $\mathfrak{m}$ a subalgebra of $\mathfrak{a}$. Then $\mathfrak{m}$ is a parabolic* (resp. *Borel*) *subalgebra of $\mathfrak{a}$ if and only if $\mathfrak{m} \otimes_k k'$ is a parabolic* (resp. *Borel*) *subalgebra of $\mathfrak{a} \otimes_k k'$.*

This follows immediately from Prop. 14.

§ 4. SPLIT SEMI-SIMPLE LIE ALGEBRA DEFINED BY A REDUCED ROOT SYSTEM

1. FRAMED SEMI-SIMPLE LIE ALGEBRAS

PROPOSITION 1. *Let $(\mathfrak{g}, \mathfrak{h})$ be a split semi-simple Lie algebra, R its root system, B a basis of R, and $(n(\alpha,\beta))_{\alpha,\beta\in\mathrm{B}}$ the corresponding Cartan matrix. For all $\alpha \in \mathrm{B}$, let $X_\alpha \in \mathfrak{g}^\alpha, X_{-\alpha} \in \mathfrak{g}^{-\alpha}$. Then, for $\alpha, \beta \in \mathrm{B}$,*

$$[H_\alpha, H_\beta] = 0 \tag{1}$$
$$[H_\alpha, X_\beta] = n(\beta,\alpha)X_\beta \tag{2}$$
$$[H_\alpha, X_{-\beta}] = -n(\beta,\alpha)X_{-\beta} \tag{3}$$
$$[X_{-\alpha}, X_\beta] = 0 \quad \text{if } \alpha \neq \beta \tag{4}$$
$$(\mathrm{ad}\, X_\alpha)^{1-n(\beta,\alpha)}X_\beta = 0 \quad \text{if } \alpha \neq \beta \tag{5}$$
$$(\mathrm{ad}\, X_{-\alpha})^{1-n(\beta,\alpha)}X_{-\beta} = 0 \quad \text{if } \alpha \neq \beta. \tag{6}$$

The family $(H_\alpha)_{\alpha\in\mathrm{B}}$ is a basis of $\mathfrak{h}$. If $X_\alpha \neq 0$ and $X_{-\alpha} \neq 0$ for all $\alpha \in \mathrm{B}$, the Lie algebra $\mathfrak{g}$ is generated by the X_α and the $X_{-\alpha}$ ($\alpha \in \mathrm{B}$).

(Recall that, if $\alpha, \beta \in \mathrm{B}$ and $\alpha \neq \beta$, $n(\beta,\alpha)$ is an integer ≤ 0, so formulas (5) and (6) make sense.)

Formulas (1), (2) and (3) are clear. If $\alpha \neq \beta$, $\beta - \alpha$ is not a root since every element of R is a linear combination of elements of B with integer coefficients all of the same sign (Chap. VI, §1, no. 6, Th. 3). This proves (4). In view of Chap. VI, §1, no. 3, Prop. 9, this also proves that the α-chain defined by β is

$$\{\beta, \beta + \alpha, \ldots, \beta - n(\beta, \alpha)\alpha\};$$

hence $\beta+(1-n(\beta,\alpha))\alpha \notin \mathrm{R}$, which proves (5). The equality (6) is established in a similar way. The family $(H_\alpha)_{\alpha\in\mathrm{B}}$ is a basis of $\mathrm{R}^\vee$, and hence of $\mathfrak{h}$. If $X_\alpha \neq 0$ and $X_{-\alpha} \neq 0$ for all $\alpha \in \mathrm{B}$, then $[X_\alpha, X_{-\alpha}] = \lambda_\alpha H_\alpha$ with $\lambda_\alpha \neq 0$, so the last assertion follows from §3, no. 3, Prop. 9 (iii).

DEFINITION 1. *Let* $(\mathfrak{g}, \mathfrak{h})$ *be a split semi-simple Lie algebra,* R *its root system. A framing of* $(\mathfrak{g}, \mathfrak{h})$ *is a pair* $(\mathrm{B}, (X_\alpha)_{\alpha\in\mathrm{B}})$*, where* B *is a basis of* R*, and where, for all* $\alpha \in \mathrm{B}$*,* X_α *is a non-zero element of* $\mathfrak{g}^\alpha$*. A framed semi-simple Lie algebra is a sequence* $(\mathfrak{g}, \mathfrak{h}, \mathrm{B}, (X_\alpha)_{\alpha\in\mathrm{B}})$ *where* $(\mathfrak{g}, \mathfrak{h})$ *is a split semi-simple Lie algebra, and where* $(\mathrm{B}, (X_\alpha)_{\alpha\in\mathrm{B}})$ *is a framing of* $(\mathfrak{g}, \mathfrak{h})$.

A framing of $\mathfrak{g}$ is a framing of $(\mathfrak{g}, \mathfrak{h})$, where $\mathfrak{h}$ is a splitting Cartan subalgebra of $\mathfrak{g}$.

Let $a_1 = (\mathfrak{g}_1, \mathfrak{h}_1, \mathrm{B}_1, (X^1_\alpha)_{\alpha\in\mathrm{B}_1})$ and $a_2 = (\mathfrak{g}_2, \mathfrak{h}_2, \mathrm{B}_2, (X^2_\alpha)_{\alpha\in\mathrm{B}_2})$ be framed semi-simple Lie algebras. An isomorphism from a_1 to a_2 is an isomorphism φ from $\mathfrak{g}_1$ to $\mathfrak{g}_2$ that takes $\mathfrak{h}_1$ to $\mathfrak{h}_2$, B_1 to B_2, and X^1_α to $X^2_{\psi\alpha}$ for all $\alpha \in \mathrm{B}_1$ (where ψ is the contragredient map of $\varphi|\mathfrak{h}_1$). In this case, φ is said to transform the framing $(\mathrm{B}_1, (X^1_\alpha)_{\alpha\in\mathrm{B}_1})$ to the framing $(\mathrm{B}_2, (X^2_\alpha)_{\alpha\in\mathrm{B}_2})$.

If $(\mathrm{B}, (X_\alpha)_{\alpha\in\mathrm{B}})$ is a framing of $(\mathfrak{g}, \mathfrak{h})$, there exists, for all $\alpha \in \mathrm{B}$, a unique element $X_{-\alpha}$ of $\mathfrak{g}^{-\alpha}$ such that $[X_\alpha, X_{-\alpha}] = -H_\alpha$ (§2, no. 2, Th. 1 (iv)). The family $(X_\alpha)_{\alpha\in\mathrm{B}\cup(-\mathrm{B})}$ is called the *generating family defined by the framing* (cf. Prop.1). This is also the generating family defined by the framing $(-\mathrm{B}, (X_\alpha)_{\alpha\in-\mathrm{B}})$. For all $\alpha \in \mathrm{B}\cup(-\mathrm{B})$, let $t_\alpha \in k^*$, and assume that $t_\alpha t_{-\alpha} = 1$ for all $\alpha \in \mathrm{B}$. Then $(t_\alpha X_\alpha)_{\alpha\in\mathrm{B}\cup(-\mathrm{B})}$ is the generating family defined by the framing $(\mathrm{B}, (t_\alpha X_\alpha)_{\alpha\in\mathrm{B}})$.

2. A PRELIMINARY CONSTRUCTION

In this number and the next, we denote by R a reduced root system in a vector space V and by B a basis of R. We denote by $(n(\alpha,\beta))_{\alpha,\beta\in\mathrm{B}}$ the Cartan matrix relative to B. Recall that $n(\alpha,\beta) = \langle\alpha, \beta^\vee\rangle$. We are going to show that R is the root system of a split semi-simple Lie algebra which is unique up to isomorphism. In the main we shall be considering the Lie algebra defined by the relations in Prop. 1.

The construction in this number applies to any square matrix $(n(\alpha,\beta))_{\alpha,\beta\in\mathrm{B}}$ over k with non-zero determinant and such that $n(\alpha,\alpha) = 2$ for all $\alpha \in \mathrm{B}$ (cf. Chap. VI, §1, no. 10, formula (14)).

Let E be the free associative algebra of the set B over k. Recall that E is **N**-graded (*Algebra*, Chap. III, §3, no. 1, Example 3). We are going to associate to each $\alpha \in \mathrm{B}$ endomorphisms $X^0_{-\alpha}, H^0_\alpha, X^0_\alpha$ of the vector space E, of degrees $1, 0, -1$ respectively. For any word $(\alpha_1, \ldots, \alpha_n)$ in elements of B, put

$$X^0_{-\alpha}(\alpha_1, \dots, \alpha_n) = (\alpha, \alpha_1, \dots, \alpha_n) \tag{7}$$

$$H^0_\alpha(\alpha_1, \dots, \alpha_n) = \left(-\sum_{i=1}^{n} n(\alpha_i, \alpha)\right)(\alpha_1, \dots, \alpha_n). \tag{8}$$

On the other hand, $X^0_\alpha(\alpha_1, \dots, \alpha_n)$ is defined by induction on n using the formula

$$X^0_\alpha(\alpha_1, \dots, \alpha_n) = (X^0_{-\alpha_1} X^0_\alpha - \delta_{\alpha,\alpha_1} H^0_\alpha)(\alpha_2, \dots, \alpha_n) \tag{9}$$

where δ_{α,α_1} is the Kronecker symbol; it is understood that $X^0_\alpha(\alpha_1, \dots, \alpha_n)$ is zero if $(\alpha_1, \dots, \alpha_n)$ is the empty word.

Lemma 1. For all $\alpha, \beta \in \mathrm{B}$, *we have*

$$[X^0_\alpha, X^0_{-\alpha}] = -H^0_\alpha \tag{10}$$

$$[H^0_\alpha, H^0_\beta] = 0 \tag{11}$$

$$[H^0_\alpha, X^0_\beta] = n(\beta, \alpha) X^0_\beta \tag{12}$$

$$[H^0_\alpha, X^0_{-\beta}] = -n(\beta, \alpha) X^0_{-\beta} \tag{13}$$

$$[X^0_\alpha, X^0_{-\beta}] = 0 \text{ if } \alpha \neq \beta. \tag{14}$$

Indeed, relation (9) can be written

$$(X^0_\alpha X^0_{-\alpha_1})(\alpha_2, \dots, \alpha_n) = (X^0_{-\alpha_1} X^0_\alpha)(\alpha_2, \dots, \alpha_n) - \delta_{\alpha,\alpha_1} H^0_\alpha(\alpha_2, \dots, \alpha_n)$$

which proves (10) and (14). Relation (11) is clear. Next

$$\begin{aligned}[H^0_\alpha, X^0_{-\beta}](\alpha_1, \dots, \alpha_n) &= H^0_\alpha(\beta, \alpha_1, \dots, \alpha_n) + \left(\sum_{i=1}^{n} n(\alpha_i, \alpha)\right)(\beta, \alpha_1, \dots, \alpha_n) \\ &= -n(\beta, \alpha)(\beta, \alpha_1, \dots, \alpha_n) \\ &= -n(\beta, \alpha) X^0_{-\beta}(\alpha_1, \dots, \alpha_n)\end{aligned}$$

hence (13). Finally,

$$\begin{aligned}0 &= [H^0_\alpha, [X^0_\beta, X^0_{-\gamma}]] \quad \text{by (10), (11), (14)} \\ &= [[H^0_\alpha, X^0_\beta], X^0_{-\gamma}] + [X^0_\beta, [H^0_\alpha, X^0_{-\gamma}]] \\ &= [[H^0_\alpha, X^0_\beta] - n(\gamma, \alpha) X^0_\beta, X^0_{-\gamma}] \quad \text{by (13)} \\ &= [[H^0_\alpha, X^0_\beta] - n(\beta, \alpha) X^0_\beta, X^0_{-\gamma}] \quad \text{by (14);}\end{aligned} \tag{15}$$

now, considering the empty word immediately gives

$$([H^0_\alpha, X^0_\beta] - n(\beta, \alpha) X^0_\beta)(\varnothing) = 0$$

so (15) implies that

$$([H^0_\alpha, X^0_\beta] - n(\beta, \alpha) X^0_\beta) X^0_{-\gamma_1} X^0_{-\gamma_2} \dots X^0_{-\gamma_n}(\varnothing) = 0$$

for all $\gamma_1, \dots, \gamma_n \in \mathrm{B}$; this proves (12).

Lemma 2. The endomorphisms X^0_α, H^0_β, $X^0_{-\gamma}$, where $\alpha, \beta, \gamma \in \mathrm{B}$, are linearly independent.

Since $X^0_{-\alpha}(\varnothing) = \alpha$, it is clear that the $X^0_{-\alpha}$ are linearly independent. Assume that $\sum_\alpha a_\alpha H^0_\alpha = 0$; then, for all $\beta \in \mathrm{B}$,

$$0 = \left[\sum_\alpha a_\alpha H^0_\alpha, X^0_{-\beta}\right] = -\sum_\alpha a_\alpha n(\beta,\alpha) X^0_{-\beta};$$

since $\det(n(\beta,\alpha)) \neq 0$, it follows that $a_\alpha = 0$ for all α. Assume that $\sum_\alpha a_\alpha X^0_\alpha = 0$. In view of formulas (7), (8), (9),

$$X^0_\alpha(\beta) = 0,$$
$$X^0_\alpha(\beta,\beta) = 2\delta_{\alpha\beta}\beta$$

for all $\beta \in \mathrm{B}$. It follows that $a_\beta = 0$ for all β. Since X^0_α, H^0_α, $X^0_{-\alpha}$ are of degree $-1, 0, 1$, respectively, the lemma follows from what has gone before.

Let I be the set $\mathrm{B} \times \{-1, 0, 1\}$. Put $x_\alpha = (\alpha, -1)$, $h_\alpha = (\alpha, 0)$, and $x_{-\alpha} = (\alpha, 1)$. Let $\mathfrak{a}$ be the Lie algebra defined by the generating family I and the following set $\mathscr{R}$ of relators:

$$[h_\alpha, h_\beta]$$
$$[h_\alpha, x_\beta] - n(\beta,\alpha)x_\beta$$
$$[h_\alpha, x_{-\beta}] + n(\beta,\alpha)x_{-\beta}$$
$$[x_\alpha, x_{-\alpha}] + h_\alpha$$
$$[x_\alpha, x_{-\beta}] \quad \text{if } \alpha \neq \beta$$

(cf. Chap. II, §2, no. 3). By Lemma 1, there exists a unique linear representation ρ of $\mathfrak{a}$ on E such that

$$\rho(x_\alpha) = X^0_\alpha, \quad \rho(h_\alpha) = H^0_\alpha, \quad \rho(x_{-\alpha}) = X^0_{-\alpha}.$$

In view of Lemma 2, this proves the following result:

Lemma 3. The canonical images in $\mathfrak{a}$ of the elements x_α, h_β, $x_{-\gamma}$, where $\alpha, \beta, \gamma \in \mathrm{B}$, are linearly independent.

In the following, we identify x_α, h_α, $x_{-\alpha}$ with their canonical images in $\mathfrak{a}$.

Lemma 4. There exists a unique involutive automorphism θ of $\mathfrak{a}$ such that

$$\theta(x_\alpha) = x_{-\alpha}, \quad \theta(x_{-\alpha}) = x_\alpha, \quad \theta(h_\alpha) = -h_\alpha$$

for all $\alpha \in \mathrm{B}$.

Indeed, there exists an involutive automorphism of the free Lie algebra L(I) satisfying these conditions. It leaves $\mathscr{R} \cup (-\mathscr{R})$ stable, and hence defines by passage to the quotient an involutive automorphism of $\mathfrak{a}$ satisfying the conditions of the lemma. The uniqueness follows from the fact that $\mathfrak{a}$ is generated by the elements x_α, h_α, $x_{-\alpha}$ ($\alpha \in \mathrm{B}$).

This automorphism is called the *canonical involutive automorphism of* $\mathfrak{a}$.

Let Q be the set of radical weights of R; this is a free **Z**-module with basis B (Chap. VI, §1, no. 9). There exists a graduation of type Q on the free Lie algebra L(I) such that x_α, h_α, $x_{-\alpha}$ are of degrees α, 0, $-\alpha$, respectively (Chap. II, §2, no. 6). Now the elements of $\mathscr{R}$ are homogeneous. Hence there exists a unique graduation of type Q on $\mathfrak{a}$ compatible with the Lie algebra structure of $\mathfrak{a}$ and such that x_α, h_α, $x_{-\alpha}$ are of degrees α, 0, $-\alpha$, respectively. For any $\mu \in \mathrm{Q}$, denote by $\mathfrak{a}^\mu$ the set of elements of $\mathfrak{a}$ homogeneous of degree μ.

Lemma 5. Let $z \in \mathfrak{a}$. Then $z \in \mathfrak{a}^\mu$ if and only if $[h_\alpha, z] = \langle \mu, \alpha^\vee \rangle z$ for all $\alpha \in \mathrm{B}$.

For $\mu \in \mathrm{Q}$, let $\mathfrak{a}^{(\mu)}$ be the set of $x \in \mathfrak{a}$ such that $[h_\alpha, x] = \langle \mu, \alpha^\vee \rangle x$ for all $\alpha \in \mathrm{B}$. The sum of the $\mathfrak{a}^{(\mu)}$ is direct. To prove the lemma, it therefore suffices to show that $\mathfrak{a}^\mu \subset \mathfrak{a}^{(\mu)}$. Let $\alpha \in \mathrm{B}$. The endomorphism u of the vector space $\mathfrak{a}$ such that $u|\mathfrak{a}^\mu = \langle \mu, \alpha^\vee \rangle .1$ is a derivation of $\mathfrak{a}$ such that $ux = (\mathrm{ad}\ h_\alpha).x$ for $x = x_\beta$, $x = h_\beta$, $x = x_{-\beta}$; hence $u = \mathrm{ad}\ h_\alpha$, which proves our assertion.

Remark. It follows from Lemma 5 that every ideal of $\mathfrak{a}$ is homogeneous, since it is stable under the ad h_α.

Denote by Q_+ (resp. Q_-) the set of linear combinations of elements of B with positive (resp. negative) integer coefficients, not all zero. Put $\mathfrak{a}_+ = \sum_{\mu \in \mathrm{Q}_+} a^\mu$ and $\mathfrak{a}_- = \sum_{\mu \in \mathrm{Q}_-} \mathfrak{a}^\mu$. Since $\mathrm{Q}_+ + \mathrm{Q}_+ \subset \mathrm{Q}_+$ and $\mathrm{Q}_- + \mathrm{Q}_- \subset \mathrm{Q}_-$, $\mathfrak{a}_+$ and $\mathfrak{a}_-$ are Lie subalgebras of $\mathfrak{a}$.

PROPOSITION 2. (i) *The Lie algebra $\mathfrak{a}_+$ is generated by the family $(x_\alpha)_{\alpha \in \mathrm{B}}$.*

(ii) *The Lie algebra $\mathfrak{a}_-$ is generated by the family $(x_{-\alpha})_{\alpha \in \mathrm{B}}$.*

(iii) *The family $(h_\alpha)_{\alpha \in \mathrm{B}}$ is a basis of the vector space $\mathfrak{a}^0$.*

(iv) *The vector space $\mathfrak{a}$ is the direct sum of $\mathfrak{a}_+$, $\mathfrak{a}^0$, $\mathfrak{a}_-$.*

Let $\mathfrak{r}$ (resp. $\mathfrak{n}$) be the Lie subalgebra of $\mathfrak{a}$ generated by $(x_\alpha)_{\alpha \in \mathrm{B}}$ (resp. $(x_{-\alpha})_{\alpha \in \mathrm{B}}$), and $\mathfrak{h}$ the vector subspace of $\mathfrak{a}$ generated by $(h_\alpha)_{\alpha \in \mathrm{B}}$. Since the x_α are homogeneous elements of $\mathfrak{a}_+$, $\mathfrak{r}$ is a graded subalgebra of $\mathfrak{a}_+$; hence, $[\mathfrak{h}, \mathfrak{r}] \subset \mathfrak{r}$, so $\mathfrak{h} + \mathfrak{r}$ is a subalgebra of $\mathfrak{a}$; since

$$[x_{-\alpha}, x_\beta] = \delta_{\alpha\beta} h_\alpha,$$

$[x_{-\alpha}, \mathfrak{r}] \subset \mathfrak{h} + \mathfrak{r}$ for all $\alpha \in \mathrm{B}$. Similarly, $\mathfrak{n}$ is a graded subalgebra of $\mathfrak{a}_-$, one has $[\mathfrak{h}, \mathfrak{n}] \subset \mathfrak{n}$, $\mathfrak{h} + \mathfrak{n}$ is a subalgebra of $\mathfrak{n}$, and $[x_\alpha, \mathfrak{n}] \subset \mathfrak{h} + \mathfrak{n}$ for all $\alpha \in \mathrm{B}$. Put

$\mathfrak{a}' = \mathfrak{r} + \mathfrak{h} + \mathfrak{n}$. The preceding shows that $\mathfrak{a}'$ is stable under ad x_α, ad h_α and ad $x_{-\alpha}$ for all $\alpha \in \mathrm{B}$, and hence is an ideal of $\mathfrak{a}$. Since $\mathfrak{a}'$ contains x_α, h_α, $x_{-\alpha}$ for all $\alpha \in \mathrm{B}$, $\mathfrak{a}' = \mathfrak{a}$. It follows from this that the inclusions $\mathfrak{r} \subset \mathfrak{a}_+$, $\mathfrak{h} \subset \mathfrak{a}^0$, $\mathfrak{n} \subset \mathfrak{a}_-$ are equalities, which proves the proposition.

PROPOSITION 3. *The Lie algebra* $\mathfrak{a}_+$ (resp. $\mathfrak{a}_-$) *is a free Lie algebra with basic family* $(x_\alpha)_{\alpha\in\mathrm{B}}$ (resp. $(x_{-\alpha})_{\alpha\in\mathrm{B}}$) (cf. Chap. II, §2, no. 3).

Let L be the Lie subalgebra of E generated by B. By Chap. II, §3, Th. 1, L can be identified with the free Lie algebra generated by B. The left regular representation of E on itself is clearly injective, and defines by restriction to L an injective representation ρ' of the Lie algebra L on E. Let φ be the unique homomorphism from L to $\mathfrak{a}_-$ which takes α to $x_{-\alpha}$ for all $\alpha \in \mathrm{B}$. Then, for all $\alpha \in \mathrm{B}$, $\rho(\varphi(\alpha))$ is the endomorphism of left multiplication by α on E, so $\rho \circ \varphi = \rho'$, which proves that φ is injective. Thus, $(x_{-\alpha})_{\alpha\in\mathrm{B}}$ is a basic family for $\mathfrak{a}_-$. Since $\theta(x_{-\alpha}) = x_\alpha$ for all α (cf. Lemma 4), $(x_\alpha)_{\alpha\in\mathrm{B}}$ is a basic family for $\mathfrak{a}_+$.

3. EXISTENCE THEOREM

We retain the hypotheses and notation of the preceding number. Recall that if $\alpha, \beta \in \mathrm{B}$ and if $\alpha \neq \beta$, then $n(\beta,\alpha) \leq 0$; moreover, if $n(\beta,\alpha) = 0$, then $n(\alpha,\beta) = 0$ (Chap. VI, §1, no. 1, formula (8)). For any pair (α,β) of distinct elements of B, put

$$x_{\alpha\beta} = (\mathrm{ad}\ x_\alpha)^{1-n(\beta,\alpha)} x_\beta \qquad y_{\alpha\beta} = (\mathrm{ad}\ x_{-\alpha})^{1-n(\beta,\alpha)} x_{-\beta}.$$

Then $x_{\alpha\beta} \in \mathfrak{a}_+, y_{\alpha\beta} \in \mathfrak{a}_-$. If θ denotes the canonical automorphism of $\mathfrak{a}$, $\theta(x_{\alpha\beta}) = y_{\alpha\beta}$.

Lemma 6. Let $\alpha, \beta \in \mathrm{B}$ *with* $\alpha \neq \beta$. *Then*

$$[\mathfrak{a}_+, y_{\alpha\beta}] = 0 \qquad [\mathfrak{a}_-, x_{\alpha\beta}] = 0.$$

The second formula follows from the first by using the automorphism θ. To prove the first, it suffices to show that $[x_\gamma, y_{\alpha\beta}] = 0$ for all $\gamma \in \mathrm{B}$. We distinguish three cases.

Case 1: $\gamma \neq \alpha$ and $\gamma \neq \beta$. In this case, x_γ commutes with $x_{-\alpha}$ and $x_{-\beta}$, and hence with $y_{\alpha\beta}$.

Case 2: $\gamma = \beta$. In this case, x_γ commutes with $x_{-\alpha}$, so

$$\begin{aligned}[x_\gamma, y_{\alpha\beta}] &= (\mathrm{ad}\ x_{-\alpha})^{1-n(\beta,\alpha)}[x_\gamma, x_{-\beta}]\\ &= -(\mathrm{ad}\ x_{-\alpha})^{1-n(\beta,\alpha)} h_\beta = -n(\alpha,\beta)(\mathrm{ad}\ x_{-\alpha})^{-n(\beta,\alpha)} x_{-\alpha}.\end{aligned}$$

If $n(\beta,\alpha) < 0$, this expression is zero since $(\mathrm{ad}\ x_{-\alpha}).x_{-\alpha} = 0$. If $n(\beta,\alpha) = 0$, then $n(\alpha,\beta) = 0$. In both cases, $[x_\gamma, y_{\alpha\beta}] = 0$.

Case 3: $\gamma = \alpha$. In the algebra of endomorphisms of $\mathfrak{a}$,

$$[-\mathrm{ad}\ h_\alpha, \mathrm{ad}\ x_{-\alpha}] = 2\,\mathrm{ad}\ x_{-\alpha}$$

and $[\mathrm{ad}\ x_\alpha, \mathrm{ad}\ x_{-\alpha}] = -\mathrm{ad}\ h_\alpha$; thus, by §1, Lemma 1,

$$[\mathrm{ad}\ x_\alpha, (\mathrm{ad}\ x_{-\alpha})^{1-n(\beta,\alpha)}] = (1-n(\beta,\alpha))(\mathrm{ad}\ x_{-\alpha})^{-n(\beta,\alpha)}(-\mathrm{ad}\ h_\alpha - n(\beta,\alpha)).$$

Consequently,

$$\begin{aligned}[x_\gamma, y_{\alpha\beta}] &= [\mathrm{ad}\ x_\alpha, (\mathrm{ad}\ x_{-\alpha})^{1-n(\beta,\alpha)}]x_{-\beta} + (\mathrm{ad}\ x_{-\alpha})^{1-n(\beta,\alpha)}(\mathrm{ad}\ x_\alpha)x_{-\beta} \\ &= -(1-n(\beta,\alpha))(\mathrm{ad}\ x_{-\alpha})^{-n(\beta,\alpha)}(\mathrm{ad}\ h_\alpha + n(\beta,\alpha))x_{-\beta} \\ &\qquad + (\mathrm{ad}\ x_{-\alpha})^{1-n(\beta,\alpha)}(\mathrm{ad}\ x_\alpha)x_{-\beta}.\end{aligned}$$

Now $[h_\alpha, x_{-\beta}] + n(\beta,\alpha)x_{-\beta} = 0$ and $[x_\alpha, x_{-\beta}] = 0$, so $[x_\gamma, y_{\alpha\beta}] = 0$.

Lemma 7. The ideal $\mathfrak{n}$ *of* $\mathfrak{a}_+$ *generated by the* $x_{\alpha\beta}$ $(\alpha,\beta \in \mathrm{B}, \alpha \neq \beta)$ *is an ideal of* $\mathfrak{a}$. *The ideal of* $\mathfrak{a}_-$ *generated by the* $y_{\alpha\beta}$ $(\alpha,\beta \in \mathrm{B}, \alpha\neq\beta)$ *is an ideal of* $\mathfrak{a}$ *and is equal to* $\theta(\mathfrak{n})$.

Let $\mathfrak{n}' = \sum_{\alpha,\beta\in\mathrm{B},\alpha\neq\beta} kx_{\alpha\beta}$. Since each $x_{\alpha\beta}$ is homogeneous in $\mathfrak{a}$, $[\mathfrak{a}^0, \mathfrak{n}'] \subset \mathfrak{n}'$ (Lemma 5 and Prop. 2). Let U (resp. V) be the enveloping algebra of $\mathfrak{a}$ (resp. $\mathfrak{a}_+$), and σ the representation of U on $\mathfrak{a}$ defined by the adjoint representation of $\mathfrak{a}$. The ideal of $\mathfrak{a}$ generated by $\mathfrak{n}'$ is $\sigma(\mathrm{U})\mathfrak{n}'$. Now $\mathfrak{a} = \mathfrak{a}_+ + \mathfrak{a}^0 + \mathfrak{a}_-$ (Prop. 2), $\sigma(\mathfrak{a}_-)\mathfrak{n}' = 0$ (Lemma 6), and $\sigma(\mathfrak{a}^0)\mathfrak{n}' \subset \mathfrak{n}'$ by the preceding. By the Poincaré-Birkhoff-Witt theorem, $\sigma(\mathrm{U})\mathfrak{n}' = \sigma(\mathrm{V})\mathfrak{n}'$, which proves the first assertion of the lemma. It follows that the ideal of $\theta(\mathfrak{a}_+) = \mathfrak{a}_-$ generated by the $\theta(x_{\alpha\beta}) = y_{\alpha\beta}$ $(\alpha,\beta\in\mathrm{B},\alpha\neq\beta)$ is the ideal $\theta(\mathfrak{n})$ of $\mathfrak{a}$. Q.E.D.

The ideal $\mathfrak{n} + \theta(\mathfrak{n})$ of $\mathfrak{a}$ is graded since it is generated by homogeneous elements. Consequently, the Lie algebra $\mathfrak{a}/(\mathfrak{n}+\theta(\mathfrak{n}))$ is a Q-graded Lie algebra; in the remainder of this paragraph, it is denoted by $\mathfrak{g}_\mathrm{B}$, or simply by $\mathfrak{g}$. By Prop. 2, if $\mathfrak{g}^\mu \neq 0$ then $\mu \in \mathrm{Q}_+$, or $\mu\in\mathrm{Q}_-$, or $\mu = 0$. Denote by X_α (resp. H_α, $X_{-\alpha}$) the canonical image of x_α (resp. h_α, $x_{-\alpha}$) in $\mathfrak{g}$. In view of the definition of $\mathfrak{a}$, $\mathfrak{n}$ and $\theta(\mathfrak{n})$, it follows that $\mathfrak{g}$ is the Lie algebra defined by the generating family $((X_\alpha, H_\alpha, X_{-\alpha}))_{\alpha\in\mathrm{B}}$ and the relations

$$[H_\alpha, H_\beta] = 0 \tag{16}$$
$$[H_\alpha, X_\beta] - n(\beta,\alpha)X_\beta = 0 \tag{17}$$
$$[H_\alpha, X_{-\beta}] + n(\beta,\alpha)X_{-\beta} = 0 \tag{18}$$
$$[X_\alpha, X_{-\alpha}] + H_\alpha = 0 \tag{19}$$
$$[X_\alpha, X_{-\beta}] = 0 \quad (\alpha\neq\beta) \tag{20}$$
$$(\mathrm{ad}\ X_\alpha)^{1-n(\beta,\alpha)}X_\beta = 0 \quad (\alpha\neq\beta) \tag{21}$$
$$(\mathrm{ad}\ X_{-\alpha})^{1-n(\beta,\alpha)}X_{-\beta} = 0 \quad (\alpha\neq\beta). \tag{22}$$

Let $z\in\mathfrak{g}$ and $\mu\in\mathrm{Q}$. Then $z\in\mathfrak{g}^\mu$ if and only if $[H_\alpha, z] = \langle\mu, \alpha^\vee\rangle z$ for all $\alpha\in\mathrm{B}$. This follows from Lemma 5.

Since $\mathfrak{a}^0 \cap (\mathfrak{n} + \theta(\mathfrak{n})) = 0$, the canonical map from $\mathfrak{a}^0$ to $\mathfrak{g}^0$ is an isomorphism. Consequently, $(H_\alpha)_{\alpha\in B}$ is a basis of the vector space $\mathfrak{g}^0$. The commutative subalgebra $\mathfrak{g}^0$ of $\mathfrak{g}_B$ will be denoted by $\mathfrak{h}_B$ or simply by $\mathfrak{h}$. There exists a unique isomorphism $\mu \mapsto \mu_B$ from V to $\mathfrak{h}^*$ such that $\langle \mu_B, H_\alpha\rangle = \langle \mu, \alpha^\vee\rangle$ for all $\mu \in V$ and all $\alpha \in B$.

The involutive automorphism θ of $\mathfrak{a}$ defines by passage to the quotient an involutive automorphism of $\mathfrak{g}$ that will also be denoted by θ. We have $\theta(X_\alpha) = X_{-\alpha}$ for $\alpha \in B \cup (-B)$, and $\theta(H_\alpha) = -H_\alpha$.

THEOREM 1. *Let* R *be a reduced root system,* B *a basis of* R. *Let* $\mathfrak{g}$ *be the Lie algebra defined by the generating family* $((X_\alpha, H_\alpha, X_{-\alpha}))_{\alpha\in B}$ *and the relations* (16) *to* (22). *Let* $\mathfrak{h} = \sum_{\alpha\in B} kH_\alpha$. *Then* $(\mathfrak{g}, \mathfrak{h})$ *is a split semi-simple Lie algebra. The isomorphism* $\mu \mapsto \mu_B$ *from* V *to* $\mathfrak{h}^*$ *maps* R *to the root system of* $(\mathfrak{g}, \mathfrak{h})$. *For all* $\mu \in R$, $\mathfrak{g}^\mu$ *is the eigenspace relative to the root* μ.

The proof follows that of Lemmas 8, 9, 10, 11.

Lemma 8. Let $\alpha \in B \cup (-B)$. *Then* $\operatorname{ad} X_\alpha$ *is locally nilpotent.*[1]

Assume that $\alpha \in B$. Let $\mathfrak{g}'$ be the set of $z \in \mathfrak{g}$ such that $(\operatorname{ad} X_\alpha)^p z = 0$ for sufficiently large p. Since $\operatorname{ad} X_\alpha$ is a derivation of $\mathfrak{g}$, $\mathfrak{g}'$ is a subalgebra of $\mathfrak{g}$. By (21), $X_\beta \in \mathfrak{g}'$ for all $\beta \in B$. By (17), (19), (20), $H_\beta \in \mathfrak{g}'$ and $X_{-\beta} \in \mathfrak{g}'$ for all $\beta \in B$. Hence $\mathfrak{g}' = \mathfrak{g}$ and $\operatorname{ad} X_\alpha$ is locally nilpotent. Since $\operatorname{ad} X_{-\alpha} = \theta(\operatorname{ad} X_\alpha)\theta^{-1}$, we see that $\operatorname{ad} X_{-\alpha}$ is locally nilpotent.

We shall see that $\mathfrak{g}$ is finite dimensional, so that $\operatorname{ad} X_\alpha$ is actually nilpotent.

Lemma 9. Let $\mu, \nu \in Q$ *and* $w \in W(R)$ *be such that* $w\mu = \nu$. *There exists an automorphism of* $\mathfrak{g}$ *that takes* $\mathfrak{g}^\mu$ *to* $\mathfrak{g}^\nu$.

For all $\alpha \in B$, let s_α be the reflection in V defined by α. Since W(R) is generated by the s_α (Chap. VI, §1, no. 5, Remark 1), it suffices to prove the lemma when $w = s_\alpha$. In view of Lemma 8, we can define

$$\theta_\alpha = e^{\operatorname{ad} X_\alpha} e^{\operatorname{ad} X_{-\alpha}} e^{\operatorname{ad} X_\alpha}.$$

It is verified as in Chap. I, §6, no. 8, that θ_α is an automorphism of $\mathfrak{g}$. We have

---

[1] An endomorphism u of a vector space V is called *locally nilpotent* (or *almost nilpotent*) if, for every $v \in V$, there exists a positive integer n such that $u^n(v) = 0$ (cf. *Commutative Algebra*, Chap. IV, §1, no. 4, Def. 2). Then $\exp(u)$, or e^u, is defined by the formula $e^u(v) = \sum_{n\geq 0}(1/n!)u^n(v)$ for all $v \in V$.

$$\begin{aligned}\theta_\alpha(H_\alpha) &= e^{\operatorname{ad} X_\alpha} e^{\operatorname{ad} X_{-\alpha}}(H_\beta - n(\alpha,\beta)X_\alpha)\\ &= e^{\operatorname{ad} X_\alpha}\left(H_\beta - n(\alpha,\beta)X_\alpha + n(\alpha,\beta)X_{-\alpha} - n(\alpha,\beta)H_\alpha - \frac{n(\alpha,\beta)}{2}2X_{-\alpha}\right)\\ &= e^{\operatorname{ad} X_\alpha}(H_\beta - n(\alpha,\beta)H_\alpha - n(\alpha,\beta)X_\alpha)\\ &= H_\beta - n(\alpha,\beta)H_\alpha - n(\alpha,\beta)X_\alpha - n(\alpha,\beta)X_\alpha - n(\alpha,\beta)(-2X_\alpha)\\ &= H_\beta - n(\alpha,\beta)H_\alpha.\end{aligned}$$

If $z \in \mathfrak{g}^\mu$,

$$\begin{aligned}[H_\beta, \theta_\alpha^{-1} z] &= \theta_\alpha^{-1}[H_\beta - n(\alpha,\beta)H_\alpha, z]\\ &= \theta_\alpha^{-1}(\langle\mu,\beta^\vee\rangle z - n(\alpha,\beta)\langle\mu,\alpha^\vee\rangle z)\\ &= \langle\mu - \langle\alpha^\vee,\mu\rangle\alpha, \beta^\vee\rangle\theta_\alpha^{-1}z = \langle s_\alpha\mu, \beta^\vee\rangle\theta_\alpha^{-1}z,\end{aligned}$$

so $\theta_\alpha^{-1}z \in \mathfrak{g}^{s_\alpha\mu}$. This shows that $\theta_\alpha^{-1}\mathfrak{g}^\mu \subset \mathfrak{g}^{s_\alpha\mu}$. Since θ_α is an automorphism and since this inclusion holds for all $\mu \in \mathrm{Q}$, we see that $\theta_\alpha^{-1}\mathfrak{g}^\mu = \mathfrak{g}^{s_\alpha\mu}$, which proves the lemma.

Lemma 10. Let $\mu \in \mathrm{Q}$, and assume that μ is not a multiple of a root. There exists $w \in \mathrm{W(R)}$ such that certain of the coordinates of $w\mu$ with respect to the basis B *are* > 0 *and certain of them are* < 0.

Let $\mathrm{V}_{\mathbf{R}}$ be the vector space $\mathrm{Q} \otimes_{\mathbf{Z}} \mathbf{R}$, in which R is a root system. By the assumption, there exists $f \in \mathrm{V}_{\mathbf{R}}^*$ such that $\langle f, \alpha\rangle \neq 0$ for all $\alpha \in \mathrm{R}$, and $\langle f, \mu\rangle = 0$. There exists a chamber C of $\mathrm{R}^\vee$ such that $f \in \mathrm{C}$. By Chap. VI, §1, no. 5, Th. 2 (i), there exists $w \in \mathrm{W(R)}$ such that wf belongs to the chamber associated to B, in other words such that $\langle wf, \alpha\rangle > 0$ for all $\alpha \in \mathrm{B}$. Write $w\mu = \sum_{\alpha\in\mathrm{B}} t_\alpha \alpha$. Then

$$0 = \langle f, \mu\rangle = \langle wf, w\mu\rangle = \sum_{\alpha\in\mathrm{B}} t_\alpha\langle wf, \alpha\rangle,$$

which proves that certain t_α are > 0 and others are < 0.

Lemma 11. Let $\mu \in \mathrm{Q}$. If $\mu \notin \mathrm{R}\cup\{0\}$, then $\mathfrak{g}^\mu = 0$. If $\mu \in \mathrm{R}$, then $\dim \mathfrak{g}^\mu = 1$.

1) If μ is not a multiple of an element of R, there exists $w \in \mathrm{W}$ such that $w\mu \notin \mathrm{Q}_+ \cup \mathrm{Q}_-$ (Lemma 10), so $\mathfrak{a}^{w\mu} = 0$, $\mathfrak{g}^{w\mu} = 0$, and hence $\mathfrak{g}^\mu = 0$ (Lemma 9).

2) Let $\alpha \in \mathrm{B}$ and let m be an integer. Since $\mathfrak{a}_+$ is a free Lie algebra with basic family $(x_\alpha)_{\alpha\in\mathrm{B}}$, we have $\dim \mathfrak{a}^\alpha = 1$ and $\mathfrak{a}^{m\alpha} = 0$ for $m > 1$ (Chap. II, §2, no. 6, Prop. 4). Hence $\dim \mathfrak{g}^\alpha \leq 1$ and $\mathfrak{g}^{m\alpha} = 0$ for $m > 1$. We cannot have $\mathfrak{g}^\alpha = 0$, as this would imply that $x_\alpha \in \mathfrak{n} + \theta\mathfrak{n}$, and hence that $\mathfrak{n} + \theta\mathfrak{n}$ contains $h_\alpha = -[x_\alpha, x_{-\alpha}]$, whereas $\mathfrak{a}^0 \cap (\mathfrak{n} + \theta\mathfrak{n}) = 0$. Consequently, $\dim \mathfrak{g}^\alpha = 1$.

3) If $\mu \in \mathrm{R}$, there exists $w \in \mathrm{W(R)}$ such that $w(\mu) \in \mathrm{B}$ (Chap. VI, §1, no. 5, Prop. 15), so $\dim \mathfrak{g}^\mu = \dim \mathfrak{g}^{w\mu} = 1$. Moreover, if n is an integer > 1 then $\mathfrak{g}^{nw(\mu)} = 0$ and so $\mathfrak{g}^{n\mu} = 0$.

Proof of Theorem 1.

Since $\dim \mathfrak{g}^0 = \operatorname{Card} \mathrm{B}$, it follows from Lemma 11 that $\mathfrak{g}$ is of *finite dimension* equal to $\operatorname{Card} \mathrm{B} + \operatorname{Card} \mathrm{R}$. We show that $\mathfrak{g}$ is semi-simple. Let $\mathfrak{k}$ be a commutative ideal of $\mathfrak{g}$. Since $\mathfrak{k}$ is stable under $\operatorname{ad}(\mathfrak{h})$, $\mathfrak{k} = (\mathfrak{k} \cap \mathfrak{h}) + \sum_{\mu \in \mathrm{R}} (\mathfrak{k} \cap \mathfrak{g}^\mu)$. It is clear that, for all $\alpha \in \mathrm{B}$, $\mathfrak{g}^\alpha + \mathfrak{g}^{-\alpha} + kH_\alpha$ is isomorphic to $\mathfrak{sl}(2, k)$. In view of Lemma 9, for all $\mu \in \mathrm{R}$, $\mathfrak{g}^\mu$ is contained in a subalgebra of $\mathfrak{g}$ isomorphic to $\mathfrak{sl}(2, k)$; consequently, $\mathfrak{k} \cap \mathfrak{g}^\mu = 0$, so $\mathfrak{k} \subset \mathfrak{h}$; hence

$$[\mathfrak{k}, \mathfrak{g}^\mu] \subset \mathfrak{k} \cap \mathfrak{g}^\mu = 0,$$

so $\mu_{\mathrm{B}}(\mathfrak{k}) = 0$ for all $\mu \in \mathrm{R}$. It follows that $\mathfrak{k} = 0$, which proves that $\mathfrak{g}$ is semi-simple.

Let $\mu \in \mathrm{R}$. There exists $\alpha \in \mathrm{B}$ such that $\langle \mu, \alpha^\vee \rangle \neq 0$, and $(\operatorname{ad} H_\alpha) | \mathfrak{g}^\mu$ is then a non-zero homothety. Consequently, $\mathfrak{h}$ is equal to its own normalizer in $\mathfrak{g}$, and hence is a Cartan subalgebra of $\mathfrak{g}$. For all $u \in \mathfrak{h}$, $\operatorname{ad} u$ is diagonalizable, so $(\mathfrak{g}, \mathfrak{h})$ is a split semi-simple Lie algebra.

For all $\mu \in \mathrm{R}$, it is clear that μ_{B} is a root of $(\mathfrak{g}, \mathfrak{h})$ and that $\mathfrak{g}^\mu$ is the corresponding eigenspace. The number of roots of $(\mathfrak{g}, \mathfrak{h})$ is $\dim \mathfrak{g} - \dim \mathfrak{h} = \operatorname{Card} \mathrm{R}$. Hence, the map $\mu \mapsto \mu_{\mathrm{B}}$ from V to $\mathfrak{h}^*$ maps R to the root system of $(\mathfrak{g}, \mathfrak{h})$.

4. UNIQUENESS THEOREM

PROPOSITION 4. *Let $(\mathfrak{g}, \mathfrak{h}, \mathrm{B}, (X_\alpha)_{\alpha \in \mathrm{B}})$ be a framed semi-simple Lie algebra. Let $(n(\alpha, \beta))_{\alpha, \beta \in \mathrm{B}}$ and $(X_\alpha)_{\alpha \in \mathrm{B} \cup (-\mathrm{B})}$ be the corresponding Cartan matrix and generating family.*

(i) *The family $((X_\alpha, H_\alpha, X_{-\alpha}))_{\alpha \in \mathrm{B}}$ and the relations* (16) *to* (22) *of* no. 3 *constitute a presentation of $\mathfrak{g}$.*

(ii) *The family $(X_\alpha)_{\alpha \in \mathrm{B}}$ and the relations* (21) *of* no. 3 *constitute a presentation of the subalgebra of $\mathfrak{g}$ generated by $(X_\alpha)_{\alpha \in \mathrm{B}}$.*

Let R be the root system of $(\mathfrak{g}, \mathfrak{h})$. Applying to R and B the constructions of nos. 2 and 3, we obtain objects that we shall denote by $\mathfrak{a}', \mathfrak{g}', X'_\alpha, H'_\alpha, \ldots$ instead of $\mathfrak{a}, \mathfrak{g}, X_\alpha, H_\alpha, \ldots$.

There exists a homomorphism φ from the Lie algebra $\mathfrak{g}'$ to the Lie algebra $\mathfrak{g}$ such that $\varphi(X'_\alpha) = X_\alpha$, $\varphi(H'_\alpha) = H_\alpha$, $\varphi(X'_{-\alpha}) = X_{-\alpha}$ for all $\alpha \in \mathrm{B}$ (Prop. 1). Since $\dim \mathfrak{g}' = \operatorname{Card} \mathrm{R} + \operatorname{Card} \mathrm{B} = \dim \mathfrak{g}$, φ is bijective. This proves (i).

The subalgebra of $\mathfrak{g}' = \mathfrak{a}'/(\mathfrak{n}' \oplus \theta' \mathfrak{n}') = (\mathfrak{a}'_+ \oplus \mathfrak{a}'^0 \oplus \mathfrak{a}'_-)/(\mathfrak{n}' \oplus \theta' \mathfrak{n}')$ generated by $(X'_\alpha)_{\alpha \in \mathrm{B}}$ can be identified with $\mathfrak{a}'_+/\mathfrak{n}'$. In view of Prop. 3 and the definition of $\mathfrak{n}'$, this proves (ii).

COROLLARY. *Every framed semi-simple Lie algebra is obtained from a framed semi-simple $\mathbf{Q}$-Lie algebra by extension of scalars from $\mathbf{Q}$ to k.*

This follows immediately from the proposition.

THEOREM 2. *Let* $(\mathfrak{g}, \mathfrak{h}, \mathrm{B}, (X_\alpha)_{\alpha\in\mathrm{B}})$ *and* $(\mathfrak{g}', \mathfrak{h}', \mathrm{B}', (X'_\alpha)_{\alpha\in\mathrm{B}'})$ *be framed semi-simple Lie algebras, let* R *and* R′ *be the root systems of* $(\mathfrak{g}, \mathfrak{h})$ *and* $(\mathfrak{g}', \mathfrak{h}')$, *let* $(n(\alpha,\beta))_{\alpha,\beta\in\mathrm{B}}$ (resp. $(n'(\alpha,\beta))_{\alpha,\beta\in\mathrm{B}'}$) *be the Cartan matrix of* R (resp. R′) *relative to* B (resp. B′), *and let* Δ (resp. Δ') *be the Dynkin graph of* R (resp. R′) *relative to* B (resp. B′).

(i) *If* φ *is an isomorphism from* $\mathfrak{h}^*$ *to* $\mathfrak{h}'^*$ *such that* $\varphi(\mathrm{R}) = \mathrm{R}'$ *and* $\varphi(\mathrm{B}) = \mathrm{B}'$, *there exists a unique isomorphism* ψ *from* $(\mathfrak{g}, \mathfrak{h}, \mathrm{B}, (X_\alpha)_{\alpha\in\mathrm{B}})$ *to* $(\mathfrak{g}', \mathfrak{h}', \mathrm{B}', (X'_\alpha)_{\alpha\in\mathrm{B}'})$ *such that* $\psi|\mathfrak{h} = {}^t\varphi^{-1}$.

(ii) *If* f *is a bijection from* B *to* B′ *such that* $n'(f(\alpha), f(\beta)) = n(\alpha,\beta)$ *for all* $\alpha,\beta \in \mathrm{B}$, *there exists an isomorphism from* $(\mathfrak{g}, \mathfrak{h}, \mathrm{B}, (X_\alpha)_{\alpha\in\mathrm{B}})$ *to* $(\mathfrak{g}', \mathfrak{h}', \mathrm{B}', (X'_\alpha)_{\alpha\in\mathrm{B}'})$.

(iii) *If there exists an isomorphism from* Δ *to* Δ', *there exists an isomorphism from* $(\mathfrak{g}, \mathfrak{h}, \mathrm{B}, (X_\alpha)_{\alpha\in\mathrm{B}})$ *to* $(\mathfrak{g}', \mathfrak{h}', \mathrm{B}', (X'_\alpha)_{\alpha\in\mathrm{B}'})$.

This follows immediately from Prop. 4 (i) (making use of Chap. VI, §4, no. 2, Prop. 1 for part (iii)).

Scholium. To any splittable semi-simple Lie algebra $\mathfrak{g}$ is associated a Dynkin graph, which determines $\mathfrak{g}$ up to isomorphism (Th. 2 (iii)). This graph is non-empty and connected if and only if $\mathfrak{g}$ is simple (§3, no. 2, Cor. 1 of Prop. 6). By Th. 1 of no. 3, and Chap. VI, §4, no. 2, Th. 3, the splittable simple Lie algebras are the algebras of type A_l $(l \geq 1)$, B_l $(l \geq 2)$, C_l $(l \geq 3)$, D_l $(l \geq 4)$, E_6, E_7, E_8, F_4, G_2. No two algebras in this list are isomorphic.

PROPOSITION 5. *Let* $(\mathfrak{g}, \mathfrak{h}, \mathrm{B}, (X_\alpha)_{\alpha\in\mathrm{B}})$ *be a framed semi-simple Lie algebra, and* $(X_\alpha)_{\alpha\in\mathrm{B}\cup(-\mathrm{B})}$ *the corresponding generating family. There exists a unique automorphism* θ *of* $\mathfrak{g}$ *such that* $\theta(X_\alpha) = X_{-\alpha}$ *for all* $\alpha \in \mathrm{B}\cup(-\mathrm{B})$. *We have* $\theta^2 = \mathrm{Id}_{\mathfrak{g}}$, *and* $\theta(h) = -h$ *for all* $h \in \mathfrak{h}$.

The uniqueness is clear since $(X_\alpha)_{\alpha\in\mathrm{B}\cup(-\mathrm{B})}$ generates the Lie algebra $\mathfrak{g}$. In view of Prop. 4, the existence of θ follows from what we said in no. 3 before Th. 1.

COROLLARY. *Let* $(\mathfrak{g}, \mathfrak{h})$ *be a split semi-simple Lie algebra. Then* $(\mathfrak{g}, \mathfrak{h})$ *possesses a Chevalley system* (§2, no. 4, Def. 3).

Let R be the root system of $(\mathfrak{g}, \mathfrak{h})$. For all $\alpha \in \mathrm{R}$, let X_α be a non-zero element of $\mathfrak{g}^\alpha$. Assume that the X_α are chosen so that $[X_\alpha, X_{-\alpha}] = -H_\alpha$ for all $\alpha \in \mathrm{R}$ (§2, no. 4, Lemma 2). Let B be a basis of R and θ the automorphism of $\mathfrak{g}$ such that $\theta(X_\alpha) = X_{-\alpha}$ for all $\alpha \in \mathrm{B}\cup(-\mathrm{B})$. We have $\theta|\mathfrak{h} = -\mathrm{Id}_{\mathfrak{h}}$. Hence, for all $\alpha \in \mathrm{R}$ there exists $t_\alpha \in k^*$ such that $\theta X_\alpha = t_\alpha X_{-\alpha}$. We have

$$\begin{aligned} t_\alpha t_{-\alpha} H_\alpha &= [t_\alpha X_{-\alpha}, t_{-\alpha} X_\alpha] = [\theta X_\alpha, \theta X_{-\alpha}] = \theta([X_\alpha, X_{-\alpha}]) \\ &= \theta(-H_\alpha) = H_\alpha \end{aligned}$$

so $t_\alpha t_{-\alpha} = 1$ for all $\alpha \in \mathrm{R}$. Introduce the $\mathrm{N}_{\alpha\beta}$ as in §2, no. 4. If $\alpha, \beta, \alpha+\beta \in \mathrm{R}$,

$$\begin{aligned} N_{-\alpha,-\beta}t_\alpha t_\beta X_{-\alpha-\beta} &= t_\alpha t_\beta [X_{-\alpha}, X_{-\beta}] = [\theta X_\alpha, \theta X_\beta] = \theta([X_\alpha, X_\beta]) \\ &= N_{\alpha\beta}\theta X_{\alpha+\beta} = N_{\alpha\beta}t_{\alpha+\beta}X_{-\alpha-\beta} \end{aligned}$$

so, in view of §2, no. 4, Lemma 4,

$$(q+1)^2 t_\alpha t_\beta = N^2_{\alpha\beta} t_{\alpha+\beta}$$

where q is an integer. It follows that if t_α and t_β are squares in k^*, so is $t_{\alpha+\beta}$. Since $t_\alpha = 1$ for all $\alpha \in B$, Prop. 19 of Chap. VI, §1, no. 6, proves that t_α is a square for all $\alpha \in R$. Choose, for all $\alpha \in R$, a $u_\alpha \in k$ such that $u_\alpha^2 = t_\alpha$. This choice can be made so that $u_\alpha u_{-\alpha} = 1$ for all $\alpha \in R$. Put $X'_\alpha = u_\alpha^{-1} X_\alpha$. Then, for all $\alpha \in R$,

$$X'_\alpha \in \mathfrak{g}^\alpha, \ [X'_\alpha, X'_{-\alpha}] = [X_\alpha, X_{-\alpha}] = -H_\alpha,$$

and $\theta(X'_\alpha) = \theta(u_\alpha^{-1}X_\alpha) = u_\alpha^{-1}t_\alpha X_{-\alpha} = u_\alpha X_{-\alpha} = u_\alpha u_{-\alpha} X'_{-\alpha} = X'_{-\alpha}$, so that $(X'_\alpha)_{\alpha\in R}$ is a Chevalley system of $(\mathfrak{g}, \mathfrak{h})$.

§ 5. AUTOMORPHISMS OF A SEMI-SIMPLE LIE ALGEBRA

In this paragraph, $\mathfrak{g}$ denotes a semi-simple Lie algebra.

1. AUTOMORPHISMS OF A FRAMED SEMI-SIMPLE LIE ALGEBRA

Recall (Chap. VII, §3, no. 1) that $\mathrm{Aut}(\mathfrak{g})$ denotes the group of automorphisms of $\mathfrak{g}$. If $\mathfrak{h}$ is a Cartan subalgebra of $\mathfrak{g}$, we denote by $\mathrm{Aut}(\mathfrak{g}, \mathfrak{h})$ the group of automorphisms of $\mathfrak{g}$ that leave $\mathfrak{h}$ stable. Assume that $\mathfrak{h}$ is splitting, and let R be the root system of $(\mathfrak{g}, \mathfrak{h})$. If $s \in \mathrm{Aut}(\mathfrak{g}, \mathfrak{h})$, the contragredient map of $s|\mathfrak{h}$ is an element of A(R) (the group of automorphisms of R) which we shall denote by $\varepsilon(s)$ in this paragraph. Thus

$$\varepsilon : \mathrm{Aut}(\mathfrak{g}, \mathfrak{h}) \to A(R)$$

is a homomorphism of groups.

For any root system R and any basis B of R, we denote by Aut(R, B) the group of automorphisms of R that leave B stable. Recall (Chap. VI, §1, no. 5, Prop. 16 and §4, no. 2, Cor. of Prop. 1) that A(R) is the semi-direct product of Aut(R, B) and W(R), and that A(R)/W(R) is canonically isomorphic to the group of automorphisms of the Dynkin graph of R.

PROPOSITION 1. *Let $(\mathfrak{g}, \mathfrak{h}, B, (X_\alpha)_{\alpha\in B})$ be a framed semi-simple Lie algebra, and R the root system of $(\mathfrak{g}, \mathfrak{h})$. Let G be the set of $s \in \mathrm{Aut}(\mathfrak{g}, \mathfrak{h})$ that leave B stable, and such that $s(X_\alpha) = X_{\varepsilon(s)\alpha}$ for all $\alpha \in B$ (in other words* *the set of automorphisms of* $(\mathfrak{g}, \mathfrak{h}, \mathrm{B}, (X_\alpha)_{\alpha \in \mathrm{B}})$. *Then the restriction of* ε *to* G *is an isomorphism from* G *to* Aut(R, B).

If $s \in \mathrm{G}$, it is clear that $\varepsilon(s) \in \mathrm{Aut(R, B)}$. On the other hand, the map

$$\varepsilon|\mathrm{G} : \mathrm{G} \to \mathrm{Aut(R, B)}$$

is bijective by Th. 2 of §4, no. 4.

2. AUTOMORPHISMS OF A SPLIT SEMI-SIMPLE LIE ALGEBRA

Let E be a commutative group, and $\mathrm{A} = \bigoplus_{\gamma \in \mathrm{E}} \mathrm{A}^\gamma$ an E-graded algebra. For any homomorphism φ from the group E to the multiplicative group k^*, let $f(\varphi)$ be the k-linear map from A to A whose restriction to each A^γ is the homothety with ratio $\varphi(\gamma)$; it is clear that $f(\varphi)$ is an automorphism of the graded algebra A, and that f is a homomorphism from the group $\mathrm{Hom}(\mathrm{E}, k^*)$ to the group of automorphisms of the graded algebra A.

Let $\mathfrak{h}$ be a splitting Cartan subalgebra of $\mathfrak{g}$, and R the root system of $(\mathfrak{g}, \mathfrak{h})$. Recall that P(R) (resp. Q(R)) denotes the group of weights (resp. radical weights) of R. Put

$$\mathrm{T_P} = \mathrm{Hom}(\mathrm{P(R)}, k^*) \quad \mathrm{T_Q} = \mathrm{Hom}(\mathrm{Q(R)}, k^*).$$

We can consider $\mathfrak{g} = \mathfrak{g}^0 + \sum_{\alpha \in \mathrm{R}} \mathfrak{g}^\alpha$ as a Q(R)-graded algebra. The preceding remarks define a canonical homomorphism from $\mathrm{T_Q}$ to $\mathrm{Aut}(\mathfrak{g}, \mathfrak{h})$, which will be denoted by f in this paragraph. In the other hand, the canonical injection from Q(R) to P(R) defines a homomorphism from $\mathrm{T_P}$ to $\mathrm{T_Q}$, which will be denoted by q:

$$\mathrm{T_P} \xrightarrow{q} \mathrm{T_Q} \xrightarrow{f} \mathrm{Aut}(\mathfrak{g}, \mathfrak{h}).$$

If $s \in \mathrm{Aut}(\mathfrak{g}, \mathfrak{h})$, let s^* be the restriction of ${}^t(s|\mathfrak{h})^{-1}$ to Q(R). Then, for all $\varphi \in \mathrm{T_Q}$,

$$f(\varphi \circ s^*) = s^{-1} \circ f(\varphi) \circ s. \tag{1}$$

Indeed, let $\gamma \in \mathrm{Q(R)}$ and $x \in \mathfrak{g}^\gamma$; then $sx \in \mathfrak{g}^{s^*\gamma}$ and

$$f(\varphi \circ s^*)x = (\varphi \circ s^*)(\gamma).x = s^{-1}(\varphi(s^*\gamma)sx) = (s^{-1} \circ f(\varphi) \circ s)(x).$$

PROPOSITION 2. *The sequence of homomorphisms*

$$1 \longrightarrow \mathrm{T_Q} \xrightarrow{f} \mathrm{Aut}(\mathfrak{g}, \mathfrak{h}) \xrightarrow{\varepsilon} \mathrm{A(R)} \longrightarrow 1$$

is exact.

a) Let $\varphi \in \operatorname{Ker} f$. Then $\varphi(\alpha) = 1$ for all $\alpha \in \mathrm{R}$. Since R generates the group $\mathrm{Q(R)}$, φ is the identity element of $\mathrm{T_Q}$.

b) Let $\varphi \in \mathrm{T_Q}$. The restriction of $f(\varphi)$ to $\mathfrak{h} = \mathfrak{g}^0$ is the identity, so

$$\operatorname{Im} f \subset \operatorname{Ker} \varepsilon.$$

c) Let $s \in \operatorname{Ker} \varepsilon$. Then $s|\mathfrak{h} = \mathrm{Id}_{\mathfrak{h}}$. For all $\alpha \in \mathrm{R}$, we have $s(\mathfrak{g}^\alpha) = \mathfrak{g}^\alpha$, and there exists a $t_\alpha \in k^*$ such that $sx = t_\alpha x$ for all $x \in \mathfrak{g}^\alpha$. Writing down the condition that $s \in \operatorname{Aut}(\mathfrak{g})$, we obtain the relations

$$\begin{aligned} t_\alpha t_{-\alpha} &= 1 && \text{for all } \alpha \in \mathrm{R} \\ t_\alpha t_\beta &= t_{\alpha+\beta} && \text{when } \alpha, \beta, \alpha+\beta \in \mathrm{R}. \end{aligned}$$

Under these conditions, there exists $\varphi \in \mathrm{T_Q}$ such that $\varphi(\alpha) = t_\alpha$ for all $\alpha \in \mathrm{R}$ (Chap. VI, §1, no. 6, Cor. 2 of Prop. 19). Then $s = f(\varphi)$. Hence, $\operatorname{Ker} \varepsilon \subset \operatorname{Im} f$.

d) The image of $\operatorname{Aut}(\mathfrak{g}, \mathfrak{h})$ under ε contains W(R) by §2, no. 2, Cor. of Th. 2, and contains Aut(R, B) by Prop. 1. Hence this image is equal to A(R).

COROLLARY 1. *Let* $(\mathrm{B}, (X_\alpha)_{\alpha \in \mathrm{B}})$ *be a framing of* $(\mathfrak{g}, \mathfrak{h})$. *Let* G *be the set of* $s \in \operatorname{Aut}(\mathfrak{g}, \mathfrak{h})$ *that leave the framing invariant. Then* $\operatorname{Aut}(\mathfrak{g}, \mathfrak{h})$ *is the semi-direct product of* G *and* $\varepsilon^{-1}(\mathrm{W(R)})$.

Indeed, $\mathrm{G} \cap \varepsilon^{-1}(\mathrm{W(R)}) = \{1\}$ by Prop. 1, and

$$\operatorname{Aut}(\mathfrak{g}, \mathfrak{h}) = \mathrm{G}.\varepsilon^{-1}(\mathrm{W(R)})$$

since ε is surjective (Prop. 2).

COROLLARY 2. *The group* $\varepsilon^{-1}(\mathrm{W(R)})$ *operates simply-transitively on the set of framings of* $(\mathfrak{g}, \mathfrak{h})$.

Indeed, $\operatorname{Aut}(\mathfrak{g}, \mathfrak{h})$ operates transitively on the set of framings of $(\mathfrak{g}, \mathfrak{h})$ by §4, no. 4, Th. 2. Cor. 2 now follows from Cor. 1.

COROLLARY 3. *Let* B *be a basis of* R. *The group* $\operatorname{Ker} \varepsilon = f(\mathrm{T_Q})$ *operates simply-transitively on the set of framings of* $(\mathfrak{g}, \mathfrak{h})$ *of the form* $(\mathrm{B}, (X_\alpha)_{\alpha \in \mathrm{B}})$.

This follows immediately from Prop. 2.

Let $\alpha \in \mathrm{R}$, $X_\alpha \in \mathfrak{g}^\alpha$, $X_{-\alpha} \in \mathfrak{g}^{-\alpha}$ be such that $[X_\alpha, X_{-\alpha}] = -H_\alpha$. We have seen (§2, no. 2, Th. 2) that, for all $t \in k^*$, the restriction of the elementary automorphism

$$\theta_\alpha(t) = e^{\operatorname{ad} tX_\alpha} e^{\operatorname{ad} t^{-1} X_{-\alpha}} e^{\operatorname{ad} tX_\alpha}$$

to $\mathfrak{h}$ is the transpose of s_α; so $\varepsilon(\theta_\alpha(t)) = s_\alpha$ and consequently $\theta_\alpha(t)\theta_\alpha(-1) \in \operatorname{Ker} \varepsilon$.

Lemma 1. Let $\alpha \in \mathrm{R}$ *and* $t \in k^*$. *Let* φ *be the homomorphism* $\lambda \mapsto t^{\lambda(H_\alpha)}$ *from* Q(R) *to* k^*. *Then* $f(\varphi) = \theta_\alpha(t)\theta_\alpha(-1)$.

Let ρ be the representation of $\mathfrak{sl}(2,k)$ on $\mathfrak{g}$ associated to X_α. Let π be the representation of $\mathbf{SL}(2,k)$ compatible with ρ. Introduce the notations $\theta(t), h(t)$ of §1, no. 5. Since $\rho(H) = \operatorname{ad} H_\alpha$, the elements of $\mathfrak{g}^\lambda$ are of weight $\lambda(H_\alpha)$ for ρ. By §2, no. 2, $\theta_\alpha(t)\theta_\alpha(-1) = \pi(\theta(t)\theta(-1)) = \pi(h(t))$. Hence the restriction of $\theta_\alpha(t)\theta_\alpha(-1)$ to $\mathfrak{g}^\lambda$ is the homothety of ratio $t^{\lambda(H_\alpha)}$ (§1, no. 5, Prop. 6), hence the lemma.

PROPOSITION 3. *The image of the composite homomorphism*

$$\mathrm{T_P} \xrightarrow{q} \mathrm{T_Q} \xrightarrow{f} \mathrm{Aut}(\mathfrak{g},\mathfrak{h})$$

is contained in $\mathrm{Aut}_e(\mathfrak{g})$.

Let B be a basis of R. Then $(H_\alpha)_{\alpha\in\mathrm{B}}$ is a basis of $\mathrm{R}^\vee$, and the dual basis of $(H_\alpha)_{\alpha\in\mathrm{B}}$ in $\mathfrak{h}^*$ is a basis of the group P(R). Hence the group $\mathrm{T_P}$ is generated by the homomorphisms $\lambda \mapsto t^{\lambda(H_\alpha)}$ ($t \in k^*, \alpha \in \mathrm{B}$). If φ is the restriction of such a homomorphism to Q(R), Lemma 1 proves that $f(\varphi) \in \mathrm{Aut}_e(\mathfrak{g})$, hence the proposition.

Let $\bar{k}$ be an algebraic closure of k. The map which associates to any automorphism s of $\mathfrak{g}$ the automorphism $s \otimes 1$ of $\mathfrak{g} \otimes_k \bar{k}$ is an injective homomorphism from $\mathrm{Aut}(\mathfrak{g})$ to $\mathrm{Aut}(\mathfrak{g}\otimes_k \bar{k})$. *We denote by* $\mathrm{Aut}_0(\mathfrak{g})$ *the normal subgroup of* $\mathrm{Aut}(\mathfrak{g})$ *which is the inverse image of* $\mathrm{Aut}_e(\mathfrak{g}\otimes_k\bar{k})$ *under this homomorphism*; this is the set of automorphisms of $\mathfrak{g}$ that become elementary on extending the base field from k to $\bar{k}$. It is clear that $\mathrm{Aut}_e(\mathfrak{g})$ is independent of the choice of $\bar{k}$, and that $\mathrm{Aut}_e(\mathfrak{g}) \subset \mathrm{Aut}_0(\mathfrak{g})$. The groups $\mathrm{Aut}_0(\mathfrak{g})$ and $\mathrm{Aut}_e(\mathfrak{g})$ can be distinct (Chap. VII, §13, no. 1). If $\mathfrak{h}$ is a Cartan subalgebra of $\mathfrak{g}$, put

$$\mathrm{Aut}_e(\mathfrak{g},\mathfrak{h}) = \mathrm{Aut}_e(\mathfrak{g}) \cap \mathrm{Aut}(\mathfrak{g},\mathfrak{h}), \qquad \mathrm{Aut}_0(\mathfrak{g},\mathfrak{h}) = \mathrm{Aut}_0(\mathfrak{g}) \cap \mathrm{Aut}(\mathfrak{g},\mathfrak{h}).$$

Lemma 2. Let $\mathfrak{h}$ *be a splitting Cartan subalgebra of* $\mathfrak{g}$, *and* $s \in \mathrm{Aut}_0(\mathfrak{g},\mathfrak{h})$. *Assume that the restriction of* s *to* $\sum_{\alpha\in\mathrm{R}} \mathfrak{g}^\alpha$ *does not have* 1 *as an eigenvalue. Then* $\varepsilon(s) = 1$.

By extension of k, we are reduced to the case where $s \in \mathrm{Aut}_e(\mathfrak{g},\mathfrak{h})$. The dimension of the nilspace of $s-1$ is at least $\dim \mathfrak{h}$ (Chap. VII, §4, no. 4, Prop. 9). Hence $(s-1)|\mathfrak{h}$ is nilpotent. Since $s|\mathfrak{h} \in \mathrm{A}(\mathrm{R}^\vee)$, $s|\mathfrak{h}$ is of finite order, and hence semi-simple (Chap. V, Appendix, Prop. 2). Consequently, $(s-1)|\mathfrak{h} = 0$, which proves that $\varepsilon(s) = 1$.

Lemma 3. (i) *Let* $m = (\mathrm{P(R)} : \mathrm{Q(R)})$. *If* φ *is the* m*th power of an element of* $\mathrm{T_Q}$, *then* $\varphi \in q(\mathrm{T_P})$.

(ii) *If* k *is algebraically closed,* $q(\mathrm{T_P}) = \mathrm{T_Q}$.

There exist a basis $(\lambda_1, \dots, \lambda_l)$ of P(R) and integers $n_1 \geq 1, \dots, n_l \geq 1$ such that $(n_1\lambda_1, \dots, n_l\lambda_l)$ is a basis of Q(R). We have $m = n_1 \dots n_l$. Let $\psi \in \mathrm{T_Q}$ and put $\psi(n_1\lambda_1) = t_1, \dots, \psi(n_l\lambda_l) = t_l$. For $i = 1, \dots, l$, put $m_i =$

$\prod_{j \neq i} n_j$. Let χ be the element of $\mathrm{T_P}$ such that $\chi(\lambda_1) = t_1^{m_1}, \ldots, \chi(\lambda_l) = t_l^{m_l}$. Then

$$\chi(n_i \lambda_i) = t_i^{m_i n_i} = t_i^m = (\psi^m)(n_i \lambda_i)$$

so $\chi | \mathrm{Q(R)} = \psi^m$. This proves (i). If k is algebraically closed, every element of k^* is the mth power of an element of k^*, so every element of $\mathrm{T_Q}$ is the mth power of an element of $\mathrm{T_Q}$; hence, (ii) follows from (i).

PROPOSITION 4. *We have* $f(\mathrm{T_Q}) \subset \mathrm{Aut_0}(\mathfrak{g}, \mathfrak{h})$ *and* $\varepsilon^{-1}(\mathrm{W(R)}) = \mathrm{Aut_0}(\mathfrak{g}, \mathfrak{h})$.

a) Let $\varphi \in \mathrm{T_Q}$ and let $\bar{k}$ be an algebraic closure of k. By Lemma 3, φ extends to an element of $\mathrm{Hom(P(R)}, k^*)$. By Prop. 3,

$$f(\varphi) \otimes 1 \in \mathrm{Aut}_e(\mathfrak{g} \otimes_k \bar{k}, \mathfrak{h} \otimes_k \bar{k}).$$

Hence $f(\varphi) \in \mathrm{Aut_0}(\mathfrak{g}, \mathfrak{h})$, and $\mathrm{Ker}\, \varepsilon \subset \mathrm{Aut_0}(\mathfrak{g}, \mathfrak{h})$.

b) The image of $\mathrm{Aut}_e(\mathfrak{g}, \mathfrak{h})$ under ε contains W(R) (§2, no. 2, Cor. of Th. 2). In view of *a*), we see that $\varepsilon^{-1}(\mathrm{W(R)}) \subset \mathrm{Aut_0}(\mathfrak{g}, \mathfrak{h})$.

c) It remains to prove that $\mathrm{Aut_0}(\mathfrak{g}, \mathfrak{h}) \subset \varepsilon^{-1}(\mathrm{W(R)})$. In view of *b*), it suffices to prove that $\varepsilon(\mathrm{Aut_0}(\mathfrak{g}, \mathfrak{h})) \cap \mathrm{Aut(R, B)}$, where B denotes a basis of R, reduces to $\{1\}$.

Let $s \in \mathrm{Aut_0}(\mathfrak{g}, \mathfrak{h})$ be such that $\varepsilon(s) \in \mathrm{Aut(R, B)}$. The subgroup of A(R) generated by $\varepsilon(s)$ has a finite number of orbits on R. Let U be such an orbit, of cardinal r, and $\mathfrak{g}^{\mathrm{U}} = \sum_{\beta \in \mathrm{U}} \mathfrak{g}^\beta$. Let $\beta_1 \in \mathrm{U}$, and put $\beta_i = \varepsilon(s)^{i-1}\beta_1$ for $1 \leq i \leq r$, so that $\mathrm{U} = \{\beta_1, \ldots, \beta_r\}$. Let X_{β_1} be a non-zero element of $\mathfrak{g}^{\beta_1}$, and put $X_{\beta_i} = s^{i-1} X_{\beta_1}$ for $1 \leq i \leq r$. There exists $c_{\mathrm{U}} \in k^*$ such that $s^r X_{\beta_1} = c_{\mathrm{U}} X_{\beta_1}$, hence $s^r X_{\beta_i} = c_{\mathrm{U}} X_{\beta_i}$ for all i, and consequently $s^r | \mathfrak{g}^{\mathrm{U}} = c_{\mathrm{U}}.1$. Let $\varphi \in \mathrm{T_Q}$, and $s' = s \circ f(\varphi)$, which by *a*) is an element of $\mathrm{Aut_0}(\mathfrak{g}, \mathfrak{h})$. We have $s'^r | \mathfrak{g}^{\mathrm{U}} = c'_{\mathrm{U}}.1$, where

$$c'_{\mathrm{U}} = c_{\mathrm{U}} \prod_{i=1}^{r} \varphi(\beta_i) = c_{\mathrm{U}} \varphi\left(\sum_{i=1}^{r} \beta_i\right).$$

Put $\mathrm{B} = \{\alpha_1, \ldots, \alpha_l\}$ and $\sum_{i=1}^{r} \beta_i = \sum_{j=1}^{l} m_j^{\mathrm{U}} \alpha_j$. Since $\varepsilon(s) \in \mathrm{Aut(R, B)}$, the m_j^{U} are integers of the same sign and not all zero. We have

$$c'_{\mathrm{U}} = c_{\mathrm{U}} \prod_{j=1}^{l} \varphi(\alpha_j)^{m_j^{\mathrm{U}}}.$$

Now φ can be chosen so that $c'_{\mathrm{U}} \neq 1$ for every orbit U; indeed, this reduces to choosing elements $\varphi(\alpha_1) = t_1, \ldots, \varphi(\alpha_l) = t_l$ of k^* which are not annihilated by a finite number of polynomials in $t_1, \ldots, t_l$, not identically zero. For such a choice of φ, $\varepsilon(s') = 1$ by Lemma 2, so

$\varepsilon(s) = \varepsilon(s')\varepsilon(f(\varphi))^{-1} = 1.$

COROLLARY. *Let* B *be a basis of* R. *The group* $\mathrm{Aut}(\mathfrak{g}, \mathfrak{h})$ *is isomorphic to the semi-direct product of the groups* $\mathrm{Aut}(\mathrm{R}, \mathrm{B})$ *and* $\mathrm{Aut}_0(\mathfrak{g}, \mathfrak{h})$.

This follows from Prop. 1, Cor. 1 of Prop. 2, and Prop. 4.

Remark. Let $\varepsilon', \varepsilon''$ be the restrictions of ε to $\mathrm{Aut}_0(\mathfrak{g}, \mathfrak{h}), \mathrm{Aut}_e(\mathfrak{g}, \mathfrak{h})$. Let f' be the homomorphism from $\mathrm{T_P}$ to $\mathrm{Aut}_e(\mathfrak{g}, \mathfrak{h})$ induced by f via the canonical injection from Q(R) to P(R). In the preceding we have established the following commutative diagram:

$$\begin{array}{ccccccccc}
1 & \longrightarrow & \mathrm{T_Q} & \xrightarrow{f} & \mathrm{Aut}(\mathfrak{g}, \mathfrak{h}) & \xrightarrow{\varepsilon} & \mathrm{A(R)} & \longrightarrow & 1 \\
 & & \uparrow & & \uparrow & & \uparrow & & \\
1 & \longrightarrow & \mathrm{T_Q} & \xrightarrow{f} & \mathrm{Aut}_0(\mathfrak{g}, \mathfrak{h}) & \xrightarrow{\varepsilon'} & \mathrm{W(R)} & \longrightarrow & 1 \\
 & & \uparrow{\scriptstyle q} & & \uparrow & & \uparrow & & \\
 & & \mathrm{T_P} & \xrightarrow{f'} & \mathrm{Aut}_e(\mathfrak{g}, \mathfrak{h}) & \xrightarrow{\varepsilon''} & \mathrm{W(R)} & \longrightarrow & 1
\end{array}$$

in which the vertical arrows other than q denote the canonical injections. We have seen (Prop. 2 and 4) that the first two rows are exact. In the third row, the homomorphism ε'' is surjective (§2, no. 2, Cor. of Th. 2); it can be shown that its kernel is $f'(\mathrm{T_P})$ (§7, Exerc. 26 *d*)).

3. AUTOMORPHISMS OF A SPLITTABLE SEMI-SIMPLE LIE ALGEBRA

PROPOSITION 5. *Assume that* $\mathfrak{g}$ *is splittable. The group* $\mathrm{Aut}_0(\mathfrak{g})$ *operates simply-transitively on the set of framings of* $\mathfrak{g}$.

Let $e_1 = (\mathfrak{g}, \mathfrak{h}_1, \mathrm{B}_1, (X^1_\alpha)_{\alpha\in\mathrm{B}_1})$, $e_2 = (\mathfrak{g}, \mathfrak{h}_2, \mathrm{B}_2, (X^2_\alpha)_{\alpha\in\mathrm{B}_2})$ be two framings of $\mathfrak{g}$. There exists at least one element of $\mathrm{Aut}_0(\mathfrak{g})$ that transforms e_1 into e_2 (Prop. 1 and Prop. 4). Let $\bar{k}$ be an algebraic closure of k. There exists an element of $\mathrm{Aut}_e(\mathfrak{g} \otimes_k \bar{k})$ that transforms $\mathfrak{h}_1 \otimes_k \bar{k}$ into $\mathfrak{h}_2 \otimes_k \bar{k}$ (Chap. VII, §3, no. 2, Th. 1). Hence, by Prop. 4 and Cor. 2 of Prop. 2, there exists an element φ of $\mathrm{Aut}_e(\mathfrak{g} \otimes_k \bar{k})$ that transforms the framing $(\mathfrak{g} \otimes_k \bar{k}, \mathfrak{h}_1 \otimes_k \bar{k}, \mathrm{B}_1, (X^1_\alpha)_{\alpha\in\mathrm{B}_1})$ of $\mathfrak{g} \otimes_k \bar{k}$ into the framing $(\mathfrak{g} \otimes_k \bar{k}, \mathfrak{h}_2 \otimes_k \bar{k}, \mathrm{B}_2, (X^2_\alpha)_{\alpha\in\mathrm{B}_2})$.Since $\mathfrak{h}_1$ and the X^1_α (resp. $\mathfrak{h}_2$ and the X^2_α) generate $\mathfrak{g}_1$ (resp. $\mathfrak{g}_2$), we have $\varphi(\mathfrak{g}_1) = \mathfrak{g}_2$, so φ is of the form $\psi \otimes 1$ where $\psi \in \mathrm{Aut}_0(\mathfrak{g})$, and ψ transforms e_1 into e_2.

COROLLARY 1. *Let* $(\mathfrak{g}, \mathfrak{h}, \mathrm{B}, (X_\alpha)_{\alpha\in\mathrm{B}})$ *be a framing of* $\mathfrak{g}$, *and* G *the group (isomorphic to* $\mathrm{Aut}(\mathrm{R}, \mathrm{B})$*) of automorphisms of* $\mathfrak{g}$ *that leave this framing invariant. Then* $\mathrm{Aut}(\mathfrak{g})$ *is the semi-direct product of* G *and* $\mathrm{Aut}_0(\mathfrak{g})$.

Indeed, every element of $\mathrm{Aut}(\mathfrak{g})$ transforms $(\mathfrak{g}, \mathfrak{h}, \mathrm{B}, (X_\alpha)_{\alpha\in\mathrm{B}})$ into a framing of $\mathfrak{g}$. By Prop. 5, every coset of $\mathrm{Aut}(\mathfrak{g})$ modulo $\mathrm{Aut}_0(\mathfrak{g})$ meets G in exactly one point. Q.E.D.

It follows from Cor. 1 that the group $\mathrm{Aut}(\mathfrak{g})/\mathrm{Aut}_0(\mathfrak{g})$ can be identified with $\mathrm{Aut}(\mathrm{R},\mathrm{B})$, and is isomorphic to the group of automorphisms of the Dynkin graph of R.

COROLLARY 2. $\mathrm{Aut}(\mathfrak{g}) = \mathrm{Aut}_0(\mathfrak{g})$ *when* $\mathfrak{g}$ *is a splittable simple Lie algebra of type* A_1, B_n $(n \geq 2)$, C_n $(n \geq 2)$, E_7, E_8, F_4, G_2. *The quotient* $\mathrm{Aut}(\mathfrak{g})/\mathrm{Aut}_0(\mathfrak{g})$ *is of order 2 when* $\mathfrak{g}$ *is of type* A_n $(n \geq 2)$, D_n $(n \geq 5)$, E_6; *it is isomorphic to* $\mathfrak{S}_3$ *when* $\mathfrak{g}$ *is of type* D_4.

This follows from Cor. 1 and Chap. VI, Plates I to IX.

Remarks. 1) Let $e_1 = (\mathfrak{g}, \mathfrak{h}_1, \mathrm{B}_1, (X^1_\alpha)_{\alpha\in\mathrm{B}_1})$, $e_2 = (\mathfrak{g}, \mathfrak{h}_2, \mathrm{B}_2, (X^2_\alpha)_{\alpha\in\mathrm{B}_2})$, $e'_2 = (\mathfrak{g}, \mathfrak{h}_2, \mathrm{B}_2, (Y^2_\alpha)_{\alpha\in\mathrm{B}_2})$ be framings of $\mathfrak{g}$, and s (resp. s') an element of $\mathrm{Aut}_0(\mathfrak{g})$ that transforms e_1 to e_2 (resp. e'_2). Then $s|\mathfrak{h}_1 = s'|\mathfrak{h}_1$. Indeed, $s'^{-1}s \in \mathrm{Aut}_0(\mathfrak{g}, \mathfrak{h}_1)$ and $s'^{-1}s(\mathrm{B}_1) = \mathrm{B}_1$, so $\varepsilon(s'^{-1}s) = 1$.

2) Let X be the set of pairs $(\mathfrak{h}, \mathrm{B})$ where $\mathfrak{h}$ is a splitting Cartan subalgebra of $\mathfrak{g}$ and B a basis of the root system of $(\mathfrak{g}, \mathfrak{h})$. If $x = (\mathfrak{h}, \mathrm{B})$ and $x' = (\mathfrak{h}', \mathrm{B}')$ are two elements of X, there exists $s \in \mathrm{Aut}_0(\mathfrak{g})$ that transforms x into x' (Prop. 5), and the restriction $s_{x',x}$ of s to $\mathfrak{h}$ does not depend on the choice of s (Remark 1). In particular, $s_{x'',x'} \circ s_{x',x} = s_{x'',x}$ if $x, x', x'' \in \mathrm{X}$, and $s_{x,x} = 1$. The set of families $(h_x)_{x\in\mathrm{X}}$ satisfying the conditions

a) $h_x \in \mathfrak{h}$ if $x = (\mathfrak{h}, \mathrm{B})$

b) $s_{x',x}(h_x) = h_{x'}$ if $x, x' \in \mathrm{X}$

is in a natural way a vector space $\mathfrak{h}(\mathfrak{g})$ which we sometimes call *the canonical Cartan subalgebra of* $\mathfrak{g}$. For $x = (\mathfrak{h}, \mathrm{B})$ and $x' = (\mathfrak{h}', \mathrm{B}')$, $s_{x',x}$ takes B to B$'$, and hence the root system of $(\mathfrak{g}, \mathfrak{h})$ to that of $(\mathfrak{g}, \mathfrak{h}')$; it follows that the dual $\mathfrak{h}(\mathfrak{g})^*$ of $\mathfrak{h}(\mathfrak{g})$ is naturally equipped with a root system $\mathrm{R}(\mathfrak{g})$ and with a basis $\mathrm{B}(\mathfrak{g})$ of $\mathrm{R}(\mathfrak{g})$. We sometimes say that $\mathrm{R}(\mathfrak{g})$ is *the canonical root system* of $\mathfrak{g}$ and that $\mathrm{B}(\mathfrak{g})$ is its *canonical basis.* The group $\mathrm{Aut}(\mathfrak{g})$ operates on $\mathfrak{h}(\mathfrak{g})$ leaving $\mathrm{R}(\mathfrak{g})$ and $\mathrm{B}(\mathfrak{g})$ stable; the elements of $\mathrm{Aut}(\mathfrak{g})$ that operate trivially on $\mathfrak{h}(\mathfrak{g})$ are those of $\mathrm{Aut}_0(\mathfrak{g})$.

PROPOSITION 6. *Let* $\mathfrak{h}$ *be a splitting Cartan subalgebra of* $\mathfrak{g}$. *We have, with the notations in* no. 1, $\mathrm{Aut}_0(\mathfrak{g}) = \mathrm{Aut}_e(\mathfrak{g}).\mathrm{Ker}\,\varepsilon = \mathrm{Aut}_e(\mathfrak{g}).f(\mathrm{T_Q})$.

By §3, no. 3, Cor. of Prop. 10, $\mathrm{Aut}_0(\mathfrak{g}) = \mathrm{Aut}_e(\mathfrak{g}).\mathrm{Aut}_0(\mathfrak{g}, \mathfrak{h})$. On the other hand, $\varepsilon(\mathrm{Aut}_e(\mathfrak{g}, \mathfrak{h})) \supset \mathrm{W}(\mathrm{R})$ by §2, no. 2, Cor. of Th. 2, so $\mathrm{Aut}_0(\mathfrak{g}, \mathfrak{h}) = \mathrm{Aut}_e(\mathfrak{g}, \mathfrak{h}).\mathrm{Ker}\,\varepsilon$.

Remark 3. Prop. 6 shows that the canonical homomorphism

$$\iota : \mathrm{T_Q}/\mathrm{Im}(\mathrm{T_P}) \to \mathrm{Aut}_0(\mathfrak{g})/\mathrm{Aut}_e(\mathfrak{g}),$$

induced by the diagram in no. 2, is *surjective.* In particular, $\mathrm{Aut}_e(\mathfrak{g})$ contains the derived group of $\mathrm{Aut}_0(\mathfrak{g})$; we shall see (§11, no. 2, Prop. 3) that they are actually equal. Moreover, it can be shown that ι is *injective,* in other words that

$$f(\mathrm{T_Q}) \cap \mathrm{Aut}_e(\mathfrak{g}) = f'(\mathrm{T_P}),$$

(cf. §7, Exerc. 26 *d*)).

PROPOSITION 7. *Let* $\mathfrak{g}$ *be a splittable semi-simple Lie algebra,* $\mathfrak{b}$ *a Borel subalgebra of* $\mathfrak{g}$*, and* $\mathfrak{p}_1$ *and* $\mathfrak{p}_2$ *distinct parabolic subalgebras of* $\mathfrak{g}$ *containing* $\mathfrak{b}$*. Then* $\mathfrak{p}_1$ *and* $\mathfrak{p}_2$ *are not conjugate under* $\mathrm{Aut}_0(\mathfrak{g})$.

We can assume that k is algebraically closed. Let $s \in \mathrm{Aut}_0(\mathfrak{g})$ be such that $s(\mathfrak{p}_1) = \mathfrak{p}_2$. Let $\mathfrak{h}$ be a Cartan subalgebra of $\mathfrak{g}$ contained in $\mathfrak{b} \cap s(\mathfrak{b})$ (§3, no. 3, Prop. 10). Since $\mathfrak{h}$ and $s(\mathfrak{h})$ are Cartan subalgebras of $s(\mathfrak{b})$, there exists $u \in [\mathfrak{b}, \mathfrak{b}]$ such that $e^{\mathrm{ad}\, u}(\mathfrak{h}) = s(\mathfrak{h})$ (Chap. VII, §3, no. 4, Th. 3). Replacing s by $e^{-\mathrm{ad}\, u}s$, we are reduced to the case in which $s(\mathfrak{h}) = \mathfrak{h}$, and s then induces on $\mathfrak{h}$ an element σ of the Weyl group W of $(\mathfrak{g}, \mathfrak{h})$ (Prop. 4). Let C be the Weyl chamber corresponding to $\mathfrak{b}$. Then $\mathfrak{p}_1$ and $\mathfrak{p}_2$ correspond to facets F_1 and F_2 of $\mathfrak{h}_{\mathbf{R}}$ contained in the closure of C. We have $\sigma(\mathrm{F}_1) = \mathrm{F}_2$. Since $\sigma \in \mathrm{W}$, this implies that $\mathrm{F}_1 = \mathrm{F}_2$ (Chap. V, §3, no. 3, Th. 2) so $\mathfrak{p}_1 = \mathfrak{p}_2$.

Remark 4. Let $\mathfrak{g}$ be a splittable semi-simple Lie algebra, $\mathscr{P}$ the set of parabolic subalgebras of $\mathfrak{g}$, a set on which $\mathrm{Aut}_0(\mathfrak{g})$ operates. Retain the notations of Remark 2. Let Σ be a subset of $\mathrm{B}(\mathfrak{g})$. Giving Σ is equivalent to giving, for every $x = (\mathfrak{h}, \mathrm{B}) \in \mathrm{X}$, a subset Σ_x of B, such that $s_{x',x}$ takes Σ_x to $\Sigma_{x'}$ for any $x, x' \in \mathrm{X}$. Let $\mathfrak{p}_x$ be the parabolic subalgebra of $\mathfrak{g}$ corresponding to Σ_x (§3, no. 4, Remark). The orbit of $\mathfrak{p}_x$ under $\mathrm{Aut}_0(\mathfrak{g})$ is the set of $\mathfrak{p}_{x'}$ for $x' \in \mathrm{X}$. This defines a map from $\mathfrak{P}(\mathrm{B}(\mathfrak{g}))$ to $\mathscr{P}/\mathrm{Aut}_0(\mathfrak{g})$. This map is surjective by the Remark of §3, no. 4, and injective by Prop. 7.

4. ZARISKI TOPOLOGY ON $\mathrm{Aut}(\mathfrak{g})$

PROPOSITION 8. *Let* V *be the set of endomorphisms of the vector space* $\mathfrak{g}$. *Then* $\mathrm{Aut}(\mathfrak{g})$ *is closed in* V *for the Zariski topology* (Chap. VII, App. I).

Let K be the Killing form of $\mathfrak{g}$. If $s \in \mathrm{Aut}(\mathfrak{g})$,

$$[sx, sy] = [x, y] \tag{2}$$

$$\mathrm{K}(sx, sy) = \mathrm{K}(x, y) \tag{3}$$

for all $x, y \in \mathfrak{g}$. Conversely, let s be an element of V satisfying (2) and (3) for all $x, y \in \mathfrak{g}$. Then $\mathrm{Ker}(s) = 0$, so s is bijective and $s \in \mathrm{Aut}(\mathfrak{g})$. But, for all $x, y \in \mathfrak{g}$, the maps $s \mapsto [sx, sy]$ and $s \mapsto \mathrm{K}(sx, sy)$ from V to $\mathfrak{g}$ and k are polynomial.

PROPOSITION 9. *Let* $\mathfrak{h}$ *be a splitting Cartan subalgebra of* $\mathfrak{g}$.

(i) *The group* $f(\mathrm{T_Q})$ *is closed in* $\mathrm{Aut}(\mathfrak{g})$ *in the Zariski topology.*

(ii) *The group* $f(q(\mathrm{T_P}))$ *is dense in* $f(\mathrm{T_Q})$ *in the Zariski topology.*

Assertion (i) follows from the equality $f(\mathrm{T_Q}) = \mathrm{Aut}(\mathfrak{g}, \mathfrak{h}) \cap \mathrm{Ker}\, \varepsilon$ (Prop. 2). Put $m = (\mathrm{P(R)} : \mathrm{Q(R)})$. Let F be a polynomial function on V; we assume that F vanishes on the mth power of every element of $f(\mathrm{T_Q})$, and show that $\mathrm{F}|f(\mathrm{T_Q}) = 0$; in view of Lemma 3, this will prove (ii).

The set V′ of elements of V inducing the identity on $\mathfrak{h}$ and leaving each $\mathfrak{g}^\alpha$ stable can be identified with k^{R}. Let F′ be the restriction of F to $\mathrm{V}' = k^{\mathrm{R}}$; this is a polynomial function. We have $f(\mathrm{T_Q}) \subset \mathrm{V}'$. Let $\mathrm{B} = (\alpha_1, \dots, \alpha_l)$ be a basis of R. For all $t = (t_1, \dots, t_l) \in k^{*\mathrm{B}}$, let $\varphi(t)$ be the homomorphism from Q(R) to the group k^* that extends t. Then $\mathrm{F}'(f(\varphi(t)))$ can be written as a finite sum

$$\sum_{n_1,\dots,n_l \in \mathbf{Z}} c_{n_1,\dots,n_l} t_1^{n_1} \dots t_l^{n_l} = \mathrm{H}(t_1, \dots, t_l).$$

By assumption,

$$0 = \mathrm{H}(t_1^m, \dots, t_l^m) = \sum_{n_1,\dots,n_l \in \mathbf{Z}} c_{n_1,\dots,n_l} t_1^{mn_1} \dots t_l^{mn_l}$$

for all $t_1, \dots, t_l \in k^*$. The $c_{n_1,\dots,n_l}$ are thus the coefficients of a polynomial in l variables which vanishes on k^{*l}; hence they are all zero.

PROPOSITION 10. *Assume that* $\mathfrak{g}$ *is splittable.*

(i) *The group* $\mathrm{Aut}_e(\mathfrak{g})$ *is dense in* $\mathrm{Aut}_0(\mathfrak{g})$ *in the Zariski topology.*

(ii) *The groups* $\mathrm{Aut}_e(\mathfrak{g})$ *and* $\mathrm{Aut}_0(\mathfrak{g})$ *are connected in the Zariski topology.*

By Prop. 3, $f(q(\mathrm{T_P})) \subset \mathrm{Aut}_e(\mathfrak{g})$. For all $s \in \mathrm{Aut}_e(\mathfrak{g})$, the closure of $s.f(q(\mathrm{T_P}))$ in the Zariski topology contains $s.f(\mathrm{T_Q})$ by Prop. 9. Hence the closure of $\mathrm{Aut}_e(\mathfrak{g})$ contains $\mathrm{Aut}_e(\mathfrak{g}).f(\mathrm{T_Q}) = \mathrm{Aut}_0(\mathfrak{g})$ (Prop. 6). This proves (i).

Let $\mathrm{Aut}_e(\mathfrak{g}) = \Omega \cup \Omega'$ be a partition of $\mathrm{Aut}_e(\mathfrak{g})$ formed by relatively open subsets in the Zariski topology, and with $\Omega \neq \varnothing$. If $\omega \in \Omega$ and if x is a nilpotent element of $\mathfrak{g}$, the map $\tau : t \mapsto \omega \exp(t \operatorname{ad} x)$ from k to $\mathrm{Aut}_e(\mathfrak{g})$ is polynomial, hence continuous in the Zariski topology; consequently, $\tau(k)$ is connected; since $\omega \in \tau(k)$, we have $\tau(k) \subset \Omega$. Thus, $\Omega.(\exp \operatorname{ad} kx) \subset \Omega$, so $\Omega.\mathrm{Aut}_e(\mathfrak{g}) \subset \Omega$ and $\Omega = \mathrm{Aut}_e(\mathfrak{g})$. This proves that $\mathrm{Aut}_e(\mathfrak{g})$ is connected. It follows, by (i), that $\mathrm{Aut}_0(\mathfrak{g})$ is connected. Q.E.D.

We shall see (§8, no. 4, Cor. of Prop. 6) that $\mathrm{Aut}_0(\mathfrak{g})$ is closed in V in the Zariski topology, and that it is the connected component of the identity element of $\mathrm{Aut}(\mathfrak{g})$. On the other hand, $\mathrm{Aut}_e(\mathfrak{g})$ is not in general closed in the Zariski topology.

Assume that $(\mathfrak{g}, \mathfrak{h})$ is split. The group $\mathrm{Aut}_0(\mathfrak{g})$ is the group $\mathrm{G}(k)$ of k-points of a connected semi-simple algebraic group G with trivial centre (adjoint group). The group $f(\mathrm{T_Q})$ is equal to $\mathrm{H}(k)$, where H is the Cartan subgroup of G with Lie algebra $\mathfrak{h}$. The inverse image $\tilde{\mathrm{H}}$ of H in the universal covering $\tilde{\mathrm{G}}$ of G (in the algebraic sense) has $\mathrm{T_P}$ as its group of k-points. The image of $\tilde{\mathrm{G}}(k)$ in $\mathrm{G}(k) = \mathrm{Aut}_0(\mathfrak{g})$ is the group $\mathrm{Aut}_e(\mathfrak{g})$.

5. LIE GROUP CASE

PROPOSITION 11. *Assume that* k *is* $\mathbf{R}$, $\mathbf{C}$ *or a non-discrete complete ultrametric field. Let* $\mathfrak{h}$ *be a splitting Cartan subalgebra of* $\mathfrak{g}$.

(i) $\mathrm{Aut}(\mathfrak{g}, \mathfrak{h})$ *is a Lie subgroup of* $\mathrm{Aut}(\mathfrak{g})$ *with Lie algebra* $\mathrm{ad}\,\mathfrak{h}$.

(ii) $f(\mathrm{T_Q})$ *and* $(q \circ f)(\mathrm{T_P})$ *are open subgroups of* $\mathrm{Aut}(\mathfrak{g}, \mathfrak{h})$.

(iii) $\mathrm{Aut}_e(\mathfrak{g})$ *is an open subgroup of* $\mathrm{Aut}(\mathfrak{g})$.

(iv) *If* $k = \mathbf{R}$ *or* $\mathbf{C}$, $\mathrm{Aut}_e(\mathfrak{g})$ *is the identity component of* $\mathrm{Aut}(\mathfrak{g})$, *in other words* $\mathrm{Int}(\mathfrak{g})$.

By Chap. III, §3, no. 8, Cor. 2 of Prop. 29, and no. 10, Prop. 36, $\mathrm{Aut}(\mathfrak{g}, \mathfrak{h})$ is a Lie subgroup of $\mathrm{Aut}(\mathfrak{g})$ whose Lie algebra is the set of $\mathrm{ad}\,x$ ($x \in \mathfrak{g}$) such that $(\mathrm{ad}\,x)\mathfrak{h} \subset \mathfrak{h}$, in other words $\mathrm{ad}\,\mathfrak{h}$.

Let $\mathrm{H} \in \mathfrak{h}$. There exists $\varepsilon > 0$ with the following properties: for $t \in k$ and $|t| < \varepsilon$, $\exp(t\gamma(\mathrm{H}))$ is defined for all $\gamma \in \mathrm{P(R)}$, and the map $\gamma \mapsto \exp(t\gamma(\mathrm{H}))$ is a homomorphism σ_t from $\mathrm{P(R)}$ to k^*. For $|t| < \varepsilon$, $\exp(t\,\mathrm{ad}\,\mathrm{H})$ is defined, induces the identity on $\mathfrak{h}$ and induces on $\mathfrak{g}^\alpha$ the homothety with ratio $\sigma_t(\alpha)$; hence $\exp t\,\mathrm{ad}\,\mathrm{H} \in (q \circ f)(\mathrm{T_P})$. This proves, in view of (i), that $(q \circ f)(\mathrm{T_P})$ contains a neighbourhood of 1 in $\mathrm{Aut}(\mathfrak{g}, \mathfrak{h})$, and consequently is an open subgroup of $\mathrm{Aut}(\mathfrak{g}, \mathfrak{h})$. *A fortiori*, $f(\mathrm{T_Q})$ is an open subgroup of $\mathrm{Aut}(\mathfrak{g}, \mathfrak{h})$.

For all $\alpha \in \mathrm{R}$, $\exp \mathrm{ad}\,\mathfrak{g}^\alpha \subset \mathrm{Aut}_e(\mathfrak{g})$. In view of (ii), $\mathrm{Aut}_e(\mathfrak{g})$ contains a neighbourhood of 1 in $\mathrm{Aut}(\mathfrak{g})$, which proves (iii).

Assume that $k = \mathbf{R}$ or $\mathbf{C}$. Then $\mathrm{Aut}_e(\mathfrak{g})$ is contained in the identity component C of $\mathrm{Aut}(\mathfrak{g})$ (Chap. VII, §3, no. 1), and is open in $\mathrm{Aut}(\mathfrak{g})$ by (iii). Thus $\mathrm{Aut}_e(\mathfrak{g}) = \mathrm{C}$. Finally, $\mathrm{C} = \mathrm{Int}(\mathfrak{g})$ by Chap. III, §9, no. 8, Prop. 30 (i).

§ 6. MODULES OVER A SPLIT SEMI-SIMPLE LIE ALGEBRA

In this paragraph, $(\mathfrak{g}, \mathfrak{h})$ *denotes a split semi-simple Lie algebra,* R *its root system,* W *its Weyl group,* B *a basis of* R, R_+ (resp. R_-) *the set of positive* (resp. *negative*) *roots relative to* B. *Put*

$$\mathfrak{n}_+ = \sum_{\alpha \in \mathrm{R}_+} \mathfrak{g}^\alpha, \quad \mathfrak{n}_- = \sum_{\alpha \in \mathrm{R}_-} \mathfrak{g}^\alpha, \quad \mathfrak{b}_+ = \mathfrak{h} + \mathfrak{n}_+ \text{ and } \mathfrak{b}_- = \mathfrak{h} + \mathfrak{n}_-.$$

We have $\mathfrak{n}_+ = [\mathfrak{b}_+, \mathfrak{b}_+]$, $\mathfrak{n}_- = [\mathfrak{b}_-, \mathfrak{b}_-]$.

For all $\alpha \in \mathrm{R}$, choose an element $X_\alpha \in \mathfrak{g}^\alpha$ such that

$$[X_\alpha, X_{-\alpha}] = -H_\alpha$$

(§2, no. 4); none of the definitions below will depend on this choice.

1. WEIGHTS AND PRIMITIVE ELEMENTS

Let V be a $\mathfrak{g}$-module. For all $\lambda \in \mathfrak{h}^*$, denote by V^λ the primary subspace, relative to λ, of V *considered as an* $\mathfrak{h}$*-module* (Chap. VII, §1, no. 1). The elements of V^λ are called the elements of weight λ of the $\mathfrak{g}$-module V. The sum of the V^λ is direct (Chap. VII, §1, no. 1, Prop. 3). For all $\alpha \in \mathfrak{h}^*$ and $\lambda \in \mathfrak{h}^*$, $\mathfrak{g}^\alpha V^\lambda \subset V^{\alpha+\lambda}$ (Chap. VII, §1, no. 3, Prop. 10 (ii)). The dimension of V^λ is called the *multiplicity* of λ in V; if it is ≥ 1, i.e. if $V^\lambda \neq 0$, λ is said to be a *weight* of V. If V is finite dimensional, the homotheties of V defined by the elements of $\mathfrak{h}$ are semi-simple, so V^λ is the set of $x \in V$ such that $Hx = \lambda(H)x$ for all $H \in \mathfrak{h}$.

Lemma 1. Let V *be a* $\mathfrak{g}$*-module and* $v \in V$*. The following conditions are equivalent:*

(i) $\mathfrak{b}_+ v \subset kv$;

(ii) $\mathfrak{h}v \subset kv$ *and* $\mathfrak{n}_+ v = 0$;

(iii) $\mathfrak{h}v \subset kv$ *and* $\mathfrak{g}^\alpha v = 0$ *for all* $\alpha \in B$.

(i) $\Longrightarrow$ (ii): Assume that $\mathfrak{b}_+ v \subset kv$. There exists $\lambda \in \mathfrak{h}^*$ such that $v \in V^\lambda$. Let $\alpha \in R_+$. Then $\mathfrak{g}^\alpha . v \subset V^\lambda \cap V^{\lambda+\alpha} = 0$. Hence $\mathfrak{n}_+ v = 0$.

(ii) $\Longrightarrow$ (iii): This is clear.

(iii) $\Longrightarrow$ (i): This follows from the fact that $(X_\alpha)_{\alpha \in B}$ generates $\mathfrak{n}_+$ (§3, no. 3, Prop. 9 (iii)).

DEFINITION 1. *Let* V *be a* $\mathfrak{g}$*-module and* $v \in V$*. Then* v *is said to be a primitive element of* V *if* $v \neq 0$ *and* v *satisfies the conditions of Lemma* 1.

A primitive element belongs to one of the V^λ. For all $\lambda \in \mathfrak{h}^*$, V^λ_π denotes the set of $v \in V^\lambda$ such that $\mathfrak{b}_+ v \subset kv$. Thus, the primitive elements of weight λ are the non-zero elements of V^λ_π.

PROPOSITION 1. *Let* V *be a* $\mathfrak{g}$*-module,* v *a primitive element of* V *and* ω *the weight of* v*. Assume that* V *is generated by* v *as a* $\mathfrak{g}$*-module.*

(i) *If* $U(\mathfrak{n}_-)$ *denotes the enveloping algebra of* $\mathfrak{n}_-$*, we have* $V = U(\mathfrak{n}_-).v$.

(ii) *For all* $\lambda \in \mathfrak{h}^*$, V^λ *is the set of* $x \in V$ *such that* $Hx = \lambda(H)x$ *for all* $H \in \mathfrak{h}$*. We have* $V = \bigoplus_{\lambda \in \mathfrak{h}^*} V^\lambda$*, and each* V^λ *is finite dimensional. The space* V^ω *is of dimension* 1*, and every weight of* V *is of the form* $\omega - \sum_{\alpha \in B} n_\alpha . \alpha$*, where the* n_α *are integers* ≥ 0.

(iii) V *is an indecomposable* $\mathfrak{g}$*-module, and its commutant reduces to the scalars.*

(iv) *Let* $U(\mathfrak{g})$ *be the enveloping algebra of* $\mathfrak{g}$*, and* $\mathscr{Z}$ *the centre of* $U(\mathfrak{g})$*. There exists a unique homomorphism* χ *from* $\mathscr{Z}$ *to* k *such that, for all* $z \in \mathscr{Z}$*,* z_V *is the homothety with ratio* $\chi(z)$.

Let $U(\mathfrak{b}_+)$ be the enveloping algebra of $\mathfrak{b}_+$. We have $U(\mathfrak{g}) = U(\mathfrak{n}_-).U(\mathfrak{b}_+)$ (Chap. I, §2, no. 7, Cor. 6 of Th. 1). Hence

$$V = U(\mathfrak{g}).v = U(\mathfrak{n}_-).U(\mathfrak{b}_+).v = U(\mathfrak{n}_-).v.$$

Denote by $\alpha_1, \ldots, \alpha_n$ the distinct elements of R_+. Then

$$(X_{-\alpha_1}^{p_1} X_{-\alpha_2}^{p_2} \ldots X_{-\alpha_n}^{p_n})_{(p_1,\ldots,p_n)\in \mathbf{N}^n}$$

is a basis of $U(\mathfrak{n}_-)$, so

$$V = \sum_{(p_1,\ldots,p_n)\in \mathbf{N}^n} kX_{-\alpha_1}^{p_1} \ldots X_{-\alpha_n}^{p_n} v. \tag{1}$$

For $\lambda \in \mathfrak{h}^*$, put

$$T_\lambda = \sum_{(p_1,\ldots,p_n)\in \mathbf{N}^n,\ \omega - p_1\alpha_1 - \cdots - p_n\alpha_n = \lambda} kX_{-\alpha_1}^{p_1} \ldots X_{-\alpha_n}^{p_n} v.$$

By Chap. VII, §1, no. 1, Prop. 2 (ii), if $h \in \mathfrak{h}$, $h_V|T_\lambda$ is the homothety with ratio $\lambda(h)$. So $T_\lambda \subset V^\lambda$. On the other hand, (1) implies that

$$V = \sum_{\lambda \in \omega - \mathbf{N}\alpha_1 - \cdots - \mathbf{N}\alpha_n} T_\lambda.$$

The sum of the V^λ is direct (Chap. VII, §1, no. 1, Prop. 3). From these observations it follows that $V^\lambda = T_\lambda$, that V is the direct sum of the V^λ, and that V^λ is the set of $x \in V$ such that $hx = \lambda(h)x$ for all $h \in \mathfrak{h}$. On the other hand, $\dim V^\lambda$ is at most the cardinal of the set of $(p_1, \ldots, p_n) \in \mathbf{N}^n$ such that $p_1\alpha_1 + \cdots + p_n\alpha_n = \omega - \lambda$. This proves that $V^\lambda = 0$ if $\omega - \lambda \notin \sum_{\alpha \in B} \mathbf{N}\alpha$, that $\dim V^\omega = 1$, and that the V^λ are all finite dimensional.

Let c be an element of the commutant of V. For all $h \in \mathfrak{h}$,

$$hc(v) = ch(v) = \omega(h)c(v),$$

so $c(v) \in V^\omega$; hence there exists $t \in k$ such that $c(v) = tv$. Now, for all $(p_1, \ldots, p_n) \in \mathbf{N}^n$,

$$cX_{-\alpha_1}^{p_1} \ldots X_{-\alpha_n}^{p_n} v = X_{-\alpha_1}^{p_1} \ldots X_{-\alpha_n}^{p_n} cv = tX_{-\alpha_1}^{p_1} \ldots X_{-\alpha_n}^{p_n} v$$

so that $c = t.1$. Hence, the commutant of V reduces to the scalars. This implies (iv) and the fact that V is indecomposable.

DEFINITION 2. *The homomorphism* χ *of Prop.* 1 (iv) *is called the central character of the* $\mathfrak{g}$*-module* V.

PROPOSITION 2. *Let* V *be a* $\mathfrak{g}$*-module generated by a primitive element* e *of weight* ω*, and* X *a semi-simple* $\mathfrak{g}$*-module. Let* Φ *be the set of homomorphisms* *from the* $\mathfrak{g}$*-module* V *to the* $\mathfrak{g}$*-module* X. *Then* $\varphi \mapsto \varphi(e)$ *is an isomorphism from the vector space* Φ *to the vector space* X_π^ω.

It is clear that $\varphi(e) \in X_\pi^\omega$ for all $\varphi \in \Phi$. If $\varphi \in \Phi$ and $\varphi(e) = 0$, then $\varphi = 0$ since e generates the $\mathfrak{g}$-module V. We show that, if f is a non-zero element of X_π^ω, there exists $\varphi \in \Phi$ such that $\varphi(e) = f$. Let X′ be the submodule of X generated by f. By Prop. 1, X′ is indecomposable, hence simple since X is semi-simple. The element (e, f) is primitive in the $\mathfrak{g}$-module $V \times X$. Let N be the submodule of $V \times X$ generated by (e, f). Then $N \cap X \subset pr_2(N) = X'$, so $N \cap X = 0$ or X′; if $N \cap X = X'$, N contains the linearly independent elements (e, f) and $(0, f)$ which are primitive of weight ω; this is absurd (Prop. 1), so $N \cap X = 0$. Thus $pr_1|N$ is an injective map h from N to V; this map is surjective since its image contains e. Thus $\varphi = pr_2 \circ h^{-1}$ is a homomorphism from the $\mathfrak{g}$-module V to the $\mathfrak{g}$-module X such that $\varphi(e) = f$.

2. SIMPLE MODULES WITH A HIGHEST WEIGHT

Recall that fixing B defines an order relation on $\mathfrak{h}_{\mathbf{Q}}^*$ (Chap. VI, §1, no. 6). The elements of $\mathfrak{h}_{\mathbf{Q}}^*$ that are ≥ 0 are the linear combinations of elements of B with rational coefficients ≥ 0.

More generally, we shall consider the following order relation between elements $\lambda, \mu \in \mathfrak{h}^*$:

$\lambda - \mu$ is a linear combination of elements of B with rational coefficients ≥ 0.

Lemma 2. Let V *be a simple* $\mathfrak{g}$*-module,* ω *a weight of* V. *The following conditions are equivalent:*

(i) *every weight of* V *is of the form* $\omega - \mu$ *where* μ *is a radical weight* ≥ 0;

(ii) ω *is the highest weight of* V;

(iii) *for all* $\alpha \in B$, $\omega + \alpha$ *is not a weight of* V;

(iv) *there exists a primitive element of weight* ω.

(i) $\Longrightarrow$ (ii) $\Longrightarrow$ (iii): This is clear.

(iii) $\Longrightarrow$ (iv): Assume that condition (iii) is satisfied. For all $h \in \mathfrak{h}$,

$$\mathrm{Ker}(h_V - \omega(h))$$

is non-zero, contained in V^ω, and stable under $\mathfrak{h}_V$. By induction on dim $\mathfrak{h}$, we see that there exists a non-zero v in V^ω such that $\mathfrak{h}v \subset kv$. Condition (iii) implies that $\mathfrak{n}_+ v = 0$, so v is primitive.

(iv) $\Longrightarrow$ (i): Let v be a primitive element of weight ω. Since V is simple, V is generated by v as a $\mathfrak{g}$-module. Assertion (i) now follows from Prop. 1.

Q.E.D.

Thus, for any simple $\mathfrak{g}$-module, the existence of a primitive element is equivalent to that of a highest weight, or to that of a maximal weight.

There exist simple $\mathfrak{sl}(2, \mathbf{C})$-modules V that have no weights for any Cartan subalgebra $\mathfrak{h}$ of $\mathfrak{sl}(2, \mathbf{C})$ (§1, Exerc. 14 *f*)). These modules are of infinite dimension over $\mathbf{C}$ (§1, no. 3, Th. 1).

PROPOSITION 3. *Let* V *be a simple* $\mathfrak{g}$*-module with a highest weight* ω.

(i) *The primitive elements of* V *are the non-zero elements of* V^ω.

(ii) V *is semi-simple as an* $\mathfrak{h}$*-module.*

(iii) *We have* $V = \bigoplus_{\lambda \in \mathfrak{h}^*} V^\lambda$. *For all* $\lambda \in \mathfrak{h}^*$, V^λ *is finite dimensional. We have* $\dim V^\omega = 1$.

(iv) *The* $\mathfrak{g}$*-module* V *is absolutely simple.*

Assertions (i), (ii) and (iii) follow from Prop. 1 and Lemma 2. Assertion (iv) follows from Prop. 1 (iii) and *Algebra*, Chap. VIII, §7, no. 3.

COROLLARY. *If* V *is finite dimensional, the canonical homomorphism* $U(\mathfrak{g}) \to \mathrm{End}(V)$ *is surjective.*

This follows from (iv), cf. *Algebra*, Chap. VIII, §3, no. 3.

PROPOSITION 4. *Let* V *be a simple* $\mathfrak{g}$*-module with a highest weight* ω, X *a semi-simple* $\mathfrak{g}$*-module, and* X′ *the isotypical component of type* V *in* X. *Then* X′ *is the submodule of* X *generated by* X^ω_π. *Its length is equal to the dimension of* X^ω_π.

Let X″ be the submodule of X generated by X^ω_π. It is clear that every submodule of X isomorphic to V is contained in X″. Hence $X' \subset X''$. On the other hand, let Φ be the set of homomorphisms from the $\mathfrak{g}$-module V to the $\mathfrak{g}$-module X. The length of X′ is $\dim_k \Phi$ (*Algebra*, Chap. VIII, §4, no. 4), that is $\dim_k X^\omega_\pi$ (Prop. 2).

3. EXISTENCE AND UNIQUENESS THEOREM

Let $\lambda \in \mathfrak{h}^*$. Since $\mathfrak{b}_+ = \mathfrak{h} \oplus \mathfrak{n}_+$ and since $\mathfrak{n}_+ = [\mathfrak{b}_+, \mathfrak{b}_+]$, the map $h + n \mapsto \lambda(h)$ (where $h \in \mathfrak{h}$, $n \in \mathfrak{n}_+$) from $\mathfrak{b}_+$ to k is a 1-dimensional representation of $\mathfrak{b}_+$. Denote by L_λ the k-vector space k equipped with the $\mathfrak{b}_+$-module structure defined by this representation. Let $U(\mathfrak{g})$, $U(\mathfrak{b}_+)$ be the enveloping algebras of $\mathfrak{g}$, $\mathfrak{b}_+$, so that $U(\mathfrak{b}_+)$ is a subalgebra of $U(\mathfrak{g})$; recall that $U(\mathfrak{g})$ is a free right $U(\mathfrak{b}_+)$-module (Chap. I, §2, no. 7, Cor. 5 of Th. 1). Put

$$Z(\lambda) = U(\mathfrak{g}) \otimes_{U(\mathfrak{b}_+)} L_\lambda. \qquad (2)$$

Then $Z(\lambda)$ is a left $\mathfrak{g}$-module. Denote by e the element $1 \otimes 1$ of $Z(\lambda)$.

PROPOSITION 5. (i) *The element* e *of* $Z(\lambda)$ *is primitive of weight* λ *and generates the* $\mathfrak{g}$*-module* $Z(\lambda)$.

(ii) *Let* $Z^+(\lambda) = \sum_{\lambda \neq \mu} Z(\lambda)^\mu$. *Every submodule of* $Z(\lambda)$ *distinct from* $Z(\lambda)$ *is contained in* $Z^+(\lambda)$.

(iii) *There exists a largest submodule* F_λ *of* $Z(\lambda)$ *distinct from* $Z(\lambda)$. *The quotient module* $Z(\lambda)/F_\lambda$ *is simple and has highest weight* λ.

It is clear that e generates the $\mathfrak{g}$-module $Z(\lambda)$. If $x \in \mathfrak{b}_+$,

$$x.e = (x.1) \otimes 1 = (1.x) \otimes 1 = 1 \otimes x.1 = \lambda(x)(1 \otimes 1) = \lambda(x)e,$$

hence (i).

The $\mathfrak{h}$-module $Z(\lambda)$ is semi-simple (Prop. 1). If G is a $\mathfrak{g}$-submodule of $Z(\lambda)$, then $G = \sum_{\mu \in \mathfrak{h}^*} (G \cap Z(\lambda)^\mu)$. The hypothesis $G \cap Z(\lambda)^\lambda \neq 0$ implies that $G = Z(\lambda)$, since $\dim Z(\lambda)^\lambda = 1$ and e generates the $\mathfrak{g}$-module $Z(\lambda)$. If $G \neq Z(\lambda)$, then $G = \sum_{\mu \neq \lambda} G \cap Z(\lambda)^\mu \subset Z^+(\lambda)$.

Let F_λ be the sum of the $\mathfrak{g}$-submodules of $Z(\lambda)$ distinct from $Z(\lambda)$. By (ii), $F_\lambda \subset Z^+(\lambda)$. Hence F_λ is the largest submodule of $Z(\lambda)$ distinct from $Z(\lambda)$. It is clear that $Z(\lambda)/F_\lambda$ is simple and that the canonical image of e in $Z(\lambda)/F_\lambda$ is primitive of weight λ.

In the remainder of this chapter, the $\mathfrak{g}$-*module* $Z(\lambda)/F_\lambda$ *of Prop.* 5 *will be denoted by* $E(\lambda)$.

THEOREM 1. *Let* $\lambda \in \mathfrak{h}^*$. *The* $\mathfrak{g}$-*module* $E(\lambda)$ *is simple and has highest weight* λ. *Every simple* $\mathfrak{g}$-*module of highest weight* λ *is isomorphic to* $E(\lambda)$.

The first assertion follows from Prop. 5 (iii). The second follows from Prop. 4.

PROPOSITION 6. *Let* V *be a* $\mathfrak{g}$-*module,* λ *an element of* $\mathfrak{h}^*$ *and* v *a primitive element of* V *of weight* λ.

(i) *There exists a unique homomorphism of* $\mathfrak{g}$-*modules* $\psi : Z(\lambda) \to V$ *such that* $\psi(e) = v$.

(ii) *Assume that* v *generates* V. *Then* ψ *is surjective. Moreover,* ψ *is bijective if and only if, for every non-zero element* u *of* $U(\mathfrak{n}_-)$, u_V *is injective.*

(iii) *The map* $u \mapsto u \otimes 1$ *from* $U(\mathfrak{n}_-)$ *to* $Z(\lambda)$ *is bijective.*

Let K be the kernel of the representation of $U(\mathfrak{b}_+)$ on L_λ; it is of codimension 1 in $U(\mathfrak{b}_+)$. Let $J = U(\mathfrak{g})K$ be the left ideal of $U(\mathfrak{g})$ generated by K; then L_λ can be identified with $U(\mathfrak{b}_+)/K$ as a left $U(\mathfrak{b}_+)$-module, and $Z(\lambda)$ can be identified with $U(\mathfrak{g})/J$ as a left $U(\mathfrak{g})$-module. We have $K.v = 0$, so $J.v = 0$, which proves (i).

Now assume that v generates V. It is clear that ψ is surjective.

By the Poincaré-Birkhoff-Witt theorem (Chap. I, §2, no. 7, Cor. 6 of Th. 1), a basis of $U(\mathfrak{n}_-)$ over k is also a basis of $U(\mathfrak{g})$ as a right $U(\mathfrak{b}_+)$-module. Hence the map $\varphi : u \mapsto u \otimes 1$ from $U(\mathfrak{n}_-)$ to $U(\mathfrak{g}) \otimes_{U(\mathfrak{b}_+)} L_\lambda$ is bijective. Let $u \in U(\mathfrak{n}_-)$. Then $\varphi^{-1} \circ u_{Z(\lambda)} \circ \varphi$ is left multiplication by u on

$\mathrm{U}(\mathfrak{n}_-)$. In view of Chap. I, §2, no. 7, Cor. 7 of Th. 1, $u_{\mathrm{Z}(\lambda)}$ is injective if $u \neq 0$. Consequently, if ψ is bijective, then u_{V} is injective for non-zero u in $\mathrm{U}(\mathfrak{n}_-)$.

Assume that ψ is not injective. There exists $u \in \mathrm{U}(\mathfrak{n}_-)$ such that $u \neq 0$ and $\psi(\varphi(u)) = 0$. Then

$$u_{\mathrm{V}}.v = u_{\mathrm{V}}.\psi(1 \otimes 1) = \psi(u \otimes 1) = \psi(\varphi(u)) = 0.$$

COROLLARY 1. *Let $\lambda \in \mathfrak{h}^*$ and $\alpha \in \mathrm{B}$ be such that $\lambda(H_\alpha) + 1 \in \mathbf{N}$. Then $\mathrm{Z}(-\alpha + s_\alpha\lambda)$ is isomorphic to a $\mathfrak{g}$-submodule of $\mathrm{Z}(\lambda)$.*

Put $m = \lambda(H_\alpha)$. Let $x = \mathrm{X}_{-\alpha}^{m+1}.e \in \mathrm{Z}(\lambda)$, and let V be the submodule of $\mathrm{Z}(\lambda)$ generated by x. Then $x \neq 0$ (Prop. 6). On the other hand, $x \in \mathrm{Z}(\lambda)^{\lambda-(m+1)\alpha}$. For $\beta \in \mathrm{B}$ and $\beta \neq \alpha$, $[\mathfrak{g}^{-\alpha}, \mathfrak{g}^\beta] = 0$ and $\mathfrak{g}^\beta.e = 0$, so $\mathfrak{g}^\beta.x = 0$. Finally, since $[X_\alpha, X_{-\alpha}] = -H_\alpha$, we have

$$[X_\alpha, X_{-\alpha}^{m+1}] = (m+1)X_{-\alpha}^m(-H_\alpha + m)$$

(§1, no. 1, Lemma 1 (ii)), so

$$X_\alpha.x = X_\alpha X_{-\alpha}^{m+1}.e = [X_\alpha, X_{-\alpha}^{m+1}].e = (m+1)X_{-\alpha}^m(me - \lambda(H_\alpha)e) = 0.$$

Thus, x is primitive of weight $\lambda - (m+1)\alpha$. In view of Prop. 6, the $\mathfrak{g}$-module V is isomorphic to $\mathrm{Z}(-\alpha + \lambda - m\alpha) = \mathrm{Z}(-\alpha + s_\alpha\lambda)$.

COROLLARY 2. *Let $\rho = \frac{1}{2}\sum_{\alpha\in\mathrm{R}_+} \alpha$, and $\lambda, \mu \in \mathfrak{h}^*$. Assume that $\lambda + \rho$ is a dominant weight in R, and that there exists $w \in \mathrm{W}$ with $\mu + \rho = w(\lambda + \rho)$. Then $\mathrm{Z}(\mu)$ is isomorphic to a submodule of $\mathrm{Z}(\lambda)$.*

The assertion is clear when $w = 1$. Assume that it is established whenever w is of length $< q$. If w is of length q, there exists $\alpha \in \mathrm{B}$ such that $w = s_\alpha w'^{-1}$, with $l(w') = q - 1$. We have $w'(\alpha) \in \mathrm{R}_+$ (Chap. VI, §1, no. 6, Cor. 2 of Prop. 17), and hence $w'^{-1}(\lambda+\rho)(H_\alpha) = (\lambda+\rho)(H_{w'\alpha})$ is an integer ≥ 0. Put

$$\mu' = w'^{-1}(\lambda + \rho) - \rho.$$

By the induction hypothesis, $\mathrm{Z}(\mu')$ is isomorphic to a submodule of $\mathrm{Z}(\lambda)$. On the other hand, by Chap. VI, §1, no. 10, Prop. 29 (ii),

$$-\alpha + s_\alpha\mu' = -\alpha + s_\alpha w'^{-1}(\lambda + \rho) - s_\alpha\rho = w(\lambda + \rho) - \rho = \mu.$$

Moreover, $\rho(H_\alpha) = 1$ (Chap. VI, §1, Prop. 29 (iii)), so $\mu'(H_\alpha)+1 \in \mathbf{N}$. Cor. 1 now implies that $\mathrm{Z}(\mu)$ is isomorphic to a submodule of $\mathrm{Z}(\mu')$, and hence also to a submodule of $\mathrm{Z}(\lambda)$.

4. COMMUTANT OF $\mathfrak{h}$ IN THE ENVELOPING ALGEBRA OF $\mathfrak{g}$

Let U be the enveloping algebra of $\mathfrak{g}$, $V \subset U$ the enveloping algebra of $\mathfrak{h}$. The algebra V can be identified with the symmetric algebra $\mathbf{S}(\mathfrak{h})$ of $\mathfrak{h}$, and also with the algebra of polynomial functions on $\mathfrak{h}^*$. Denote by $\alpha_1, \dots, \alpha_n$ the pairwise distinct positive roots. Let $(H_1, \dots, H_l)$ be a basis of $\mathfrak{h}$. By the Poincaré-Birkhoff-Witt theorem, the elements

$$u((q_i), (m_i), (p_i)) = X_{-\alpha_1}^{q_1} \dots X_{-\alpha_n}^{q_n} H_1^{m_1} \dots H_l^{m_l} X_{\alpha_1}^{p_1} \dots X_{\alpha_n}^{p_n}$$

(q_i, m_i, p_i integers ≥ 0) form a basis of the vector space U. For all $h \in \mathfrak{h}$, we have

$$[h, u((q_i),(m_i),(p_i))] = ((p_1 - q_1)\alpha_1 + \dots + (p_n - q_n)\alpha_n)(h) u((q_i),(m_i),(p_i)). \quad (3)$$

The vector space U is a $\mathfrak{g}$-module (hence also an $\mathfrak{h}$-module) under the adjoint representation. If $\lambda \in \mathfrak{h}^*$, the subspaces U^λ and U_λ are defined (Chap. VII, §1, no. 3); formula (3) shows that $U^\lambda = U_\lambda$ and that $U = \bigoplus_{\lambda \in Q} U^\lambda$ (where Q is the group of radical weights of R). In particular, U^0 is the commutant of $\mathfrak{h}$, or of V, in U.

Lemma 3. Put $L = (\mathfrak{n}_- U) \cap U^0$.

(i) *We have* $L = (U\mathfrak{n}_+) \cap U^0$, *and* L *is a two-sided ideal of* U^0.

(ii) *We have* $U^0 = V \oplus L$.

It is clear that $\mathfrak{n}_- U$ (resp. $U\mathfrak{n}_+$) is the set of linear combinations of the elements $u((q_i), (m_i), (p_i))$ such that $\sum q_i > 0$ (resp. $\sum p_i > 0$). On the other hand

$$u((q_i), (m_i), (p_i)) \in U^0 \iff p_1\alpha_1 + \dots + p_n\alpha_n = q_1\alpha_1 + \dots + q_n\alpha_n.$$

This implies that $(\mathfrak{n}_- U) \cap U^0 = (U\mathfrak{n}_+) \cap U^0$. Finally, $(\mathfrak{n}_- U) \cap U^0$ (resp. $(U\mathfrak{n}_+) \cap U^0$) is a right (resp. left) ideal of U^0, hence (i). Further, an element $u((q_i), (m_i), (p_i))$ that is in U^0 belongs to V (resp. to L) if and only if $p_1 = \dots = p_n = q_1 = \dots = q_n = 0$ (resp. $p_1 + \dots + p_n + q_1 + \dots + q_n > 0$), hence (ii). Q.E.D.

In view of Lemma 3, the projection of U^0 onto V with kernel L is a homomorphism of algebras. It is called the *Harish-Chandra homomorphism* from U^0 to V (relative to B). Recall that V can be identified with the algebra of polynomial functions on $\mathfrak{h}^*$.

PROPOSITION 7. *Let* $\lambda \in \mathfrak{h}^*$, E *a* $\mathfrak{g}$-*module generated by a primitive element of weight* λ, χ *the central character of* E, *and* φ *the Harish-Chandra homomorphism from* U^0 *to* V. *Then,* $\chi(z) = (\varphi(z))(\lambda)$ *for all* z *in the centre of* U.

Let v be a primitive element of E of weight λ, and z an element of the centre of U. There exist $u_1, \dots, u_p \in \mathrm{U}$ and $n_1, \dots, n_p \in \mathfrak{n}_+$ such that $z = \varphi(z) + u_1 n_1 + \dots + u_p n_p$. Then

$$\chi(z)v = zv = \varphi(z)v + u_1 n_1 v + \dots + u_p n_p v = \varphi(z)v = (\varphi(z))(\lambda)v.$$

COROLLARY. *Let $\langle \cdot, \cdot \rangle$ be a non-degenerate invariant symmetric bilinear form on $\mathfrak{g}$, C the Casimir element associated to $\langle \cdot, \cdot \rangle$. Denote also by $\langle \cdot, \cdot \rangle$ the inverse form on $\mathfrak{h}^*$ of the restriction of $\langle \cdot, \cdot \rangle$ to $\mathfrak{h}$* (§2, no. 3, Prop. 5). *Then $\chi(\mathrm{C}) = \langle \lambda, \lambda + 2\rho \rangle$, where $\rho = \frac{1}{2} \sum_{\alpha \in \mathrm{R}_+} \alpha$.*

We recall the notations of §2, no. 3, Prop. 6. We have

$$\mathrm{C} = \sum_{\alpha \in \mathrm{R}_-} \frac{1}{\langle X_\alpha, X_{-\alpha} \rangle} X_\alpha X_{-\alpha} + \sum_{\alpha \in \mathrm{R}_+} \frac{1}{\langle X_\alpha, X_{-\alpha} \rangle} X_{-\alpha} X_\alpha$$
$$+ \sum_{\alpha \in \mathrm{R}_+} \frac{1}{\langle X_\alpha, X_{-\alpha} \rangle} [X_\alpha, X_{-\alpha}] + \sum_{i \in I} H_i H_i'$$

so

$$\varphi(\mathrm{C}) = \sum_{\alpha \in \mathrm{R}_+} \frac{1}{\langle X_\alpha, X_{-\alpha} \rangle} [X_\alpha, X_{-\alpha}] + \sum_{i \in I} H_i H_i'.$$

By Prop. 7,

$$\chi(\mathrm{C}) = \sum_{\alpha \in \mathrm{R}_+} \frac{1}{\langle X_\alpha, X_{-\alpha} \rangle} \lambda([X_\alpha, X_{-\alpha}]) + \sum_{i \in I} \lambda(H_i)\lambda(H_i').$$

Let h_λ be the element of $\mathfrak{h}$ such that $\langle h_\lambda, h \rangle = \lambda(h)$ for all $h \in \mathfrak{h}$. By §2, no. 2, Prop. 1,

$$\lambda\left(\frac{1}{\langle X_\alpha, X_{-\alpha} \rangle}[X_\alpha, X_{-\alpha}]\right) = \left\langle h_\lambda, \frac{1}{\langle X_\alpha, X_{-\alpha} \rangle}[X_\alpha, X_{-\alpha}] \right\rangle = \alpha(h_\lambda) = \langle \lambda, \alpha \rangle.$$

Hence

$$\chi(\mathrm{C}) = \left(\sum_{\alpha \in \mathrm{R}_+} \langle \lambda, \alpha \rangle \right) + \langle \lambda, \lambda \rangle = \langle \lambda, \lambda + 2\rho \rangle.$$

§ 7. FINITE DIMENSIONAL MODULES OVER A SPLIT SEMI-SIMPLE LIE ALGEBRA

In this paragraph, we retain the general notations of §6. *We denote by* P (resp. Q) *the group of weights of* R (resp. *radical weights of* R). *We denote by* P_+ (resp. Q_+) *the set of elements of* P (resp. Q) *that are positive for the order relation defined by* B. *We denote by* P_{++} *the set of dominant weights of* R *relative to* B (Chap. VI, §1, no. 10). *An element* λ *of* $\mathfrak{h}^*$ *belongs to* P (resp. *to* P_{++}) *if and only if all the* $\lambda(H_\alpha)$, $\alpha \in \mathrm{B}$, *are integers* (resp. *integers* ≥ 0). *We have* $\mathrm{P}_{++} \subset \mathrm{P}_+$ (Chap. VI, §1, no. 6). *If* $w \in \mathrm{W}$, *we denote by* $\varepsilon(w)$ *the determinant of* w, *which is equal to* 1 *or* -1. *We put* $\rho = \frac{1}{2} \sum_{\alpha \in \mathrm{R}_+} \alpha$.

1. WEIGHTS OF A FINITE DIMENSIONAL SIMPLE $\mathfrak{g}$-MODULE

PROPOSITION 1. *Let* V *be a finite dimensional* $\mathfrak{g}$*-module.*

(i) *All the weights of* V *belong to* P.

(ii) $\mathrm{V} = \bigoplus_{\mu \in \mathrm{P}} \mathrm{V}^\mu$.

(iii) *For all* $\mu \in \mathfrak{h}^*$, V^μ *is the set of* $x \in \mathrm{V}$ *such that* $h.x = \mu(h)x$ *for all* $h \in \mathfrak{h}$.

For all $\alpha \in \mathrm{B}$, there exists a homomorphism from $\mathfrak{sl}(2, k)$ to $\mathfrak{g}$ that takes H to H_α. Thus, by §1, no. 2, Cor. of Prop. 2, $(H_\alpha)_\mathrm{V}$ is diagonalizable and its eigenvalues are integers. Hence, the set of $(H_\alpha)_\mathrm{V}$, for $\alpha \in \mathrm{B}$, is diagonalizable (*Algebra*, Chap. VII, §5, no. 6, Prop. 13). Consequently, for all $h \in \mathfrak{h}$, h_V is diagonalizable. By Chap. VII, §1, no. 3, Prop. 9, $\mathrm{V} = \bigoplus_{\mu \in \mathfrak{h}^*} \mathrm{V}^\mu$. On the other hand, if $\mathrm{V}^\mu \neq 0$, the preceding shows that $\mu(H_\alpha) \in \mathbf{Z}$ for all $\alpha \in \mathrm{B}$, so $\mu \in \mathrm{P}$. This proves (i) and (ii). We see in the same way that $\mathfrak{h}_\mathrm{V}$ is diagonalizable, hence (iii).

COROLLARY. *Let* ρ *be a finite dimensional representation of* $\mathfrak{g}$ *and* Φ *the bilinear form associated to* ρ.

(i) *If* $x, y \in \mathfrak{h}_\mathbf{Q}$, *then* $\Phi(x, y) \in \mathbf{Q}$ *and* $\Phi(x, x) \in \mathbf{Q}_+$.

(ii) *If* ρ *is injective, the restriction of* Φ *to* $\mathfrak{h}$ *is non-degenerate.*

Assertion (i) follows from Prop. 1 since the elements of P have rational values on $\mathfrak{h}_\mathbf{Q}$. If ρ is injective, Φ is non-degenerate (Chap. I, §6, no. 1, Prop. 1), so the restriction of Φ to $\mathfrak{h}$ is non-degenerate (Chap. VII, §1, no. 3, Prop. 10 (iii)).

Lemma 1. Let V *be a* $\mathfrak{g}$*-module and* ρ *the corresponding representation of* $\mathfrak{g}$.

(i) *If* a *is a nilpotent element of* $\mathfrak{g}$, *and if* $\rho(a)$ *is locally nilpotent,*

$$\rho(e^{\operatorname{ad} a}b) = e^{\rho(a)}\rho(b)e^{-\rho(a)}$$

for all $b \in \mathfrak{g}$.

(ii) *If* $\alpha \in \mathrm{R}$ *and if the images under* ρ *of the elements of* $\mathfrak{g}^{\alpha}$ *and* $\mathfrak{g}^{-\alpha}$ *are locally nilpotent, the set of weights of* V *is stable under the reflection* s_α.

With the assumptions in (i), we have $\rho((\operatorname{ad} a)^n b) = (\operatorname{ad}\rho(a))^n\rho(b)$ for all $n \geq 0$, so $\rho(e^{\operatorname{ad} a}b) = e^{\operatorname{ad}\rho(a)}\rho(b)$. On the other hand,

$$e^{\operatorname{ad}\rho(a)}\rho(b) = e^{\rho(a)}\rho(b)e^{-\rho(a)}$$

is assertion (ii) of Chap. VII, §3, no. 1, Lemma 1.

We now adopt the assumptions in (ii). Let $\theta_\alpha = e^{\operatorname{ad} X_\alpha}e^{\operatorname{ad} X_{-\alpha}}e^{\operatorname{ad} X_\alpha}$. By (i), there exists $\mathrm{S} \in \mathbf{GL}(\mathrm{V})$ such that $\rho(\theta_\alpha b) = \mathrm{S}\rho(b)\mathrm{S}^{-1}$ for all $b \in \mathfrak{g}$. Now $\theta_\alpha|\mathfrak{h}$ is the transpose of s_α (§2, no. 2, Lemma 1). Let λ be a weight of V. There exists a non-zero element x of V such that $\rho(h)x = \lambda(h)x$ for all $h \in \mathfrak{h}$. Then

$$\rho(h)\mathrm{S}^{-1}x = \mathrm{S}^{-1}\rho({}^t s_\alpha h)x = \mathrm{S}^{-1}\lambda({}^t s_\alpha h)x = (s_\alpha\lambda)(h)\mathrm{S}^{-1}x$$

for all $h \in \mathfrak{h}$. Consequently, $s_\alpha\lambda$ is a weight of V.

PROPOSITION 2. *Let* V *be a finite dimensional* $\mathfrak{g}$*-module and* $s \in \mathrm{Aut}_0(\mathfrak{g})$.

(i) *There exists* $\mathrm{S} \in \mathbf{GL}(\mathrm{V})$ *such that* $(s(x))_{\mathrm{V}} = \mathrm{S}x_{\mathrm{V}}\mathrm{S}^{-1}$ *for all* $x \in \mathfrak{g}$.

(ii) *If* $s \in \mathrm{Aut}_e(\mathfrak{g})$, S *can be chosen to be an element of* $\mathbf{SL}(\mathrm{V})$ *leaving stable all the* $\mathfrak{g}$*-submodules of* V.

Assertion (ii) follows from Lemma 1 (i). Now let $s \in \mathrm{Aut}_0(\mathfrak{g})$ and denote by ρ the representation of $\mathfrak{g}$ defined by V. By (ii), the representations ρ and $\rho \circ s$ become equivalent after extension of scalars. They are therefore equivalent (Chap. I, §3, no. 8, Prop. 13), hence the existence of S.

Remark 1. Let S satisfy the condition in Prop. 2 (i), and let $\mathfrak{h}' = s(\mathfrak{h})$; denote by s^* the isomorphism $\lambda \mapsto \lambda \circ s^{-1}$ from $\mathfrak{h}^*$ to $\mathfrak{h}'^*$. It is clear that

$$\mathrm{S}(\mathrm{V}^{\lambda}) = \mathrm{V}^{s^*\lambda}.$$

In particular:

COROLLARY 1. *The isomorphism* s^* *takes the weights of* V *with respect to* $\mathfrak{h}$ *to those of* V *with respect to* $\mathfrak{h}'$; *corresponding weights have the same multiplicity.*

COROLLARY 2. *Let* $w \in \mathrm{W}$. *For all* $\lambda \in \mathfrak{h}^*$, *the vector subspaces* V^{λ} *and* $\mathrm{V}^{w\lambda}$ *have the same dimension. The set of weights of* V *is stable under* W.

Indeed, w is of the form s^* with $s \in \mathrm{Aut}_e(\mathfrak{g}, \mathfrak{h})$ (§2, no. 2, Cor. of Th. 2).

Remark 2. By Cor. 1 of Prop. 2 and §5, no. 3, Remark 2, it makes sense to speak of the weights of V with respect to the canonical Cartan subalgebra of $\mathfrak{g}$, and of their multipicities.

Remark 3. Lemma 1 (i) and Prop. 2 remain valid, with the same proof, even if $\mathfrak{g}$ is not assumed to be splittable.

2. HIGHEST WEIGHT OF A FINITE DIMENSIONAL SIMPLE $\mathfrak{g}$-MODULE

THEOREM 1. *A simple $\mathfrak{g}$-module is finite dimensional if and only if it has a highest weight belonging to* P_{++}.

We denote by V a simple $\mathfrak{g}$-module and by $\mathscr{X}$ its set of weights.

a) Assume that V is finite dimensional. Then $\mathscr{X}$ is finite and non-empty (Prop. 1) and so has a maximal element ω. Let $\alpha \in \mathrm{B}$. Then $\omega + \alpha \notin \mathscr{X}$, which proves that ω is the highest weight of V (§6, no. 2, Lemma 2). On the other hand, there exists a homomorphism from $\mathfrak{sl}(2,k)$ to $\mathfrak{g}$ that takes H to H_α; by §1, Prop. 2 (i), $\omega(H_\alpha)$ is an integer ≥ 0, so $\omega \in \mathrm{P}_{++}$.

b) Assume that V has a highest weight $\omega \in \mathrm{P}_{++}$. Let $\alpha \in \mathrm{B}$ and let e be a primitive element of weight ω in V. Put $e_j = X_{-\alpha}^j e$ for $j \geq 0$, $m = \omega(H_\alpha) \in \mathbf{N}$, and $\mathrm{N} = \sum_{j=0}^{m} ke_j$. By §1, no. 2, Prop. 1, $X_\alpha e_{m+1} = 0$. If $\beta \in \mathrm{B}$ and $\beta \neq \alpha$, then $[X_\beta, X_{-\alpha}] = 0$ so $X_\beta e_{m+1} = X_\beta X_{-\alpha}^{m+1} e = X_{-\alpha}^{m+1} X_\beta e = 0$. If $e_{m+1} \neq 0$, we conclude that e_{m+1} is primitive, which is absurd (§6, Prop. 3 (i)); so $e_{m+1} = 0$. Thus, by §1, no. 2, Cor. of Prop. 1, N is stable under the subalgebra $\mathfrak{s}_\alpha$ generated by H_α, X_α and $X_{-\alpha}$. Now $\mathfrak{s}_\alpha$ is reductive in $\mathfrak{g}$, so the sum of the finite dimensional subspaces of V that are stable under $\mathfrak{s}_\alpha$ is a $\mathfrak{g}$-submodule of V (Chap. I, §6, no. 6, Prop. 7); since this sum is non-zero, it is equal to V. It follows from this that $(X_\alpha)_\mathrm{V}$ and $(X_{-\alpha})_\mathrm{V}$ are locally nilpotent. In view of Lemma 1 (ii), $\mathscr{X}$ is stable under s_α, and this holds for all α. Hence $\mathscr{X}$ is stable under W. Now every orbit of W on P meets P_{++} (Chap. VI, §1, no. 10). On the other hand, if $\lambda \in \mathscr{X} \cap \mathrm{P}_{++}$, then $\lambda = \omega - \sum_{\alpha\in\mathrm{B}} n_\alpha \alpha = \sum_{\alpha\in\mathrm{B}} n'_\alpha \alpha$ with $n_\alpha \in \mathbf{N}$ and $n'_\alpha \geq 0$ for all $\alpha \in \mathrm{B}$ (Chap. V, §3, no. 5, Lemma 6). So $\mathscr{X} \cap \mathrm{P}_{++}$ is finite and hence so is $\mathscr{X}$. Since each weight has finite multiplicity (§6, no. 1, Prop. 1 (ii)), V is finite dimensional.

COROLLARY 1. *If $\lambda \in \mathfrak{h}^*$ and $\lambda \notin \mathrm{P}_{++}$, the $\mathfrak{g}$-module* $\mathrm{E}(\lambda)$ (§6, no. 3) *is infinite dimensional.*

COROLLARY 2. *The $\mathfrak{g}$-modules $\mathrm{E}(\lambda)$ for $\lambda \in \mathrm{P}_{++}$ constitute a set of representatives of the classes of finite dimensional simple $\mathfrak{g}$-modules.*

The $\mathfrak{g}$-modules $\mathrm{E}(\lambda)$, where λ is a fundamental weight, are called the *fundamental $\mathfrak{g}$-modules*; the corresponding representations are called the *fundamental representations of* $\mathfrak{g}$; they are absolutely irreducible (§6, no. 2, Prop. 3 (iv)).

If V is a finite dimensional $\mathfrak{g}$-module and $\lambda \in \mathrm{P}_{++}$, the isotypical component of V of type $\mathrm{E}(\lambda)$ is called *the isotypical component of highest weight* λ *of* V.

Remark 1. Let $\lambda \in \mathrm{P}_{++}$, ρ_λ the representation of $\mathfrak{g}$ on $\mathrm{E}(\lambda)$, $s \in \mathrm{Aut}(\mathfrak{g})$, and σ the canonical image of s in $\mathrm{Aut}(\mathrm{R}, \mathrm{B})$ (§5, no. 3, Cor. 1 of Prop. 5). Then $\rho_\lambda \circ s$ is equivalent to $\rho_{\sigma\lambda}$; indeed, if $s \in \mathrm{Aut}_0(\mathfrak{g})$, $\rho_\lambda \circ s$ and $\rho_{\sigma\lambda}$ are equivalent to ρ_λ (Prop. 2); and, if s leaves $\mathfrak{h}$ and B stable, $\rho_\lambda \circ s$ is simple of highest weight $\sigma\lambda$.

In particular, the fundamental representations are permuted by s, and this permutation is the identity if and only if $s \in \mathrm{Aut}_0(\mathfrak{g})$.

PROPOSITION 3. *Let* V *be a finite dimensional* $\mathfrak{g}$*-module and* $\mathscr{X}$ *its set of weights. Let* $\lambda \in \mathscr{X}$, $\alpha \in \mathrm{R}$, I *the set of* $t \in \mathbf{Z}$ *such that* $\lambda + t\alpha \in \mathscr{X}$, p (resp. $-q$) *the largest* (resp. *smallest*) *element of* I. *Let* m_t *be the multiplicity of* $\lambda + t\alpha$.

(i) $\mathrm{I} = [\![-q, p]\!]$ *and* $q - p = \lambda(H_\alpha)$.

(ii) *For any integer* $u \in [\![0, p+q]\!]$, $\lambda + (p-u)\alpha$ *and* $\lambda + (-q+u)\alpha$ *are conjugate under* s_α, *and* $m_{-q+u} = m_{p-u}$.

(iii) *If* $t \in \mathbf{Z}$ *and* $t < (p-q)/2$, $(X_\alpha)_{\mathrm{V}}$ *maps* $\mathrm{V}^{\lambda+t\alpha}$ *injectively into* $\mathrm{V}^{\lambda+(t+1)\alpha}$.

(iv) *The function* $t \mapsto m_t$ *is increasing on* $[\![-q, (p-q)/2]\!]$ *and decreasing on* $[\![(p-q)/2, p]\!]$.

Let $\alpha \in \mathrm{B}$. Give V the $\mathfrak{sl}(2, k)$-module structure defined by the elements $X_\alpha, X_{-\alpha}, H_\alpha$ of $\mathfrak{g}$. Every non-zero element of $\mathrm{V}^{\lambda+p\alpha}$ is then primitive. Consequently, $(\lambda + p\alpha)(H_\alpha) \geq 0$ and $(X_{-\alpha})^r \mathrm{V}^{\lambda+p\alpha} \neq 0$ for

$$0 \leq r \leq (\lambda + p\alpha)(H_\alpha) = \lambda(H_\alpha) + 2p$$

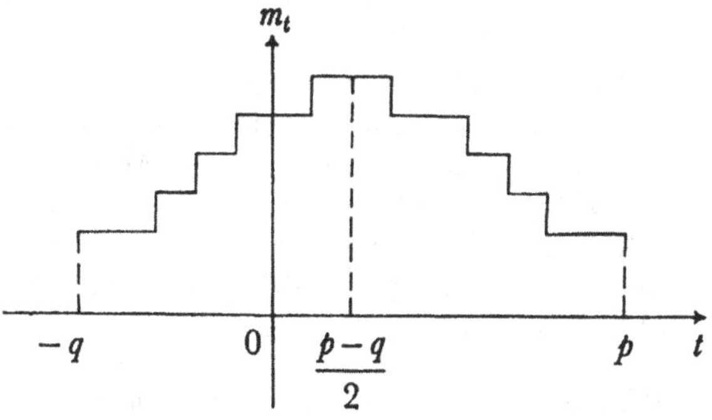

(§1, no. 2, Prop. 2). It follows that $V^{\lambda+t\alpha} \neq 0$ for $p \geq t \geq p - (\lambda(H_\alpha) + 2p)$, so $p + q \geq \lambda(H_\alpha) + 2p$. Applying this result to $-\alpha$ gives

$$p + q \geq \lambda(H_{-\alpha}) + 2q = -\lambda(H_\alpha) + 2q.$$

Hence $q - p = \lambda(H_\alpha)$ and $\lambda + t\alpha \in \mathscr{X}$ for $p \geq t \geq -q$, which proves (i).

We have $s_\alpha(\alpha) = -\alpha$, and $s_\alpha(\mu) \in \mu + k\alpha$ for all $\mu \in \mathfrak{h}^*$. Since W leaves $\mathscr{X}$ stable (Cor. 2 of Prop. 2), s_α leaves $\{\lambda - q\alpha, \lambda - q\alpha + \alpha, \ldots, \lambda + p\alpha\}$ stable and takes $\lambda - q\alpha + u\alpha$ to $\lambda + p\alpha - u\alpha$ for all $u \in k$. Using Cor. 2 of Prop. 2 again, we see that $m_{-q+u} = m_{p-u}$ for every integer $u \in [0, p+q]$. This proves (ii).

By §1, Cor. of Prop. 2, $(X_\alpha)_V | V^{\lambda+t\alpha}$ is injective for $t < (p-q)/2$. Now $(X_\alpha)_V$ maps $V^{\lambda+t\alpha}$ to $V^{\lambda+(t+1)\alpha}$. Hence $m_{t+1} \geq m_t$ for $t < (p-q)/2$. Changing α to $-\alpha$, we see that $m_{t+1} \leq m_t$ for $t > (p-q)/2$. This proves (iii) and (iv).

COROLLARY 1. *If $\lambda \in \mathscr{X}$ and $\lambda(H_\alpha) \geq 1$, then $\lambda - \alpha \in \mathscr{X}$. If $\lambda + \alpha \in \mathscr{X}$ and $\lambda(H_\alpha) = 0$, then $\lambda \in \mathscr{X}$ and $\lambda - \alpha \in \mathscr{X}$.*

This follows immediately from Prop. 3 (i).

COROLLARY 2. *Let $\mu \in P_{++}$ and $\nu \in Q_+$. If $\mu + \nu \in \mathscr{X}$, then $\mu \in \mathscr{X}$.*

Write $\nu = \sum_{\alpha \in B} c_\alpha . \alpha$, where $c_\alpha \in \mathbf{N}$ for all $\alpha \in B$. The corollary is clear when $\sum_{\alpha \in B} c_\alpha = 0$; assume that $\sum_{\alpha \in B} c_\alpha > 0$ and argue by induction on $\sum_{\alpha \in B} c_\alpha$. Let $(\cdot | \cdot)$ be a W-invariant non-degenerate positive symmetric bilinear form on $\mathfrak{h}^*_{\mathbf{R}}$. Then $(\nu | \sum_{\alpha \in B} c_\alpha . \alpha) > 0$, so there exists $\beta \in B$ such that $c_\beta \geq 1$ and $(\nu | \beta) > 0$, hence $\nu(H_\beta) \geq 1$. Since $\mu \in P_{++}$, it follows that $(\mu + \nu)(H_\beta) \geq 1$. By Cor. 1, $\mu + (\nu - \beta) \in \mathscr{X}$, and it suffices to apply the induction hypothesis.

COROLLARY 3. *Let $v \in V$ be primitive of weight ω. Let Σ be the set of $\alpha \in B$ such that $\omega(H_\alpha) = 0$. The stabilizer in $\mathfrak{g}$ of the line kv is the parabolic subalgebra $\mathfrak{p}_\Sigma$ associated to Σ* (§3, no. 4, Remark).

Replacing V by the $\mathfrak{g}$-submodule generated by v, if necessary, we can assume that V is simple. Let $\mathfrak{s}$ be the stabilizer. We have $(\mathfrak{n}_+)_V v = 0$, $(\mathfrak{h})_V v \subset kv$. Let $\alpha \in B$ be such that $\omega(H_\alpha) = 0$. We have $\omega + \alpha \notin \mathscr{X}$, hence $\omega - \alpha \notin \mathscr{X}$ (Prop. 3 (i)) and consequently $(\mathfrak{g}^{-\alpha})_V v = 0$. The preceding proves that $\mathfrak{p}_\Sigma \subset \mathfrak{s}$. If $\mathfrak{p}_\Sigma \neq \mathfrak{s}$, then $\mathfrak{s} = \mathfrak{p}_{\Sigma'}$, where Σ' is a subset of B strictly containing Σ. Let $\beta \in \Sigma' - \Sigma$. Then $\mathfrak{g}^{-\beta}$ stabilizes kv, and hence annihilates v. But, since $\omega(H_\beta) > 0$, this contradicts Prop. 3 (iii). Q.E.D.

A subset $\mathscr{X}$ of P is called R-*saturated* if it satisfies the following condition: for all $\lambda \in \mathscr{X}$ and all $\alpha \in R$, we have $\lambda - t\alpha \in \mathscr{X}$ for all integers t between 0 and $\lambda(H_\alpha)$. Since $s_\alpha(\lambda) = \lambda - \lambda(H_\alpha)\alpha$, we see that an R-saturated subset of P is stable under W. Let $\mathscr{Y} \subset P$. An element λ of $\mathscr{Y}$ is called R-*extremal in* $\mathscr{Y}$ if, for all $\alpha \in R$, either $\lambda + \alpha \notin \mathscr{Y}$ or $\lambda - \alpha \notin \mathscr{Y}$.

PROPOSITION 4. *Let* V *be a finite dimensional* $\mathfrak{g}$*-module and* d *an integer* ≥ 1. *The set of weights of* V *of multiplicity* $\geq d$ *is* R*-saturated.*

This follows immediately from Prop. 3.

PROPOSITION 5. *Let* V *be a finite dimensional simple* $\mathfrak{g}$*-module,* ω *its highest weight,* $\mathscr{X}$ *its set of weights. Choose a* W*-invariant non-degenerate positive symmetric bilinear form* $(\cdot|\cdot)$ *on* $\mathfrak{h}_{\mathbf{R}}^{*}$, *and let* $\lambda \mapsto \| \lambda \| = (\lambda|\lambda)^{1/2}$ *be the corresponding norm.*

(i) $\mathscr{X}$ *is the smallest* R*-saturated subset of* P *containing* ω.

(ii) *The* R*-extremal elements of* $\mathscr{X}$ *are the* W*-transforms of* ω.

(iii) *If* $\mu \in \mathscr{X}$, *we have* $\| \mu \| \leq \| \omega \|$. *If, in addition,* $\mu \neq \omega$, *we have* $\| \mu + \rho \| < \| \omega + \rho \|$. *If* μ *is not* R*-extremal in* $\mathscr{X}$, *then* $\|\mu\| < \|\omega\|$.

(iv) *We have* $\mathscr{X} = \mathrm{W}.(\mathscr{X} \cap \mathrm{P}_{++})$. *An element* λ *of* P_{++} *belongs to* $\mathscr{X} \cap \mathrm{P}_{++}$ *if and only if* $\omega - \lambda \in \mathrm{Q}_{+}$.

(i) Let $\mathscr{X}'$ be the smallest R-saturated subset of P containing ω. We have $\mathscr{X}' \subset \mathscr{X}$ (Prop. 4). Assume that $\mathscr{X} \neq \mathscr{X}'$. Let λ be a maximal element of $\mathscr{X} - \mathscr{X}'$. Since $\lambda \neq \omega$, there exists $\alpha \in \mathrm{B}$ such that $\lambda + \alpha \in \mathscr{X}$. Introduce p and q as in Prop. 3. Since λ is maximal in $\mathscr{X} - \mathscr{X}'$, $\lambda + p\alpha \in \mathscr{X}'$. By Prop. 3 (ii), $\lambda - q\alpha \in \mathscr{X}'$ since $\mathscr{X}'$ is stable under W. Hence $\lambda + u\alpha \in \mathscr{X}'$ for every integer u in the interval $[-q, p]$. This contradicts $\lambda \notin \mathscr{X}'$ and proves (i).

(ii) It is clear that ω is an R-extremal element of $\mathscr{X}$; its W-transforms are therefore also R-extremal in $\mathscr{X}$. Let λ be an R-extremal element of $\mathscr{X}$; we shall prove that $\lambda \in \mathrm{W}.\omega$. Since there exists $w \in \mathrm{W}$ such that $w\lambda \in \mathrm{P}_{++}$ (Chap. VI, §1, no. 10), we can assume that $\lambda \in \mathrm{P}_{++}$. Let $\alpha \in \mathrm{B}$; introduce p and q as in Prop. 3. Since λ is R-extremal, either $p = 0$ or $q = 0$. Since

$$q - p = \lambda(H_\alpha) \geq 0,$$

we cannot have $p > 0$. Hence $p = 0$ and $\lambda = \omega$.

(iii) Let $\mu \in \mathscr{X} \cap \mathrm{P}_{++}$. Then $\omega + \mu \in \mathrm{P}_{++}$ and $\omega - \mu \in \mathrm{Q}_{+}$ (§6, no. 1, Prop. 1), so $0 \leq (\omega - \mu|\omega + \mu) = (\omega|\omega) - (\mu|\mu)$; hence, $(\mu|\mu) \leq (\omega|\omega)$, and this extends to all $\mu \in \mathscr{X}$ by using the Weyl group. If $\mu \in \mathscr{X} - \{\omega\}$,

$$\begin{aligned}(\mu + \rho|\mu + \rho) &= (\mu|\mu) + 2(\mu|\rho) + (\rho|\rho) \leq (\omega|\omega) + 2(\mu|\rho) + (\rho|\rho)\\ &= (\omega + \rho|\omega + \rho) - 2(\omega - \mu|\rho).\end{aligned}$$

Now $\omega - \mu = \sum_{\alpha \in \mathrm{B}} n_\alpha \alpha$ with integers $n_\alpha \geq 0$ not all zero, so $(\omega - \mu|\rho) > 0$ since $(\rho|\alpha) > 0$ for all $\alpha \in \mathrm{B}$ (Chap. VI, §1, no. 10, Prop. 29 (iii)). If μ is not R-extremal in $\mathscr{X}$, there exists $\alpha \in \mathrm{R}$ such that $\mu + \alpha \in \mathscr{X}$ and $\mu - \alpha \in \mathscr{X}$; then

$$\|\mu\| < \sup(\|\mu + \alpha\|, \|\mu - \alpha\|) \leq \sup_{\lambda \in \mathscr{X}} \|\lambda\|$$

and this last upper bound is $\|\omega\|$ by the preceding.

(iv) We have $\mathscr{X} = \mathrm{W}.(\mathscr{X} \cap \mathrm{P}_{++})$ by Chap. VI, §1, no. 10. If $\lambda \in \mathscr{X}$, then $\omega - \lambda \in \mathrm{Q}_+$ (§6, no. 1, Prop. 1). If $\lambda \in \mathrm{P}_{++}$ and $\omega - \lambda \in \mathrm{Q}_+$, then $\lambda \in \mathscr{X}$ (Cor. 2 of Prop. 3).

COROLLARY. *Let* $\mathscr{X}$ *be a finite* R*-saturated subset of* P. *There exists a finite dimensional* $\mathfrak{g}$*-module whose set of weights is* $\mathscr{X}$.

Since $\mathscr{X}$ is stable under W, $\mathscr{X}$ is the smallest R-saturated set containing $\mathscr{X} \cap \mathrm{P}_{++}$. By Prop. 5 (i), $\mathscr{X}$ is the set of weights of $\bigoplus_{\lambda \in \mathscr{X} \cap \mathrm{P}_{++}} \mathrm{E}(\lambda)$.

Remark 2. Recall (Chap. VI, §1, no. 6, Cor. 3 of Prop. 17) that there exists a unique element w_0 of W that transforms B into $-\mathrm{B}$; we have $w_0^2 = 1$. and $-w_0$ respects the order relation on P. With this in mind, let V be a finite dimensional simple $\mathfrak{g}$-module, ω its highest weight. Then $w_0(\omega)$ is the lowest weight of V, and its multiplicity is 1.

3. MINUSCULE WEIGHTS

PROPOSITION 6. *Let* $\lambda \in \mathrm{P}$, *and* $\mathscr{X}$ *the smallest* R*-saturated subset of* P *containing* λ. *Choose a norm* $\|\cdot\|$ *as in Prop.* 5. *The following conditions are equivalent:*

(i) $\mathscr{X} = \mathrm{W}.\lambda$;

(ii) *all the elements of* $\mathscr{X}$ *have the same norm;*

(iii) *for all* $\alpha \in \mathrm{R}$, *we have* $\lambda(H_\alpha) \in \{0, 1, -1\}$.

Every non-empty R*-saturated subset of* P *contains an element* λ *satisfying the above conditions.*

Introduce the condition:

(ii′) for all $\alpha \in \mathrm{R}$ and for every integer t between 0 and $\lambda(H_\alpha)$,

$\|\lambda - t\alpha\| \geq \|\lambda\|$.

(i) $\Longrightarrow$ (ii) $\Longrightarrow$ (ii′): This is clear.

(ii′) $\Longrightarrow$ (iii): Assume that condition (ii′) is satisfied. Let $\alpha \in \mathrm{R}$. We have $\|\lambda\| = \|\lambda - \lambda(H_\alpha)\alpha\|$, so $\|\lambda - t\alpha\| < \|\lambda\|$ for every integer t strictly between 0 and $\lambda(H_\alpha)$; hence, there can be no such integers, so $|\lambda(H_\alpha)| \leq 1$.

(iii) $\Longrightarrow$ (i): Assume that condition (iii) is satisfied. Let $w \in \mathrm{W}$ and $\alpha \in \mathrm{R}$. Then $(w\lambda)(H_\alpha) = \lambda(H_{w^{-1}\alpha}) \in \{0, 1, -1\}$; thus, if t is an integer between 0 and $(w\lambda)(H_\alpha)$, $w\lambda - t\alpha$ is equal to $w\lambda$ or $s_\alpha(w\lambda)$. This proves that $\mathrm{W}.\lambda$ is R-saturated, so $\mathscr{X} = \mathrm{W}.\lambda$.

Let $\mathscr{Y}$ be a non-empty R-saturated subset of P. There exists in $\mathscr{Y}$ an element λ of minimum norm. It is clear that λ satisfies condition (ii′), hence the last assertion of the proposition.

PROPOSITION 7. *Let* V *be a finite dimensional simple* $\mathfrak{g}$*-module,* $\mathscr{X}$ *the set of weights of* V, *and* λ *the highest element of* $\mathscr{X}$ (cf. Prop. 5 (i)). *Conditions* (i), (ii) *and* (iii) *of Prop.* 6 *are equivalent to:*

(iv) *for all* $\alpha \in \mathrm{R}$ *and all* $x \in \mathfrak{g}^\alpha$, *we have* $(x_\mathrm{V})^2 = 0$.

If these conditions are satisfied, all the weights of V *have multiplicity* 1.

If (i) is satisfied, then $\mathscr{X} = \mathrm{W}.\lambda$ and the weights all have the same multiplicity as λ (Cor. 2 of Prop. 2), in other words, multiplicity 1. Moreover, if $w \in \mathrm{W}$ and $\alpha \in \mathrm{R}$, $w\lambda + t\alpha$ cannot be a weight of V unless $|t| \leq 1$; thus, if $x \in \mathfrak{g}^\alpha$,

$$(x_\mathrm{V})^2(\mathrm{V}^{w(\lambda)}) \subset \mathrm{V}^{w(\lambda)+2\alpha} = 0,$$

so $(x_\mathrm{V})^2 = 0$, which proves that (i) $\Longrightarrow$ (iv).

Conversely, assume that (iv) is satisfied. Let $\alpha \in \mathrm{R}$, and give V the $\mathfrak{sl}(2,k)$-module structure defined by the elements $X_\alpha, X_{-\alpha}, H_\alpha$ of $\mathfrak{g}$. Condition (iv), applied to $x = X_\alpha$, implies that the weights of the $\mathfrak{sl}(2,k)$-module V belong to $\{0, 1, -1\}$ (cf. §1, no. 2, Cor. of Prop. 2). In particular, $\lambda(H_\alpha) \in \{0, 1, -1\}$, so (iv) $\Longrightarrow$ (iii).

PROPOSITION 8. *Assume that* $\mathfrak{g}$ *is simple. Denote by* $\alpha_1, \ldots, \alpha_l$ *the elements of* B. *Let* $\varpi_1, \ldots, \varpi_l$ *be the corresponding fundamental weights. Let* $H = n_1 H_{\alpha_1} + \cdots + n_l H_{\alpha_l}$ *be the highest root of* $\mathrm{R}^\vee$, *and* J *the set of* $i \in \{1, \ldots, l\}$ *such that* $n_i = 1$. *Let* $\lambda \in \mathrm{P}_{++} - \{0\}$. *Then conditions* (i), (ii) *and* (iii) *of Prop.* 6 *are equivalent to each of the following conditions*:

(v) $\lambda(H) = 1$;

(vi) *there exists* $i \in \mathrm{J}$ *such that* $\lambda = \varpi_i$.

The ϖ_i, *for* $i \in \mathrm{J}$, *form a system of representatives in* P(R) *of the non-zero elements of* P(R)/Q(R).

Let $\lambda = u_1\varpi_1 + \cdots + u_l\varpi_l$, where $u_1, \ldots, u_l$ are integers ≥ 0 and not all zero. Then $\lambda(H) = u_1 n_1 + \cdots + u_l n_l$ and $n_1 \geq 1, \ldots, n_l \geq 1$, which gives the equivalence of (v) and (vi) immediately. On the other hand, $\lambda(H) = \sup_{\alpha \in \mathrm{R}_+} \lambda(H_\alpha)$, and $\lambda(H) > 0$ since λ is a non-zero element of P_{++}. Hence condition (v) is equivalent to the condition $\lambda(H_\alpha) \in \{0, 1\}$ for all $\alpha \in \mathrm{R}$, in other words to condition (iii) of Prop. 6.

The last assertion of the proposition follows from Chap. VI, §2, no. 3, Cor. of Prop. 6.

DEFINITION 1. *Assume that* $\mathfrak{g}$ *is simple. A minuscule weight of* $(\mathfrak{g}, \mathfrak{h})$ *is an element of* $\mathrm{P}_{++} - \{0\}$ *which satisfies the equivalent conditions* (i), (ii), (iii), (iv), (v) *and* (vi) *of Prop.* 6, 7 *and* 8.

Remark. Assume that $\mathfrak{g}$ is simple. Let $\Sigma'^\vee$ be the Coxeter graph of the affine Weyl group $\mathrm{W}_a(\mathrm{R}^\vee)$. Recall that the vertices of $\Sigma'^\vee$ are the vertices of the Coxeter graph $\Sigma^\vee$ of $\mathrm{W}(\mathrm{R}^\vee)$, together with a supplementary vertex 0. The group $\mathrm{A}(\mathrm{R}^\vee)$ operates on $\Sigma'^\vee$ leaving 0 fixed. The group $\mathrm{Aut}(\Sigma'^\vee)$ is canonically isomorphic to the semi-direct product of $\mathrm{A}(\mathrm{R}^\vee)/\mathrm{W}(\mathrm{R}^\vee)$ with a group Γ_C (cf. Chap. VI, §2, no. 3, and Chap. VI, §4, no. 3); clearly

$(\mathrm{Aut}\,\Sigma'^{\vee})(0) = \Gamma_C(0)$; and $\Gamma_C(0)$ consists of 0 and the vertices of $\Sigma^{\vee}$ corresponding to the ϖ_i for $i \in \mathrm{J}$ (cf. Chap. VI, §2, Prop. 5 and Remark 1 of no. 3). In summary, *the minuscule weights are the fundamental weights corresponding to the vertices of $\Sigma^{\vee}$ which can be obtained from 0 by the operation of an element of* $\mathrm{Aut}(\Sigma'^{\vee})$.

With the notations of Chap. VI, Plates I to IX, we deduce from the preceding that the minuscule weights are the following:

For type A_l $(l \geq 1)$: $\varpi_1, \ldots, \varpi_l$.
For type B_l $(l \geq 2)$: ϖ_l.
For type C_l $(l \geq 2)$: ϖ_1.
For type D_l $(l \geq 3)$: $\varpi_1, \varpi_{l-1}, \varpi_l$.
For type E_6: ϖ_1, ϖ_6.
For type E_7: ϖ_7.
For types $\mathrm{E}_8, \mathrm{F}_4, \mathrm{G}_2$ there are no minuscule weights.

4. TENSOR PRODUCTS OF $\mathfrak{g}$-MODULES

Let E, F be $\mathfrak{g}$-modules. For all $\lambda, \mu \in \mathfrak{h}^*$, $\mathrm{E}^{\lambda} \otimes \mathrm{F}^{\mu} \subset (\mathrm{E} \otimes \mathrm{F})^{\lambda+\mu}$ (Chap. VII, §1, no. 1, Prop. 2 (ii)). If E and F are finite dimensional, then $\mathrm{E} = \sum_{\lambda \in \mathrm{P}} \mathrm{E}^{\lambda}$ and $\mathrm{F} = \sum_{\mu \in \mathrm{P}} \mathrm{F}^{\mu}$; consequently,

$$(\mathrm{E} \otimes \mathrm{F})^{\nu} = \sum_{\lambda, \mu \in \mathrm{P}, \lambda+\mu=\nu} \mathrm{E}^{\lambda} \otimes \mathrm{F}^{\mu}.$$

In other words, equipped with its graduation of type P, $\mathrm{E} \otimes \mathrm{F}$ is the graded tensor product of the graded vector spaces E and F.

PROPOSITION 9. *Let* E, F *be finite dimensional simple $\mathfrak{g}$-modules, with highest weights λ, μ, respectively.*

(i) *The component of* $\mathrm{E} \otimes \mathrm{F}$ *of highest weight $\lambda + \mu$ is a simple submodule, generated by* $(\mathrm{E} \otimes \mathrm{F})^{\lambda+\mu} = \mathrm{E}^{\lambda} \otimes \mathrm{F}^{\mu}$.

(ii) *The highest weight of any simple submodule of* $\mathrm{E} \otimes \mathrm{F}$ *is* $\leq \lambda + \mu$ (cf. §9, Prop. 2).

If $\alpha, \beta \in \mathrm{P}$ and if $\mathrm{E}^{\alpha} \otimes \mathrm{F}^{\beta} \neq 0$, then $\alpha \leq \lambda$ and $\beta \leq \mu$. Consequently, $(\mathrm{E} \otimes \mathrm{F})^{\lambda+\mu}$ is equal to $\mathrm{E}^{\lambda} \otimes \mathrm{F}^{\mu}$, and hence is of dimension 1, and $\lambda + \mu$ is the highest weight of $\mathrm{E} \otimes \mathrm{F}$. Every non-zero element of $\mathrm{E}^{\lambda} \otimes \mathrm{F}^{\mu}$ is primitive. By Prop. 4 of §6, no. 2, the length of the isotypical component of $\mathrm{E} \otimes \mathrm{F}$ of highest weight $\lambda + \mu$ is 1.

Remark. Retain the notations of Prop. 9. Let C be the isotypical component of $\mathrm{E} \otimes \mathrm{F}$ of highest weight $\lambda + \mu$. Then C depends only on E and F and not on the choice of $\mathfrak{h}$ and the basis B. In other words, let $\mathfrak{h}'$ be a splitting Cartan subalgebra of $\mathfrak{g}$, R′ the root system of $(\mathfrak{g}, \mathfrak{h}')$, and B′ a basis of R′; let λ', μ' be the highest weights of E, F relative to $\mathfrak{h}'$ and B'; let C' be the isotypical component of $E \otimes F$ of highest weight $\lambda' + \mu'$; then $C' = C$. Indeed, to prove this we can assume, by extension of the base field, that k is algebraically closed. Then there exists $s \in \mathrm{Aut}_e(\mathfrak{g})$ that takes $\mathfrak{h}$ to $\mathfrak{h}'$, R to R', B to B'. Let $S \in \mathbf{SL}(E \otimes F)$ have the properties in Prop. 2 of no. 1. Then $S((E \otimes F)^{\lambda+\mu}) = (E \otimes F)^{\lambda'+\mu'}$ and $S(C) = C$. Hence $(E \otimes F)^{\lambda'+\mu'} \subset C' \cap S(C) = C' \cap C$, so $C' = C$. Thus, to 2 classes of finite dimensional simple $\mathfrak{g}$-modules we can associate canonically a third; in other words, we have defined on the set $\mathfrak{S}_{\mathfrak{g}}$ of classes of finite dimensional simple $\mathfrak{g}$-modules a composition law. With this structure, $\mathfrak{S}_{\mathfrak{g}}$ is canonically isomorphic to the additive monoid P_{++}.

COROLLARY 1. *Let* $(\varpi_\alpha)_{\alpha \in B}$ *be the family of fundamental weights relative to* B. *Let* $\lambda = \sum_{\alpha \in B} m_\alpha \varpi_\alpha \in P_{++}$. *For all* $\alpha \in B$, *let* E_α *be a simple* $\mathfrak{g}$-*module of highest weight* ϖ_α. *In the* $\mathfrak{g}$-*module* $\bigotimes_{\alpha \in B} (\bigotimes^{m_\alpha} E_\alpha)$, *the isotypical component of highest weight* λ *is of length* 1.

This follows from Prop. 9 by induction on $\sum_{\alpha \in B} m_\alpha$.

COROLLARY 2. *Assume that* k *is* **R** *or* **C** *or a non-discrete complete ultrametric field. Let* G *be a Lie group with Lie algebra* $\mathfrak{g}$. *Assume that, for any fundamental representation* ρ *of* $\mathfrak{g}$, *there exists an analytic linear representation* ρ' *of* G *such that* $\rho = L(\rho')$. *Then, for any finite dimensional linear representation* π *of* $\mathfrak{g}$, *there exists an analytic linear representation* π' *of* G *such that* $\pi = L(\pi')$.

We use the notations of Cor. 1. There exists a representation σ of G on $X = \bigotimes_{\alpha \in B} (\bigotimes^{m_\alpha} E_\alpha)$ such that $L(\sigma)$ corresponds to the $\mathfrak{g}$-module structure of X (Chap. III, §3, no. 11, Cor. 3 of Prop. 41). Let C be the isotypical component of X of highest weight λ. In view of Chap. III, §3, no. 11, Prop. 40, it suffices to prove that C is stable under $\sigma(G)$. Let $g \in G$ and $\varphi = \mathrm{Ad}(g)$. Then $\sigma(g) a_X \sigma(g)^{-1} = (\varphi(a))_X$ for all $a \in \mathfrak{g}$. On the other hand, φ is an automorphism of $\mathfrak{g}$ that takes $\mathfrak{h}$ to $\mathfrak{h}'$, R to $R' = R(\mathfrak{g}, \mathfrak{h}')$, B to a basis B' of R', and ϖ_α to the highest weight ϖ'_α of E_α relative to $\mathfrak{h}'$ and B' (since φ transforms E_α into a $\mathfrak{g}$-module isomorphic to E_α). Hence φ takes λ to $\sum m_\alpha \varpi'_\alpha$. By the *Remark* above, $\sigma(g)(C) = C$.

PROPOSITION 10. *Let* $\lambda, \mu \in P_{++}$. *Let* E, F, G *be simple* $\mathfrak{g}$-*modules with highest weights* $\lambda, \mu, \lambda + \mu$. *Let* $\mathscr{X}$ (resp. $\mathscr{X}', \mathscr{X}''$) *be the set of weights of* E (resp. F, G). *Then* $\mathscr{X}'' = \mathscr{X} + \mathscr{X}'$.

We have $E = \bigoplus_{\nu \in P} E^\nu$, $F = \bigoplus_{\sigma \in P} F^\sigma$, so $E \otimes F$ is the direct sum of the

$$(E \otimes F)^\tau = \sum_{\nu + \sigma = \tau} E^\nu \otimes F^\sigma.$$

By Prop. 9, G can be identified with a $\mathfrak{g}$-submodule of $E \otimes F$, so $\mathscr{X}'' \subset \mathscr{X} + \mathscr{X}'$. We have $G^\tau = G \cap (E \otimes F)^\tau$, and it is enough to show that, for $\nu \in \mathscr{X}$ and

$\sigma \in \mathscr{X}'$, we have $G \cap (E \otimes F)^{\nu+\sigma} \neq 0$. Let $(e_1, \ldots, e_n)$ (resp. $(f_1, \ldots, f_p)$) be a basis of E (resp. F) consisting of elements each of which belong to some E^ν (resp. F^σ), and such that $e_1 \in E^\lambda$ (resp. $f_1 \in F^\mu$). The $e_i \otimes f_j$ form a basis of $E \otimes F$. Suppose that the result to be proved is false. Then there exists a pair (i, j) such that the coordinate of index (i, j) of every element of G is zero. Let U be the enveloping algebra of $\mathfrak{g}$, U' the dual of U, c the coproduct of U. For all $u \in U$, let $x_i(u)$ (resp. $y_j(u)$) be the coordinate of $u(e_1)$ (resp. $u(f_1)$) of index i (resp. j); let $z_{ij}(u)$ be the coordinate of index (i, j) of $u(e_1 \otimes f_1)$. Then $x_i, y_j, z_{ij} \in U'$. Now e_1 generates the $\mathfrak{g}$-module E, so $x_i \neq 0$, and similarly $y_j \neq 0$. By the definition of the $\mathfrak{g}$-module $E \otimes F$ (Chap. I, §3, no. 2), if $c(u) = \sum u_s \otimes u'_s$, we have

$$z_{ij}(u) = \sum_s x_i(u_s).y_j(u'_s) = \langle c(u), x_i \otimes y_j \rangle.$$

In other words, z_{ij} is the product of x_i and y_j in the algebra U'. But this algebra is an integral domain (Chap. II, §1, no. 5, Prop. 10), so $z_{ij} \neq 0$. Since $u(e_1 \otimes f_1) \in G$ for all $u \in U$, this is a contradiction.

5. DUAL OF A $\mathfrak{g}$-MODULE

Let E, F be $\mathfrak{g}$-modules. Recall (Chap. I, §3, no. 3) that $\mathrm{Hom}_k(E, F)$ has a canonical $\mathfrak{g}$-module structure. Let φ be an element of weight λ in $\mathrm{Hom}_k(E, F)$. If $\mu \in \mathfrak{h}^*$, then $\varphi(E^\mu) \subset F^{\lambda+\mu}$ (Chap. VII, §1, no. 1, Prop. 2 (ii)). Thus, if E and F are finite dimensional, the elements of weight λ in $\mathrm{Hom}_k(E, F)$ are the graded homomorphisms of degree λ in the sense of *Algebra*, Chap. II, §11, no. 2, Def. 4.

PROPOSITION 11. *Let* E *be a finite dimensional* $\mathfrak{g}$*-module, and consider the* $\mathfrak{g}$*-module* $E^* = \mathrm{Hom}_k(E, k)$.

(i) *An element* $\lambda \in P$ *is a weight of* E^* *if and only if* $-\lambda$ *is a weight of* E, *and the multiplicity of* λ *in* E^* *is equal to that of* $-\lambda$ *in* E.

(ii) *If* E *is simple and has highest weight* ω, E^* *is simple and has highest weight* $-w_0(\omega)$ (cf. no. 2, Remark 2).

Consider k as a trivial $\mathfrak{g}$-module whose elements are of weight 0. By what was said above, the elements of E^* of weight λ are the homomorphisms from E to k which vanish on E^μ if $\mu \neq -\lambda$. This proves (i). If E is simple, E^* is simple (Chap. I, §3, no. 3), and the last assertion follows from Remark 2 of no. 2.

Remarks. 1) Let E, E^* be as in Prop. 11, and $\sigma \in \mathrm{Aut}(\mathfrak{g}, \mathfrak{h})$ be such that $\varepsilon(\sigma) = -w_0$ in the notations of §5, no. 1 (§5, no. 2, Prop. 2). Let ρ, ρ' be the representations of $\mathfrak{g}$ associated to E, E^*. Then $\rho \circ \sigma$ is a simple representation of $\mathfrak{g}$ with highest weight $-w_0(\omega)$, so $\rho \circ \sigma$ is equivalent to ρ'.

2) Assume that $w_0 = -1$. Then, for any finite dimensional $\mathfrak{g}$-module E, E is isomorphic to E^*. Recall that, if $\mathfrak{g}$ is simple, $w_0 = -1$ in the following cases: $\mathfrak{g}$ of type A_1, B_l ($l \geq 2$), C_l ($l \geq 2$), D_l (l even ≥ 4), E_7, E_8, F_4, G_2 (Chap. VI, Plates).

Lemma 2. Let $h^0 = \sum_{\alpha \in R_+} H_\alpha$. *Then* $h^0 = \sum_{\alpha \in B} a_\alpha H_\alpha$, *where the* a_α *are integers* ≥ 1. *Let* $(b_\alpha)_{\alpha \in B}, (c_\alpha)_{\alpha \in B}$ *be families of scalars such that* $b_\alpha c_\alpha = a_\alpha$ *for all* $\alpha \in B$. *Put* $x = \sum_{\alpha \in B} b_\alpha X_\alpha, y = \sum_{\alpha \in B} c_\alpha X_{-\alpha}$. *There exists a homomorphism* φ *from* $\mathfrak{sl}(2, k)$ *to* $\mathfrak{g}$ *such that* $\varphi(H) = h^0, \varphi(X_+) = x, \varphi(X_-) = y$.

The fact that the a_α are integers ≥ 1 follows from the fact that $(H_\alpha)_{\alpha \in B}$ is a basis of the root system $(H_\alpha)_{\alpha \in B}$ (cf. Chap. VI, §1, no. 5, Remark 5). We have:

$$\alpha(h^0) = 2 \tag{1}$$

for all $\alpha \in B$ (Chap. VI, §1, no. 10, Cor. of Prop. 29), so

$$[h^0, x] = \sum_{\alpha \in B} b_\alpha \alpha(h^0) X_\alpha = 2x \tag{2}$$

$$[h^0, y] = \sum_{\alpha \in B} c_\alpha(-\alpha(h^0)) X_{-\alpha} = -2y. \tag{3}$$

On the other hand,

$$[x, y] = \sum_{\alpha,\beta \in B} b_\alpha c_\beta [X_\alpha, X_{-\beta}] = \sum_{\alpha \in B} b_\alpha c_\alpha [X_\alpha, X_{-\alpha}] = -\sum_{\alpha \in B} a_\alpha H_\alpha = -h^0, \tag{4}$$

hence the existence of the homomorphism φ.

PROPOSITION 12. *Let* E *be a finite dimensional simple* $\mathfrak{g}$*-module,* ω *its highest weight, and* $\mathscr{B}$ *the vector space of* $\mathfrak{g}$*-invariant bilinear forms on* E. *Let* m *be the integer* $\sum_{\alpha \in R_+} \omega(H_\alpha)$, *so that* $m/2$ *is the sum of the coordinates of* ω *with respect to* B (Chap. VI, §1, no. 10, Cor. of Prop. 29). *Let* w_0 *be the element of* W *such that* $w_0(B) = -B$.

(i) *If* $w_0(\omega) \neq -\omega$, *then* $\mathscr{B} = 0$.

(ii) *Assume that* $w_0(\omega) = -\omega$. *Then* $\mathscr{B}$ *is of dimension* 1, *and every non-zero element of* $\mathscr{B}$ *is non-degenerate. If* m *is even* (resp. *odd*), *every element of* $\mathscr{B}$ *is symmetric* (resp. *alternating*).

a) Let $\Phi \in \mathscr{B}$. The map φ from E to E^* defined, for $x, y \in E$, by $\varphi(x)(y) = \Phi(x, y)$ is a homomorphism of $\mathfrak{g}$-modules. If $\Phi \neq 0$, then $\varphi \neq 0$, so φ is an isomorphism by Schur's lemma, and hence Φ is non-degenerate. Consequently, the $\mathfrak{g}$-module E is isomorphic to the $\mathfrak{g}$-module E^*, so that $w_0(\omega) = -\omega$. We have thus proved (i).

b) Assume from now on that $w_0(\omega) = -\omega$. Then E is isomorphic to E^*. The vector space $\mathscr{B}$ is isomorphic to $\mathrm{Hom}_{\mathfrak{g}}(E, E^*)$, and hence to $\mathrm{Hom}_{\mathfrak{g}}(E, E)$ which is of dimension 1 (§6, no. 1, Prop. 1 (iii)). Hence $\dim \mathscr{B} = 1$. Every non-zero element Φ of $\mathscr{B}$ is non-degenerate by *a*). Put $\Phi_1(x, y) = \Phi(y, x)$ for $x, y \in E$. By the preceding, there exists $\lambda \in k$ such that $\Phi_1(x, y) = \lambda\Phi(x, y)$ for all $x, y \in E$. Then $\Phi(y, x) = \lambda\Phi(x, y) = \lambda^2\Phi(y, x)$, so $\lambda^2 = 1$ and $\lambda = \pm 1$. Thus, Φ is either symmetric or alternating.

c) By Chap. VII, §1, no. 3, Prop. 9 (v), E^λ and E^μ are orthogonal with respect to Φ if $\lambda + \mu \neq 0$. Since Φ is non-degenerate, it follows that $E^\omega, E^{-\omega}$ are not orthogonal with respect to Φ.

d) There exists a homomorphism φ from $\mathfrak{sl}(2, k)$ onto a subalgebra of $\mathfrak{g}$ that takes H to $\sum_{\alpha \in R_+} H_\alpha$ (Lemma 2). Consider E as an $\mathfrak{sl}(2, k)$-module via this homomorphism. Then the elements of E^λ are of weight $\lambda\left(\sum_{\alpha \in R_+} H_\alpha\right)$. If $\lambda \in P$ is such that $E^\lambda \neq 0$ and $\lambda \neq \omega, \lambda \neq -\omega$, then $-\omega < \lambda < \omega$, so

$$-m = -\omega\left(\sum_{\alpha \in R_+} H_\alpha\right) < \lambda\left(\sum_{\alpha \in R_+} H_\alpha\right) < \omega\left(\sum_{\alpha \in R_+} H_\alpha\right) = m.$$

Let G be the isotypical component of type $V(m)$ of the $\mathfrak{sl}(2, k)$-module E. By the preceding, G is of length 1 and contains $E^\omega, E^{-\omega}$. By *c*), the restriction of Φ to G is non-zero. By §1, no. 3, Remark 3, m is even or odd according as this restriction is symmetric or alternating. In view of *b*), this completes the proof.

DEFINITION 2. *A finite dimensional irreducible representation ρ of $\mathfrak{g}$ is said to be orthogonal* (resp. *symplectic*) *if there exists on* E *a non-degenerate symmetric* (resp. *alternating*) *bilinear form invariant under ρ.*

6. REPRESENTATION RING

Let $\mathfrak{a}$ be a finite dimensional Lie algebra. Let $\mathscr{F}_{\mathfrak{a}}$ (resp. $\mathfrak{S}_{\mathfrak{a}}$) be the set of classes of finite dimensional (resp. finite dimensional simple) $\mathfrak{a}$-modules. Let $\mathscr{R}(\mathfrak{a})$ be the free abelian group $\mathbf{Z}^{(\mathfrak{S}_{\mathfrak{a}})}$. For any finite dimensional simple $\mathfrak{a}$-module E, denote its class by [E]. Let F be a finite dimensional $\mathfrak{a}$-module; let $(F_n, F_{n-1}, \ldots, F_0)$ be a Jordan-Hölder series for F; the element $\sum_{i=1}^{n} [F_i/F_{i-1}]$ of $\mathscr{R}(\mathfrak{a})$ depends only on F and not on the choice of Jordan-Hölder series; we denote it by [F]. If

$$0 \longrightarrow F' \longrightarrow F \longrightarrow F'' \longrightarrow 0$$

is an exact sequence of finite dimensional $\mathfrak{a}$-modules, then $[F] = [F'] + [F'']$.

Let F be a finite dimensional semi-simple $\mathfrak{a}$-module; for all $\mathrm{E} \in \mathfrak{S}_\mathfrak{a}$, let n_E be the length of the isotypical component of F of type E; then $[\mathrm{F}] = \sum_{\mathrm{E}\in\mathfrak{S}_\mathfrak{a}} n_\mathrm{E}.\mathrm{E}$. If F, F' are finite dimensional semi-simple $\mathfrak{a}$-modules, and if $[\mathrm{F}] = [\mathrm{F}']$, then F and F′ are isomorphic.

Lemma 3. Let G *be an abelian group written additively, and* $\varphi : \mathscr{F}_\mathfrak{a} \to \mathrm{G}$ *a map; by abuse of notation, we denote by* $\varphi(\mathrm{F})$ *the image under* φ *of the class of any finite dimensional* $\mathfrak{a}$*-module* F. *Assume that, for any exact sequence*

$$0 \longrightarrow \mathrm{F}' \longrightarrow \mathrm{F} \longrightarrow \mathrm{F}'' \longrightarrow 0$$

of finite dimensional $\mathfrak{a}$*-modules, we have* $\varphi(\mathrm{F}) = \varphi(\mathrm{F}') + \varphi(\mathrm{F}'')$. *Then, there exists a unique homomorphism* $\theta : \mathscr{R}(\mathfrak{a}) \to \mathrm{G}$ *such that* $\theta([\mathrm{F}]) = \varphi(\mathrm{F})$ *for every finite dimensional* $\mathfrak{a}$*-module* F.

There exists a unique homomorphism θ from $\mathscr{R}(\mathfrak{a})$ to G such that $\theta([\mathrm{E}]) = \varphi(\mathrm{E})$ for every finite dimensional simple $\mathfrak{a}$-module E. Let F be a finite dimensional $\mathfrak{a}$-module, and $(\mathrm{F}_n, \mathrm{F}_{n-1}, \ldots, \mathrm{F}_0)$ a Jordan-Hölder series of F; if $n > 0$, we have, by induction on n,

$$\theta([\mathrm{F}]) = \sum_{i=1}^{n} \theta([\mathrm{F}_i/\mathrm{F}_{i-1}]) = \sum_{i=1}^{n} \varphi(\mathrm{F}_i/\mathrm{F}_{i-1}) = \varphi(\mathrm{F}).$$

If $n = 0$ then $[\mathrm{F}] = 0$ so $\theta([\mathrm{F}]) = 0$; on the other hand, by considering the exact sequence $0 \longrightarrow 0 \longrightarrow 0 \longrightarrow 0 \longrightarrow 0$ we see that $\varphi(0) = 0$.

Example. Take $\mathrm{G} = \mathbf{Z}$ and $\varphi(\mathrm{F}) = \dim \mathrm{F}$. The corresponding homomorphism from $\mathscr{R}(\mathfrak{a})$ to $\mathbf{Z}$ is denoted by dim. Let c be the class of a trivial $\mathfrak{a}$-module of dimension 1, and let ψ be the homomorphism $n \mapsto nc$ from $\mathbf{Z}$ to $\mathscr{R}(\mathfrak{a})$. It is immediate that

$$\dim \circ \psi = \mathrm{Id}_\mathbf{Z},$$

so that $\mathscr{R}(\mathfrak{a})$ is the direct sum of $\operatorname{Ker}\dim$ and $\mathbf{Z}c$.

Lemma 4. There exists on the additive group $\mathscr{R}(\mathfrak{a})$ *a unique multiplication distributive over addition such that* $[\mathrm{E}][\mathrm{F}] = [\mathrm{E}\otimes\mathrm{F}]$ *for all finite dimensional* $\mathfrak{a}$*-modules* E, F. *In this way* $\mathscr{R}(\mathfrak{a})$ *is given the structure of a commutative ring. The class of the trivial* $\mathfrak{a}$*-module of dimension* 1 *is the unit element of this ring.*

The uniqueness is clear. There exists a commutative multiplication on $\mathscr{R}(\mathfrak{a}) = \mathbf{Z}^{(\mathfrak{S}_\mathfrak{a})}$ that is distributive over addition and such that $[\mathrm{E}][\mathrm{F}] = [\mathrm{E}\otimes\mathrm{F}]$ for all $\mathrm{E}, \mathrm{F} \in \mathfrak{S}_\mathfrak{a}$. Let $\mathrm{E}_1, \mathrm{E}_2$ be finite dimensional $\mathfrak{a}$-modules, l_1 and l_2 their lengths; we show that $[\mathrm{E}_1][\mathrm{E}_2] = [\mathrm{E}_1 \otimes \mathrm{E}_2]$ by induction on $l_1 + l_2$. This is clear if $l_1 + l_2 \leq 2$. On the other hand, let F_1 be a submodule of E_1 distinct from 0 and E_1. Then

$$[\mathrm{F}_1][\mathrm{E}_2] = [\mathrm{F}_1 \otimes \mathrm{E}_2] \quad \text{and} \quad [\mathrm{E}_1/\mathrm{F}_1][\mathrm{E}_2] = [(\mathrm{E}_1/\mathrm{F}_1) \otimes \mathrm{E}_2]$$

by the induction hypothesis. On the other hand, $(E_1 \otimes E_2)/(F_1 \otimes E_2)$ is isomorphic to $(E_1/F_1) \otimes E_2$. Hence

$$[E_1][E_2] = ([E_1/F_1] + [F_1]).[E_2] = [(E_1/F_1) \otimes E_2] + [F_1 \otimes E_2] = [E_1 \otimes E_2],$$

which proves our assertion. It follows immediately that the multiplication defined above is associative, so $\mathscr{R}(\mathfrak{a})$ has the structure of a commutative ring. Finally, it is clear that the class of the trivial $\mathfrak{a}$-module of dimension 1 is the unit element of this ring.

Lemma 5. There exists a unique involutive automorphism $X \mapsto X^*$ *of the ring* $\mathscr{R}(\mathfrak{a})$ *such that* $[E]^* = [E^*]$ *for every finite dimensional* $\mathfrak{a}$*-module* E.

The uniqueness is clear. By Lemma 3, there exists a homomorphism $X \mapsto X^*$ from the additive group $\mathscr{R}(\mathfrak{a})$ to itself such that $[E]^* = [E^*]$ for every finite dimensional $\mathfrak{a}$-module E. We have $(X^*)^* = X$, so this homomorphism is involutive. It is an automorphism of the ring $\mathscr{R}(\mathfrak{a})$ since $(E \otimes F)^*$ is isomorphic to $E^* \otimes F^*$ for all finite dimensional $\mathfrak{a}$-modules E and F. Q.E.D.

Let $U(\mathfrak{a})$ be the enveloping algebra of $\mathfrak{a}$, $U(\mathfrak{a})^*$ the vector space dual of $U(\mathfrak{a})$. Recall (Chap. II, §1, no. 5) that the coalgebra structure of $U(\mathfrak{a})$ defines on $U(\mathfrak{a})^*$ a commutative, associative algebra structure with unit element. For any finite dimensional $\mathfrak{a}$-module E, the map $u \mapsto \mathrm{Tr}(u_E)$ from $U(\mathfrak{a})$ to k is an element τ_E of $U(\mathfrak{a})^*$. If $0 \longrightarrow E' \longrightarrow E \longrightarrow E'' \longrightarrow 0$ is an exact sequence of finite dimensional $\mathfrak{a}$-modules, then $\tau_E = \tau_{E'} + \tau_{E''}$. Hence, by Lemma 3 there exists a unique homomorphism, which we denote by Tr, from the additive group $\mathscr{R}(\mathfrak{a})$ to the group $U(\mathfrak{a})^*$ such that $\mathrm{Tr}[E] = \tau_E$ for every finite dimensional $\mathfrak{a}$-module E. If k denotes the trivial $\mathfrak{a}$-module of dimension 1, it is easy to check that $\mathrm{Tr}[k]$ is the unit element of $U(\mathfrak{a})^*$. Finally, let E and F be finite dimensional $\mathfrak{a}$-modules. Let $u \in U(\mathfrak{a})$ and let c be the coproduct of $U(\mathfrak{a})$. By definition of the U-module $E \otimes F$ (Chap. I, §3, no. 2), if $c(u) = \sum_i u_i \otimes u'_i$,

$$u_{E\otimes F} = \sum_i (u_i)_E \otimes (u'_i)_F.$$

Consequently

$$\tau_{E\otimes F}(u) = \sum_i \mathrm{Tr}(u_i)_E \mathrm{Tr}(u'_i)_F = \sum_i \tau_E(u_i)\tau_F(u'_i)$$
$$= (\tau_E \otimes \tau_F)(c(u)).$$

This means that $\tau_E\tau_F = \tau_{E\otimes F}$. Thus, $\mathrm{Tr} : \mathscr{R}(\mathfrak{a}) \to U(\mathfrak{a})^*$ is a *homomorphism of rings.*

Let $\mathfrak{a}_1$ and $\mathfrak{a}_2$ be Lie algebras, f a homomorphism from $\mathfrak{a}_1$ to $\mathfrak{a}_2$. Every finite dimensional $\mathfrak{a}_2$-module E defines by means of f an $\mathfrak{a}_1$-module, hence elements of $\mathscr{R}(\mathfrak{a}_2)$ and $\mathscr{R}(\mathfrak{a}_1)$ that we denote provisionally by $[E]_2$ and $[E]_1$.

By Lemma 3, there exists a unique homomorphism, denoted by $\mathscr{R}(f)$, from the group $\mathscr{R}(\mathfrak{a}_2)$ to the group $\mathscr{R}(\mathfrak{a}_1)$ such that $\mathscr{R}(f)[E]_2 = [E]_1$ for every finite dimensional $\mathfrak{a}_2$-module E. Moreover, $\mathscr{R}(f)$ is a homomorphism of rings. If $U(f)$ is the homomorphism from $U(\mathfrak{a}_1)$ to $U(\mathfrak{a}_2)$ extending f, the following diagram is commutative

$$\begin{array}{ccc} \mathscr{R}(\mathfrak{a}_2) & \xrightarrow{\mathscr{R}(f)} & \mathscr{R}(\mathfrak{a}_1) \\ {\scriptstyle \mathrm{Tr}}\downarrow & & \downarrow{\scriptstyle \mathrm{Tr}} \\ U(\mathfrak{a}_2)^* & \xrightarrow{{}^tU(f)} & U(\mathfrak{a}_1)^*. \end{array}$$

In what follows we take for $\mathfrak{a}$ the splittable semi-simple Lie algebra $\mathfrak{g}$. The ring $\mathscr{R}(\mathfrak{g})$ is called the *representation ring* of $\mathfrak{g}$. For all $\lambda \in P_{++}$, we denote by $[\lambda]$ the class of the simple $\mathfrak{g}$-module $E(\lambda)$ of highest weight λ.

7. CHARACTERS OF $\mathfrak{g}$-MODULES

Let Δ be a commutative monoid written additively, and $\mathbf{Z}[\Delta] = \mathbf{Z}^{(\Delta)}$ the algebra of the monoid Δ over $\mathbf{Z}$ (*Algebra*, Chap. III, §2, no. 6). Denote by $(e^\lambda)_{\lambda\in\Delta}$ the canonical basis of $\mathbf{Z}[\Delta]$. For all $\lambda, \mu \in \Delta$, we have $e^{\lambda+\mu} = e^\lambda e^\mu$. If 0 is the neutral element of Δ, then e^0 is the unit element of $\mathbf{Z}[\Delta]$; it is denoted by 1.

Let E be a Δ-graded vector space over a field κ, and let $(E^\lambda)_{\lambda\in\Delta}$ be its graduation. If each E^λ is finite dimensional, the *character* of E, denoted by ch(E), is the element $(\dim E^\lambda)_{\lambda\in\Delta}$ of $\mathbf{Z}^\Delta$. If E itself is finite dimensional,

$$\mathrm{ch}(E) = \sum_{\lambda\in\Delta} (\dim E^\lambda) e^\lambda \in \mathbf{Z}[\Delta]. \tag{5}$$

Let E', E, E'' be Δ-graded vector spaces such that the $E'^\lambda, E^\lambda, E''^\lambda$ are finite dimensional over κ, and $0 \longrightarrow E' \longrightarrow E \longrightarrow E'' \longrightarrow 0$ an exact sequence of graded homomorphisms of degree 0. It is immediate that

$$\mathrm{ch}(E) = \mathrm{ch}(E') + \mathrm{ch}(E''). \tag{6}$$

In particular, if F_1, F_2 are Δ-graded vector spaces such that the F_1^λ and the F_2^λ are finite dimensional over κ, then

$$\mathrm{ch}(F_1 \oplus F_2) = \mathrm{ch}(F_1) + \mathrm{ch}(F_2). \tag{7}$$

If F_1 and F_2 are finite dimensional, we also have

$$\mathrm{ch}(F_1 \otimes F_2) = \mathrm{ch}(F_1).\mathrm{ch}(F_2). \tag{8}$$

Example. Assume that $\Delta = \mathbf{N}$. Let T be an indeterminate. There exists a unique isomorphism from the algebra $\mathbf{Z}[\mathbf{N}]$ to the algebra $\mathbf{Z}[T]$ that takes e^n to T^n for all $n \in \mathbf{N}$. For any finite dimensional $\mathbf{N}$-graded vector space E, the image of ch(E) in $\mathbf{Z}[T]$ is the Poincaré polynomial of E (Chap. V, §5, no. 1).

Let E be a $\mathfrak{g}$-module such that $E = \sum_{\lambda \in \mathfrak{h}^*} E^\lambda$ and such that each E^λ is finite dimensional. We know that $(E^\lambda)_{\lambda \in \mathfrak{h}^*}$ is a graduation of the vector space E. In what follows we shall reserve the notation ch(E) for the character of E considered as a $\mathfrak{h}^*$-graded vector space. Thus, the character ch(E) is an element of $\mathbf{Z}^{\mathfrak{h}^*}$. If E is finite dimensional, $\mathrm{ch}(E) \in \mathbf{Z}[P]$. By formula (6) and Lemma 3 of no. 6, there exists a unique homomorphism from the group $\mathscr{R}(\mathfrak{g})$ to $\mathbf{Z}[P]$ that takes E to ch(E), for any finite dimensional $\mathfrak{g}$-module E; this homomorphism will be denoted by ch. Relation (8) shows that ch *is a homomorphism from the ring* $\mathscr{R}(\mathfrak{g})$ *to the ring* $\mathbf{Z}[P]$.

Remark. Every element of P defines a simple $\mathfrak{h}$-module of dimension 1, hence a homomorphism from the group $\mathbf{Z}[P]$ to the group $\mathscr{R}(\mathfrak{h})$, which is an injective homomorphism of rings. It is immediate that the composite

$$\mathscr{R}(\mathfrak{g}) \longrightarrow \mathbf{Z}[P] \longrightarrow \mathscr{R}(\mathfrak{h})$$

is the homomorphism defined by the canonical injection of $\mathfrak{h}$ into $\mathfrak{g}$ (no. 6).

The Weyl group W operates by automorphisms on the group P, and hence operates on $\mathbf{Z}^P$. For all $\lambda \in P$ and all $w \in W$, we have $we^\lambda = e^{w\lambda}$. Let $\mathbf{Z}[P]^W$ be the subring of $\mathbf{Z}[P]$ consisting of the elements invariant under W.

Lemma 6. If $\lambda \in P_{++}$, *then* $\mathrm{ch}[\lambda] \in \mathbf{Z}[P]^W$. *The unique maximal term of* ch[λ] (Chap. VI, §3, no. 2, Def. 1) *is* e^λ.

The first assertion follows from no. 1, Cor. 2 of Prop. 2, and the second from §6, no. 1, Prop. 1 (ii).

THEOREM 2. (i) *Let* $(\varpi_\alpha)_{\alpha \in B}$ *be the family of fundamental weights relative to* B. *Let* $(T_\alpha)_{\alpha \in B}$ *be a family of indeterminates. The map* $f \mapsto f(([\varpi_\alpha])_{\alpha \in B})$ *from* $\mathbf{Z}[(T_\alpha)_{\alpha \in B}]$ *to* $\mathscr{R}(\mathfrak{g})$ *is an isomorphism of rings.*

(ii) *The homomorphism* ch *from* $\mathscr{R}(\mathfrak{g})$ *to* $\mathbf{Z}[P]$ *induces an isomorphism from the ring* $\mathscr{R}(\mathfrak{g})$ *to the ring* $\mathbf{Z}[P]^W$.

(iii) *Let* E *be a finite dimensional* $\mathfrak{g}$*-module. If* $\mathrm{ch}\,E = \sum_{\lambda \in P_{++}} m_\lambda \mathrm{ch}[\lambda]$, *the isotypical component of* E *of highest weight* λ *has length* m_λ.

The family $([\lambda])_{\lambda \in P_{++}}$ is a basis of the **Z**-module $\mathscr{R}(\mathfrak{g})$, and the family $(\mathrm{ch}[\lambda])_{\lambda \in P_{++}}$ is a basis of the **Z**-module $\mathbf{Z}[P]^W$ (Lemma 6, and Chap. VI, §3, no. 4, Prop. 3). This proves (ii) and (iii). Assertion (i) follows from (ii), Lemma 6 and Chap. VI, §3, no. 4, Th. 1.

COROLLARY. *Let* E, E′ *be finite dimensional* $\mathfrak{g}$*-modules. Then* E *is isomorphic to* E′ *if and only if* ch E = ch E′.

This follows from Th. 2 (ii) and the fact that E, E′ are semi-simple.

§8. SYMMETRIC INVARIANTS

In this paragraph, we denote by $(\mathfrak{g}, \mathfrak{h})$ *a split semi-simple Lie algebra, by* R *its root system, by* W *its Weyl group, and by* P *its group of weights.*

1. EXPONENTIAL OF A LINEAR FORM

Let V be a finite dimensional vector space, $\mathbf{S}(V)$ its symmetric algebra. The coalgebra structure of $\mathbf{S}(V)$ defines on $\mathbf{S}(V)^*$ a commutative and associative algebra structure (*Algebra*, Chap. III, §11, pp. 579 to 582). The vector space $\mathbf{S}(V)^*$ can be identified canonically with $\prod_{m\geq 0} \mathbf{S}^m(V)^*$, and $\mathbf{S}^m(V)^*$ can be identified canonically with the space of symmetric m-linear forms on V. The canonical injection of $V^* = \mathbf{S}^1(V)^*$ into $\mathbf{S}(V)^*$ defines an injective homomorphism from the algebra $\mathbf{S}(V^*)$ to the algebra $\mathbf{S}(V)^*$, whose image is $\mathbf{S}(V)^{*gr} = \sum_{m\geq 0} \mathbf{S}^m(V)^*$ (*Algebra*, Chap. III, §11, no. 5, Prop. 8). We identify the algebras $\mathbf{S}(V^*)$ and $\mathbf{S}(V)^{*gr}$ by means of this homomorphism; we also identify $\mathbf{S}(V^*)$ with the algebra of polynomial functions on V (Chap. VII, App. I, no. 1).

The elements $(u_m) \in \prod_{m\geq 0} \mathbf{S}^m(V)^*$ such that $u_0 = 0$ form an ideal J of $\mathbf{S}(V)^*$; we give $\mathbf{S}(V)^*$ the J-adic topology (*Commutative Algebra*, Chap. III, §2, no. 5), in which $\mathbf{S}(V)^*$ is complete and $\mathbf{S}(V^*)$ is dense in $\mathbf{S}(V)^*$. If $(e_i^*)_{1\leq i\leq n}$ is a basis of V^*, and if $T_1, \dots, T_n$ are indeterminates, the homomorphism from $k[T_1, \dots, T_n]$ to $\mathbf{S}(V^*)$ that takes T_i to e_i^* $(1 \leq i \leq n)$ is an isomorphism of algebras, and extends to a continuous isomorphism from the algebra $k[[T_1, \dots, T_n]]$ to the algebra $\mathbf{S}(V)^*$.

For all $\lambda \in V^*$, the family $\lambda^n/n!$ is summable in $\mathbf{S}(V)^*$. Its sum is called the *exponential of* λ and is denoted by $\exp(\lambda)$ (conforming to Chap. II, §6, no. 1). Let $x_1, \dots, x_n \in V$; we have

$$\langle \exp \lambda, x_1 \dots x_n \rangle = \frac{1}{n!}\langle \lambda^n, x_1 \dots x_n \rangle = \langle \lambda, x_1 \rangle \dots \langle \lambda, x_n \rangle$$

by *Algebra*, Chap. III, §11, no. 5, formula (29). It follows immediately that $\exp(\lambda)$ *is the unique homomorphism from the algebra* $\mathbf{S}(V)$ *to* k *that extends* λ.

We have $\exp(\lambda+\mu) = \exp(\lambda)\exp(\mu)$ for all $\lambda, \mu \in V^*$ (Chap. II, §6, no. 1, Remark). Thus, *the map* $\exp : V^* \to \mathbf{S}(V)^*$ *is a homomorphism from the additive group* V^* *to the multiplicative group of invertible elements of* $\mathbf{S}(V)^*$. The family $(\exp \lambda)_{\lambda \in V^*}$ is a free family in the vector space $\mathbf{S}(V)^*$ (*Algebra*, Chap. V, §7, no. 3, Th. 1).

Lemma 1. Let Π *be a subgroup of* V^* *that generates the vector space* V^*, *and* m *an integer* ≥ 0. *Then* $\mathrm{pr}_m(\exp \Pi)$ *generates the vector space* $\mathbf{S}^m(V^*)$.

By *Algebra*, Chap. I, §8, no. 2, Prop. 2, any product of m elements of V^* is a k-linear combination of elements of the form x^m where $x \in \Pi$. But $x^m = m!\,\mathrm{pr}_m(\exp x)$. Q.E.D.

By transport of structure, every automorphism of V defines automorphisms of the algebras $\mathbf{S}(V)$ and $\mathbf{S}(V)^*$; this gives linear representations of $\mathbf{GL}(V)$ on $\mathbf{S}(V)$ and $\mathbf{S}(V)^*$.

2. INJECTION OF $k[P]$ INTO $\mathbf{S}(\mathfrak{h})^*$

The map $p \mapsto \exp p$ from P to $\mathbf{S}(\mathfrak{h})^*$ is a homomorphism from the additive group P to $\mathbf{S}(\mathfrak{h})^*$ equipped with its multiplicative structure (no. 1). Consequently, there exists a unique homomorphism ψ from the algebra $k[P]$ of the monoid P to the algebra $\mathbf{S}(\mathfrak{h})^*$ such that

$$\psi(e^\lambda) = \exp(\lambda) \qquad (\lambda \in P)$$

(in the notations of §7, no. 7). By no. 1, ψ is injective. By transport of structure, $\psi(w(e^\lambda)) = w(\psi(e^\lambda))$ for all $\lambda \in P$ and all $w \in W$. Hence, if $k[P]^W$ (resp. $\mathbf{S}(\mathfrak{h})^{*W}$) denotes the set of elements of $k[P]$ (resp. $\mathbf{S}(\mathfrak{h})^*$) invariant under W, we have $\psi(k[P]^W) \subset \mathbf{S}(\mathfrak{h})^{*W}$.

PROPOSITION 1. *Let* $S^m(\mathfrak{h}^*)^W$ *be the set of elements of* $S^m(\mathfrak{h}^*)$ *invariant under* W. *Then* $\mathrm{pr}_m(\psi(k[P]^W)) = \mathbf{S}^m(\mathfrak{h}^*)^W$.

It is clear from the preceding that $\mathrm{pr}_m(\psi(k[P]^W)) \subset \mathbf{S}^m(\mathfrak{h}^*)^W$. Every element of $\mathbf{S}^m(\mathfrak{h}^*)$ is a k-linear combination of elements of the form

$$\mathrm{pr}_m(\exp \lambda) = (\mathrm{pr}_m \circ \psi)(e^\lambda)$$

where $\lambda \in P$ (Lemma 1). Hence every element of $\mathbf{S}^m(\mathfrak{h}^*)^W$ is a linear combination of elements of the form

$$\sum_{w \in W} w((\mathrm{pr}_m \circ \psi)(e^\lambda)) = (\mathrm{pr}_m \circ \psi)\left(\sum_{w \in W} w(e^\lambda)\right),$$

each of which belongs to $\mathrm{pr}_m(\psi(k[P]^W))$.

PROPOSITION 2. *Let* E *be a finite dimensional* $\mathfrak{g}$*-module. Let* $U(\mathfrak{h}) = \mathbf{S}(\mathfrak{h})$ *be the enveloping algebra of* $\mathfrak{h}$. *If* $u \in U(\mathfrak{h})$, *then*

$$\mathrm{Tr}(u_E) = \langle \psi(\mathrm{ch}\, E), u \rangle.$$

It suffices to treat the case in which $u = h_1 \dots h_m$ with $h_1, \dots, h_m \in \mathfrak{h}$. For all $\lambda \in P$, let $d_\lambda = \dim E^\lambda$. Then $\mathrm{ch}\, E = \sum_\lambda d_\lambda e^\lambda$, so $\psi(\mathrm{ch}\, E) = \sum_\lambda d_\lambda \exp(\lambda)$ and hence

$$\begin{aligned}\langle\psi(\operatorname{ch} E), u\rangle &= \sum_\lambda d_\lambda \langle \exp\lambda, h_1 \dots h_m\rangle \\ &= \sum_\lambda d_\lambda \lambda(h_1)\dots\lambda(h_m) \quad \text{(no. 1)} \\ &= \operatorname{Tr} u_{\mathrm{E}}.\end{aligned}$$

COROLLARY 1. *Let* $\mathrm{U}(\mathfrak{g})$ *be the enveloping algebra of* $\mathfrak{g}$. *Let the homomorphism* $\zeta : \mathrm{U}(\mathfrak{g})^* \to \mathrm{U}(\mathfrak{h})^* = \mathbf{S}(\mathfrak{h})^*$ *be the transpose of the canonical injection* $\mathrm{U}(\mathfrak{h}) \to \mathrm{U}(\mathfrak{g})$. *The following diagram commutes*

$$\begin{array}{ccc} \mathscr{R}(\mathfrak{g}) & \xrightarrow{\operatorname{ch}} & \mathbf{Z}[\mathrm{P}] \\ {\scriptstyle\operatorname{Tr}}\downarrow & & \downarrow{\scriptstyle\psi} \\ \mathrm{U}(\mathfrak{g})^* & \xrightarrow{\zeta} & \mathbf{S}(\mathfrak{h})^*. \end{array}$$

This is simply a reformulation of Prop. 2.

COROLLARY 2. *Let* m *be an integer* ≥ 0. *Every element of* $\mathbf{S}^m(\mathfrak{h}^*)^{\mathrm{W}}$ *is a linear combination of polynomial functions on* $\mathfrak{h}$ *of the form* $x \mapsto \operatorname{Tr}(\rho(x)^m)$, *where* ρ *is a finite dimensional linear representation of* $\mathfrak{g}$.

By Prop. 1, $\mathbf{S}^m(\mathfrak{h}^*)^{\mathrm{W}} = (\mathrm{pr}_m \circ \psi)(k[\mathrm{P}]^{\mathrm{W}})$. Now $\mathbf{Z}[\mathrm{P}]^{\mathrm{W}} = \operatorname{ch}\mathscr{R}(\mathfrak{g})$ (§7, no. 7, Th. 2 (ii)). Thus, by Chap. VI, §3, no. 4, Lemma 3, $\psi(k[\mathrm{P}]^{\mathrm{W}})$ is the k-vector subspace of $\mathbf{S}(\mathfrak{h})^*$ generated by $\psi(\operatorname{ch}\mathscr{R}(\mathfrak{g})) = \zeta(\operatorname{Tr}\mathscr{R}(\mathfrak{g}))$. Consequently, $\mathbf{S}^m(\mathfrak{h}^*)^{\mathrm{W}}$ is the vector subspace of $\mathbf{S}^m(\mathfrak{h}^*)$ generated by $(\mathrm{pr}_m \circ \zeta \circ \operatorname{Tr})(\mathscr{R}(\mathfrak{g}))$. But, if ρ is a finite dimensional linear representation of $\mathfrak{g}$,

$$((\mathrm{pr}_m \circ \zeta \circ \operatorname{Tr})(\rho))(x) = \left\langle (\zeta \circ \operatorname{Tr})(\rho), \frac{x^m}{m!} \right\rangle = \frac{1}{m!}\operatorname{Tr}(\rho(x)^m)$$

for all $x \in \mathfrak{h}$.

3. INVARIANT POLYNOMIAL FUNCTIONS

Let $\mathfrak{a}$ be a finite dimensional Lie algebra. In accordance with the conventions of no. 1, we identify the algebra $\mathbf{S}(\mathfrak{a}^*)$, the algebra $\mathbf{S}(\mathfrak{a})^{*gr}$, and the algebra of polynomial functions on $\mathfrak{a}$. For all $a \in \mathfrak{a}$, let $\theta(a)$ be the derivation of $\mathbf{S}(\mathfrak{a})$ such that $\theta(a)x = [a, x]$ for all $x \in \mathfrak{a}$. We know (Chap. I, §3, no. 2) that θ is a representation of $\mathfrak{a}$ on $\mathbf{S}(\mathfrak{a})$. Let $\theta^*(a)$ be the restriction of $-{}^t\theta(a)$ to $\mathbf{S}(\mathfrak{a}^*)$. Then θ^* is a representation of $\mathfrak{a}$. If $f \in \mathbf{S}^n(\mathfrak{a}^*)$, then $\theta^*(a)f \in \mathbf{S}^n(\mathfrak{a}^*)$ and, for $x_1, \dots, x_n \in \mathfrak{a}$,

$$(\theta^*(a)f)(x_1, \dots, x_n) = -\sum_{1 \leq i \leq n} f(x_1, \dots, x_{i-1}, [a, x_i], x_{i+1}, \dots, x_n). \quad (1)$$

We deduce easily from (1) that $\theta^*(a)$ is a derivation of $\mathbf{S}(\mathfrak{a}^*)$. An element of $\mathbf{S}(\mathfrak{a})$ (resp. $\mathbf{S}(\mathfrak{a}^*)$) that is invariant under the representation θ (resp. θ^*) of $\mathfrak{a}$ is called an *invariant element* of $\mathbf{S}(\mathfrak{a})$ (resp. $\mathbf{S}(\mathfrak{a}^*)$).

Lemma 2. Let ρ be a finite dimensional linear representation of $\mathfrak{a}$, and m an integer ≥ 0. The function $x \mapsto \mathrm{Tr}(\rho(x)^m)$ on $\mathfrak{a}$ is an invariant polynomial function.

Put $g(x_1, \ldots, x_m) = \mathrm{Tr}(\rho(x_1) \ldots \rho(x_m))$ for $x_1, \ldots, x_m \in \mathfrak{a}$. If $x \in \mathfrak{a}$, we have

$$\begin{aligned} -(\theta^*(x)g)(x_1, \ldots, x_m) \\ &= \sum_{1 \leq i \leq m} \mathrm{Tr}(\rho(x_1) \ldots \rho(x_{i-1})[\rho(x), \rho(x_i)]\rho(x_{i+1}) \ldots \rho(x_m)) \\ &= \mathrm{Tr}(\rho(x)\rho(x_1) \ldots \rho(x_m)) - \mathrm{Tr}(\rho(x_1) \ldots \rho(x_m)\rho(x)) = 0, \end{aligned}$$

so $\theta^*(x)g = 0$. Let h be the symmetric multilinear form defined by

$$h(x_1, \ldots, x_m) = \frac{1}{m!} \sum_{\sigma \in \mathfrak{S}_m} g(x_{\sigma(1)}, \ldots, x_{\sigma(m)}).$$

For all $x \in \mathfrak{a}$, we have $\theta^*(x)h = 0$ and $\mathrm{Tr}(\rho(x)^m) = h(x, \ldots, x)$, hence the lemma.

Lemma 3. Let E be a finite dimensional $\mathfrak{g}$-module, and $x \in \mathrm{E}$. Then x is an invariant element of the $\mathfrak{g}$-module E if and only if $(\exp a_{\mathrm{E}}).x = x$ for every nilpotent element a of $\mathfrak{g}$.

The condition is clearly necessary. Assume now that it is satisfied. Let a be a nilpotent element of $\mathfrak{g}$. There exists an integer n such that $a_{\mathrm{E}}^n = 0$. For all $t \in k$, we have

$$0 = \exp(ta_{\mathrm{E}}).x - x = ta_{\mathrm{E}}x + \frac{1}{2!}t^2 a_{\mathrm{E}}^2 x + \cdots + \frac{1}{(n-1)!}t^{n-1}a_{\mathrm{E}}^{n-1}x,$$

so $a_{\mathrm{E}}x = 0$. But the Lie algebra $\mathfrak{g}$ is generated by its nilpotent elements (§4, no. 1, Prop. 1). Hence x is an invariant element of the $\mathfrak{g}$-module E. Q.E.D.

For any $\xi \in \mathbf{GL}(\mathfrak{g})$, let $\mathbf{S}(\xi)$ be the automorphism of $\mathbf{S}(\mathfrak{g})$ that extends ξ, and $\mathbf{S}^*(\xi)$ the restriction to $\mathbf{S}(\mathfrak{g}^*)$ of the contragredient automorphism of $\mathbf{S}(\xi)$. Then $\mathbf{S}$ and $\mathbf{S}^*$ are representations of $\mathbf{GL}(\mathfrak{g})$. If a is a nilpotent element of $\mathfrak{g}$, $\theta(a)$ is locally nilpotent on $\mathbf{S}(\mathfrak{g})$ and $\mathbf{S}(\exp \mathrm{ad}\ a) = \exp \theta(a)$, so

$$\mathbf{S}^*(\exp \mathrm{ad}\ a) = \exp \theta^*(a). \tag{2}$$

PROPOSITION 3. *Let f be a polynomial function on $\mathfrak{g}$. The following conditions are equivalent:*

(i) *$f \circ s = f$ for all $s \in \mathrm{Aut}_e(\mathfrak{g})$;*

(ii) $f \circ s = f$ *for all* $s \in \mathrm{Aut}_0(\mathfrak{g})$;

(iii) f *is invariant.*

The equivalence of (i) and (iii) follows from formula (2) and Lemma 3. It follows from this that (iii) implies (ii) by extension of the base field. The implication (ii) $\Longrightarrow$ (i) is clear.

Note carefully that, if f satisfies the conditions of Prop. 3, f is not in general invariant under $\mathrm{Aut}(\mathfrak{g})$ (Exerc. 1 and 2).

THEOREM 1. *Let* $\mathrm{I}(\mathfrak{g}^*)$ *be the algebra of invariant polynomial functions on* $\mathfrak{g}$. *Let* $i : \mathbf{S}(\mathfrak{g}^*) \to \mathbf{S}(\mathfrak{h}^*)$ *be the restriction homomorphism.*

(i) *The map* $i|\mathrm{I}(\mathfrak{g}^*)$ *is an isomorphism from the algebra* $\mathrm{I}(\mathfrak{g}^*)$ *to the algebra* $\mathbf{S}(\mathfrak{h}^*)^{\mathrm{W}}$.

(ii) *For any integer* $n \geq 0$, *let* $\mathrm{I}^n(\mathfrak{g}^*)$ *be the set of homogeneous elements of* $\mathrm{I}(\mathfrak{g}^*)$ *of degree* n. *Then* $\mathrm{I}^n(\mathfrak{g}^*)$ *is the set of linear combinations of functions on* $\mathfrak{g}$ *of the form* $x \mapsto \mathrm{Tr}(\rho(x)^n)$, *where* ρ *is a finite dimensional linear representation of* $\mathfrak{g}$.

(iii) *Let* $l = \mathrm{rk}(\mathfrak{g})$. *There exist* l *algebraically independent homogeneous elements of* $\mathrm{I}(\mathfrak{g}^*)$ *that generate the algebra* $\mathrm{I}(\mathfrak{g}^*)$.

a) Let $f \in \mathrm{I}(\mathfrak{g}^*)$ and $w \in \mathrm{W}$. There exists $s \in \mathrm{Aut}_e(\mathfrak{g}, \mathfrak{h})$ such that $s|\mathfrak{h} = w$ (§2, no. 2, Cor. of Th. 2). Since f is invariant under s (Prop. 3), $i(f)$ is invariant under w. Hence $i(\mathrm{I}(\mathfrak{g}^*)) \subset \mathbf{S}(\mathfrak{h}^*)^{\mathrm{W}}$.

b) We prove that, if $f \in \mathrm{I}(\mathfrak{g}^*)$ is such that $i(f) = 0$, then $f = 0$. Extending the base field if necessary, we can assume that k is algebraically closed. By Prop. 3, f vanishes on $s(\mathfrak{h})$ for all $s \in \mathrm{Aut}_e(\mathfrak{g})$. Hence f vanishes on every Cartan subalgebra of $\mathfrak{g}$ (Chap. VII, §3, no. 2, Th. 1), and in particular on the set of regular elements of $\mathfrak{g}$. But this set is dense in $\mathfrak{g}$ for the Zariski topology (Chap. VII, §2, no. 2).

c) Let n be an integer ≥ 0. Let L^n be the set of linear combinations of functions of the form $x \mapsto \mathrm{Tr}(\rho(x)^n)$ on $\mathfrak{g}$, where ρ is a finite dimensional linear representation of $\mathfrak{g}$. By Lemma 2, $\mathrm{L}^n \subset \mathrm{I}^n(\mathfrak{g}^*)$. Thus

$$i(\mathrm{L}^n) \subset i(\mathrm{I}^n(\mathfrak{g}^*)) \subset \mathbf{S}^n(\mathfrak{h}^*)^{\mathrm{W}}.$$

By Cor. 2 of Prop. 2, $\mathbf{S}^n(\mathfrak{h}^*)^{\mathrm{W}} \subset i(\mathrm{L}^n)$. Hence $i(\mathrm{I}^n(\mathfrak{g}^*)) = \mathbf{S}^n(\mathfrak{h}^*)^{\mathrm{W}}$, which proves (i), and $i(\mathrm{L}^n) = i(\mathrm{I}^n(\mathfrak{g}^*))$ so $\mathrm{L}^n = \mathrm{I}^n(\mathfrak{g}^*)$ by *b*). Thus (ii) is proved.

d) Assertion (iii) follows from (i) and Chap. V, §5, no. 3, Th. 3.

COROLLARY 1. *Assume that* $\mathfrak{g}$ *is simple. Let* $m_1, \ldots, m_l$ *be the exponents of the Weyl group of* $\mathfrak{g}$. *There exist elements* $\mathrm{P}_1, \ldots, \mathrm{P}_l$ *of* $\mathrm{I}(\mathfrak{g}^*)$, *homogeneous of degrees*

$$m_1 + 1, \ldots, m_l + 1,$$

which are algebraically independent and generate the algebra $\mathrm{I}(\mathfrak{g}^*)$.

This follows from Th. 2 (i) and Chap. V, §6, no. 2, Prop. 3.

COROLLARY 2. *Let* B *be a basis of* R, R_+ (resp. R_-) *the set of positive* (resp. *negative*) *roots of* $(\mathfrak{g}, \mathfrak{h})$ *relative to* B, $\mathfrak{n}_+ = \sum_{\alpha\in R_+} \mathfrak{g}^\alpha$, $\mathfrak{n}_- = \sum_{\alpha\in R_-} \mathfrak{g}^\alpha$, $\mathbf{S}(\mathfrak{h})$ *the symmetric algebra of* $\mathfrak{h}$, *and* J *the ideal of* $\mathbf{S}(\mathfrak{g})$ *generated by* $\mathfrak{n}_+ \cup \mathfrak{n}_-$.

(i) $\mathbf{S}(\mathfrak{g}) = \mathbf{S}(\mathfrak{h}) \oplus J$.

(ii) *Let* j *be the homomorphism from the algebra* $\mathbf{S}(\mathfrak{g})$ *to the algebra* $\mathbf{S}(\mathfrak{h})$ *defined by the preceding decomposition of* $\mathbf{S}(\mathfrak{g})$. *Let* $I(\mathfrak{g})$ *be the set of invariant elements of* $\mathbf{S}(\mathfrak{g})$. *Let* $\mathbf{S}(\mathfrak{h})^W$ *be the set of elements of* $\mathbf{S}(\mathfrak{h})$ *invariant under the operation of* W. *Then* $j|I(\mathfrak{g})$ *is an isomorphism from* $I(\mathfrak{g})$ *to* $\mathbf{S}(\mathfrak{h})^W$.

Assertion (i) is clear. The Killing form defines an isomorphism from the vector space $\mathfrak{g}^*$ to the vector space $\mathfrak{g}$, which extends to an isomorphism ξ from the $\mathfrak{g}$-module $\mathbf{S}(\mathfrak{g}^*)$ to the $\mathfrak{g}$-module $\mathbf{S}(\mathfrak{g})$. We have $\xi(I(\mathfrak{g}^*)) = I(\mathfrak{g})$. The orthogonal complement of $\mathfrak{h}$ with respect to the Killing form is $\mathfrak{n}_+ + \mathfrak{n}_-$ (§2, no. 2, Prop. 1). If we identify $\mathfrak{h}^*$ with the orthogonal complement of $\mathfrak{n}_+ + \mathfrak{n}_-$ in $\mathfrak{g}^*$, then $\xi(\mathfrak{h}^*) = \mathfrak{h}$, so $\xi(\mathbf{S}(\mathfrak{h}^*)) = \mathbf{S}(\mathfrak{h})$ and $\xi(\mathbf{S}(\mathfrak{h}^*)^W) = \mathbf{S}(\mathfrak{h})^W$. Finally, $\xi^{-1}(J)$ is the set of polynomial functions on $\mathfrak{g}$ that vanish on $\mathfrak{h}$. This proves that ξ transforms the homomorphism i of Th. 1 into the homomorphism j of Cor. 2. Thus assertion (ii) follows from Th. 1 (i).

PROPOSITION 4. *Let* $\mathfrak{a}$ *be a semi-simple Lie algebra,* l *its rank. Let* I (resp. I′) *be the set of elements of* $\mathbf{S}(\mathfrak{a}^*)$ (resp. $\mathbf{S}(\mathfrak{a})$) *invariant under the representation induced by the adjoint representation of* $\mathfrak{a}$. *Let* Z *be the centre of the enveloping algebra of* $\mathfrak{a}$.

(i) I *and* I′ *are graded polynomial algebras* (Chap. V, § 5, no. 1) *of transcendance degree* l.

(ii) Z *is isomorphic to the algebra of polynomials in* l *indeterminates over* k.

The canonical filtration of the enveloping algebra of $\mathfrak{a}$ induces a filtration of Z. By Chap. I, §2, no. 7, Th. 1 and p. 25, gr Z is isomorphic to I′. In view of *Commutative Algebra*, Chap. III, §2, no. 9, Prop. 10, it follows that (i) $\Longrightarrow$ (ii).

On the other hand, Th. 1 and its Cor. 2 show that (i) is true whenever $\mathfrak{a}$ is split. The general case reduces to that case in view of the following lemma:

Lemma 4.[2] *Let* $A = \bigoplus_{n\geq 0} A^n$ *be a graded* k*-algebra,* k' *an extension of* k, *and* $A' = A \otimes_k k'$. *Assume that* A′ *is a graded polynomial algebra over* k'. *Then* A *is a graded polynomial algebra over* k.

We have $A'^0 = k'$, so $A^0 = k$. Put $A_+ = \bigoplus_{n\geq 1} A^n$ and $P = A_+/A_+^2$. Then P is a graded vector space, and there is a graded linear map $f : P \to A_+$ of

[2] In Lemmas 4, 5 and 6, k can be any (commutative) field.

degree zero such that the composite with the canonical projection $A_+ \to P$ is the identity on P. Give $\mathbf{S}(P)$ the graded structure induced by that of P (*Algebra*, Chap. III, p. 506). The homomorphism of k-algebras $g : \mathbf{S}(P) \to A$ that extends f (*Algebra*, Chap. III, p. 497) is a graded homomorphism of degree 0; an immediate induction on the degree shows that g is surjective.

Lemma 5. A is a graded polynomial algebra if and only if P is finite dimensional and g is bijective.

If P is finite dimensional, $\mathbf{S}(P)$ is clearly a graded polynomial algebra, and so is A if g is bijective. Conversely, assume that A is generated by algebraically independent homogeneous elements $x_1, \dots, x_m$ of degrees $d_1, \dots, d_m$. Let $\bar{x}_i$ be the image of x_i in P. It is immediate that the $\bar{x}_i$ form a basis of P; since $\bar{x}_i$ is of degree d_i, it follows that $\mathbf{S}(P)$ and A are isomorphic; in particular, $\dim \mathbf{S}(P)^n = \dim A^n$ for all n. Since g is surjective, it is necessarily bijective.

Lemma 4 is now immediate. Indeed, Lemma 5, applied to the k'-algebra A', shows that $g \otimes 1 : \mathbf{S}(P) \otimes k' \to A \otimes k'$ is bijective, and hence so is g.

PROPOSITION 5. *We retain the notations of Prop. 4, and denote by* $\mathfrak{p}$ *the ideal of* $\mathbf{S}(\mathfrak{a}^*)$ *generated by the homogeneous elements of* I *of degree* ≥ 1. *Let* $x \in \mathfrak{a}$. *Then* x *is nilpotent if and only if* $f(x) = 0$ *for all* $x \in \mathfrak{p}$.[3]

Extending the base field if necessary, we can assume that $\mathfrak{a} = \mathfrak{g}$ is splittable. Assume that x is nilpotent. For any finite dimensional linear representation ρ of $\mathfrak{g}$, and any integer $n \geq 1$, we have $\mathrm{Tr}(\rho(x)^n) = 0$, so $f(x) = 0$ for all homogeneous $f \in I(\mathfrak{g}^*)$ of degree ≥ 1 (Th. 1 (ii)), and hence $f(x) = 0$ for all $f \in \mathfrak{p}$. Conversely, if $f(x) = 0$ for all $f \in \mathfrak{p}$, then $\mathrm{Tr}((\mathrm{ad}\ x)^n) = 0$ for all $n \geq 1$ (Th. 1 (ii)), so x is nilpotent.

**Remarks* 1) Let $P_1, \dots, P_l$ be algebraically independent homogeneous elements of I that generate the algebra I. *Then* $(P_1, \dots, P_l)$ *is an* $\mathbf{S}(\mathfrak{a}^*)$*-regular sequence* (Chap. V, §5, no. 5). Indeed, extending the base field if necessary, we can assume that $\mathfrak{a} = \mathfrak{g}$ is splittable. Now let $N = \dim \mathfrak{g}$, and let

$$(Q_1, \dots, Q_{N-l})$$

be a basis of the orthogonal complement of $\mathfrak{h}$ in $\mathfrak{g}^*$. Let $\mathfrak{m}$ be the ideal of $\mathbf{S}(\mathfrak{g}^*)$ generated by $P_1, \dots, P_l, Q_1, \dots, Q_{N-l}$. Then $\mathbf{S}(\mathfrak{g}^*)\mathfrak{m}$ is isomorphic to $\mathbf{S}(\mathfrak{h}^*)/J$, where J is the ideal of $\mathbf{S}(\mathfrak{h}^*)$ generated by $i(P_1), \dots, i(P_l)$. By Th. 1 and Chap. V, §5, no. 2, Th. 2, $\mathbf{S}(\mathfrak{h}^*)/J$ is a finite dimensional vector space, and hence so is $\mathbf{S}(\mathfrak{g}^*)/\mathfrak{m}$. By a result of *Commutative Algebra*, it follows that $(P_1, \dots, P_l, Q_1, \dots, Q_{N-l})$ is an $\mathbf{S}(\mathfrak{g}^*)$-regular sequence, and *a fortiori* so is $(P_1, \dots, P_l)$.

[3] It can be shown (B. KOSTANT, Lie group representations on polynomial rings, *Amer. J. Math.*, Vol. LXXXV (1963), pp. 327-404, Th. 10 and 15) that $\mathfrak{p}$ is a prime ideal of $\mathbf{S}(\mathfrak{a}^*)$ and that $\mathbf{S}(\mathfrak{a}^*)/\mathfrak{p}$ is integrally closed.

2) *The algebra* $\mathbf{S}(\mathfrak{a}^*)$ *is a graded free module over* I. Indeed, this follows from Prop. 4, Remark 1, and Chap. V, §5, no. 5, Lemma 5.$_*$

4. PROPERTIES OF Aut_0

Lemma 6. Let V *be a finite dimensional vector space,* G *a finite group of automorphisms of* V*, and* v *and* v' *elements of* V *such that* $v' \notin \mathrm{G}v$*. There exists a* G*-invariant polynomial function* f *on* V *such that* $f(v') \neq f(v)$*.*

Indeed, for each $s \in \mathrm{G}$ there exists a polynomial function g_s on V equal to 1 at v and to 0 at sv'. Then the function $g = 1 - \prod_{s\in\mathrm{G}} g_s$ is equal to 0 at v and to 1 on $\mathrm{G}v'$. The polynomial function $f = \prod_{t\in g} t.g$ is G-invariant, equal to 0 at v and to 1 at v'.

PROPOSITION 6. *Let* $\mathfrak{a}$ *be a semi-simple Lie algebra and* $s \in \mathrm{Aut}(\mathfrak{a})$*. The following conditions are equivalent:*

(i) $s \in \mathrm{Aut}_0(\mathfrak{a})$;

(ii) *for any invariant polynomial function* f *on* $\mathfrak{a}$*, we have* $f \circ s = f$*.*

By extending scalars if necessary, we can assume that k is algebraically closed. The implication (i) $\Longrightarrow$ (ii) follows from Prop. 3. We assume that condition (ii) is satisfied and prove (i). In view of Prop. 3, and §5, no. 3, Cor. 1 of Prop. 5, we can assume that $s \in \mathrm{Aut}(\mathfrak{g}, \mathfrak{h})$ and that s leaves stable a Weyl chamber C. Let $x \in \mathrm{C} \cap \mathfrak{h}_{\mathbf{Q}}$. We have $sx \in \mathrm{C}$. If g is a W-invariant polynomial function on $\mathfrak{h}$, we have $g(x) = g(sx)$ (Th. 1 (i)). By Lemma 6, it follows that $sx \in \mathrm{W}x$. Since $sx \in \mathrm{C}$, we have $x = sx$ (Chap. V, §3, no. 3, Th. 2). Then $s|\mathfrak{h} = \mathrm{Id}_{\mathfrak{h}}$, and $s \in \mathrm{Aut}_0(\mathfrak{g}, \mathfrak{h})$ (§5, no. 2, Prop. 4).

COROLLARY. *The group* $\mathrm{Aut}_0(\mathfrak{a})$ *is open and closed in* $\mathrm{Aut}(\mathfrak{a})$ *in the Zariski topology.*

Prop. 6 shows that $\mathrm{Aut}_0(\mathfrak{a})$ is closed. Let $\bar{k}$ be an algebraic closure of k. The group $\mathrm{Aut}(\mathfrak{a} \otimes \bar{k})/\mathrm{Aut}_0(\mathfrak{a} \otimes \bar{k})$ is finite (§5, no. 3, Cor. 1 of Prop. 5); *a fortiori*, the group $\mathrm{Aut}(\mathfrak{a})/\mathrm{Aut}_0(\mathfrak{a})$ is finite. Since the cosets of $\mathrm{Aut}(\mathfrak{a})$ in $\mathrm{Aut}(\mathfrak{a})$ are closed, it follows that $\mathrm{Aut}_0(\mathfrak{a})$ is open in $\mathrm{Aut}(\mathfrak{a})$.

5. CENTRE OF THE ENVELOPING ALGEBRA

In this number, we choose a basis B of R. Let R_+ be the set of positive roots relative to B. Let $\rho = \frac{1}{2} \sum_{\alpha\in\mathrm{R}_+} \alpha$, and δ the automorphism of the algebra $\mathbf{S}(\mathfrak{h})$ that takes every $x \in \mathfrak{h}$ to $x - \rho(x)$, and hence the polynomial function p on $\mathfrak{h}^*$ to the function $\lambda \mapsto p(\lambda - \rho)$.

THEOREM 2. *Let* U *be the enveloping algebra of* $\mathfrak{g}$*,* Z *its centre,* $\mathrm{V} \subset \mathrm{U}$ *the enveloping algebra of* $\mathfrak{h}$ *(identified with* $\mathbf{S}(\mathfrak{h})$*),* U^0 *the commutant of* V *in* U*,*

φ the Harish-Chandra homomorphism (§6, no. 4) *from* U^0 *to* V *relative to* B. *Let* $\mathbf{S}(\mathfrak{h})^{\mathrm{W}}$ *be the set of elements of* $\mathbf{S}(\mathfrak{h})$ *invariant under the action of* W. *Then* $(\delta\circ\varphi)|\mathrm{Z}$ *is an isomorphism from* Z *to* $\mathbf{S}(\mathfrak{h})^{\mathrm{W}}$, *independent of the choice of* B.

a) Let P_{++} be the set of dominant weights of R, $w \in \mathrm{W}$, $\lambda \in \mathrm{P}_{++}$, $\mu = w\lambda$. Then $\mathrm{Z}(\mu-\rho)$ is isomorphic to a submodule of $\mathrm{Z}(\lambda-\rho)$ (§6, no. 3, Cor. 2 of Prop. 6), and $\varphi(u)(\lambda-\rho) = \varphi(u)(\mu-\rho)$ for all $u \in \mathrm{Z}$ (§6, no. 4, Prop. 7). Thus, the polynomial functions $(\delta\circ\varphi)(u)$ and $(\delta\circ\varphi)(u)\circ w$ on $\mathfrak{h}^*$ coincide on P_{++}. But P_{++} is *dense* in $\mathfrak{h}^*$ in the Zariski topology: this can be seen by identifying $\mathfrak{h}^*$ with k^{B} by means of the basis consisting of the fundamental weights ϖ_α, and by applying Prop. 9 of *Algebra*, Chap. IV, §2, no. 3. Hence

$$(\delta\circ\varphi)(u) = (\delta\circ\varphi)(u)\circ w,$$

which proves that $(\delta\circ\varphi)(\mathrm{Z}) \subset \mathbf{S}(\mathfrak{h})^{\mathrm{W}}$.

b) Let η be the isomorphism from $\mathbf{I}(\mathfrak{g})$ to $\mathbf{S}(\mathfrak{h})^{\mathrm{W}}$ defined in no. 3, Cor. 2 of Th. 1. Consider the canonical isomorphism from the $\mathfrak{g}$-module U to the $\mathfrak{g}$-module $\mathbf{S}(\mathfrak{g})$ (Chap. I, §2, no. 8), and let θ be its restriction to Z. Then $\theta(\mathrm{Z}) = \mathbf{I}(\mathfrak{g})$. Let z be an element of Z with filtration $\leq f$ in U.

$$\begin{array}{ccc} \mathrm{Z} & \xrightarrow{\theta} & \mathbf{I}(\mathfrak{g}) \\ {\scriptstyle\varphi}\downarrow & & \downarrow{\scriptstyle\eta} \\ \mathbf{S}(\mathfrak{h}) & \xrightarrow{\delta} & \mathbf{S}(\mathfrak{h}). \end{array}$$

Introduce the notations of §6, no. 4, and put

$$z = \sum_{\sum q_i + \sum m_i + \sum p_i \leq f} \lambda_{(q_i),(m_i),(p_i)} u((q_i),(m_i),(p_i)).$$

Let $v((q_i),(m_i),(p_i))$ be the monomial $X_{-\alpha_1}^{q_1}\dots X_{-\alpha_n}^{q_n} H_1^{m_1}\dots H_l^{m_l} X_{\alpha_1}^{p_1}\dots X_{\alpha_n}^{p_n}$ calculated in $\mathbf{S}(\mathfrak{g})$. Denoting by $\mathbf{S}_d(\mathfrak{g})$ the sum of the homogeneous components of $\mathbf{S}(\mathfrak{g})$ of degrees $0, 1, \dots, d$, we have

$$\theta(z) \equiv \sum_{\sum q_i + \sum m_i + \sum p_i = f} \lambda_{(q_i),(m_i),(p_i)} v((q_i),(m_i),(p_i)) \pmod{\mathbf{S}_{f-1}(\mathfrak{g})}$$

so

$$(\eta\circ\theta)(z) \equiv \sum_{\sum m_i = f} \lambda_{(0),(m_i),(0)} v((0),(m_i),(0)) \pmod{\mathbf{S}_{f-1}(\mathfrak{h})}$$

and consequently

$$(\eta\circ\theta)(z) \equiv \varphi(z) \pmod{\mathbf{S}_{f-1}(\mathfrak{h})}. \tag{3}$$

c) We show that $\delta\circ\varphi : \mathrm{Z} \to \mathbf{S}(\mathfrak{h})^{\mathrm{W}}$ is bijective. The canonical filtrations on U and $\mathbf{S}(\mathfrak{g})$ induce filtrations on Z, $\mathbf{I}(\mathfrak{g})$ and $\mathbf{S}(\mathfrak{h})^{\mathrm{W}}$, and θ, η are compatible with these filtrations, so that $\mathrm{gr}(\eta \circ \theta)$ is an isomorphism from the vector space $\mathrm{gr}(Z)$ to the vector space $\mathrm{gr}(\mathbf{S}(\mathfrak{h})^{W})$. By (3), $\mathrm{gr}(\varphi) = \mathrm{gr}(\eta \circ \theta)$, and it is clear that $\mathrm{gr}(\delta)$ is the identity. Hence $\mathrm{gr}(\delta \circ \varphi)$ is bijective, so

$$\delta \circ \varphi : Z \to \mathbf{S}(\mathfrak{h})^{W}$$

is bijective (*Commutative Algebra*, Chap. III, §2, no. 8, Cor. 1 and 2 of Th. 1).

d) Recall the notations in *a*). Let E be a simple $\mathfrak{g}$-module of highest weight λ, and χ its central character (§6, no. 1, Def. 2). Let φ' and δ' be the homomorphisms analogous to φ and δ relative to the basis $w(B)$. The highest weight of E relative to $w(B)$ is $w(\lambda)$. By §6, no. 4, Prop. 7,

$$\varphi(u)(\lambda) = \chi(u) = \varphi'(u)(w\lambda)$$

for all $u \in Z$, so, by *a*),

$$\begin{aligned}(\delta \circ \varphi)(u)(w\lambda + w\rho) = (\delta \circ \varphi)(u)(\lambda + \rho) = \varphi(u)(\lambda) = \varphi'(u)(w\lambda)\\ = (\delta' \circ \varphi')(u)(w\lambda + w\rho).\end{aligned}$$

Thus, the polynomial functions $(\delta \circ \varphi)(u)$ and $(\delta' \circ \varphi')(u)$ coincide on $w(P_{++}) + w\rho$, and hence are equal.

COROLLARY 1. *For all* $\lambda \in \mathfrak{h}^*$, *let* χ_λ *be the homomorphism* $z \mapsto (\varphi(z))(\lambda)$ *from* Z *to* k.

(i) *If* k *is algebraically closed, every homomorphism from* Z *to* k *is of the form* χ_λ *for some* $\lambda \in \mathfrak{h}^*$.

(ii) *Let* $\lambda, \mu \in \mathfrak{h}^*$. *Then* $\chi_\lambda = \chi_\mu$ *if and only if* $\mu + \rho \in W(\lambda + \rho)$.

If k is algebraically closed, every homomorphism from $\mathbf{S}(\mathfrak{h})^{W}$ to k extends to a homomorphism from $\mathbf{S}(\mathfrak{h})$ to k (*Commutative Algebra*, Chap. V, §1, no. 9, Prop. 22, and §2, no. 1, Cor. 4 of Th. 1), and every homomorphism from $\mathbf{S}(\mathfrak{h})$ to k is of the form $f \mapsto f(\lambda)$ for some $\lambda \in \mathfrak{h}^*$ (Chap. VII, App. I, Prop. 1). Hence, if χ is a homomorphism from Z to k, there exists (Th. 2) a $\mu \in \mathfrak{h}^*$ such that, for all $z \in Z$,

$$\chi(z) = ((\delta \circ \varphi)(z))(\mu) = (\varphi(z))(\mu - \rho)$$

hence (i).

Let $\lambda, \mu \in \mathfrak{h}^*$ and assume that $\chi_\lambda = \chi_\mu$. Then, for all $z \in Z$,

$$((\delta \circ \varphi)(z))(\lambda + \rho) = (\varphi(z))(\lambda) = \chi_\lambda(z) = \chi_\mu(z) = ((\delta \circ \varphi)(z))(\mu + \rho);$$

in other words, the homomorphisms from $\mathbf{S}(\mathfrak{h})$ to k defined by $\lambda + \rho$ and $\mu + \rho$ coincide on $\mathbf{S}(\mathfrak{h})^{W}$; thus, assertion (ii) follows from *Commutative Algebra*, Chap. V, §2, no. 2, Cor. of Th. 2.

COROLLARY 2. *Let* E, E′ *be finite dimensional simple* $\mathfrak{g}$-*modules, and* χ, χ' *their central characters. If* $\chi = \chi'$, E *and* E′ *are isomorphic.*

Let λ, λ' be the highest weights of E, E'. By §6, no. 4, Prop. 7, $\chi_\lambda = \chi = \chi' = \chi_{\lambda'}$, so there exists $w \in \mathrm{W}$ such that $\lambda' + \rho = w(\lambda + \rho)$. Since $\lambda + \rho$ and $\lambda' + \rho$ belong to the chamber defined by B, we have $w = 1$. Thus, $\lambda = \lambda'$, hence the corollary.

PROPOSITION 7. *For any class γ of finite dimensional simple $\mathfrak{g}$-modules, let U_γ be the isotypical component of type γ of the $\mathfrak{g}$-module* U *(for the adjoint representation of $\mathfrak{g}$ on* U*). Let γ_0 be the class of the trivial $\mathfrak{g}$-module of dimension* 1. *Let* [U, U] *be the vector subspace of* U *generated by the brackets of pairs of elements of* U.

(i) U *is the direct sum of the* U_γ.

(ii) $\mathrm{U}_{\gamma_0} = \mathrm{Z}$, *and* $\sum_{\gamma \neq \gamma_0} \mathrm{U}_\gamma = [\mathrm{U}, \mathrm{U}]$.

(iii) *Let $u \mapsto u^\natural$ be the projection of* U *onto* Z *defined by the decomposition* $\mathrm{U} = \mathrm{Z} \oplus [\mathrm{U}, \mathrm{U}]$. *If $u \in \mathrm{U}$ and $v \in \mathrm{U}$, we have $(uv)^\natural = (vu)^\natural$. If $u \in \mathrm{U}$ and $z \in \mathrm{Z}$, we have $(uz)^\natural = u^\natural z$.*

(iv) *Let φ be the Harish-Chandra homomorphism. Let $\lambda \in \mathrm{P}_{++}$, and let* E *be a finite dimensional simple $\mathfrak{g}$-module of highest weight λ. For all $u \in \mathrm{U}$, we have*

$$\frac{1}{\dim \mathrm{E}} \mathrm{Tr}(u_\mathrm{E}) = (\varphi(u^\natural))(\lambda).$$

The $\mathfrak{g}$-module U is a direct sum of finite dimensional submodules. This implies (i).

It is clear that $\mathrm{U}_{\gamma_0} = \mathrm{Z}$. Let U′ be a vector subspace of U defining a subrepresentation of class γ of the adjoint representation. Then either $[\mathfrak{g}, \mathrm{U}'] = \mathrm{U}'$ or $[\mathfrak{g}, \mathrm{U}'] = 0$. Thus, if $\gamma \neq \gamma_0$ then $[\mathfrak{g}, \mathrm{U}'] = \mathrm{U}'$, so $\sum_{\gamma \neq \gamma_0} \mathrm{U}_\gamma \subset [\mathrm{U}, \mathrm{U}]$. On the other hand, if $u \in \mathrm{U}$ and $x_1, \dots, x_n \in \mathfrak{g}$, then

$$\begin{aligned}[x_1 \dots x_n, u] = (x_1 \dots x_n u - x_2 \dots x_n u x_1) + (x_2 \dots x_n u x_1 - x_3 \dots x_n u x_1 x_2) \\ + \dots + (x_n u x_1 \dots x_{n-1} - u x_1 \dots x_n) \in [\mathfrak{g}, \mathrm{U}].\end{aligned}$$

Hence $[\mathrm{U}, \mathrm{U}] \subset \left[\mathfrak{g}, \sum_\gamma \mathrm{U}_\gamma\right] = \left[\mathfrak{g}, \sum_{\gamma \neq \gamma_0} \mathrm{U}_\gamma\right] \subset \sum_{\gamma \neq \gamma_0} \mathrm{U}_\gamma$. This proves (ii). Under these conditions, (iii) follows from Chap. I, §6, no. 9, Lemma 5.

Finally, let E, λ be as in (iv). Then

$$\begin{aligned}\mathrm{Tr}(u_\mathrm{E}) &= \mathrm{Tr}((u^\natural)_\mathrm{E}) \qquad \text{since } u - u^\natural \in [\mathrm{U}, \mathrm{U}] \\ &= \mathrm{Tr}(\varphi(u^\natural)(\lambda).1) \qquad (\S 6,\ \text{no. } 4,\ \text{Prop. } 7) \\ &= (\dim \mathrm{E}).\varphi(u^\natural)(\lambda).\end{aligned}$$

§9. THE FORMULA OF HERMANN WEYL

In this paragraph, we retain the general notations of §6 *and* §7.

1. CHARACTERS OF FINITE DIMENSIONAL $\mathfrak{g}$-MODULES

Let $(e^\lambda)_{\lambda\in\mathfrak{h}^*}$ be the canonical basis of the ring $\mathbf{Z}[\mathfrak{h}^*]$. Give the space $\mathbf{Z}^{\mathfrak{h}^*}$ of all maps from $\mathfrak{h}^*$ to $\mathbf{Z}$ the product topology of the discrete topologies on the factors. If $\varphi \in \mathbf{Z}^{\mathfrak{h}^*}$, the family $(\varphi(\nu)e^\nu)_{\nu\in\mathfrak{h}^*}$ is summable, and

$$\varphi = \sum_{\nu\in\mathfrak{h}^*} \varphi(\nu)e^\nu.$$

Let $\mathbf{Z}\langle\mathrm{P}\rangle$ be the set of $\varphi \in \mathbf{Z}^{\mathfrak{h}^*}$ whose support is contained in a finite union of sets of the form $\nu - \mathrm{P}_+$, where $\nu \in \mathfrak{h}^*$. Then $\mathbf{Z}[\mathrm{P}] \subset \mathbf{Z}\langle\mathrm{P}\rangle \subset \mathbf{Z}^{\mathfrak{h}^*}$. Define on $\mathbf{Z}\langle\mathrm{P}\rangle$ a ring structure extending that of $\mathbf{Z}[\mathrm{P}]$ by putting, for $\varphi, \psi \in \mathbf{Z}\langle\mathrm{P}\rangle$ and $\nu \in \mathfrak{h}^*$,

$$(\varphi\psi)(\nu) = \sum_{\mu\in\mathfrak{h}^*} \varphi(\mu)\psi(\nu-\mu)$$

(the family $(\varphi(\mu)\psi(\nu-\mu))_{\mu\in\mathfrak{h}^*}$ has finite support, in view of the condition satisfied by the supports of φ and ψ). If $\varphi = \sum_\nu x_\nu e^\nu$ and $\psi = \sum_\nu y_\nu e^\nu$, then $\varphi\psi = \sum_{\nu,\mu} x_\nu y_\mu e^{\nu+\mu}$.

Let $\nu \in \mathfrak{h}^*$. A *partition of* ν *into positive roots* is a family $(n_\alpha)_{\alpha\in\mathrm{R}_+}$, where the n_α are integers ≥ 0 such that $\nu = \sum_{\alpha\in\mathrm{R}_+} n_\alpha\alpha$. *We denote by* $\mathfrak{P}(\nu)$ *the number of partitions of* ν *into positive roots.* We have

$$\mathfrak{P}(\nu) > 0 \Longleftrightarrow \nu \in \mathrm{Q}_+.$$

In this paragraph, we denote by K the following element of $\mathbf{Z}\langle\mathrm{P}\rangle$:

$$\mathrm{K} = \sum_{\gamma\in\mathrm{Q}_+} \mathfrak{P}(\gamma)e^{-\gamma}.$$

Now recall (Chap. VI, §3, no. 3, Prop. 2) that

$$d = \prod_{\alpha\in\mathrm{R}_+} (e^{\alpha/2} - e^{-\alpha/2}) = \sum_{w\in\mathrm{W}} \varepsilon(w)e^{w\rho}$$

is an anti-invariant element of $\mathbf{Z}[\mathrm{P}]$.

Lemma 1. In the ring $\mathbf{Z}\langle\mathrm{P}\rangle$, *we have* $\mathrm{K}.\prod_{\alpha\in\mathrm{R}_+}(1-e^{-\alpha}) = \mathrm{K}e^{-\rho}d = 1.$

Indeed,

$$\mathrm{K} = \prod_{\alpha \in \mathrm{R}_+} (e^0 + e^{-\alpha} + e^{-2\alpha} + \cdots)$$

so

$$Ke^{-\rho}d = \prod_{\alpha \in \mathrm{R}_+} (1 + e^{-\alpha} + e^{-2\alpha} + \cdots) \prod_{\alpha \in \mathrm{R}_+} (1 - e^{-\alpha}) = 1.$$

Lemma 2. Let $\lambda \in \mathfrak{h}^*$. *The module* $\mathrm{Z}(\lambda)$ (§6, no. 3) *admits a character that is an element of* $\mathbf{Z}\langle \mathrm{P} \rangle$, *and we have* $d.\operatorname{ch} \mathrm{Z}(\lambda) = e^{\lambda+\rho}$.

Let $\alpha_1, \ldots, \alpha_q$ be distinct elements of R_+. The $X_{-\alpha_1}^{n_1} X_{-\alpha_2}^{n_2} \ldots X_{-\alpha_q}^{n_q} \otimes 1$ form a basis of $\mathrm{Z}(\lambda)$ (§6, Prop. 6 (iii)). For $h \in \mathfrak{h}$, we have

$$\begin{aligned} h.(X_{-\alpha_1}^{n_1} X_{-\alpha_2}^{n_2} \ldots X_{-\alpha_q}^{n_q} \otimes 1) & \\ &= [h, X_{-\alpha_1}^{n_1} \ldots X_{-\alpha_q}^{n_q}] \otimes 1 + (X_{-\alpha_1}^{n_1} \ldots X_{-\alpha_q}^{n_q}) \otimes h.1 \\ &= (\lambda - n_1\alpha_1 - \cdots - n_q\alpha_q)(h)(X_{-\alpha_1}^{n_1} \ldots X_{-\alpha_q}^{n_q} \otimes 1). \end{aligned}$$

Thus, the dimension of $\mathrm{Z}(\lambda)^{\lambda-\mu}$ is $\mathfrak{P}(\mu)$. This proves that $\operatorname{ch} \mathrm{Z}(\lambda)$ is defined, is an element of $\mathbf{Z}\langle \mathrm{P} \rangle$, and that

$$\operatorname{ch} \mathrm{Z}(\lambda) = \sum_{\mu} \mathfrak{P}(\mu) e^{\lambda-\mu} = \mathrm{K}e^{\lambda}.$$

It now suffices to apply Lemma 1.

Lemma 3. Let M *be a* $\mathfrak{g}$-*module which admits a character* ch(M) *whose support is contained in a finite union of the sets* $\mu - \mathrm{P}_+$. *Let* U *be the enveloping algebra of* $\mathfrak{g}$, Z *the centre of* U, $\lambda_0 \in \mathfrak{h}^*$, *and* χ_{λ_0} *the corresponding homomorphism from* Z *to* k (§8, Cor. 1 of Th. 2). *Assume that, for all* $z \in \mathrm{Z}$, z_{M} *is the homothety with ratio* $\chi_{\lambda_0}(z)$. *Let* $\mathrm{D_M}$ *be the set of* $\lambda \in \mathrm{W}(\lambda_0 + \rho) - \rho$ *such that* $\lambda + \mathrm{Q}_+$ *meets* Supp(ch M). *Then* ch(M) *is a* $\mathbf{Z}$-*linear combination of the* $\operatorname{ch} \mathrm{Z}(\lambda)$ *for* $\lambda \in \mathrm{D_M}$.

If Supp(ch M) is empty, the lemma is clear. Assume that Supp(ch M) $\neq \varnothing$. Let λ be a maximal element of this support, and put $\dim \mathrm{M}^{\lambda} = m$. There exists a $\mathfrak{g}$-homomorphism φ from $(\mathrm{Z}(\lambda))^m$ to M which maps $(\mathrm{Z}(\lambda)^{\lambda})^m$ bijectively onto M^{λ} (§6, no. 3, Prop. 6 (i)). Thus, the central character of $\mathrm{Z}(\lambda)$ is χ_{λ_0}, so $\lambda \in \mathrm{W}(\lambda_0+\rho)-\rho$ (§8, no. 5, Cor. 1 of Th. 2). This proves that $\mathrm{D_M} \neq \varnothing$, and allows us to argue by induction on $\operatorname{Card} \mathrm{D_M}$. Let L and N be the kernel and cokernel of φ. Then we have an exact sequence of $\mathfrak{g}$-homomorphisms:

$$0 \to \mathrm{L} \to (\mathrm{Z}(\lambda))^m \to \mathrm{M} \to \mathrm{N} \to 0$$

so

$$\operatorname{ch}(\mathrm{M}) = -\operatorname{ch}(\mathrm{L}) + m \operatorname{ch} \mathrm{Z}(\lambda) + \operatorname{ch}(\mathrm{N})$$

(§7, no. 7, formula (6)). The sets Supp(ch L) and Supp(ch N) are contained in a finite union of sets $\mu - \mathrm{P}_+$. For $z \in \mathrm{Z}$, z_{L} and z_{N} are homotheties with ratio

$\chi_{\lambda_0}(z)$. Clearly, $D_N \subset D_M$. On the other hand, $(\lambda + Q_+) \cap \mathrm{Supp}(\mathrm{ch}\, M) = \{\lambda\}$, and $\lambda \notin \mathrm{Supp}(\mathrm{ch}\, N)$, so $\lambda \notin D_N$ and

$$\mathrm{Card}\, D_N < \mathrm{Card}\, D_M.$$

On the other hand, L is a submodule of $(Z(\lambda))^m$; if $\lambda' \in D_L$, then $\lambda' + Q_+$ meets $\mathrm{Supp}(\mathrm{ch}\, L) \subset \mathrm{Supp}\,\mathrm{ch}\, Z(\lambda)$, so $\lambda \in \lambda' + Q_+$ (§6, no. 1, Prop. 1 (ii)); it follows that $D_L \subset D_M$. Since $L \cap (Z(\lambda)^\lambda)^m = 0$, we have $\lambda \notin D_L$, so

$$\mathrm{Card}\, D_L < \mathrm{Card}\, D_M.$$

It now suffices to apply the induction hypothesis.

THEOREM 1 (*Character Formula of H. Weyl*). *Let* M *be a finite dimensional simple* $\mathfrak{g}$*-module, and* λ *its highest weight. Then*

$$\left(\sum_{w \in W} \varepsilon(w) e^{w\rho}\right) . \mathrm{ch}\, M = \sum_{w \in W} \varepsilon(w) e^{w(\lambda+\rho)}.$$

With the notations of Lemma 3, the central character of M is χ_λ (§6, no. 4, Prop. 7). Hence, by Lemmas 2 and 3, $d.\mathrm{ch}\, M$ is a $\mathbf{Z}$-linear combination of the $e^{\mu+\rho}$ such that

$$\mu + \rho \in W(\lambda + \rho).$$

On the other hand, by §7, no. 7, Lemma 7, $d.\mathrm{ch}\, M$ is anti-invariant, and its unique maximal term is $e^{\lambda+\rho}$, hence the theorem.

Example. Take $\mathfrak{g} = \mathfrak{sl}(2, k)$, $\mathfrak{h} = kH$. Let α be the root of $(\mathfrak{g}, \mathfrak{h})$ such that $\alpha(H) = 2$. The $\mathfrak{g}$-module $V(m)$ has highest weight $(m/2)\alpha$. Hence

$$\begin{aligned}
\mathrm{ch}(V(m)) &= (e^{(m/2)\alpha + \frac{1}{2}\alpha} - e^{-(m/2)\alpha - \frac{1}{2}\alpha})/(e^{\frac{1}{2}\alpha} - e^{-\frac{1}{2}\alpha}) \\
&= e^{-(m/2)\alpha}.(e^{(m+1)\alpha} - 1)/(e^\alpha - 1) \\
&= e^{-(m/2)\alpha}(e^{m\alpha} + e^{(m-1)\alpha} + \cdots + 1) \\
&= e^{(m/2)\alpha} + e^{(m-2)\alpha/2} + \cdots + e^{-(m/2)\alpha}
\end{aligned}$$

which also follows easily from §1, no. 2, Prop. 2.

2. DIMENSIONS OF SIMPLE $\mathfrak{g}$-MODULES

If $\mu \in \mathfrak{h}^*$, put $J(e^\mu) = \sum_{w \in W} \varepsilon(w) e^{w\mu}$, cf. Chap. VI, §3, no. 3.

THEOREM 2. *Let* E *be a finite dimensional simple* $\mathfrak{g}$*-module,* λ *its highest weight and* $(\cdot|\cdot)$ *a* W*-invariant non-degenerate positive symmetric bilinear form on* $\mathfrak{h}^*$*. Then:*

$$\dim \mathrm{E} = \prod_{\alpha \in \mathrm{R}_+} \frac{\langle \lambda + \rho, H_\alpha \rangle}{\langle \rho, H_\alpha \rangle} = \prod_{\alpha \in \mathrm{R}_+} \left(1 + \frac{(\lambda|\alpha)}{(\rho|\alpha)} \right).$$

Let T be an indeterminate. For all $\nu \in \mathrm{P}$, denote by f_ν the homomorphism from $\mathbf{Z}[\mathrm{P}]$ to $\mathbf{R}[[\mathrm{T}]]$ that takes e^μ to $e^{(\nu|\mu)\mathrm{T}}$ for all $\mu \in \mathrm{P}$. Then dim E is the constant term of the series $f_\nu(\mathrm{ch}\,\mathrm{E})$.

For all $\mu, \nu \in \mathrm{P}$, we have

$$\begin{aligned} f_\nu(\mathrm{J}(e^\mu)) &= \sum_{w \in \mathrm{W}} \varepsilon(w) e^{(\nu|w\mu)\mathrm{T}} \\ &= \sum_{w \in \mathrm{W}} \varepsilon(w) e^{(w^{-1}\nu|\mu)\mathrm{T}} = f_\mu(\mathrm{J}(e^\nu)). \end{aligned}$$

In particular, in view of Chap. VI, §3, no. 3, formula (3),

$$f_\rho(\mathrm{J}(e^\mu)) = f_\mu(\mathrm{J}(e^\rho)) = e^{(\mu|\rho)\mathrm{T}} \prod_{\alpha \in \mathrm{R}_+} (1 - e^{-(\mu|\alpha)\mathrm{T}}).$$

Hence, setting $\mathrm{Card}(\mathrm{R}_+) = \mathrm{N}$,

$$f_\rho(\mathrm{J}(e^\mu)) \equiv \mathrm{T}^\mathrm{N} \prod_{\alpha \in \mathrm{R}} (\mu|\alpha) \qquad (\mathrm{mod}\ \mathrm{T}^{\mathrm{N}+1}\mathbf{R}[[\mathrm{T}]]).$$

The equality $\mathrm{J}(e^{\lambda+\rho}) = \mathrm{ch}(\mathrm{E}).\mathrm{J}(e^\rho)$ (Th. 1) thus implies that

$$\mathrm{T}^\mathrm{N} \prod_{\alpha \in \mathrm{R}_+} (\lambda + \rho|\alpha) \equiv f_\rho(\mathrm{ch}\,\mathrm{E}).\mathrm{T}^\mathrm{N} \prod_{\alpha \in \mathrm{R}_+} (\rho|\alpha) \qquad (\mathrm{mod}\ \mathrm{T}^{\mathrm{N}+1}\mathbf{R}[[\mathrm{T}]])$$

so

$$\dim \mathrm{E} = \Big(\prod_{\alpha \in \mathrm{R}_+} (\lambda + \rho|\alpha) \Big) \Big/ \Big(\prod_{\alpha \in \mathrm{R}_+} (\rho|\alpha) \Big) = \prod_{\alpha \in \mathrm{R}_+} \left(1 + \frac{(\lambda|\alpha)}{(\rho|\alpha)} \right).$$

Now, if $\alpha \in \mathrm{R}_+$, α can be identified with an element of $\mathfrak{h}_\mathbf{R}$ proportional to H_α, so

$$(\lambda + \rho|\alpha)/(\rho|\alpha) = \langle \lambda + \rho, H_\alpha \rangle / \langle \rho, H_\alpha \rangle.$$

Examples. 1) In the Example of no. 1, we find that

$$\dim \mathrm{V}(m) = \left(\frac{m}{2}\alpha + \frac{\alpha}{2} \right)(H_\alpha) / \frac{\alpha}{2}(H_\alpha) = m + 1,$$

which we knew in §1.

2) Take $\mathfrak{g}$ to be the splittable simple Lie algebra of type G_2 and adopt the notations of Chap. VI, Plate IX. Give $\mathfrak{h}^*_\mathbf{R}$ the W-invariant positive symmetric form $(\cdot|\cdot)$ such that $(\alpha_1|\alpha_1) = 1$. Then $\rho = \varpi_1 + \varpi_2$ and

$$(\varpi_1|\alpha_1) = \frac{1}{2}, \quad (\varpi_1|\alpha_2) = 0, \quad (\varpi_1|\alpha_2+\alpha_1) = \frac{1}{2},$$
$$(\varpi_1|\alpha_2+2\alpha_1) = 1, \quad (\varpi_1|\alpha_2+3\alpha_1) = \frac{3}{2}, \quad (\varpi_1|2\alpha_2+3\alpha_1) = \frac{3}{2},$$
$$(\varpi_2|\alpha_1) = 0, \quad (\varpi_2|\alpha_2) = \frac{3}{2}, \quad (\varpi_2|\alpha_2+\alpha_1) = \frac{3}{2},$$
$$(\varpi_2|\alpha_2+2\alpha_1) = \frac{3}{2}, \quad (\varpi_2|\alpha_2+3\alpha_1) = \frac{3}{2}, \quad (\varpi_2|2\alpha_2+3\alpha_1) = 3.$$

Thus, if n_1, n_2 are integers ≥ 0, the dimension of the simple representation of highest weight $n_1\varpi_1 + n_2\varpi_2$ is

$$\left(1+\frac{n_1/2}{\frac{1}{2}}\right)\left(1+\frac{3n_2/2}{\frac{3}{2}}\right)\left(1+\frac{n_1/2+3n_2/2}{\frac{1}{2}+\frac{3}{2}}\right)\left(1+\frac{n_1+3n_2/2}{1+\frac{3}{2}}\right)$$
$$\times\left(1+\frac{3n_1/2+3n_2/2}{\frac{3}{2}+\frac{3}{2}}\right)\left(1+\frac{3n_1/2+3n_2}{\frac{3}{2}+3}\right)$$
$$= (1+n_1)(1+n_2)\left(1+\frac{n_1+3n_2}{4}\right)\left(1+\frac{2n_1+3n_2}{5}\right)\left(1+\frac{n_1+n_2}{2}\right)$$
$$\times\left(1+\frac{n_1+2n_2}{3}\right)$$
$$= \frac{(1+n_1)(1+n_2)(2+n_1+n_2)(3+n_1+2n_2)(4+n_1+3n_2)(5+2n_1+3n_2)}{5!}.$$

In particular, the fundamental representation of highest weight ϖ_1 (resp. ϖ_2) is of dimension 7 (resp. 14).

3. MULTIPLICITIES OF WEIGHTS OF SIMPLE $\mathfrak{g}$-MODULES

PROPOSITION 1. *Let* $\omega \in \mathrm{P}_{++}$. *For all* $\lambda \in \mathrm{P}$, *the multiplicity of* λ *in* $\mathrm{E}(\omega)$ *is*

$$m_\lambda = \sum_{w\in\mathrm{W}} \varepsilon(w)\mathfrak{P}(w(\omega+\rho)-(\lambda+\rho)).$$

By Th. 1 and Lemma 1,

$$\mathrm{ch}\,\mathrm{E}(\omega) = \mathrm{K}\,e^{-\rho}d\,\mathrm{ch}\,\mathrm{E}(\omega) = \mathrm{K}\,e^{-\rho}\sum_{w\in\mathrm{W}}\varepsilon(w)e^{w(\omega+\rho)}$$

so

$$\mathrm{ch}\,\mathrm{E}(\omega) = \sum_{w\in\mathrm{W},\gamma\in\mathrm{Q}_+} \varepsilon(w)\mathfrak{P}(\gamma)e^{-\rho+w(\omega+\rho)-\gamma}$$

and

$$m_\lambda = \sum_{w\in\mathrm{W},\gamma\in\mathrm{Q}_+,\gamma=-\lambda-\rho+w(\omega+\rho)} \varepsilon(w)\mathfrak{P}(\gamma).$$

COROLLARY. *If* λ *is a weight of* $\mathrm{E}(\omega)$ *distinct from* ω,

$$m_\lambda = - \sum_{w \in \mathrm{W}, w \neq 1} \varepsilon(w) m_{\lambda+\rho-w\rho}.$$

Apply Prop. 1 with $\omega = 0$. If $\mu \in \mathrm{P} - \{0\}$, we find that

$$0 = \sum_{w \in \mathrm{W}} \varepsilon(w) \mathfrak{P}(w\rho + \mu - \rho)$$

hence

$$\mathfrak{P}(\mu) = - \sum_{w \in \mathrm{W}, w \neq 1} \varepsilon(w) \mathfrak{P}(\mu + w\rho - \rho). \tag{1}$$

Prop. 1 also gives

$$m_\lambda = - \sum_{w \in \mathrm{W}} \varepsilon(w) \sum_{w' \in \mathrm{W}, w' \neq 1} \varepsilon(w') \mathfrak{P}(w(\omega + \rho) - (\lambda + \rho) + w'\rho - \rho)$$

since $w(\omega + \rho) \neq \lambda + \rho$ for all $w \in \mathrm{W}$ (§7, Prop. 5 (iii)). Hence,

$$\begin{aligned} m_\lambda &= - \sum_{w' \in \mathrm{W}, w' \neq 1} \varepsilon(w') \sum_{w \in \mathrm{W}} \varepsilon(w) \mathfrak{P}(w(\omega + \rho) - (\lambda + \rho - w'\rho + \rho)) \\ &= - \sum_{w' \in \mathrm{W}, w' \neq 1} \varepsilon(w') m_{\lambda+\rho-w'\rho} \qquad \text{(Prop. 1).} \end{aligned}$$

4. DECOMPOSITION OF TENSOR PRODUCTS OF SIMPLE $\mathfrak{g}$-MODULES

PROPOSITION 2. *Let* $\lambda, \mu \in \mathrm{P}_{++}$. *In* $\mathscr{R}(\mathfrak{g})$, *we have*

$$[\lambda].[\mu] = \sum_{\nu \in \mathrm{P}_{++}} m(\lambda, \mu, \nu)[\nu]$$

with

$$m(\lambda, \mu, \nu) = \sum_{w, w' \in \mathrm{W}} \varepsilon(ww') \mathfrak{P}(w(\lambda + \rho) + w'(\mu + \rho) - (\nu + 2\rho)).$$

Let E, F be finite dimensional simple $\mathfrak{g}$-modules of highest weights λ, μ. Let l_ν be the length of the isotypical component of $\mathrm{E} \otimes \mathrm{F}$ of highest weight ν. It suffices to show that

$$l_\nu = \sum_{w, w' \in \mathrm{W}} \varepsilon(ww') \mathfrak{P}(w(\lambda + \rho) + w'(\mu + \rho) - (\nu + 2\rho)). \tag{2}$$

Put $c_1 = \mathrm{ch}(\mathrm{E}) = \sum_{\sigma \in \mathrm{P}} m_\sigma e^\sigma$, $c_2 = \mathrm{ch}(\mathrm{F})$, and $d = \mathrm{J}(e^\rho)$, where J is defined as in no. 2. We have

$$\sum_{\xi\in P_{++}} l_\xi \operatorname{ch}[\xi] = \operatorname{ch}(E\otimes F) = c_1c_2$$

so, after multiplying by d and using Th. 1,

$$\sum_{\xi\in P_{++}} l_\xi J(e^{\xi+\rho}) = c_1 J(e^{\mu+\rho}) = \left(\sum_{\sigma\in P} m_\sigma e^\sigma\right)\left(\sum_{w\in W}\varepsilon(w)e^{w(\mu+\rho)}\right) \tag{3}$$

$$= \sum_{\tau\in P}\left(\sum_{w\in W}\varepsilon(w)m_{\tau+\rho-w(\mu+\rho)}\right)e^{\tau+\rho}.$$

Now, if $\xi \in P_{++}$, $\xi+\rho$ belongs to the chamber defined by B (Chap. VI, §1, no. 10); thus, for all $w \in W$ distinct from 1, we have $w(\xi+\rho) \notin P_{++}$. Consequently, the coefficient of $e^{\nu+\rho}$ in $\sum_{\xi\in P_{++}} l_\xi J(e^{\xi+\rho})$ is equal to l_ν. In view of (3), we obtain

$$l_\nu = \sum_{w\in W}\varepsilon(w)m_{\nu+\rho-w(\mu+\rho)},$$

that is, by Prop. 1,

$$l_\nu = \sum_{w,w'\in W}\varepsilon(w)\varepsilon(w')\mathfrak{P}(w'(\lambda+\rho)-(\nu+\rho-w(\mu+\rho)+\rho))$$

which proves (2).

Example. We return to the Example of no. 1. Let $\lambda = (n/2)\alpha, \mu = (p/2)\alpha$, $\nu = (q/2)\alpha$ with $n \geq p$. We have

$$\begin{aligned} m(\lambda,\mu,\nu) &= \mathfrak{P}\left(\frac{n}{2}\alpha+\frac{\alpha}{2}+\frac{p}{2}\alpha+\frac{\alpha}{2}-\frac{q}{2}\alpha-\alpha\right)\\ &\quad -\mathfrak{P}\left(\frac{n}{2}\alpha+\frac{\alpha}{2}-\frac{p}{2}\alpha-\frac{\alpha}{2}-\frac{q}{2}\alpha-\alpha\right)\\ &\quad -\mathfrak{P}\left(-\frac{n}{2}\alpha-\frac{\alpha}{2}+\frac{p}{2}\alpha+\frac{\alpha}{2}-\frac{q}{2}\alpha-\alpha\right)\\ &\quad +\mathfrak{P}\left(-\frac{n}{2}\alpha-\frac{\alpha}{2}-\frac{p}{2}\alpha-\frac{\alpha}{2}-\frac{q}{2}\alpha-\alpha\right)\\ &= \mathfrak{P}\left(\frac{n+p-q}{2}\alpha\right)-\mathfrak{P}\left(\frac{n-p-q-2}{2}\alpha\right). \end{aligned}$$

This is zero if $n+p+q$ is not divisible by 2, or if $q \geq n+p$. If

$$q = n+p-2r$$

with r an integer ≥ 0, we have

$$m(\lambda,\mu,\nu) = \mathfrak{P}(r\alpha)-\mathfrak{P}((r-p-1)\alpha)$$

hence $m(\lambda, \mu, \nu) = 1$ if $r \leq p$ and $m(\lambda, \mu, \nu) = 0$ if $r > p$. Finally, the $\mathfrak{g}$-module $\mathrm{V}(n) \otimes \mathrm{V}(p)$ is isomorphic to

$$\mathrm{V}(n+p) \oplus \mathrm{V}(n+p-2) \oplus \mathrm{V}(n+p-4) \oplus \cdots \oplus \mathrm{V}(n-p)$$

(*Clebsch-Gordan formula*).

§10. MAXIMAL SUBALGEBRAS OF SEMI-SIMPLE LIE ALGEBRAS

THEOREM 1. *Let* V *be a finite dimensional vector space,* $\mathfrak{g}$ *a reductive Lie subalgebra in* $\mathfrak{gl}(\mathrm{V})$, $\mathfrak{q}$ *a Lie subalgebra of* $\mathfrak{g}$ *and* Φ *the bilinear form* $(x, y) \mapsto \mathrm{Tr}(xy)$ *on* $\mathfrak{g} \times \mathfrak{g}$. *Assume that the orthogonal complement* $\mathfrak{n}$ *of* $\mathfrak{q}$ *with respect to* Φ *is a Lie subalgebra of* $\mathfrak{g}$ *consisting of nilpotent endomorphisms of* V. *Then* $\mathfrak{q}$ *is a parabolic subalgebra of* $\mathfrak{g}$.

a) $\mathfrak{q}$ *is the normalizer of* $\mathfrak{n}$ *in* $\mathfrak{g}$: let $\mathfrak{p}$ be this normalizer. Let $x \in \mathfrak{q}$ and $y \in \mathfrak{n}$; for all $z \in \mathfrak{q}$, we have $[z, x] \in \mathfrak{q}$, so

$$\Phi([x, y], z) = \Phi(y, [z, x]) = 0;$$

in other words, $[x, y] \in \mathfrak{n}$. Hence $\mathfrak{q} \subset \mathfrak{p}$. Since $\mathfrak{n}$ is an ideal of $\mathfrak{p}$ consisting of nilpotent endomorphisms of V, P is orthogonal to $\mathfrak{n}$ with respect to Φ (Chap. I, no. 3, Prop. 4 *d*)). Since Φ is non-degenerate[4], $\mathfrak{p} \subset \mathfrak{q}$, hence our assertion.

b) There exists a reductive Lie subalgebra $\mathfrak{m}$ *in* $\mathfrak{gl}(\mathrm{V})$ *such that* $\mathfrak{q}$ *is the semi-direct product of* $\mathfrak{m}$ *and* $\mathfrak{n}$: let $\mathfrak{n}_{\mathrm{V}}(\mathfrak{q})$ be the largest ideal of $\mathfrak{q}$ consisting of nilpotent endomorphisms of V. Then $\mathfrak{n}_{\mathrm{V}}(\mathfrak{q})$ contains $\mathfrak{n}$, and it is orthogonal to $\mathfrak{q}$ (*loc. cit.*); hence $\mathfrak{n} = \mathfrak{n}_{\mathrm{V}}(\mathfrak{q})$. Moreover, $\mathfrak{g}$ is reductive in $\mathfrak{gl}(\mathrm{V})$ by hypothesis, hence decomposable (Chap. VII, §5, no. 1, Prop. 2); since $\mathfrak{q}$ is the intersection of $\mathfrak{g}$ with the normalizer of $\mathfrak{n}$ in $\mathfrak{gl}(\mathrm{V})$, it is a decomposable Lie algebra (*loc. cit.*, Cor. 1 of Prop. 3). Thus, our assertion follows from Prop. 7 of Chap. VII, §5, no. 3.

Choose a Cartan subalgebra $\mathfrak{h}$ of $\mathfrak{m}$; denote by $\mathfrak{g}_1$ the commutant of $\mathfrak{h}$ in $\mathfrak{g}$, and put $\mathfrak{q}_1 = \mathfrak{q} \cap \mathfrak{g}_1, \mathfrak{n}_1 = \mathfrak{n} \cap \mathfrak{g}_1$.

c) The Lie algebras $\mathfrak{g}_1, \mathfrak{q}_1$ *and* $\mathfrak{n}_1$ *satisfy the same hypotheses as* $\mathfrak{g}, \mathfrak{q}$ *and* $\mathfrak{n}$: since $\mathfrak{m}$ is reductive in $\mathfrak{gl}(\mathrm{V})$, $\mathfrak{h}$ is commutative and is composed of semi-simple endomorphisms of V (Chap. VII, §2, no. 4, Cor. 3 of Th. 2). Thus $\mathfrak{g}_1 = \mathfrak{g}^0(\mathfrak{h})$ is reductive in $\mathfrak{g}$ (Chap. VII, §1, no. 3, Prop. 11), hence also in $\mathfrak{gl}(\mathrm{V})$ (Chap. I, §6, no. 6, Cor. 2 of Prop. 7). It is clear that $\mathfrak{n}_1$ is composed

[4] Let $\mathfrak{z}$ be the orthogonal complement of $\mathfrak{g}$ with respect to Φ; this is an ideal of $\mathfrak{g}$ contained in $\mathfrak{n}$, so every element of $\mathfrak{z}$ is nilpotent. The identity representation of $\mathfrak{g}$ is semi-simple (Chap. I, §6, Cor. 1 of Prop. 7). Hence $z = 0$ (Chap. I, §4, no. 3, Lemma 2).

of nilpotent endomorphisms of V. Since $\mathfrak{h}$ is a subalgebra of $\mathfrak{q}$, reductive in $\mathfrak{gl}(V)$, the adjoint representation of $\mathfrak{h}$ on $\mathfrak{q}$ is semi-simple; by construction, $\mathfrak{q}_1$ is the set of invariants of $\mathrm{ad}_{\mathfrak{q}}(\mathfrak{h})$, so $\mathfrak{q} = \mathfrak{q}_1 + [\mathfrak{h}, \mathfrak{q}]$ (Chap. I, §3, no. 5, Prop. 6). Since

$$\Phi(\mathfrak{g}_1, [\mathfrak{h}, \mathfrak{q}]) = \Phi([\mathfrak{h}, \mathfrak{g}_1], \mathfrak{q}) = 0,$$

an element of $\mathfrak{g}_1$ is orthogonal to $\mathfrak{q}_1$ if and only if it is orthogonal to $\mathfrak{q}$; consequently, $\mathfrak{n}_1 = \mathfrak{g}_1 \cap \mathfrak{n}$ is the orthogonal complement of $\mathfrak{q}_1$ in $\mathfrak{g}_1$.

d) The Cartan subalgebra $\mathfrak{h}$ *of* $\mathfrak{m}$ *is a Cartan subalgebra of* $\mathfrak{g}$: We have $\mathfrak{q} = \mathfrak{m} \oplus \mathfrak{n}$ and $\mathfrak{h} = \mathfrak{m} \cap \mathfrak{g}_1$, so it is immediate that $\mathfrak{q}_1 = \mathfrak{h} \oplus \mathfrak{n}_1$. Moreover, $[\mathfrak{h}, \mathfrak{n}_1] = 0$, $\mathfrak{h}$ is commutative and $\mathfrak{n}_1$ is nilpotent, so the Lie algebra $\mathfrak{q}_1$ is nilpotent. By *a*) and *c*), $\mathfrak{q}_1$ is the normalizer of $\mathfrak{n}_1$ in $\mathfrak{g}_1$; *a fortiori*, $\mathfrak{q}_1$ is equal to its normalizer in $\mathfrak{g}_1$, hence is a Cartan subalgebra of $\mathfrak{g}_1$. Since $\mathfrak{g}_1$ is reductive in $\mathfrak{gl}(V)$, it follows from Cor. 3 of Th. 2 of Chap. VII, §2, no. 4, that $\mathfrak{q}_1$ is composed of semi-simple endomorphisms of V; thus, since $\mathfrak{n}_1$ is composed of nilpotent endomorphisms of V, we have $\mathfrak{n}_1 = 0$. Consequently, $\mathfrak{h} = \mathfrak{q}_1$ is a Cartan subalgebra of $\mathfrak{g}_1$, and since $\mathfrak{g}_1$ normalizes $\mathfrak{h}$, we have $\mathfrak{h} = \mathfrak{g}_1$. Thus, we have proved that every element of $\mathfrak{h}$ is a semi-simple element of $\mathfrak{g}$, and that the commutant of $\mathfrak{h}$ in $\mathfrak{g}$ is equal to $\mathfrak{h}$; it follows that $\mathfrak{h} = \mathfrak{g}^0(\mathfrak{h})$, so $\mathfrak{h}$ is a Cartan subalgebra of $\mathfrak{g}$.

e) $\mathfrak{q}$ *is a parabolic subalgebra of* $\mathfrak{g}$: by the preceding, $\mathfrak{h}$ is a Cartan subalgebra of $\mathfrak{g}$, $\mathfrak{n}$ consists of nilpotent elements of $\mathfrak{g}$, and $[\mathfrak{h}, \mathfrak{n}] \subset \mathfrak{n}$. Let $\bar{k}$ be an algebraic closure of k; by definition, $\mathfrak{q}$ is parabolic in $\mathfrak{g}$ if and only if $\bar{k} \otimes_k \mathfrak{q}$ is a parabolic subalgebra of $\bar{k} \otimes_k \mathfrak{g}$. The properties stated above being preserved by extension of scalars, for the proof we can restrict ourselves to the case in which $\mathfrak{h}$ is splitting. Let R be the root system of $(\mathfrak{g}, \mathfrak{h})$; by Prop. 2 (v) of §3, no. 1, there exists a subset P of R such that $P \cap (-P) = \varnothing$ and $\mathfrak{n} = \sum_{\alpha \in P} \mathfrak{g}^\alpha$. Let P′ be the set of roots α such that $-\alpha \notin P$; we have $P' \cup (-P') = R$, and the orthogonal complement $\mathfrak{q}$ of $\mathfrak{n}$ in $\mathfrak{g}$ is equal to $\mathfrak{h} + \sum_{\alpha \in P'} \mathfrak{g}^\alpha$. We have proved that $\mathfrak{q}$ is parabolic. Q.E.D.

Lemma 1. Let $\mathfrak{g}$ *be a semi-simple Lie algebra,* V *a finite dimensional vector space,* ρ *a linear representation of* $\mathfrak{g}$ *on* V, D *a vector subspace of* V, $\mathfrak{h}$ *a Cartan subalgebra of* $\mathfrak{g}$, $\mathfrak{s}$ (resp. $\mathfrak{s}'$) *the set of* $x \in \mathfrak{h}$ *such that* $\rho(x)D \subset D$ (resp. $\rho(x)D = 0$), *and* Φ *the bilinear form on* $\mathfrak{g}$ *associated*[5] *to* ρ.

(i) *If* $\mathfrak{h}$ *is splitting, the vector subspaces* $\mathfrak{s}$ *and* $\mathfrak{s}'$ *of* $\mathfrak{h}$ *are rational over* $\mathbf{Q}$.

(ii) *If* ρ *is injective, the restriction of* Φ *to* $\mathfrak{s}$ (resp. $\mathfrak{s}'$) *is non-degenerate.*

Assume that the Cartan subalgebra $\mathfrak{h}$ is splitting. Let d be the dimension of D; put $W = \bigwedge^d(V)$ and $\sigma = \bigwedge^d(\rho)$; denote also by $(e_1, \dots, e_d)$ a basis of D and $e = e_1 \wedge \dots \wedge e_d$ a decomposable d-vector associated to D. Let P be the set of weights of σ with respect to $\mathfrak{h}$; denote by W^μ the subspace of

[5] In other words, $\Phi(x, y) = \mathrm{Tr}(\rho(x)\rho(y))$ for $x, y \in \mathfrak{g}$.

W associated to the weight μ, and put $e = \sum_{\mu \in P} e^{\mu}$ (with $e^{\mu} \in W^{\mu}$ for all $\mu \in P$); finally, let P′ be the set of weights μ such that $e^{\mu} \neq 0$ and let P″ be the set of differences of elements of P′. Let x be in $\mathfrak{h}$; then x belongs to $\mathfrak{s}$ if and only if there exists c in k such that $\rho(x).e = c.e$ (Chap. VII, §5, no. 4, Lemma 2 (i)). Since $\rho(x).e^{\mu} = \mu(x).e^{\mu}$, we see that $x \in \mathfrak{s}$ is equivalent to the relation "$\mu(x) = 0$ for all $\mu \in P''$". Now, the **Q**-structure of $\mathfrak{h}$ is the **Q**-vector subspace $\mathfrak{h}_{\mathbf{Q}}$ of $\mathfrak{h}$ generated by the coroots H_α and all μ in P″ take rational values on $\mathfrak{h}_{\mathbf{Q}}$; it follows (*Algebra*, Chap. II, §8, no. 4, Prop. 5) that $\mathfrak{s}$ is a subspace of $\mathfrak{h}$ rational over **Q**.

For any weight $\mu \in P$, let p_μ be the projection onto V^{μ} associated to the decomposition $V = \bigoplus_{\mu \in P} V^{\mu}$; denote by P_1 the set of $\mu \in P$ such that $p_\mu(D) \neq 0$. It is immediate that $\mathfrak{s}'$ is the intersection of the kernels (in $\mathfrak{h}$) of the elements of P_1; it follows, in the same way as for $\mathfrak{s}$, that $\mathfrak{s}'$ is a subspace of $\mathfrak{h}$ rational over **Q**. This proves (i).

By extension of scalars, it suffices to prove (ii) when k is algebraically closed, hence when $\mathfrak{h}$ is splitting. Let $\mathfrak{m}$ be a vector subspace of $\mathfrak{h}$ rational over **Q**; for all non-zero x in $\mathfrak{m}_{\mathbf{Q}} = \mathfrak{m} \cap \mathfrak{h}_{\mathbf{Q}}$, we have $\Phi(x, x) > 0$ by the Cor. of Prop. 1 of §7, no. 1. The restriction of Φ to $\mathfrak{m}_{\mathbf{Q}}$ is non-degenerate, and hence so is the restriction of Φ to $\mathfrak{m}$ since $\mathfrak{m}$ is canonically isomorphic to $k \otimes_{\mathbf{Q}} \mathfrak{m}_{\mathbf{Q}}$.

DEFINITION 1. *Let $\mathfrak{q}$ be a Lie subalgebra of the semi-simple Lie algebra $\mathfrak{g}$. Then $\mathfrak{q}$ is said to be decomposable in $\mathfrak{g}$ if, for all $x \in \mathfrak{q}$, the semi-simple and nilpotent components of x in $\mathfrak{g}$ belong to $\mathfrak{q}$. Denote by $\mathfrak{n}_{\mathfrak{g}}(\mathfrak{q})$ the set of elements x of the radical of $\mathfrak{q}$ such that* $\mathrm{ad}_{\mathfrak{g}} x$ *is nilpotent.*

Let ρ be an injective representation of $\mathfrak{g}$ on a finite dimensional vector space V. We know (Chap. I, §6, no. 3, Th. 3) that an element x of $\mathfrak{g}$ is semi-simple (resp. nilpotent) if and only if the endomorphism $\rho(x)$ of V is semi-simple (resp. nilpotent). It follows immediately that the algebra $\mathfrak{q}$ is decomposable in $\mathfrak{g}$ if and only if $\rho(\mathfrak{q})$ is a decomposable subalgebra of $\mathfrak{gl}(V)$ in the sense of Definition 1 of Chap. VII, §5, no. 1. With the notations of Chap. VII, §5, no. 3, we also have

$$\rho(\mathfrak{n}_{\mathfrak{g}}(\mathfrak{q})) = \mathfrak{n}_V(\rho(\mathfrak{q})).$$

THEOREM 2. *Let $\mathfrak{g}$ be a semi-simple Lie algebra, $\mathfrak{n}$ a subalgebra of $\mathfrak{g}$ consisting of nilpotent elements, $\mathfrak{q}$ the normalizer of $\mathfrak{n}$ in $\mathfrak{g}$. Assume that $\mathfrak{n}$ is the set of nilpotent elements of the radical of $\mathfrak{q}$. Then $\mathfrak{q}$ is parabolic.*

Note first of all that $\mathfrak{q}$ is decomposable (Chap. VII, §5, no. 1, Cor. 1 of Prop. 3). By Th. 1, it suffices to prove that $\mathfrak{q}$ is the orthogonal complement $\mathfrak{n}^0$ of $\mathfrak{n}$ with respect to the Killing form Φ of $\mathfrak{g}$. We know that $\mathfrak{q} \subset \mathfrak{n}^0$ (Chap. I, §4, no. 3, Prop. 4 *d*)). By Chap. VII, §5, no. 3, Prop. 7, there exists a subalgebra $\mathfrak{m}$ of $\mathfrak{q}$, reductive in $\mathfrak{g}$, such that $\mathfrak{q}$ is the semi-direct product of $\mathfrak{m}$ and $\mathfrak{n}$. We show that the restriction of Φ to $\mathfrak{m}$ is non-degenerate. Let $\mathfrak{c}$ be the centre of

$\mathfrak{m}$. We have $\Phi([\mathfrak{m},\mathfrak{m}],\mathfrak{c}) = 0$ by Chap. I, §5, no. 5, Prop. 5, and the restriction of Φ to $[\mathfrak{m},\mathfrak{m}]$ is non-degenerate by Chap. I, §6, no. 1, Prop. 1. It remains to see that the restriction of Φ to $\mathfrak{c}$ is non-degenerate. Let $\mathfrak{k}$ be a Cartan subalgebra of $[\mathfrak{m},\mathfrak{m}]$; then $\mathfrak{k} \oplus \mathfrak{c}$ is commutative and reductive in $\mathfrak{g}$. Let $\mathfrak{h}$ be a Cartan subalgebra of $\mathfrak{g}$ containing $\mathfrak{k} \oplus \mathfrak{c}$ (Chap. VII, §2, no. 3, Prop. 10). Then $\mathfrak{h} \cap \mathfrak{q}$ is a commutative subalgebra of $\mathfrak{q}$ containing $\mathfrak{k} \oplus \mathfrak{c}$, and $\mathrm{ad}_{\mathfrak{q}} x$ is semi-simple for all $x \in \mathfrak{h} \cap \mathfrak{q}$; hence $\mathfrak{h} \cap \mathfrak{q}$ is contained in a Cartan subalgebra $\mathfrak{h}'$ of $\mathfrak{q}$ (Chap. VII, §2, no. 3, Prop. 10); let f be the projection of $\mathfrak{q}$ onto $\mathfrak{m}$ with kernel $\mathfrak{n}$; then $f(\mathfrak{h}')$ is a Cartan subalgebra of $\mathfrak{m}$ (Chap. VII, §2, no. 1, Cor. 2 of Prop. 4) containing $\mathfrak{k}\oplus\mathfrak{c}$, and consequently equal to $\mathfrak{k}\oplus\mathfrak{c}$; this proves that $f(\mathfrak{h} \cap \mathfrak{q}) = \mathfrak{k} \oplus \mathfrak{c}$, and since every element of $\mathfrak{h}$ is semi-simple in $\mathfrak{g}$, we have $\mathfrak{h} \cap \mathfrak{q} = \mathfrak{k} \oplus \mathfrak{c}$. Thus,

$$\mathfrak{c} = \{x \in \mathfrak{h} \mid [x,\mathfrak{n}] \subset \mathfrak{n} \text{ and } [x,[\mathfrak{m},\mathfrak{m}]] = 0\}.$$

By Lemma 1, the restriction of Φ to $\mathfrak{c}$ is non-degenerate.

Let $\mathfrak{q}^0$ be the orthogonal complement of $\mathfrak{q}$ in $\mathfrak{g}$ relative to Φ. The preceding proves that $\mathfrak{q} \cap \mathfrak{q}^0 = \mathfrak{n}$. Assume that $\mathfrak{q} \neq \mathfrak{q}^0$, so $\mathfrak{q}^0 \neq \mathfrak{n}$ (and $\mathfrak{q}^0 \supset \mathfrak{n}$). Since $\mathrm{ad}_{\mathfrak{g}}\mathfrak{n}$ leaves $\mathfrak{q}$ stable, $\mathrm{ad}_{\mathfrak{g}}\mathfrak{n}$ leaves $\mathfrak{q}^0$ stable; Engel's theorem proves that there exists $x \in \mathfrak{q}^0$ such that $x \notin \mathfrak{n}$ and $[x,\mathfrak{n}] \subset \mathfrak{n}$. But then $x \in \mathfrak{q}^0 \cap \mathfrak{q} = \mathfrak{n}$, a contradiction. Hence $\mathfrak{q} = \mathfrak{n}^0$.

COROLLARY 1. *Let $\mathfrak{q}$ be a maximal element of the set of subalgebras of $\mathfrak{g}$ distinct from $\mathfrak{g}$. Then $\mathfrak{q}$ is either parabolic or reductive in $\mathfrak{g}$.*

We can assume that $\mathfrak{g}$ is a Lie subalgebra of $\mathfrak{gl}(V)$ for some finite dimensional vector space V. Let $\mathfrak{e}(\mathfrak{q}) \subset \mathfrak{g}$ be the decomposable envelope of $\mathfrak{q}$. If $\mathfrak{e}(\mathfrak{q}) = \mathfrak{g}$, $\mathfrak{q}$ is an ideal of $\mathfrak{g}$ (Chap. VII, §5, no. 2, Prop. 4), hence is semi-simple, and consequently $\mathfrak{q}$ is reductive in $\mathfrak{g}$. Assume that $\mathfrak{e}(\mathfrak{q}) \neq \mathfrak{g}$. Then $\mathfrak{e}(\mathfrak{q}) = \mathfrak{q}$, so $\mathfrak{q}$ is decomposable. Assume that $\mathfrak{q}$ is not reductive in $\mathfrak{g}$. Let $\mathfrak{n}$ be the set of nilpotent elements of the radical of $\mathfrak{q}$. Then $\mathfrak{n} \neq 0$ (Chap. VII, §5, no. 3, Prop. 7 (i)). Let $\mathfrak{p}$ be the normalizer of $\mathfrak{n}$ in $\mathfrak{g}$. Then $\mathfrak{p} \supset \mathfrak{q}$, and $\mathfrak{p} \neq \mathfrak{g}$ since $\mathfrak{g}$ is semi-simple. Hence $\mathfrak{p} = \mathfrak{q}$. Thus $\mathfrak{q}$ is parabolic (Th. 1).

COROLLARY 2. *Let $\mathfrak{n}$ be a subalgebra of $\mathfrak{g}$ consisting of nilpotent elements. There exists a parabolic subalgebra $\mathfrak{q}$ of $\mathfrak{g}$ with the following properties:*

(i) $\mathfrak{n} \subset \mathfrak{n}_{\mathfrak{g}}(\mathfrak{q})$;

(ii) *the normalizer of $\mathfrak{n}$ in $\mathfrak{g}$ is contained in $\mathfrak{q}$;*

(iii) *every automorphism of $\mathfrak{g}$ leaving $\mathfrak{n}$ invariant leaves $\mathfrak{q}$ invariant.*

If $\mathfrak{g}$ is splittable, $\mathfrak{n}$ is contained in a Borel subalgebra of $\mathfrak{g}$.

Let $\mathfrak{q}_1$ be the normalizer of $\mathfrak{n}$ in $\mathfrak{g}$. This is a decomposable subalgebra of $\mathfrak{g}$. Let $\mathfrak{n}_1 = \mathfrak{n}_{\mathfrak{g}}(\mathfrak{q}_1)$. Define inductively $\mathfrak{q}_i$ to be the normalizer of $\mathfrak{n}_{i-1}$ in $\mathfrak{g}$, and $\mathfrak{n}_i$ to be equal to $\mathfrak{n}_{\mathfrak{g}}(\mathfrak{q}_i)$. The sequences $(\mathfrak{n},\mathfrak{n}_1,\mathfrak{n}_2,\ldots)$ and $(\mathfrak{q}_1,\mathfrak{q}_2,\ldots)$ are increasing. There exists j such that $\mathfrak{q}_j = \mathfrak{q}_{j+1}$, in other words $\mathfrak{q}_j$ is the normalizer of $\mathfrak{n}_{\mathfrak{g}}(\mathfrak{q}_j)$ in $\mathfrak{g}$. Thus $\mathfrak{q}_j$ is parabolic (Th. 1). We have $\mathfrak{n} \subset \mathfrak{n}_j =$

$\mathfrak{n}_{\mathfrak{g}}(\mathfrak{q}_j)$, and $\mathfrak{q}_1 \subset \mathfrak{q}_j$; every automorphism of $\mathfrak{g}$ leaving $\mathfrak{n}$ invariant evidently leaves $\mathfrak{n}_1, \mathfrak{n}_2, \ldots$ and $\mathfrak{q}_1, \mathfrak{q}_2, \ldots$ invariant. If $\mathfrak{g}$ is splittable, $\mathfrak{q}_j$ contains a Borel subalgebra $\mathfrak{b}$, and consequently (§3, no. 4, Prop. 13), we have $\mathfrak{b} \supset \mathfrak{n}_{\mathfrak{g}}(\mathfrak{q}_j) \supset \mathfrak{n}$.

THEOREM 3. *Assume that k is algebraically closed. Let $\mathfrak{g}$ be a semi-simple Lie algebra. Let $\mathfrak{a}$ be a solvable subalgebra of $\mathfrak{g}$. There exists a Borel subalgebra of $\mathfrak{g}$ containing $\mathfrak{a}$.*

By Chap. VII, §5, no. 2, Cor. 1 (ii) of Prop. 4, we can assume that $\mathfrak{a}$ is decomposable. There exists a commutative subalgebra $\mathfrak{t}$ of $\mathfrak{g}$, consisting of semi-simple elements, such that $\mathfrak{a}$ is the semi-direct product of $\mathfrak{t}$ and $\mathfrak{n}_{\mathfrak{g}}(\mathfrak{a})$ (Chap. VII, §5, no. 3, Cor. 2 of Prop. 6). There exists (Cor. 2 of Th. 2) a parabolic subalgebra $\mathfrak{q}$ of $\mathfrak{g}$ such that $\mathfrak{n}_{\mathfrak{g}}(\mathfrak{a}) \subset \mathfrak{n}_{\mathfrak{g}}(\mathfrak{q})$, and such that the normalizer of $\mathfrak{n}_{\mathfrak{g}}(\mathfrak{a})$ in $\mathfrak{g}$ is contained in $\mathfrak{q}$; *a fortiori*, $\mathfrak{a} \subset \mathfrak{q}$. Let $\mathfrak{b}$ be a Borel subalgebra of $\mathfrak{g}$ contained in $\mathfrak{q}$ and $\mathfrak{h}$ a Cartan subalgebra of $\mathfrak{g}$ contained in $\mathfrak{b}$. Then $\mathfrak{h}$ is a Cartan subalgebra of $\mathfrak{q}$, so there exists $s \in \mathrm{Aut}_e(\mathfrak{q})$ such that $s(\mathfrak{t}) \subset \mathfrak{h}$ (Chap. VII, §2, no. 3, Prop. 10 and Chap. VII, §3, no. 2, Th. 1). We have $s(\mathfrak{n}_{\mathfrak{g}}(\mathfrak{q})) = \mathfrak{n}_{\mathfrak{g}}(\mathfrak{q})$ (Chap. VII, §3, no. 1, Remark 1), so

$$s(\mathfrak{a}) = s(\mathfrak{t}) + s(\mathfrak{n}_{\mathfrak{g}}(\mathfrak{a})) \subset \mathfrak{h} + s(\mathfrak{n}_{\mathfrak{g}}(\mathfrak{q})) = \mathfrak{h} + \mathfrak{n}_{\mathfrak{g}}(\mathfrak{q}) \subset \mathfrak{b}.$$

COROLLARY. *If k is algebraically closed, every maximal solvable subalgebra of $\mathfrak{g}$ is a Borel subalgebra.*

§11. CLASSES OF NILPOTENT ELEMENTS AND $\mathfrak{sl}_2$-TRIPLETS

In this paragraph, $\mathfrak{g}$ denotes a finite dimensional Lie algebra.

1. DEFINITION OF $\mathfrak{sl}_2$-TRIPLETS

DEFINITION 1. *An $\mathfrak{sl}_2$-triplet in $\mathfrak{g}$ is a sequence (x, h, y) of elements of $\mathfrak{g}$, distinct from $(0, 0, 0)$, such that*

$$[h, x] = 2x, \quad [h, y] = -2y, \quad [x, y] = -h.$$

Let (x, h, y) be an $\mathfrak{sl}_2$-triplet in $\mathfrak{g}$. The linear map τ from $\mathfrak{sl}(2, k)$ to $\mathfrak{g}$ such that $\tau(X_+) = x, \tau(H) = h, \tau(X_-) = y$ is a homomorphism which is non-zero and hence injective (since $\mathfrak{sl}(2, k)$ is simple), and with image $kx + kh + ky$. We thus obtain a canonical bijection from the set of $\mathfrak{sl}_2$-triplets in $\mathfrak{g}$ to the set of injective homomorphisms from $\mathfrak{sl}(2, k)$ to $\mathfrak{g}$. If $\mathfrak{g}$ is semi-simple and if (x, h, y) is an $\mathfrak{sl}_2$-triplet in $\mathfrak{g}$, then x and y are nilpotent elements of $\mathfrak{g}$ and h is a semi-simple element of $\mathfrak{g}$ (Chap. I, §6, no. 3, Prop. 4).

Lemma 1. Let $x, h, y, y' \in \mathfrak{g}$. *If* (x, h, y) *and* (x, h, y') *are* $\mathfrak{sl}_2$*-triplets in* $\mathfrak{g}$, *then* $y = y'$.

Indeed, $y - y' \in \operatorname{Ker}(\operatorname{ad}_{\mathfrak{g}} x)$ and $(\operatorname{ad}_{\mathfrak{g}} h)(y - y') = -2(y - y')$. But $\operatorname{ad}_{\mathfrak{g}} x$ is injective on $\operatorname{Ker}(p + \operatorname{ad}_{\mathfrak{g}} h)$ for every integer $p > 0$ (§1, no. 2, Cor. of Prop. 2).

Lemma 2. Let $\mathfrak{n}$ *be a subalgebra of* $\mathfrak{g}$ *such that, for all* $n \in \mathfrak{n}$, $\operatorname{ad}_{\mathfrak{g}}(n)$ *is nilpotent. Let* $h \in \mathfrak{g}$ *be such that* $[h, \mathfrak{n}] = \mathfrak{n}$. *Then* $e^{\operatorname{ad}_{\mathfrak{g}} \mathfrak{n}}.h = h + \mathfrak{n}$.

It is clear that $e^{\operatorname{ad}_{\mathfrak{g}}(\mathfrak{n})}.h \subset h + \mathfrak{n}$. We shall prove that, if $v \in \mathfrak{n}$, then $h + v \in e^{\operatorname{ad}_{\mathfrak{g}}(\mathfrak{n})}.h$. It suffices to prove that $h + v \in e^{\operatorname{ad}_{\mathfrak{g}}(\mathfrak{n})}.h + \mathscr{C}^p\mathfrak{n}$ for all $p \geq 1$ (since $\mathscr{C}^p\mathfrak{n} = 0$ for sufficiently large p). This is clear for $p = 1$ since $\mathscr{C}^1\mathfrak{n} = \mathfrak{n}$. Assume now that we have proved the existence of $y_p \in \mathfrak{n}$ and $z_p \in \mathscr{C}^p\mathfrak{n}$ such that $h + v = e^{\operatorname{ad}_{\mathfrak{g}} y_p}.h + z_p$. Since $(\operatorname{ad}_{\mathfrak{g}} h)(\mathfrak{n}) = \mathfrak{n}$, $(\operatorname{ad}_{\mathfrak{g}} h)|\mathfrak{n}$ is a bijection from $\mathfrak{n}$ to $\mathfrak{n}$, hence its restriction to $\mathscr{C}^p\mathfrak{n}$, which leaves $\mathscr{C}^p\mathfrak{n}$ stable, is also bijective; consequently, there exists $z \in \mathscr{C}^p\mathfrak{n}$ such that $z_p = [z, h]$. Then

$$e^{\operatorname{ad}_{\mathfrak{g}}(y_p + z)}h - e^{\operatorname{ad}_{\mathfrak{g}} y_p}h \in [z, h] + \mathscr{C}^{p+1}\mathfrak{n}$$

so

$$e^{\operatorname{ad}_{\mathfrak{g}}(y_p + z)}h \in h + v - z_p + [z, h] + \mathscr{C}^{p+1}\mathfrak{n} = h + v + \mathscr{C}^{p+1}\mathfrak{n}$$

which establishes our assertion by induction on p.

Lemma 3. Let $x \in \mathfrak{g}$, $\mathfrak{p} = \operatorname{Ker}(\operatorname{ad} x)$, $\mathfrak{q} = \operatorname{Im}(\operatorname{ad} x)$. *Then* $[\mathfrak{p}, \mathfrak{q}] \subset \mathfrak{q}$, *and* $\mathfrak{p} \cap \mathfrak{q}$ *is a subalgebra of* $\mathfrak{g}$.

If $u \in \mathfrak{p}$ and $v \in \mathfrak{q}$, there exists $w \in \mathfrak{g}$ such that $v = [x, w]$, so

$$[u, v] = [u, [x, w]] = [x, [u, w]] - [[x, u], w] = [x, [u, w]] \in \mathfrak{q}.$$

On the other hand, $\mathfrak{p}$ is a subalgebra of $\mathfrak{g}$, so $[\mathfrak{p} \cap \mathfrak{q}, \mathfrak{p} \cap \mathfrak{q}] \subset \mathfrak{p} \cap \mathfrak{q}$.

Lemma 4. Let (x, h, y) *and* (x, h', y') *be* $\mathfrak{sl}_2$*-triplets in* $\mathfrak{g}$. *There exists* $z \in \mathfrak{g}$ *such that* $\operatorname{ad}_{\mathfrak{g}} z$ *is nilpotent and such that*

$$e^{\operatorname{ad}_{\mathfrak{g}} z}x = x, \quad e^{\operatorname{ad}_{\mathfrak{g}} z}h = h', \quad e^{\operatorname{ad}_{\mathfrak{g}} z}y = y'.$$

Let $\mathfrak{n} = \operatorname{Ker}(\operatorname{ad} x) \cap \operatorname{Im}(\operatorname{ad} x)$. For all $p \in \mathbf{Z}$, let $\mathfrak{g}_p = \operatorname{Ker}(\operatorname{ad} h - p)$. By §1, no. 3 (applied to the adjoint representation of $kx + ky + kh$ on $\mathfrak{g}$), we have that $\mathfrak{n} \subset \sum_{p>0} \mathfrak{g}_p$, so $\operatorname{ad}_{\mathfrak{g}} n$ is nilpotent for all $n \in \mathfrak{n}$, and $[h, \mathfrak{n}] = \mathfrak{n}$. We have $[x, h' - h] = 0$ and $[x, y - y'] = h' - h$, so $h' - h \in \mathfrak{n}$. By Lemmas 2 and 3, there exists $z \in \mathfrak{n}$ such that $e^{\operatorname{ad}_{\mathfrak{g}} z}h = h'$. Since $z \in \operatorname{Ker} \operatorname{ad}_{\mathfrak{g}} x$, we have $e^{\operatorname{ad}_{\mathfrak{g}} z}x = x$. Lemma 1 now proves that $e^{\operatorname{ad}_{\mathfrak{g}} z}y = y'$. Q.E.D.

Let G be a group of automorphisms of $\mathfrak{g}$. Then two $\mathfrak{sl}_2$-triplets (x, h, y), (x', h', y') are said to be G-conjugate if there exists $g \in \mathrm{G}$ such that $gx = x'$, $gh = h'$, $gy = y'$.

PROPOSITION 1. *Let* G *be a group of automorphisms of* $\mathfrak{g}$ *containing* $\mathrm{Aut}_e(\mathfrak{g})$. *Let* (x, h, y) *and* (x', h', y') *be* $\mathfrak{sl}_2$*-triplets in* $\mathfrak{g}$. *Let*

$$\mathfrak{t} = kx + kh + ky, \quad \mathfrak{t}' = kx' + kh' + ky'.$$

Consider the following conditions:

(i) x *and* x' *are* G*-conjugate*;

(ii) (x, h, y) *and* (x', h', y') *are* G*-conjugate*;

(iii) $\mathfrak{t}$ *and* $\mathfrak{t}'$ *are* G*-conjugate.*

We have (i) $\Longleftrightarrow$ (ii) $\Longrightarrow$ (iii). *If* k *is algebraically closed, the three conditions are equivalent.*

(i) $\Longleftrightarrow$ (ii): This follows from Lemma 4.

(ii) $\Longrightarrow$ (iii): This is clear.

We assume that k is algebraically closed and prove that (iii) $\Longrightarrow$ (i). We treat first the case in which $\mathfrak{t} = \mathfrak{t}' = \mathfrak{g} = \mathfrak{sl}(2, k)$. Since $\mathrm{ad}_{\mathfrak{g}}\, x$ is nilpotent, the endomorphism x of k^2 is nilpotent (Chap. I, §6, Th. 3), so there exists a matrix $A \in \mathbf{GL}(2, k)$ such that $AxA^{-1} = X$, and consequently an automorphism α of $\mathfrak{sl}(2, k)$ such that $\alpha(x) = x'$; now $\alpha \in \mathrm{Aut}_e(\mathfrak{g})$ (§5, no. 3, Cor. 2 of Prop. 5). We now pass to the general case; we assume that $\mathfrak{t}$ and $\mathfrak{t}'$ are G-conjugate and prove that x and x' are G-conjugate. We can assume that $\mathfrak{t} = \mathfrak{t}'$. By the preceding, there exists $\beta \in \mathrm{Aut}_e(\mathfrak{t})$ such that $\beta x = x'$. Now, if $t \in \mathfrak{t}$ is such that $\mathrm{ad}_{\mathfrak{t}}\, t$ is nilpotent, then $\mathrm{ad}_{\mathfrak{g}}\, t$ is nilpotent; so β extends to an element of $\mathrm{Aut}_e(\mathfrak{g})$.

Remark. The three conditions of Prop. 1 are equivalent if we assume only that $k = k^2$ (cf. Exerc. 1).

2. $\mathfrak{sl}_2$-TRIPLETS IN SEMI-SIMPLE LIE ALGEBRAS

Lemma 5. Let V *be a finite dimensional vector space,* A *and* B *endomorphisms of* V. *Assume that* A *is nilpotent and that* $[\mathrm{A}, [\mathrm{A}, \mathrm{B}]] = 0$. *Then* AB *is nilpotent.*

Put $\mathrm{C} = [\mathrm{A}, \mathrm{B}]$. Since $[\mathrm{A}, \mathrm{C}] = 0$,

$$[\mathrm{A}, \mathrm{BC}^p] = [\mathrm{A}, \mathrm{B}]\mathrm{C}^p = \mathrm{C}^{p+1}$$

for every integer $p \geq 0$. Consequently, $\mathrm{Tr}(\mathrm{C}^p) = 0$ for $p \geq 1$, which proves that C is nilpotent (*Algebra*, Chap. VII, §3, no. 5, Cor. 4 of Prop. 13). Now let $\bar{k}$ be an algebraic closure of k, and let $\lambda \in \bar{k}$, $x \in \mathrm{V} \otimes_k \bar{k}$ be such that $\mathrm{AB}x = \lambda x$, $x \neq 0$. The relation $[[\mathrm{B}, \mathrm{A}], \mathrm{A}] = 0$ shows that $[\mathrm{B}, \mathrm{A}^p] = p[\mathrm{B}, \mathrm{A}]\mathrm{A}^{p-1}$ for every integer $p \geq 0$. Let r be the smallest integer such that $\mathrm{A}^r x = 0$. Then

$$\lambda \mathrm{A}^{r-1}x = \mathrm{A}^{r-1}\mathrm{AB}x = \mathrm{A}^r\mathrm{B}x = \mathrm{BA}^r x - [\mathrm{B}, \mathrm{A}^r]x = -r[\mathrm{B}, \mathrm{A}]\mathrm{A}^{r-1}x.$$

Since $[\mathrm{B}, \mathrm{A}]$ is nilpotent and since $\mathrm{A}^{r-1}x \neq 0$, this proves that $\lambda = 0$. Thus, all the eigenvalues of AB are zero, hence the lemma.

Lemma 6. Let $h, x \in \mathfrak{g}$ be such that $[h, x] = 2x$ and $h \in (\operatorname{ad} x)(\mathfrak{g})$. Then there exists $y \in \mathfrak{g}$ such that (x, h, y) is either $(0, 0, 0)$ or an $\mathfrak{sl}_2$-triplet.

Let $\mathfrak{g}'$ be the solvable Lie algebra $kh + kx$. Since $x \in [\mathfrak{g}', \mathfrak{g}']$, $\operatorname{ad}_{\mathfrak{g}} x$ is nilpotent (Chap. I, §5, no. 3, Th. 1); let $\mathfrak{n}$ be its kernel. Since $[\operatorname{ad} h, \operatorname{ad} x] = 2 \operatorname{ad} x$, we have $(\operatorname{ad} h)\mathfrak{n} \subset \mathfrak{n}$. Let $z \in \mathfrak{g}$ be such that $h = -[x, z]$. For any integer $n \geq 0$, put $\mathrm{M}_n = (\operatorname{ad} x)^n \mathfrak{g}$. If $n > 0$, we have (§1, no. 1, Lemma 1)

$$[\operatorname{ad} z, (\operatorname{ad} x)^n] = n((\operatorname{ad} h) - n + 1)(\operatorname{ad} x)^{n-1}$$

so, if $u \in \mathrm{M}_{n-1}$,

$$n((\operatorname{ad} h) - n + 1)u \in (\operatorname{ad} z)(\operatorname{ad} x)u + \mathrm{M}_n.$$

Since $(\operatorname{ad} h)\mathfrak{n} \subset \mathfrak{n}$, it follows that

$$((\operatorname{ad} h) - n + 1)(\mathfrak{n} \cap \mathrm{M}_{n-1}) \subset \mathfrak{n} \cap \mathrm{M}_n.$$

Since $\operatorname{ad} x$ is nilpotent, $\mathrm{M}_n = 0$ for sufficiently large n. Consequently, the eigenvalues of $\operatorname{ad} h|\mathfrak{n}$ are integers ≥ 0. Thus, the restriction of $\operatorname{ad} h + 2$ to $\mathfrak{n}$ is invertible.

Now $[h, z] + 2z \in \mathfrak{n}$ since

$$\begin{aligned}[x, [h, z] + 2z] &= [[x, h], z] + [h, [x, z]] + 2[x, z] \\ &= [-2x, z] + [h, -h] + 2[x, z] = 0.\end{aligned}$$

Hence there exists $z' \in \mathfrak{n}$ such that $[h, z'] + 2z' = [h, z] + 2z$, that is, $[h, y] = -2y$, putting $y = z - z'$. Since $[x, y] = [x, z] = -h$, this completes the proof.

PROPOSITION 2 (*Jacobson-Morozov*). *Assume that $\mathfrak{g}$ is semi-simple. Let x be a non-zero nilpotent element of $\mathfrak{g}$. There exist $h, y \in \mathfrak{g}$ such that (x, h, y) is an $\mathfrak{sl}_2$-triplet.*

Let $\mathfrak{n} = \operatorname{Ker}(\operatorname{ad} x)^2$. If $z \in \mathfrak{n}$, then $[\operatorname{ad} x, [\operatorname{ad} x, \operatorname{ad} z]] = \operatorname{ad}([x, [x, z]]) = 0$. By Lemma 5, $\operatorname{ad} x \circ \operatorname{ad} z$ is nilpotent, so $\operatorname{Tr}(\operatorname{ad} x \circ \operatorname{ad} z) = 0$. This shows that x is orthogonal to $\mathfrak{n}$ with respect to the Killing form Φ of $\mathfrak{g}$. Since

$$\Phi((\operatorname{ad} x)^2 y, y') = \Phi(y, (\operatorname{ad} x)^2 y')$$

for all $y, y' \in \mathfrak{g}$, and since Φ is non-degenerate, the orthogonal complement of $\mathfrak{n}$ is the image of $(\operatorname{ad} x)^2$. Hence there exists $y' \in \mathfrak{g}$ such that $x = (\operatorname{ad} x)^2 y'$. Put

$$h = -2[x, y'];$$

we have $[h, x] = 2x$ and $h \in (\operatorname{ad} x)(\mathfrak{g})$. It now suffices to apply Lemma 6.

COROLLARY. *Assume that $\mathfrak{g}$ is semi-simple. Let G be a group of automorphisms of $\mathfrak{g}$ containing $\operatorname{Aut}_e(\mathfrak{g})$. The map which associates to any $\mathfrak{sl}_2$-triplet*

(x, h, y) in $\mathfrak{g}$ the nilpotent element x defines, by passage to the quotient, a bijection from the set of G*-conjugacy classes of $\mathfrak{sl}_2$-triplets to the set of* G*-conjugacy classes of non-zero nilpotent elements.*

This follows from Prop. 1 and 2.

Lemma 7. Let K *be a commutative field with at least* 4 *elements. Let* G *be the group of matrices* $\begin{pmatrix} \alpha & \beta \\ 0 & \alpha^{-1} \end{pmatrix}$ *where* $\alpha \in \mathrm{K}^*$, $\beta \in \mathrm{K}$. *Let* G' *be the group of such matrices such that* $\alpha = 1$. *Then* $\mathrm{G}' = (\mathrm{G}, \mathrm{G})$.

If $\alpha, \alpha' \in \mathrm{K}^*$ and $\beta, \beta' \in \mathrm{K}$,

$$\begin{pmatrix} \alpha & \beta \\ 0 & \alpha^{-1} \end{pmatrix} \begin{pmatrix} \alpha' & \beta' \\ 0 & \alpha'^{-1} \end{pmatrix} \begin{pmatrix} \alpha & \beta \\ 0 & \alpha^{-1} \end{pmatrix}^{-1} \begin{pmatrix} \alpha' & \beta' \\ 0 & \alpha'^{-1} \end{pmatrix}^{-1} = \begin{pmatrix} 1 & -\alpha'\beta' - \alpha\beta\alpha'^2 + \alpha^2\alpha'\beta' + \alpha\beta \\ 0 & 1 \end{pmatrix}.$$

In particular,

$$\begin{pmatrix} 1 & \beta \\ 0 & 1 \end{pmatrix} \begin{pmatrix} \alpha' & 0 \\ 0 & \alpha'^{-1} \end{pmatrix} \begin{pmatrix} 1 & \beta \\ 0 & 1 \end{pmatrix}^{-1} \begin{pmatrix} \alpha' & 0 \\ 0 & \alpha'^{-1} \end{pmatrix}^{-1} = \begin{pmatrix} 1 & \beta(1-\alpha'^2) \\ 0 & 1 \end{pmatrix}.$$

But there exists $\alpha'_0 \in \mathrm{K}^*$ such that $\alpha'_0 \neq 1$ and $\alpha'_0 \neq -1$, and then $k.(1-{\alpha'_0}^2) = k$, hence the lemma.

PROPOSITION 3. *Assume that $\mathfrak{g}$ is semi-simple. The group $\mathrm{Aut}_e(\mathfrak{g})$ is equal to its derived group. If $\mathfrak{g}$ is splittable, $\mathrm{Aut}_e(\mathfrak{g})$ is the derived group of $\mathrm{Aut}_0(\mathfrak{g})$.*

Let x be a non-zero nilpotent element of $\mathfrak{g}$. Choose $h, y \in \mathfrak{g}$ be such that (x, h, y) is an $\mathfrak{sl}_2$-triplet (Prop. 2). The subalgebra $\mathfrak{s}$ of $\mathfrak{g}$ generated by (x, h, y) can be identified with $\mathfrak{sl}(2, k)$. Let ρ be the representation $z \mapsto \mathrm{ad}_{\mathfrak{g}}\, z$ of $\mathfrak{s} = \mathfrak{sl}(2, k)$ on $\mathfrak{g}$, and let π be the representation of $\mathbf{SL}(2, k)$ compatible with ρ (§1, no. 4). The image of π is generated by the $\exp(t\,\mathrm{ad}_{\mathfrak{g}}\, x)$ and the $\exp(t\,\mathrm{ad}_{\mathfrak{g}}\, y)$ with $t \in k$ (*Algebra*, Chap. III, §8, no. 9, Prop. 17), hence is contained in $\mathrm{Aut}_e(\mathfrak{g})$. Since $\mathbf{SL}(2, k)$ is equal to its derived group (Lemma 7 and *loc. cit.*), $\exp(\mathrm{ad}_{\mathfrak{g}}\, x)$ belongs to the derived group G of $\mathrm{Aut}_e(\mathfrak{g})$. Hence $\mathrm{Aut}_e(\mathfrak{g})$ is equal to G. Assume now that $\mathfrak{g}$ is splittable. Since $\mathrm{Aut}_0(\mathfrak{g})/\mathrm{Aut}_e(\mathfrak{g})$ is commutative (§5, no. 3, Remark 3), the preceding proves that the derived group of $\mathrm{Aut}_0(\mathfrak{g})$ is $\mathrm{Aut}_e(\mathfrak{g})$.

3. SIMPLE ELEMENTS

DEFINITION 2. *An element h of $\mathfrak{g}$ is said to be simple if there exist $x, y \in \mathfrak{g}$ such that (x, h, y) is an $\mathfrak{sl}_2$-triplet in $\mathfrak{g}$.*

We also say that h is the simple element of the $\mathfrak{sl}_2$-triplet (x, h, y).

PROPOSITION 4. *Let h be a non-zero element of $\mathfrak{g}$. Then h is simple if and only if there exists $x \in \mathfrak{g}$ such that $[h, x] = 2x$ and $h \in (\mathrm{ad}\, x)(\mathfrak{g})$.*

The condition is clearly necessary. It is sufficient by Lemma 6.

PROPOSITION 5. *Assume that $\mathfrak{g}$ is splittable semi-simple. Let $\mathfrak{h}$ be a splitting Cartan subalgebra of $\mathfrak{g}$, R the set of roots of $(\mathfrak{g}, \mathfrak{h})$, and B a basis of R. Let h be a simple element of $\mathfrak{g}$ belonging to $\mathfrak{h}$. Then h is conjugate under $\mathrm{Aut}_e(\mathfrak{g}, \mathfrak{h})$ to an element h' of $\mathfrak{h}$ such that $\alpha(h') \in \{0, 1, 2\}$ for all $\alpha \in \mathrm{B}$.*

The eigenvalues of $\mathrm{ad}_{\mathfrak{g}} h$ belong to $\mathbf{Z}$ (§1, no. 2, Cor. of Prop. 2). Hence $h \in \mathfrak{h}_{\mathbf{Q}}$. There exists an element w of the Weyl group of $(\mathfrak{g}, \mathfrak{h})$ such that $\alpha(wh) \geq 0$ for all $\alpha \in \mathrm{B}$ (Chap. VI, §1, no. 5, Th. 2 (i)). In view of §2, no. 2, Cor. of Th. 2, we are reduced to the case in which $\alpha(h) \in \mathbf{N}$ for all $\alpha \in \mathrm{B}$. Let R_+ be the set of positive roots relative to B, and $\mathrm{R}_- = -\mathrm{R}_+$. There exists an $\mathfrak{sl}_2$-triplet in $\mathfrak{g}$ of the form (x, h, y). Let T be the set of roots β such that $\beta(h) = -2$. Then $\mathrm{T} \subset \mathrm{R}_-$ and $y \in \sum_{\beta \in \mathrm{T}} \mathfrak{g}^\beta$. Assume that there exists $\alpha \in \mathrm{B}$ such that $\alpha(h) > 2$. For all $\beta \in \mathrm{T}$, we have $(\alpha + \beta)(h) > 0$, so $\alpha + \beta \notin \mathrm{R}_-$ and $\alpha + \beta \neq 0$; on the other hand, since $\beta \in \mathrm{R}_-$ and $\alpha \in \mathrm{B}$, we have $\alpha + \beta \notin \mathrm{R}_+$; hence $\alpha + \beta \notin \mathrm{R} \cup \{0\}$, so $[\mathfrak{g}^\alpha, \mathfrak{g}^\beta] = 0$. Thus, $[y, \mathfrak{g}^\alpha] = 0$. But $\mathrm{ad}_{\mathfrak{g}} y | \mathfrak{g}^\alpha$ is injective since $\alpha(h) > 0$ (§1, no. 2, Cor. of Prop. 2). This contradiction proves that $\alpha(h) \leq 2$ for all $\alpha \in \mathrm{B}$.

COROLLARY. *If k is algebraically closed and if $\mathfrak{g}$ is semi-simple of rank l, the number of conjugacy classes of simple elements of $\mathfrak{g}$, relative to $\mathrm{Aut}_e(\mathfrak{g})$, is at most 3^l.*

Indeed, every semi-simple element of $\mathfrak{g}$ is conjugate under $\mathrm{Aut}_e(\mathfrak{g})$ to an element of $\mathfrak{h}$.

Lemma 8. Assume that k is algebraically closed and that $\mathfrak{g}$ is semi-simple. Let h be a semi-simple element of $\mathfrak{g}$ such that the eigenvalues of $\mathrm{ad}\, h$ are rational. Let $\mathfrak{g}^0 = \mathrm{Ker}(\mathrm{ad}\, h)$, $\mathfrak{g}^2 = \mathrm{Ker}(\mathrm{ad}\, h - 2)$. Let G_h be the set of elementary automorphisms of $\mathfrak{g}$ leaving h fixed. Let $x \in \mathfrak{g}^2$ be such that $[x, \mathfrak{g}^0] = \mathfrak{g}^2$. Then $\mathrm{G}_h x$ contains a subset of $\mathfrak{g}^2$ that is dense and open in the Zariski topology.

Let $\mathfrak{h}$ be a Cartan subalgebra of $\mathfrak{g}^0$. This is a Cartan subalgebra of $\mathfrak{g}$ containing h (Chap. VII, §2, no. 3, Prop. 10). We have $h \in \mathfrak{h}_{\mathbf{Q}}$. Let R be the root system of $(\mathfrak{g}, \mathfrak{h})$, Q the group of radical weights. There exists a basis B of R such that $\alpha(h) \geq 0$ for all $\alpha \in \mathrm{B}$.

Let U be the set of $z \in \mathfrak{h}$ such that $\alpha(z) \neq 0$ for all $\alpha \in \mathrm{B}$. Let $(H'_\alpha)_{\alpha \in \mathrm{B}}$ be the basis of $\mathfrak{h}$ dual to B. If $z \in \mathrm{U}$, there exists a homomorphism from Q to k^* that takes any $\gamma \in \mathrm{Q}$ to $\prod_{\alpha \in \mathrm{B}} \alpha(z)^{\gamma(H'_\alpha)}$. By §5, Prop. 2 and 4, the endomorphism $\varphi(z)$ of the vector space $\mathfrak{g}$ which induces on $\mathfrak{g}^\gamma$ the homothety with ratio $\prod_{\alpha \in \mathrm{B}} \alpha(z)^{\gamma(H'_\alpha)}$ is an elementary automorphism of $\mathfrak{g}$, which clearly belongs to G_h.

Let $s \in \mathfrak{h}$. If $\gamma \in \mathrm{R}$ is such that $\mathfrak{g}^\gamma \cap \mathfrak{g}^2 \neq 0$,

$$2 = \gamma(h) = \gamma\left(\sum_{\alpha\in\mathrm{B}} \alpha(h)H'_\alpha\right) = \sum_{\alpha\in\mathrm{B}} \alpha(h)\gamma(H'_\alpha);$$

since $\alpha(h) \geq 0$ for all $\alpha \in \mathrm{B}$, and since the $\gamma(H'_\alpha)$ are integers either all ≥ 0 or all ≤ 0, we have $\gamma(H'_\alpha) \in \mathbf{N}$ for all $\alpha \in \mathrm{B}$. Thus, we can consider (for $z \in \mathfrak{h}$) the endomorphism $\psi(z)$ of the vector space $\mathfrak{g}^2$ that induces on $\mathfrak{g}^\gamma \cap \mathfrak{g}^2$ the homothety with ratio $\prod_{\alpha\in\mathrm{B}} \alpha(z)^{\gamma(H'_\alpha)}$. The map $z \mapsto \psi(z)$ from $\mathfrak{h}$ to $\mathrm{End}(\mathfrak{g}^2)$ is polynomial. For $z \in \mathrm{U}$, we have $\psi(z) = \varphi(z)|\mathfrak{g}^2$.

Let $\gamma_1, \dots, \gamma_r$ be the distinct roots of $(\mathfrak{g}, \mathfrak{h})$ vanishing on h. If $y_1 \in \mathfrak{g}^{\gamma_1}, \dots, y_r \in \mathfrak{g}^{\gamma_r}$, we have $e^{\mathrm{ad}\, y_1} \dots e^{\mathrm{ad}\, y_r} \in \mathrm{G}_h$. We can thus define a map ρ from $\mathfrak{h} \times \mathfrak{g}^{\gamma_1} \times \dots \times \mathfrak{g}^{\gamma_r}$ to $\mathfrak{g}^2$ by putting

$$\rho(z, y_1, \dots, y_r) = \psi(z)e^{\mathrm{ad}\, y_1} \dots e^{\mathrm{ad}\, y_r}x$$

for $z \in \mathfrak{h}$, $y_1 \in \mathfrak{g}^{\gamma_1}, \dots, y_r \in \mathfrak{g}^{\gamma_r}$. This map is polynomial, and $\rho(\mathrm{U}, \mathfrak{g}^{\gamma_1}, \dots, \mathfrak{g}^{\gamma_r}) \subset \mathrm{G}_h x$. By Chap. VII, App. I, Prop. 3 and 4, it suffices to prove that the tangent linear map of ρ is surjective at some point.

Now let T be the tangent linear map of $z \mapsto \psi(z)$ at $h_0 = \sum_{\alpha\in\mathrm{B}} H'_\alpha$. Then $\mathrm{T}(z)$ is the endomorphism of $\mathfrak{g}^2$ that induces on $\mathfrak{g}^\gamma \cap \mathfrak{g}^2$ the homothety with ratio

$$\sum_{\alpha\in\mathrm{B}} \gamma(H'_\alpha)\alpha(h_0)^{\gamma(H'_\alpha)-1}\alpha(z) \prod_{\beta\in\mathrm{B},\beta\neq\alpha} \beta(h_0)^{\gamma(H'_\alpha)} = \sum_{\alpha\in\mathrm{B}} \gamma(H'_\alpha)\alpha(z) = \gamma(z).$$

Thus, the tangent linear map of $z \mapsto \rho(z, 0, \dots, 0)$ at h_0 is the map $z \mapsto [z, x]$; its image is $[x, \mathfrak{h}]$. The tangent linear map at 0 of the map $y_1 \mapsto \rho(h_0, y_1, 0, \dots, 0)$ is the map $y_1 \mapsto \psi(h_0)[y_1, x]$; this last map has image $\psi(h_0)[x, \mathfrak{g}^{\gamma_1}] = [x, \mathfrak{g}^{\gamma_1}]$. Similarly, the tangent linear map at 0 of the map $y_i \mapsto \rho(h_0, 0, \dots, 0, y_i, 0, \dots, 0)$ has image $[x, \mathfrak{g}^{\gamma_i}]$. Finally, the tangent linear map of ρ at $(h_0, 0, \dots, 0)$ has image

$$[x, \mathfrak{h} + \mathfrak{g}^{\gamma_1} + \dots + \mathfrak{g}^{\gamma_r}] = [x, \mathfrak{g}^0] = \mathfrak{g}^2. \qquad \text{Q.E.D.}$$

The group G_h is an algebraic group with Lie algebra ad $\mathfrak{g}^0$.

PROPOSITION 6. *Assume that k is algebraically closed and that $\mathfrak{g}$ is semi-simple. Let* G *be a group of automorphisms of $\mathfrak{g}$ containing* $\mathrm{Aut}_e(\mathfrak{g})$. *Let (x, h, y) and (x', h', y') be $\mathfrak{sl}_2$-triplets in $\mathfrak{g}$. The following conditions are equivalent:*

(i) *h and h' are* G*-conjugate;*

(ii) *(x, h, y) and (x', h', y') are* G*-conjugate.*

We only have to prove the implication (i) $\Longrightarrow$ (ii), and we are reduced immediately to the case in which $h = h'$. Introduce $\mathfrak{g}^2$ and G_h as in Lemma 8. We have $x \in \mathfrak{g}^2$, and $[x, \mathfrak{g}^0] = \mathfrak{g}^2$ by §1, no. 2, Cor. of Prop. 2. Hence $\mathrm{G}_h x$ contains a subset of $\mathfrak{g}^2$ that is dense and open in the Zariski topology, and so does $G_h x'$. So there exists $a \in G_h$ such that $a(x) = x'$. We have $a(h) = h$, and consequently $a(y) = y'$ (no. 1, Lemma 1).

COROLLARY 1. *The map which associates to any* $\mathfrak{sl}_2$*-triplet its simple element defines by passage to the quotients a bijection from the set of* G*-conjugacy classes of* $\mathfrak{sl}_2$*-triplets to the set of* G*-conjugacy classes of simple elements.*

COROLLARY 2. *If* $\operatorname{rk}(\mathfrak{g}) = l$*, the number of conjugacy classes, relative to* $\operatorname{Aut}_e(\mathfrak{g})$*, of non-zero nilpotent elements of* $\mathfrak{g}$ *is at most* 3^l*.*

This follows from Cor. 1, the Cor. of Prop. 2, and the Cor. of Prop. 5.

COROLLARY 3. *If* $\operatorname{rk}(\mathfrak{g}) = l$*, the number of conjugacy classes, relative to* $\operatorname{Aut}_e(\mathfrak{g})$*, of subalgebras of* $\mathfrak{g}$ *isomorphic to* $\mathfrak{sl}(2, k)$ *is at most* 3^l*.*

This follows from Cor. 1, Prop. 1, and the Cor. of Prop. 5.

4. PRINCIPAL ELEMENTS

DEFINITION 3. *Assume that* $\mathfrak{g}$ *is semi-simple.*

(i) *A nilpotent element* x *of* $\mathfrak{g}$ *is said to be principal if the dimension of* $\operatorname{Ker}\operatorname{ad} x$ *is the rank of* $\mathfrak{g}$*.*

(ii) *A simple element* h *of* $\mathfrak{g}$ *is said to be principal if* h *is regular and the eigenvalues of* $\operatorname{ad} h$ *in an algebraic closure of* k *belong to* $2\mathbf{Z}$*.*

(iii) *An* $\mathfrak{sl}_2$*-triplet* (x, h, y) *of* $\mathfrak{g}$ *is said to be principal if the length of* $\mathfrak{g}$*, considered as a module over* $kx + kh + ky$*, is equal to the rank of* $\mathfrak{g}$*.*

PROPOSITION 7. *Assume that* $\mathfrak{g}$ *is semi-simple. Let* (x, h, y) *be an* $\mathfrak{sl}_2$*-triplet in* $\mathfrak{g}$*. The following conditions are equivalent:*

(i) x *is principal;*

(ii) h *is principal;*

(iii) (x, h, y) *is principal.*

For $p \in \mathbf{Z}$, let $\mathfrak{g}^p = \operatorname{Ker}(\operatorname{ad} h - p)$. Let $\mathfrak{g}' = \sum_{p \in \mathbf{Z}} \mathfrak{g}^{2p}$. If $\mathfrak{g}$ is considered as a module over $\mathfrak{a} = kx + kh + ky$, $\mathfrak{g}'$ is the sum of the simple submodules of odd dimension (§1, no. 2, Cor. of Prop. 2). Let l (resp. l') be the length of $\mathfrak{g}$ (resp. $\mathfrak{g}'$) considered as an $\mathfrak{a}$-module. By §1, no. 2,

$$\dim(\operatorname{Ker}\operatorname{ad} x) = l \geq l' = \dim(\operatorname{Ker}\operatorname{ad} h) \geq \operatorname{rk}(\mathfrak{g}).$$

The equivalence of (i) and (iii) follows immediately. On the other hand, condition (ii) means that $\dim(\operatorname{Ker}\operatorname{ad} h) = \operatorname{rk}(\mathfrak{g})$ and $\mathfrak{g}' = \mathfrak{g}$, in other words that

$$\dim(\operatorname{Ker}\operatorname{ad} h) = \operatorname{rk}(\mathfrak{g})$$

and $l = l'$. The equivalence of (ii) and the other conditions follows.

PROPOSITION 8. *Assume that* $\mathfrak{g}$ *is semi-simple* $\neq 0$. *Let* $\mathfrak{h}$ *be a splitting Cartan subalgebra of* $\mathfrak{g}$, R *the root system of* $(\mathfrak{g}, \mathfrak{h})$, B *a basis of* R, h^0 *the element of* $\mathfrak{h}$ *such that* $\alpha(h^0) = 2$ *for all* $\alpha \in \mathrm{B}$.

(i) *The element* h^0 *is simple and principal.*

(ii) *The elements* x *of* $\mathfrak{g}$ *such that there exists an* $\mathfrak{sl}_2$*-triplet of the form* (x, h^0, y) *are the elements of* $\sum_{\alpha \in \mathrm{B}} \mathfrak{g}^\alpha$ *that have a non-zero component in each* $\mathfrak{g}^\alpha$.

The element h^0 is that considered in §7, no. 5, Lemma 2 (cf. *loc. cit.*, formula (1)). It follows from this lemma that h^0 is simple principal and that, if $x \in \sum_{\alpha \in \mathrm{B}} \mathfrak{g}^\alpha$ has a non-zero component in each $\mathfrak{g}^\alpha$, there exists an $\mathfrak{sl}_2$-triplet of the form (x, h^0, y). Conversely, let (x, h^0, y) be a $\mathfrak{sl}_2$-triplet. We have $[h^0, x] = 2x$, so $x \in \sum_{\gamma \in \mathrm{R}, \gamma(h^0)=2} \mathfrak{g}^\gamma = \sum_{\alpha \in \mathrm{B}} \mathfrak{g}^\alpha$. Similarly, $y \in \sum_{\alpha \in \mathrm{B}} \mathfrak{g}^{-\alpha}$. Write

$$h^0 = \sum_{\alpha \in \mathrm{B}} a_\alpha H_\alpha \quad \text{where } a_\alpha > 0 \text{ for all } \alpha \in \mathrm{B},$$

$$x = \sum_{\alpha \in \mathrm{B}} X_\alpha \quad \text{where } X_\alpha \in \mathfrak{g}^\alpha \text{ for all } \alpha \in \mathrm{B},$$

$$y = \sum_{\alpha \in \mathrm{B}} X_{-\alpha} \quad \text{where } X_{-\alpha} \in \mathfrak{g}^{-\alpha} \text{ for all } \alpha \in \mathrm{B}.$$

Then

$$\sum_{\alpha \in \mathrm{B}} a_\alpha H_\alpha = h^0 = [y, x] = \sum_{\alpha, \beta \in \mathrm{B}} [X_{-\beta}, X_\alpha] = \sum_{\alpha \in \mathrm{B}} [X_{-\alpha}, X_\alpha]$$

so $[X_{-\alpha}, X_\alpha] \neq 0$ for all $\alpha \in \mathrm{B}$.

COROLLARY. *In a splittable semi-simple Lie algebra, there exist principal nilpotent elements.*

In a non-splittable semi-simple Lie algebra, 0 may be the only nilpotent element.

PROPOSITION 9. *Assume that* k *is algebraically closed and that* $\mathfrak{g}$ *is semi-simple. All the principal simple (resp. nilpotent) elements of* $\mathfrak{g}$ *are conjugate under* $\mathrm{Aut}_e(\mathfrak{g})$.

We retain the notations of Prop. 8. Let h be a principal simple element. It is conjugate under $\mathrm{Aut}_e(\mathfrak{g})$ to an $h' \in \mathfrak{h}$ such that $\alpha(h') \in \{0, 1, 2\}$ for all $\alpha \in \mathrm{B}$ (no. 3, Prop. 5). Since h' is principal simple, $\alpha(h') \neq 0$ and $\alpha(h') \in 2\mathbf{Z}$ for all $\alpha \in \mathrm{B}$, so $\alpha(h') = 2$ for all $\alpha \in \mathrm{B}$, and hence $h' = h^0$. This proves the assertion for principal simple elements.

Let x, x' be principal nilpotent elements. There exist $\mathfrak{sl}_2$-triplets (x, h, y), (x', h', y'). By Prop. 7, h and h' are principal simple, hence conjugate under $\mathrm{Aut}_e(\mathfrak{g})$ by the preceding. So x and x' are conjugate under $\mathrm{Aut}_e(\mathfrak{g})$ (Prop. 6).

Lemma 9. With the notations of Prop. 8, put $\mathfrak{g}^p = \mathrm{Ker}(\mathrm{ad}\, h^0 - p)$ *for* $p \in \mathbf{Z}$. *Let* $\mathfrak{g}^2_*$ *be the set of elements of* $\mathfrak{g}^2 = \sum_{\alpha \in \mathrm{B}} \mathfrak{g}^\alpha$ *that have a non-zero component in each* $\mathfrak{g}^\alpha$. *Let* R_+ *be the set of positive roots relative to* B, $\mathfrak{n}_+ = \sum_{\alpha \in \mathrm{R}_+} \mathfrak{g}^\alpha$, *and* $x \in \mathfrak{g}^2_*$. *Then* $e^{\mathrm{ad}\, \mathfrak{n}_+}.x = x + [\mathfrak{n}_+, \mathfrak{n}_+]$.

It is clear that $e^{\mathrm{ad}\, \mathfrak{n}_+}.x \subset x + [\mathfrak{n}_+, \mathfrak{n}_+]$. We prove that, if $v \in [\mathfrak{n}_+, \mathfrak{n}_+]$, then $x + v \in e^{\mathrm{ad}\, \mathfrak{n}_+}.x$. Put $\mathfrak{n}^{(p)} = \sum_{r \geq p} \mathfrak{g}^{2r}$; it suffices to prove that

$$x + v \in e^{\mathrm{ad}\, \mathfrak{n}_+}.x + \mathfrak{n}^{(p)}$$

for all $p \geq 2$. This is clear for $p = 2$ since $\mathfrak{n}^{(2)} = [\mathfrak{n}_+, \mathfrak{n}_+]$ (§3, no. 3, Prop. 9 (iii)). Assume that we have found $z \in \mathfrak{n}_+$ such that $v + x - e^{\mathrm{ad}\, z}.x \in \mathfrak{n}^{(p)}$. Since there exists an $\mathfrak{sl}_2$-triplet of the form (x, h^0, y) (Prop. 8), §1, no. 2, Cor. of Prop. 2 proves that $[x, \mathfrak{g}^{2p-2}] = \mathfrak{g}^{2p}$; hence, there exists $z' \in \mathfrak{g}^{2p-2} \subset \mathfrak{n}_+$ such that

$$v + x - e^{\mathrm{ad}\, z}.x \in [z', x] + \mathfrak{n}^{(p+1)}.$$

So $v + x \in e^{\mathrm{ad}(z+z')}.x + \mathfrak{n}^{(p+1)}$, and our assertion is established by induction.

PROPOSITION 10. *Assume that* $\mathfrak{g}$ *is semi-simple. Let* $\mathfrak{h}$ *be a splitting Cartan subalgebra of* $\mathfrak{g}$, R *the root system of* $(\mathfrak{g}, \mathfrak{h})$, B *a basis of* R, R_+ *the set of positive roots relative to* B, *and* $\mathfrak{n}_+ = \sum_{\alpha \in \mathrm{R}_+} \mathfrak{g}^\alpha$. *The principal nilpotent elements belonging to* $\mathfrak{n}_+$ *are the elements of* $\mathfrak{n}_+$ *having a non-zero component in* $\mathfrak{g}^\alpha$ *for all* $\alpha \in \mathrm{B}$.

Prop. 8 and Lemma 9 prove that such elements are principal nilpotent. We prove the converse. Evidently we can assume that $\mathfrak{g}$ is simple. Let h^0 and $\mathfrak{g}^p$ be as in Prop. 8 and Lemma 9. Let ω be the highest root, and put $\omega(h^0) = 2q$; we have $q = h - 1$, where h is the Coxeter number of R, cf. Chap. VI, §1, no. 11, Prop. 31. Then $\mathfrak{g}^{2q} = \mathfrak{g}^\omega$, $\mathfrak{g}^{-2q} = \mathfrak{g}^{-\omega}$, and $\mathfrak{g}^{2k} = 0$ for $|k| > q$. There exists a principal $\mathfrak{sl}_2$-triplet (x^0, h^0, y^0). By §1, no. 2, Cor. of Prop. 2, $(\mathrm{ad}\, x^0)^{2q}(\mathfrak{g}^{-\omega}) = \mathfrak{g}^\omega$, so $(\mathrm{ad}\, x^0)^{2q} \neq 0$. Let x be a principal nilpotent element of $\mathfrak{g}$ belonging to $\mathfrak{n}_+$. If $\bar{k}$ is an algebraic closure of k, $x \otimes 1$ and $x^0 \otimes 1$ are conjugate under an automorphism of $\mathfrak{g} \otimes_k \bar{k}$ (Prop. 9), so $(\mathrm{ad}\, x)^{2q} \neq 0$. There exists $\lambda \in \mathrm{R}$ such that $(\mathrm{ad}\, x)^{2q}\mathfrak{g}^\lambda \neq 0$. Put $x = \sum_{n \geq 1} x_n$, where $x_n \in \mathfrak{g}^{2n}$. Then

$$(\mathrm{ad}\, x)^{2q}\mathfrak{g}^\lambda \subset (\mathrm{ad}\, x_1)^{2q}\mathfrak{g}^\lambda + \sum_{k > 4q + \lambda(h^0)} \mathfrak{g}^k = (\mathrm{ad}\, x_1)^{2q}\mathfrak{g}^\lambda,$$

since $4q + \lambda(h^0) \geq 4q - 2q = 2q$. Now $(\operatorname{ad} x_1)^{2q}\mathfrak{g}^\lambda \subset \mathfrak{g}^{4q+\lambda(h^0)}$, where $\lambda = -\omega$. Thus, $(\operatorname{ad} x_1)^{2q}\mathfrak{g}^{-\omega} = \mathfrak{g}^\omega$. We have $\omega = \sum_{\alpha \in \mathrm{B}} n_\alpha \alpha$ with $n_\alpha > 0$ for all $\alpha \in \mathrm{B}$ (Chap. VI, §1, no. 8, Remark). If there exists $\alpha_0 \in \mathrm{B}$ such that $x_1 \in \sum_{\alpha \in \mathrm{B}, \alpha \neq \alpha_0} \mathfrak{g}^\alpha$, the relation

$$\omega \notin -\omega + \sum_{\alpha \in \mathrm{B}, \alpha \neq \alpha_0} k\alpha$$

implies that $\mathfrak{g}^\omega \not\subset (\operatorname{ad} x_1)^p \mathfrak{g}^{-\omega}$ for all p; this is absurd, so the component of x_1 in $\mathfrak{g}^\alpha$ is non-zero for all $\alpha \in \mathrm{B}$.

§12. CHEVALLEY ORDERS

1. LATTICES AND ORDERS

Let V be a **Q**-vector space. A *lattice* in V is a free **Z**-submodule $\mathscr{V}$ of V such that the **Q**-linear map $\alpha_{\mathscr{V},\mathrm{V}} : \mathscr{V} \otimes_{\mathbf{Z}} \mathbf{Q} \to \mathrm{V}$ induced by the injection of $\mathscr{V}$ into V is bijective. When V is finite dimensional, this is the same as saying that $\mathscr{V}$ is a **Z**-submodule of finite type which generates the **Q**-vector space V (recall that a torsion-free **Z**-module of finite type is free by *Algebra*, Chap. VII, §4, no. 4, Cor. 2); moreover, in this case our definition is a special case of that of *Commutative Algebra*, Chap. VII, §4, no. 1, Def. 1 (*loc. cit.*, Example 3). If W is a vector subspace of V, and $\mathscr{V}$ is a lattice in V, then $\mathscr{V} \cap \mathrm{W}$ is a lattice in W.

If V is a **Q**-algebra, an *order* in V is a lattice $\mathscr{V}$ in the underlying vector space that is a **Z**-subalgebra of V; the map $\alpha_{\mathscr{V},\mathrm{V}}$ is then an isomorphism of **Q**-algebras. If V is a unital **Q**-algebra, a *unital order* in V is an order in V containing the unit element.

Assume that V is a **Q**-bigebra, with coproduct c and counit γ. If $\mathscr{V}$ is a lattice in the vector space V, the canonical map $i : \mathscr{V} \otimes_{\mathbf{Z}} \mathscr{V} \to \mathrm{V} \otimes_{\mathbf{Q}} \mathrm{V}$ is injective; a *biorder* in V is a unital order $\mathscr{V}$ in the unital algebra V such that $\gamma(\mathscr{V}) \subset \mathbf{Z}$ and $c(\mathscr{V}) \subset i(\mathscr{V} \otimes_{\mathbf{Z}} \mathscr{V})$; the maps

$$\gamma_{\mathscr{V}} : \mathscr{V} \to \mathbf{Z} \quad \text{and} \quad c_{\mathscr{V}} : \mathscr{V} \to \mathscr{V} \otimes_{\mathbf{Z}} \mathscr{V}$$

induced by γ and c give $\mathscr{V}$ the structure of a **Z**-bigebra, and the map $\alpha_{\mathscr{V},\mathrm{V}}$ is then an isomorphism of **Q**-bigebras.

2. DIVIDED POWERS IN A BIGEBRA

Let A be a unital k-algebra, $x \in \mathrm{A}$, $d \in k$, $n \in \mathbf{N}$. Put

$$x^{(n,d)} = \frac{x(x-d)\dots(x-d(n-1))}{n!} = \prod_{i=0}^{n-1}(x-id)/(i+1). \tag{1}$$

In particular, $x^{(0,d)} = 1$, $x^{(1,d)} = x$. We agree that $x^{(n,d)} = 0$ for n an integer < 0. Put

$$x^{(n)} = x^{(n,0)} = \frac{x^n}{n!} \tag{2}$$

$$\binom{x}{n} = x^{(n,1)} = \frac{x(x-1)\dots(x-n+1)}{n!}. \tag{3}$$

PROPOSITION 1. *Let* A *be a bigebra, with coproduct* c, *and* x *a primitive element* (Chap. II, §1, no. 2) *of* A. *Then*

$$c(x^{(n,d)}) = \sum_{p\in\mathbf{N},q\in\mathbf{N},p+q=n} x^{(p,d)} \otimes x^{(q,d)}. \tag{4}$$

The proposition is trivial for $n \leq 0$. We argue by induction on n. If formula (4) is true for n, then

$$\begin{aligned}(n+1)c(x^{(n+1,d)}) &= c(x-dn)c(x^{(n,d)})\\ &= (x\otimes 1 + 1\otimes x - dn\,1\otimes 1)c(x^{(n,d)})\\ &= \sum_{p+q=n}[xx^{(p,d)}\otimes x^{(q,d)} + x^{(p,d)}\otimes xx^{(q,d)} - (p+q)dx^{(p,d)}\otimes x^{(q,d)}]\\ &= \sum_{p+q=n}(x-pd)x^{(p,d)}\otimes x^{(q,d)} + \sum_{p+q=n} x^{(p,d)}\otimes(x-qd)x^{(q,d)}\\ &= \sum_{p+q=n}(p+1)x^{(p+1,d)}\otimes x^{(q,d)} + \sum_{p+q=n}(q+1)x^{(p,d)}\otimes x^{(q+1,d)}\\ &= \sum_{r+s=n+1} rx^{(r,d)}\otimes x^{(s,d)} + \sum_{r+s=n+1} sx^{(r,d)}\otimes x^{(s,d)}\\ &= (n+1)\sum_{r+s=n+1} x^{(r,d)}\otimes x^{(s,d)},\end{aligned}$$

hence formula (4) for $n+1$.

3. INTEGRAL VARIANT OF THE POINCARÉ-BIRKHOFF-WITT THEOREM

Let $\mathfrak{g}$ be a finite dimensional $\mathbf{Q}$-Lie algebra, $\mathrm{U}(\mathfrak{g})$ its enveloping bigebra. If I is a totally ordered set, $\mathbf{x} = (x_i)_{i\in\mathrm{I}}$ a family of elements of $\mathfrak{g}$, and $\mathbf{n} = (n_i)_{i\in\mathrm{I}} \in \mathbf{N}^{(\mathrm{I})}$ a multi-index, put

$$\mathbf{x}^{(\mathbf{n})} = \prod_{i\in\mathrm{I}} \frac{x_i^{n_i}}{n_i!}, \tag{5}$$

the product being calculated in $\mathrm{U}(\mathfrak{g})$ in accordance with the ordered set I.

THEOREM 1. *Let $\mathscr{U}$ be a biorder in the bigebra* $\mathrm{U}(\mathfrak{g})$. *Let $\mathscr{G} = \mathscr{U} \cap \mathfrak{g}$, which is an order in the Lie algebra $\mathfrak{g}$. Let $(x_i)_{i \in \mathrm{I}}$ be a basis of $\mathscr{G}$. Give* I *a total order, and assume that we are given, for all $\mathbf{n} \in \mathbf{N}^{\mathrm{I}}$, an element $[\mathbf{n}]$ of $\mathscr{U}$ such that $[\mathbf{n}] - \mathbf{x}^{(\mathbf{n})}$ has filtration $< |\mathbf{n}|$ in $\mathrm{U}(\mathfrak{g})$. Then, the family of the $[\mathbf{n}]$ for $\mathbf{n} \in \mathbf{N}^{\mathrm{I}}$ is a basis of the $\mathbf{Z}$-module $\mathscr{U}$.*

For $p \in \mathbf{N}$, let $\mathrm{U}_p(\mathfrak{g})$ be the set of elements of $\mathrm{U}(\mathfrak{g})$ of filtration $\leq p$; then the images in $\mathrm{U}_p(\mathfrak{g})/\mathrm{U}_{p-1}(\mathfrak{g})$ of the $\mathbf{x}^{(\mathbf{n})}$ such that $|\mathbf{n}| = p$ form a basis of this $\mathbf{Q}$-vector space (Chap. I, §2, no. 7, Th. 1); hence the $[\mathbf{n}]$ form a basis of the $\mathbf{Q}$-vector space $\mathrm{U}(\mathfrak{g})$. It remains to prove the following assertion (in which we put $\mathrm{M} = \mathbf{N}^{\mathrm{I}}$):

(*) if $u \in \mathscr{U}$, $(a_{\mathbf{n}}) \in \mathbf{Z}^{(\mathrm{M})}$, and $d \in \mathbf{N} - \{0\}$ are such that

$$du = \sum_{\mathbf{n} \in \mathrm{M}} a_{\mathbf{n}}[\mathbf{n}], \tag{6}$$

then d divides each $a_{\mathbf{n}}$.

For each integer $r \geq 0$, introduce the *iterated coproduct*

$$c_i : \mathscr{U} \to \mathbf{T}^r(\mathscr{U}) = \mathscr{U} \otimes \mathscr{U} \otimes \cdots \otimes \mathscr{U};$$

by definition, c_0 is the counit of $\mathscr{U}$, $c_1 = \mathrm{Id}_{\mathscr{U}}$, $c_2 = c$ (the coproduct of $\mathscr{U}$), and, for $r \geq 2$, c_{r+1} is defined as the composite $p \circ (c_r \otimes 1) \circ c$:

$$\mathscr{U} \xrightarrow{c} \mathscr{U} \otimes_{\mathbf{Z}} \mathscr{U} \xrightarrow{c_r \otimes 1} \mathbf{T}^r(\mathscr{U}) \otimes_{\mathbf{Z}} \mathscr{U} \xrightarrow{p} \mathbf{T}^{r+1}(\mathscr{U})$$

where p is defined by using the multiplication in the algebra $\mathbf{T}(\mathscr{U})$. Further, consider the canonical projection π of $\mathscr{U}$ onto $\mathscr{U}^+ = \operatorname{Ker} c_0$, and the composite

$$c_r^+ = \mathbf{T}^r(\pi) \circ c_r : \mathscr{U} \to \mathbf{T}^r(\mathscr{U}^+).$$

Lemma 1. Let $\mathbf{n} \in \mathbf{N}^{\mathrm{I}}$. If $|\mathbf{n}| < r$, then $c_r^+([\mathbf{n}]) = 0$. If $|\mathbf{n}| = r$, then

$$c_r^+([\mathbf{n}]) = \sum_{\varphi} x_{\varphi(1)} \otimes x_{\varphi(2)} \otimes \cdots \otimes x_{\varphi(r)}, \tag{7}$$

where φ belongs to the set of maps from $\{1, 2, \ldots, r\}$ to I *which take each value $i \in \mathrm{I}$ n_i times.*

By Prop. 1,

$$c_r(\mathbf{x}^{(\mathbf{n})}) = \sum \mathbf{x}^{(\mathbf{p}_1)} \otimes \cdots \otimes \mathbf{x}^{(\mathbf{p}_r)}$$

where the summation extends over the set of sequences $(\mathbf{p}_1, \ldots, \mathbf{p}_r)$ of r elements of M such that $\mathbf{p}_1 + \cdots + \mathbf{p}_r = \mathbf{n}$. In view of Chap. II, §1, no. 3, Prop. 6, the map c_r^+, extended by linearity to a map from $\mathrm{U}(\mathfrak{g})$ to $\mathbf{T}^r(\mathrm{U}^+(\mathfrak{g}))$, vanishes on $\mathrm{U}_{r-1}(\mathfrak{g})$. It follows that, for $r \geq |\mathbf{n}|$,

$$c_r^+([\mathbf{n}]) = c_r^+(\mathbf{x}^{(\mathbf{n})}) = \sum \pi(\mathbf{x}^{(\mathbf{p}_1)}) \otimes \cdots \otimes \pi(\mathbf{x}^{(\mathbf{p}_r)}). \tag{8}$$

For $r > |\mathbf{n}|$, the relation $\mathbf{p}_1 + \cdots + \mathbf{p}_r = \mathbf{n}$ implies that at least one of the $\mathbf{p}_i$ is zero, so $c_r^+([\mathbf{n}]) = 0$. For $r = |\mathbf{n}|$, the only non-zero terms of the third member of (8) are those for which $|\mathbf{p}_1| = \cdots = |\mathbf{p}_r| = 1$, hence (7).

We return to the proof of Th. 1. We retain the notations of (*) and prove, by descending induction on $|\mathbf{n}|$, that d divides $a_\mathbf{n}$, which is clear when $|\mathbf{n}|$ is sufficiently large. If d divides $a_\mathbf{n}$ for $|\mathbf{n}| > r$ then, putting

$$u' = u - \sum_{|\mathbf{n}|>r} (a_\mathbf{n}/d)[\mathbf{n}] \in \mathscr{U},$$

we have

$$du' = \sum_{|\mathbf{n}|\le r} a_\mathbf{n}[\mathbf{n}]. \tag{9}$$

For any map φ from $\{1, 2, \ldots, r\}$ to I, put

$$e_\varphi = x_{\varphi(1)} \otimes \cdots \otimes x_{\varphi(r)}$$

and $a_\varphi = a_\mathbf{n}$ where $\mathbf{n} = (\operatorname{Card} \varphi^{-1}(i))_{i\in \mathrm{I}}$. By Lemma 1, (9) implies that

$$dc_r^+(u') = \sum_{\varphi\in \mathrm{I}^r} a_\varphi e_\varphi \tag{10}$$

so $c_r^+(u') \in \mathbf{T}^r(\mathscr{U}^+) \cap \mathbf{Q}\mathbf{T}^r(\mathscr{G})$. But the submodule $\mathscr{G}$ of $\mathscr{U}^+$ is a direct factor (*Algebra*, Chap. VII, §4, no. 3, Cor. of Th. 1), so the submodule $\mathbf{T}^r(\mathscr{G})$ is a direct factor of $\mathbf{T}^r(\mathscr{U}^+)$, and hence $c_r^+(u') \in \mathbf{T}^r(\mathscr{G})$. On the other hand, the x_i form a basis of $\mathscr{G}$ by hypothesis, so the e_φ form a basis of $\mathbf{T}^r(\mathscr{G})$. Then (10) proves that d divides the a_φ, that is, the $a_\mathbf{n}$ for $|\mathbf{n}| = r$. This proves (*).

4. EXAMPLE: POLYNOMIALS WITH INTEGER VALUES

Let V be a finite dimensional $\mathbf{Q}$-vector space, V^* its dual, $\mathscr{V}$ a lattice in V, $\mathscr{V}^*$ the dual $\mathbf{Z}$-module of $\mathscr{V}$, which can be identified canonically with a lattice in V^*, $\mathbf{S}(\mathrm{V})$ the symmetric algebra of V, and

$$\lambda : \mathbf{S}(\mathrm{V}) \to \mathrm{A}(\mathrm{V}^*)$$

the canonical bijection from $\mathbf{S}(\mathrm{V})$ to the algebra of polynomial functions on V^* (*Algebra*, Chap. IV, §5, no. 11, Remark 1). If we identify $\mathrm{A}(\mathrm{V}^* \times \mathrm{V}^*)$ with $\mathrm{A}(\mathrm{V}^*) \otimes_\mathbf{Q} \mathrm{A}(\mathrm{V}^*)$, then λ transforms the coproduct of $\mathbf{S}(\mathrm{V})$ into the map $\mathrm{A}(\mathrm{V}^*) \to \mathrm{A}(\mathrm{V}^* \times \mathrm{V}^*)$ which associates to the polynomial function φ on V^* the polynomial function

$$(x, y) \mapsto \varphi(x + y)$$

on $\mathrm{V}^* \times \mathrm{V}^*$ (*Algebra*, Chap. IV, §5, no. 11, Remark 2).

Denote by $\binom{\mathscr{V}}{\mathbf{Z}}$ the subset of $\mathbf{S}(\mathrm{V})$ consisting of the elements which correspond to polynomial maps from V^* to $\mathbf{Q}$ that take integer values on $\mathscr{V}^*$.

PROPOSITION 2. (i) $\binom{\mathscr{V}}{\mathbf{Z}}$ *is a biorder in the bigebra* $\mathbf{S}(\mathrm{V})$, *and* $\binom{\mathscr{V}}{\mathbf{Z}} \cap \mathrm{V} = \mathscr{V}$.

(ii) *The* $\mathbf{Z}$*-algebra* $\binom{\mathscr{V}}{\mathbf{Z}}$ *is generated by the* $\binom{h}{n}$ *for* $h \in \mathscr{V}, n \in \mathbf{N}$.

(iii) *If* $(h_1, \ldots, h_r)$ *is a basis of* $\mathscr{V}$, *the elements*

$$\binom{h}{\mathbf{n}} = \binom{h_1}{n_1} \cdots \binom{h_r}{n_r},$$

where $\mathbf{n} = (n_1, \ldots, n_r)$ *belongs to* $\mathbf{N}^r$, *form a basis of the* $\mathbf{Z}$*-module* $\binom{\mathscr{V}}{\mathbf{Z}}$.

For $m \in \mathbf{N}$, put $\mathbf{S}_m(\mathrm{V}) = \sum_{i \leq m} \mathbf{S}^i(\mathrm{V})$, $\mathbf{S}_m(\mathscr{V}) = \sum_{i \leq m} \mathbf{S}^i(\mathscr{V})$. By *Algebra*, Chap. IV, §5, no. 9, Prop. 15 and Remark,

$$\mathbf{S}_m(\mathscr{V}) \subset \mathbf{S}_m(\mathrm{V}) \cap \binom{\mathscr{V}}{\mathbf{Z}} \subset \frac{1}{m!}\mathbf{S}_m(\mathscr{V})$$

so $\binom{\mathscr{V}}{\mathbf{Z}} \cap \mathrm{V} = \mathscr{V}$. Since $\mathbf{S}_m(\mathscr{V})$ is a lattice in $\mathbf{S}_m(\mathrm{V})$, $\mathbf{S}_m(\mathrm{V}) \cap \binom{\mathscr{V}}{\mathbf{Z}}$ is also a lattice in $\mathbf{S}_m(\mathrm{V})$. On the other hand, $\mathbf{S}_m(\mathrm{V}) \cap \binom{\mathscr{V}}{\mathbf{Z}}$ is a direct factor of $\mathbf{S}_{m+1}(\mathrm{V}) \cap \binom{\mathscr{V}}{\mathbf{Z}}$ (since the quotient is torsion-free), hence it admits a complement which is a free $\mathbf{Z}$-module. It follows that $\binom{\mathscr{V}}{\mathbf{Z}}$ is a free $\mathbf{Z}$-module. It is clear that this is a unital order in the algebra $\mathbf{S}(\mathrm{V})$. Let $(u_n)_{n \in \mathbf{N}}$ be a basis of the $\mathbf{Z}$-module $\binom{\mathscr{V}}{\mathbf{Z}}$. This is also a basis of the $\mathbf{Q}$-module $\mathbf{S}(\mathrm{V})$ and, for all

$$\varphi \in \mathbf{S}(\mathrm{V} \times \mathrm{V}) = \mathbf{S}(\mathrm{V}) \otimes_{\mathbf{Q}} \mathbf{S}(\mathrm{V}),$$

there exists a unique sequence (v_n) of elements of $\mathbf{S}(\mathrm{V})$ such that $\varphi = \sum u_n \otimes v_n$. As above, identify $\mathbf{S}(\mathrm{V})$ with $\mathrm{A}(\mathrm{V}^*)$ and $\mathbf{S}(\mathrm{V}) \otimes \mathbf{S}(\mathrm{V})$ with $\mathrm{A}(\mathrm{V}^* \times \mathrm{V}^*)$. If $\varphi \in \binom{\mathscr{V} \times \mathscr{V}}{\mathbf{Z}}$, the polynomial function $x \mapsto \varphi(x, y)$ belongs to $\binom{\mathscr{V}}{\mathbf{Z}}$ for all $y \in \mathscr{V}^*$. It follows that $v_n(y) \in \mathbf{Z}$ for all n and all $y \in \mathscr{V}^*$, in other words that $v_n \in \binom{\mathscr{V}}{\mathbf{Z}}$. This proves that the coproduct maps $\binom{\mathscr{V}}{\mathbf{Z}}$ to $\binom{\mathscr{V}}{\mathbf{Z}} \otimes_{\mathbf{Z}} \binom{\mathscr{V}}{\mathbf{Z}}$. If $h \in \mathscr{V}$ and $n \in \mathbf{N}$, then $\binom{h}{n}$ maps $u \in \mathscr{V}^*$ to the integer $\binom{u(h)}{n}$, so $\binom{h}{n} \in \binom{\mathscr{V}}{\mathbf{Z}}$. Assertion (iii) is now obtained by applying Th. 1 to the commutative Lie algebra V, and (ii) follows.

COROLLARY. *Let* X *be an indeterminate. The polynomials* $\binom{\mathrm{X}}{n}$, *where* $n \in \mathbf{N}$, *form a basis of the* $\mathbf{Z}$*-module consisting of the polynomials* $\mathrm{P} \in k[\mathrm{X}]$ *such that* $\mathrm{P}(\mathbf{Z}) \subset \mathbf{Z}$.

If $\mathrm{P}(\mathbf{Z}) \subset \mathbf{Z}$, the Lagrange interpolation formula (*Algebra*, Chap. IV, §2, no. 1, Prop. 6) shows that the coefficients of P belong to $\mathbf{Q}$; thus, we can assume that $k = \mathbf{Q}$ and apply Prop. 2 with $\mathrm{V} = \mathbf{Q}$, $\mathscr{V} = \mathbf{Z}$.

5. SOME FORMULAS

In this number, A *denotes a unital associative algebra. If* $x \in \mathrm{A}$, *we write* $\mathrm{ad}\, x$ *instead of* $\mathrm{ad}_{\mathrm{A}} x$.

Lemma 2. If $x, y \in \mathrm{A}$ *and* $n \in \mathbf{N}$,

$$\frac{(\mathrm{ad}\, x)^n}{n!} y = \sum_{p+q=n} (-1)^q \frac{x^p}{p!} y \frac{x^q}{q!} = \sum_{p+q=n} (-1)^q x^{(p)} y x^{(q)}. \tag{11}$$

Indeed, if we denote by L_x and R_x the maps $z \mapsto xz$ and $z \mapsto zx$ from A to A, we have, since L_x and R_x commute,

$$\frac{1}{n!}(\mathrm{ad}\, x)^n = \frac{1}{n!}(\mathrm{L}_x - \mathrm{R}_x)^n = \sum_{p+q=n} (-1)^q \frac{1}{p!} \mathrm{L}_x^p \frac{1}{q!} \mathrm{R}_x^q.$$

Lemma 3. Let $x, h \in \mathrm{A}$ *and* $\lambda \in k$ *be such that* $(\mathrm{ad}\, h)x = \lambda x$. *For all* $n \in \mathbf{N}$, *and all* $\mathrm{P} \in k[\mathrm{X}]$, *we have*

$$\mathrm{P}(h) x^{(n)} = x^{(n)} \mathrm{P}(h + n\lambda). \tag{12}$$

Since $\mathrm{ad}\, h$ is a derivation of A and since $(\mathrm{ad}\, h)x$ commutes with x, we have

$$(\mathrm{ad}\, h) x^n = n x^{n-1}((\mathrm{ad}\, h)x) = n\lambda x^n, \tag{13}$$

so

$$(\mathrm{ad}\, h) x^{(n)} = n\lambda x^{(n)}.$$

Thus, formula (12) follows from the special case

$$\mathrm{P}(h) x = x \mathrm{P}(h + \lambda) \tag{14}$$

by replacing x by $x^{(n)}$ and λ by $n\lambda$. It suffices to prove (14) when $\mathrm{P} = \mathrm{X}^m$, by induction on m. It is clear when $m = 0, 1$. If (14) is true for $\mathrm{P} = \mathrm{X}^m$, then

$$h^{m+1} x = h.h^m x = hx(h + \lambda)^m = x(h + \lambda)^{m+1}$$

which proves (12).

Lemma 4. Let $x, y, h \in \mathrm{A}$ *be such that*

$$[y, x] = h, \quad [h, x] = 2x, \quad [h, y] = -2y. \tag{15}$$

(i) *For $m, n \in \mathbf{N}$, we have*

$$x^{(n)}y^{(m)} = \sum_{p\geq 0} y^{(m-p)} \binom{m+n-p-1-h}{p} x^{(n-p)}. \tag{16}$$

(ii) *Let* A′ *be the* **Z**-*subalgebra of* A *generated by the* $x^{(m)}$ *and the* $y^{(m)}$ *for* $m \in \mathbf{N}$. *Then* $\binom{h}{n} \in$ A′ *for all* $n \in \mathbf{N}$.

Formula (16) can be written in the equivalent form

$$(\operatorname{ad} x^{(n)})y^{(m)} = \sum_{p\geq 1} y^{(m-p)} \binom{m+n-p-1-h}{p} x^{(n-p)}. \tag{17_m}$$

This is trivial for $m = 0$. We argue by induction on m. From (17_m), we obtain

$$\begin{aligned}(m+1)(\operatorname{ad} x^{(n)})y^{(m+1)} &= (\operatorname{ad} x^{(n)})y^{(m)}.y + y^{(m)}.(\operatorname{ad} x^{(n)})y &(18)\\ &= \sum_{p\geq 1} y^{(m-p)} \binom{m+n-p-1-h}{p} x^{(n-p)}y + y^{(m)}(n-1-h)x^{(n-1)}\end{aligned}$$

(§1, no. 1, Lemma 1). Now, applying the same lemma, and then Lemma 3, we have

$$\begin{aligned}&\binom{m+n-p-1-h}{p} x^{(n-p)}y\\ &\quad= \binom{m+n-p-1-h}{p}(yx^{(n-p)} + (n-p-1-h)x^{(n-p-1)})\\ &\quad= y\binom{m+n-p+1-h}{p} x^{(n-p)}\\ &\qquad\qquad + \binom{m+n-p-1-h}{p}(n-p-1-h)x^{(n-p-1)}.\end{aligned}$$

Inserting this into (18), we obtain

$$\begin{aligned}&(m+1)(\operatorname{ad} x^{(n)})y^{(m+1)}\\ &\quad= \sum_{p\geq 1}(m-p+1)y^{(m-p+1)}\binom{m+n-p+1-h}{p} x^{(n-p)}\\ &\qquad+ \sum_{p\geq 1} y^{(m-p)}\binom{m+n-p-1-h}{p}(n-p-1-h)x^{(n-p-1)}\\ &\qquad+ y^{(m)}(n-1-h)x^{(n-1)}\\ &\quad= \sum_{p\geq 1}(m-p+1)y^{(m-p+1)}\binom{m+n-p+1-h}{p} x^{(n-p)}\\ &\qquad+ \sum_{p\geq 0} y^{(m-p)}\binom{m+n-p-1-h}{p}(n-p-1-h)x^{(n-p-1)}.\end{aligned}$$

Changing p to $p-1$ in the second sum, and regrouping the terms, we obtain

$$(m+1)(\operatorname{ad} x^{(n)})y^{(m+1)} = \sum_{p\geq 1} y^{(m-p+1)} A_p x^{(n-p)} \tag{19}$$

with

$$A_p = (m-p+1)\binom{m+n-p+1-h}{p} + (n-p-h)\binom{m+n-p-h}{p-1}.$$

Putting $z = m+n-p-h$, this can also be written as

$$\begin{aligned} A_p &= \frac{1}{p!}(m-p+1)(z+1)z(z-1)\dots(z-p+2) \\ &\quad + \frac{1}{(p-1)!}(z-m)z(z-1)\dots(z-p+2) \\ &= \frac{1}{p!}z(z-1)\dots(z-p+2)[(m-p+1)(z+1)+p(z-m)] \\ &= (m+1)\binom{z}{p} = (m+1)\binom{(m+1)+n-p-1-h}{p}. \end{aligned}$$

Inserting this into (19), we obtain (17_{m+1}), hence (i).

Assume that $\binom{h}{p} \in \mathrm{A}'$ for $p < n$. Then, for all $\mathrm{P} \in \mathbf{Q}[\mathrm{T}]$ of degree $< n$ such that $\mathrm{P}(\mathbf{Z}) \subset \mathbf{Z}$, we have $\mathrm{P}(h) \in \mathrm{A}'$ (no. 4, Cor. of Prop. 2). Hence, in view of (16) with $m = n$,

$$\begin{aligned} (-1)^n\binom{h}{n} &= \binom{n-1-h}{n} \\ &= -x^{(n)}y^{(n)} + \sum_{p=0}^{n-1} y^{(n-p)}\binom{2n-p-1-h}{p}x^{(n-p)} \in \mathrm{A}'; \end{aligned}$$

hence (ii) by induction on n.

6. BIORDERS IN THE ENVELOPING ALGEBRA OF A SPLIT REDUCTIVE LIE ALGEBRA

Let $\mathfrak{g}$ be a reductive Lie algebra over $\mathbf{Q}$, $\mathfrak{h}$ a splitting Cartan subalgebra of $\mathfrak{g}$, and $\mathrm{R} = \mathrm{R}(\mathfrak{g}, \mathfrak{h})$ (§2, no. 1, Remark 5).

DEFINITION 1. *A lattice $\mathscr{H}$ in $\mathfrak{h}$ is said to be permissible (relative to $\mathfrak{g}$) if, for all $\alpha \in \mathrm{R}$, we have $H_\alpha \in \mathscr{H}$ and $\alpha(\mathscr{H}) \subset \mathbf{Z}$.*

Remarks. 1) Let B be a basis of R. A lattice $\mathscr{H}$ in $\mathfrak{h}$ is permissible if and only if $H_\alpha \in \mathscr{H}$ and $\alpha(\mathscr{H}) \subset \mathbf{Z}$ for all $\alpha \in \mathrm{B}$.

2) Let $\mathfrak{c}$ be the centre of $\mathfrak{g}$. Then, a lattice $\mathscr{H}$ in $\mathfrak{h}$ is permissible if and only if $\mathrm{Q}(\mathrm{R}^\vee) \subset \mathscr{H} \subset \mathrm{P}(\mathrm{R}^\vee) \oplus \mathfrak{c}$. The lattice $\mathscr{H} \cap \mathscr{D}\mathfrak{g}$ is then permissible in

the Cartan subalgebra $\mathfrak{h} \cap \mathscr{D}\mathfrak{g}$ of $\mathscr{D}\mathfrak{g}$. There may exist permissible lattices $\mathscr{H}$ such that $\mathscr{H} \neq (\mathscr{H} \cap \mathscr{D}\mathfrak{g}) \oplus (\mathscr{H} \cap \mathfrak{c})$ (cf. §13, no. 1.IX).

3) If $\mathfrak{g}$ is semi-simple, the permissible lattices in $\mathfrak{h}$ are the subgroups $\mathscr{H}$ of $\mathfrak{h}$ such that $Q(R^\vee) \subset \mathscr{H} \subset P(R^\vee)$.

In the remainder of this number, we assume fixed a split reductive Lie algebra $(\mathfrak{g}, \mathfrak{h})$, a basis B of $R = R(\mathfrak{g}, \mathfrak{h})$ and, for each $\alpha \in B$, a pair (x_α, y_α) with

$$y_\alpha \in \mathfrak{g}^{-\alpha}, \quad x_\alpha \in \mathfrak{g}^\alpha, \quad [y_\alpha, x_\alpha] = H_\alpha. \tag{20}$$

If we denote by $\mathfrak{n}_+$ (resp. $\mathfrak{n}_-$) the subalgebra of $\mathfrak{g}$ generated by the x_α (resp. the y_α), we know (§3, no. 3, Prop. 9 (iii)) that

$$\mathfrak{g} = \mathfrak{n}_- \oplus \mathfrak{h} \oplus \mathfrak{n}_+ \tag{21}$$

$$U(\mathfrak{g}) = U(\mathfrak{n}_-) \otimes_{\mathbf{Q}} U(\mathfrak{h}) \otimes_{\mathbf{Q}} U(\mathfrak{n}_+) \tag{22}$$

(where $U(\mathfrak{g}), \ldots$ are the enveloping algebras of $\mathfrak{g}, \ldots$).

Denote by $\mathscr{U}_+$ the $\mathbf{Z}$-subalgebra of $U(\mathfrak{n}_+)$ generated by the $x_\alpha^{(n)}$ for $\alpha \in B$ and $n \in \mathbf{N}$. Let W be the Weyl group of R, R_+ the set of positive roots relative to B.

Lemma 5. (i) *$\mathscr{U}_+$ is a lattice in the vector space* $U(\mathfrak{n}_+)$.

(ii) *For all $\alpha \in B$, we have* $\mathscr{U}_+ \cap U(\mathfrak{g}^\alpha) = \bigoplus_{n \in \mathbf{N}} \mathbf{Z} x_\alpha^{(n)}$.

By definition, $\mathscr{U}_+$ is generated as a $\mathbf{Z}$-module by the elements

$$x_\varphi^{(\mathbf{n})} = \prod_{1 \leq i \leq r} x_{\varphi(i)}^{(n(i))}$$

where $r \in \mathbf{N}$, $\varphi = (\varphi(i)) \in B^r$, and $\mathbf{n} = (n(i)) \in \mathbf{N}^r$. Give the algebra $U(\mathfrak{n}_+)$ the graduation of type Q(R) for which each $\mathfrak{g}^\alpha$ ($\alpha \in R_+$) is homogeneous of degree α. A monomial $x_\varphi^{(\mathbf{n})}$ of the preceding type is homogeneous of degree

$$\sum_{1 \leq i \leq r} n(i)\varphi(i) \in Q(R).$$

The monomials of this kind having a given degree q are finite in number, and generate over $\mathbf{Q}$ the homogeneous component of $U(\mathfrak{n}_+)$ of degree q. This proves (i).

If $\alpha \in B$, $\mathscr{U}_+ \cap U(\mathfrak{g}^\alpha)$ is contained in the sum of the homogeneous components of degrees which are multiples of α; thus, by the preceding, $\mathscr{U}_+ \cap U(\mathfrak{g}^\alpha)$ is generated by the $x_\varphi^{(\mathbf{n})}$ such that $\sum n(i)\varphi(i) \in \mathbf{N}\alpha$, which forces $\varphi(i) = \alpha$ for all i (since B is a basis of R), so

$$x_\varphi^{(\mathbf{n})} = x_\alpha^{(n(1))} \ldots x_\alpha^{(n(r))} = \frac{(n(1) + \cdots + n(r))!}{n(1)! \ldots n(r)!} x_\alpha^{(n(1)+\cdots+n(r))}.$$

Thus, $\mathscr{U}_+ \cap \mathscr{U}(\mathfrak{g}^\alpha) \subset \bigoplus_n \mathbf{Z} x_\alpha^{(n)}$, hence (ii). Q.E.D.

In the remainder of this paragraph, if E and F are $\mathbf{Z}$-submodules of $U(\mathfrak{g})$, we denote by E.F the $\mathbf{Z}$-submodule of $U(\mathfrak{g})$ generated by the products ab, where $a \in E, b \in F$.

PROPOSITION 3. *Let $\mathscr{H}$ be a permissible lattice in $\mathfrak{h}$. Let $\mathscr{U}_+, \mathscr{U}_-, \mathscr{U}_0$ be the $\mathbf{Z}$-subalgebras of $U(\mathfrak{g})$ generated respectively by the elements $x_\alpha^{(n)}$ $(\alpha \in B, n \in \mathbf{N})$, $y_\alpha^{(n)}$ $(\alpha \in B, n \in \mathbf{N})$, $\binom{h}{n}$ $(h \in \mathscr{H}, n \in \mathbf{N})$. Let $\mathscr{U}$ be the $\mathbf{Z}$-subalgebra of $U(\mathfrak{g})$ generated by $\mathscr{U}_+, \mathscr{U}_-, \mathscr{U}_0$.*

(i) *$\mathscr{U}$ is a biorder in the bigebra $U(\mathfrak{g})$.*

(ii) *We have $\mathscr{U} = \mathscr{U}_-.\mathscr{U}_0.\mathscr{U}_+$, $\mathscr{U} \cap \mathfrak{h} = \mathscr{H}$ and, for all $\alpha \in B$,*

$$\mathscr{U} \cap \mathfrak{g}^\alpha = \mathbf{Z}x_\alpha, \quad V \cap \mathfrak{g}^{-\alpha} = \mathbf{Z}y_\alpha.$$

By Lemma 5 and Prop. 2, $\mathscr{U}_+, \mathscr{U}_-, \mathscr{U}_0$ are orders in the $\mathbf{Q}$-algebras $U(\mathfrak{n}_+), U(\mathfrak{n}_-), U(\mathfrak{h})$, respectively, and

$$\binom{\pm h + q}{p} \in \mathscr{U}_0 \quad \text{for } h \in \mathscr{H}, q \in \mathbf{Z}, p \in \mathbf{N}. \tag{23}$$

Put $\mathscr{L} = \mathscr{U}_-.\mathscr{U}_0.\mathscr{U}_+ \subset U(\mathfrak{g})$. By (22), $\mathscr{L}$ is a lattice in $U(\mathfrak{g})$. By construction,

$$\mathscr{U}_-.\mathscr{L} \subset \mathscr{L} \tag{24}$$

$$\mathscr{L}.\mathscr{U}_+ \subset \mathscr{L} \tag{25}$$

while Lemma 3 and (23) imply that

$$\mathscr{U}_0.\mathscr{L} \subset \mathscr{L} \tag{26}$$

$$\mathscr{L}.\mathscr{U}_0 \subset \mathscr{L}. \tag{27}$$

Let $\alpha \in B, n \in \mathbf{N}, r \in \mathbf{N}, \varphi = (\varphi(i)) \in B^r$, and

$(m(1), \dots, m(r)) \in \mathbf{N}^r$.

We show that

$$x_\alpha^{(n)} y_{\varphi(1)}^{(m(1))} \dots y_{\varphi(r)}^{(m(r))} \in \mathscr{L} \tag{28}$$

or equivalently, in view of (25), that

$$[x_\alpha^{(n)}, y_{\varphi(1)}^{(m(1))} \dots y_{\varphi(r)}^{(m(r))}] \in \mathscr{L}. \tag{29}$$

We argue by induction on r. The bracket to be studied is the sum of the terms

$$y_{\varphi(1)}^{(m(1))} \dots y_{\varphi(k)}^{(m(k))} [x_\alpha^{(n)}, y_{\varphi(k+1)}^{(m(k+1))}] y_{\varphi(k+2)}^{(m(k+2))} \dots y_{\varphi(r)}^{(m(r))}. \tag{30}$$

For $\alpha \neq \varphi(k+1)$, x_α and $y_{\varphi(k+1)}$ commute, so $[x_\alpha^{(n)}, y_{\varphi(k+1)}^{(m(k+1))}] = 0$. If $\alpha = \varphi(k+1)$, the expression (30) is, by (17), the sum of expressions of the form

$$y_{\varphi(1)}^{(m(1))} \dots y_{\varphi(k)}^{(m(k))} y_{\varphi(k+1)}^{(m(k+1)-p)} \binom{q-h}{p} x_\alpha^{(n-p)} y_{\varphi(k+2)}^{(m(k+2))} \dots y_{\varphi(r)}^{(m(r))} \qquad (31)$$

where $q \in \mathbf{Z}, p \in \mathbf{N} - \{0\}, h \in \mathscr{H}$. The induction hypothesis, together with (24) and (26), proves that the expression (31) belongs to $\mathscr{L}$. We have thus proved (28).

By (28), $x_\alpha^{(n)} \mathscr{U}_- \subset \mathscr{L}$; thus, by (25) and (27), $x_\alpha^{(n)} \mathscr{L} \subset \mathscr{L}$, so $\mathscr{U}_+ . \mathscr{L} \subset \mathscr{L}$ and

$$\mathscr{L}.\mathscr{L} \subset \mathscr{U}_-.\mathscr{U}_0.\mathscr{L} \subset \mathscr{U}_-.\mathscr{L} \subset \mathscr{L}.$$

Thus, $\mathscr{L}$ is a **Z**-subalgebra of $\mathrm{U}(\mathfrak{g})$, so $\mathscr{U} = \mathscr{L}$. If c is the coproduct of $\mathrm{U}(\mathfrak{g})$, $c(\mathscr{U}) \subset \mathscr{U} \otimes_{\mathbf{Z}} \mathscr{U}$ (no. 2, Prop. 1). Let γ be the counit of $\mathrm{U}(\mathfrak{g})$. Since $\gamma(x_\alpha^{(n)}) = \gamma(y_\alpha^{(n)}) = \gamma\left(\binom{h}{n}\right) = 0$ for $n > 0$, we have $\gamma(\mathscr{U}) \subset \mathbf{Z}$. This proves (i). On the other hand,

$$\mathscr{U} \cap \mathfrak{h} = \mathscr{L} \cap \mathfrak{h} = \mathscr{U}_0 \cap \mathfrak{h} = \mathscr{H}$$

by Prop. 2 of no. 4; similarly,

$$\mathscr{U} \cap \mathfrak{g}^\alpha = \mathscr{U}_+ \cap \mathfrak{g}^\alpha = \mathbf{Z} x_\alpha$$

by Lemma 5. This proves (ii).

Remark 4. By Prop. 5 of §4, no. 4, there exists a unique automorphism θ of $\mathfrak{g}$ such that $\theta(x_\alpha) = y_\alpha$ and $\theta(y_\alpha) = x_\alpha$ for all $\alpha \in \mathrm{B}$, and $\theta(h) = -h$ for all $h \in \mathfrak{h}$; we have $\theta^2 = 1$. By construction of $\mathscr{U}$, we see that the automorphism of $\mathrm{U}(\mathfrak{g})$ that extends θ leaves $\mathscr{U}$ stable.

COROLLARY 1. *Put* $\mathscr{G} = \mathscr{U} \cap \mathfrak{g}$. *Then* $\mathscr{G}$ *is an order in the Lie algebra* $\mathfrak{g}$, *stable under* θ. *We have* $\mathscr{G} = \mathscr{H} + \sum_{\alpha \in \mathrm{R}} (\mathscr{G} \cap \mathfrak{g}^\alpha)$. *For all* $\alpha \in \mathrm{B}$ *and all* $n \in \mathbf{N}$, *the maps* $(\operatorname{ad} x_\alpha)^n/n!$, $(\operatorname{ad} y_\alpha)^n/n!$ *leave* $\mathscr{U}$ *and* $\mathscr{G}$ *stable.*

The first assertion is clear. The second follows by considering the graduation of type $\mathrm{Q}(\mathrm{R})$ on $\mathrm{U}(\mathfrak{g})$ and $\mathscr{U}$. The third follows from Lemma 2 of no. 5.

COROLLARY 2. *Let* $w \in \mathrm{W}$. *There exists an elementary automorphism* φ *of* $\mathfrak{g}$ *that commutes with* θ, *leaves* $\mathscr{G}$ *and* $\mathscr{U}$ *stable, and extends* w.

It suffices to treat the case in which w is of the form s_α $(\alpha \in \mathrm{B})$. Note first of all that $\operatorname{ad} x_\alpha$ and $\operatorname{ad} y_\alpha$ are locally nilpotent on $\mathrm{U}(\mathfrak{g})$, in other words that for all $u \in \mathrm{U}(\mathfrak{g})$ there exists an integer n such that $(\operatorname{ad} x_\alpha)^n u = (\operatorname{ad} y_\alpha)^n u = 0$. This enables us to define the automorphisms $e^{\operatorname{ad} x_\alpha} = \sum_{n=0}^{\infty} \frac{1}{n!} (\operatorname{ad} x_\alpha)^n$ and

$e^{\operatorname{ad} y_\alpha}$ of $\mathrm{U}(\mathfrak{g})$; we verify immediately that these automorphisms of $\mathrm{U}(\mathfrak{g})$ leave $\mathscr{U}$ stable. Put $\varphi_1 = e^{\operatorname{ad} x_\alpha} e^{\operatorname{ad} y_\alpha} e^{\operatorname{ad} x_\alpha}$, $\varphi_2 = e^{\operatorname{ad} y_\alpha} e^{\operatorname{ad} x_\alpha} e^{\operatorname{ad} y_\alpha}$. We have $\varphi_1|\mathfrak{g} = \varphi_2|\mathfrak{g}$ (§2, no. 2, formula (1)), so $\varphi_1 = \varphi_2$. Put $\varphi_1 = \varphi_2 = \varphi$. We have $\theta\varphi\theta^{-1} = \varphi$, so θ and φ commute. On the other hand, $\varphi|\mathfrak{h} = w$ by §2, no. 2, Lemma 1.

COROLLARY 3. *Let $\alpha \in \mathrm{R}$. If $x \in \mathscr{G} \cap \mathfrak{g}^\alpha$ and $n \in \mathbf{N}$, we have $x^{(n)} \in \mathscr{U}$, and $(\operatorname{ad} x)^n/n!$ leaves $\mathscr{G}$ and $\mathscr{U}$ stable.*

This is clear if $\alpha \in \mathrm{B}$, by construction of $\mathscr{U}$ and Cor. 1. In the general case, there exists $w \in \mathrm{W}$ such that $w(\alpha) \in \mathrm{B}$ (Chap. VI, §1, no. 5, Prop. 15). By Cor. 2, there exists an automorphism φ of $\mathfrak{g}$ that leaves $\mathscr{G}$ and $\mathscr{U}$ stable and takes $\mathfrak{g}^\alpha$ to $\mathfrak{g}^{w(\alpha)}$, hence the corollary by transport of structure.

COROLLARY 4. *There exists a Chevalley system $(X_\alpha)_{\alpha\in\mathrm{R}}$ in $(\mathfrak{g}, \mathfrak{h})$* (§2, no. 4, Def. 3) *such that $X_\alpha = x_\alpha$ and $X_{-\alpha} = y_\alpha$ for $\alpha \in \mathrm{B}$. For every Chevalley system $(X'_\alpha)_{\alpha\in\mathrm{R}}$ having these properties, and for all $\alpha \in \mathrm{R}$, X'_α is a basis of $\mathscr{G} \cap \mathfrak{g}^\alpha$.*

For $\alpha \in \mathrm{B}$, put $X_\alpha = x_\alpha, X_{-\alpha} = y_\alpha$. For $\alpha \in \mathrm{R}_+ - \mathrm{B}$, choose a $w \in \mathrm{W}$ such that $w(\alpha) \in \mathrm{B}$ and an automorphism φ of $\mathfrak{g}$ such that $\theta\varphi = \varphi\theta$, $\varphi(\mathscr{G}) = \mathscr{G}$ and $\varphi(h) = w^{-1}(h)$ for $h \in \mathfrak{h}$ (Cor. 2); put $X_\alpha = \varphi(x_{w(\alpha)}), X_{-\alpha} = \varphi(y_{w(\alpha)})$. Then

$$[X_{-\alpha}, X_\alpha] = \varphi([y_{w(\alpha)}, x_{w(\alpha)}]) = \varphi(H_{w(\alpha)}) = w^{-1}(H_{w(\alpha)}) = H_\alpha$$
$$\theta(X_\alpha) = \theta\varphi(x_{w(\alpha)}) = \varphi\theta(x_{w(\alpha)}) = \varphi(y_{w(\alpha)}) = X_{-\alpha}$$

so $(X_\alpha)_{\alpha\in\mathrm{R}}$ is a Chevalley system. Moreover,

$$\mathscr{G} \cap \mathfrak{g}^\alpha = \varphi(\mathscr{G} \cap \mathfrak{g}^{w(\alpha)}) = \varphi(\mathbf{Z}\, x_{w(\alpha)}) = \mathbf{Z}\, X_\alpha \tag{32}$$
$$\mathscr{G} \cap \mathfrak{g}^{-\alpha} = \varphi(\mathscr{G} \cap \mathfrak{g}^{-w(\alpha)}) = \varphi(\mathbf{Z}\, y_{w(\alpha)}) = \mathbf{Z}\, X_{-\alpha}. \tag{33}$$

Let $(X'_\alpha)_{\alpha\in\mathrm{R}}$ be a Chevalley system such that $X'_\alpha = x_\alpha, X'_{-\alpha} = y_\alpha$ for $\alpha \in \mathrm{B}$. Let S be the set of $\alpha \in \mathrm{R}$ such that $X'_\alpha = \pm X_\alpha$. By §2, no. 4, Prop. 7, S is a closed set of roots. Since $\mathrm{S} \supset \mathrm{B} \cup (-\mathrm{B})$, we have $\mathrm{S} = \mathrm{R}$ (Chap. VI, §1, no. 6, Prop. 19). Thus, by (32) and (33), we have $\mathscr{G} \cap \mathfrak{g}^\alpha = \mathbf{Z}\, X'_\alpha$ for all $\alpha \in \mathrm{R}$.

Remarks. 5) Let $(X_\alpha)_{\alpha\in\mathrm{R}}$ be the Chevalley system constructed above. If $\alpha, \beta, \alpha+\beta \in \mathrm{R}$ and if we put $[X_\alpha, X_\beta] = \mathrm{N}_{\alpha,\beta} X_{\alpha+\beta}$, we have $[X_\alpha, X_\beta] \in \mathscr{G} \cap \mathfrak{g}^{\alpha+\beta}$, and we recover the fact that $\mathrm{N}_{\alpha,\beta} \in \mathbf{Z}$ (cf. §2, no. 4, Prop. 7).

6) We have obtained in passing a new proof of the existence of Chevalley systems (cf. §4, no. 4, Cor. of Prop. 5), independent of Lemma 4, §2.

7. CHEVALLEY ORDERS

Let $(\mathfrak{g}, \mathfrak{h})$ be a split reductive Lie algebra over $\mathbf{Q}$, R its root system. Choose:

a) a permissible lattice $\mathscr{H}$ in $\mathfrak{h}$ (no. 6, Def. 1);

b) for all $\alpha \in \mathrm{R}$, a lattice $\mathscr{G}^\alpha$ in $\mathfrak{g}^\alpha$.

Put $\mathscr{G} = \mathscr{H} \oplus \sum_{\alpha \in \mathrm{R}} \mathscr{G}^\alpha$. This is a lattice in $\mathfrak{g}$. Denote by $\mathscr{U}$ the $\mathbf{Z}$-subalgebra of $\mathrm{U}(\mathfrak{g})$ generated by the $\binom{h}{n}$ $(h \in \mathscr{H}, n \in \mathbf{N})$ and the $x^{(n)}$ $(x \in \mathscr{G}^\alpha, \alpha \in \mathrm{R}, n \in \mathbf{N})$. Finally, for $\alpha \in \mathrm{R}$ and $x \in \mathfrak{g}^\alpha - \{0\}$, put

$$w_\alpha(x) = (\exp \operatorname{ad} x)(\exp \operatorname{ad} y)(\exp \operatorname{ad} x),$$

where y is the unique element of $\mathfrak{g}^{-\alpha}$ such that $[y, x] = H_\alpha$. With these notations:

THEOREM 2. *The following conditions are equivalent:*

(i) *There exists a Chevalley system* $(X_\alpha)_{\alpha \in \mathrm{R}}$ *of* $(\mathfrak{g}, \mathfrak{h})$ *such that* $\mathscr{G}_\alpha = \mathbf{Z}\, X_\alpha$ *for all* $\alpha \in \mathrm{R}$.

(ii) $\mathscr{U} \cap \mathfrak{g} = \mathscr{G}$ *and* $[\mathscr{G}^\alpha, \mathscr{G}^{-\alpha}] = \mathbf{Z}\, H_\alpha$ *for all* $\alpha \in \mathrm{R}$.

(iii) *For all* $\alpha \in \mathrm{R}, x \in \mathscr{G}^\alpha, n \in \mathbf{N}$, *the endomorphism* $(\operatorname{ad} x)^n/n!$ *of* $\mathfrak{g}$ *maps* $\mathscr{G}$ *to* $\mathscr{G}$, *and* $[\mathscr{G}^\alpha, \mathscr{G}^{-\alpha}] = \mathbf{Z}\, H_\alpha$.

(iv) *For all* $\alpha \in \mathrm{R}$ *and every basis* x *of* $\mathscr{G}^\alpha$, $w_\alpha(x)$ *maps* $\mathscr{G}$ *to* $\mathscr{G}$ *(that is, maps* $\mathscr{G}^\beta$ *to* $\mathscr{G}^{s_\alpha(\beta)}$ *for all* $\beta \in \mathrm{R}$*).*

(i) $\Longrightarrow$ (ii): let $(X_\alpha)_{\alpha \in \mathrm{R}}$ be a Chevalley system in $(\mathfrak{g}, \mathfrak{h})$ such that $\mathscr{G}^\alpha = \mathbf{Z}\, X_\alpha$ for all $\alpha \in \mathrm{R}$, and let B be a basis of R. For $\alpha \in \mathrm{B}$, put $x_\alpha = X_\alpha, y_\alpha = X_{-\alpha}$. Let $\mathscr{U}'$ be the biorder associated by Prop. 3 of no. 6 to $\mathscr{H}$, the x_α and the y_α. It is clear that $\mathscr{U}' \subset \mathscr{U}$. By Cor. 3 and 4 of Prop. 3, $x^{(n)} \in \mathscr{U}'$ for all $\alpha \in \mathrm{R}$, $x \in \mathscr{G}^\alpha$ and $n \in \mathbf{N}$. Thus $\mathscr{U} = \mathscr{U}'$, which proves (ii).

(ii) $\Longrightarrow$ (iii): this is clear by Lemma 2 of no. 5.

(iii) $\Longrightarrow$ (iv): let $\alpha \in \mathrm{R}$ and let x be a basis of $\mathscr{G}^\alpha$. Since $[\mathscr{G}^\alpha, \mathscr{G}^{-\alpha}] = \mathbf{Z}\, H_\alpha$, the unique $y \in \mathfrak{g}^{-\alpha}$ such that $[y, x] = H_\alpha$ belongs to $\mathscr{G}^{-\alpha}$. Since $\exp \operatorname{ad} x$ and $\exp \operatorname{ad} y$ leave $\mathscr{G}$ stable by (iii), so does $w_\alpha(x)$.

(iv) $\Longrightarrow$ (i): let B be a basis of R. Choose a basis x_α of $\mathscr{G}^\alpha$ for all $\alpha \in \mathrm{B}$. Let $y_\alpha \in \mathscr{G}^{-\alpha}$ be such that $[y_\alpha, x_\alpha] = H_\alpha$. By §1, no. 5, formulas (5), we have $y_\alpha = w_\alpha(x_\alpha).x_\alpha$ so y_α is a basis of $\mathscr{G}^{-\alpha}$ by (iv). Let $\mathscr{G}$ be the order in $\mathfrak{g}$ defined by $\mathscr{H}$, the x_α and the y_α (no. 6, Cor. 1 of Prop. 3). Then $\mathscr{G}$ is stable under the $(\operatorname{ad} x_\alpha)^n/n!, (\operatorname{ad} y_\alpha)^n/n!$ (*loc. cit.*), and hence under the $w_\alpha(x_\alpha)$.

Now let $\beta \in \mathrm{R}$. There exist $\alpha_0, \alpha_1, \ldots, \alpha_r \in \mathrm{B}$ such that

$$\beta = s_{\alpha_r} s_{\alpha_{r-1}} \cdots s_{\alpha_1}(\alpha_0)$$

(Chap. VI, §1, no. 5, Prop. 15). Then $w_{\alpha_r}(x_{\alpha_r}).w_{\alpha_{r-1}}(x_{\alpha_{r-1}}) \ldots w_{\alpha_1}(x_{\alpha_1})$ maps $\mathscr{G}^{\alpha_0}$ to $\mathscr{G}^\beta$ by (iv), and maps $\mathscr{G} \cap \mathfrak{g}^{\alpha_0}$ to $\mathscr{G} \cap \mathfrak{g}^\beta$ by the preceding. Since $\mathscr{G} \cap \mathfrak{g}^{\alpha_0} = \mathscr{G}^{\alpha_0}$ (Prop. 3 (ii)), we have $\mathscr{G} \cap \mathfrak{g}^\beta = \mathscr{G}^\beta$. Thus

$$\mathscr{G} = \mathscr{H} \oplus \sum_{\beta \in R} (\mathscr{G} \cap \mathfrak{g}^\beta) = \mathscr{H} \oplus \sum_{\beta \in R} \mathscr{G}^\beta = \mathscr{G}$$

and Cor. 4 of Prop. 3 concludes the proof.

DEFINITION 2. *When conditions* (i) *to* (iv) *of Th.* 2 *are satisfied,* $\mathscr{G}$ *is said to be a Chevalley order in* $(\mathfrak{g}, \mathfrak{h})$.

Remark. Chevalley orders in $(\mathfrak{g}, \mathfrak{h})$ always exist. Indeed, the Chevalley orders are the sets of the form $\mathscr{H} \oplus \sum_{\alpha \in R} \mathbf{Z} X_\alpha$, where $(X_\alpha)_{\alpha \in R}$ is a Chevalley system in $(\mathfrak{g}, \mathfrak{h})$ and $\mathscr{H}$ is a lattice in $\mathfrak{h}$ such that

$$Q(R^\vee) \subset \mathscr{H} \subset P(R^\vee) \oplus \mathfrak{c}$$

($\mathfrak{c}$ being the centre of $\mathfrak{g}$).

THEOREM 3. *We retain the notations at the beginning of no.* 7, *and assume that* $\mathscr{G}$ *is a Chevalley order in* $(\mathfrak{g}, \mathfrak{h})$.

(i) $\mathscr{U}$ *is a biorder in* $U(\mathfrak{g})$.

(ii) *Let* B *be a basis of* R, *and* $(X_\alpha)_{\alpha \in B \cup (-B)}$ *a family of elements of* $\mathfrak{g}$ *such that* $\mathscr{G}^\alpha = \mathbf{Z} X_\alpha$ *for* $\alpha \in B \cup (-B)$. *The* $\mathbf{Z}$*-algebra* $\mathscr{U}$ *is generated by the* $\binom{h}{n}$ *and the* $X_\alpha^{(n)}$ $(h \in \mathscr{H}, \alpha \in B \cup (-B), n \in \mathbf{N})$. *If* $\mathfrak{g}$ *is semi-simple and* $\mathscr{H} = Q(R^\vee)$, *the* $\mathbf{Z}$*-algebra* $\mathscr{U}$ *is generated by the* $X_\alpha^{(n)}$ $(\alpha \in B \cup (-B), n \in \mathbf{N})$.

(iii) *Let* B *be a basis of* R, R_+ *the corresponding set of positive roots,* $R_- = -R_+$, $\mathfrak{n}_+ = \sum_{\alpha \in R_+} \mathfrak{g}^\alpha$, $\mathfrak{n}_- = \sum_{\alpha \in R_-} \mathfrak{g}^\alpha$. *Then,*

$$\mathscr{U} = (\mathscr{U} \cap U(\mathfrak{n}_-)).(\mathscr{U} \cap U(\mathfrak{h})).(\mathscr{U} \cap U(\mathfrak{n}_+)).$$

Let $(h_i)_{i \in I}$ *be a basis of* $\mathscr{H}$. *For all* $\alpha \in R$, *let* X_α *be a basis of* $\mathscr{G}^\alpha$. *Give the set* $I \cup R$ *a total order (we assume that* $I \cap R = \varnothing$). *For* $\lambda \in I \cup R$ *and* $n \in \mathbf{N}$, *put* $e_\lambda^{\langle n \rangle} = \binom{h_\lambda}{n}$ *if* $\lambda \in I$, $e_\lambda^{\langle n \rangle} = X_\lambda^{(n)}$ *if* $\lambda \in R$. *Then the products* $\prod_{\lambda \in I \cup R} e_\lambda^{\langle n_\lambda \rangle}$, *where* (n_λ) *belongs to* $\mathbf{N}^{I \cup R}$, *form a basis of the* $\mathbf{Z}$*-module* $\mathscr{U}$. *The products* $\prod_{\lambda \in I} \binom{h_\lambda}{n_\lambda}$, *where* (n_λ) *belongs to* $\mathbf{N}^I$, *form a basis of the* $\mathbf{Z}$*-module* $\mathscr{U} \cap U(\mathfrak{h})$. *The products* $\prod_{\lambda \in R_+} X_\lambda^{(n_\lambda)}$, *where* (n_λ) *belongs to* $\mathbf{N}^{R_+}$, *form a basis of the* $\mathbf{Z}$*-module* $\mathscr{U} \cap U(\mathfrak{n}_+)$.

Let B and $(X_\alpha)_{\alpha \in B \cup (-B)}$ be as in (ii), and such that $[X_{-\alpha}, X_\alpha] = H_\alpha$. Let $\mathscr{U}'$ be the $\mathbf{Z}$-subalgebra of $U(\mathfrak{g})$ generated by the $\binom{h}{n}$ and the $X_\alpha^{(n)}$ $(h \in \mathscr{H}, \alpha \in B \cup (-B), n \in \mathbf{N})$. We have seen in the proof of Th. 2, (i) $\Longrightarrow$ (ii), that $\mathscr{U}'$ is equal to $\mathscr{U}$ and is a biorder in $U(\mathfrak{g})$. This proves (i) and the first assertion of (ii); the second follows from Lemma 4 (ii). Assertion (iii) follows from Th. 1 (no. 3) and Prop. 3 (no. 6).

8. ADMISSIBLE LATTICES

Generalizing the terminology adopted for vector spaces, an endomorphism u of a module M is said to be *diagonalizable* if there exists a basis of M such that the matrix of u relative to this basis is diagonal.

Lemma 6. Let M *be a free* **Z**-*module of finite type, u an endomorphism of* M, *and v the endomorphism $u \otimes 1$ of* $\mathrm{M} \otimes_{\mathbf{Z}} \mathbf{Q}$. *Assume that* $\binom{v}{n}(\mathrm{M}) \subset \mathrm{M}$ *for all* $n \in \mathbf{N}$. *Then u is diagonalizable.*

a) For any polynomial $\mathrm{P} \in \mathbf{Q}[\mathrm{T}]$ such that $\mathrm{P}(\mathbf{Z}) \subset \mathbf{Z}$, we have $\mathrm{P}(v)(\mathrm{M}) \subset \mathrm{M}$ (no. 4, Cor. of Prop. 2), so $\det \mathrm{P}(v) \in \mathbf{Z}$.

b) Denote by $\chi_v(t) = t^d + \alpha_1 t^{d-1} + \cdots$ the characteristic polynomial of v. Let $k \in \mathbf{Z}, n \in \mathbf{N}$. Applying a) to the polynomial $\binom{\mathrm{T}-k}{n}$, we see that the number

$$a_n = \det \binom{v-k}{n} = \frac{1}{(n!)^d} \det(v-k)\det(v-k-1)\ldots\det(v-k-n+1)$$
$$= \frac{(-1)^n}{(n!)^4}\chi_v(k)\chi_v(k+1)\ldots\chi_v(k+n-1)$$

is an integer. Take $k-1 < -\alpha_1/d$. Then

$$\chi_v(k+n-1) = n^d + (\alpha_1 + (k-1)d)n^{d-1} + \cdots$$

and

$$|a_n| = \frac{|\chi_v(k+n-1)|}{n^d}|a_{n-1}|;$$

hence, if $a_n \neq 0$ for all $n \in \mathbf{N}$, the sequence of the $|a_n|$ is strictly decreasing for n sufficiently large, which is absurd. It follows that v has an integer eigenvalue λ. Put $\mathrm{M}' = \mathrm{Ker}(u - \lambda.1)$ and $\mathrm{M}'' = \mathrm{M}/\mathrm{M}'$. Then M' is the intersection with M of a vector subspace of $\mathrm{M} \otimes_{\mathbf{Z}} \mathbf{Q}$, so the **Z**-module M'' is torsion-free of finite type, and consequently free of rank $< d$. Arguing by induction on d and applying the induction hypothesis to the endomorphism of M'' induced by u, we conclude that all the eigenvalues of v in an algebraically closed extension of **Q** are integers.

c) We show that v is diagonalizable. Let λ be an eigenvalue of v and let $x \in \mathrm{M} \otimes_{\mathbf{Z}} \mathbf{Q}$ be such that $(v-\lambda)^2 x = 0$. We have $v(vx - \lambda x) = \lambda(vx - \lambda x)$, so

$$\frac{1}{n!}(v-\lambda-n+1)(v-\lambda-n+2)\ldots(v-\lambda-1)(v-\lambda)x$$
$$= \frac{(-1)^{n-1}}{n}(vx - \lambda x).$$

By a), this implies that $vx - \lambda x \in n\mathrm{M}$ for all $n \in \mathbf{N}$, so $(v-\lambda)x = 0$.

d) Let λ be an eigenvalue of v and let $(\lambda - a, \lambda + b)$ be an interval in $\mathbf{Z}$ containing all the eigenvalues of v. Consider the polynomial

$$\mathrm{P(T)} = (-1)^b \frac{(\mathrm{T} - \lambda - 1)(\mathrm{T} - \lambda - 2) \ldots (\mathrm{T} - \lambda - b)}{b!} \times \frac{(\mathrm{T} - \lambda + 1)(\mathrm{T} - \lambda + 2) \ldots (\mathrm{T} - \lambda + a)}{a!}.$$

We have $\mathrm{P}(\mathbf{Z}) \subset \mathbf{Z}, \mathrm{P}(\lambda) = 1, \mathrm{P}(\mu) = 0$ for $\mu \in \mathbf{Z} \cap (\lambda - a, \lambda + b)$ and $\mu \neq \lambda$. By *a*), $\mathrm{P}(v)(\mathrm{M}) \subset \mathrm{M}$. By *c*), $\mathrm{P}(v)$ is a projection of $\mathrm{M} \otimes_{\mathbf{Z}} \mathbf{Q}$ onto the eigenspace corresponding to λ. Q.E.D.

Remark 1. If we only assume that v is diagonalizable with integer eigenvalues, u is not necessarily diagonalizable (for example, take $\mathrm{M} = \mathbf{Z}^2$ and $u(x, y) = (y, x)$ for all $(x, y) \in \mathrm{M}$).

Let $\mathfrak{g}, \mathfrak{h}, \mathrm{R}, \mathscr{H}, \mathscr{G}^\alpha, \mathscr{G}, \mathscr{U}$ be as in no. 7, and assume that $\mathscr{G}$ is a Chevalley order in $(\mathfrak{g}, \mathfrak{h})$.

DEFINITION 3. *Let* E *be a* $\mathfrak{g}$*-module. A lattice* $\mathscr{E}$ *in* E *is said to be admissible (relative to* $\mathscr{G}$*) if the following conditions are satisfied:*

(i) $\mathscr{U}$ *maps* $\mathscr{E}$ *to* $\mathscr{E}$;

(ii) $\mathscr{E}$ *is stable under* $\binom{h}{n}$ *and* $x^{(n)}$ *for all* $\alpha \in \mathrm{R}, x \in \mathscr{G}^\alpha, n \in \mathbf{N}, h \in \mathscr{H}$.

Remarks. 2) Let ρ be the adjoint representation of $\mathfrak{g}$ on $\mathrm{U}(\mathfrak{g})$. Let α, x, n, h be as in (ii) above. We have $\rho(x^{(n)}).\mathscr{U} \subset \mathscr{U}$ by Lemma 2. On the other hand, if $p \in \mathbf{N}$,

$$\rho\left(\binom{h}{p}\right) x^{(n)} = \binom{\mathrm{ad}\, h}{p} x^{(n)} = \binom{n\alpha(h)}{p} x^{(n)}$$

(no. 5, formula (13)), so $\rho\left(\binom{h}{p}\right).\mathscr{U} \subset \mathscr{U}$. This proves that $\mathscr{U}$ is an admissible lattice in $\mathrm{U}(\mathfrak{g})$, and it follows that $\mathscr{G}$ is an admissible lattice in $\mathfrak{g}$ (for the adjoint representation).

3) Let E be a finite dimensional $\mathfrak{g}$-module, $\mathscr{E}$ an admissible lattice in E, $\mathfrak{c}$ the centre of $\mathfrak{g}$. By Lemma 6, every element of $\mathfrak{c}$ defines a diagonalizable endomorphism of E. Hence E is semi-simple (Chap. I, §6, no. 5, Th. 4). Thus, E is a direct sum of simple $\mathscr{D}\mathfrak{g}$-modules on which $\mathfrak{c}$ induces homotheties. By Lemma 6, $\mathscr{E} = \oplus(\mathscr{E} \cap \mathrm{E}^\lambda)$ and, for all weights λ of E, we have

$$\lambda(\mathscr{H}) \subset \mathbf{Z}.$$

4) If $\mathfrak{g}$ is semi-simple and $\mathscr{H} = \mathrm{Q}(\mathrm{R}^\vee)$, conditions (i) and (ii) of Def. 3 are equivalent, by Th. 3 (ii), to

(iii) $\mathscr{E}$ is stable under $x^{(n)}$ for all $\alpha \in \mathrm{R}, x \in \mathscr{G}^\alpha, n \in \mathbf{N}$.

5) Let B be a basis of R; in conditions (i) and (ii) above, "$\alpha \in \mathrm{R}$" can be replaced by "$\alpha \in \mathrm{B} \cup (-\mathrm{B})$" (*loc. cit*).

THEOREM 4. *Let* E *be a finite dimensional* $\mathfrak{g}$*-module. The following conditions are equivalent:*

(i) E *has an admissible lattice;*

(ii) *every element of* $\mathscr{H}$ *defines a diagonalizable endomorphism of* E *with integer eigenvalues.*

(i) $\Longrightarrow$ (ii): this follows from Remark 3.

(ii) $\Longrightarrow$ (i): we assume that condition (ii) is satisfied and prove (i). By Th. 4 of Chap. I, §6, no. 5, we can assume that the elements of $\mathfrak{c}$ define homotheties of E, and that E is a simple $\mathscr{D}\mathfrak{g}$-module. Let B be a basis of R, and $\mathfrak{g} = \mathfrak{n}_- \oplus \mathfrak{h} \oplus \mathfrak{n}_+$ the corresponding decomposition of $\mathfrak{g}$. Let λ be the highest weight of the $\mathscr{D}\mathfrak{g}$-module E, and let $e \in \mathrm{E}^\lambda - \{0\}$. Put $\mathscr{E} = \mathscr{U}.e$. It is clear that $\mathscr{U}.\mathscr{E} \subset \mathscr{E}$. Since E is simple, $\mathrm{U}(\mathfrak{g}).e = \mathrm{E}$ and hence $\mathscr{E}$ generates E as a $\mathbf{Q}$-vector space. For $h \in \mathscr{H}$ and $n \in \mathbf{N}$, we have $\binom{h}{n} e = \binom{\lambda(h)}{n} e \in \mathbf{Z}\, e$, so

$$(\mathscr{U} \cap \mathrm{U}(\mathfrak{h})).e = \mathbf{Z}\, e.$$

Since $\mathrm{U}(\mathfrak{n}_+).e = 0$, we have $\mathscr{E} = (\mathscr{U} \cap \mathrm{U}(\mathfrak{n}_-)).e$ by Prop. 3. It now follows from Th. 3 (iii) that $\mathscr{E}$ is a $\mathbf{Z}$-module of finite type.

COROLLARY. *If* $\mathfrak{g}$ *is semi-simple and* $\mathscr{H} = \mathrm{Q}(\mathrm{R}^\vee)$, *every finite dimensional* $\mathfrak{g}$*-module has an admissible lattice.*

§13. CLASSICAL SPLITTABLE SIMPLE LIE ALGEBRAS

In this paragraph we describe explicitly, for each type of classical splittable simple Lie algebra:

(I) an algebra of this type, its dimension and its splitting Cartan subalgebras;

(II) its coroots;

(III) its Borel subalgebras and its parabolic subalgebras;

(IV) its fundamental simple representations;

(V) those of its fundamental simple representations which are orthogonal or symplectic;

(VI) the algebra of invariant polynomial functions;

(VII) certain properties of the groups $\mathrm{Aut}\,\mathfrak{g}$, $\mathrm{Aut}_0\mathfrak{g}$, $\mathrm{Aut}_e\mathfrak{g}$;

(VIII) the restriction of the Killing form to a Cartan subalgebra;

(IX) the Chevalley orders.

1. ALGEBRAS OF TYPE A_l ($l \geq 1$)

(I) Let V be a vector space of dimension $l+1$ over k, and let $\mathfrak{g}$ be the algebra $\mathfrak{sl}(V)$ of endomorphisms of V of trace zero. Let $(e_i)_{1\leq i\leq l+1}$ be a basis of V; the map which associates to an element of $\mathfrak{g}$ its matrix with respect to this basis is an identification of $\mathfrak{g}$ with the algebra $\mathfrak{sl}(l+1,k)$ of matrices of trace zero. We know that $\mathfrak{g}$ is semi-simple (Chap. I, §6, no. 7, Prop. 8).

Recall (*Algebra*, Chap. II, §10, no. 3) that E_{ij} denotes the matrix (α_{mp}) such that $\alpha_{ij} = 1$ and $\alpha_{mp} = 0$ for $(m,p) \neq (i,j)$. The matrices

$$\begin{array}{ll} E_{ij} & (1 \leq i,j \leq l+1,\ i \neq j) \\ E_{i,i} - E_{i+1,i+1} & (1 \leq i \leq l) \end{array}$$

form a basis of $\mathfrak{g}$. Hence

$$\dim \mathfrak{g} = l(l+2).$$

Let $\hat{\mathfrak{h}}$ be the set of diagonal elements of $\mathfrak{gl}(l+1,k)$; the sequence $(E_{ii})_{1\leq i\leq l+1}$ is a basis of the vector space $\hat{\mathfrak{h}}$; let $(\hat{\varepsilon}_i)_{1\leq i\leq l+1}$ be the basis of $\hat{\mathfrak{h}}^*$ dual to $(E_{ii})_{1\leq i\leq l+1}$. For all $h \in \hat{\mathfrak{h}}$,

$$[h, E_{ij}] = (\hat{\varepsilon}_i(h) - \hat{\varepsilon}_j(h))E_{ij} \tag{1}$$

by Chap. I, §1, no. 2, formulas (5). Let $\mathfrak{h}$ be the set of elements of $\hat{\mathfrak{h}}$ of trace zero, and put $\varepsilon_i = \hat{\varepsilon}_i|\mathfrak{h}$. Then $\mathfrak{h}$ is a Cartan subalgebra of $\mathfrak{g}$ (Chap. VII, §2, no. 1, Example 4). Relation (1) proves that this Cartan subalgebra is splitting, and that the roots of $(\mathfrak{g},\mathfrak{h})$ are the $\varepsilon_i - \varepsilon_j$ $(i \neq j)$. Let $\hat{\mathfrak{h}}_0^*$ be the set of elements of $\hat{\mathfrak{h}}^*$ the sum of whose coordinates with respect to $(\hat{\varepsilon}_i)$ is zero. The map $\lambda \mapsto \lambda|\mathfrak{h}$ from $\hat{\mathfrak{h}}_0^*$ to $\mathfrak{h}^*$ is bijective. Thus, the root system R of $(\mathfrak{g},\mathfrak{h})$ is of type A_l (Chap. VI, §4, no. 7). Consequently, $\mathfrak{g}$ is simple (§3, no. 2, Cor. 1 of Prop. 6). Thus, *$\mathfrak{g}$ is a splittable simple Lie algebra of type* A_l.

Every splitting Cartan subalgebra $\mathfrak{h}'$ of $\mathfrak{g}$ is a transform of $\mathfrak{h}$ under an elementary automorphism (§3, no. 3, Cor. of Prop. 10). Since $\mathrm{Aut}_e\mathfrak{g}$ is the set of automorphisms $x \mapsto sxs^{-1}$ of $\mathfrak{g}$ with $s \in \mathbf{SL}(V)$ (Chap. VII, §3, no. 1, Remark 2; cf. also (VII)), there exists a basis β of V such that $\mathfrak{h}'$ is the set $\mathfrak{h}_\beta$ of elements of $\mathfrak{g}$ whose matrix with respect to the basis β is diagonal. Since $\mathfrak{h}_\beta$ contains an element with distinct eigenvalues, the only vector subspaces of V stable under the elements of $\mathfrak{h}_\beta$ are those generated by a subset of β. It follows that the map $\beta \mapsto \mathfrak{h}_\beta$ induces by passage to the quotient a bijection from the set of decompositions of V into the direct sum of $l+1$ subspaces of dimension 1 to the set of splitting Cartan subalgebras of $\mathfrak{g}$.

(II) Let $\alpha = \varepsilon_i - \varepsilon_j$ $(i \neq j)$ be a root. We have $\mathfrak{g}^\alpha = kE_{ij}$. Since

$$[E_{ij}, E_{ji}] = E_{ii} - E_{jj}$$

and since $\alpha(E_{ii} - E_{jj}) = 2$, we have (§2, no. 2, Th. 1 (ii))

$$H_\alpha = E_{ii} - E_{jj}.$$

(III) Put $\alpha_1 = \varepsilon_1 - \varepsilon_2, \alpha_2 = \varepsilon_2 - \varepsilon_3, \ldots, \alpha_l = \varepsilon_l - \varepsilon_{l+1}$. By Chap. VI, §4, no. 7.I, $(\alpha_1, \ldots, \alpha_l)$ is a basis B of R; the positive roots relative to B are the $\varepsilon_i - \varepsilon_j$ for $i < j$. The corresponding Borel subalgebra $\mathfrak{b}$ is the set of upper triangular matrices of trace zero.

A *flag* in V is a set of vector subspaces of V, distinct from $\{0\}$ and V, totally ordered by inclusion. Order the set of flags of V by inclusion. The maximal flags are the sets $\{W_1, \ldots, W_l\}$, where W_i is an i-dimensional vector subspace and

$$W_1 \subset \cdots \subset W_l.$$

For example, if V_i denotes the subspace of V generated by $e_1, \ldots, e_i$, then $\{V_1, \ldots, V_l\}$ is a maximal flag.

It is immediate that $\mathfrak{b}$ is the set of elements of $\mathfrak{g}$ leaving stable the elements of the maximal flag $\{V_1, \ldots, V_l\}$. Conversely, since $\mathfrak{b}$ contains $\mathfrak{h}$ and the matrices E_{ij} for $i < j$, we see that the V_i are the only non-trivial vector subspaces stable under $\mathfrak{b}$.

Now let δ be a maximal flag in V. It follows from the preceding that the set $\mathfrak{b}_\delta$ of elements of $\mathfrak{g}$ leaving stable all the elements of δ is a Borel subalgebra of $\mathfrak{g}$. Since every Borel subalgebra of $\mathfrak{g}$ is a transform of $\mathfrak{b}$ under an elementary automorphism, we see that the map $\delta \mapsto \mathfrak{b}_\delta$ is a bijection from the set of maximal flags to the set of Borel subalgebras of $\mathfrak{g}$.

Let β be a basis of V. By (I) and the preceding, the Borel subalgebras containing $\mathfrak{h}_\beta$ are those corresponding to the maximal flags each of whose elements is generated by a subset of β. These flags correspond bijectively to the total orders on β in the following way: to a total order ω on β is associated the flag $\{W_1, \ldots, W_l\}$, where W_i is the vector subspace generated by the first i elements of β for the order ω. Since there are $(l+1)!$ total orders on β, we recover the fact that there exist $(l+1)!$ Borel subalgebras of $(\mathfrak{sl}(V), \mathfrak{h}_\beta)$ (§3, no. 3, Remark).

Let γ be a flag in V. Since γ is contained in a maximal flag, the set $\mathfrak{p}_\gamma$ of elements of $\mathfrak{g}$ leaving stable the elements of γ is a parabolic subalgebra of $\mathfrak{g}$. We show that the only non-trivial vector subspaces stable under $\mathfrak{p}_\gamma$ are the elements of γ. For this, we can assume that $\gamma = \{V_{i_1}, \ldots, V_{i_q}\}$ with $1 \leq i_1 < \cdots < i_q \leq l$. Put $i_0 = 0, i_{q+1} = l+1$. The non-empty intervals

$$[\![i_0 + 1, i_1]\!], [\![i_1 + 1, i_2]\!], \ldots, [\![i_q + 1, i_{q+1}]\!]$$

form a partition of $\{1, \ldots, l+1\}$, so that any square matrix of order $l+1$ can be written as a block matrix $(X_{ab})_{1 \leq a,b \leq q+1}$. The algebra $\mathfrak{p}_\gamma$ is then the set $\mathfrak{p}_{i_1,\ldots,i_q}$ of elements $(X_{ab})_{1\leq a,b\leq q+1}$ of $\mathfrak{sl}(l+1,k)$ such that $X_{ab}=0$ for $a>b$. Since $\mathfrak{p}_{i_1,\ldots,i_q}\supset\mathfrak{b}$, a non-trivial vector subspace stable under $\mathfrak{p}_{i_1,\ldots,i_q}$ is one of the V_i; if $i_k<i<i_{k+1}$, the algebra $\mathfrak{p}_{i_1,\ldots,i_q}$ contains $E_{i_{k+1},i}$ and V_i is not stable, hence our assertion.

Consequently, the 2^l flags contained in the maximal flag $\{V_1,\ldots,V_l\}$ give rise to 2^l distinct parabolic subalgebras containing $\mathfrak{b}$; since there are exactly 2^l parabolic subalgebras containing $\mathfrak{b}$ (§3, no. 4, Remark), it follows that the map $\gamma\mapsto\mathfrak{p}_\gamma$ is a bijection from the set of flags of V to the set of parabolic subalgebras of $\mathfrak{g}$. Moreover, $\mathfrak{p}_\gamma\supset\mathfrak{p}_{\gamma'}$ if and only if $\gamma\subset\gamma'$.

Recall the parabolic subalgebra $\mathfrak{p}=\mathfrak{p}_{i_1,\ldots,i_q}$ $(1\leq i_1<\cdots<i_q\leq l)$. Let $\mathfrak{s}$ (resp. $\mathfrak{n}$) be the set of $(X_{ab})_{1\leq a,b\leq q+1}$ in $\mathfrak{sl}(l+1,k)$ such that $X_{ab}=0$ for $a\neq b$ (resp. $a\geq b$). In view of Prop. 13 of §3, no. 4, we have $\mathfrak{p}=\mathfrak{s}\oplus\mathfrak{n}$, the subalgebra $\mathfrak{s}$ is reductive in $\mathfrak{g}$ and $\mathfrak{n}$ is both the largest nilpotent ideal and the nilpotent radical of $\mathfrak{p}$.

(IV) For $r=1,2,\ldots,l$, let $\varpi_r=\varepsilon_1+\cdots+\varepsilon_r$. We have $\varpi_i(H_{\alpha_j})=\delta_{ij}$, so ϖ_r is the fundamental weight corresponding to α_r.

Let σ be the identity representation of $\mathfrak{g}$ on V. The exterior power $\bigwedge^r\sigma$ of σ is a representation on $E=\bigwedge^r(V)$. Let $(e_1,\ldots,e_{l+1})$ be the chosen basis of V. The $e_{i_1}\wedge\cdots\wedge e_{i_r}$, where $i_1<\cdots<i_r$, form a basis of E. If $h\in\mathfrak{h}$,

$$(\textstyle\bigwedge^r\sigma)(h).e_{i_1}\wedge\cdots\wedge e_{i_r}=(\varepsilon_{i_1}+\cdots+\varepsilon_{i_r})(h)e_{i_1}\wedge\cdots\wedge e_{i_r}.$$

Thus, every weight is of multiplicity 1, ϖ_r is a weight of $\bigwedge^r\sigma$, and every other weight is of the form $\varpi_r-\mu$, where μ is a positive radical weight. Consequently, ϖ_r is the highest weight of $\bigwedge^r\sigma$, and $e_1\wedge\cdots\wedge e_r$ is a primitive element. By Chap. VI, §4, no. 7.IX, the Weyl group can be identified with the symmetric group of

$$\{\varepsilon_1,\ldots,\varepsilon_{l+1}\}.$$

The orbit of ϖ_r under the Weyl group thus contains all the $\varepsilon_{i_1}+\cdots+\varepsilon_{i_r}$ with $i_1<\cdots<i_r$. The simple submodule generated by the primitive element $e_1\wedge\cdots\wedge e_r$ thus admits all the $\varepsilon_{i_1}+\cdots+\varepsilon_{i_r}$ as weights and consequently is equal to E. Thus, *$\bigwedge^r\sigma$ is irreducible with highest weight ϖ_r.*

Thus, the representations $\bigwedge^r\sigma$ $(1\leq r\leq l)$ are the fundamental representations. We have $\dim(\bigwedge^r\sigma)=\binom{l+1}{r}$.

(V) We have $w_0(\alpha_1)=-\alpha_l, w_0(\alpha_2)=-\alpha_{l-1},\ldots$ (Chap. VI, §4, no. 7, XI), so

$$-w_0(\varpi_1)=\varpi_l,\quad -w_0(\varpi_2)=\varpi_{l-1},\ldots.$$

Let

$$\omega=n_1\varpi_1+\cdots+n_l\varpi_l\qquad(n_1,\ldots,n_l\in\mathbf{N})$$

be a dominant weight. Then, the simple representation with highest weight ω is orthogonal or symplectic if and only if

$$n_1 = n_l, \quad n_2 = n_{l-1}, \ldots$$

(§7, no. 5, Prop. 12). In particular, if l is even, none of the fundamental representations of $\mathfrak{sl}(l+1, k)$ is orthogonal or symplectic. If l is odd, the representation $\bigwedge^i \sigma$ for $i \neq (l+1)/2$ is neither orthogonal nor symplectic; by Chap. VI, §4, no. 7.VI, the sum of the coordinates of $\varpi_{(l+1)/2}$ with respect to $(\alpha_1, \ldots, \alpha_l)$ is

$$\frac{1}{l+1}\left[\frac{l+1}{2}\left(1+2+\cdots+\frac{l-1}{2}\right)+\frac{l+1}{2}\left(1+2+\cdots+\frac{l+1}{2}\right)\right]$$
$$= 1+2+\cdots+\frac{l-1}{2}+\frac{l+1}{4}$$

so $\bigwedge^{(l+1)/2} \sigma$ is orthogonal if $l \equiv -1 \pmod{4}$ and symplectic if $l \equiv 1 \pmod{4}$ (§7, no. 5, Prop. 12). This last result can be made more precise as follows. Choose a non-zero element e in $\bigwedge^{l+1}(\mathrm{V})$. The multiplication in the exterior algebra V defines a bilinear map from

$$\bigwedge^{(l+1)/2}(\mathrm{V}) \times \bigwedge^{(l+1)/2}(\mathrm{V})$$

to $\bigwedge^{l+1}(\mathrm{V})$, which can be written $(u, v) \mapsto \Phi(u, v)e$, where Φ is a bilinear form on $\bigwedge^{(l+1)/2}(\mathrm{V})$. It is immediately verified that Φ is non-zero, invariant under $\mathfrak{g}$ (and hence non-degenerate), symmetric if $(l+1)/2$ is even, and alternating if $(l+1)/2$ is odd.

(VI) For all $x \in \mathfrak{g}$, the characteristic polynomial of $\sigma(x) = x$ can be written

$$\mathrm{T}^{l+1} + f_2(x)\mathrm{T}^{l-1} + f_3(x)\mathrm{T}^{l-2} + \cdots + f_{l+1}(x)$$

where $f_2, \ldots, f_{l+1}$ are polynomial functions invariant under $\mathfrak{g}$ (§8, no. 3, Lemma 2).

If $x = \xi_1 E_{11} + \cdots + \xi_{l+1} E_{l+1\, l+1} \in \mathfrak{h}$, the $f_i(x)$ are, up to sign, the elementary symmetric functions of $\xi_1, \ldots, \xi_{l+1}$ of degree $2, \ldots, l+1$. Thus, by Chap. VI, §4, no. 7.IX, the $f_i|\mathfrak{h}$ generate the algebra of elements of $\mathbf{S}(\mathfrak{h}^*)$ invariant under the Weyl group, and are algebraically independent. Hence (§8, no. 3, Prop. 3) $f_2, f_3, \ldots, f_{l+1}$ generate the algebra of polynomial functions invariant under $\mathfrak{g}$, and are algebraically independent.

(VII) For all $g \in \mathbf{GL}(l+1, k)$, let $\varphi_k(g) = \varphi(g)$ be the automorphism $x \mapsto gxg^{-1}$ of $\mathfrak{g}$. Then φ is a homomorphism from $\mathbf{GL}(l+1, k)$ to $\mathrm{Aut}(\mathfrak{g})$. We have

$$\varphi(\mathbf{SL}(l+1, k)) = \mathrm{Aut}_e(\mathfrak{g})$$

(Chap. VII, §3, no. 1, Remark 2). Let $\bar{k}$ be an algebraic closure of k. We have

$$\mathbf{GL}(l+1, \bar{k}) = \bar{k}^*.\mathbf{SL}(l+1, \bar{k}),$$

so $\varphi_{\bar{k}}(\mathbf{GL}(l+1,\bar{k})) = \varphi_{\bar{k}}(\mathbf{SL}(l+1,\bar{k})) = \mathrm{Aut}_e(\mathfrak{g}\otimes_k \bar{k})$; it follows that $\varphi(\mathbf{GL}(l+1,k)) \subset \mathrm{Aut}_0(\mathfrak{g})$. On the other hand, $\mathrm{Aut}_0(\mathfrak{g}) \subset \varphi(\mathbf{GL}(l+1,k))$, by Prop. 2 of §7, no. 1, applied to the identity representation of $\mathfrak{g}$. Hence,

$$\mathrm{Aut}_0(\mathfrak{g}) = \varphi(\mathbf{GL}(l+1,k)).$$

The kernel of φ is the set of elements of $\mathbf{GL}(l+1,k)$ that commute with every matrix of order $l+1$, that is the set k^* of invertible scalar matrices. Thus, $\mathrm{Aut}_0(\mathfrak{g})$ can be identified with the group $\mathbf{GL}(l+1,k)/k^* = \mathbf{PGL}(l+1,k)$. The kernel of $\varphi' = \varphi|\mathbf{SL}(l+1,k)$ is $\mu_{l+1}(k)$, where $\mu_{l+1}(k)$ denotes the set of $(l+1)$th roots of unity in k. Thus, $\mathrm{Aut}_e(\mathfrak{g})$ can be identified with the group $\mathbf{SL}(l+1,k)/\mu_{l+1}(k) = \mathbf{PSL}(l+1,k)$. On the other hand, we have the exact sequence

$$1 \longrightarrow \mathbf{SL}(l+1,k) \longrightarrow \mathbf{GL}(l+1,k) \xrightarrow{\det} k^* \longrightarrow 1$$

and the image of k^* under det is $k^{*\,l+1}$. It follows that there are canonical isomorphisms

$$\begin{aligned}\mathrm{Aut}_0(\mathfrak{g})/\mathrm{Aut}_e(\mathfrak{g}) &\longrightarrow \mathbf{PGL}(l+1,k)/\mathbf{PSL}(l+1,k)\\ &\longrightarrow \mathbf{GL}(l+1,k)/k^*.\mathbf{SL}(l+1,k) \longrightarrow k^*/k^{*\,l+1}.\end{aligned}$$

If $k = \mathbf{R}$, we see that $\mathrm{Aut}_0(\mathfrak{g}) = \mathrm{Aut}_e(\mathfrak{g})$ if $l+1$ is odd, and that $\mathrm{Aut}_0(\mathfrak{g})/\mathrm{Aut}_e(\mathfrak{g})$ is isomorphic to $\mathbf{Z}/2\mathbf{Z}$ if $l+1$ is even.

With the notations of §5, $f(\mathrm{T_Q})$ is the set of automorphisms of $\mathfrak{g}$ that induce the identity on $\mathfrak{h}$, and hence is equal to $\varphi(\mathrm{D})$, where D is the set of diagonal elements of $\mathbf{GL}(l+1,k)$ (§5, Prop. 4). Let D′ be the set of diagonal elements of $\mathbf{SL}(l+1,k)$. By Prop. 3 of §5, and the determination of $\mathrm{Aut}_e(\mathfrak{g})$, we have $f(q(\mathrm{T_P})) \subset \varphi(\mathrm{D}')$. We show that $f(q(\mathrm{T_P})) = \varphi(\mathrm{D}')$. Let

$$d = \begin{pmatrix} \lambda_1 & & 0 \\ & \ddots & \\ 0 & & \lambda_{l+1} \end{pmatrix}.$$

be an element of D′. There exists a $\zeta \in \mathrm{Hom}(\mathrm{Q(R)},k^*) = \mathrm{T_Q}$ such that $\zeta(\varepsilon_i - \varepsilon_j) = \lambda_i\lambda_j^{-1}$ for all i and j. It is easy to verify that $f(\zeta) = \varphi(d)$. By Chap. VI, §4, no. 7.VIII, P(R) is generated by Q(R) and the element $\varepsilon = \varepsilon_1$, whose image in P(R)/Q(R) is of order $l+1$; but

$$\begin{aligned}\zeta((l+1)\varepsilon) &= \zeta((\varepsilon_1-\varepsilon_2)+(\varepsilon_1-\varepsilon_3)+\cdots+(\varepsilon_1-\varepsilon_{l+1}))\\ &= \lambda_1^l\lambda_2^{-1}\lambda_3^{-1}\ldots\lambda_{l+1}^{-1} = \lambda_1^{l+1}\end{aligned}$$

so ζ extends to a homomorphism from P(R) to k^*. This proves that $\zeta \in q(\mathrm{T_P})$, so $\varphi(d) \in f(q(\mathrm{T_P}))$.

Recall (§5, no. 3, Cor. 2 of Prop. 5) that $\mathrm{Aut}(\mathfrak{g}) = \mathrm{Aut}_0(\mathfrak{g})$ for $l = 1$, and that $\mathrm{Aut}(\mathfrak{g})/\mathrm{Aut}_0(\mathfrak{g})$ is isomorphic to $\mathbf{Z}/2\mathbf{Z}$ for $l \geq 2$. The map $\theta : x \mapsto -{}^t x$ is an automorphism of $\mathfrak{sl}(l+1,k)$ and $a_0 = \theta|\mathfrak{h} \notin \mathrm{W}$ if $l \geq 2$ (Chap. VI, §4, no. 7.XI), so the class of a_0 in $\mathrm{Aut}(\mathfrak{g})/\mathrm{Aut}_0(\mathfrak{g})$ is the non-trivial element of this group (§5, no. 2, Prop. 4).

(VIII) The restriction to $\mathfrak{h}$ of the Killing form is

$$\begin{aligned}
&\Phi(\xi_1 E_{11} + \cdots + \xi_{l+1}E_{l+1,l+1}, \xi'_1 E_{11} + \cdots + \xi'_{l+1}E_{l+1,l+1}) \\
&\quad = \sum_{i\neq j}(\xi_i - \xi_j)(\xi'_i - \xi'_j) = \sum_{i,j}(\xi_i - \xi_j)(\xi'_i - \xi'_j) \\
&\quad = (l+1)\sum_i \xi_i\xi'_i + (l+1)\sum_j \xi_j\xi'_j - 2\left(\sum_i \xi_i\right)\left(\sum_j \xi'_j\right) \\
&\quad = 2(l+1)\sum_i \xi_i\xi'_i.
\end{aligned}$$

(IX) For $1 \leq i < j \leq l+1$, put

$$X_{\varepsilon_i-\varepsilon_j} = E_{ij} \qquad X_{\varepsilon_j-\varepsilon_i} = -E_{ji}.$$

Then, for all $\alpha \in \mathrm{R}$, we have $[X_\alpha, X_{-\alpha}] = -H_\alpha$ and $\theta(X_\alpha) = X_{-\alpha}$ (where θ is the automorphism $x \mapsto -{}^t x$ introduced in (VII)). Consequently, $(X_\alpha)_{\alpha\in\mathrm{R}}$ is a Chevalley system in $(\mathfrak{g}, \mathfrak{h})$.

Take $k = \mathbf{Q}$. The permissible lattices in $\mathfrak{h}$ (§12, no. 6, Def. 1) are those lying between the $\mathbf{Z}$-module $\mathrm{Q}(\mathrm{R}^\vee)$ generated by the $E_{ii} - E_{i+1,i+1}$, that is, consisting of the diagonal matrices belonging to $\mathfrak{sl}(l+1, \mathbf{Z})$, and the $\mathbf{Z}$-module $\mathrm{P}(\mathrm{R}^\vee)$ generated by $\mathrm{Q}(\mathrm{R}^\vee)$ and $E_{11} - (l+1)^{-1}\sum E_{ii}$ (Chap. VI, §4, no. 7.VIII), that is, consisting of the diagonal matrices of trace zero of the form $x + (l+1)^{-1}a.1$, where x has integer entries and $a \in \mathbf{Z}$. It follows that $\mathfrak{sl}(l+1, \mathbf{Z})$ is the Chevalley order in $(\mathfrak{g}, \mathfrak{h})$ associated to the permissible lattice $\mathrm{Q}(\mathrm{R}^\vee)$ and the Chevalley system (X_α). It is easy to verify that $\bigwedge^r \mathbf{Z}^{l+1}$ is an admissible lattice in $\bigwedge^r \mathbf{Q}^{l+1}$ relative to $\mathfrak{sl}(l+1, \mathbf{Z})$ (§12, no. 8, Def. 3).

On the other hand, $\mathfrak{gl}(l+1, \mathbf{Z})$ is a Chevalley order in the split reductive algebra $\mathfrak{gl}(l+1, \mathbf{Q})$; its projection onto $\mathfrak{sl}(l+1, \mathbf{Q})$ parallel to the centre $\mathbf{Q}.1$ of $\mathfrak{gl}(l+1, \mathbf{Q})$ is the Chevalley order in $(\mathfrak{g}, \mathfrak{h})$ defined by the permissible lattice $\mathrm{P}(\mathrm{R}^\vee)$ in $\mathfrak{h}$ and the Chevalley system (X_α). We remark that $\mathfrak{gl}(l+1, \mathbf{Z})$ is not the direct sum of its intersections with $\mathfrak{sl}(l+1, \mathbf{Q})$ and the centre of $\mathfrak{gl}(l+1, \mathbf{Q})$.

2. ALGEBRAS OF TYPE B_l ($l \geq 1$)

(I) Let V be a finite dimensional vector space, and Ψ a non-degenerate symmetric bilinear form on V. The set of endomorphisms x of V such that $\Psi(xv, v') + \Psi(v, xv') = 0$ for all $v, v' \in \mathrm{V}$ is a Lie subalgebra of $\mathfrak{sl}(\mathrm{V})$, semi-simple for $\dim \mathrm{V} \neq 2$ (Chap. I, §6, no. 7, Prop. 9). *We denote it by* $\mathfrak{o}(\Psi)$ *and call it the orthogonal Lie algebra associated to* Ψ.

Assume that V is of odd dimension $2l+1 \geq 3$ and that Ψ is of maximum index l. Denote by Q the quadratic form such that Ψ is associated to Q. We have $Q(x) = \frac{1}{2}\Psi(x, x)$ for $x \in V$. By *Algebra*, Chap. IX, §4, no. 2, V can be written as the direct sum of two maximal totally isotropic subspaces F and F′ and the orthogonal complement G of F + F′, which is non-isotropic and 1-dimensional. Up to multiplying Ψ by a non-zero constant, we can assume that there exists $e_0 \in G$ such that $\Psi(e_0, e_0) = -2$. On the other hand, F and F′ are in duality via Ψ; let $(e_i)_{1\leq i\leq l}$ be a basis of F and $(e_{-i})_{1\leq i\leq l}$ the dual basis of F′. Then

$$(e_1, \dots, e_l, e_0, e_{-l}, \dots, e_{-1})$$

is a basis of V; we have

$$Q\left(\sum x_i e_i\right) = -x_0^2 + \sum_{i=1}^{i=l} x_i x_{-i}$$

and the matrix of Ψ with respect to this basis is the square matrix of order $2l+1$

$$S = \begin{pmatrix} 0 & 0 & s \\ 0 & -2 & 0 \\ s & 0 & 0 \end{pmatrix}, \qquad s = \begin{pmatrix} 0 & 0 & \cdots & 0 & 1 \\ 0 & 0 & \cdots & 1 & 0 \\ \vdots & \vdots & \ddots & \vdots & \vdots \\ 0 & 1 & \cdots & 0 & 0 \\ 1 & 0 & \cdots & 0 & 0 \end{pmatrix},$$

where s is the square matrix of order l all of whose entries are zero except those on the second diagonal[6] which are equal to 1. A basis of V with the preceding properties will be called a *Witt basis* of V. The algebra $\mathfrak{g} = \mathfrak{o}(\Psi)$ can then be identified with the algebra $\mathfrak{o}_S(2l+1, k)$ of square matrices a of order $2l+1$ such that $a = -S^{-1}aS$ (*Algebra*, Chap. IX, §1, no. 10, formulas (50)). An easy calculation shows that $\mathfrak{g}$ is the set of matrices of the form

$$\begin{pmatrix} A & 2s^t x & B \\ y & 0 & x \\ C & 2s^t y & D \end{pmatrix} \tag{2}$$

where x and y are matrices with 1 row and l columns and A, B, C, D are square matrices of order l such that $B = -s^t Bs$, $C = -s^t Cs$, and $D = -s^t As$. Since the map $A \mapsto s^t As$ from $\mathbf{M}_l(k)$ to itself is the symmetry with respect to the second diagonal, it follows that

$$\dim \mathfrak{g} = 2l + l^2 + 2\frac{l(l-1)}{2} = l(2l+1).$$

Let $\mathfrak{h}$ be the set of diagonal elements of $\mathfrak{g}$. This is a commutative subalgebra of $\mathfrak{g}$, with basis the elements

[6] The second diagonal of a square matrix $(a_{ij})_{1\leq i,j\leq n}$ is the family of a_{ij} such that $i + j = n + 1$.

$$H_i = E_{i,i} - E_{-i,-i} \quad (1 \le i \le l).$$

Let (ε_i) be the basis of $\mathfrak{h}^*$ dual to (H_i). Put

$$\begin{cases} X_{\varepsilon_i} &= 2E_{i,0} + E_{0,-i} & (1 \le i \le l) \\ X_{-\varepsilon_i} &= -2E_{-i,0} - E_{0,i} & (1 \le i \le l) \\ X_{\varepsilon_i-\varepsilon_j} &= E_{i,j} - E_{-j,-i} & (1 \le i < j \le l) \\ X_{\varepsilon_j-\varepsilon_i} &= -E_{j,i} + E_{-i,-j} & (1 \le i < j \le l) \\ X_{\varepsilon_i+\varepsilon_j} &= E_{i,-j} - E_{j,-i} & (1 \le i < j \le l) \\ X_{-\varepsilon_i-\varepsilon_j} &= -E_{-j,i} + E_{-i,j} & (1 \le i < j \le l). \end{cases} \tag{3}$$

It is easy to verify that these elements form a basis of a complement of $\mathfrak{h}$ in $\mathfrak{g}$ and that, for $h \in \mathfrak{h}$,

$$[h, X_\alpha] = \alpha(h)X_\alpha \tag{4}$$

for all $\alpha \in \mathrm{R}$, where R is the set of the $\pm\varepsilon_i$ and the $\pm\varepsilon_i \pm \varepsilon_j$ $(1 \le i < j \le l)$. It follows that $\mathfrak{h}$ is equal to its normalizer in $\mathfrak{g}$, and hence is a Cartan subalgebra of $\mathfrak{g}$, that $\mathfrak{h}$ is splitting, and that the roots of $(\mathfrak{g}, \mathfrak{h})$ are the elements of R. The root system R of $(\mathfrak{g}, \mathfrak{h})$ is of type B_l for $l \ge 2$, and of type A_1 (also said to be of type B_1) for $l = 1$ (Chap. VI, §4, no. 5.I, extended to the case $l = 1$). Consequently, $\mathfrak{g}$ *is a splittable simple Lie algebra of type* B_l.

Every splitting Cartan subalgebra of $\mathfrak{o}(\Psi)$ is a transform of $\mathfrak{h}$ by an elementary automorphism of $\mathfrak{o}(\Psi)$, and hence by an element of $\mathbf{O}(\Psi)$ (cf. (VII)), and consequently is the set $\mathfrak{h}_\beta$ of elements of $\mathfrak{g}$ whose matrix with respect to a Witt basis β of V is diagonal. We verify immediately that the only vector subspaces invariant under $\mathfrak{h}_\beta$ are those generated by a subset of β.

If $l = 1$, the algebras $\mathfrak{o}(\Psi)$ and $\mathfrak{sl}(2, k)$ have the same root systems, and are thus isomorphic (cf. also §1, Exerc. 16). From now on, we assume that $l \ge 2$.

(II) The root system $\mathrm{R}^\vee$ is determined by means of Chap. VI, §4, no. 5.V, and we find that

$$H_{\varepsilon_i} = 2H_i, \quad H_{\varepsilon_i-\varepsilon_j} = H_i - H_j, \quad H_{\varepsilon_i+\varepsilon_j} = H_i + H_j.$$

(III) Put $\alpha_1 = \varepsilon_1 - \varepsilon_2, \dots, \alpha_{l-1} = \varepsilon_{l-1} - \varepsilon_l, \alpha_l = \varepsilon_l$. By Chap. VI, §4, no. 5.II, $(\alpha_1, \dots, \alpha_l)$ is a basis B of R; the positive roots relative to B are the ε_i and the $\varepsilon_i \pm \varepsilon_j$ $(i < j)$. The corresponding Borel subalgebra $\mathfrak{b}$ is the set of upper triangular matrices in $\mathfrak{g}$.

It is immediately verified that the only vector subspaces of V distinct from $\{0\}$ and V stable under $\mathfrak{b}$ are the elements of the maximal flag corresponding to the basis (e_i), that is, the totally isotropic subspaces $\mathrm{V}_1, \dots, \mathrm{V}_l$, where V_i is generated by $e_1, \dots, e_i$, together with their orthogonal complements $\mathrm{V}_{-1}, \dots, \mathrm{V}_{-i}$: the orthogonal complement V_{-i} of V_i is generated by $e_1, \dots, e_l, e_0, e_{-l}, \dots, e_{-i-1}$ and is not totally isotropic. On the other hand, if an element of $\mathfrak{g}$ leaves stable a vector subspace, it leaves stable its orthogonal complement. Consequently, $\mathfrak{b}$ is the set of elements of $\mathfrak{g}$ leaving stable the elements of the flag $\{V_1, \ldots, V_l\}$.

A flag is said to be *isotropic* if each of its elements is totally isotropic. The flag $\{V_1, \ldots, V_l\}$ is a maximal isotropic flag. Since the group $\mathbf{O}(\Psi)$ operates transitively both on the Borel subalgebras of $\mathfrak{g}$ (cf. (VII)) and on the maximal isotropic flags (*Algebra*, Chap. IX, §4, no. 3, Th. 1), we see that, for any maximal isotropic flag δ in V, the set $\mathfrak{b}_\delta$ of elements of $\mathfrak{g}$ leaving stable the elements of δ is a Borel subalgebra of $\mathfrak{g}$ and that the map $\delta \mapsto \mathfrak{b}_\delta$ is a bijection from the set of maximal isotropic flags to the set of Borel subalgebras of $\mathfrak{g}$.

Let δ be an isotropic flag and let $\mathfrak{p}_\delta$ be the set of elements of $\mathfrak{g}$ leaving stable the elements of δ. If $\delta \subset \{V_1, \ldots, V_l\}$, then $\mathfrak{p}_\delta$ is a parabolic subalgebra of $\mathfrak{g}$ containing $\mathfrak{b}$, and it is easy to verify that the only totally isotropic subspaces $\neq \{0\}$ stable under $\mathfrak{p}_\delta$ are the elements of δ. This gives 2^l parabolic subalgebras of $\mathfrak{g}$ containing $\mathfrak{b}$. We see as above that the map $\delta \mapsto \mathfrak{p}_\delta$ is a bijection from the set of isotropic flags in V to the set of parabolic subalgebras of $\mathfrak{g}$. Moreover, $\mathfrak{p}_\delta \subset \mathfrak{p}_{\delta'}$ if and only if $\delta \supset \delta'$.

(IV) The fundamental weights corresponding to $\alpha_1, \ldots, \alpha_l$ are, by Chap. VI, §4, no. 5.VI,

$$\varpi_i = \varepsilon_1 + \cdots + \varepsilon_i \quad (1 \leq i \leq l-1)$$
$$\varpi_l = \frac{1}{2}(\varepsilon_1 + \cdots + \varepsilon_l).$$

Let σ be the identity representation of $\mathfrak{g}$ on V. The exterior power $\bigwedge^r \sigma$ operates on $E = \bigwedge^r V$. If $h \in \mathfrak{h}$,

$$\sigma(h).e_i = \varepsilon_i(h)e_i \quad \text{for } 1 \leq i \leq l$$
$$\sigma(h).e_0 = 0$$
$$\sigma(h).e_{-i} = -\varepsilon_i(h)e_{-i} \quad \text{for } 1 \leq i \leq l.$$

It follows that, for $1 \leq r \leq l$, $\varepsilon_1 + \cdots + \varepsilon_r$ is the highest weight of $\bigwedge^r \sigma$, the elements of weight $\varepsilon_1 + \cdots + \varepsilon_r$ being those proportional to $e_1 \wedge \cdots \wedge e_r$. We shall show that for $1 \leq r \leq l-1$, *the representation $\bigwedge^r \sigma$ is a fundamental representation of $\mathfrak{g}$ of highest weight ϖ_r*. For this, it is enough to show that $\bigwedge^r \sigma$ is irreducible for $0 \leq r \leq 2l+1$. But the bilinear form Φ on $\bigwedge^r V \times \bigwedge^{2l+1-r} V$ defined by

$$x \wedge y = \Phi(x, y)e_1 \wedge \cdots \wedge e_l \wedge e_0 \wedge e_{-l} \wedge \cdots \wedge e_{-1}$$

is invariant under $\mathfrak{g}$ and puts $\bigwedge^r V$ and $\bigwedge^{2l+1-r} V$ in duality. Thus, the representation $\bigwedge^{2l+1-r} \sigma$ is the dual of $\bigwedge^r \sigma$ and it suffices to prove the irreducibility of $\bigwedge^r \sigma$ for $0 \leq r \leq l$, or that the smallest subspace T_r of $\bigwedge^r V$ containing $e_1 \wedge \cdots \wedge e_r$ and stable under $\mathfrak{g}$ is the whole of $\bigwedge^r V$. This is immediate for $r = 0$ and $r = 1$ (cf. formula (2)). For $r = 2$ (and hence $l \geq 2$), the representation $\bigwedge^2 \sigma$ and the adjoint representation of $\mathfrak{g}$ (which is irreducible) have the same dimension $l(2l+1)$ and the same highest weight $\varepsilon_1+\varepsilon_2$ (Chap. VI, §4, no. 5.IV). We conclude that $\bigwedge^2 \sigma$ is *equivalent to the adjoint representation*, and hence is irreducible. This proves our assertion for $l=1$ and $l=2$.

We now argue by induction on l, and assume that $l \geq r \geq 3$. We remark first of all that if W is a non-isotropic subspace of V of odd dimension, with orthogonal complement W′, the restriction $\Psi_{\rm W}$ of Ψ to W is non-degenerate and $\mathfrak{o}(\Psi_{\rm W})$ can be identified with the subalgebra of $\mathfrak{g}$ consisting of the elements vanishing on W′. If $\dim \mathrm{W} < \dim \mathrm{V}$, and if $\Psi_{\rm W}$ is of maximal index, the induction hypothesis implies that if T_r contains a non-zero element of the form $w' \wedge w$, with $w' \in \bigwedge^{r-k} \mathrm{W}'$ and $w \in \bigwedge^k \mathrm{W}$ $(0 \leq k \leq r)$, then T_r contains $w' \wedge \bigwedge^k \mathrm{W}$: indeed, we have $a.(w' \wedge w) = w' \wedge a.w$ for all $a \in \mathfrak{o}(\Psi_{\rm W})$. We show by induction on $p \in [0, r]$ that T_r contains the elements

$$x = e_{i_1} \wedge \cdots \wedge e_{i_{r-p}} \wedge e_{j_1} \wedge \cdots \wedge e_{j_p}$$

for $1 \leq i_1 < \cdots < i_{r-p} \leq l$ and $-l \leq j_1 < \cdots < j_p \leq 0$. For $p=0$, this follows from the irreducibility of the operation of $\mathfrak{gl}(\mathrm{F})$ on $\bigwedge^r \mathrm{F}$ (no. 1), since $\mathfrak{g}$ contains the elements leaving $\mathrm{F} = \mathrm{V}_l = \sum_{i=1}^{l} ke_i$ fixed and inducing on it any endomorphism (cf. formula (2)). If $p=1$, let $q \in (1, l)$ be such that $q \neq -j_1$ and such that there exists $\lambda \in [1, r-p]$ with $q = i_\lambda$; if $p \geq 2$, let $q \in [1, l]$ be such that $-q \in \{j_1, \ldots, j_p\}$. Permuting the e_i if necessary, we can assume that $q=1$. Now take for W the orthogonal complement of $\mathrm{W}' = ke_1 + ke_{-1}$. If $p=1$, we have $x \in e_1 \wedge \bigwedge^{r-1} \mathrm{W}$; since T_r contains $e_1 \wedge \cdots \wedge e_r$, we see that T_r contains x. If $p \geq 2$, either $x \in e_{-1} \wedge \bigwedge^{r-1} \mathrm{W}$ or $x \in e_1 \wedge e_{-1} \wedge \bigwedge^{r-2} \mathrm{W}$; since T_r contains $e_{-1} \wedge e_2 \wedge \cdots \wedge e_{r-1}$ and $e_{-1} \wedge e_1 \wedge e_2 \wedge \cdots \wedge e_{r-2}$ by the induction hypothesis, we see that T_r contains x, which completes the proof.

For another proof of the irreducibility of $\bigwedge^r \sigma$, see Exerc. 6.

We shall now determine the fundamental representation with highest weight ϖ_l.

Lemma 1. Let V *be a finite dimensional vector space,* Q *a non-degenerate quadratic form on* V, Ψ *the symmetric bilinear form associated to* Q, C(Q) *the Clifford algebra of* V *relative to* Q, f_0 *the composite of the canonical maps*

$$\mathfrak{o}(\Psi) \longrightarrow \mathfrak{gl}(\mathrm{V}) \longrightarrow \mathrm{V} \otimes \mathrm{V}^* \longrightarrow \mathrm{V} \otimes \mathrm{V} \longrightarrow \mathrm{C}^+(\mathrm{Q})$$

(*the* 1*st is the canonical injection, the* 3*rd is defined by the canonical isomorphism from* V* *to* V *corresponding to* Ψ, *the* 4*th is defined by the multiplication in* C(Q), cf. *Algebra*, Chap. IX, § 9, no. 1). *Put* $f = \frac{1}{2} f_0$.

(i) *If* $(e_r), (e'_r)$ *are bases of* V *such that* $\Psi(e_r, e'_s) = \delta_{rs}$, *we have* $f_0(a) = \sum_r (ae_r)e'_r$ *for all* $a \in \mathfrak{o}(\Psi)$.

(ii) *If* $a, b \in \mathfrak{o}(\Psi)$, *we have* $\sum_r (ae_r)(be'_r) = -\sum_r (abe_r)e'_r$.

(iii) *If* $a \in \mathfrak{o}(\Psi)$ *and* $v \in \mathrm{V}$, *we have* $[f(a), v] = av$.

(iv) *If* $a, b \in \mathfrak{o}(\Psi)$, *we have* $[f(a), f(b)] = f([a, b])$.

(v) $f(\mathfrak{o}(\Psi))$ *generates the associative algebra* $\mathrm{C}^+(\mathrm{Q})$.

(vi) *Let* N *be a left* $\mathrm{C}^+(\mathrm{Q})$*-module and* ρ *the corresponding homomorphism from* $\mathrm{C}^+(\mathrm{Q})$ *to* $\mathrm{End}_k(\mathrm{N})$. *Then* $\rho \circ f$ *is a representation of* $\mathfrak{o}(\Psi)$ *on* N. *If* N *is simple,* $\rho \circ f$ *is irreducible.*

Assertion (i) is clear. If $a, b \in \mathfrak{o}(\Psi)$, we have (putting $\Psi(x, y) = \langle x, y\rangle$):

$$\begin{aligned}\sum_r (ae_r)(be'_r) &= \sum_{r,s,t} \langle ae_r, e'_s\rangle\langle be'_r, e_t\rangle e_s e'_t = \sum_{r,s,t} \langle e_r, ae'_s\rangle\langle e'_r, be_t\rangle e_s e'_t \\ &= \sum_{s,t} \langle ae'_s, be_t\rangle e_s e'_t = -\sum_{s,t} \langle e'_s, abe_t\rangle e_s e'_t = -\sum_t (abe_t)e'_t\end{aligned}$$

which proves (ii). Next, for all $v \in \mathrm{V}$, we have by (i),

$$\begin{aligned}[f(a), v] &= \frac{1}{2}\sum_r ((ae_r)e'_r v - v(ae_r)e'_r) \\ &= \frac{1}{2}\sum_r ((ae_r)e'_r v + (ae_r)ve'_r - (ae_r)ve'_r - v(ae_r)e'_r) \\ &= \frac{1}{2}\sum_r ((ae_r)\langle e'_r, v\rangle - \langle ae_r, v\rangle e'_r) \\ &= \frac{1}{2}a\left(\sum_r \langle e'_r, v\rangle e_r\right) + \frac{1}{2}\sum_r \langle e_r, av\rangle e'_r = \frac{1}{2}av + \frac{1}{2}av = av,\end{aligned}$$

which proves (iii). Then

$$\begin{aligned}[f(a), f(b)] &= \left[f(a), \frac{1}{2}\sum_r (be_r)e'_r\right] \qquad \text{by (i)} \\ &= \frac{1}{2}\sum_r ([f(a), be_r]e'_r + (be_r)[f(a), e'_r]) \\ &= \frac{1}{2}\sum_r ((abe_r)e'_r + (be_r)(ae'_r)) \qquad \text{by (iii)} \\ &= \frac{1}{2}\sum_r ((abe_r)e'_r - (bae_r)e'_r) \qquad \text{by (ii)} \\ &= f([a, b]) \qquad \text{by (i)}\end{aligned}$$

which proves (iv). To prove (v), we can, by extending scalars, assume that k is algebraically closed. Choose then a basis (e_r) of V such that $\Psi(e_r, e_s) = \delta_{rs}$, so $e'_r = e_r$. If $i \neq j$, then $\mathrm{E}_{ij} - \mathrm{E}_{ji} \in \mathfrak{o}(\Psi)$ and

$$f(\mathrm{E}_{ij} - \mathrm{E}_{ji}) = \frac{1}{2}(e_i e_j - e_j e_i) = e_i e_j;$$

but the $e_i e_j$ generate $\mathrm{C}^+(\mathrm{Q})$.

Assertion (vi) follows from (iv) and (v). Q.E.D.

Recall now the notations used at the beginning of this number. Put $\tilde{\mathrm{V}} = \mathrm{F} + \mathrm{F}'$ and let $\tilde{\mathrm{Q}}$ (resp. $\tilde{\Psi}$) be the restriction of Q (resp. Ψ) to $\tilde{\mathrm{V}}$. Then $\tilde{\mathrm{Q}}$ is a non-degenerate quadratic form of maximum index l on the space $\tilde{\mathrm{V}}$ of dimension $2l$ and the Clifford algebra $\mathrm{C}(\tilde{\mathrm{Q}})$ is a central simple algebra of dimension 2^{2l} (*Algebra*, Chap. IX, §9, no. 4, Th. 2). Let N be the exterior algebra of the maximal isotropic subspace F′ generated by $e_{-1}, \ldots, e_{-l}$. Identify F with the dual of F′ by means of Ψ and for $x \in \mathrm{F}'$ (resp. $y \in \mathrm{F}$) denote by $\lambda(x)$ (resp. $\lambda(y)$) the left exterior product with x (resp. the left interior product with y) in N; if $a_1, \ldots, a_k \in \mathrm{F}'$, then

$$\lambda(x).(a_1 \wedge \cdots \wedge a_k) = x \wedge a_1 \wedge \cdots \wedge a_k$$

$$\lambda(y).(a_1 \wedge \cdots \wedge a_k) = \sum_{i=1}^{k} (-1)^{i-1} \Psi(a_i, y) a_1 \wedge \cdots \wedge a_{i-1} \wedge a_{i+1} \wedge \cdots \wedge a_k.$$

It is easily verified that $\lambda(x)^2 = \lambda(y)^2 = 0$ and that

$$\lambda(x)\lambda(y) + \lambda(y)\lambda(x) = \Psi(x, y).1.$$

It follows (*Algebra*, Chap. IX, §9, no. 1) that there exists a unique homomorphism (again denoted by λ) from $\mathrm{C}(\tilde{\mathrm{Q}})$ to End(N) extending the map $\lambda : \mathrm{F} \cup \mathrm{F}' \to \mathrm{End}(\mathrm{N})$. Since $\dim \mathrm{N} = 2^l$ and since $\mathrm{C}(\tilde{\mathrm{Q}})$ has a unique class of simple modules, of dimension 2^l (*Algebra*, Chap. IX, §9, no. 4, Th. 2), the representation of $\mathrm{C}(\tilde{\mathrm{Q}})$ on N defined by λ is irreducible and is a spinor representation of $\mathrm{C}(\tilde{\mathrm{Q}})$ (*loc. cit.*).

Consider now the map $\mu : v \mapsto e_0 v$ from $\tilde{\mathrm{V}}$ to $\mathrm{C}^+(\mathrm{Q})$. For $v \in \tilde{\mathrm{V}}$, we have

$$(e_0 v)^2 = -e_0^2 v^2 = -\mathrm{Q}(e_0)\mathrm{Q}(v) = \mathrm{Q}(v) = \tilde{\mathrm{Q}}(v)$$

and μ extends uniquely to a homomorphism, again denoted by μ, from $\mathrm{C}(\tilde{\mathrm{Q}})$ to $\mathrm{C}^+(\mathrm{Q})$. Since $\mathrm{C}(\tilde{\mathrm{Q}})$ is simple and since

$$\dim \mathrm{C}^+(\mathrm{Q}) = \dim \mathrm{C}(\tilde{\mathrm{Q}}) = 2^{2l},$$

we see that μ is an *isomorphism*. Consequently, $\lambda \circ \mu^{-1}$ defines a simple $\mathrm{C}^+(\mathrm{Q})$-module structure on N and $\rho = \lambda \circ \mu^{-1} \circ f$ is an irreducible representation of $\mathfrak{g}$ on N (Lemma 1 (vi)).

On the other hand, in view of Lemma 1 (i), we have

$$f(H_i) = \frac{1}{2}(e_i e_{-i} - e_{-i} e_i).$$

Since $e_i e_{-i} = -e_0^2 e_i e_{-i} = e_0 e_i e_0 e_{-i}$ and $e_i e_{-i} + e_{-i} e_i = 1$, we have

$$\mu^{-1} \circ f(H_i) = \frac{1}{2} - e_{-i} e_i = -\frac{1}{2} + e_i e_{-i}.$$

We deduce that, for $1 \leq i_1 < \cdots < i_k \leq l$:

$$\rho(H_i)(e_{-i_1} \wedge \cdots \wedge e_{-i_k}) = \begin{cases} -\frac{1}{2}e_{-i_1} \wedge \cdots \wedge e_{-i_k} & \text{if } i \in \{i_1, \dots, i_k\} \\ \frac{1}{2}e_{-i_1} \wedge \cdots \wedge e_{-i_k} & \text{if } i \notin \{i_1, \dots i_k\} \end{cases}$$

and for $h \in \mathfrak{h}$

$$\rho(h)(e_{-i_1} \wedge \cdots \wedge e_{-i_k}) \qquad (5)$$
$$= (\frac{1}{2}(\varepsilon_1 + \cdots + \varepsilon_l) - (\varepsilon_{i_1} + \cdots + \varepsilon_{i_k}))(h)(e_{-i_1} \wedge \cdots \wedge e_{-i_k}).$$

This shows that the highest weight of ρ is ϖ_l. We call ρ the *spinor representation* of $\mathfrak{g}$. Note that its weights are all simple (moreover, ϖ_l is a minuscule weight).

(V) We have $w_0 = -1$, so every finite dimensional simple representation of $\mathfrak{g}$ is orthogonal or symplectic. By Chap. VI, §4, no. 5.VI, the sum of the coordinates of ϖ_r with respect to $(\alpha_1, \dots, \alpha_l)$ is integral for $1 \leq r \leq l-1$: thus, the representation $\bigwedge^r \sigma$ is orthogonal. Moreover, it leaves invariant the extension $\Psi_{(r)}$ of Ψ to $\bigwedge^r \mathrm{V}$.

For the spinor representation, the sum of the coordinates of ϖ_l with respect to $(\alpha_1, \dots, \alpha_l)$ is $\frac{1}{2}(1 + \cdots + l) = \frac{l(l+1)}{4}$ (*loc. cit.*). Thus, it is orthogonal for $l \equiv 0$ or -1 (mod. 4) and symplectic for $l \equiv 1$ or 2 (mod. 4). In fact, consider the bilinear form Φ on $\mathrm{N} = \bigwedge \mathrm{F}'$ defined as follows: if $x \in \bigwedge^p \mathrm{F}'$ and $y \in \bigwedge^q \mathrm{F}'$, put $\Phi(x, y) = 0$ if $p + q \neq l$ and

$$x \wedge y = (-1)^{\frac{p(p+1)}{2}} \Phi(x, y) e_{-1} \wedge \cdots \wedge e_{-r}$$

if $p + q = l$. It is easily verified that Φ is non-degenerate and is orthogonal for $l \equiv 0, -1$ (mod. 4) and alternating for $l \equiv 1, 2$ (mod. 4). On the other hand, in view of Lemma 1 (i),

$$f(X_{\varepsilon_i}) = e_0 e_i, \quad f(X_{-\varepsilon_i}) = -e_0 e_{-i}$$

for $1 \leq i \leq l$ and

$$f(X_{\varepsilon_i - \varepsilon_j}) = \frac{1}{2}(e_i e_{-j} - e_{-j} e_i) = e_i e_{-j} = e_0 e_i e_0 e_{-j}$$

for $1 \leq i < j \leq l$, and similarly

$$f(X_{\varepsilon_j - \varepsilon_i}) = -e_0 e_j e_0 e_{-i}, \; f(X_{\varepsilon_i + \varepsilon_j}) = e_0 e_i e_0 e_j, \; f(X_{-\varepsilon_i - \varepsilon_j}) = e_0 e_{-i} e_0 e_{-j};$$

hence

$$\mu^{-1} \circ f(X_{\varepsilon_i}) = e_i, \quad \mu^{-1} f(X_{-\varepsilon_i}) = -e_{-i}$$

and

$$\mu^{-1} \circ f(X_{\pm\varepsilon_i \pm \varepsilon_j}) = c e_{\pm i} e_{\pm j} \quad \text{for } 1 \leq i, j \leq l, \; i \neq j \text{ with } c \in \{1, -1\}.$$

It is now painless to verify that Φ is actually $\mathfrak{g}$-invariant (cf. Exerc. 18).

(VI) For $x \in \mathfrak{g}$, the characteristic polynomial of $\sigma(x)$ takes the form

$$T^{2l+1} + f_1(x)T^{2l} + f_2(x)T^{2l-1} + \cdots + f_{2l+1}(x)$$

where $f_1, \ldots, f_{2l+1}$ are invariant polynomial functions on $\mathfrak{g}$.

If $x = \xi_1 H_1 + \cdots + \xi_l H_l \in \mathfrak{h}$, the $f_i(x)$ are, up to sign, the elementary symmetric functions of $\xi_1, \ldots, \xi_l, -\xi_1, \ldots, -\xi_l$; these symmetric functions are zero in odd degrees, and

$$T^{2l+1} + f_2(x)T^{2l-1} + f_4(x)T^{2l-3} + \cdots + f_{2l}(x)T = T(T^2 - \xi_1^2) \ldots (T^2 - \xi_l^2)$$

so that $f_2, \ldots, f_{2l}$ are, up to sign, the elementary symmetric functions of $\xi_1^2, \ldots, \xi_l^2$, which are algebraically independent generators of $\mathbf{S}(\mathfrak{h}^*)^W$ (Chap. VI, §4, no. 5.IX). In view of §8, no. 3, Th. 1 (i), we see that $f_1 = f_3 = f_5 = \cdots = 0$ and that $(f_2, f_4, \ldots, f_{2l})$ is an algebraically free family generating the algebra of invariant polynomial functions on $\mathfrak{g}$.

(VII) Since the only automorphism of the Dynkin graph is the identity, we have $\mathrm{Aut}(\mathfrak{g}) = \mathrm{Aut}_0(\mathfrak{g})$.

Let Σ be the group of similarities of V relative to Ψ. For all $g \in \Sigma$, let $\varphi(g)$ be the automorphism $x \mapsto gxg^{-1}$ of $\mathfrak{g}$. Then φ is a homomorphism from Σ to $\mathrm{Aut}(\mathfrak{g})$. We show that it is surjective. Let $\alpha \in \mathrm{Aut}(\mathfrak{g}) = \mathrm{Aut}_0(\mathfrak{g})$. By Prop. 2 of §7, no. 1, there exists $s \in \mathbf{GL}(V)$ such that $\alpha(x) = sxs^{-1}$ for all $x \in \mathfrak{g}$. Then s transforms Ψ into a bilinear form Ψ' on V that is invariant under $\mathfrak{g}$, and hence proportional to Ψ (§7, no. 5, Prop. 12). This proves that $s \in \Sigma$.

Since the identity representation of $\mathfrak{g}$ is irreducible, its commutant reduces to the scalars (§6, no. 1, Prop. 1), so the kernel of φ is k^*. Thus, the group $\mathrm{Aut}(\mathfrak{g}) = \mathrm{Aut}_0(\mathfrak{g})$ can be identified with Σ/k^*. But, it follows from *Algebra*, Chap. IX, §6, no. 5, that the group Σ is the product of the groups k^* and $\mathbf{SO}(\Psi)$; hence $\mathrm{Aut}(\mathfrak{g}) = \mathrm{Aut}_0(\mathfrak{g})$ can be identified with $\mathbf{SO}(\Psi)$.

Let $\mathbf{O}_0^+(\Psi)$ be the reduced orthogonal group of Ψ (*Algebra*, Chap. IX, §9, no. 5). Since $\mathbf{SO}(\Psi)/\mathbf{O}_0^+(\Psi)$ is commutative (*loc. cit.*), the group $\mathrm{Aut}_e(\mathfrak{g})$ is contained in $\mathbf{O}_0^+(\Psi)$ (§11, no. 2, Prop. 3); in fact, it is equal to it (Exerc. 7).

(VIII) The canonical bilinear form Φ_R on $\mathfrak{h}^*$ is given by

$$\Phi_R(\xi_1\varepsilon_1 + \cdots + \xi_l\varepsilon_l, \xi_1'\varepsilon_1 + \cdots + \xi_l'\varepsilon_l) = \frac{1}{4l-2}(\xi_1\xi_1' + \cdots + \xi_l\xi_l')$$

(Chap. VI, §4, no. 5.V). The isomorphism from $\mathfrak{h}$ to $\mathfrak{h}^*$ defined by Φ_R takes H_i to $(4l-2).\varepsilon_i$. Thus, the inverse form of Φ_R, that is the restriction to $\mathfrak{h}$ of the Killing form, is

$$\Phi(\xi_1 H_1 + \cdots + \xi_l H_l, \xi_1' H_1 + \cdots + \xi_l' H_l) = (4l-2)(\xi_1\xi_1' + \cdots + \xi_l\xi_l').$$

(IX) Recall the X_α ($\alpha \in \mathrm{R}$) defined by the formulas (3). It is easy to verify that $[X_\alpha, X_{-\alpha}] = -H_\alpha$ for $\alpha \in \mathrm{R}$. On the other hand, let M be the matrix $I + E_{0,0}$; since $M = S^t M^{-1} S$, the map

$$\theta : g \mapsto -M^{-1t} g M$$

is an automorphism of $\mathfrak{g}$ and $\theta(X_\alpha) = X_{-\alpha}$ for all $\alpha \in \mathrm{R}$. Consequently, (X_α) *is a Chevalley system in* $(\mathfrak{g}, \mathfrak{h})$.

Assume that $k = \mathbf{Q}$. The Cartan subalgebra $\mathfrak{h}$ has two permissible lattices: the lattice $\mathrm{Q}(\mathrm{R}^\vee)$ generated by the H_α and the lattice $\mathrm{P}(\mathrm{R}^\vee)$ that is generated by the H_i and consists of the diagonal matrices in $\mathfrak{h}$ with integer entries. It follows that $\mathfrak{o}_S(2l+1, \mathbf{Z})$ (the set of matrices in $\mathfrak{g}$ with integer entries) is the Chevalley order $\mathrm{P}(\mathrm{R}^\vee) + \sum \mathbf{Z}.X_\alpha$ in $\mathfrak{g}$. Since $(X_{\pm\varepsilon_i})^2 = 2E_{\pm i, \mp i}$, $(X_{\pm\varepsilon_i})^3 = 0$ and $(X_{\pm\varepsilon_i \pm\varepsilon_j})^2 = 0$, we see that the lattice $\mathscr{V}$ generated by the Witt basis $(e_i)_{-l \le i \le l}$ is an admissible lattice for $\mathfrak{o}_S(2l+1, \mathbf{Z})$ in V. The same is true for $\bigwedge^r \mathscr{V}$ in $\bigwedge^r \mathrm{V}$.

Now consider the spinor representation ρ of $\mathfrak{g}$ on $\mathrm{N} = \bigwedge \mathrm{F}'$. As its weights do not map $\mathrm{P}(\mathrm{R}^\vee)$ to $\mathbf{Z}$, it has no admissible lattice for $\mathfrak{o}_S(2l+1, \mathbf{Z})$. On the other hand, the lattice $\mathscr{N}$ generated by the canonical basis $(e_{-i_1} \wedge \cdots \wedge e_{-i_k})$ of N (for $1 \le i_1 < \cdots < i_k \le l$) is an admissible lattice for the Chevalley order $\mathscr{G} = \mathrm{Q}(\mathrm{R}^\vee) + \sum_{\alpha \in \mathrm{R}} \mathbf{Z}.X_\alpha$. Indeed, it is immediate that $\mathscr{N}$ is stable under the exterior product with the e_{-i} and the interior product with the e_i (for $1 \le i \le l$). The formulas of (V) then show that $\mathscr{N}$ is stable under $\rho(\mathscr{G})$. Moreover, since $\rho(X_\alpha)^2 = 0$ for all $\alpha \in \mathrm{R}$, it follows that $\mathscr{N}$ is admissible.

3. ALGEBRAS OF TYPE C_l ($l \ge 1$)

(I) Let Ψ be a non-degenerate alternating bilinear form on a vector space V of finite dimension $2l \ge 2$; the set of endomorphisms x of V such that $\Psi(xv, v') + \Psi(v, xv') = 0$ for all $v, v' \in \mathrm{V}$ is a semi-simple Lie subalgebra of $\mathfrak{sl}(\mathrm{V})$ (Chap. I, §6, no. 7, Prop. 9). *We denote it by* $\mathfrak{sp}(\Psi)$ *and call it the symplectic Lie algebra associated to* Ψ.

By *Algebra*, Cap. IX, §4, no. 2, V can be written as the direct sum of two maximal totally isotropic subspaces F and F′, which are in duality relative to Ψ. Let $(e_i)_{1 \le i \le l}$ be a basis of F, and $(e_{-i})_{1 \le i \le l}$ the dual basis of F′. Then

$$(e_1, \ldots, e_l, e_{-l}, \ldots, e_{-1})$$

is a basis of V; we say that it is a *Witt basis* (or symplectic basis) of V. The matrix of Ψ with respect to this basis is the square matrix of order $2l$

$$J = \begin{pmatrix} 0 & s \\ -s & 0 \end{pmatrix}$$

where s is the square matrix of order l all of whose entries are zero except those on the second diagonal which are equal to 1, cf. no. 2.I.

The algebra $\mathfrak{g} = \mathfrak{sp}(\Psi)$ can be identified with the algebra $\mathfrak{sp}(2l, k)$ of square matrices a of order $2l$ such that $a = -J^{-1t}aJ = J^t aJ$ (*Algebra*, Chap. IX, §1, no. 10, formulas (50)), that is of the form

$$a = \begin{pmatrix} A & B \\ C & -s^t As \end{pmatrix}$$

where A, B, C are square matrices of order l such that $B = s^t Bs$ and $C = c^t Cs$; in other words, B and C are symmetric with respect to the second diagonal. It follows that

$$\dim \mathfrak{g} = l^2 + 2\frac{l(l+1)}{2} = l(2l+1).$$

Let $\mathfrak{h}$ be the set of diagonal matrices in $\mathfrak{g}$. This is a commutative subalgebra of $\mathfrak{g}$, with basis the elements $H_i = E_{i,i} - E_{-i,-i}$ for $1 \leq i \leq l$. Let $(\varepsilon_i)_{1 \leq i \leq l}$ be the dual basis of (H_i). For $1 \leq i < j \leq l$, put

$$\begin{cases} X_{2\varepsilon_i} & = E_{i,-i} \\ X_{-2\varepsilon_i} & = -E_{-i,i} \\ X_{\varepsilon_i - \varepsilon_j} & = E_{i,j} - E_{-j,-i} \\ X_{-\varepsilon_i + \varepsilon_j} & = -E_{j,i} + E_{-i,-j} \\ X_{\varepsilon_i + \varepsilon_j} & = E_{i,-j} + E_{j,-i} \\ X_{-\varepsilon_i - \varepsilon_j} & = -E_{-i,j} - E_{-j,i}. \end{cases} \tag{6}$$

It is easily verified that these elements form a basis of a complement of $\mathfrak{h}$ in $\mathfrak{g}$ and that, for $h \in \mathfrak{h}$,

$$[h, X_\alpha] = \alpha(h)X_\alpha \tag{7}$$

for all $\alpha \in \mathrm{R}$, where R is the set of the $\pm 2\varepsilon_i$ and the $\pm\varepsilon_i \pm \varepsilon_j$ $(i < j)$. It follows that $\mathfrak{h}$ is equal to its own normalizer in $\mathfrak{g}$, and hence is a Cartan subalgebra of $\mathfrak{g}$, that $\mathfrak{h}$ is splitting, and that the roots of $(\mathfrak{g}, \mathfrak{h})$ are the elements of R. The root system R of $(\mathfrak{g}, \mathfrak{h})$ is of type C_l for $l \geq 2$, and of type A_1 (in other words of type C_1) for $l = 1$ (Chap. VI, §4, no. 6.I extended to the case $l = 1$). Consequently, $\mathfrak{g}$ *is a splittable simple Lie algebra of type* C_l.

Every splitting Cartan subalgebra of $\mathfrak{g}$ is transformed into $\mathfrak{h}$ by an elementary automorphism, hence by an element of the symplectic group $\mathbf{Sp}(\Psi)$ (cf. (VII)), and consequently is the set $\mathfrak{h}_\beta$ of elements of $\mathfrak{g}$ whose matrix with respect to a Witt basis β of $\mathfrak{g}$ is diagonal. It is immediately verified that the only vector subspaces of V stable under $\mathfrak{h}_\beta$ are those generated by a subset of β.

We have $\mathfrak{sp}(2, k) = \mathfrak{sl}(2, k)$. On the other hand, the algebras $\mathfrak{sp}(4, k)$ and $\mathfrak{o}_S(5, k)$ have the same root system, and hence are isomorphic (cf. Exerc. 3). *From now on, we assume that* $l \geq 2$.

(II) The root system $\mathrm{R}^\vee$ is determined by means of Chap. VI, §4, no. 6.I and 6.V; we find that

$$H_{2\varepsilon_i} = H_i, \quad H_{\varepsilon_i - \varepsilon_j} = H_i - H_j, \quad H_{\varepsilon_i + \varepsilon_j} = H_i + H_j.$$

(III) Put $\alpha_1 = \varepsilon_1 - \varepsilon_2, \ldots, \alpha_{l-1} = \varepsilon_{l-1} - \varepsilon_l, \alpha_l = 2\varepsilon_l$. By Chap. VI, §4, no. 6.II, $\{\alpha_1, \ldots, \alpha_l\}$ is a basis B of R; the positive roots relative to B are the $2\varepsilon_i$ and the $\varepsilon_i \pm \varepsilon_j$ $(i < j)$. The corresponding Borel subalgebra $\mathfrak{b}$ is the set of upper triangular matrices in $\mathfrak{g}$.

Let δ be an isotropic flag in V (that is, whose elements are all totally isotropic subspaces for Ψ), and let $\mathfrak{p}_\delta$ be the subalgebra consisting of the elements of $\mathfrak{g}$ leaving stable the elements of δ. We show as in no. 2.III that the map $\delta \mapsto \mathfrak{p}_\delta$ is a bijection from the set of isotropic flags (resp. the maximum isotropic flags) to the set of parabolic (resp. Borel) subalgebras of $\mathfrak{g}$; we have $\mathfrak{p}_\delta \supset \mathfrak{p}_{\delta'}$ if and only if $\delta \subset \delta'$.

(IV) The fundamental weights corresponding to $\alpha_1, \ldots, \alpha_l$ are, by Chap. VI, §4, no. 6.VI, the $\varpi_i = \varepsilon_1 + \cdots + \varepsilon_i$ $(1 \leq i \leq l)$.

We are going to show how the fundamental representation σ_r of weight ϖ_r can be realised as a subrepresentation of $\bigwedge^r \sigma$, where σ is the identity representation of $\mathfrak{g}$ on V, and for this we shall study the decomposition of the representation $\bigwedge \sigma$ of $\mathfrak{g}$ on the exterior algebra $\bigwedge$ V.

Let (e_i^*) be the basis of V* dual to (e_i). The alternating bilinear form Ψ can be identified with an element $\Gamma^* \in \bigwedge^2 \mathrm{V}^*$ (*Algebra*, Chap. III, §7, no. 4, Prop. 7 and §11, no. 10) and it is easy to verify that

$$\Gamma^* = -\sum_{i=1}^{l} e_i^* \wedge e_{-i}^*.$$

Let Ψ^* be the inverse form of Ψ (*Algebra*, Chap. IX, §1, no. 7); it is immediate that

$$\Psi^*(e_i^*, e_j^*) = 0$$

for $i \neq -j$ and $\Psi^*(e_i^*, e_{-i}^*) = -1$ for $1 \leq i \leq l$. If we identify Ψ^* with an element $\Gamma \in \bigwedge^2 \mathrm{V}$, then

$$\Gamma = \sum_{i=1}^{l} e_i \wedge e_{-i}.$$

Denote by X_- the endomorphism of $\bigwedge$ V given by the left exterior product with Γ and by X_+ the endomorphism of $\bigwedge$ V given by the left interior product with $-\Gamma^*$:

$$X_- u = \left(\sum_{i=1}^{l} e_i \wedge e_{-i}\right) \wedge u,$$

$$X_+ u = \left(\sum_{i=1}^{l} e_i^* \wedge e_{-i}^*\right) \wedge u.$$

To calculate X_+ and X_-, introduce a basis of $\bigwedge V$ in the following way: for any triplet (A, B, C) formed by three *disjoint* subsets of $[1, l]$, put

$$e_{A,B,C} = e_{a_1} \wedge \cdots \wedge e_{a_m} \wedge e_{-b_1} \wedge \cdots \wedge e_{-b_n} \wedge e_{c_1} \wedge e_{-c_1} \wedge \cdots \wedge e_{c_p} \wedge e_{-c_p}$$

where $(a_1, \ldots, a_m)$ (resp. $(b_1, \ldots, b_n), (c_1, \ldots, c_p)$) are the elements of A (resp. B, C) arranged in increasing order. We obtain in this way a basis of $\bigwedge V$ and simple calculations show that

$$X_-.e_{A,B,C} = \sum_{j \in [1,l], j \notin A \cup B \cup C} e_{A,B,C \cup \{j\}} \tag{8}$$

$$X_+.e_{A,B,C} = -\sum_{j \in C} e_{A,B,C-\{j\}}. \tag{9}$$

Let H be the endomorphism of $\bigwedge V$ that reduces to multiplication by $(l - r)$ on $\bigwedge^r V$ $(0 \leq r \leq 2l)$. It is painless to verify (cf. Exerc. 19) that

$$[X_+, X_-] = -H$$
$$[H, X_+] = 2X_+$$
$$[H, X_-] = -2X_-.$$

In other words, the vector subspace $\mathfrak{s}$ generated by X_+, X_- and H is a Lie subalgebra of $\mathrm{End}(\bigwedge V)$, isomorphic to $\mathfrak{sl}(2, k)$, and $\bigwedge^r V$ is the subspace of elements of weight $l - r$. Denote by E_r the subspace of $\bigwedge^r V$ consisting of the primitive elements, that is, $E_r = (\bigwedge^r V) \cap \mathrm{Ker} X_+$. It follows from §1 that, for $r < l$, the restriction of X_- to $\bigwedge^r V$ is injective and that, for $r \leq l$, $\bigwedge^r V$ decomposes as a direct sum

$$\bigwedge^r V = E_r \oplus X_-(E_{r-2}) \oplus X_-^2(E_{r-4}) \oplus \cdots$$
$$= E_r \oplus X_-(\bigwedge^{r-2} V).$$

This shows in particular that $\dim E_r = \binom{2l}{r} - \binom{2l}{r-2}$ for $0 \leq r \leq l$.

On the other hand, the very definition of $\mathfrak{sp}(\Psi)$ shows that Γ^* is annihilated by the second exterior power of the dual of σ. Similarly, Γ is annihilated by $\bigwedge^2 \sigma$. We deduce immediately that X_+ and X_-, and hence also H, commute with the endomorphisms $\bigwedge \sigma(g)$ for $g \in \mathfrak{g}$. Consequently, the subspaces E_r for $0 \leq r \leq l$ are stable under $\bigwedge^r \sigma$; we shall show that *the restriction of* $\bigwedge^r \sigma$ *to* E_r *is a fundamental representation* σ_r *of weight* ϖ_r $(1 \leq r \leq l)$.

We remark first of all that the weights of $\bigwedge^r \sigma$ relative to $\mathfrak{h}$ are the

$$\varepsilon_{i_1} + \cdots + \varepsilon_{i_k} - (\varepsilon_{j_1} + \cdots + \varepsilon_{j_{r-k}}),$$

where $i_1, \ldots, i_k$ (resp. $j_1, \ldots, j_{r-k}$) are distinct elements of $[1, l]$; thus, the highest weight of $\bigwedge^r \sigma$ is indeed

$$\varpi_r = \varepsilon_1 + \cdots + \varepsilon_r$$

and the vectors of weight ϖ_r are those proportional to $e_1 \wedge \cdots \wedge e_r = e_{\{1,\ldots,r\},\varnothing,\varnothing}$. Formula (9) shows that $e_1 \wedge \cdots \wedge e_r \in \mathrm{E}_r$. Thus, it suffices to prove that the restriction of $\bigwedge^r \sigma$ to E_r is irreducible.

If $s \in \mathbf{Sp}(\Psi)$, the extension of s to $\bigwedge \mathrm{V}$ (resp. $\bigwedge \mathrm{V}^*$) fixes Γ (resp. Γ^*), and hence commutes with X_+ and X_- and leaves E_r stable. Consequently, E_r contains the vector subspace F_r generated by the transforms of $e_1 \wedge \cdots \wedge e_r$ by $\mathbf{Sp}(\Psi)$. The theorem of Witt shows that these are the non-zero decomposable r-vectors such that corresponding vector subspace of V is a totally isotropic subspace, r-vectors that we shall call isotropic.

Lemma 2. For $1 \leq r \leq l$, let F_r be the subspace of $\bigwedge^r \mathrm{V}$ generated by the isotropic r-vectors. Then

$$\bigwedge^r \mathrm{V} = \mathrm{F}_r + X_-(\bigwedge^{r-2} \mathrm{V}) = \mathrm{F}_r + \left(\sum_{i=1}^{l} e_i \wedge e_{-i}\right) \wedge \bigwedge^{r-2} \mathrm{V}.$$

We show first of all how Lemma 2 implies our assertion. Since $\mathrm{F}_r \subset \mathrm{E}_r$ and $\mathrm{E}_r \cap X_-(\bigwedge^{r-2} \mathrm{V}) = \{0\}$, the lemma implies that $\mathrm{F}_r = \mathrm{E}_r$. On the other hand, let $s \in \mathbf{Sp}(\Psi)$; the automorphism $a \mapsto sas^{-1}$ of $\mathrm{End}(\mathrm{V})$ preserves $\mathfrak{g}$ and induces on it an element of $\mathrm{Aut}_0(\mathfrak{g})$ (cf. (VII)), and hence transforms every irreducible representation of $\mathfrak{g}$ into an equivalent representation (§7, no. 1, Prop. 2). Since $e_1 \wedge \cdots \wedge e_r$ belongs to an irreducible component of $\bigwedge^r \sigma$, and since $\mathrm{E}_r = \mathrm{F}_r$ is generated by the transforms of $e_1 \wedge \cdots \wedge e_r$ by $\mathbf{Sp}(\Psi)$, it follows that the representation of $\mathfrak{g}$ on E_r is isotypical. But the multiplicity of its highest weight ϖ_r is 1, so it is irreducible.

It remains to prove the lemma. It is clear for $r = 1$. We argue by induction, and assume that $r \geq 2$. By the induction hypothesis, we are reduced to proving that

$$\mathrm{F}_{r-1} \wedge \mathrm{V} \subset \mathrm{F}_r + \Gamma \wedge \bigwedge^{r-2} \mathrm{V},$$

or that, if y is a decomposable $(r-1)$-vector and $x \in \mathrm{V}$, then

$$z = y \wedge x \in \mathrm{F}_r + \Gamma \wedge \bigwedge^{r-2} \mathrm{V}.$$

Let $(f_i)_{1 \leq \pm i \leq l}$ be a Witt basis of V such that $y = f_1 \wedge \cdots \wedge f_{r-1}$. It suffices to carry out the proof when $x = f_i$. If $i \notin [\![1-r, -1]\!]$, the r-vector $f_1 \wedge \cdots \wedge f_{r-1} \wedge f_i$ is isotropic. Otherwise, we can assume, renumbering the f_i if necessary, that $i = 1 - r$. Then $\Gamma = \sum_{j=1}^{l} f_j \wedge f_{-j}$, so

$$f_{r-1} \wedge f_{1-r} = \frac{1}{l-r+2}\left(\Gamma - \sum_{i=1}^{r-2} f_i \wedge f_{-i} + \sum_{j=r}^{l} (f_{r-1} \wedge f_{1-r} - f_j \wedge f_{-j})\right),$$

$$z = \frac{1}{l-r+2}\Gamma \wedge f_1 \wedge \cdots \wedge f_{r-2}$$

$$+ \frac{1}{l-r+2}\sum_{j=r}^{l} (f_1 \wedge \cdots \wedge f_{r-2}) \wedge (f_{r-1} \wedge f_{1-r} - f_j \wedge f_{-j}).$$

But

$$f_{r-1} \wedge f_{1-r} - f_j \wedge f_{-j} = (f_{r-1} + f_j) \wedge (f_{1-r} - f_{-j}) - f_j \wedge f_{1-r} + f_{r-1} \wedge f_{-j}$$

and we verify immediately that the r-vectors

$$f_1 \wedge \cdots \wedge f_{r-2} \wedge (f_{r-1} + f_j) \wedge (f_{1-r} - f_{-j}),$$
$$f_1 \wedge \cdots \wedge f_{r-2} \wedge f_j \wedge f_{1-r} \quad \text{and} \quad f_1 \wedge \cdots \wedge f_{r-1} \wedge f_{-j}$$

are isotropic for $r \leq j \leq l$. Consequently, $z \in \mathrm{F}_r + \Gamma \wedge \bigwedge^{r-2} \mathrm{V}$, which completes the proof.

(V) We have $w_0 = -1$, so every finite dimensional simple representation of $\mathfrak{g}$ is orthogonal or symplectic. By Chap. VI, §4, no. 6.VI, the sum of the coordinates of ϖ_r with respect to $(\alpha_1, \ldots, \alpha_l)$ is

$$1 + 2 + \cdots + (r-1) + r + r + \cdots + r + \frac{r}{2}$$

so σ_r is orthogonal for r even and symplectic for r odd.

Since $e_1 \wedge \cdots \wedge e_r$ and $e_{-1} \wedge \cdots \wedge e_{-r}$ belong to E_r and since

$$\Psi_{(r)}(e_1 \wedge \cdots \wedge e_r\,,\, e_{-1} \wedge \cdots \wedge e_{-r}) = 1,$$

we see that the restriction of $\Psi_{(r)}$ to E_r is non-zero: this is, up to a constant factor, the bilinear form; it is symmetric if r is even, alternating if r is odd, and invariant under σ_r.

(VI) For all $x \in \mathfrak{g}$, the characteristic polynomial of $\sigma(x)$ takes the form

$$\mathrm{T}^{2l} + f_1(x)\mathrm{T}^{2l-1} + \cdots + f_{2l}(x)$$

where $f_1, \ldots, f_{2l}$ are invariant polynomial functions on $\mathfrak{g}$.

If $x = \xi_1 H_1 + \cdots + \xi_l H_l \in \mathfrak{h}$, the $f_i(x)$ are, up to sign, the elementary symmetric functions of $\xi_1, \ldots, \xi_l, -\xi_1, \ldots, -\xi_l$; these symmetric functions are zero in odd degrees, and

$$\mathrm{T}^{2l} + f_2(x)\mathrm{T}^{2l-2} + \cdots + f_{2l}(x) = (\mathrm{T}^2 - \xi_1^2) \ldots (\mathrm{T}^2 - \xi_l^2).$$

As in no. 2.VI, it follows that $f_1 = f_3 = f_5 = \cdots = 0$, and that

$$(f_2, f_4, \ldots, f_{2l})$$

is an algebraically free family generating the algebra of invariant polynomial functions on $\mathfrak{g}$.

(VII) Since the only automorphism of the Dynkin graph is the identity, we have $\mathrm{Aut}(\mathfrak{g}) = \mathrm{Aut}_0(\mathfrak{g})$.

Let Σ be the group of similarities of V relative to Ψ (*Algebra*, Chap. IX, §6, end of no. 5). One proves as in no. 2.VII that the automorphisms of $\mathfrak{g}$

are the maps $x \mapsto sxs^{-1}$ where $s \in \Sigma$, so that $\mathrm{Aut}(\mathfrak{g}) = \mathrm{Aut}_0(\mathfrak{g})$ can be identified with Σ/k^*.

For all $s \in \Sigma$, let $\mu(s)$ be the multiplier of s. The map $s \mapsto \mu(s) \bmod k^{*2}$ from Σ to k^*/k^{*2} is a homomorphism whose kernel contains $k^*.1$, and hence gives a homomorphism λ from Σ/k^* to k^*/k^{*2}. We have $\mathbf{Sp}(\Psi) \cap k^* = \{1, -1\}$. Consider the sequence of homomorphisms

$$1 \longrightarrow \mathbf{Sp}(\Psi)/\{1,-1\} \xrightarrow{\iota} \Sigma/k^* \xrightarrow{\lambda} k^*/k^{*2} \longrightarrow 1. \tag{10}$$

The map ι is injective, and $\mathrm{Im}(\iota) \subset \mathrm{Ker}\lambda$ since the multiplier of an element of $\mathbf{Sp}(\Psi)$ is 1. If the multiplier of $s \in \Sigma$ is an element of k^{*2}, there exists $\nu \in k^*$ such that $\nu s \in \mathbf{Sp}(\Psi)$; thus, $\mathrm{Im}(\iota) = \mathrm{Ker}(\lambda)$. In summary, the sequence (10) is exact. We identify $\mathbf{Sp}(\Psi)/\{1,-1\}$ with a subgroup of Σ/k^*. Since k^*/k^{*2} is commutative, $\mathbf{Sp}(\Psi)/\{1,-1\}$ contains the derived group of Σ/k^*. Thus, $\mathrm{Aut}_e(\mathfrak{g})$ is contained in $\mathbf{Sp}(\Psi)/\{1,-1\}$ (§11, no. 2, Prop. 3). In fact, it is equal to it, and $\mathrm{Aut}(\mathfrak{g})/\mathrm{Aut}_e(\mathfrak{g})$ is identified with k^*/k^{*2} (Exerc. 9).

(VIII) The canonical bilinear form Φ_{R} on $\mathfrak{h}^*$ is given by

$$\Phi_{\mathrm{R}}(\xi_1\varepsilon_1 + \cdots + \xi_l\varepsilon_l, \xi'_1\varepsilon_1 + \cdots + \xi'_l\varepsilon_l) = \frac{1}{4(l+1)}(\xi_1\xi'_1 + \cdots + \xi_l\xi'_l)$$

(Chap. VI, §4, no. 6.V). Thus, the inverse form of Φ_{R}, that is, the restriction to $\mathfrak{h}$ of the Killing form, is

$$\Phi(\xi_1 H_1 + \cdots + \xi_l H_l, \xi'_1 H_1 + \cdots + \xi'_l H_l) = 4(l+1)(\xi_1\xi'_1 + \cdots + \xi_l\xi'_l).$$

(IX) Recall the X_α defined by formulas (6) ($\alpha \in \mathrm{R}$). It is easily verified that $[X_\alpha, X_{-\alpha}] = -H_\alpha$ for $\alpha \in \mathrm{R}$. On the other hand, the map $\theta : a \mapsto -{}^t a$ is an automorphism of $\mathfrak{g}$ and $\theta(X_\alpha) = X_{-\alpha}$ for all $\alpha \in \mathrm{R}$. Consequently, $(X_\alpha)_{\alpha \in \mathrm{R}}$ is a Chevalley system in $(\mathfrak{g}, \mathfrak{h})$.

Assume that $k = \mathbf{Q}$. The Cartan subalgebra $\mathfrak{h}$ has two permissible lattices $\mathrm{Q}(\mathrm{R}^\vee) = \sum_{i=1}^{l} \mathbf{Z}.H_i$ and $\mathrm{P}(\mathrm{R}^\vee) = \mathrm{Q}(\mathrm{R}^\vee) + \frac{1}{2}\mathbf{Z}.\sum_{i=1}^{l} H_i$ (Chap. VI, §4, no. 5.VIII). We see that $\mathrm{Q}(\mathrm{R}^\vee)$ is the set of matrices with integer entries belonging to $\mathfrak{h}$. It follows that the Chevalley order $\mathrm{Q}(\mathrm{R}^\vee) + \sum_{\alpha \in \mathrm{R}} \mathbf{Z}.X_\alpha$ is the set $\mathfrak{sp}(2l, \mathbf{Z})$ of matrices in $\mathfrak{g}$ with integer entries.

Consider the reductive Lie algebra $\mathfrak{sp}(\Psi) + \mathbf{Q}.1$. It is easy to see that the set of its elements with integer entries is a Chevalley order, whose projection onto $\mathfrak{sp}(\Psi)$ parallel to $\mathbf{Q}.1$ is the Chevalley order $\mathrm{P}(\mathrm{R}^\vee) + \sum \mathbf{Z}.X_\alpha$.

Finally, $X_\alpha^2 = 0$ for all $\alpha \in \mathrm{R}$. It follows that the lattice $\mathscr{V}$ in V generated by the e_i is admissible for the Chevalley order $\mathfrak{sp}(2l, \mathbf{Z})$. The same holds for the lattice $\mathrm{E}_r \cap \bigwedge^r \mathscr{V}$ in E_r.

Finally, E_r has an admissible lattice for the Chevalley order

$$\mathrm{P}(\mathrm{R}^\vee) + \sum \mathbf{Z}.X_\alpha$$

only if r is even; then $\mathrm{E}_r \cap \bigwedge^r \mathscr{V}$ is such a lattice.

4. ALGEBRAS OF TYPE D_l ($l \geq 2$)

(I) Let V be a vector space of even dimension $2l \geq 4$ and let Ψ be a non-degenerate symmetric bilinear form of maximum index l on V. By *Algebra*, Chap. IX, §4, no. 2, V can be written as the direct sum of two maximal totally isotropic subspaces F and F′. Let $(e_i)_{1\leq i\leq l}$ be a basis of F and $(e_{-i})_{1\leq i\leq l}$ the dual basis of F′ (for the duality between F and F′ defined by Ψ). Then $e_1, \ldots, e_l, e_{-l}, \ldots, e_{-1}$ is a basis of V; we shall call it a *Witt basis* of V. The matrix of Ψ with respect to this basis is the square matrix S of order $2l$ all of whose entries are zero, except those situated on the second diagonal, which are equal to 1. The algebra $\mathfrak{g} = \mathfrak{o}(\Psi)$ can be identified with the algebra $\mathfrak{o}_S(2l, k)$ of square matrices g of order $2l$ such that $g = -S^t gS$. It has dimension $l(2l-1)$. An easy calculation shows that $\mathfrak{g}$ is the set of matrices of the form

$$\begin{pmatrix} A & B \\ C & D \end{pmatrix}$$

where A, B, C, D are square matrices of order l such that $B = -s^t Bs, C = -s^t Cs$ and $D = -s^t As$ (s is the matrix of order l all of whose entries are zero except those situated on the second diagonal which are equal to 1).

Let $\mathfrak{h}$ be the set of diagonal matrices belonging to $\mathfrak{g}$. This is a commutative subalgebra of $\mathfrak{g}$, with basis formed by the elements $H_i = E_{i,i} - E_{-i,-i}$ for $1 \leq i \leq l$. Let (ε_i) be the basis of $\mathfrak{h}^*$ dual to (H_i). Put, for $1 \leq i < j \leq l$,

$$\begin{cases} X_{\varepsilon_i-\varepsilon_j} &= E_{i,j} - E_{-j,-i} \\ X_{-\varepsilon_i+\varepsilon_j} &= -E_{j,i} + E_{-i,-j} \\ X_{\varepsilon_i+\varepsilon_j} &= E_{i,-j} - E_{j,-i} \\ X_{-\varepsilon_i-\varepsilon_j} &= -E_{-j,i} + E_{-i,j}. \end{cases} \tag{11}$$

These elements form a basis of a complement of $\mathfrak{h}$ in $\mathfrak{g}$. For $h \in \mathfrak{h}$,

$$[h, X_\alpha] = \alpha(h) X_\alpha$$

for all $\alpha \in$ R, where R is the set of the $\pm\varepsilon_i \pm \varepsilon_j$ ($i < j$). Thus, $\mathfrak{h}$ is a splitting Cartan subalgebra of $\mathfrak{g}$, and the roots of $(\mathfrak{g}, \mathfrak{h})$ are the elements of R. The root system R of $(\mathfrak{g}, \mathfrak{h})$ is thus of type D_l for $l \geq 3$, of type $A_1 \times A_1$ (in other words of type D_2) for $l = 2$ (Chap. VI, §4, no. 8.I extended to the case $l = 2$). Consequently, $\mathfrak{g}$ *is a splittable simple Lie algebra of type* D_l *if* $l \geq 3$.

Every splitting Cartan subalgebra of $\mathfrak{g}$ is transformed into $\mathfrak{h}$ by an elementary automorphism of $\mathfrak{g}$, and hence by an element of $\mathbf{O}(\Psi)$ (cf. (VII)) and consequently is the set $\mathfrak{h}_\beta$ of elements of $\mathfrak{g}$ whose matrix with respect to a Witt basis β of V is diagonal. We verify immediately that the only subspaces invariant under $\mathfrak{h}_\beta$ are those generated by a subset of β.

Since the algebras $\mathfrak{o}_S(4, k)$ and $\mathfrak{sl}(2, k) \times \mathfrak{sl}(2, k)$ have the same root systems, they are isomorphic. Similarly, $\mathfrak{o}_S(6, k)$ and $\mathfrak{sl}(4, k)$ are isomorphic (cf. also Exerc. 3). *From now on, we assume that* $l \geq 3$.

(II) We determine $R^\vee$ by means of Chap. VI, §4, no. 8.V. We find that

$$H_{\varepsilon_i-\varepsilon_j} = H_i - H_j, \quad H_{\varepsilon_i+\varepsilon_j} = H_i + H_j.$$

(III) Put $\alpha_1 = \varepsilon_1 - \varepsilon_2, \alpha_2 = \varepsilon_2 - \varepsilon_3, \dots, \alpha_{l-1} = \varepsilon_{l-1} - \varepsilon_l, \alpha_l = \varepsilon_{l-1} + \varepsilon_l$. By Chap. VI, §4, no. 8.II, $(\alpha_1, \dots, \alpha_l)$ is a basis B of R; the positive roots relative to B are the $\varepsilon_i \pm \varepsilon_j$ $(i < j)$. The corresponding Borel subalgebra $\mathfrak{b}$ is the set of upper triangular matrices belonging to $\mathfrak{g}$.

It is easily verified that the only non-trivial vector subspaces invariant under $\mathfrak{b}$ are the totally isotropic subspaces $V_1, \dots, V_l, V'_l$, where V_i is generated by $e_1, \dots, e_i$ and V'_l by $e_1, \dots, e_{l-1}, e_{-l}$, and the orthogonal complements $V_{-1}, \dots, V_{-l+1}$ of $V_1, \dots, V_{l-1}$; the orthogonal complement V_{-i} of V_i is generated by $e_1, \dots, e_l, e_{-l}, \dots, e_{-(i+1)}$. But an immediate calculation shows that, if an element $a \in \mathfrak{g}$ leaves V_{l-1} stable, its matrix is of the form

$$\begin{pmatrix} A & x & B \\ 0 & \begin{pmatrix} \lambda & 0 \\ 0 & -\lambda \end{pmatrix} & y \\ 0 & 0 & D \end{pmatrix}$$

where A, B, D are square matrices of order $l-1$, x (resp. y) is a matrix with 2 columns and $l-1$ rows (resp. 2 rows and $l-1$ columns), and $\lambda \in k$. It follows that a leaves V_l and V'_l stable. Consequently, $\mathfrak{b}$ is the set of $a \in \mathfrak{g}$ leaving all the elements of the isotropic flag $(V_1, \dots, V_{l-1})$ stable. Note that the preceding and Witt's theorem (*Algebra*, Chap. IX, §4, no. 3, Th. 1) imply that V_l and V'_l are the only maximal totally isotropic subspaces containing V_{l-1}.

We say that an isotropic flag is *quasi-maximal* if it is composed of $l-1$ totally isotropic subspaces of dimensions $1, \dots, l-1$. We then see as in no. 2 that, for any quasi-maximal isotropic flag δ, the set $\mathfrak{b}_\delta$ of $a \in \mathfrak{g}$ leaving the elements of δ stable is a Borel subalgebra of $\mathfrak{g}$ and that the map $\delta \mapsto \mathfrak{b}_\delta$ is a bijection from the set of quasi-maximal isotropic flags to the set of Borel subalgebras.

We say that an isotropic flag is proper if it does not contain both a subspace of dimension l and a subspace of dimension $l-1$. Let δ be such an isotropic flag and let $\mathfrak{p}_\delta$ be the set of $a \in \mathfrak{g}$ leaving stable the elements of δ. If $\delta \subset \{V_1, \dots, V_l, V'_l\}$, then $\mathfrak{p}_\delta$ is a parabolic subalgebra of $\mathfrak{g}$, containing $\mathfrak{b}$, and it is easy to verify that the only totally isotropic subspaces $\neq \{0\}$ stable under $\mathfrak{p}_\delta$ are the elements of δ. Since there are 2^{l-2} proper isotropic flags contained in $\{V_1, \dots, V_l, V'_l\}$ and containing V_{l-1} (resp. V_l, resp. V'_l, resp. containing neither V_{l-1}, nor V_l, nor V'_l), this gives 2^l parabolic subalgebras containing $\mathfrak{b}$. It follows as above that the map $\delta \mapsto \mathfrak{p}_\delta$ is a bijection from the set of proper isotropic flags to the set of parabolic subalgebras of $\mathfrak{g}$.

(IV) The fundamental weights corresponding to $\alpha_1, \dots, \alpha_l$ are, by Chap. VI, §4, no. 8.VI,

$$\varpi_i = \varepsilon_1 + \varepsilon_2 + \cdots + \varepsilon_i \qquad (1 \leq i \leq l-2)$$
$$\varpi_{l-1} = \frac{1}{2}(\varepsilon_1 + \varepsilon_2 + \ldots + \varepsilon_{l-2} + \varepsilon_{l-1} - \varepsilon_l)$$
$$\varpi_l = \frac{1}{2}(\varepsilon_1 + \varepsilon_2 + \ldots + \varepsilon_{l-2} + \varepsilon_{l-1} + \varepsilon_l).$$

Let σ be the identity representation of $\mathfrak{g}$ on V. The exterior power $\bigwedge^r \sigma$ operates on $\mathrm{E} = \bigwedge^r(\mathrm{V})$. If $h \in \mathfrak{h}$, we have

$$\sigma(h)e_i = \varepsilon_i(h)e_i, \qquad \sigma(h)e_{-i} = -\varepsilon_i(h)e_{-i}$$

for $1 \leq i \leq l$. It follows that, for $1 \leq r \leq l$, $\varepsilon_1 + \cdots + \varepsilon_r$ is the highest weight of $\bigwedge^r \sigma$, the elements of weight $\varepsilon_1 + \cdots + \varepsilon_r$ being those proportional to $e_1 \wedge \cdots \wedge e_r$.

We shall show that, for $1 \leq r \leq l-2$, *the representation* $\bigwedge^r \sigma$ *is a fundamental representation of weight* ϖ_r.

For this, it suffices to show that $\bigwedge^r \sigma$ is irreducible for $1 \leq r \leq l-1$ (note that the representation $\bigwedge^l \sigma$ is not irreducible, cf. Exerc. 10), or that the smallest subspace T_r of $\bigwedge^r \mathrm{V}$ containing $e_1 \wedge \cdots \wedge e_r$ and stable under $\mathfrak{g}$ is the whole of $\bigwedge^r \mathrm{V}$. This is immediate for $r = 1$. For $r = 2$, we see as in no. 2 that $\bigwedge^2 \sigma$ is equivalent to the adjoint representation of $\mathfrak{g}$, which is irreducible since $\mathfrak{g}$ is simple. The proof is completed by arguing by induction on l, as in no. 2, but assuming that $l-1 \geq r \geq 3$.

We are now going to determine the fundamental representations of highest weight ϖ_{l-1} and ϖ_l. Let Q be the quadratic form $x \mapsto \frac{1}{2}\Psi(x,x)$. We have defined in no. 2.IV the spinor representation λ of the Clifford algebra C(Q) on $\mathrm{N} = \bigwedge \mathrm{F}'$. We verify immediately that the subspace N_+ (resp. N_-) of N given by the sum of the $\bigwedge^p \mathrm{F}'$ for p even (resp. odd) is stable under the restriction of λ to $\mathrm{C}^+(\mathrm{Q})$. Consequently, the representations λ_+ and λ_- of $\mathrm{C}^+(\mathrm{Q})$ on N_+ and N_- respectively are the semi-spinor representations of $\mathrm{C}^+(\mathrm{Q})$ (*Algebra*, Chap. IX, §9, no. 4); they are irreducible, of dimension 2^{l-1} and inequivalent. Let $\rho_+ = \lambda_+ \circ f$ and $\rho_- = \lambda_- \circ f$ be the corresponding irreducible representations of $\mathfrak{g}$ (no. 2, Lemma 1 (vi)). In view of Lemma 1 (i), we have

$$f(H_i) = \frac{1}{2}(e_i e_{-i} - e_{-i}e_i) = e_i e_{-i} - \frac{1}{2} = \frac{1}{2} - e_{-i}e_i$$

and we see, as in no. 2.IV, that, for $h \in \mathfrak{h}$ and $1 \leq i_1 < \cdots < i_k \leq l$,

$$\lambda \circ f(h)(e_{-i_1} \wedge \cdots \wedge e_{-i_k})$$
$$= (\frac{1}{2}(\varepsilon_1 + \cdots + \varepsilon_l) - (\varepsilon_{i_1} + \cdots + \varepsilon_{i_k}))(h)(e_{-i_1} \wedge \cdots \wedge e_{-i_k}).$$

Consequently, the highest weight of ρ_+ (resp. ρ_-) is ϖ_l (resp. ϖ_{l-1}).

We say that ρ_+ and ρ_- are the *semi-spinor representations* of $\mathfrak{g}$. All their weights are simple. We also say that $\rho = \lambda \circ f = \rho_+ \oplus \rho_-$ is the *spinor representation* of $\mathfrak{g}$.

(V) For $1 \leq r \leq l-2$, the fundamental representation $\bigwedge^r \sigma$ is orthogonal: it leaves invariant the extension of Ψ to $\bigwedge^r \mathrm{V}$.

Consider now the spinor representation ρ of $\mathfrak{g}$. We show as in no. 2 that, for $1 \leq i,j \leq l, i \neq j$,

$$f(X_{\varepsilon_i - \varepsilon_j}) = \pm e_i e_{-j}$$
$$f(X_{\varepsilon_i + \varepsilon_j}) = \pm e_i e_j$$
$$f(X_{-\varepsilon_i - \varepsilon_j}) = \pm e_{-i} e_{-j}.$$

It follows that the non-degenerate bilinear form Φ introduced in no. 2.V is invariant under $\rho(\mathfrak{g})$. Thus, the spinor representation ρ leaves invariant a non-degenerate form that is symmetric for $l \equiv 0, -1$ (mod. 4) and alternating for $l \equiv 1, 2$ (mod. 4).

If l is even, the restrictions of Φ to N_+ and N_- are non-degenerate and the semi-spinor representations are orthogonal for $l \equiv 0$ (mod. 4) and symplectic for $l \equiv 2$ (mod. 4). Moreover, we remark that $w_0 = -1$ (Chap. VI, §4, no. 8.XI).

On the other hand, if l is odd, N_+ and N_- are totally isotropic for Φ. Moreover, $-w_0(\alpha_i) = \alpha_i$ for $1 \leq i \leq l-2$, $-w_0(\alpha_l) = \alpha_{l-1}$ and $-w_0(\alpha_{l-1}) = \alpha_l$ (Chap. VI, §4, no. 8.XI), so $-w_0(\varpi_l) = \varpi_{l-1}$ and the semi-spinor representations are neither orthogonal nor symplectic; each of them is isomorphic to the dual of the other.

(VI) For all $x \in \mathfrak{g}$, the characteristic polynomial of $\sigma(x)$ takes the form

$$\mathrm{T}^{2l} + f_1(x)\mathrm{T}^{2l-1} + \cdots + f_{2l}(x).$$

We see as in no. 3 that $f_1 = f_3 = f_5 = \cdots = 0$. By Chap. VI, §4, no. 8.IX and §8, no. 3, Th. 1, there exists a polynomial function $\tilde{f}$ on $\mathfrak{g}$ such that $f_2, f_4, \ldots, f_{2l-2}, \tilde{f}$ generate the algebra $\mathrm{I}(\mathfrak{g}^*)$ of invariant polynomial functions on $\mathfrak{g}$, are algebraically independent, and further $\tilde{f}^2 = (-1)^l f_{2l}$.

For all $x \in \mathfrak{g}$, we have ${}^t(Sx) = {}^t x S = -Sx$, so we can consider $\mathrm{Pf}(Sx)$, which is a polynomial function of x. Now:

$$f_{2l}(x) = \det(x) = (-1)^l \det(Sx) = (-1)^l (\mathrm{Pf}(Sx))^2.$$

Thus, we can take $\tilde{f}(x) = \mathrm{Pf}(Sx)$.

(VII) Recall (§5, no. 3, Cor. 1 of Prop. 5) that $\mathrm{Aut}(\mathfrak{g})/\mathrm{Aut}_0(\mathfrak{g})$ can be identified with the group $\mathrm{Aut}(\mathrm{D})$ of automorphisms of the Dynkin graph D of $(\mathfrak{g}, \mathfrak{h})$. When $l \neq 4$, $\mathrm{Aut}(\mathrm{D})$ is the group of order 2 consisting of the permutations of $\alpha_1, \ldots, \alpha_l$ that leave $\alpha_1, \ldots, \alpha_{l-2}$ fixed. When $l = 4$, $\mathrm{Aut}(\mathrm{D})$ consists of the permutations of $\alpha_1, \ldots, \alpha_4$ that leave α_2 fixed; it is isomorphic to $\mathfrak{S}_3$ (cf. Chap. VI, §4, no. 8.XI). In all cases, the subgroup of $\mathrm{Aut}(\mathrm{D})$ consisting of the elements that leave α_1 fixed is of order 2. We denote by $\mathrm{Aut}'(\mathfrak{g})$ the corresponding subgroup of $\mathrm{Aut}(\mathfrak{g})$; we have $\mathrm{Aut}'(\mathfrak{g}) = \mathrm{Aut}(\mathfrak{g})$ if $l \neq 4$ and $(\mathrm{Aut}(\mathfrak{g}) : \mathrm{Aut}'(\mathfrak{g})) = 3$ if $l = 4$; moreover,

$$(\mathrm{Aut}'(\mathfrak{g}) : \mathrm{Aut}_0(\mathfrak{g})) = 2.$$

An element $s \in \mathrm{Aut}(\mathfrak{g})$ belongs to $\mathrm{Aut}'(\mathfrak{g})$ if and only if $\sigma \circ s$ is equivalent to σ (this follows from the fact that ϖ_1 is the highest weight of σ). We conclude as in no. 2.VII that $\mathrm{Aut}'(\mathfrak{g})$ can be identified with Σ/k^*, where Σ is the group of similarities of V relative to Ψ.

Let $s \in \Sigma$, and let $\lambda(s)$ be the multiplier of s. We have $\det(s) = \lambda(s)^l$ if s is *direct*, and $\det(s) = -\lambda(s)^l$ if s is *inverse* (*Algebra*, Chap. IX, §6, no. 5). The direct similarities form a subgroup Σ_0 of index 2 in Σ; we have $\Sigma_0 \supset k^*$. *The group Σ_0/k^* is equal to the subgroup* $\mathrm{Aut}_0(\mathfrak{g})$ *of* $\mathrm{Aut}'(\mathfrak{g}) = \Sigma/k^*$. Indeed, it suffices to verify this when k is algebraically closed: in that case $\mathrm{Aut}_0(\mathfrak{g}) = \mathrm{Aut}_e(\mathfrak{g})$ is equal to its derived group (§11, no. 2, Prop. 3), hence is contained in Σ_0/k^*, and since they are both of index 2 in Σ/k^*, they are equal.

On the other hand, as in no. 3.VII there is an exact sequence

$$1 \longrightarrow \mathbf{SO}(\Psi)/\{1,-1\} \longrightarrow \Sigma_0/k^* \longrightarrow k^*/k^{*2} \longrightarrow 1.$$

Identify $\mathbf{SO}(\Psi)/\{1,-1\}$ with a subgroup of $\Sigma_0/k^* = \mathrm{Aut}_0(\mathfrak{g})$. Since k^*/k^{*2} is commutative, we have $\mathrm{Aut}_e(\mathfrak{g}) \subset \mathbf{SO}(\Psi)/\{1,-1\}$. In fact, it can be shown (Exerc. 11) that $\mathrm{Aut}_e(\mathfrak{g})$ is equal to the image in $\mathbf{SO}(\Psi)/\{1,-1\}$ of the reduced orthogonal group $\mathbf{O}_0^+(\Psi)$ of Ψ (*Algebra*, Chap. IX, §9, no. 5).

(VIII) The canonical bilinear form Φ_{R} on $\mathfrak{h}^*$ is given by

$$\Phi_{\mathrm{R}}(\xi_1\varepsilon_1 + \cdots + \xi_l\varepsilon_l, \xi'_1\varepsilon_1 + \cdots + \xi'_l\varepsilon_l) = \frac{1}{4(l-1)}(\xi_1\xi'_1 + \cdots + \xi_l\xi'_l)$$

(Chap. VI, §4, no. 8.V). Thus, the restriction of the Killing form to $\mathfrak{h}$ is

$$\Phi(\xi_1 H_1 + \cdots + \xi_l H_l, \xi'_1 H_1 + \cdots + \xi'_l H_l) = 4(l-1)(\xi_1\xi'_1 + \cdots + \xi_l\xi'_l).$$

(IX) Recall the X_α ($\alpha \in \mathrm{R}$) defined by formulas (11). We verify easily that $[X_\alpha, X_{-\alpha}] = -H_\alpha$ for $\alpha \in \mathrm{R}$. On the other hand, the map $\theta : a \mapsto -{}^t a$ is an automorphism of $\mathfrak{g}$ and $\theta(X_\alpha) = X_{-\alpha}$ for all $\alpha \in \mathrm{R}$. Consequently $(X_\alpha)_{\alpha\in\mathrm{R}}$ is a Chevalley system in $(\mathfrak{g}, \mathfrak{h})$.

Assume that $k = \mathbf{Q}$. By Chap. VI, §4, no. 8.VIII, the subalgebra $\mathfrak{h}$ has three permissible lattices if l is odd and four permissible lattices if l is even. In particular, the lattice $\mathscr{H}$ generated by the H_i is permissible. But this lattice is the set of diagonal matrices in $\mathfrak{g}$ with integer entries. It follows that $\mathfrak{o}_S(2l, \mathbf{Z})$ is the Chevalley order $\mathscr{H} + \sum \mathbf{Z}.X_\alpha$ in $\mathfrak{g}$. Since $X_\alpha^2 = 0$ for all $\alpha \in \mathrm{R}$, we see that the lattice $\mathscr{V}$ in V generated by the Witt basis (e_i) is an admissible lattice in V for $\mathfrak{o}_S(2l, \mathbf{Z})$. The same holds for $\bigwedge^r \mathscr{V}$ in $\bigwedge^r \mathrm{V}$.

On the other hand, if we take $\mathrm{P}(\mathrm{R}^\vee) = \mathbf{Z}.\frac{1}{2}\sum_{i=1}^{l} H_i + \mathscr{H}$ as permissible lattice and $\mathscr{G} = \mathrm{P}(\mathrm{R}^\vee) + \sum \mathbf{Z}.X_\alpha$ as Chevalley order, we see that $\bigwedge^r \mathrm{V}$ has an admissible lattice only if r is even; $\bigwedge^r \mathscr{V}$ is then admissible.

Consider the reductive Lie algebra $\mathfrak{o}(\Psi) + \mathbf{Q}.1$; we see immediately that the lattice $\tilde{\mathscr{G}} = (\mathfrak{o}(\Psi) + \mathbf{Q}.1) \cap \mathfrak{gl}(2l, \mathbf{Z})$ is a Chevalley order. The Chevalley order $\mathscr{G}$ is the projection of $\tilde{\mathscr{G}}$ onto $\mathfrak{o}(\Psi)$ parallel to the centre $\mathbf{Q}.1$.

Finally, we see as in no. 2 that the lattice $\mathscr{N}_+$ (resp. $\mathscr{N}_-$) generated by the $e_{-i_1} \wedge \cdots \wedge e_{-i_{2k}}$ (resp. $e_{-i_1} \wedge \cdots \wedge e_{-i_{2k+1}}$) is admissible for the semi-spinor representation of the Chevalley order $\mathrm{Q}(\mathrm{R}^\vee) + \sum_{\alpha\in\mathrm{R}} \mathbf{Z}.X_\alpha$. On the other hand, $\mathscr{N}_+$ and $\mathscr{N}_-$ have no admissible lattice for $\mathfrak{o}_S(2l, \mathbf{Z})$.

TABLE 1

We associate to each fundamental weight the number 1 (resp. $-1, 0$) if the corresponding simple representation is orthogonal (resp. symplectic, resp. neither orthogonal nor symplectic). The calculation of this number has in essence been given in §13 for types A_l, B_l, C_l, D_l. The results are also indicated below for types E_6, E_7, E_8, F_4, G_2 (it suffices to apply §7, Prop. 12, and Chap. VI, §4, nos. 9.VI, 9.XI, 10.VI, 10.XI, 11.VI, 11.XI, 12.VI, 12.XI, 13.VI, 13.XI).

A_l $(l \geq 1)$

$$\varpi_r \begin{cases} 0 & r \neq \frac{l+1}{2} \\ (-1)^r & r = \frac{l+1}{2} \end{cases}$$

B_l $(l \geq 2)$

$$\begin{array}{lll} \varpi_r & 1 & r \neq l \\ \varpi_l & (-1)^{l(l+1)/2} & \end{array}$$

C_l $(l \geq 2)$

$$\varpi_r \quad (-1)^r$$

D_l $(l \geq 2)$

$$\begin{array}{ll} \varpi_r & 1 \quad r \neq l-1, l \\ \varpi_l \text{ and } \varpi_{l-1} & \begin{cases} 0 & \text{if } l \text{ is odd} \\ (-1)^{l/2} & \text{if } l \text{ is even} \end{cases} \end{array}$$

E_6		E_7		E_8		F_4		G_2	
ϖ_1	0	ϖ_1	1	ϖ_1	1	ϖ_1	1	ϖ_1	1
ϖ_2	1	ϖ_2	-1	ϖ_2	1	ϖ_2	1	ϖ_2	1
ϖ_3	0	ϖ_3	1	ϖ_3	1	ϖ_3	1		
ϖ_4	1	ϖ_4	1	ϖ_4	1	ϖ_4	1		
ϖ_5	0	ϖ_5	-1	ϖ_5	1				
ϖ_6	0	ϖ_6	1	ϖ_6	1				
		ϖ_7	-1	ϖ_7	1				
				ϖ_8	1				

TABLE 2

We associate to each fundamental weight the dimension of the corresponding simple representation, calculated by means of Th. 2 of §9.

A_l $(l \geq 1)$

ϖ_r $(1 \leq r \leq l)$	$\binom{l+1}{r}$

B_l $(l \geq 2)$

ϖ_r $(1 \leq r \leq l-1)$	$\binom{2l+1}{r}$
ϖ_l	2^l

C_l $(l \geq 2)$

ϖ_r $(1 \leq r \leq l)$	$\binom{2l}{r} - \binom{2l}{r-2}$

D_l $(l \geq 2)$

ϖ_r $(1 \leq r \leq l-2)$	$\binom{2l}{r}$
ϖ_{l-1}	2^{l-1}
ϖ_l	2^{l-1}

E_6

ϖ_1	$27 = 3^3$
ϖ_2	$78 = 2.3.13$
ϖ_3	$351 = 3^3.13$
ϖ_4	$2925 = 3^2.5^2.13$
ϖ_5	$351 = 3^3.13$
ϖ_6	$27 = 3^3$

E_7

ϖ_1	$133 = 7.19$
ϖ_2	$912 = 2^4.3.19$
ϖ_3	$8645 = 5.7.13.19$
ϖ_4	$365750 = 2.5^3.7.11.19$
ϖ_5	$27664 = 2^4.7.13.19$
ϖ_6	$1539 = 3^4.19$
ϖ_7	$56 = 2^3.7$

E_8

ϖ_1	$3875 = 5^3.31$
ϖ_2	$147250 = 2.5^3.19.31$
ϖ_3	$6696000 = 2^6.3^3.5^3.31$
ϖ_4	$6899079264 = 2^5.3.7^2.11^2.17.23.31$
ϖ_5	$146325270 = 2.3.5.7^2.13^2.19.31$
ϖ_6	$2450240 = 2^6.5.13.19.31$
ϖ_7	$30380 = 2^2.5.7^2.31$
ϖ_8	$248 = 2^3.31$

F_4

ϖ_1	$52 = 2^2.13$
ϖ_2	$1274 = 2.7^2.13$
ϖ_3	$273 = 3.7.13$
ϖ_4	$26 = 2.13$

G_2

ϖ_1	7
ϖ_2	$14 = 2.7$

EXERCISES

The base field k is assumed to be of characteristic zero.

Unless explicitly stated otherwise, Lie algebras are assumed to be finite dimensional.

§1

We denote by $\mathfrak{s}$ the Lie algebra $\mathfrak{sl}(2, k)$.

1) Let U be the enveloping algebra of $\mathfrak{s}$. Show that the element

$$\mathrm{C} = H^2 - 2(X_+X_- + X_-X_+) = H^2 + 2H - 4X_-X_+$$

belongs to the centre of U, and that its image in the representation associated to $\mathrm{V}(m)$ is the homothety with ratio $m(m+2)$.

2) Let $\lambda \in k$, let $\mathrm{Z}(\lambda)$ be a vector space having a basis (e_n), with $n = 0, 1, \ldots$, and let X_+, X_-, H be the endomorphisms of $\mathrm{Z}(\lambda)$ defined by formulas (2) of Prop. 1.

a) Verify that this gives an $\mathfrak{s}$-module structure on $\mathrm{Z}(\lambda)$.

b) Assume that λ is not an integer ≥ 0. Show that the $\mathfrak{s}$-module $\mathrm{Z}(\lambda)$ is simple.

c) Assume that λ is an integer ≥ 0. Let Z' be the subspace of $\mathrm{Z}(\lambda)$ generated by the e_n, $n > \lambda$. Show that Z' is an $\mathfrak{s}$-submodule of $\mathrm{Z}(\lambda)$, isomorphic to $\mathrm{Z}(-\lambda-2)$, and that the quotient $\mathrm{Z}(\lambda)/\mathrm{Z}'$ is isomorphic to the simple $\mathfrak{s}$-module $\mathrm{V}(\lambda)$. Show that the only $\mathfrak{s}$-submodules of $\mathrm{Z}(\lambda)$ are $0, \mathrm{Z}'$ and $\mathrm{Z}(\lambda)$.

3) Let E be a vector space having a basis $(e_n)_{n\in\mathbf{Z}}$. Let

$$a(n) = a_0 + a_1 n, \quad b(n) = b_0 + b_1 n, \quad c(n) = c_0 + c_1 n$$

be three affine functions with coefficients in k. Define endomorphisms X_+, X_-, H of E by the formulas

$$X_+e_n = a(n)e_{n-1}, \quad X_-e_n = b(n)e_{n+1}, \quad He_n = c(n)e_n.$$

This gives an $\mathfrak{s}$-module structure on E if and only if:

$$a_1b_1 = 1, \quad c_1 = -2, \quad c_0 = -a_0b_1 - a_1b_0.$$

Under what condition is this $\mathfrak{s}$-module simple?

¶ 4) Let $\mathfrak{g}$ be a Lie algebra. A $\mathfrak{g}$-module E is said to be *locally finite* if it is a union of finite dimensional $\mathfrak{g}$-submodules; equivalently, every $\mathfrak{g}$-submodule of E of finite type (as a $U(\mathfrak{g})$-module) is finite dimensional.

a) Let $0 \to E' \to E \to E'' \to 0$ be an exact sequence of $\mathfrak{g}$-modules. Show that, if E′ and E″ are locally finite, so is E (reduce to the case in which E is of finite type, and use the fact that $U(\mathfrak{g})$ is noetherian, cf. Chap. I, §2, no. 6).

b) Assume that $\mathfrak{g}$ is semi-simple. Show that E is locally finite if and only if E is a direct sum of finite dimensional simple $\mathfrak{g}$-modules.

c) Assume that $\mathfrak{g} = \mathfrak{s}$. Show that E is locally finite if and only if the following conditions are satisfied:
c_1) E has a basis consisting of eigenvectors of H,
c_2) the endomorphisms X_{+E} and X_{-E} are locally nilpotent.

(Begin by showing that, if E satisfies c_1) and c_2), and is not reduced to 0, it contains a primitive element e of integer weight $m \geq 0$; prove next that $X^{m+1}e = 0$, and hence that E contains $V(m)$. Conclude by applying *a*) to the largest locally finite submodule E′ of E.)

5) Let E be a locally finite $\mathfrak{s}$-module (Exerc. 4). For all $m \geq 0$, denote by $L(m)$ the vector space of $\mathfrak{s}$-homomorphisms from $V(m)$ to E.

a) Define an isomorphism from E to $\bigoplus_m L_m \otimes V(m)$.

b) Denote by Φ_m the invariant bilinear form on $V(m)$ defined in no. 3, Remark 3. For all $m \in \mathbf{N}$, let b_m be a bilinear form on L_m; let b be the bilinear form on E which corresponds, *via* the isomorphism in *a*), to the direct sum of the forms $b_m \otimes \Phi_m$. Show that b is invariant, and that every invariant bilinear form on E is obtained in this way in a unique manner. The form b is symmetric (resp. alternating) if and only if the b_m, m even, are symmetric (resp. alternating), and the b_m, m odd, are alternating (resp. symmetric). The form b is non-degenerate if and only if the b_m are non-degenerate.

c) Assume that E is finite dimensional. Show that E is monogenic (as a $U(\mathfrak{s})$-module) if and only if $\dim L_m \leq m+1$ for all $m \geq 0$.

6) If E is a finite dimensional $\mathfrak{s}$-module, and n an integer, denote by a_n the dimension of the eigenspace of H_E relative to the eigenvalue n. Denote by $c_E(T)$ the element of $\mathbf{Z}[T, T^{-1}]$ defined by $c_E(T) = \sum_{n \in \mathbf{Z}} a_n T^n$.

a) Define L_m as in Exerc. 5. Show that

$$\dim L_m = a_m - a_{m+2} \quad \text{for every integer } m \geq 0.$$

Deduce that $c_E(T) = c_{E'}(T)$ if and only if E and E′ are isomorphic. Recover this result by using Exerc. 18 *e*) of Chap. VII, §3.

b) Show that $c_{E \oplus F} = c_E + c_F$ and $c_{E \otimes F} = c_E . c_F$.

c) We have $c_{V(m)}(T) = (T^{m+1} - T^{-m-1})/(T - T^{-1})$.

d) Deduce from *a*), *b*), *c*) that, if $m \geq m' \geq 0$, the $\mathfrak{s}$-module $\mathrm{V}(m) \otimes \mathrm{V}(m')$ is isomorphic to

$$\mathrm{V}(m+m') \oplus \mathrm{V}(m+m'-2) \oplus \mathrm{V}(m+m'-4) \oplus \cdots \oplus \mathrm{V}(m-m').$$

If m is ≥ 0, the $\mathfrak{s}$-module $\mathbf{S}^2\mathrm{V}(m)$ is isomorphic to

$$\mathrm{V}(2m) \oplus \mathrm{V}(2m-4) \oplus \mathrm{V}(2m-8) \oplus \cdots \oplus \begin{cases} \mathrm{V}(0) & \text{if } m \text{ is even} \\ \mathrm{V}(2) & \text{if } m \text{ is odd} \end{cases}$$

and the $\mathfrak{s}$-module $\bigwedge^2 \mathrm{V}(m)$ is isomorphic to

$$\mathrm{V}(2m-2) \oplus \mathrm{V}(2m-6) \oplus \mathrm{V}(2m-10) \oplus \cdots \oplus \begin{cases} \mathrm{V}(2) & \text{if } m \text{ is even} \\ \mathrm{V}(0) & \text{if } m \text{ is odd.} \end{cases}$$

7) Let E be a finite dimensional $\mathfrak{s}$-module. Show that the dual of E is isomorphic to E.

8) Show that the $\mathfrak{s}$-module $\mathrm{V}(m)$ can be realized as the space of homogeneous polynomials $f(u, v)$ of degree m in two variables, the operators X_+, X_- and H being given by:

$$X_+ f = u\frac{\partial f}{\partial v}, \quad X_- f = -v\frac{\partial f}{\partial u}, \quad Hf = u\frac{\partial f}{\partial u} - v\frac{\partial f}{\partial v}.$$

¶ 9) Consider the operation of $\mathfrak{s}$, via the adjoint representation, on its enveloping algebra U and on its symmetric algebra $\mathbf{S} = \bigoplus \mathbf{S}^n$.

a) Determine the weights of the $\mathfrak{s}$-module $\mathbf{S}^n$. Deduce the following isomorphisms of $\mathfrak{s}$-modules:

$$\mathbf{S}^n \longrightarrow \mathrm{V}(2n) \oplus \mathrm{V}(2n-4) \oplus \mathrm{V}(2n-8) \oplus \cdots \oplus \mathrm{V}(0) \quad n \text{ even}$$
$$\mathbf{S}^n \longrightarrow \mathrm{V}(2n) \oplus \mathrm{V}(2n-4) \oplus \mathrm{V}(2n-8) \oplus \cdots \oplus \mathrm{V}(2) \quad n \text{ odd.}$$

In particular, the elements of $\mathbf{S}^n$ invariant under $\mathfrak{s}$ form a space of dimension 1 (resp. 0) if n is even (resp. odd).

b) Show that the subalgebra of $\mathbf{S}$ consisting of the elements invariant under $\mathfrak{s}$ is the polynomial algebra $k[\Gamma]$, where $\Gamma = H^2 - 4X_-X_+$. The module $\mathbf{S}^n/\Gamma.\mathbf{S}^{n-2}$ is isomorphic to $\mathrm{V}(2n)$.

c) Show that the centre of U is the polynomial algebra $k[\mathrm{C}]$, where C is the element defined in Exerc. 1 (use *b*) and the Poincaré-Birkhoff-Witt theorem).

10) Let m be an integer > 0, $\mathbf{S}$ the graded algebra $k[\mathrm{X}_1, \ldots, \mathrm{X}_m]$, and $a_1, \ldots, a_m$ elements of k^*. Let $\Phi = \sum_{i,j=1}^{m} a_{ij} X_i X_j$ be a quadratic form; put $\mathrm{D}_i = \frac{\partial}{\partial X_i}$ and $\mathrm{D} = \sum_{i,j=1}^{m} b_{ij}\mathrm{D}_i\mathrm{D}_j$ where (b_{ij}) is the inverse matrix of (a_{ij}).

a) Show that there exists a unique $\mathfrak{s}$-module structure on $\mathbf{S}$ such that $X_+ f = \frac{1}{2}\mathrm{D}(f), X_- f = \frac{1}{2}\Phi f$ and $Hf = -(m/2+n)f$ if f is homogeneous of degree n.
b) Put $\mathrm{A}^n = \mathbf{S}^n \cap \mathrm{Ker(D)}$. Show that $\mathbf{S}^n$ is the direct sum of the subspaces $\Phi^p.\mathrm{A}^q$ for $2p+q=n$. Deduce the identity

$$\sum_{n=0}^{\infty} \dim(\mathrm{A}^n)\mathrm{T}^n = (1+\mathrm{T})/(1-\mathrm{T})^{m-1}.$$

c) If f is a non-zero element of A^n, the $\Phi^p f$, $p \geq 0$, form a basis of a simple $\mathfrak{s}$-submodule of $\mathbf{S}$, isomorphic to the module $\mathrm{Z}\left(-\frac{m}{2}-n\right)$ of Exerc. 2.
d) Make the cases $m=1$ and $m=2$ explicit. Use the case $m=3$ to recover the results of Exerc. 9.

11) Assume that k is algebraically closed. Let $\mathfrak{g}$ be a Lie algebra, V a simple $\mathfrak{g}$-module, and D the field of endomorphisms of the $\mathfrak{g}$-module V. Show that, if k is uncountable,[7] then $\mathrm{D} = k$. (If not, D would contain a subfield isomorphic to $k(\mathrm{X})$, X being an indeterminate, and we would have $\dim_k \mathrm{D} > \aleph_0$, hence also $\dim_k \mathrm{V} > \aleph_0$, which is absurd since V is a monogenic $\mathrm{U}(\mathfrak{g})$-module.)

¶ 12) Let $q \in k$, and let W be a vector space with basis $(e_0, e_1, e_2, \ldots)$.
a) Show that there exists a unique representation ρ_q of $\mathfrak{s}$ on W such that

$$\rho_q(H)e_n = 2e_{n+1}, \quad \rho_q(X_+)e_n = (\frac{1}{2}\rho_q(H)-1)^n e_0,$$
$$\rho_q(X_-)e_n = (\frac{1}{2}\rho_q(H)+1)^n(-qe_0+e_1+e_2).$$

We have $\rho_q(\mathrm{C}) = 4q$, where $\mathrm{C} = H^2 + 2H - 4X_-X_+$ (cf. Exerc. 1). The representation ρ_q is simple. The elements $x \in \mathfrak{s}$ such that $\rho_q(x)$ admits an eigenvalue are the multiples of X_+. The endomorphism $\rho_q(X_+)$ admits 1 as an eigenvalue, with multiplicity 1.
b) Let ρ be a simple representation of $\mathfrak{s}$ such that $\rho(\mathrm{C}) = 4q$ and such that $\rho(X_+)$ has 1 as an eigenvalue. Show that ρ is equivalent to ρ_q.

¶ 13) Assume that $k = \mathbf{C}$. Put $\mathrm{C} = H^2 + 2H - 4X_-X_+$, cf. Exerc. 1. A representation ρ of $\mathfrak{s} = \mathfrak{sl}(2, \mathbf{C})$ is said to be H-diagonalizable if the underlying space of ρ has a basis consisting of eigenvectors of $\rho(H)$. Let $q \in \mathbf{C}$ and $v \in \mathbf{Q}/\mathbf{Z}$.
a) Let S be a vector space with a basis $(e_w)_{w\in\mathbf{C}}$ indexed by the elements of $\mathbf{C}$. Let $\mathrm{S}_v = \sum_{w \in v} \mathbf{C}e_w$. There exists a unique representation $\rho_{v,q}$ of $\mathfrak{s}$ on S_v such that

[7] The conclusion remains true even if k is countable, cf. D. QUILLEN, On the endomorphism ring of a simple module over an enveloping algebra, *Proc. Amer. Math. Soc.*, Vol. XXI (1969), pp. 171-172.

$$\rho_{v,q}(X_+)e_w = (q - w^2 - w)^{1/2} e_{w+1}$$
$$\rho_{v,q}(X_-)e_w = (q - w^2 + w)^{1/2} e_{w-1}$$
$$\rho_{v,q}(H)e_w = 2we_w.$$

(We agree that, for all $z \in \mathbf{C}$, $z^{1/2}$ is the square root of z whose amplitude belongs to $[0, \pi)$.) Denote by $S_{v,q}$ the $\mathfrak{s}$-module defined by $\rho_{v,q}$. We have $\rho_{v,q}(C) = 4q$.

b) If $q \neq u^2 + u$ for all $u \in v$, $S_{v,q}$ is simple.

c) Assume that $2v \neq 0$, and that q is of the form $u^2 + u$, where $u \in v, u \geq 0$ (which defines u uniquely). Let $S^-_{v,q}$ (resp. $S^+_{v,q}$) be the vector subspace of S_v generated by the e_w for $w \leq u$ (resp. $w > u$). Then $S^-_{v,q}$ and $S^+_{v,q}$ are simple $\mathfrak{s}$-submodules of $S_{v,q}$.

d) Assume that $2v = 0$, and that q is of the form $u^2 + u$, where $u \in v, u \geq 0$ (which defines u uniquely). Let $S^-_{v,q}$ (resp. $S^0_{v,q}, S^+_{v,q}$) be the vector subspace of S_v generated by the e_w for $w < -u$ (resp. $-u \leq w \leq u, w > u$). Then $S^-_{v,q}, S^0_{v,q}$ and $S^+_{v,q}$ are simple $\mathfrak{s}$-submodules of $S_{v,q}$.

e) Assume that $v = -\frac{1}{2} + \mathbf{Z}$ and $q = -\frac{1}{4}$. Let $S^-_{-1/2,-1/4}$ (resp. $S^+_{-1/2,-1/4}$) be the vector subspace of $S_{-1/2}$ generated by the e_w for $w \leq -\frac{1}{2}$ (resp. $w > -\frac{1}{2}$). Then $S^-_{-1/2,-1/4}$ and $S^+_{-1/2,-1/4}$ are simple $\mathfrak{s}$-submodules of $S_{-1/2,-1/4}$.

f) Denote by $\rho^\pm_{v,q}, \rho^0_{v,q}$ the representations corresponding to $S^\pm_{v,q}, S^0_{v,q}$. In case *b*), the elements $x \in \mathfrak{s}$ such that $\rho^-_{v,q}(x)$ admits an eigenvalue are those in $\mathbf{C}H$. In cases *c*), *d*), *e*), the elements $x \in \mathfrak{s}$ such that $\rho^-_{v,q}(x)$ admits an eigenvalue are those in $\mathbf{C}H + \mathbf{C}X_+$; if, in addition, x is nilpotent (and hence proportional to X_+) and non-zero, $\rho^-_{v,q}(x)$ admits 0 as its only eigenvalue, and this of multiplicity 1; on the other hand, if x is semi-simple, the underlying space of $\rho^-_{v,q}$ has a basis consisting of eigenvectors of $\rho^-_{v,q}(x)$. There are analogous results for $\rho^+_{v,q}$, replacing X_+ by X_-.

g) Let V be a simple $\mathfrak{s}$-module and ρ the corresponding representation. Then $\rho(C)$ is a homothety (use Exerc. 11). Assume that $\rho(C) = 4q$. Show that, if ρ is H-diagonalizable, V is isomorphic to one of the modules $S_{v,q}, S^\pm_{v,q}, S^0_{v,q}$ considered in *b*), *c*), *d*), *e*). Moreover, ρ is H-diagonalizable if and only if $\rho(H)$ admits an eigenvalue; it suffices that $\rho(X_+)$ admits the eigenvalue 0.

¶ 14) The notations are those of the preceding exercise. Denote by B_q the quotient of $U(\mathfrak{s})$ by the two-sided ideal generated by $C - 4q$, and denote by $u \mapsto u^\bullet$ the canonical map $U(\mathfrak{s}) \to B_q$. Consider the representations of Exerc. 12 and 13 as representations of B_q.

a) Every element of B_q can be expressed uniquely in the form

$$\sum_{r \geq 0} X_+^{\bullet\, r} p_r(H^\bullet) + \sum_{s > 0} q_s(H^\bullet) X_-^{\bullet\, s},$$

where the p_r and the q_s are polynomials. If two elements of B_q generate the same left ideal, they are proportional.

b) Consider case *b*) of Exerc. 13. Let a be a non-zero element of $\mathrm{S}_{v,q}$. If a is an eigenvector of $\rho_{v,q}(H)$, with eigenvalue λ, the annihilator of a in B_q is the left ideal generated by $H^\bullet - \lambda$; if a is not an eigenvector of $\rho_{v,q}(H)$, its annihilator is a non-monogenic left ideal.

c) Consider the case of the representations $\rho^{\pm}_{v,q}$ of Exerc. 13. If a is a non-zero element of the space of such a representation, its annihilator in B_q is a non-monogenic left ideal.

d) Consider the representation ρ_q of Exerc. 12. The annihilator of e_0 in B_q is generated by $X^\bullet_+ - 1$. If $a \in \mathrm{W}$ is not proportional to e_0, its annihilator is a non-monogenic left ideal.

e) Every automorphism of $\mathfrak{s}$ extends to $\mathrm{U}(\mathfrak{s})$ and defines by passage to the quotient an automorphism of B_q; let A' be the subgroup of $\mathrm{A} = \mathrm{Aut}(\mathrm{B}_q)$ thus obtained. Show that $\mathrm{A}' \neq \mathrm{A}$. (Let φ be the endomorphism $x \mapsto [X^{\bullet 2}_+, x]$ of B_q; this automorphism is locally nilpotent, and $e^\varphi \in \mathrm{A}, e^\varphi \notin \mathrm{A}'$.)

f) The group A operates by transport of structure on the set of classes of simple representations of B_q. Let π_1 (resp. π_2, π_3) be a representation of type $\rho_{v,q}$ of Exerc. 13 *b*) (resp. of type $\rho^{\pm}_{v,q}$ of Exerc. 13, resp. of type ρ_q of Exerc. 12). Then $\mathrm{A}\pi_1, \mathrm{A}\pi_2$ and $\mathrm{A}\pi_3$ are pairwise disjoint. (Use *a*), *b*), *c*), *d*).) If $\psi \in \mathrm{A}$ is such that $\psi\pi_1$ is of type $\rho_{v,q}$ of Exerc. 13 *b*), then $\psi \in \mathrm{A}'$ (use *a*) and *b*)). Deduce that, if ψ is the automorphism e^φ of *e*), the representation $\sigma = \pi_1 \circ \psi$ has the following property: for all $x \in \mathfrak{s} - \{0\}$, $\sigma(x)$ has no eigenvalue.[8]

15) Let $\mathfrak{g}$ be a Lie algebra of dimension 3. Show the equivalence of the following conditions (cf. Chap. I, §6, Exerc. 23).

(i) $\mathfrak{g} = [\mathfrak{g}, \mathfrak{g}]$.
(ii) The Killing form of $\mathfrak{g}$ is non-degenerate.
(iii) $\mathfrak{g}$ is semi-simple.
(iv) $\mathfrak{g}$ is simple.

¶ 16) Let $\mathfrak{g}$ be a simple Lie algebra of dimension 3, and let Φ be its Killing form. Denote by $\mathfrak{o}(\Phi)$ the orthogonal algebra of Φ, i.e. the subalgebra of $\mathfrak{gl}(\mathfrak{g})$ consisting of the elements leaving Φ invariant.

a) Show that $\mathrm{ad} : \mathfrak{g} \to \mathfrak{o}(\Phi)$ is an isomorphism.

b) Prove the equivalence of the following properties:

(i) $\mathfrak{g}$ contains a non-zero isotropic vector (for Φ).
(ii) $\mathfrak{g}$ contains a non-zero nilpotent element.
(iii) $\mathfrak{g}$ is isomorphic to $\mathfrak{s}$.

c) Show that there exists an extension k_1 of k, of degree ≤ 2, such that $\mathfrak{g}_{(k_1)} = k_1 \otimes_k \mathfrak{g}$ is isomorphic to $\mathfrak{s}_{(k_1)}$.

[8] For more details on Exerc. 12, 13 and 14, see: D. ARNAL and G. PINCZON, Sur les représentations algébriquement irréductibles de l'algèbre de Lie $\mathfrak{sl}(2)$, *J. Math. Phys.*, Vol. 15 (1974), pp. 350-359.

d) Give the vector space $A = k \oplus \mathfrak{g}$ the unique algebra structure admitting 1 as unit element, and such that the product $\mathfrak{g} \times \mathfrak{g} \to A$ is given by the formula

$$x.y = \frac{1}{8}\Phi(x,y).1 + \frac{1}{2}[x,y] \quad (x, y \in \mathfrak{g}).$$

Show that A is a quaternion algebra over k, and that $\mathfrak{g}$ is the Lie subalgebra of A consisting of the elements of trace zero (reduce, by extension of the base field, to the case where $\mathfrak{g} = \mathfrak{s}$, and show that A can then be identified with the matrix algebra $\mathbf{M}_2(k)$).

Conversely, if D is a quaternion algebra over k, the elements of D of trace zero form a simple Lie algebra of dimension 3, and the corresponding algebra A can be identified with D.

e) Prove the formulas

$$\Phi(x,x) = -8\mathrm{Nrd}_A(x), \qquad \Phi(x,y) = 4\mathrm{Trd}_A(xy)$$
$$2[x,[y,z]] = \Phi(x,y)z - \Phi(x,z)y \qquad (x, y, z \in \mathfrak{g}).$$

f) Show that the discriminant of Φ (with respect to any basis of $\mathfrak{g}$) is of the form $-2\lambda^2$, with $\lambda \in k^*$.

g) Let n be an integer ≥ 0. Show that the $\mathfrak{g}$-module $\mathbf{S}^n(\mathfrak{g})$ has a unique simple submodule of dimension $2n+1$, and that this module is absolutely simple (reduce to the case where $\mathfrak{g} = \mathfrak{s}$, and use Exerc. 9).

Show that $\mathfrak{g}$ has an absolutely simple module of dimension $2n$ only if $\mathfrak{g}$ is isomorphic to $\mathfrak{s}$, i.e. only if A is isomorphic to $\mathbf{M}_2(k)$. (Let V be such a module. If $n \geq 2$, show by means of Exerc. 6 *c*) that the $\mathfrak{g}$-module $V \otimes \mathfrak{g}$ has a unique absolutely simple submodule of dimension $2n-2$. Then reduce to the case $n = 1$, which is trivial.)

¶ 17) We retain the notations of the preceding exercise. Show that, for all $n \geq 1$, the algebra $U(\mathfrak{g})$ has a unique two-sided ideal $\mathfrak{m}_n$ such that $U(\mathfrak{g})/\mathfrak{m}_n$ is a central simple algebra of dimension n^2 (extend scalars to reduce to the case in which $\mathfrak{g} = \mathfrak{s}$ and show that $\mathfrak{m}_n$ is then the kernel of the homomorphism $U(\mathfrak{g}) \to \mathrm{End}(V(n-1))$). Every two-sided ideal of $U(\mathfrak{g})$ of finite codimension is of the form $\mathfrak{m}_{n_1} \cap \mathfrak{m}_{n_2} \cap \cdots \cap \mathfrak{m}_{n_h}$, where $n_1, \ldots, n_h$ are distinct; its codimension is $n_1^2 + \cdots + n_h^2$ (apply the density theorem). The $\mathfrak{m}_n$ are the only maximal two-sided ideals of $U(\mathfrak{g})$ of finite codimension.

Show that $\mathfrak{m}_2$ is generated (as a two-sided ideal) by the elements $x^2 - \frac{1}{8}\Phi(x,x)$ $(x \in \mathfrak{g})$, and that the quotient $U(\mathfrak{g})/\mathfrak{m}_2$ can be identified with the quaternion algebra A of Exerc. 16.

When $\mathfrak{g}$ is isomorphic to $\mathfrak{s}$, $U(\mathfrak{g})/\mathfrak{m}_n$ is isomorphic to $\mathbf{M}_n(k)$. Show that, when $\mathfrak{g}$ is not isomorphic to $\mathfrak{s}$, $U(\mathfrak{g})/\mathfrak{m}_n$ is isomorphic to $\mathbf{M}_n(k)$ if and only if n is odd (use Exerc. 16 *g*)).[9]

[9] It can be shown that, when n is even, $U(\mathfrak{g})/\mathfrak{m}_n$ is isomorphic to $\mathbf{M}_{n/2}(A)$.

¶ 18) Assume that $k = \mathbf{R}, \mathbf{C}$ or a field complete for a discrete valuation and with residue field of characteristic $p \neq 0$ (for example a finite extension of the p-adic field $\mathbf{Q}_p$).

a) Let $\mathfrak{n}$ be a nilpotent Lie algebra, N the Lie group obtained by giving $\mathfrak{n}$ the Hausdorff law (Chap. III, §9, no. 5), and $\rho : \mathfrak{n} \to \mathfrak{gl}(\mathrm{E})$ a linear representation of $\mathfrak{n}$ on a finite dimensional vector space E. Assume that $\rho(x)$ is nilpotent for all $x \in \mathfrak{n}$, and put $\pi(x) = \exp(\rho(x))$. Show that π is the only homomorphism of Lie groups $\varphi : \mathrm{N} \to \mathbf{GL}(\mathrm{E})$ such that $\mathrm{L}(\varphi) = \rho$.

(When $k = \mathbf{R}$ or $\mathbf{C}$, use the fact that N is connected. When k is ultrametric, show that the eigenvalues of $\varphi(x)$, $x \in \mathrm{N}$, are equal to 1; for this, prove first of all that, if k' is a finite extension of k, and if $(\lambda_1, \ldots, \lambda_n, \ldots)$ is a sequence of elements of k' such that $\lambda_n = \lambda_{n+1}^p$ for all n and $\lambda_1 = 1$, then $\lambda_n = 1$ for all n.)

b) Let $\rho : \mathfrak{s} \to \mathfrak{gl}(\mathrm{E})$ be a finite dimensional linear representation of $\mathfrak{s}$, and let π be the homomorphism from $\mathbf{SL}(2, k)$ to $\mathbf{GL}(\mathrm{E})$ compatible with ρ (no. 4). Show that π is the unique homomorphism of Lie groups $\varphi : \mathbf{SL}(2, k) \to \mathbf{GL}(\mathrm{E})$ such that $\mathrm{L}(\varphi) = \rho$. (Use *a*) to prove that π and ρ coincide on the $\exp(n)$, with n nilpotent in $\mathfrak{s}$, and remark that $\mathbf{SL}(2, k)$ is generated by the $\exp(n)$.)

§2

1) Let $\mathfrak{g}$ be a simple Lie algebra of dimension 3, and Φ its Killing form (cf. §1, Exerc. 15, 16, 17).

a) An element $x \in \mathfrak{g}$ is regular if and only if $\Phi(x, x) \neq 0$. Let $\mathfrak{h}_x = kx$ be the Cartan subalgebra generated by such an element. Show that $\mathfrak{h}_x$ is splitting if and only if $2\Phi(x, x)$ is a square in k.

b) Show that

$$\mathfrak{g} \text{ is splittable} \Longleftrightarrow \mathfrak{g} \text{ is isomorphic to } \mathfrak{sl}(2, k).$$

2) Let k_1 be an extension of k of finite degree $n \geq 2$.

a) Show that the semi-simple k-algebra $\mathfrak{sl}(2, k_1)$ is not splittable.

b) Show that the splittable simple k-algebra $\mathfrak{sl}(n, k)$ contains a Cartan subalgebra $\mathfrak{h}_1$ that is not splitting. (Choose an embedding of the algebra k_1 into $\mathbf{M}_n(k)$, and take $\mathfrak{h}_1 = k_1 \cap \mathfrak{sl}(n, k)$.)

3) Let $(\mathfrak{g}, \mathfrak{h})$ be a split semi-simple Lie algebra, R its root system, and K the restriction of the Killing form of $\mathfrak{g}$ to $\mathfrak{h}$. With the notations of Chap. VI, §1, no. 12, we have $\mathrm{K} = \mathrm{B}_{\mathrm{R}^\vee}$ and $\mathrm{K} = 4\gamma(\mathrm{R})\Phi_{\mathrm{R}^\vee}$ if R is irreducible; moreover, if all the roots of R have the same length, then

$$\mathrm{K}(H_\alpha, H_\alpha) = 4h \quad \text{for all } \alpha \in \mathrm{R},$$

where h is the Coxeter number of $\mathrm{W}(\mathrm{R})$.

If for example $\mathfrak{g}$ is simple of type E_6, E_7 or E_8, $K(H_\alpha, H_\alpha)$ is equal to 48, 72 or 120.

4) Let $(\mathfrak{g}, \mathfrak{h})$ be a split semi-simple Lie algebra, and $(X_\alpha)_{\alpha\in R}$ a family of elements satisfying the conditions of Lemma 2. If $\alpha, \beta \in R$ and $\alpha + \beta \in R$, define $N_{\alpha\beta}$ by the formula $[X_\alpha, X_\beta] = N_{\alpha\beta}X_{\alpha+\beta}$; if $\alpha + \beta \notin R$, put $N_{\alpha\beta} = 0$. Prove the following formulas:

a) $N_{\alpha\beta} = -N_{\beta\alpha}$.

b) If $\alpha, \beta, \gamma \in R$ are such that $\alpha + \beta + \gamma = 0$, then

$$\frac{N_{\alpha\beta}}{\langle\gamma,\gamma\rangle} = \frac{N_{\beta\gamma}}{\langle\alpha,\alpha\rangle} = \frac{N_{\gamma\alpha}}{\langle\beta,\beta\rangle}.$$

c) If $\alpha, \beta, \gamma, \delta \in R$ are such that $\alpha + \beta + \gamma + \delta = 0$, and if no pair has sum zero, we have:

$$\frac{N_{\alpha\beta}N_{\gamma\delta}}{\langle\gamma+\delta,\gamma+\delta\rangle} + \frac{N_{\beta\gamma}N_{\alpha\delta}}{\langle\alpha+\delta,\alpha+\delta\rangle} + \frac{N_{\gamma\alpha}N_{\beta\delta}}{\langle\beta+\delta,\beta+\delta\rangle} = 0.$$

5) Let $(\mathfrak{g}, \mathfrak{h})$ be a split semi-simple Lie algebra, and $(X_\alpha)_{\alpha\in R}$ a family of elements satisfying the conditions of Lemma 2. Let B be a basis of R. The H_α ($\alpha \in B$) and the X_α ($\alpha \in R$) form a basis of the vector space $\mathfrak{g}$. Show that the discriminant of the Killing form of $\mathfrak{g}$ with respect to this basis (*Algebra*, Chap. IX, §2) is a rational number, independent of the choice of $\mathfrak{h}$, of B, and of the X_α.[10] Deduce that, if $n = \dim \mathfrak{g}$, the element of $\bigwedge^n \mathfrak{g}$ defined by the exterior product of the H_α ($\alpha \in B$) and the X_α ($\alpha \in R$) is independent, up to sign, of the choice of $\mathfrak{h}$, of B, and of the X_α.

6) Let $(X_\alpha)_{\alpha\in R}$ be a Chevalley system in the split semi-simple Lie algebra $(\mathfrak{g}, \mathfrak{h})$. Let $\alpha, \beta \in R$, and let p (resp. q) be the largest integer j such that $\beta + j\alpha \in R$ (resp. $\beta - j\alpha \in R$), cf. Lemma 4. Show that

$$\mathrm{ad}(X_\alpha)^k(X_{\beta-q\alpha}) = \pm k! X_{\beta+(k-q)\alpha} \quad \text{for} \quad 0 \leq k \leq p+q.$$

Deduce that the algebra $\mathfrak{g}_{\mathbf{Z}}$ of Prop. 8 is stable under the $\mathrm{ad}(X_\alpha)^k/k!$ and under the $e^{\mathrm{ad}(X_\alpha)}$ (cf. §12).

7) Assume that k is an ordered field. Let $(\mathfrak{g}, \mathfrak{h})$ be a split semi-simple Lie algebra of rank l; put $\dim \mathfrak{g} = l + 2m$.

a) Show that the Killing form Φ of $\mathfrak{g}$ is the direct sum of a neutral form of rank $2m$ and a positive non-degenerate form of rank l; in particular, its index is m.

b) Let φ be an involutive automorphism of $\mathfrak{g}$ whose restriction to $\mathfrak{h}$ is $-\mathrm{Id}$. Show that the form

[10]For an explicit calculation of this discriminant, see: T. A. SPRINGER and R. STEINBERG, Conjugacy Classes (no. 4.8), Seminar on Algebraic Groups and Related Finite Groups, *Lect. Notes in Math.* 131, Springer-Verlag (1970).

$$(x, y) \mapsto -\Phi(x, \varphi(y)), \quad x, y \in \mathfrak{g},$$

is symmetric, non-degenerate, and positive.

c) Let $k' = k(\sqrt{\alpha})$, where α is an element < 0 in k. Denote by c the non-trivial k-automorphism of k'. Let $\mathfrak{g}_c$ be the k-subspace of $\mathfrak{g}_{(k')} = k' \otimes_k \mathfrak{g}$ formed by the elements y such that

$$(1 \otimes \varphi).y = (c \otimes 1).y.$$

Show that $\mathfrak{g}_c$ is a k-Lie subalgebra of $\mathfrak{g}_{(k')}$ and that the injection of $\mathfrak{g}_c$ into $\mathfrak{g}_{(k')}$ extends to an isomorphism from $k' \otimes_k \mathfrak{g}_c$ to $\mathfrak{g}_{(k')}$. The algebra $\mathfrak{g}_c$ is semi-simple, and $\sqrt{\alpha}\,\mathfrak{h}$ is a Cartan subalgebra of it.

d) Show that the Killing form of $\mathfrak{g}_c$ is negative. Deduce that $\mathfrak{g}_c$ is not splittable (unless $\mathfrak{g} = 0$).

e) When $k = \mathbf{R}$, show that $\mathrm{Int}(\mathfrak{g}_c)$ is compact.

8) *a*) Let $\mathfrak{g}$ be a Lie algebra and n an integer ≥ 0. Let $\Sigma_n(\mathfrak{g})$ be the subset of $\mathfrak{g}^n$ consisting of the families $(x_1, \dots, x_n)$ that generate $\mathfrak{g}$ as a k-algebra. Show that $\Sigma_n(\mathfrak{g})$ is open in $\mathfrak{g}^n$ in the Zariski topology (Chap. VII, App. I). If k' is an extension of k, then $\Sigma_n(\mathfrak{g}_{(k')}) \cap \mathfrak{g}^n = \Sigma_n(\mathfrak{g})$. Deduce that, if $\mathfrak{g}_{(k')}$ can be generated by n elements, so can $\mathfrak{g}$.

b) Let $(\mathfrak{g}, \mathfrak{h})$ be a split semi-simple Lie algebra. Let x be an element of $\mathfrak{h}$ such that $\alpha(x) \neq 0$ for all $\alpha \in \mathrm{R}$ and $\alpha(x) \neq \beta(x)$ for any pair of distinct elements $\alpha, \beta \in \mathrm{R}$. For all $\alpha \in \mathrm{R}$, let y_α be a non-zero element of $\mathfrak{g}^\alpha$, and let $y = \sum_{\alpha \in \mathrm{R}} y_\alpha$. Show that, for all $\alpha \in \mathrm{R}$, there exists a polynomial $\mathrm{P}_\alpha(\mathrm{T}) \in k[\mathrm{T}]$, without constant term, such that $y_\alpha = \mathrm{P}_\alpha(\mathrm{ad}\, x).y$. Deduce that $\mathfrak{g}$ is generated by $\{x, y\}$.

c) Show, by means of *a*) and *b*), that every semi-simple Lie algebra can be generated by two elements.

¶ 9) Let G be a connected finite dimensional real Lie group. Let $\mathfrak{g}$ be its Lie algebra, and let $\{x_1, \dots, x_n\}$ be a generating family of $\mathfrak{g}$. For $m \geq 0$, denote by Γ_m the subgroup of G generated by the $\exp(2^{-m} x_i)$, $1 \leq i \leq n$. We have $\Gamma_0 \subset \Gamma_1 \subset \cdots$.

a) Show that the union of the Γ_m is dense in G.

b) Let H_m be the identity component of the closure $\overline{\Gamma}_m$ of Γ_m. We have $\mathrm{H}_0 \subset \mathrm{H}_1 \subset \cdots$, and the family (H_m) is stationary; let H be the common value of the H_m for m sufficiently large. Show that H is normal in G (observe that H is normalized by all the Γ_m), and that the image of Γ_m in G/H is a discrete subgroup of G/H. The union of these subgroups being dense in G/H, deduce (cf. Chap. III, §6, Exerc. 23 *d*)) that G/H is nilpotent.

c) Assume that $\mathfrak{g} = \mathscr{D}\mathfrak{g}$. Show that G = H, in other words, that Γ_m is dense in G if m is sufficiently large.

10) Let G be a connected semi-simple real Lie group. Show, by using Exerc. 8 and 9, that there exists a dense subgroup of G generated by two elements.

11) Let R be a root system of rank l, and Φ_{R} the corresponding canonical bilinear form (Chap. VI, §1, no. 12). The matrix $\Phi = (\Phi_{\mathrm{R}}(\alpha,\beta))_{\alpha,\beta\in\mathrm{R}}$ is of rank l, and $\Phi^2 = \Phi$. Deduce the formula $\sum_{\alpha\in\mathrm{R}} \Phi_{\mathrm{R}}(\alpha,\alpha) = \mathrm{Tr}\,\Phi = l$.

§3

1) Let Γ be a subgroup of Q(R) of finite index, and $\mathrm{P} = \Gamma \cap \mathrm{R}$. The algebra $\mathfrak{h} + \mathfrak{g}^{\mathrm{P}}$ is reductive, and any reductive subalgebra of $\mathfrak{g}$ containing $\mathfrak{h}$ can be obtained in this way (use Chap. VI, §1, Exerc. 6 b)).

2) Let X be the set of reductive subalgebras of $\mathfrak{g}$, distinct from $\mathfrak{g}$, and containing $\mathfrak{h}$. Determine the minimal elements of X by means of Chap. VI, §4, Exerc. 4. Show that the centre of such a maximal subalgebra is of dimension 0 or 1, according to whether we are in case *a*) or case *b*) of the exercise in question.

3) Let $\mathfrak{b}$ be a Borel subalgebra of $\mathfrak{g}$, and let $l = \mathrm{rk}(\mathfrak{g})$. Show that the minimum number of generators of the algebra $\mathfrak{b}$ is l if $l \neq 1$, and 2 if $l = 1$.

4) Assume that $k = \mathbf{R}$ or $\mathbf{C}$. Put $\mathrm{G} = \mathrm{Int}(\mathfrak{g})$, and identify the Lie algebra of G with $\mathfrak{g}$. Let $\mathfrak{b}$ be a Borel subalgebra of $(\mathfrak{g},\mathfrak{h})$, and $\mathfrak{n}$ the set of its nilpotent elements. Denote by H, B, N the integral subgroups of G with Lie algebras $\mathfrak{h}, \mathfrak{b}, \mathfrak{n}$, respectively. Show that H, B, N are Lie subgroups of G, that N is simply-connected, and that B is the semi-direct product of H by N.

5) Let $\mathfrak{m}$ be a parabolic subalgebra of a semi-simple Lie algebra $\mathfrak{a}$.

a) Let $\mathfrak{p}$ be a subalgebra of $\mathfrak{m}$. Then $\mathfrak{p}$ is a parabolic subalgebra of $\mathfrak{a}$ if and only if $\mathfrak{p}$ contains the radical $\mathfrak{r}$ of $\mathfrak{m}$ and $\mathfrak{p}/\mathfrak{r}$ is a parabolic subalgebra of the semi-simple algebra $\mathfrak{m}/\mathfrak{r}$.

b) If $\mathfrak{m}'$ is a parabolic subalgebra of $\mathfrak{a}$, every Cartan subalgebra of $\mathfrak{m} \cap \mathfrak{m}'$ is a Cartan subalgebra of $\mathfrak{a}$. (Reduce to the split case, and apply Prop. 10 to the Borel subalgebras contained in $\mathfrak{m}$ and $\mathfrak{m}'$.)

6) Two Borel subalgebras of a semi-simple Lie algebra $\mathfrak{a}$ are said to be *opposite* if their intersection is a Cartan subalgebra. Show that, if $\mathfrak{b}$ is a Borel subalgebra of $\mathfrak{a}$, and $\mathfrak{h}$ is a Cartan subalgebra of $\mathfrak{b}$, there exists a unique Borel subalgebra of $\mathfrak{a}$ that is opposite to $\mathfrak{b}$ and contains $\mathfrak{h}$. (Reduce to the split case.)

7) Let $\mathfrak{a}$ be a semi-simple Lie algebra, $\mathfrak{h}$ a Cartan subalgebra of $\mathfrak{a}$, and $\mathfrak{s}$ a semi-simple subalgebra of $\mathfrak{a}$ containing $\mathfrak{h}$. Show that:

$$(\mathfrak{a},\mathfrak{h}) \text{ is split} \iff (\mathfrak{s},\mathfrak{h}) \text{ is split.}$$

Construct an example in which $\mathfrak{a}$ is splittable but $\mathfrak{s}$ is not. (Take $\mathfrak{a} = \mathfrak{sp}(4, k)$ and $\mathfrak{s} = \mathfrak{sl}(2, k')$, where k' is a quadratic extension of k.)

8) Choose a basis of R, and hence a set of positive roots R_+. Assume that R is irreducible. Let $\tilde{\alpha}$ be the highest root, S the set of roots orthogonal to $\tilde{\alpha}$, and $S_+ = S \cap R_+$.

a) Let

$$\mathfrak{g}' = \mathfrak{g}^S \oplus \mathfrak{h}_S, \quad \mathfrak{m}_+ = \sum_{\alpha \in S_+} \mathfrak{g}^\alpha, \quad \mathfrak{m}_- = \sum_{\alpha \in S_+} \mathfrak{g}^{-\alpha}.$$

Then $\mathfrak{g}'$ is semi-simple and $\mathfrak{g}' = \mathfrak{h}_S \oplus \mathfrak{m}_+ \oplus \mathfrak{m}_-$.

b) Put:

$$\mathfrak{n}_+ = \sum_{\alpha \in R_+} \mathfrak{g}^\alpha, \quad \mathfrak{p} = \sum_{\alpha \in R_+ - S_+} \mathfrak{g}^\alpha, \quad \mathfrak{p}_0 = \sum_{\alpha \in R_+ - S_+ - \{\tilde{\alpha}\}} \mathfrak{g}^\alpha.$$

Then

$$\mathfrak{n}_+ = \mathfrak{m}_+ \oplus \mathfrak{p}, \quad [\mathfrak{m}_+, \mathfrak{p}_0] \subset \mathfrak{p}_0, \quad [\mathfrak{n}_+, \mathfrak{g}^{\tilde{\alpha}}] = 0.$$

In particular, $\mathfrak{p}$ is an ideal of $\mathfrak{n}_+$.

c) For all $\alpha \in R_+ - S_+ - \{\tilde{\alpha}\}$, there exists a unique $\alpha' \in R_+ - S_+ - \{\tilde{\alpha}\}$ such that $\alpha + \alpha' = \tilde{\alpha}$. For all $\alpha \in R_+ - S_+ - \{\tilde{\alpha}\}$, a non-zero element X_α of $\mathfrak{g}^\alpha$ can be chosen so that

$$\begin{aligned} &[X_\alpha, X_{\alpha'}] = \pm X_{\tilde{\alpha}} && \text{if } \alpha \in R_+ - S_+ - \{\tilde{\alpha}\} \\ &[X_\alpha, X_\beta] = 0 && \text{if } \alpha, \beta \in R_+ - S_+,\ \beta \neq \alpha'. \end{aligned}$$ [11]

9) Construct examples of semi-simple Lie algebras $\mathfrak{a}$ such that:

i) $\mathfrak{a}$ has no Borel subalgebra;

ii) $\mathfrak{a}$ has a Borel subalgebra, but is not splittable.

¶ 10) Let $\mathfrak{a}$ be a semi-simple Lie algebra, and x an element of $\mathfrak{a}$. We say that x is *diagonalizable* if ad x is diagonalizable (*Algebra*, Chap. VII), in other words, if there exists a basis of $\mathfrak{a}$ consisting of eigenvectors of ad x.

a) Let $\mathfrak{c}$ be a commutative subalgebra of $\mathfrak{a}$ consisting of diagonalizable elements, and let L be the set of weights of $\mathfrak{c}$ in the representation $\text{ad}_{\mathfrak{a}}$. The set L is a finite subset of $\mathfrak{c}^*$ containing 0 (unless $\mathfrak{a} = 0$) and $\mathfrak{a} = \bigoplus_{\lambda \in L} \mathfrak{a}^\lambda(\mathfrak{c})$. Show that there exists a subset M of L such that $L - \{0\}$ is the disjoint union of M and $-M$, and such that $(M + M) \cap L \subset M$. If M has these properties, put $\mathfrak{a}^M = \bigoplus_{\lambda \in M} \mathfrak{a}^\lambda(\mathfrak{c})$, and $\mathfrak{p}^M = \mathfrak{a}^0(\mathfrak{c}) \oplus \mathfrak{a}^M$. The algebra $\mathfrak{a}^0(\mathfrak{c})$ is the commutant of $\mathfrak{c}$ in $\mathfrak{a}$; it is reductive in $\mathfrak{a}$ (Chap. VII, §1, no. 5), and its Cartan subalgebras are the Cartan subalgebras of $\mathfrak{a}$ (Chap. VII, §2, no. 3). Show that $\mathfrak{p}^M$ is a

[11]This exercise was communicated to us by A. JOSEPH.

parabolic subalgebra of $\mathfrak{a}$, and that $\mathfrak{a}^{\mathrm{M}}$ is the set of nilpotent elements of the radical of $\mathfrak{p}^{\mathrm{M}}$. (Use the fact that $\mathfrak{c}$ is contained in a Cartan subalgebra of $\mathfrak{a}$ and reduce, by extension of scalars, to the case in which this subalgebra is splitting.)

b) Retain the hypotheses and notations of *a*), and assume in addition that $\mathfrak{a}$ is splittable. Show that $\mathfrak{c}$ is contained in a splitting Cartan subalgebra of $\mathfrak{a}$. (Take a splitting Cartan subalgebra $\mathfrak{h}$ of $\mathfrak{a}$ contained in $\mathfrak{p}^{\mathrm{M}}$; then there exists a unique Cartan subalgebra $\mathfrak{h}'$ of $\mathfrak{a}^0(\mathfrak{c})$ such that $\mathfrak{h}$ is contained in $\mathfrak{h}' \oplus \mathfrak{a}^{\mathrm{M}}$; show that $\mathfrak{h}'$ is a splitting Cartan subalgebra of $\mathfrak{a}$, and that $\mathfrak{h}'$ contains $\mathfrak{c}$.)

11) Let $\mathfrak{a} = \prod \mathfrak{a}_i$ be a finite product of semi-simple Lie algebras. A subalgebra $\mathfrak{q}$ of $\mathfrak{a}$ is a parabolic (resp. Borel) subalgebra if and only if it is of the form $\mathfrak{q} = \prod \mathfrak{q}_i$ where, for all i, $\mathfrak{q}_i$ is a parabolic (resp. Borel) subalgebra of $\mathfrak{a}_i$.

¶ 12) Let k' be a finite extension of k, $\mathfrak{a}'$ a semi-simple k'-Lie algebra, and $\mathfrak{a}$ the underlying k-Lie algebra. Show that the parabolic (resp. Borel) subalgebras of $\mathfrak{a}$ are the same as those of $\mathfrak{a}'$. (Extend scalars to an algebraic closure of k, and use Exerc. 11.)

13) Let $\mathfrak{p}$ and $\mathfrak{q}$ be two parabolic subalgebras of a semi-simple Lie algebra $\mathfrak{a}$, and let $\mathfrak{n}$ be the nilpotent radical of $\mathfrak{p}$. Show that $\mathfrak{m} = (\mathfrak{p} \cap \mathfrak{q}) + \mathfrak{n}$ is a parabolic subalgebra of $\mathfrak{a}$. (Reduce to the case where $\mathfrak{a}$ is split and $\mathfrak{q}$ is a Borel subalgebra; choose a Cartan subalgebra $\mathfrak{h}$ contained in $\mathfrak{p} \cap \mathfrak{q}$, and determine the subset P of the corresponding root system such that $\mathfrak{m} = \mathfrak{h} + \mathfrak{g}^{\mathrm{P}}$.)

14) Retain the notations of Prop. 9.

a) Let $\alpha \in \mathrm{B}$. Show that $\mathfrak{n} \cap \operatorname{Ker} \operatorname{ad} X_\alpha$ is the direct sum of the $\mathfrak{g}^\beta$, where β belongs to the set of elements of R_+ such that $\alpha + \beta \notin \mathrm{R}_+$.

b) Deduce that, if $\mathfrak{g}$ is simple, the centre of $\mathfrak{n}$ is equal to $\mathfrak{g}^{\tilde{\alpha}}$, where $\tilde{\alpha}$ is the highest root of R_+. In the general case, the dimension of the centre of $\mathfrak{n}$ is equal to the number of simple components of $\mathfrak{g}$.

§4

The Lie algebras considered in this paragraph are not necessarily finite dimensional.

1) Retain the notations of no. 2. Let $\lambda \in k^{\mathrm{B}}$. Associate to any $\alpha \in \mathrm{B}$ the endomorphisms $X^\lambda_{-\alpha}, H^\lambda_\alpha, X^\lambda_\alpha$ of the vector space E such that

$$X^\lambda_{-\alpha}(\alpha_1, \ldots, \alpha_n) = (\alpha, \alpha_1, \ldots, \alpha_n)$$

$$H^\lambda_\alpha(\alpha_1, \ldots, \alpha_n) = (\lambda(\alpha) - \sum_{i=1}^{n} n(\alpha_i, \alpha))(\alpha_1, \ldots, \alpha_n).$$

The vector $X^\lambda_\alpha(\alpha_1, \ldots, \alpha_n)$ is defined by induction on n by the formula

$$X^\lambda_\alpha(\alpha_1, \ldots, \alpha_n) = (X^\alpha_{-\alpha_1} X^\lambda_\alpha - \delta_{\alpha,\alpha_1} H^\lambda_\alpha)(\alpha_2, \ldots, \alpha_n),$$

where δ_{α,α_1} is the Kronecker symbol; we agree that, if $(\alpha_1, \ldots, \alpha_n)$ is the empty word, then $X_\alpha^\lambda(\alpha_1, \ldots, \alpha_n)$ is zero.

Show that Lemmas 1 and 2 remain true for these endomorphisms. We thus obtain a representation $\rho_\lambda : \mathfrak{a} \to \mathfrak{gl}(\mathrm{E})$ such that

$$\rho_\lambda(x_\alpha) = X_\alpha^\lambda, \quad \rho_\lambda(h_\alpha) = H_\alpha^\lambda, \quad \rho_\lambda(x_{-\alpha}) = X_{-\alpha}^\lambda.$$

¶ 2) Retain the notations of no. 3. Let $\mathfrak{m}$ be an ideal of $\mathfrak{a}$.

a) Let α and β be two distinct elements of B. Assume that $(\operatorname{ad} x_\alpha)^{\mathrm{N}} x_\beta$ belongs to $\mathfrak{m}$ for N sufficiently large. Show that $x_{\alpha\beta} \in \mathfrak{m}$. (Apply the results of §1, no. 2 to $\mathfrak{a}/\mathfrak{m}$, provided with a suitable $\mathfrak{sl}(2,k)$-module structure.)

b) Show that $\mathfrak{n} + \theta\mathfrak{n}$ is the smallest finite codimensional ideal of $\mathfrak{a}$. Show that this is also the smallest ideal containing the $(\operatorname{ad} x_\alpha)^4 x_\beta$ and the $(\operatorname{ad} x_{-\alpha})^4 x_{-\beta}$.

3) Let $(\mathfrak{g}, \mathfrak{h})$ be a split semi-simple Lie algebra, R the corresponding root system, and B a basis of R. For any $\alpha \in \mathrm{B}$ (resp. for any pair $(\alpha, \beta) \in \mathrm{B}^2$), let $\mathrm{R}(\alpha)$ (resp. $\mathrm{R}(\alpha, \beta)$) be the closed subset of R formed by the $\pm\alpha$ (resp. the smallest closed subset of R containing $\pm\alpha$ and $\pm\beta$). Let $\mathfrak{g}(\alpha)$ (resp. $\mathfrak{g}(\alpha, \beta)$) be the derived algebra of the algebra $\mathfrak{h} + \mathfrak{g}^{\mathrm{R}(\alpha)}$ (resp. $\mathfrak{h} + \mathfrak{g}^{\mathrm{R}(\alpha,\beta)}$), cf. §3.

a) Show that $\mathfrak{g}(\alpha) = kH_\alpha \oplus \mathfrak{g}^\alpha \oplus \mathfrak{g}^{-\alpha}$; it is isomorphic to $\mathfrak{sl}(2, k)$.

b) Show that $\mathfrak{g}(\alpha, \beta)$ is semi-simple, and that it is generated by $\mathfrak{g}(\alpha)$ and $\mathfrak{g}(\beta)$. Its root system can be identified with $\mathrm{R}(\alpha, \beta)$.

c) Let $\mathfrak{s}$ be a Lie algebra (not necessarily finite dimensional). For all $\alpha \in \mathrm{B}$, let f_α be a homomorphism from $\mathfrak{g}(\alpha)$ to $\mathfrak{s}$. Assume that, for any pair (α, β), there exists a homomorphism $f_{\alpha\beta} : \mathfrak{g}(\alpha, \beta) \to \mathfrak{s}$ that extends both f_α and f_β. Show that, in that case, there exists a unique homomorphism $f : \mathfrak{g} \to \mathfrak{s}$ that extends the f_α. (Use Prop. 4 (i).)

4) Let $\mathfrak{g}$ be a splittable semi-simple Lie algebra, and σ an automorphism of k. Let $\mathfrak{g}_\sigma$ be the Lie algebra obtained from $\mathfrak{g}$ by extending scalars by means of σ. Show that $\mathfrak{g}_\sigma$ is isomorphic to $\mathfrak{g}$. (Use the Cor. of Prop. 4.)

5) *a*) Let $\mathfrak{g}$ be a simple Lie algebra, and k_1 the commutant of the adjoint representation of $\mathfrak{g}$. Show that k_1 is a commutative field, a finite extension of k, and that $\mathfrak{g}$ is an absolutely simple k_1-Lie algebra.

Conversely, if k_1 is a finite extension of k, and $\mathfrak{g}$ an absolutely simple k_1-Lie algebra, then $\mathfrak{g}$ is a simple k-Lie algebra, and the commutant of its adjoint representation can be identified with k_1.

b) Let k' be a Galois extension of k containing k_1. Show that $\mathfrak{g}_{(k')}$ is a product of $[k_1 : k]$ absolutely simple algebras. When $\mathfrak{g}_{(k')}$ is split, these algebras are mutually isomorphic. (Use Exerc. 4.)

6) Let A be a commutative ring, and let $\mathfrak{u}$ be the A-Lie algebra defined by the family of generators $\{x, y\}$ and the relations

$$[x,[x,y]]=0, \quad [y,[y,[y,x]]]=0.$$

Show that $\mathfrak{u}$ is a free A-module with basis

$$\{x, y, [x,y], [y,[x,y]]\}.$$

Show that, when $A = k$, $\mathfrak{u}$ is isomorphic to the algebra $\mathfrak{a}_+/\mathfrak{n}$ corresponding to a root system of type B_2.

¶ 7) Let A be a commutative ring in which 2 is invertible, and let $\mathfrak{u}$ be the A-Lie algebra defined by the family of generators $\{x, y\}$ and the relations

$$[x,[x,y]]=0, \quad [y,[y,[y,[y,x]]]]=0.$$

Show that $\mathfrak{u}$ is a free A-module with basis

$$\{x, y, [x,y], [y,[x,y]], [y,[y,[x,y]]], [x,[y,[y,[x,y]]]]\}.$$

Show that, when $A = k$, $\mathfrak{u}$ is isomorphic to the algebra $\mathfrak{a}_+/\mathfrak{n}$ corresponding to a root system of type G_2.

§5

1) The index of $\mathrm{Aut}_0(\mathfrak{g})$ in $\mathrm{Aut}(\mathfrak{g})$ is finite.

2) We have $\mathrm{Aut}_e(\mathfrak{g}) = \mathrm{Aut}(\mathfrak{g})$ if $\mathfrak{g}$ is splittable, simple, and of type G_2, F_4 or E_8.

3) Let $\mathfrak{h}$ be a splitting Cartan subalgebra of $\mathfrak{g}$, $\mathfrak{b}$ a Borel subalgebra of $(\mathfrak{g}, \mathfrak{h})$, $\mathfrak{n} = [\mathfrak{b}, \mathfrak{b}]$ and $N = \exp \mathrm{ad}_{\mathfrak{g}} \mathfrak{n}$. Then

$$\mathrm{Aut}(\mathfrak{g}) = N.\mathrm{Aut}(\mathfrak{g}, \mathfrak{h}).N.$$

(Let $s \in \mathrm{Aut}(\mathfrak{g})$. Apply Prop. 10 of §3, no. 3 to $\mathfrak{b} \cap s(\mathfrak{b})$, then apply Chap. VII, §3, no. 4, Th. 3.)

4) Let $\mathfrak{h}$ be a Cartan subalgebra of $\mathfrak{g}$, and s an element of $\mathrm{Aut}(\mathfrak{g}, \mathfrak{h})$ such that $sH \neq H$ for all non-zero H in $\mathfrak{h}$. Show that s is of finite order. (Reduce to the case in which $\mathfrak{h}$ is splitting, and choose an integer $n \geq 1$ such that $\varepsilon(s)^n = 1$. Then there exists $\varphi \in T_Q$ such that $f(\varphi) = s^n$. Let σ be the transpose of $s|\mathfrak{h}$. Show that $1 + \sigma + \sigma^2 + \cdots + \sigma^{n-1} = 0$, and deduce that $s^{n^2} = 1$.)

¶ 5) *a)* Let $a \in \mathrm{Aut}(\mathfrak{g})$, and $\mathfrak{n}$ the nilspace of $a - 1$. Show that the following conditions are equivalent:
(i) $\mathrm{Ker}(a-1)$ is a Cartan subalgebra of $\mathfrak{g}$.
(ii) $\mathfrak{n}$ is a Cartan subalgebra of $\mathfrak{g}$, and $a \in \mathrm{Aut}_0(\mathfrak{g})$.
(iii) $\dim \mathfrak{n} = \mathrm{rk}(\mathfrak{g})$ and $a \in \mathrm{Aut}_0(\mathfrak{g})$.

b) Assume from now on that k is algebraically closed. Let V be a vector space, R a root system in V, T_Q the group $\mathrm{Hom}(Q(R), k^*)$, n an integer ≥ 1 and T_n the subgroup of T_Q consisting of the elements whose order divides n. Let

ζ be a primitive nth root of unity in k. For all $H \in \mathrm{P}(\mathrm{R}^\vee)$, let $\psi(H)$ be the element $\gamma \mapsto \zeta^{\gamma(H)}$ of $\mathrm{T_Q}$. The map ψ is a homomorphism from $\mathrm{P}(\mathrm{R}^\vee)$ to T_n, with kernel $n\mathrm{P}(\mathrm{R}^\vee)$. Let $t \in \mathrm{T}_n$, and let C be an alcove in $\mathrm{P}(\mathrm{R}^\vee) \otimes \mathbf{R}$. There exists $w \in \mathrm{W}(\mathrm{R})$ and $H \in \mathrm{P}(\mathrm{R}^\vee)$ such that $\frac{1}{n}H \in \overline{\mathrm{C}}$ and $\psi(wH) = t$. (Use Chap. VI, §2, no. 1.)

c) Let $\mathfrak{h}$ be a Cartan subalgebra of $\mathfrak{g}$, and R the root system of $(\mathfrak{g}, \mathfrak{h})$. Let n, ζ, ψ be as in *b*), and f as in no. 2. Let $H \in \mathrm{P}(\mathrm{R}^\vee)$. The set of elements of $\mathfrak{g}$ invariant under $f(\psi(H))$ is $\mathfrak{h} \oplus \sum_{\alpha \in \mathrm{R}'} \mathfrak{g}^\alpha$, where R′ is the set of $\alpha \in \mathrm{R}$ such that $\alpha(H) \in n\mathbf{Z}$, and $f(\psi(H))$ satisfies the conditions of *a*) if and only if $\frac{1}{n}H$ belongs to an alcove.

d) Assume from now on that $\mathfrak{g}$ is simple. Let $\mathfrak{h}$ and R be as in *c*), $(\alpha_1, \ldots, \alpha_l)$ a basis of R, h the Coxeter number of R, ζ a primitive hth root of unity in k, and H the element of $\mathfrak{h}$ such that $\alpha_i(H) = 1$ for $i = 1, \ldots, l$. Prove the following properties:
(i) The homomorphism $\gamma \mapsto \zeta^{\gamma(H)}$ from $\mathrm{Q}(\mathrm{R})$ to k^* defines an element of $\mathrm{Aut}(\mathfrak{g}, \mathfrak{h})$ which satisfies the conditions in *a*), and has order h. (Use *c*), Chap. VI, §2, Prop. 5 and Chap. VI, §1, Prop. 31.)
(ii) Every automorphism of $\mathfrak{g}$ of finite order satisfies the conditions in *a*) and is of order $\geq h$.
(iii) The automorphisms of $\mathfrak{g}$ of order h satisfying the conditions of *a*) form a conjugacy class in $\mathrm{Aut}_e(\mathfrak{g})$. (Use Prop. 5 of no. 3.)

e) Let $\mathfrak{h}$ and R be as in *c*), and let w be a Coxeter transformation in W(R). Let $s \in \mathrm{Aut}(\mathfrak{g}, \mathfrak{h})$ be such that $\varepsilon(s) = w$. Show that s satisfies the conditions in *a*), and is of order h. (Use Chap. VI, §1, Prop. 33, Chap. V, §6, no. 2 and Chap. VII, §4, Prop. 9.)

f) If $s \in \mathrm{Aut}(\mathfrak{g})$, the following conditions are equivalent:
(i) s satisfies the conditions in *a*), and is of order h;
(ii) there exists a Cartan subalgebra $\mathfrak{h}$ of $\mathfrak{g}$ stable under s such that $s|\mathfrak{h}$ is a Coxeter transformation in the Weyl group of $(\mathfrak{g}, \mathfrak{h})$. (Use *d*) and *e*).)

g) The characteristic polynomial of the automorphism in *d*) (i) is

$$\mathrm{A}(\mathrm{T}) = (\mathrm{T}-1)^l \prod_{\alpha \in \mathrm{R}} (\mathrm{T} - \zeta^{\alpha(H)}).$$

That of the automorphism s in *e*) is

$$\mathrm{B}(\mathrm{T}) = (\mathrm{T}^h - 1)^l \prod_{i=1}^{l} (\mathrm{T} - \zeta^{m_i}) \quad (m_i \text{ being the exponents of R}).$$

(Use Prop. 33 (iv) of Chap. VI, §1, no. 11.) Deduce from the relation $\mathrm{A}(\mathrm{T}) = \mathrm{B}(\mathrm{T})$ that, for all $j \geq 1$, the number of i such that $m_i \geq j$ is equal to the number of $\alpha \in \mathrm{R}_+$ such that $\alpha(H) = j$; hence recover the result of Exerc. 6 *c*) of Chap. VI, §4.[12]

6) Assume that $k = \mathbf{R}$ or $\mathbf{C}$. Let G be the Lie group $\mathrm{Aut}_0(\mathfrak{g})$. Show that an element $a \in \mathrm{G}$ is regular (in the sense of Chap. VII, §4, no. 2) if and only if it satisfies the conditions of Exerc. 5 *a*).

7) Assume that $\mathfrak{g}$ is splittable. Let $\mathrm{B}(\mathfrak{g})$ be the canonical basis of the canonical Cartan subalgebra of $\mathfrak{g}$ (no. 3, Remark 2). If $s \in \mathrm{Aut}(\mathfrak{g})$, s induces a permutation of $\mathrm{B}(\mathfrak{g})$; denote by $\mathrm{sgn}(s)$ the sign of this permutation. Show that s operates on $\bigwedge^n \mathfrak{g}$ (with $n = \dim \mathfrak{g}$) by

$$x \mapsto \mathrm{sgn}(s).x.$$

¶ 8) Let $\bar{k}$ be an algebraic closure of k, and $\bar{\mathfrak{g}} = \bar{k} \otimes_k \mathfrak{g}$. The Galois group $\mathrm{Gal}(\bar{k}/k)$ operates naturally on $\bar{\mathfrak{g}}$, on the canonical Cartan subalgebra of $\bar{\mathfrak{g}}$ (no. 3, Remark 2), as well as on its root system $\bar{\mathrm{R}}$ and its canonical basis $\bar{\mathrm{B}}$. We thus obtain a continuous homomorphism (i.e. with open kernel)

$$\pi : \mathrm{Gal}(\bar{k}/k) \to \mathrm{Aut}(\bar{\mathrm{R}}, \bar{\mathrm{B}}).$$

Show that $\mathfrak{g}$ is splittable if and only if the following two conditions are satisfied:
(i) The homomorphism π is trivial.
(ii) $\mathfrak{g}$ has a Borel subalgebra $\mathfrak{b}$.
(Show that a Cartan subalgebra $\mathfrak{h}$ contained in $\mathfrak{b}$ is splitting if and only if π is trivial.)

9) Let R be a reduced root system, B a basis of R, and $(\mathfrak{g}_0, \mathfrak{h}_0, \mathrm{B}, (X_\alpha)_{\alpha\in\mathrm{B}})$ a corresponding framed semi-simple Lie algebra (§4). Let $\bar{k}$ be an algebraic closure of k, and $\rho : \mathrm{Gal}(\bar{k}/k) \to \mathrm{Aut}(\mathrm{R}, \mathrm{B})$ a continuous homomorphism (cf. Exerc. 8); if $\sigma \in \mathrm{Gal}(\bar{k}/k)$, denote by ρ_σ the k-linear automorphism of $\bar{\mathfrak{g}}_0 = \bar{k} \otimes_k \mathfrak{g}_0$ such that $\rho_\sigma(X_\alpha) = X_{\rho(\sigma)\alpha}$. On the other hand, the natural operation of $\mathrm{Gal}(\bar{k}/k)$ on $\bar{k}$ can be extended to an operation on $\bar{\mathfrak{g}}_0$. Let $\mathfrak{g}$ be the subset of $\bar{\mathfrak{g}}_0$ consisting of the elements x such that $\rho_\sigma(x) = \sigma^{-1}.x$ for all $\sigma \in \mathrm{Gal}(\bar{k}/k)$.

a) Show that $\mathfrak{g}$ is a k-Lie subalgebra of $\bar{\mathfrak{g}}_0$, and that the injection of $\mathfrak{g}$ into $\bar{\mathfrak{g}}_0$ extends to an isomorphism from $\bar{k} \otimes_k \mathfrak{g}$ to $\bar{\mathfrak{g}}_0$. In particular, $\mathfrak{g}$ is semi-simple.

b) Let $\mathfrak{b}_0$ be the subalgebra of $\mathfrak{g}_0$ generated by $\mathfrak{h}_0$ and the X_α. Put

$$\bar{\mathfrak{b}}_0 = \bar{k} \otimes_k \mathfrak{b}_0, \quad \mathfrak{h} = \mathfrak{g} \cap \bar{\mathfrak{b}}_0, \quad \bar{\mathfrak{h}}_0 = \bar{k} \otimes_k \mathfrak{h}_0, \quad \mathfrak{h} = \mathfrak{g} \cap \bar{\mathfrak{h}}_0.$$

Show that $\bar{k} \otimes_k \mathfrak{b} = \bar{\mathfrak{b}}_0$ and that $\bar{k} \otimes_k \mathfrak{h} = \bar{\mathfrak{h}}_0$, so that $\mathfrak{b}$ is a Borel subalgebra of $\mathfrak{g}$ and $\mathfrak{h}$ is a Cartan subalgebra contained in it.

[12] For more details on this exercise, see: B. KOSTANT, The principal three-dimensional subgroup and the Betti numbers of a complex simple Lie group, *Amer. J. Math.*, Vol. LXXXI (1959), pp. 973-1032.

c) Show that the homomorphism π associated to the algebra $\mathfrak{g}$ (cf. Exerc. 8) is equal to ρ.

¶ 10) Let $\mathfrak{h}$ be a splitting Cartan subalgebra of $\mathfrak{g}$, and $(X_\alpha)_{\alpha\in R}$ a Chevalley system in $(\mathfrak{g}, \mathfrak{h})$, cf. §2, no. 4. If $\alpha \in R$, put

$$\theta_\alpha = e^{\mathrm{ad}\ X_\alpha} e^{\mathrm{ad}\ X_{-\alpha}} e^{\mathrm{ad}\ X_\alpha},$$

and denote by $\overline{W}$ the subgroup of $\mathrm{Aut}_e(\mathfrak{g}, \mathfrak{h})$ generated by the θ_α.

a) Show that $\varepsilon(\overline{W}) = W(R)$.

b) Let $s \in \overline{W}$, and let $w = \varepsilon(s)$. Show that

$$s(X_\alpha) = \pm X_{w(\alpha)} \qquad \text{for all } \alpha \in R.$$

(Use Exerc. 5 of §2.)

c) Let M be the kernel of $\varepsilon : \overline{W} \to W(R)$. Show that M is contained in the subgroup of $f(T_Q)$ consisting of the elements $f(\varphi)$ such that $\varphi^2 = 1$. Show that M contains the elements $f(\varphi_\alpha)$ defined by $\varphi_\alpha(\beta) = (-1)^{\langle\beta,\alpha^\vee\rangle}$ (remark that $\theta_\alpha^2 = f(\varphi_\alpha)$).

d) *Let $\varphi \in \mathrm{Hom}(Q, \{\pm 1\})$. Show that $f(\varphi)$ belongs to M if and only if φ extends to a homomorphism from P to $\{\pm 1\}$. (Sufficiency follows from the fact that M contains the $f(\varphi_\alpha)$. To prove necessity, reduce to the case in which $k = \mathbf{Q}$, and use the fact that M is contained in $f(T_Q) \cap \mathrm{Aut}_e(\mathfrak{g}) = \mathrm{Im}(T_P)$, cf. §7, Exerc. 26 *d*).) Deduce that M is isomorphic to the dual of the group $Q/(Q \cap 2P)$.[13]*

11) With the notations of no. 2, assume that k is non-discrete and *locally compact*, hence isomorphic to $\mathbf{R}, \mathbf{C}$ or a finite extension of $\mathbf{Q}_p$ (*Commutative Algebra*, Chap. VI, §9, no. 3). For all $n \geq 1$, the quotient k^*/k^{*n} is finite (cf. *Commutative Algebra*, Chap. VI, §9, Exerc. 3 for the ultrametric case). Deduce that the quotients $T_Q/\mathrm{Im}(T_P)$ and $\mathrm{Aut}(\mathfrak{g})/\mathrm{Aut}_e(\mathfrak{g})$ are finite.

When $k = \mathbf{R}$, show that $T_Q/\mathrm{Im}(T_P)$ is isomorphic to the dual of the $\mathbf{F}_2$-vector space $(Q \cap 2P)/2Q$. When $k = \mathbf{C}$, we have $T_Q = \mathrm{Im}(T_P)$; this is the integral subgroup of the Lie group $\mathrm{Aut}(\mathfrak{g})$ with Lie algebra $\mathfrak{h}$.

¶ 12) Let $\mathfrak{h}$ be a splitting Cartan subalgebra of $\mathfrak{g}$, A a subset of $\mathfrak{h}$, and $s \in \mathrm{Aut}(\mathfrak{g})$ such that $sA = A$. Show that there exists $t \in \mathrm{Aut}(\mathfrak{g}, \mathfrak{h})$ such that $t|A = s|A$ and $ts^{-1} \in \mathrm{Aut}_e(\mathfrak{g})$. (Let $\mathfrak{a}$ be the commutant of A in $\mathfrak{g}$; this is a reductive subalgebra of $\mathfrak{g}$, of which $s\mathfrak{h}$ and $\mathfrak{h}$ are splitting Cartan subalgebras; deduce the existence of $u \in \mathrm{Aut}_e(\mathfrak{a})$ such that $us\mathfrak{h} = \mathfrak{h}$; show that there exists $v \in \mathrm{Aut}_e(\mathfrak{g})$ extending u such that $v|A = \mathrm{Id}_A$; take $t = vs$.) Deduce that, if $s \in \mathrm{Aut}_0(\mathfrak{g})$, there exists $w \in W(R)$ such that $w|A = s|A$.

¶ 13) Let $(\mathfrak{g}, \mathfrak{h}, B, (X_\alpha)_{\alpha\in B})$ be a framed semi-simple Lie algebra, R the root system of $(\mathfrak{g}, \mathfrak{h})$, Δ the corresponding Dynkin graph, and Φ a subgroup of

[13] For more details on this exercise, see: J. TITS, Normalisateurs de tores. I. Groupes de Coxeter étendus, *J. Alg.*, Vol. IV (1966), pp. 96-116.

$\mathrm{Aut}(\mathrm{R},\mathrm{B}) = \mathrm{Aut}(\Delta)$.

If $s \in \Phi$, extend s to an automorphism of $\mathfrak{g}$ by the conditions

$$s(X_\alpha) = X_{s\alpha}, \quad \text{and} \quad s(H_\alpha) = H_{s\alpha} \quad \text{for all } \alpha \in \mathrm{B}, \text{ cf. Prop. 1.}$$

We thus identify Φ with a subgroup of $\mathrm{Aut}(\mathfrak{g},\mathfrak{h})$; denote by $\tilde{\mathfrak{g}}$ (resp. $\tilde{\mathfrak{h}}$) the subalgebra of $\mathfrak{g}$ (resp. $\mathfrak{h}$) consisting of the elements invariant under Φ.

a) Let $\alpha \in \mathrm{B}$, and let $\mathrm{X} = \Phi.\alpha$. Show, by using the Plates in Chap. VI, that only two cases are possible:
(i) every element of X distinct from α is orthogonal to α;
(ii) there exists a unique element α' of $\mathrm{X} - \{\alpha\}$ that is not orthogonal to α, and $n(\alpha,\alpha') = n(\alpha',\alpha) = -1$.

b) Let i be the restriction map $\mathfrak{h}^* \to \tilde{\mathfrak{h}}^*$, and let $\tilde{\mathrm{B}} = i(\mathrm{B})$; the map $\mathrm{B} \to \tilde{\mathrm{B}}$ identifies $\tilde{\mathrm{B}}$ with B/Φ. Show that $\tilde{\mathfrak{g}}$ is a semi-simple Lie algebra, that $\tilde{\mathfrak{h}}$ is a splitting Cartan subalgebra of it, and that $\tilde{\mathrm{B}}$ is a basis of $\mathrm{R}(\tilde{\mathfrak{g}},\tilde{\mathfrak{h}})$. (Observe that $\tilde{\mathrm{B}}$ is contained in $\mathrm{R}(\tilde{\mathfrak{g}},\tilde{\mathfrak{h}})$, and that every element of $\mathrm{R}(\tilde{\mathfrak{g}},\tilde{\mathfrak{h}})$ is a linear combination of elements of $\tilde{\mathrm{B}}$ with integer coefficients of the same sign.) If $\tilde{\alpha} \in \tilde{\mathrm{B}}$, the corresponding inverse root $H_{\tilde{\alpha}} \in \tilde{\mathfrak{h}}$ is given by

$$H_{\tilde{\alpha}} = \sum_{i(\alpha)=\tilde{\alpha}} H_\alpha \quad \text{in case (i) of } a)$$

$$H_{\tilde{\alpha}} = 2\sum_{i(\alpha)=\tilde{\alpha}} H_\alpha \quad \text{in case (ii) of } a),$$

where the summation is over those elements $\alpha \in \mathrm{B}$ such that $i(\alpha) = \tilde{\alpha}$. If $\beta \in \mathrm{B}$ has image $\tilde{\beta} \in \tilde{\mathrm{B}}$, then

$$n(\tilde{\beta},\tilde{\alpha}) = \sum_{i(\alpha)=\tilde{\alpha}} n(\beta,\alpha) \quad \text{in case (i)}$$

$$n(\tilde{\beta},\tilde{\alpha}) = 2\sum_{i(\alpha)=\tilde{\alpha}} n(\beta,\alpha) \quad \text{in case (ii).}$$

Deduce how the Dynkin graph of $\mathrm{R}(\tilde{\mathfrak{g}},\tilde{\mathfrak{h}})$ is determined starting from the pair (Δ,Φ).

c) Show that, if $\mathfrak{g}$ is simple, so is $\tilde{\mathfrak{g}}$.

If $\mathfrak{g}$ is of type A_l, $l \geq 2$, and Φ is of order 2, then $\tilde{\mathfrak{g}}$ is of type $\mathrm{B}_{l/2}$ if l is even, and of type $\mathrm{C}_{(l+1)/2}$ if l is odd.

If $\mathfrak{g}$ is of type D_l, $l \geq 4$, and Φ is of order 2, then $\tilde{\mathfrak{g}}$ is of type B_{l-1}.

If $\mathfrak{g}$ is of type D_4, and Φ is of order 3 or 6, then $\tilde{\mathfrak{g}}$ is of type G_2.

If $\mathfrak{g}$ is of type E_6, and Φ is of order 2, then $\tilde{\mathfrak{g}}$ is of type F_4.

§6

1) Show that Z(λ) can be defined as a quotient of the representation ρ_λ of §4, Exerc. 1.

2) Let μ be a weight of Z(λ) (resp. E(λ)). Show that there exists a sequence of weights $\mu_0, \dots, \mu_n$ of Z(λ) (resp. of E(λ)) such that $\mu_0 = \lambda, \mu_n = \mu$, and $\mu_{i-1} - \mu_i \in \mathrm{B}$ for $1 \leq i \leq n$.

3) Assume that $\mathfrak{g}$ is simple, and denote by $\tilde{\alpha}$ the highest root of R. The module E($\tilde{\alpha}$) is isomorphic to $\mathfrak{g}$, equipped with the adjoint representation. If C is the Casimir element associated to the Killing form of $\mathfrak{g}$, the image of C in End E($\tilde{\alpha}$) is the identity (cf. Chap. I, §3, no. 7, Prop. 12). Deduce, by using the Cor. of Prop. 7, that

$$\Phi_{\mathrm{R}}(\tilde{\alpha}, \tilde{\alpha} + 2\rho) = 1,$$

where Φ_{R} is the canonical bilinear form on $\mathfrak{h}^*$ (Chap. VI, §1, no. 12).

4) We use the notations of no. 4.

a) Let $m \in \mathbf{N}$. Give $\mathbf{N}^m$ the product order. Show that, for any subset S of $\mathbf{N}^m$, the set of minimal elements of S is finite.

b) Let $\alpha_1, \dots, \alpha_m$ be distinct elements of R, $X_i \in \mathfrak{g}^{\alpha_i} - \{0\}$, S the set of non-zero sequences $(p_i) \in \mathbf{N}^m$ such that $\sum p_i \alpha_i = 0$, M the set of minimal elements of S. Then $\mathfrak{h}$ and the $X_1^{p_1} \dots X_m^{p_m}$, where $(p_1, \dots, p_m) \in \mathrm{M}$, generate the algebra U^0.

c) Show that U^0 is both a left- and right-noetherian algebra. (Give U^0 the filtration induced by that of U($\mathfrak{g}$), and show, by using *a*) and *b*), that gr U^0 is commutative of finite type.)

d) Show that, for all $\lambda \in \mathfrak{h}^*$, U^λ is a left (resp. right) U^0-module of finite type.

e) Let V be a simple $\mathfrak{g}$-module such that $\mathrm{V} = \bigoplus_{\lambda \in \mathfrak{h}^*} \mathrm{V}_\lambda$. If one of the $\mathrm{V}_\lambda \neq 0$ is finite dimensional, then all of the V_λ are finite dimensional. (Use *d*).)

5) Show that, if $\mathfrak{g} = \mathfrak{sl}(2, k)$, the modules Z($\lambda$) of this paragraph are isomorphic to the modules Z(λ) of §1, Exerc. 2.

§7

All the $\mathfrak{g}$-modules considered (except those in Exerc. 14 and 15) are assumed to be finite dimensional.

1) Let $\omega \in \mathrm{P}_{++}$; denote by S($\omega$) the set of weights of E(ω), in other words the smallest R-saturated subset of P containing ω (Prop. 5). If $\lambda \in \mathrm{S}(\omega)$, we have $\lambda \equiv \omega$ (mod. Q). Conversely, let $\lambda \in \mathrm{P}$ be such that $\lambda \equiv \omega$ (mod. Q); prove the equivalence of the following properties:

(i) $\lambda \in S(\omega)$;
(ii) $\omega - w\lambda \in Q_+$ for all $w \in W$;
(iii) λ belongs to the convex hull of $W.\omega$ in $\mathfrak{h}^*_{\mathbf{R}}$.
(To prove that (iii) $\Longrightarrow$ (ii), remark that $\omega - w\omega$ is a linear combination of elements of R_+ with coefficients ≥ 0; deduce, by convexity, that $\omega - w\lambda$ has the same property; since $\omega - w\lambda$ belongs to Q, this implies that $\omega - w\lambda \in Q_+$. To prove that (ii) $\Longrightarrow$ (i), choose w such that $w\lambda \in P_{++}$, and apply Cor. 2 of Prop. 3. The implication (i) $\Longrightarrow$ (iii) is immediate.)

2) Let $(R_i)_{i \in I}$ be the family of irreducible components of R, and $\mathfrak{g} = \prod_{i \in I} \mathfrak{g}_i$ the corresponding decomposition of $\mathfrak{g}$ into a product of simple algebras. Identify P with the product of the $P(R_i)$, and give each $P(R_i)$ the order relation defined by the basis $B_i = B \cap R_i$.

a) Let $\omega = (\omega_i)_{i \in I}$ be an element of $P_{++} = \prod_{i \in I} P_{++}(R_i)$. Show that the simple $\mathfrak{g}$-module $E(\omega)$ is isomorphic to the tensor product of the simple $\mathfrak{g}_i$-modules $E(\omega_i)$.

b) Let $\mathscr{M}$ (resp. $\mathscr{M}_i$) be the set of elements of P_{++} (resp. of $P_{++}(R_i)$) having the equivalent properties (i), (ii), (iii), (iv) of Prop. 6 and 7. Show that $\mathscr{M} = \prod_{i \in I} \mathscr{M}_i$, in other words that $\omega \in \mathscr{M}$ if and only if, for all $i \in I$, ω_i is either zero or a minuscule weight of R_i. Deduce that $\mathscr{M}$ is a system of representatives in P of the elements of P/Q.

c) Let E be a simple $\mathfrak{g}$-module, and $\mathscr{X}$ its set of weights. Show that $\mathscr{X}$ contains a unique element of $\mathscr{M}$, and that the multiplicity of this element is equal to the upper bound of the multiplicities of the elements of $\mathscr{X}$.

3) *a*) Let E be a $\mathfrak{g}$-module. Show the equivalence of the conditions:
(i) The rank of the semi-direct product of $\mathfrak{g}$ by E is strictly larger than that of $\mathfrak{g}$.
(ii) 0 is a weight of E.
(iii) There exists a weight of E that is radical (i.e. belongs to Q).

b) Assume that E is simple. Show that (i), (ii), (iii) are equivalent to
(iv) The highest weight of E is radical.

If these conditions are satisfied, there exists no non-zero invariant alternating bilinear form. (Use Prop. 12 and Prop. 1 (ii) of §6.)

4) Let k' be an extension of k, and $\mathfrak{g}' = \mathfrak{g}_{(k')}$. Show that every $\mathfrak{g}'$-module arises, by extension of scalars, from a $\mathfrak{g}$-module that is unique up to isomorphism.

5) Let E be a $\mathfrak{g}$-module. Show the equivalence of the conditions:
(i) E is faithful (i.e. the canonical map from $\mathfrak{g}$ to $\mathfrak{gl}(E)$ is injective).
(ii) Every root of $\mathfrak{g}$ is the difference between two weights of E.

¶6) Let φ be an involutive automorphism of $\mathfrak{g}$ whose restriction to $\mathfrak{h}$ is $-\mathrm{Id}$. Show that, if E is a $\mathfrak{g}$-module, there exists a non-degenerate symmetric bilinear form Ψ on E such that

$$\Psi(x.a, b) + \Psi(a, \varphi(x).b) = 0 \quad \text{for } x \in \mathfrak{g},\ a, b \in \mathrm{E}.$$

(Reduce to the case in which E is simple. Show that the transform of E by φ is isomorphic to the dual E^* of E, and deduce the existence of a non-degenerate bilinear form Ψ satisfying the conditions above. Show next that, if e is a primitive vector in E, then $\Psi(e, e) \neq 0$. Deduce that Ψ is symmetric.)

7) If $\lambda \in \mathrm{P}_{++}$, denote by $\rho_\lambda : \mathrm{U}(\mathfrak{g}) \to \mathrm{End}(\mathrm{E}(\lambda))$ the representation defined by the simple module $\mathrm{E}(\lambda)$. We have $\mathrm{Im}(\rho_\lambda) = \mathrm{End}(\mathrm{E}(\lambda))$; put $\mathfrak{m}_\lambda = \mathrm{Ker}(\rho_\lambda)$.

a) Show that the $\mathfrak{m}_\lambda$ are pairwise distinct, and that they are the only two-sided ideals $\mathfrak{m}$ of $\mathrm{U}(\mathfrak{g})$ such that $\mathrm{U}(\mathfrak{g})/\mathfrak{m}$ is a finite dimensional simple k-algebra.

b) If I is a finite subset of P_{++}, put $\mathfrak{m}_\mathrm{I} = \bigcap_{\lambda \in \mathrm{I}} \mathfrak{m}_\lambda$. Show that the canonical map $\mathrm{U}(\mathfrak{g})/\mathfrak{m}_\mathrm{I} \to \prod_{\lambda \in \mathrm{I}} \mathrm{U}(\mathfrak{g})/\mathfrak{m}_\lambda$ is an isomorphism, and that every two-sided ideal of $\mathrm{U}(\mathfrak{g})$ of finite codimension is equal to exactly one of the $\mathfrak{m}_\mathrm{I}$.

c) Show that the principal anti-automorphism of $\mathrm{U}(\mathfrak{g})$ transforms $\mathfrak{m}_\lambda$ to $\mathfrak{m}_{\lambda^*}$, where $\lambda^* = -w_0\lambda$ (cf. Prop. 11).

¶ 8) Let $\mathfrak{g}$ be a semi-simple Lie algebra; exceptionally, we do not assume in this exercise that $\mathfrak{g}$ is split. Let $\bar{k}$ be an algebraic closure of k and let

$$\pi : \mathrm{Gal}(\bar{k}/k) \to \mathrm{Aut}(\bar{\mathrm{R}}, \bar{\mathrm{B}})$$

be the homomorphism defined in §5, Exerc. 8. Let $\mathrm{Gal}(\bar{k}/k)$ operate, via π, on the set $\bar{\mathrm{P}}_{++}$ of dominant weights of $\bar{\mathrm{R}}$ relative to $\bar{\mathrm{B}}$; let Ω be a system of representatives of the elements of the quotient $\bar{\mathrm{P}}_{++}/\mathrm{Gal}(\bar{k}/k)$.

a) Put $\bar{\mathfrak{g}} = \bar{k} \otimes_k \mathfrak{g}$. If I is a finite subset of $\bar{\mathrm{P}}_{++}$ stable under $\mathrm{Gal}(\bar{k}/k)$, the two-sided ideal $\bar{\mathfrak{m}}_\mathrm{I}$ of $\mathrm{U}(\bar{\mathfrak{g}})$ associated to I (cf. Exerc. 7) is of the form $\bar{k} \otimes_k \mathfrak{m}_\mathrm{I}$, where $\mathfrak{m}_\mathrm{I}$ is a two-sided ideal of $\mathrm{U}(\mathfrak{g})$. Show that every two-sided ideal of $\mathrm{U}(\mathfrak{g})$ of finite codimension is obtained in this way exactly once.

b) Let $\omega \in \Omega$, $\mathrm{I}(\omega)$ its orbit under $\mathrm{Gal}(\bar{k}/k)$, and G_ω the stabilizer of ω; let k_ω be the sub-extension of $\bar{k}$ corresponding to G_ω by Galois theory. Show that $\mathrm{U}(\mathfrak{g})/\mathfrak{m}_{\mathrm{I}(\omega)}$ is a simple algebra whose centre is isomorphic to k_ω. Every two-sided ideal $\mathfrak{m}$ of $\mathrm{U}(\mathfrak{g})$ such that $\mathrm{U}(\mathfrak{g})/\mathfrak{m}$ is a finite dimensional simple algebra is equal to exactly one of the $\mathfrak{m}_{\mathrm{I}(\omega)}$.

c) The group $\mathrm{Gal}(\bar{k}/k)$ operates, via π, on the ring $\mathrm{R}(\bar{\mathfrak{g}})$; denote by $\mathrm{R}(\bar{\mathfrak{g}})^{\mathrm{inv}}$ the subring of $\mathrm{R}(\bar{\mathfrak{g}})$ consisting of the elements invariant under $\mathrm{Gal}(\bar{k}/k)$. Show that the map $[\mathrm{E}] \to [\bar{k} \otimes_k \mathrm{E}]$ extends to an injective homomorphism from $\mathrm{R}(\mathfrak{g})$ to $\mathrm{R}(\bar{\mathfrak{g}})^{\mathrm{inv}}$ whose cokernel is a torsion group; this is an isomorphism if and only if, for all $\omega \in \Omega$, $\mathrm{U}(\mathfrak{g})/\mathfrak{m}_{\mathrm{I}(\omega)}$ is an algebra of matrices over k_ω; show that this is the case when $\mathfrak{g}$ has a Borel subalgebra. [14]

[14] For more details on this exercise, see: J. TITS, Représentations linéaires irréductibles d'un groupe réductif sur un corps quelconque, *J. für die reine und angewandte Math.* Vol. CCXLVII (1971), pp. 196-220.

9) Let $\mathfrak{a}_1$ and $\mathfrak{a}_2$ be Lie algebras. Show that there exists a unique homomorphism

$$f : \mathrm{R}(\mathfrak{a}_1) \otimes_{\mathbf{Z}} \mathrm{R}(\mathfrak{a}_2) \to \mathrm{R}(\mathfrak{a}_1 \times \mathfrak{a}_2)$$

such that $f([\mathrm{E}_1] \otimes [\mathrm{E}_2]) = [\mathrm{E}_1 \otimes \mathrm{E}_2]$ if E_i is an $\mathfrak{a}_i$-module ($i = 1, 2$). Show that f is injective, and that it is bijective if $\mathfrak{a}_1$ and $\mathfrak{a}_2$ are splittable semi-simple.

10) Let Γ be a subgroup of P containing Q. Such a subgroup is stable under W.

a) Show that, if $\lambda \in \mathrm{P}_{++} \cap \Gamma$, every weight of $\mathrm{E}(\lambda)$ belongs to Γ.

b) Let $\mathrm{R}_\Gamma(\mathfrak{g})$ be the subgroup of $\mathrm{R}(\mathfrak{g})$ with basis the $[\lambda]$, with $\lambda \in \mathrm{P}_{++} \cap \Gamma$. If E is a $\mathfrak{g}$-module, $[\mathrm{E}] \in \mathrm{R}_\Gamma(\mathfrak{g})$ if and only if the weights of E belong to Γ. Deduce that $\mathrm{R}_\Gamma(\mathfrak{g})$ is a subring of $\mathrm{R}(\mathfrak{g})$.

c) Show that the homomorphism $\mathrm{ch} : \mathrm{R}_\Gamma(\mathfrak{g}) \to \mathbf{Z}[\Gamma]$ is an isomorphism from $\mathrm{R}_\Gamma(\mathfrak{g})$ to the subring of $\mathbf{Z}[\Gamma]$ consisting of the elements invariant under W. (Use Th. 2 (ii).)

d) Describe $\mathrm{R}(\mathfrak{g})$ and $\mathrm{R}_\Gamma(\mathfrak{g})$ explicitly when $\mathfrak{g} = \mathfrak{sl}(2, k)$ and $\Gamma = \mathrm{Q}$.

¶ 11) The notations are those of no. 7. For any integer $m \geq 1$, denote by Ψ^m the endomorphism of $\mathbf{Z}[\Delta]$ that takes e^λ to $e^{m\lambda}$. We have $\Psi^1 = \mathrm{Id}$ and $\Psi^m \circ \Psi^n = \Psi^{mn}$.

Let E be a finite dimensional Δ-graded vector space. For all $n \geq 0$, denote by $a_n\mathrm{E}$ (resp. $s_n\mathrm{E}$) the nth exterior (resp. symmetric) power of E, equipped with its natural grading.

a) Show that

$$n\,\mathrm{ch}(s_n\mathrm{E}) = \sum_{m=1}^{n} \Psi^m(\mathrm{ch}(\mathrm{E}))\mathrm{ch}(s_{n-m}\mathrm{E})$$

and

$$n\,\mathrm{ch}(a_n\mathrm{E}) = \sum_{m=1}^{n} (-1)^{m-1}\Psi^m(\mathrm{ch}(\mathrm{E}))\mathrm{ch}(a_{n-m}\mathrm{E}).$$

Deduce that $\mathrm{ch}(s_n\mathrm{E})$ and $\mathrm{ch}(a_n\mathrm{E})$ can be expressed as polynomials, with rational coefficients, in the $\Psi^m(\mathrm{ch}(\mathrm{E}))$, $1 \leq m \leq n$. For example:

$$\mathrm{ch}(s_2\mathrm{E}) = \frac{1}{2}\mathrm{ch}(\mathrm{E})^2 + \frac{1}{2}\Psi^2(\mathrm{ch}(\mathrm{E}))$$

$$\mathrm{ch}(a_2\mathrm{E}) = \frac{1}{2}\mathrm{ch}(\mathrm{E})^2 - \frac{1}{2}\Psi^2(\mathrm{ch}(\mathrm{E})).$$

b) Prove the following identities (in the algebra of formal power series in a variable T, with coefficients in $\mathbf{Q}[\Delta]$):

$$\sum_{n=0}^{\infty} \mathrm{ch}(s_n \mathrm{E})\mathrm{T}^n = \exp\left\{\sum_{m=1}^{\infty} \Psi^m(\mathrm{ch}(\mathrm{E}))\mathrm{T}^m/m\right\}$$

and

$$\sum_{n=0}^{\infty} \mathrm{ch}(a_n \mathrm{E})\mathrm{T}^n = \exp\left\{\sum_{m=1}^{\infty} (-1)^{m-1}\Psi^m(\mathrm{ch}(\mathrm{E}))\mathrm{T}^m/m\right\}.$$

c) Assume that Δ is a group, so that Ψ^m can be defined for all $m \in \mathbf{Z}$. Show that, if E^* is the graded dual of E,

$$\mathrm{ch}(\mathrm{E}^*) = \Psi^{-1}(\mathrm{ch}(\mathrm{E})).$$

d) Identify $\mathrm{R}(\mathfrak{g})$, by means of ch, with a subring of $\mathbf{Z}[\mathrm{P}]$. Show that $\mathrm{R}(\mathfrak{g})$ is stable under the Ψ^m, $m \in \mathbf{Z}$, and that so is the subring $\mathrm{R}_\Gamma(\mathfrak{g})$ defined in the preceding exercise.

12) Let $\lambda \in \mathrm{P}_{++}$ and let $\mathscr{X}_\lambda$ be the set of weights of $\mathrm{E}(\lambda)$. Show that

$$\mathscr{X}_\lambda \subset \lambda - \mathrm{P}_{++}$$

does not hold in general. (Consider, for example, the adjoint representation of $\mathfrak{sl}(3, k)$.)

13) Let $\lambda = \sum_{\alpha \in \mathrm{B}} a_\alpha \alpha$ be an element of P_{++}. For $n = 0, 1, \ldots$, denote by $\mathscr{X}_n$ the set of weights μ of $\mathrm{E}(\lambda)$ such that $\lambda - \mu$ is the sum of n elements of B. Let s_n be the sum of the multiplicities of the elements of $\mathscr{X}_n$ (as weights of $\mathrm{E}(\lambda)$). Let $\mathrm{T} = 2\sum_{\alpha \in \mathrm{B}} a_\alpha$. Show that:

a) T is an integer ≥ 0.

b) $s_n = 0$ for $n > \mathrm{T}$, and $s_{\mathrm{T}-n} = s_n$.

c) If r is the integer part of $\mathrm{T}/2$, then $s_1 \leq s_2 \leq \cdots \leq s_{r+1}$.

¶ 14) Let $\lambda \in \mathrm{P}_{++}$, F_λ be the largest proper submodule of $\mathrm{Z}(\lambda)$ and v a primitive element of $\mathrm{Z}(\lambda)$ of weight λ, cf. §6, no. 3. Show that

$$\mathrm{F}_\lambda = \sum_{\alpha \in \mathrm{B}} \mathrm{U}(\mathfrak{g}) X_{-\alpha}^{\lambda(H_\alpha)+1} v = \sum_{\alpha \in \mathrm{B}} \mathrm{U}(\mathfrak{n}_-) X_{-\alpha}^{\lambda(H_\alpha)+1} v.$$

15) Let $\lambda \in \mathfrak{h}^*$ and let v (resp. v') be a primitive element of $\mathrm{Z}(\lambda)$ (resp. of $\mathrm{E}(\lambda)$) of weight λ. Let I (resp. I$'$) be the annihilator of v (resp. v') in $\mathrm{U}(\mathfrak{g})$.

a) $\mathrm{I} = \mathrm{U}(\mathfrak{g})\mathfrak{n}_+ + \sum_{h \in \mathfrak{h}} \mathrm{U}(\mathfrak{g})(h - \lambda(h))$.

b) I$'$ is the largest left ideal of $\mathrm{U}(\mathfrak{g})$ distinct from $\mathrm{U}(\mathfrak{g})$ and containing I.

c) If $\lambda \in \mathrm{P}_{++}$, then

$$\mathrm{I}' = \mathrm{I} + \sum_{\alpha \in \mathrm{B}} \mathrm{U}(\mathfrak{g}) X_{-\alpha}^{\lambda(H_\alpha)+1} = \mathrm{I} + \sum_{\alpha \in \mathrm{B}} \mathrm{U}(\mathfrak{n}_-) X_{-\alpha}^{\lambda(H_\alpha)+1}.$$

(Use the preceding exercise.)

¶ 16) Let V and V′ be two $\mathfrak{g}$-modules. Then V′ is said to be *subordinate* to V if there exists a linear map $f : \mathrm{V} \to \mathrm{V}'$ such that:

α) f is surjective; β) the image under f of a primitive element of V is either 0 or a primitive element of V′; γ) f is an $\mathfrak{n}_-$-homomorphism.

a) Let $f : \mathrm{V} \to \mathrm{V}'$ satisfy α), β), γ). Let v be a primitive element of V. Then the image under f of the $\mathfrak{g}$-submodule generated by v is the $\mathfrak{g}$-submodule generated by $f(v)$.

b) Let $f : \mathrm{V} \to \mathrm{V}'$ satisfy α), β), γ). Let W be a $\mathfrak{g}$-submodule of V. Then $f(\mathrm{W})$ is a $\mathfrak{g}$-submodule of V′ that is subordinate to W.

c) Let $\mathrm{V} = \mathrm{E}_1 \oplus \cdots \oplus \mathrm{E}_s$ and $\mathrm{V}' = \mathrm{E}'_1 \oplus \cdots \mathrm{E}'_{s'}$ be decompositions of V and V′ into sums of simple modules. Then V′ is subordinate to V if and only if $s' \leq s$ and there exists $\sigma \in \mathfrak{S}_s$ such that E'_i is subordinate to $\mathrm{E}_{\sigma(i)}$ for $i = 1, \ldots, s'$.

d) If V′ is subordinate to V and if V is simple, then V′ is simple or reduced to 0.

e) Assume that V and V′ are simple. Let λ and λ' be their highest weights. Then V′ is subordinate to V if and only if $\lambda'(H_\alpha) \leq \lambda(H_\alpha)$ for all $\alpha \in \mathrm{B}$. (For the sufficiency, use the preceding exercise.)

17) Let $\lambda, \mu \in \mathrm{P}_{++}$ and $\alpha \in \mathrm{B}$ be such that $\lambda(H_\alpha) \geq 1$ and $\mu(H_\alpha) \geq 1$. Let $\mathrm{F} = \mathrm{E}(\lambda) \otimes \mathrm{E}(\mu)$.

a) Show that $\dim \mathrm{F}^{\lambda+\mu} = 1$ and $\dim \mathrm{F}^{\lambda+\mu-\alpha} = 2$.

b) Show that $X_\alpha : \mathrm{F}^{\lambda+\mu-\alpha} \to \mathrm{F}^{\lambda+\mu}$ is surjective, and that the non-zero elements of its kernel are primitive (remark that, if $\beta \in \mathrm{B}$ is distinct from α, $\lambda + \mu - \alpha + \beta$ is not a weight of F).

c) Deduce that $\mathrm{E}(\lambda) \otimes \mathrm{E}(\mu)$ contains a unique submodule isomorphic to

$$\mathrm{E}(\lambda + \mu - \alpha).$$

d) Show that $\mathbf{S}^2(\mathrm{E}(\lambda))$ (resp. $\bigwedge^2 \mathrm{E}(\lambda)$) contains a unique submodule isomorphic to $\mathrm{E}(2\lambda)$ (resp. to $\mathrm{E}(2\lambda - \alpha)$).

¶ 18) Choose a positive non-degenerate symmetric bilinear form $(\cdot|\cdot)$ on $\mathfrak{h}^*_{\mathbf{R}}$ invariant under W. Let $\lambda \in \mathrm{P}_{++}$.

a) Let μ be a weight of $\mathrm{E}(\lambda)$. Write $\lambda - \mu$ in the form $\sum_{\alpha \in \mathrm{B}} k_\alpha \alpha$. Let $\alpha \in \mathrm{B}$ be such that $k_\alpha \neq 0$. Show that there exists $\alpha_1, \ldots, \alpha_n \in \mathrm{B}$ such that $(\lambda|\alpha_1) \neq 0$, $(\alpha_1|\alpha_2) \neq 0, \ldots, (\alpha_{n-1}|\alpha_n) \neq 0, (\alpha_n|\alpha) \neq 0$.

b) Let v be a primitive element of $\mathrm{E}(\lambda)$, and let $\alpha_1, \ldots, \alpha_n \in \mathrm{B}$ satisfy the following conditions:
(i) $(\alpha_i|\alpha_{i+1}) \neq 0$ for $i = 1, 2, \ldots, n-1$;
(ii) $(\alpha_i|\alpha_j) = 0$ for $j > i + 1$;
(iii) $\lambda(H_{\alpha_1}) \neq 0$ and $\lambda(H_{\alpha_2}) = \cdots = \lambda(H_{\alpha_n}) = 0$.
Show that $X_{-\alpha_n} X_{-\alpha_{n-1}} \cdots X_{-\alpha_1} v \neq 0$. (Observe that, for $1 \leq s \leq n$,

$$\lambda - \alpha_1 - \cdots - \alpha_{s-1} + \alpha_n$$

is not a weight of $\mathrm{E}(\lambda)$, and deduce that $X_{\alpha_s} X_{-\alpha_s} X_{-\alpha_{s-1}} \dots X_{-\alpha_1} v \neq 0$ by induction on s.) If $\sigma \in \mathfrak{S}_n$ and $\sigma \neq 1$, then $X_{-\alpha_{\sigma(n)}} X_{-\alpha_{\sigma(n-1)}} \dots X_{-\alpha_{\sigma(1)}} v = 0$. (Let r be the smallest integer such that $\sigma(r) \neq r$. Use *a*) to show that $\lambda - \alpha_{\sigma(1)} - \dots - \alpha_{\sigma(r)}$ is not a weight of $\mathrm{E}(\lambda)$.)

c) Let $\lambda' \in \mathrm{P}_{++}$. A *chain* joining λ to λ' is a sequence $(\alpha_1, \dots, \alpha_n)$ of elements of B such that $n \geq 1$, $(\lambda|\alpha_1) \neq 0, (\alpha_1|\alpha_2) \neq 0, \dots, (\alpha_{n-1}|\alpha_n) \neq 0$, $(\alpha_n|\lambda') \neq 0$. Such a chain is called *minimal* if no strict subsequence of $(\alpha_1, \dots, \alpha_n)$ joins λ to λ'. In that case, we have $(\alpha_i|\alpha_j) = 0$ if $|i-j| \geq 2$, $(\lambda|\alpha_i) = 0$ if $i \geq 2$ and $(\lambda'|\alpha_i) = 0$ if $i \leq n-1$.

d) Let $(\alpha_1, \dots, \alpha_n)$ be a minimal chain joining λ to λ'. If v' is a primitive vector in $\mathrm{E}(\lambda')$, put:

$$v_s = X_{-\alpha_s} X_{-\alpha_{s-1}} \dots X_{-\alpha_1} v \quad (s = 0, 1, \dots, n)$$
$$v'_s = X_{-\alpha_{s+1}} X_{-\alpha_{s+2}} \dots X_{-\alpha_n} v' \quad (s = 0, 1, \dots, n)$$

$a_0 = (\lambda|\alpha_1), a_n = (-1)^n(\lambda'|\alpha_n), a_s = (-1)^{s+1}(\alpha_s|\alpha_{s+1}), 1 \leq s \leq n-1$. Show that

$$\sum_{s=0}^{n} a_s v_s \otimes v'_s$$

is a primitive element of $\mathrm{E}(\lambda) \otimes \mathrm{E}(\lambda')$ of weight $\lambda + \lambda' - \alpha_1 - \cdots - \alpha_n$, and that it is unique up to homothety.

(Use *b*) and *c*) to show that every element of $\mathrm{E}(\lambda) \otimes \mathrm{E}(\lambda')$ of weight

$$\lambda + \lambda' - \alpha_1 - \cdots - \alpha_n$$

is a linear combination of the $v_s \otimes v'_s$. Then write down the condition for such a linear combination to be a primitive vector.)

Deduce that $\mathrm{E}(\lambda) \otimes \mathrm{E}(\lambda')$ contains a unique $\mathfrak{g}$-submodule isomorphic to

$$\mathrm{E}(\lambda + \lambda' - \alpha_1 - \cdots - \alpha_n).$$

(When $n = 1$ we recover Exerc. 17.)

e) Let w be a primitive element of $\mathrm{E}(\lambda) \otimes \mathrm{E}(\lambda')$. Assume that the weight ν of w is distinct from $\lambda + \lambda'$. Show that there exists a chain $(\alpha_1, \dots, \alpha_n)$ joining λ to λ' such that

$$\nu \leq \lambda + \lambda' - \alpha_1 - \cdots - \alpha_n.$$

(Let C be the set of $\alpha \in \mathrm{B}$ such that the coordinate of index α of $\lambda + \lambda' - \nu$ is $\neq 0$. Let D (resp. D′) be the set of $\alpha \in \mathrm{C}$ such that there exists $\gamma_1, \dots, \gamma_t \in \mathrm{C}$ satisfying $(\lambda|\gamma_1) \neq 0$ (resp. $(\lambda'|\gamma_1) \neq 0$), $(\gamma_1|\gamma_2) \neq 0, \dots, (\gamma_{t-1}|\gamma_t) \neq 0$, $(\gamma_t|\alpha) \neq 0$. Let Y (resp. Y′) be the set of weights of $\mathrm{E}(\lambda)$ (resp. $\mathrm{E}(\lambda')$) of the form $\lambda - \sum_{\alpha \in \mathrm{C}} k_\alpha \alpha$ (resp. $\lambda' - \sum_{\alpha \in \mathrm{C}} k_\alpha \alpha$), with $k_\alpha \in \mathbf{N}$. Show that w belongs to $\left(\sum_{\mu\in Y} E(\lambda)^{\mu}\right) \otimes \left(\sum_{\mu'\in Y'} E(\lambda')^{\mu'}\right)$. Using *a*) and the fact that $\nu \neq \lambda + \lambda'$, show that $D \cap D' \neq \varnothing$.)

f) Show the equivalence of the following properties:
(i) $E(\lambda) \otimes E(\lambda')$ is isomorphic to $E(\lambda + \lambda')$.
(ii) $E(\lambda) \otimes E(\lambda')$ is a simple module.
(iii) There is no chain joining λ to λ'.
(iv) R is the direct sum of two root systems R_1 and R'_1 such that $\lambda \in P(R_1)$ and $\lambda' \in P(R'_1)$.
(v) $\mathfrak{g}$ is the product of two ideals $\mathfrak{s}$ and $\mathfrak{s}'$ such that $\mathfrak{s}'.E(\lambda) = 0$ and $\mathfrak{s}.E(\lambda') = 0$.

(Use *d*) to prove the equivalence of (ii) and (iii).)[15]

19) We recall the notations of Prop. 10. Let $\bar{k}$ be an algebraic closure of k. If $x \in \mathfrak{g}$, denote by $\mathscr{X}_E(x)$ (resp. $\mathscr{X}_F(x)$, resp. $\mathscr{X}_G(x)$) the set of eigenvalues of x_E (resp. x_F, resp. x_G) in $\bar{k}$. Show that $\mathscr{X}_G(x) = \mathscr{X}_E(x) + \mathscr{X}_F(x)$ for all $x \in \mathfrak{g}$ and that, when E and F are given, this property characterizes the simple $\mathfrak{g}$-module G up to isomorphism.

20) Assume that $k = \mathbf{R}$ or $\mathbf{C}$. Let Γ be a simply-connected Lie group with Lie algebra $\mathfrak{g}$. Let λ, μ, E, F, G be as in Prop. 10. Let $(e_1, \ldots, e_n)$ (resp. $(f_1, \ldots, f_p)$) be a basis of E (resp. F) consisting of eigenvectors of $\mathfrak{h}$, with $e_1 \in E^{\lambda}$ and $f_1 \in F^{\mu}$. We can consider E, F, G as Γ-modules. If $\gamma \in \Gamma$, denote by $a_i(\gamma)$ the coordinate of $\gamma.e_1$ with index i, and $b_j(\gamma)$ the coordinate of $\gamma.f_1$ with index j. Show that the function $a_i b_j$ on Γ is not identically zero. Deduce that, for all (i, j), there exists an element of $G \subset E \otimes F$ whose coordinate of index (i, j) is $\neq 0$; for $k = \mathbf{R}$ or $\mathbf{C}$, this gives a new proof of Prop. 10. Pass from this to the case $k = \mathbf{Q}$, and then to the case of an arbitrary field, cf. Exerc. 4.

21) Let $\lambda, \mu \in P_{++}$. Let E, F, G be simple $\mathfrak{g}$-modules of highest weights $\lambda, \mu, \lambda + \mu$, and let n be an integer ≥ 1. If ω is a weight of E of multiplicity n, then $\omega + \mu$ is a weight of G of multiplicity $\geq n$. (We have $G^{\omega+\mu} \subset \bigoplus_{\nu+\sigma=\omega+\mu} E^{\nu} \otimes F^{\sigma}$. If $\dim G^{\omega+\mu} < n$, the projection of $G^{\omega+\mu}$ onto $E^{\omega} \otimes F^{\mu}$ is of the form $E' \otimes F^{\mu}$, with E' strictly contained in E^{ω}. Derive a contradiction from this by choosing adapted bases of E and F and imitating the proof of Prop. 10.)

22) Assume that $\mathfrak{g}$ is simple, in other words that R is irreducible. Show that there exists a unique dominant weight $\lambda \neq 0$ such that the set of weights of $E(\lambda)$ is $W.\lambda \cup \{0\}$: we have $\lambda = \alpha$, where $\alpha \in R$ is such that H_{α} is the highest root of $R^{\vee}$. When all the roots are of the same length (cases A_l, D_l, E_6, E_7, E_8), we have $\lambda = \tilde{\alpha}$; this is the only root that is a dominant

[15] For more details, cf. E. B. DYNKIN, Maximal subgroups of classical groups [in Russian], *Trudy Moskov. Mat. Obšč.*, Vol. 1 (1952), pp. 39-166 (= Amer. Math. Soc. Transl., Vol. 6 (1957), pp. 245-374).

weight; the corresponding representation is the adjoint representation of $\mathfrak{g}$. In the other cases, λ is the only root of minimum length that is a dominant weight; with the notations of Chap. VI, *Plates*, we have: $\lambda = \varpi_1$ (type B_l); $\lambda = \varpi_2$ (type C_l); $\lambda = \varpi_4$ (type F_4); $\lambda = \varpi_1$ (type G_2).

23) Let U^0 be the commutant of $\mathfrak{h}$ in $U(\mathfrak{g})$ (cf. §6, no. 4).

a) Let V be a (finite dimensional) $\mathfrak{g}$-module. Show that the V^λ, $\lambda \in P$, are stable under U^0, and that, if V is simple and $V^\lambda \neq 0$, V^λ is a simple U^0-module (use the decomposition $U(\mathfrak{g}) = \oplus U^\lambda$, *loc. cit.*).

b) Show that, for every element $c \neq 0$ of U^0, there exists a finite dimensional simple representation ρ of U^0 such that $\rho(c) \neq 0$. (Use *a*) and Chap. I, §7, Exerc. 3 *a*).)

24) Let A be a unital associative algebra and M the set of its finite codimensional two-sided ideals.

a) Let U^* be the vector space dual of U. If $\theta \in U^*$, show the equivalence of the following properties:
(i) there exists $\mathfrak{m} \in M$ such that $\theta(\mathfrak{m}) = 0$;
(ii) there exist two finite families (θ'_i) and (θ''_i) of elements of U^* such that

$$\theta(xy) = \sum_i \theta'_i(x)\theta''_i(y) \quad \text{for all } x, y \in U.$$

The elements θ with these properties form a subspace U' of U^*, which coincides with that denoted by B' in *Algebra*, Chap. III, §11, Exerc. 27. There exists a unique coalgebra structure on U' whose coproduct $c : U' \to U' \otimes U'$ is given by

$$c(\theta) = \sum_i \theta'_i \otimes \theta''_i,$$

where $\theta'_i, \theta''_i \in U'$ are such that $\theta(xy) = \sum_i \theta'_i(x)\theta''_i(y)$ for all $x, y \in U$, cf. (ii).

The coalgebra U' is the union of the increasing filtration by the subspaces $(U/\mathfrak{m})^*$, $\mathfrak{m} \in M$, which are finite dimensional. The dual of U' can be identified with the algebra $\hat{U} = \varprojlim U/\mathfrak{m}$; if $\hat{U}$ is given the projective limit of the discrete topologies on the $U/\mathfrak{m}$, $\mathfrak{m} \in M$, the continuous linear forms on $\hat{U}$ are given by the elements of U'.

b) Let E be a finite dimensional left U-module. Its annihilator $\mathfrak{m}_E$ belongs to M; the composite $\hat{U} \to U/\mathfrak{m}_E \to \mathrm{End}(E)$ gives E the structure of a left $\hat{U}$-module. If F is a finite dimensional left U-module, a linear map $f : E \to F$ is a U-homomorphism if and only if it is a $\hat{U}$-homomorphism. If $a \in E, b \in E^*$, the linear form $\theta_{a,b} : x \mapsto \langle xa, b\rangle$ belongs to U', and

$$\langle x, \theta_{a,b}\rangle = \langle xa, b\rangle \quad \text{for all } x \in \hat{U}.$$

The $\theta_{a,b}$ (for varying E, a, b) generate the k-vector space U'.

c) Let X_U be the set of isomorphism classes of finite dimensional left U-modules. For all $E \in X_U$, let u_E be a k-linear endomorphism of E; assume that

$$f \circ u_E = u_F \circ f \quad \text{for all } E, F \in X_U \text{ and } f \in \text{Hom}_U(E, F).$$

Show that there exists a unique element $x \in \hat{U}$ such that $x_E = u_E$ for all $E \in X_U$. (Reduce to the case in which U is finite dimensional.)

¶ 25) Let $\mathfrak{a}$ be a Lie algebra and U its enveloping algebra. Apply the definitions and results of Exerc. 24 to the algebra U. In particular, $U' \subset U^*$ and the dual of U' can be identified with the algebra $\hat{U} = \varprojlim U/\mathfrak{m}$; the canonical map $U \to \hat{U}$ is injective (Chap. I, §7, Exerc. 3).

a) The coalgebra structure of U (Chap. II, §1, no. 4) defines an algebra structure on the dual U^* (cf. Chap. II, §1, no. 5, Prop. 10, and *Algebra*, Chap. III, §11, no. 2). If E, F are finite dimensional $\mathfrak{a}$-modules,

$$\theta_{a,b} \circ \theta_{c,d} = \theta_{a \otimes c, b \otimes d} \quad \text{for } a \in E, b \in E^*, c \in F, d \in F^*.$$

Deduce that U' is a subalgebra of U^*. The coalgebra and algebra structures of U' make it a commutative *bigebra* (*Algebra*, Chap. III, §11, no. 4).

b) Let x be an element of $\hat{U}$; identify x with a linear form $U' \to k$. Prove the equivalence of the following properties:

(i) x is a homomorphism of algebras from U' to k.

(ii) $x_{E \otimes F} = x_E \otimes x_F$ for all finite dimensional $\mathfrak{a}$-modules E and F.

(Prove first that (ii) is equivalent to

(ii′) $x(\theta_{a\otimes c, b\otimes d}) = x(\theta_{a,b})x(\theta_{c,d})$ if $a \in E, b \in E^*, c \in F, d \in F^*$,

and use the fact that the $\theta_{a,b}$ generate U'.)

c) Let x be an element of $\hat{U}$ satisfying conditions (i) and (ii) of *b*). Show the equivalence of the following conditions:

(iii) x takes the unit element of U' to the unit element of k.

(iv) $x \neq 0$.

(v) If k is given the trivial $\mathfrak{a}$-module structure, we have $x_k = \text{Id}$.

(vi) x_E is invertible for all E.

(The equivalence (iii) $\Longleftrightarrow$ (iv) follows from the fact that x is a homomorphism of algebras. On the other hand, $x_k = \lambda\,\text{Id}$, with $\lambda \in k$. Using the $\mathfrak{a}$-isomorphism $k \otimes E \to E$, deduce that $\lambda x_E = x_E$ for all E, and, in particular, that $\lambda^2 = \lambda$ by taking $E = k$. The case $\lambda = 1$ corresponds to $x \neq 0$, hence (iv) $\Longleftrightarrow$ (v), and (vi) $\Longrightarrow$ (v). To prove that (v) $\Longrightarrow$ (vi), show that, if F is the dual of E, then ${}^t x_F \circ x_E = \lambda\,\text{Id}_E$.)

d) Let G be the set of elements of $\hat{U}$ satisfying conditions (i) to (vi) above. Show that G is a subgroup of the group of invertible elements of $\hat{U}$.

Let $x \in G$. If E is a finite dimensional $\mathfrak{a}$-module, then $x_E \in \mathbf{GL}(E)$. This applies in particular to $E = \mathfrak{a}$, equipped with the adjoint representation; this gives an element $x_{\mathfrak{a}} \in \mathbf{GL}(\mathfrak{a})$. Show that $x_{\mathfrak{a}}$ is an automorphism of $\mathfrak{a}$

(use the $\mathfrak{a}$-homomorphism $\mathfrak{a} \otimes \mathfrak{a} \to \mathfrak{a}$ given by the bracket). This gives a homomorphism $v : \mathrm{G} \to \mathrm{Aut}(\mathfrak{a})$. If E is a finite dimensional $\mathfrak{a}$-module,

$$x_{\mathrm{E}}(y.e) = v(x)(y).x_{\mathrm{E}}(e) \quad \text{if } y \in \mathfrak{a}, e \in \mathrm{E}$$

(use the $\mathfrak{a}$-homomorphism $\mathfrak{a} \otimes \mathrm{E} \to \mathrm{E}$ given by the operation of $\mathfrak{a}$ on E).

e) The principal anti-automorphism σ of U extends by continuity to $\hat{\mathrm{U}}$. Its transpose leaves U′ stable and induces an inversion on U′ (*Algebra*, Chap. III, §11, Exerc. 4). If $x \in \mathrm{G}$, then $\sigma(x) = x^{-1}$.

f) Assume that $\mathfrak{a}$ is semi-simple[16]. Let n be a nilpotent element of $\mathfrak{a}$. Then there exists a unique element e^n of G such that $(e^n)_{\mathrm{E}} = \exp(n_{\mathrm{E}})$ for every finite dimensional $\mathfrak{a}$-module E. We have $v(e^n) = \exp(\mathrm{ad}\, n) \in \mathrm{Aut}(\mathfrak{a})$, so $\mathrm{Aut}_e(\mathfrak{a}) \subset v(\mathrm{G})$.

Show that, if $\mathfrak{b}$ is a subalgebra of $\mathfrak{a}$ consisting of nilpotent elements, then

$$e^n.e^m = e^{\mathrm{H}(n,m)} \quad \text{for } n, m \in \mathfrak{b},$$

where H denotes the Hausdorff series (Chap. II, §6).

¶ 26) Apply the notations and results of Exerc. 25 to the case in which $\mathfrak{a} = \mathfrak{g}$ (split case).

a) Let $x \in \mathrm{G}$ and let $\sigma = v(x)$ be its image in $\mathrm{Aut}(\mathfrak{g})$. If ρ is a representation of $\mathfrak{g}$, ρ and $\rho \circ \sigma$ are equivalent. Deduce (cf. no. 2, Remark 1) that σ belongs to $\mathrm{Aut}_0(\mathfrak{g})$. Extend this result to arbitrary semi-simple algebras.

b) Let $\varphi \in \mathrm{T_P} = \mathrm{Hom}(\mathrm{P}, k^*)$, where $\mathrm{P} = \mathrm{P}(\mathrm{R})$. If E is a $\mathfrak{g}$-module, let φ_{E} be the endomorphism of E whose restriction to each E^λ ($\lambda \in \mathrm{P}$) is the homothety of ratio $\varphi(\lambda)$. Show that there exists a unique element $t(\varphi) \in \mathrm{G}$ such that $t(\varphi)_{\mathrm{E}} = \varphi_{\mathrm{E}}$ for all E (Use Exerc. 24 *c*), and the characterizations (ii) and (vi) of Exerc. 25.) This gives a homomorphism $t : \mathrm{T_P} \to \mathrm{G}$. Show that t is injective. We use this to identify $\mathrm{T_P}$ with a subgroup of G. The composite $\mathrm{T_P} \to \mathrm{G} \to \mathrm{Aut}(\mathfrak{g})$ is the homomorphism denoted by $f \circ q$ in §5, no. 2.

c) Let $x \in \mathrm{G}$ be such that $\sigma = v(x)$ belongs to the subgroup $f(\mathrm{T_Q})$ of $\mathrm{Aut}_0(\mathfrak{g})$ (§5, no. 2), in other words, such that it operates trivially on $\mathfrak{h}$; denote the element of $\mathrm{T_Q} = \mathrm{Hom}(\mathrm{Q}, k^*)$ corresponding to x by ψ. We are going to show that x belongs to $\mathrm{T_P}$. Prove successively:

c_1) If E is a $\mathfrak{g}$-module, x_{E} is an $\mathfrak{h}$-endomorphism of E.

(Use the $\mathfrak{g}$-homomorphism $\mathfrak{g} \otimes \mathrm{E} \to \mathrm{E}$, and the fact that x operates trivially on $\mathfrak{h}$.) In particular, the E^μ are stable under x_{E}.

c_2) There exists $\varphi \in \mathrm{T_P}$ such that, for every $\mathfrak{g}$-module E, and every primitive element e of E of weight λ, we have $x_{\mathrm{E}}e = \varphi(\lambda)e$.

(Choose φ such that this relation is true when E is a fundamental module $\mathrm{E}(\varpi_\alpha)$. Deduce the case of the $\mathrm{E}(\lambda)$, $\lambda \in \mathrm{P}_{++}$, by using the embedding of

[16]*In this case, it can be shown that U′ is the bigebra of the simply-connected semi-simple algebraic group A with Lie algebra $\mathfrak{a}$, and that G is the group of k-points of A.*

such a module in a tensor product of the $E(\varpi_\alpha)$. Pass from this to the general case.)

c_3) Choose φ as in c_2). Let E be a simple $\mathfrak{g}$-module of highest weight λ, and let μ be a weight of E; then $\lambda - \mu \in Q$. Show that the restriction of x_E to E^μ is the homothety of ratio $\varphi(\lambda)\psi(\mu - \lambda)$. (Same method as for c_1).)

c_4) If $\lambda, \mu \in P_{++}$, and if $\alpha \in B$ is not orthogonal to either λ or μ, then

$$\varphi(\lambda + \mu - \alpha) = \varphi(\lambda + \mu)\psi(-\alpha),$$

so $\varphi(\alpha) = \psi(\alpha)$. (Use c_2), c_3), and the embedding of $E(\lambda + \mu - \alpha)$ in $E(\lambda) \otimes E(\mu)$, cf. Exerc. 17.)

c_5) Deduce from c_4) that $\varphi|Q = \psi$, and use c_3) to deduce that $x = t(\varphi)$.

d) Identify T_Q with a subgroup of $\mathrm{Aut}_0(\mathfrak{g})$ by means of f. By *a*), we have

$$\mathrm{Aut}_e(\mathfrak{g}) \subset v(G) \subset \mathrm{Aut}_0(\mathfrak{g})$$

and, by *c*), $v(G) \cap T_Q = \mathrm{Im}(T_P)$. Deduce (cf. §5, no. 3) that $\mathrm{Aut}_e(\mathfrak{g}) \cap T_Q = \mathrm{Im}(T_P)$ and that $f(G) = \mathrm{Aut}_e(\mathfrak{g})$. The canonical map

$$\iota : T_Q/\mathrm{Im}(T_P) \to \mathrm{Aut}_0(\mathfrak{g})/\mathrm{Aut}_e(\mathfrak{g})$$

is therefore an isomorphism.

e) The kernel of $v : G \to \mathrm{Aut}_e(\mathfrak{g})$ is equal to the kernel of $T_P \to T_Q$; it is isomorphic to

$$\mathrm{Hom}(P/Q, k^*);$$

this is a finite abelian group contained in the centre of G, and its order divides $(P : Q)$; if k is algebraically closed, it is isomorphic to the dual of P/Q (*Algebra*, Chap. VII, §4, no. 8).

f) Let $\alpha \in R, X_\alpha \in \mathfrak{g}^\alpha$ and $X_{-\alpha} \in \mathfrak{g}^{-\alpha}$ be such that $[X_\alpha, X_{-\alpha}] = -H_\alpha$, and let ρ_α be the corresponding representation of $\mathfrak{sl}(2, k)$ on $\mathfrak{g}$. If E is a $\mathfrak{g}$-module, deduce (§1, no. 4) a representation of $\mathbf{SL}(2, k)$ on E, and hence (Exerc. 25 *b*), *c*)) a homomorphism

$$\varphi_\alpha : \mathbf{SL}(2, k) \to G.$$

Show that $\mathrm{Im}(\varphi_\alpha)$ contains the elements of T_P of the form $\lambda \mapsto t^{\lambda(H_\alpha)}, t \in k^*$. Deduce that the $\mathrm{Im}(\varphi_\alpha)$, $\alpha \in B$, generate G (show first that the group they generate contains T_P). In particular, G is generated by the e^n, with $n \in \mathfrak{g}^\alpha$, $\alpha \in B \cup -B$. The derived group of G is equal to G.

g) If a subgroup G′ of G is such that $v(G') = \mathrm{Aut}_e(\mathfrak{g})$, then $G' = G$ (use f)).

h) Let E be a faithful $\mathfrak{g}$-module, and let Γ be the subgroup of P generated by the weights of E. Then $P \supset \Gamma \supset Q$, cf. Exerc. 5. Show that the kernel of the canonical homomorphism $G \to \mathbf{GL}(E)$ is equal to the subgroup of T_P consisting of the elements φ whose restriction to Γ is trivial. In particular, if $\Gamma = P$ the homomorphism $G \to \mathbf{GL}(E)$ is injective. If $\Gamma = Q$, this homomorphism factorizes as $G \xrightarrow{v} \mathrm{Aut}_e(\mathfrak{g}) \longrightarrow \mathbf{GL}(E)$, and the homomorphism $\mathrm{Aut}_e(\mathfrak{g}) \to \mathbf{GL}(E)$ is injective.

27) Let $\Omega = P/Q$. If $\omega \in \Omega$, and if E is a $\mathfrak{g}$-module, denote by E_ω the direct sum of the E^λ, for $\lambda \in \omega$. We have $E = \bigoplus_{\omega \in \Omega} E_\omega$.

a) Show that E_ω is a $\mathfrak{g}$-submodule of E. We have $(E^*)_\omega = (E_{-\omega})^*$ and

$$(E \otimes F)_\omega = \bigoplus_{\alpha+\beta=\omega} E_\alpha \otimes F_\beta$$

if F is another $\mathfrak{g}$-module.

b) Let $\chi \in \mathrm{Hom}(\Omega, k^*) = \mathrm{Ker}(T_P \to T_Q)$. Identify χ with an element of the kernel of $f : G \to \mathrm{Aut}_e(\mathfrak{g})$, cf. Exerc. 26 *e*). Show that the operation of χ on E_ω is the homothety of ratio $\chi(\omega)$.

c) What are the E_ω when $\mathfrak{g} = \mathfrak{sl}(2, k)$?

§8

1) Let f be an invariant polynomial function on $\mathfrak{g}$. Show that f is invariant under $\mathrm{Aut}(\mathfrak{g})$ if and only if $f|\mathfrak{h}$ is invariant under $\mathrm{Aut}(R)$. Deduce that, if the Dynkin graph of R has a non-trivial automorphism, there exists an invariant polynomial function on $\mathfrak{g}$ that is not invariant under $\mathrm{Aut}(\mathfrak{g})$.

2) Take $\mathfrak{g} = \mathfrak{sl}(3, k)$. Show that $x \mapsto \det(x)$ is an invariant polynomial function on $\mathfrak{g}$ that is not invariant under $\mathrm{Aut}(\mathfrak{g})$ (use the automorphism $x \mapsto -{}^t x$).

3) Let $\mathfrak{a}$ be a semi-simple Lie algebra, and $s \in \mathrm{Aut}(\mathfrak{a})$. Show the equivalence of:

(i) $s \in \mathrm{Aut}_0(\mathfrak{a})$.

(ii) s operates trivially on the centre of $U(\mathfrak{a})$.

(iii) For all $x \in \mathfrak{a}$, there exists $t \in \mathrm{Aut}_0(\mathfrak{a})$ such that $tx = sx$.

(Use Prop. 6 to show that (iii) $\Longrightarrow$ (i).)

4) Show that, in Cor. 2 of Prop. 2, and in Th. 1 (ii), we can restrict ourselves to representations ρ whose weights are radical weights (remark that Prop. 1 remains valid when $k[P]^W$ is replaced by $k[Q]^W$, where Q is the group of radical weights).

5) We retain the notations of §6 and §7. Let $\lambda \in \mathfrak{h}^*$. If, for any $w \in W$, $w \neq 1$, we have $(\lambda + \rho) - w(\lambda + \rho) \notin Q_+$, then $Z(\lambda)$ is simple.

(Use Cor. 1 (ii) of Th. 2.)

6) Let $\mathfrak{a}$ be a Lie algebra, f a polynomial function on $\mathfrak{a}$, and x, y two elements of $\mathfrak{a}$. Put $f_y = \theta^*(y)f$ (cf. no. 3), and denote by $D_x f$ the tangent linear map

of f at x (Chap. VII, App. I, no. 2). Show that $f_y(x) = (\mathrm{D}_x f)([x, y])$. Deduce that $\mathrm{D}_x f$ vanishes on $\mathrm{Im}(\mathrm{ad}\ x)$ when f is invariant.

7) *Let $d_1, \ldots, d_l$ be the characteristic degrees of the algebra $\mathrm{I}(\mathfrak{g})$, cf. Chap. V, §5, no. 1. For any integer $n \geq 0$, denote by r_n the number of elements of degree n in a homogeneous basis of $\mathbf{S}(\mathfrak{g})$ over $\mathrm{I}(\mathfrak{g})$ (cf. no. 3, Remark 2), and put $r(\mathrm{T}) = \sum_{n=0}^{\infty} r_n \mathrm{T}^n$. Show that

$$r(\mathrm{T}) = (1 - \mathrm{T})^{-\mathrm{N}} \prod_{i=1}^{l} (1 - \mathrm{T}^{d_i}), \quad \text{where } \mathrm{N} = \dim(\mathfrak{g}).*$$

8) Put $l = \mathrm{rk}(\mathfrak{g}), \mathrm{N} = \dim(\mathfrak{g})$. If $x \in \mathfrak{g}$, define $a_i(x)$, $0 \leq i \leq \mathrm{N}$, by the formula

$$\det(\mathrm{T} + \mathrm{ad}\ x) = \sum_{i=0}^{\mathrm{N}} \mathrm{T}^{\mathrm{N}-i} a_i(x).$$

The function a_i thus defined is homogeneous polynomial of degree i, and invariant under $\mathrm{Aut}(\mathfrak{g})$. If $x \in \mathfrak{h}$ and $i \leq \mathrm{N} - l$, $a_i(x)$ is the ith elementary symmetric function of the $\alpha(x)$, $\alpha \in \mathrm{R}$; in particular, $a_{\mathrm{N}-l}(x) = \prod_{\alpha \in \mathrm{R}} \alpha(x)$. Construct an example in which the a_i do not generate the algebra of polynomial functions on $\mathfrak{g}$ invariant under $\mathrm{Aut}(\mathfrak{g})$.

9) The notations are those of no. 5. Let $\lambda \in \mathfrak{h}^*, z \in \mathrm{Z}$, z' the image of z under the principal anti-automorphism of $\mathrm{U}(\mathfrak{g})$, and w_0 the element of W that transforms B into $-$B. Show that $\chi_\lambda(z) = \chi_{-w_0\lambda}(z')$.

(It suffices to prove this for $\lambda \in \mathrm{P}_{++}$. Consider the operation of z on $\mathrm{E}(\lambda)$ and $\mathrm{E}(\lambda)^*$; use Prop. 11 of §7.)

10) (In this exercise, and in the following three, we retain the notations of §6.) Let $\lambda \in \mathfrak{h}^*$.

a) Let N, N' be $\mathfrak{g}$-submodules of $\mathrm{Z}(\lambda)$ such that $\mathrm{N}' \subset \mathrm{N}$ and N/N' is simple. Show that there exists $\mu \in \lambda - \mathrm{Q}_+$ such that N/N' is isomorphic to $\mathrm{E}(\mu)$ (apply Th. 1 of §6), and such that $\mu + \rho \in \mathrm{W}.(\lambda + \rho)$ (apply Cor. 1 of Th. 2).

b) Show that $\mathrm{Z}(\lambda)$ admits a Jordan-Hölder sequence. (Apply *a*) and the fact that the weights of $\mathrm{Z}(\lambda)$ are of finite multiplicity.)

11) Let $\lambda \in \mathfrak{h}^*$ and let V be a non-zero $\mathfrak{g}$-submodule of $\mathrm{Z}(\lambda)$. Show that there exists $\mu \in \mathfrak{h}^*$ such that V contains a simple $\mathfrak{g}$-submodule isomorphic to $\mathrm{Z}(\mu)$.

(Let A be the set of $\nu \in \mathfrak{h}^*$ such that V contains a $\mathfrak{g}$-submodule isomorphic to $\mathrm{Z}(\nu)$. By using Prop. 6 of no. 6, show first that $\mathrm{A} \neq \varnothing$. Then show that A is finite, and consider an element μ of A such that $(\mu - \mathrm{Q}_+) \cap \mathrm{A} = \{\mu\}$.)

¶ 12) *a*) Let $\lambda, \mu \in \mathfrak{h}^*$. Every non-zero $\mathfrak{g}$-homomorphism from $\mathrm{Z}(\mu)$ to $\mathrm{Z}(\lambda)$ is injective. (Use Prop. 6 of no. 6.)

b) Let $r \in \mathbf{N}$, A a finite subset of $\mathbf{N}^r$, $m = \mathrm{Card}(\mathrm{A})$. For all $\xi \in \mathbf{N}^r$, let $s(\xi)$ be the sum of the coordinates of ξ, and $\mathfrak{P}_{\mathrm{A}}(\xi)$ the number of families $(n_\alpha)_{\alpha\in\mathrm{A}}$ of integers ≥ 0 such that $\xi = \sum_{\alpha\in\mathrm{A}} n_\alpha\alpha$. Then $\mathfrak{P}_{\mathrm{A}}(\xi) \leq (s(\xi)+1)^m$ for all $\xi \in \mathbf{N}^r$. (Argue by induction on m.)

c) Let $\lambda, \mu \in \mathfrak{h}^*$. Show that $\dim \mathrm{Hom}_{\mathfrak{g}}(\mathrm{Z}(\mu), \mathrm{Z}(\lambda)) \leq 1$. (Let φ_1 and φ_2 be non-zero $\mathfrak{g}$-homomorphisms from $\mathrm{Z}(\mu)$ to $\mathrm{Z}(\lambda)$. If $\mathrm{Im}(\varphi_1) = \mathrm{Im}(\varphi_2)$, φ_1 and φ_2 are linearly dependent by Prop. 1 (iii) of §6. Assume that $\mathrm{Im}(\varphi_1) \neq \mathrm{Im}(\varphi_2)$. If $\mathrm{Z}(\mu)$ is simple, the sum $\mathrm{Im}(\varphi_1) + \mathrm{Im}(\varphi_2)$ is direct; deduce that $\mathfrak{P}(\xi+\lambda-\mu) \geq 2\mathfrak{P}(\xi)$ for all $\xi \in \mathfrak{h}^*$, contradicting *b*). In the general case, use Exerc. 11.)

When $\dim \mathrm{Hom}_{\mathfrak{g}}(\mathrm{Z}(\mu), \mathrm{Z}(\lambda)) = 1$, write $\mathrm{Z}(\mu) \subset \mathrm{Z}(\lambda)$ by abuse of notation.

d) Let $\nu \in \mathfrak{h}^*$. The set of $\lambda \in \mathfrak{h}^*$ such that $\mathrm{Z}(\lambda-\nu) \subset \mathrm{Z}(\lambda)$ is closed in $\mathfrak{h}^*$ in the Zariski topology.

¶ 13) *a*) Let $\mathfrak{a}$ be a nilpotent Lie algebra, $x \in \mathfrak{a}, n \in \mathbf{N}, p \in \mathbf{N}$. There exists $l \in \mathbf{N}$ such that $x^l y_1 \ldots y_n \in \mathrm{U}(\mathfrak{a})x^p$ for all $y_1, \ldots, y_n \in \mathfrak{a}$.

b) Let $\lambda, \mu \in \mathfrak{h}^*$, and $\alpha \in \mathrm{B}$ be such that

$$\mathrm{Z}(s_\alpha\mu - \rho) \subset \mathrm{Z}(\mu-\rho) \subset \mathrm{Z}(\lambda-\rho).$$

Assume that $\lambda \in \mathrm{P}$. Let $p = \lambda(H_\alpha) \in \mathbf{Z}$. Show that:

b_1) If $p \leq 0$, then $\mathrm{Z}(\lambda-\rho) \subset \mathrm{Z}(s_\alpha\lambda-\rho)$.

b_2) If $p > 0$, then $\mathrm{Z}(s_\alpha\mu-\rho) \subset \mathrm{Z}(s_\alpha\lambda-\rho) \subset \mathrm{Z}(\lambda-\rho)$.

(Use *a*), and §6, Cor. 1 of Prop. 6.)

c) Let $\lambda \in \mathfrak{h}^*, \alpha \in \mathrm{R}$, and $m = \lambda(H_\alpha)$. Assume that $m \in \mathbf{N}$. Show that

$$\mathrm{Z}(s_\alpha\lambda-\rho) \subset \mathrm{Z}(\lambda-\rho).$$

(Prove this first for $\lambda \in \mathrm{P}$ by using *b*), and then in the general case by using Exerc. 12 *d*).)[17]

14) *Let $\mathfrak{a}$ be a semi-simple Lie algebra, and $\mathrm{Z}(\mathfrak{a})$ the centre of $\mathrm{U}(\mathfrak{a})$. Show that $\mathrm{U}(\mathfrak{a})$ is a free $\mathrm{Z}(\mathfrak{a})$-module. (Remark that $\mathrm{gr}\,\mathrm{U}(\mathfrak{a})$ is isomorphic to $\mathbf{S}(\mathfrak{a})$ and $\mathbf{S}(\mathfrak{a}^*)$, and use Remark 2 of no. 3.)$_*$

15) Let x be a diagonalizable element of $\mathfrak{g}$ (§3, Exerc. 10), and y a semi-simple element of $\mathfrak{g}$ such that $f(x) = f(y)$ for every invariant polynomial function f on $\mathfrak{g}$. Show that there exists $s \in \mathrm{Aut}_e(\mathfrak{g})$ such that $sy = x$. (Remark that $\mathrm{ad}\,x$ and $\mathrm{ad}\,y$ have the same characteristic polynomial, cf. Exerc. 8, hence

[17] For more details on Exercises 10 to 13, see: I. N. BERNSTEIN, I. M. GELFAND and S. I. GELFAND, Structure of representations generated by highest weight vectors [in Russian], *Funct. Anal. i evo prilojenie*, Vol. V (1971), pp. 1-9.
In this memoir, it is further proved that, if $\lambda, \lambda' \in \mathfrak{h}^*$ are such that $\mathrm{Z}(\lambda-\rho) \subset \mathrm{Z}(\lambda'-\rho)$, there exist $\gamma_1, \ldots, \gamma_n \in \mathrm{R}_+$ such that $\lambda = s_{\gamma_n}\ldots s_{\gamma_2}s_{\gamma_1}\lambda'$ and $(s_{\gamma_i}\ldots s_{\gamma_1}\lambda')(H_{\gamma_{i+1}}) \in \mathbf{N}$ for $0 \leq i \leq n$. It follows that $\mathrm{Z}(\lambda-\rho)$ is simple if and only if $\lambda(H_\alpha) \in \mathbf{N}^*$ for all $\alpha \in \mathrm{R}_+$.

the fact that y is diagonalizable. Next, reduce to the case in which x and y are contained in $\mathfrak{h}$ by using Exerc. 10 of §3, and use Th. 1 (i) and Lemma 6 to prove that x and y are conjugate under W.)

16) Let $\mathfrak{a}$ be a semi-simple Lie algebra, x an element of $\mathfrak{a}$, and x_s the semi-simple component of x. Show that, if f is an invariant polynomial function on $\mathfrak{a}$, then $f(x) = f(x_s)$. (Reduce to the case in which f is of the form $x \mapsto \operatorname{Tr} \rho(x)^n$.)

17) Assume that k is algebraically closed, and put $\mathrm{G} = \mathrm{Aut}_e(\mathfrak{g})$. Let $x, y \in \mathfrak{g}$. Show the equivalence of:
(i) The semi-simple components of x and y are G-conjugate.
(ii) For every invariant polynomial function f on $\mathfrak{g}$, we have $f(x) = f(y)$.
(Use Exerc. 15 and 16.)

18) Let $\mathfrak{a}$ be a semi-simple Lie algebra, $l = \mathrm{rk}(\mathfrak{a})$, I the algebra of invariant polynomial functions on $\mathfrak{a}$, and $\mathrm{P}_1, \dots, \mathrm{P}_l$ homogeneous elements of I generating I as an algebra. The P_i define a polynomial map $\mathrm{P} : \mathfrak{a} \to k^l$. If $x \in \mathfrak{a}$, denote by $\mathrm{D}_x\mathrm{P} : \mathfrak{a} \to k^l$ the tangent linear map of P at x (Chap. VII, App. I, no. 2).

a) Let $\mathfrak{h}$ be a Cartan subalgebra of $\mathfrak{a}$, and let $x \in \mathfrak{h}$. Prove the equivalence of:
(i) $\mathrm{D}_x\mathrm{P}|\mathfrak{h}$ is an isomorphism from $\mathfrak{h}$ to k^l;
(ii) x is regular.
(Reduce to the split case. Choose a basis of $\mathfrak{h}$, and denote by $d(x)$ the determinant of the matrix of $\mathrm{D}_x\mathrm{P}|\mathfrak{h}$ relative to this basis. Show, by means of Prop. 5 of Chap. V, §5, no. 4, that there exists $c \in k^*$ such that $d(x)^2 = c \prod_{\alpha \in \mathrm{R}} \alpha(x)$, where α belongs to the set R of roots of $(\mathfrak{a}, \mathfrak{h})$.)

If these conditions are satisfied, $\mathfrak{a} = \mathfrak{h} \oplus \mathrm{Im}\,\mathrm{ad}(x)$, and $\mathrm{Ker}\,\mathrm{D}_x\mathrm{P} = \mathrm{Im}\,\mathrm{ad}(x)$ (use Exerc. 6 to show that $\mathrm{D}_x\mathrm{P}$ vanishes on $\mathrm{Im}\,\mathrm{ad}(x)$).

b) Show that the set of $x \in \mathfrak{a}$ such that $\mathrm{D}_x\mathrm{P}$ is of rank l is a dense open subset of $\mathfrak{a}$ in the Zariski topology.

§9

All the $\mathfrak{g}$-modules considered are assumed to be finite dimensional.

1) If m is an integer ≥ 0, we have $\dim \mathrm{E}(m\rho) = (m+1)^{\mathrm{N}}$, where $\mathrm{N} = \mathrm{Card}(\mathrm{R}_+)$.

2) Show that there exists a unique polynomial function d on $\mathfrak{h}^*$ such that $d(\lambda) = \dim \mathrm{E}(\lambda)$ for all $\lambda \in \mathrm{P}_{++}$; its degree is $\mathrm{Card}(\mathrm{R}_+)$. We have

$$d(w\lambda - \rho) = \varepsilon(w) d(\lambda - \rho) \quad \text{if } w \in \mathrm{W}, \lambda \in \mathfrak{h}^*.$$

In particular, the function $\lambda \mapsto d(\lambda - \rho)^2$ is invariant under W. Deduce that there exists a unique element u of the centre of $\mathrm{U}(\mathfrak{g})$ such that $\chi_\lambda(u) = d(\lambda)^2$

for all $\lambda \in \mathfrak{h}^*$ (apply Th. 2 of §8, no. 5). When $\mathfrak{g} = \mathfrak{sl}(2, k)$, we have $u = \mathrm{C}+1$, where C is the element defined in Exerc. 1 of §1.

3) Let $k_1, \ldots, k_l$ be the characteristic degrees of the algebra of invariants of W (cf. Chap. V, §5).

a) Show that, for all $j \geq 1$, the number of i such that $k_i > j$ is equal to the number of $\alpha \in \mathrm{R}_+$ such that $\langle \rho, H_\alpha \rangle = j$. (Reduce to the case in which R is irreducible, and use Chap. VI, §4, Exerc. 6 *c*).) (Cf. §5, Exerc. 5 *g*).)

b) Deduce the formula

$$\prod_{\alpha \in \mathrm{R}_+} \langle \rho, H_\alpha \rangle = \prod_{i=1}^{l} (k_i - 1)!.$$

¶ 4) Assume that $\mathfrak{g}$ is simple, and denote by γ the element of R_+ such that H_γ is the highest root of $\mathrm{R}^\vee$; write $H_\gamma = \sum_{\alpha \in \mathrm{B}} n_\alpha H_\alpha$. We have $\langle \rho, H_\gamma \rangle = \sum n_\alpha = h - 1$, where h is the Coxeter number of R (Chap. VI, §1, no. 11, Prop. 31).

a) Let $\alpha \in \mathrm{B}$. Show that, for all $\beta \in \mathrm{R}_+$,

$$\langle \varpi_\alpha + \rho, H_\beta \rangle \leq h + n_\alpha - 1,$$

and that equality holds if $\beta = \gamma$. Deduce that every prime factor of $\dim \mathrm{E}(\varpi_\alpha)$ is $\leq h + n_\alpha - 1$.

b) Assume that ϖ_α is not minuscule, i.e. $n_\alpha \geq 2$. Let $m \in (2, n_\alpha)$ and $p = h + m - 1$. Verify (cf. Chap. VI, *Plates*) that there exists $\beta \in \mathrm{R}_+$ such that $\langle \varpi_\alpha, H_\beta \rangle = n_\alpha$ and $\langle \rho, H_\beta \rangle = h - 1 - (n_\alpha - m)$, hence $\langle \varpi_\alpha + \rho, H_\beta \rangle = p$. Deduce that, if p is prime, p divides $\dim \mathrm{E}(\varpi_\alpha)$. (Remark that p does not divide any of the $\langle \rho, H_\beta \rangle$ for $\beta \in \mathrm{R}_+$, cf. Exerc. 3.)

c) When $\mathfrak{g}$ is of type G_2 (resp. $\mathrm{F}_4, \mathrm{E}_8$), we have $h = 6$ (resp. 12, 30), and $\dim \mathrm{E}(\varpi_\alpha)$ is divisible by 7 (resp. 13, 31), When $\mathfrak{g}$ is of type E_6 (resp. E_7), and ϖ_α is not minuscule, $\dim \mathrm{E}(\varpi_\alpha)$ is divisible by 13 (resp. 19).

¶ 5) *a*) Let $\alpha \in \mathrm{R}, x \in \mathfrak{g}^\alpha, y \in \mathfrak{g}^{-\alpha}$, and let E be a $\mathfrak{g}$-module. Show that, for all $\lambda \in \mathrm{P}$,

$$\mathrm{Tr}((xy)_\mathrm{E} | \mathrm{E}^\lambda) = \mathrm{Tr}((xy)_\mathrm{E} | \mathrm{E}^{\lambda+\alpha}) + \lambda([x, y]) \dim \mathrm{E}^\lambda.$$

Deduce that

$$\sum_{\lambda \in \mathrm{P}} \lambda([x, y]) \dim \mathrm{E}^\lambda . e^\lambda = (1 - e^{-\alpha}) \sum_{\lambda \in \mathrm{P}} \mathrm{Tr}((xy)_\mathrm{E} | \mathrm{E}^\lambda) . e^\lambda.$$

b) Give $\mathfrak{h}^*$ a non-degenerate W-invariant symmetric bilinear form $\langle \, , \, \rangle$. Let Δ be the endomorphism of the vector space $k[\mathrm{P}]$ such that $\Delta(e^\mu) = \langle \mu, \mu \rangle e^\mu$ for all $\mu \in \mathrm{P}$; if $a, b \in k[\mathrm{P}]$, put

$$\Delta'(a, b) = \Delta(ab) - a\Delta(b) - b\Delta(a).$$

Prove that

$$\begin{aligned}\Delta(\mathrm{J}(e^{\mu})) &= \langle\mu,\mu\rangle \mathrm{J}(e^{\mu}) \quad \text{for } \mu \in \mathrm{P},\\ \Delta'(e^{\lambda},e^{\mu}) &= 2\langle\lambda,\mu\rangle e^{\lambda+\mu} \quad \text{for } \lambda,\mu \in \mathrm{P},\\ \Delta'(ab,c) &= a\Delta'(b,c)+b\Delta'(a,c) \quad \text{for } a,b,c \in k[\mathrm{P}].\end{aligned}$$

c) Let $\lambda \in \mathrm{P}_{++}$, $c_\lambda = \mathrm{ch}(\mathrm{E}(\lambda))$ and $d = \mathrm{J}(e^{\rho})$. Prove that

$$\Delta(c_\lambda d) = \langle\lambda+\rho,\lambda+\rho\rangle c_\lambda d.$$

(Use *a*), *b*), §6, Cor. of Prop. 7, and Chap. VI, §3, no. 3, formula (3).)

d) Deduce another proof of the formula of H. Weyl from the above and §7, no. 2, Prop. 5 (iii).

e) For all $\lambda \in \mathfrak{h}^*$, put $\dim \mathrm{E}^{\lambda} = m(\lambda)$. Deduce from *a*) that

$$\mathrm{Tr}((xy)_{\mathrm{E}}|\mathrm{E}^{\lambda}) = \sum_{i=0}^{+\infty}(\lambda+i\alpha)([x,y])m(\lambda+i\alpha)$$

$$\sum_{i=-\infty}^{+\infty}(\lambda+i\alpha)([x,y])m(\lambda+i\alpha) = 0.$$

f) Let $\langle\cdot,\cdot\rangle$ be a non-degenerate invariant symmetric bilinear form on $\mathfrak{g}$ whose restriction to $\mathfrak{h}$ is the inverse of the form chosen above. Let Γ be the corresponding Casimir element. Assume that E is simple; put $\Gamma_{\mathrm{E}} = \gamma.1$, where $\gamma \in k$. By using *e*) and §2, no. 3, Prop. 6, show that

$$\gamma m(\lambda) = \langle\lambda,\lambda\rangle m(\lambda) + \sum_{\alpha\in\mathrm{R}}\sum_{i=0}^{+\infty}\langle\lambda+i\alpha,\alpha\rangle m(\lambda+i\alpha)$$

for all $\lambda \in \mathfrak{h}^*$, and then that

$$\begin{aligned}\gamma m(\lambda) &= \langle\lambda,\lambda\rangle m(\lambda) + \sum_{\alpha\in\mathrm{R}_+} m(\lambda)\langle\lambda,\alpha\rangle + 2\sum_{\alpha\in\mathrm{R}_+}\sum_{i=1}^{+\infty} m(\lambda+i\alpha)\langle\lambda+i\alpha,\alpha\rangle\\ &= \langle\lambda,\lambda+2\rho\rangle m(\lambda) + 2\sum_{\alpha\in\mathrm{R}_+}\sum_{i=1}^{+\infty} m(\lambda+i\alpha)\langle\lambda+i\alpha,\alpha\rangle.\end{aligned}$$

g) Continue to assume that E is simple; let ω be its highest weight. Deduce from *f*) that, for all $\lambda \in \mathfrak{h}^*$,

$$(\langle\omega+\rho,\omega+\rho\rangle - \langle\lambda+\rho,\lambda+\rho\rangle)m(\lambda) = 2\sum_{\alpha\in\mathrm{R}_+}\sum_{i=1}^{+\infty} m(\lambda+i\alpha)\langle\lambda+i\alpha,\alpha\rangle.$$

(Recall that, by Prop. 5 of §7, $\langle\omega+\rho,\omega+\rho\rangle > \langle\lambda+\rho,\lambda+\rho\rangle$ if λ is a weight of E distinct from ω. The preceding formula thus gives a procedure for calculating $m(\lambda)$ step by step.)[18]

6) Let $x \mapsto x^*$ be the involution of $k[\mathrm{P}]$ that takes e^p to e^{-p} for all $p \in \mathrm{P}$.

a) Put $\mathrm{D} = d^*d = \prod_{\alpha\in\mathrm{R}}(1-e^\alpha)$. Show that $d^* = (-1)^{\mathrm{N}}d$, where $\mathrm{N} = \mathrm{Card}(\mathrm{R}_+)$, and hence that $\mathrm{D} = (-1)^{\mathrm{N}}d^2$.

b) Define two linear forms ε and I on $k[\mathrm{P}]$ by the formulas:

$$\varepsilon(1) = 1, \quad \varepsilon(e^p) = 0 \quad \text{if } p \in \mathrm{P} - \{0\}$$

and

$$\mathrm{I}(f) = \frac{1}{m}\varepsilon(\mathrm{D}.f), \qquad \text{where } m = \mathrm{Card}(\mathrm{W}).$$

Show, by using the formula $d = \mathrm{J}(e^\rho)$, that $\mathrm{I}(1) = 1$.

c) Let $\lambda \in \mathrm{P}_{++}$ and $c_\lambda = \mathrm{ch}\,\mathrm{E}(\lambda) = \mathrm{J}(e^{\lambda+\rho})/d$. Show that $\mathrm{I}(c_\lambda) = 0$ if $\lambda \neq 0$. (Same method as for *b*).)

d) Show that I takes integer values on the subalgebra $\mathbf{Z}[\mathrm{P}]^{\mathrm{W}} = \mathrm{ch}\,\mathrm{R}(\mathfrak{g})$ of $k[\mathrm{P}]$. If E is a $\mathfrak{g}$-module, the dimension of the space of invariants of $\mathfrak{g}$ in E is equal to $\mathrm{I}(\mathrm{ch}\,\mathrm{E})$. (Reduce to the case in which E is simple and use *b*) and *c*).)

e) We have $\dim \mathrm{E} = \sum_{\lambda\in\mathrm{P}_{++}} \mathrm{I}(c_\lambda^*\mathrm{ch}\,\mathrm{E})d(\lambda)$, where $d(\lambda) = \dim \mathrm{E}(\lambda)$. In particular:

$$\mathrm{I}(c_\lambda^* c_\mu) = \delta_{\lambda\mu} \quad \text{if } \lambda, \mu \in \mathrm{P}_{++}.$$

f) If $\lambda, \mu, \nu \in \mathrm{P}_{++}$, the integer $m(\lambda,\mu,\nu)$ of Prop. 2 is equal to $\mathrm{I}(c_\lambda c_\mu c_\nu^*)$. Deduce the identity

$$d(\lambda)d(\mu) = \sum_{\nu\in\mathrm{P}_{++}} m(\lambda,\mu,\nu)d(\nu) \qquad \lambda,\mu,\nu \in \mathrm{P}_{++}.$$

(Apply *e*) to the $\mathfrak{g}$-module $\mathrm{E} = \mathrm{E}(\lambda) \otimes \mathrm{E}(\mu)$.)

¶ 7) We retain the notations of the proof of Th. 2.

a) Show that

$$f_\rho(\mathrm{J}(e^\mu)) = \prod_{\alpha\in\mathrm{R}_+} (e^{(\mu|\alpha)\mathrm{T}/2} - e^{-(\mu|\alpha)\mathrm{T}/2}).$$

b) Take $(\cdot|\cdot)$ to be the canonical bilinear form Φ_{R} (Chap. VI, §1, no. 12). Show that

[18] For more details on this exercise, see: H. FREUDENTHAL, Zur Berechnung der Charaktere der halbeinfachen Lieschen Gruppen, *Proc. Kon. Akad. Wet. Amsterdam*, Vol. LVII (1954), pp. 369-376.

$$f_\rho(\mathrm{J}(e^\mu)) \equiv d_\mu \mathrm{T}^{\mathrm{N}} \left(1 + \frac{\mathrm{T}^2}{48}(\mu|\mu)\right) \quad (\text{mod. } \mathrm{T}^{\mathrm{N}+3}\mathbf{R}[[\mathrm{T}]]),$$

where $d_\mu = \prod_{\alpha \in \mathrm{R}_+} (\mu|\alpha)$.

c) From *b*) and the equality $\mathrm{J}(e^{\lambda+\rho}) = \mathrm{ch}(\mathrm{E}).\mathrm{J}(e^\rho)$, deduce the formula

$$\sum_{\mu \in \mathrm{P}} (\mu|\rho)^2 \dim \mathrm{E}^\mu = \frac{\dim \mathrm{E}}{24}(\lambda|\lambda + 2\rho).$$

d) Assume that $\mathfrak{g}$ is simple. Show that $(\rho|\rho) = \dim \mathfrak{g}/24$. (Apply *c*) with λ the highest root of R, and use Exerc. 3 of §6.)

¶ 8) Let ψ be a polynomial function on $\mathfrak{h}^*$ of degree r. Show that there exists a unique polynomial function Ψ on $\mathfrak{h}^*$ that is invariant under W, of degree $\leq r$, and such that

$$\sum_{\alpha \in \mathrm{P}} \psi(\mu) \dim \mathrm{E}^\mu = \Psi(\lambda + \rho) \dim \mathrm{E}$$

for any simple $\mathfrak{g}$-module E of highest weight λ.

(Treat first the case in which $\psi(\mu) = (\mu|\nu)^r$, where $\nu \in \mathrm{P}$ is not orthogonal to any root; for this use the homomorphism f_ν in the proof of Th. 2 and Chap. V, §5, no. 4, Prop. 5 (i).)

9) We use the notations of Chap. VI, Plate I, in the case of an algebra $\mathfrak{g}$ of type A_2. Let n, p be integers ≥ 0.

a) $\mathfrak{P}(n\alpha_1 + p\alpha_2) = 1 + \inf(n, p)$.

b) Let $\lambda = n\varpi_1 + p\varpi_2$. Then $\dim \mathrm{E}(\lambda) = \frac{1}{2}(n+1)(p+1)(n+p+2)$. The multiplicity of the weight 0 of $\mathrm{E}(\lambda)$ is 0 if λ is not radical, i.e. if $n \not\equiv p$ (mod. 3); if λ is radical, it is $1 + \inf(n, p)$.

10) We use the notations of Chap. VI, Plate II, in the case of an algebra of type B_2. Let n, p be integers ≥ 0.

a) We have

$$\mathfrak{P}(n\alpha_1 + p(\alpha_1 + 2\alpha_2)) = 1 + \frac{1}{2}p(p+3)$$

$$\mathfrak{P}(n\alpha_2 + p(\alpha_1 + \alpha_2)) = [p^2/4] + p + 1$$

$$\mathfrak{P}(n(\alpha_1 + \alpha_2) + p(\alpha_1 + 2\alpha_2)) = [n^2/4] + n + 1 + np + \frac{1}{2}p(p+3).$$

b) Let $\lambda = n(\alpha_1 + \alpha_2) + p(\alpha_1 + 2\alpha_2)$. The multiplicity of the weight 0 in $\mathrm{E}(\lambda)$ is

$$[n/2] + 1 + np + p.$$

11) Assume that $\mathfrak{g}$ is not a product of algebras of rank 1. Let $n \in \mathbf{N}$. Show that there exists a simple $\mathfrak{g}$-module one of whose weights is of multiplicity

$\geq n$. (Suppose not. Let E_λ be the simple $\mathfrak{g}$-module of highest weight $\lambda \in P_{++}$. Compare $\dim E_\lambda$ and the number of distinct weights of E_λ when $(\lambda|\lambda) \to \infty$ (with the notations of Th. 2).)

12) Let U^0 be the commutant of $\mathfrak{h}$ in $U(\mathfrak{g})$. If $\mathfrak{g}$ is of rank 1, U^0 is commutative. If $\mathfrak{g}$ is of rank ≥ 2, U^0 admits simple representations of arbitrarily large finite dimension. (Use Exerc. 11 and Exerc. 23 *a*) of §7.)

13) Let R be a root system in a vector space V. Two elements v_1, v_2 of V are said to be *disjoint* if R is the direct sum of two root systems R_1 and R_2 (Chap. VI, §1, no. 2) such that v_i belongs to the vector subspace of V generated by R_i, $i = 1, 2$. Show that two elements of $V_{\mathbf{R}}$ that belong to the same chamber of R are disjoint if and only if they are orthogonal.

¶ 14) *a*) Let $\mu, \nu \in P_{++}$ and let γ be a weight of $E(\mu)$. Let ρ_μ be the representation of $\mathfrak{g}$ on $E(\mu)$. Let $X_\alpha \in \mathfrak{g}^\alpha - \{0\}, Y_\alpha \in \mathfrak{g}^{-\alpha} - \{0\}$. If $\alpha \in B$, put $\nu_\alpha = \nu(H_\alpha)$, and

$$E^+(\mu, \gamma, \nu) = E(\mu)^\gamma \cap \bigcap_{\alpha \in B} \operatorname{Ker} \rho_\mu(X_\alpha)^{\nu_\alpha + 1},$$

$$E^-(\mu, \gamma, \nu) = E(\mu)^\gamma \cap \bigcap_{\alpha \in B} \operatorname{Ker} \rho_\mu(Y_\alpha)^{\nu_\alpha + 1},$$

$$d^+(\mu, \gamma, \nu) = \dim E^+(\mu, \gamma, \nu), \quad d^-(\mu, \gamma, \nu) = \dim E^-(\mu, \gamma, \nu).$$

For all $\lambda \in \mathfrak{h}^*$, put $\lambda^* = -w_0\lambda$, where w_0 is the element of W that takes B to $-B$. Show that

$$d^+(\mu, \gamma, \nu) = d^-(\mu, -\gamma^*, \nu^*).$$

b) Let $\lambda_1, \lambda_2 \in P_{++}$, V the $\mathfrak{g}$-module $\operatorname{Hom}_k(E(\lambda_1^*), E(\lambda_2))$, U the set of $\varphi \in V$ such that $Y_\alpha.\varphi = 0$ for all $\alpha \in B$, and w a primitive vector in $E(\lambda_1^*)$. Show that $\varphi \mapsto \varphi(w)$ is an isomorphism from U to the set of $v \in E(\lambda_2)$ such that $Y_\alpha^{\lambda_1^*(H_\alpha)+1}.v = 0$ for all $\alpha \in B$. (Use Exerc. 15 of §7 to prove surjectivity.)

c) With the notations of Prop. 2, prove that, for $\lambda_1, \lambda_2, \lambda \in P_{++}$,

$$\begin{aligned} m(\lambda_1, \lambda_2, \lambda) &= d^+(\lambda, \lambda_2 - \lambda_1^*, \lambda_1^*) = d^-(\lambda, \lambda_1 - \lambda_2^*, \lambda_1) \\ &= d^+(\lambda_1, \lambda - \lambda_2, \lambda_2) = d^-(\lambda_1, \lambda_2^* - \lambda^*, \lambda_2^*). \end{aligned}$$

(Observe that $m(\lambda_1, \lambda_2, \lambda)$ is the dimension of the space of $\mathfrak{g}$-invariant elements of

$$E(\lambda)^* \otimes E(\lambda_1) \otimes E(\lambda_2),$$

and hence is equal to $m(\lambda_1^*, \lambda, \lambda_2)$.)

d) Let $\lambda_1, \lambda_2 \in P_{++}$, and let λ be the unique element of $P_{++} \cap W.(\lambda_1 - \lambda_2^*)$. We have

$$m(\lambda_1, \lambda_2, \lambda) = 1.$$

(Use *c*).) Deduce that $E(\lambda_1) \otimes E(\lambda_2)$ contains a unique submodule isomorphic to $E(\lambda)$.

e) We retain the notations of *d*). Show the equivalence of the following conditions:
(i) $\lambda = \lambda_1 + \lambda_2$
(ii) $\| \lambda \| = \| \lambda_1 + \lambda_2 \|$
(iii) λ_1 and λ_2^* are orthogonal
(iv) λ_1 and λ_2^* are disjoint (Exerc. 13)
(v) λ_1 and λ_2 are disjoint.
Conclude that $E(\lambda_1) \otimes E(\lambda_2)$ is not a simple module unless λ_1 and λ_2 are disjoint (and hence another proof of Exerc. 18 *f*) of §7).

15) Put $N = \mathrm{Card}(R_+)$, $c = \prod_{\alpha \in R_+} \langle \rho, H_\alpha \rangle$, and $d(\lambda) = \dim E(\lambda)$ if $\lambda \in P_{++}$. Show that, for any real number $s > 0$,

$$\sum_{\lambda \in P_{++}} d(\lambda)^{-s} \leq \frac{1}{c} \left(\sum_{m=1}^{\infty} m^{-s} \right)^{N}.$$

Deduce that $\sum_{\lambda \in P_{++}} d(\lambda)^{-s} < +\infty$ if $s > 1$.

¶ 16) *a*) Take $\mathfrak{g}$ to be of type F_4 and use the notations of Chap. VI, Plate VIII. If $i = 1, 2, 3, 4$, put

$$\mathscr{X}_i = P_{++} \cap (\varpi_i - Q_+).$$

The set of weights of $E(\varpi_i)$ is the disjoint union of the $W\omega$, where ω belongs to $\mathscr{X}_i$ (cf. §7, Prop. 5 (iv)). We have:

$$\mathscr{X}_1 = \{0, \varpi_1, \varpi_4\};$$
$$\mathscr{X}_2 = \{0, \varpi_1, \varpi_2, \varpi_3, \varpi_4, \varpi_1 + \varpi_4, 2\varpi_4\};$$
$$\mathscr{X}_3 = \{0, \varpi_1, \varpi_3, \varpi_4\};$$
$$\mathscr{X}_4 = \{0, \varpi_4\}.$$

b) Show, by means of Th. 2, that:

$$\dim E(\varpi_1) = 52, \quad \dim E(\varpi_2) = 1274, \quad \dim E(\varpi_3) = 273, \quad \dim E(\varpi_4) = 26.$$

c) By using Chap. V, §3, Prop. 1, and the plates of Chap. VI, show that

$$\mathrm{Card}(W\varpi_1) = 2^7 3^2 2^{-3} (3!)^{-1} = 24.$$

Calculate $\mathrm{Card}(W\varpi_2), \ldots, \mathrm{Card}(W.2\varpi_4)$ similarly. Deduce that the number of weights of $E(\varpi_2)$ is 553; since this number is strictly less than $\dim E(\varpi_2)$, one of these weights is of multiplicity ≥ 2.

d) Make analogous calculations for $\varpi_1, \varpi_3, \varpi_4$. By using Exerc. 21 of §7, deduce that, if ρ is a non-zero simple representation of $\mathfrak{g}$, ρ admits a weight of multiplicity ≥ 2.

e) Prove the same result for a simple algebra of type E_8.

f) Let $\mathfrak{a}$ be a splittable simple Lie algebra. Deduce from *d*), *e*) and Prop. 7 and 8 of §7 the equivalence of the following properties:
(i) $\mathfrak{a}$ admits a non-zero simple representation all of whose weights are of multiplicity 1;
(ii) $\mathfrak{a}$ is neither of type F_4 nor of type E_8.

§10

1) Let $\mathfrak{s} = \mathfrak{sl}(2, k)$ and $\mathfrak{g} = \mathfrak{sl}(3, k)$. Identify $\mathfrak{s}$ with a subalgebra of $\mathfrak{g}$ by means of an irreducible representation of $\mathfrak{s}$ of degree 3. Show that every subspace of $\mathfrak{g}$ containing $\mathfrak{s}$ and stable under $\mathrm{ad}_{\mathfrak{g}}\mathfrak{s}$ is equal to either $\mathfrak{s}$ or $\mathfrak{g}$; deduce that $\mathfrak{s}$ is maximal among the subalgebras of $\mathfrak{g}$ distinct from $\mathfrak{g}$.

2) Let $m = \frac{1}{2}(\dim(\mathfrak{g}) + \mathrm{rk}(\mathfrak{g}))$. Every solvable subalgebra of $\mathfrak{g}$ is of dimension $\leq m$; if it is of dimension m, it is a Borel subalgebra. (Reduce to the algebraically closed case, and use Th. 2.)

3) Assume that k is $\mathbf{R}, \mathbf{C}$, or a non-discrete complete ultrametric field. Give the grassmannian $\mathbf{G}(\mathfrak{g})$ of vector subspaces of $\mathfrak{g}$ its natural structure of analytic manifold over k (*Differentiable and Analytic Manifolds, Results*, 5.2.6). Consider the subsets of $\mathbf{G}(\mathfrak{g})$ formed by:
(i) the subalgebras
(ii) the solvable subalgebras
(iii) the nilpotent subalgebras
(iv) the subalgebras consisting of nilpotent elements
(v) the Borel subalgebras.
Show that these subsets are closed (for (v), use Exerc. 2). Deduce that these subsets are compact when k is locally compact.

Show by examples that the subsets of $\mathbf{G}(\mathfrak{g})$ formed by:
(vi) the Cartan subalgebras
(vii) the subalgebras reductive in $\mathfrak{g}$
(viii) the semi-simple subalgebras
(ix) the decomposable subalgebras
are not necessarily closed, even when $k = \mathbf{C}$.

4) Assume that $k = \mathbf{C}$. Let $G = \mathrm{Int}(\mathfrak{g}) = \mathrm{Aut}_0(\mathfrak{g})$, and let B be an integral subgroup of G whose Lie algebra $\mathfrak{b}$ is a Borel subalgebra of $\mathfrak{g}$. Show that B is the normalizer of $\mathfrak{b}$ in G (use Exerc. 4 of §3 and Exerc. 11 of §5). Deduce by means of Exerc. 3 that G/B is compact.

¶ 5) Assume that $\mathfrak{g}$ is splittable. If $\mathfrak{h}$ is a splitting Cartan subalgebra of $\mathfrak{g}$, denote by $E(\mathfrak{h})$ the subgroup of $\mathrm{Aut}_e(\mathfrak{g})$ generated by the $e^{\mathrm{ad}\, x}$, $x \in \mathfrak{g}^{\alpha}(\mathfrak{h})$, $\alpha \in R(\mathfrak{g}, \mathfrak{h})$, cf. Chap. VII, §3, no. 2.

a) Let $\mathfrak{b}$ be a Borel subalgebra of $\mathfrak{g}$ containing $\mathfrak{h}$, and let $\mathfrak{h}_1$ be a Cartan subalgebra of $\mathfrak{b}$. Show that $\mathfrak{h}_1$ is conjugate to $\mathfrak{h}$ by an element of $\mathrm{E}(\mathfrak{h})$. Deduce that $\mathrm{E}(\mathfrak{h}) = \mathrm{E}(\mathfrak{h}_1)$.

b) Let $\mathfrak{h}'$ be a splitting Cartan subalgebra of $\mathfrak{g}$. Show that $\mathrm{E}(\mathfrak{g}) = \mathrm{E}(\mathfrak{h}')$. (If $\mathfrak{b}'$ is a Borel subalgebra containing $\mathfrak{h}'$, choose a Cartan subalgebra $\mathfrak{h}_1$ of $\mathfrak{b} \cap \mathfrak{b}'$ and apply *a*) to show that $\mathrm{E}(\mathfrak{h}) = \mathrm{E}(\mathfrak{h}_1) = \mathrm{E}(\mathfrak{h}')$.)

c) Let x be a nilpotent element of $\mathfrak{g}$. Show that $e^{\mathrm{ad}\, x} \in \mathrm{E}(\mathfrak{h})$. (Use *b*) and Cor. 2 of Th. 1 to reduce to the case in which $x \in [\mathfrak{b}, \mathfrak{b}]$.) Deduce that $\mathrm{E}(\mathfrak{h}) = \mathrm{Aut}_e(\mathfrak{g})$.

6) *a*) Show the equivalence of the properties:
(i) $\mathfrak{g}$ has no nilpotent element $\neq 0$.
(ii) $\mathfrak{g}$ has no parabolic subalgebra $\neq \mathfrak{g}$.
(Use Cor. 2 of Th. 1.)

Such an algebra is called *anisotropic.*

b) Let $\mathfrak{p}$ be a minimal parabolic subalgebra of $\mathfrak{g}$, $\mathfrak{r}$ the radical of $\mathfrak{p}$, and $\mathfrak{s} = \mathfrak{p}/\mathfrak{r}$. Show that $\mathfrak{s}$ is anisotropic. (Remark that, if $\mathfrak{q}$ is a parabolic subalgebra of $\mathfrak{s}$, the inverse image of $\mathfrak{q}$ in $\mathfrak{p}$ is a parabolic subalgebra of $\mathfrak{g}$, cf. §3, Exerc. 5 *a*).)

¶ 7) *a*) Show that the following properties of k are equivalent:
(i) Every anisotropic semi-simple k-Lie algebra reduces to 0.
(ii) Every semi-simple k-Lie algebra has a Borel subalgebra.
(Use Exerc. 6 to prove that (i) $\Longrightarrow$ (ii).)

b) Show that (i) and (ii) imply[19]:
(iii) Every finite dimensional k-algebra that is a field is commutative. (Or again: the Brauer group of every algebraic extension of k reduces to 0.)

(Use the Lie algebra of elements of trace zero in such an algebra.)

c) Show that (i) and (ii) are implied by:
(iv) For any finite family of homogeneous polynomials $f_\alpha \in k[(X_i)_{i \in \mathrm{I}}]$ of degrees ≥ 1 such that $\sum\limits_\alpha \deg f_\alpha < \mathrm{Card}(\mathrm{I})$, there exist elements $x_i \in k$, not all zero, such that $f_\alpha((x_i)_{i \in \mathrm{I}}) = 0$ for all α.

(Use Prop. 5 of §8.)

§11

1) Let $\mathfrak{g} = \mathfrak{sl}(2, k)$. Put $\mathrm{G} = \mathrm{Aut}_e(\mathfrak{g})$; this group can be identified with $\mathbf{PSL}_2(k)$, cf. Chap. VII, §3, no. 1, Remark 2.

a) Every nilpotent element of $\mathfrak{g}$ is G-conjugate to $\begin{pmatrix} 0 & \lambda \\ 0 & 0 \end{pmatrix}$ for some $\lambda \in k$. Such an element is principal if and only if it is non-zero.

[19]In fact, (iii) is *equivalent* to (i) and (ii). For this, see: R. STEINBERG, Regular elements of semi-simple algebraic groups, *Publ. Math. I.H.E.S.*, Vol. XXV (1965), pp. 49-80.

b) The elements $\begin{pmatrix} 0 & \lambda \\ 0 & 0 \end{pmatrix}$, $\begin{pmatrix} 0 & \mu \\ 0 & 0 \end{pmatrix}$, where $\lambda, \mu \in k^*$, are G-conjugate if and only if $\lambda^{-1}\mu$ is a square in k.

c) Every simple element of $\mathfrak{g}$ is G-conjugate to $\begin{pmatrix} 1 & 0 \\ 0 & -1 \end{pmatrix}$.

2) Let $A = \begin{pmatrix} 0 & 1 \\ 0 & 0 \end{pmatrix}, B = \begin{pmatrix} 0 & 0 \\ 1 & 0 \end{pmatrix}$. Then A is nilpotent, AB is not, and

$$[A,[A,B]] = \begin{pmatrix} 0 & -2 \\ 0 & 0 \end{pmatrix}.$$

Deduce that Lemma 5 does not extend to fields of characteristic 2.

3) Let $\mathfrak{r}$ be the radical of $\mathfrak{g}$, and $\mathfrak{s} = \mathfrak{g}/\mathfrak{r}$. Show the equivalence of:
(i) $\mathfrak{g}$ contains no $\mathfrak{sl}_2$-triplet;
(ii) $\mathfrak{s}$ contains no $\mathfrak{sl}_2$-triplet;
(iii) $\mathfrak{s}$ is anisotropic (§10, Exerc. 6);
(iv) $\mathfrak{s}$ contains no non-zero diagonalizable element (§3, Exerc. 10).
(Use Prop. 2 and Exerc. 10 *a*) of §3.)

4) Let V be a vector space of dimension $n \geq 2$, $\mathfrak{g} = \mathfrak{sl}(\mathrm{V})$ and $\mathrm{G} = \mathbf{PGL}(\mathrm{V})$, identified with a group of automorphisms of $\mathfrak{g}$. An $\mathfrak{sl}_2$-triplet in $\mathfrak{g}$ gives V a faithful $\mathfrak{sl}(2,k)$-module structure, and conversely any such structure arises from an $\mathfrak{sl}_2$-triplet; an $\mathfrak{sl}_2$-triplet is principal if and only if the corresponding $\mathfrak{sl}(2,k)$-module is simple; two $\mathfrak{sl}_2$-triplets are G-conjugate if and only if the corresponding $\mathfrak{sl}(2,k)$-modules are isomorphic. Deduce that the G-conjugacy classes of $\mathfrak{sl}_2$-triplets correspond bijectively with the families $(m_1, m_2, \ldots)$ of integers ≥ 0 such that

$$m_1 + 2m_2 + 3m_3 + \cdots = n \quad \text{and} \quad m_1 < n.$$

¶ 5) Assume that $\mathfrak{g}$ is semi-simple. Let $\mathfrak{a}$ be a subalgebra of $\mathfrak{g}$, reductive in $\mathfrak{g}$, of the same rank as $\mathfrak{g}$, and containing a principal $\mathfrak{sl}_2$-triplet of $\mathfrak{g}$. Show that $\mathfrak{a} = \mathfrak{g}$.

¶ 6) Assume that $\mathfrak{g}$ is absolutely simple, and denote its Coxeter number by h. Let x be a nilpotent element of $\mathfrak{g}$. Show that $(\mathrm{ad}\ x)^{2h-1} = 0$ and that $(\mathrm{ad}\ x)^{2h-2} \neq 0$ if and only if x is principal. (Reduce to the case in which $\mathfrak{g}$ is split, and x is contained in the subalgebra $\mathfrak{n}_+$ of Prop. 10. Repeat the proof of Prop. 10.)

¶ 7) Assume that $\mathfrak{g}$ is semi-simple. Let x be a nilpotent element of $\mathfrak{g}$. Then x is principal if and only if x is contained in a unique Borel subalgebra of $\mathfrak{g}$. (Reduce to the case in which k is algebraically closed. Use Prop. 10 and Prop. 10 of §3, no. 3.)

8) A semi-simple Lie algebra has an $\mathfrak{sl}_2$-triplet if and only if it is $\neq 0$ and has a Borel subalgebra.

9) Assume that $\mathfrak{g}$ is semi-simple. Let N (resp. P) be the set of nilpotent (resp. principal nilpotent) elements of $\mathfrak{g}$.

a) Show that P is an open subset of N in the Zariski topology (use Exerc. 6).

b) Assume that $\mathfrak{g}$ is splittable. Show that P is dense in N. (Use Prop. 10 and Cor. 2 of Th. 1 of §10.)

10) Assume that $\mathfrak{g}$ is splittable semi-simple. Let (x, h, y) be an $\mathfrak{sl}_2$-triplet in $\mathfrak{g}$.

a) Show that there exists a splitting Cartan subalgebra $\mathfrak{h}$ of $\mathfrak{g}$ containing h. (Use Exerc. 10 *b*) of §3.)

b) Choose $\mathfrak{h}$ as in *a*). Show that $h \in \mathfrak{h}_{\mathbf{Q}}$ and that there exists a basis B of $\mathrm{R}(\mathfrak{g}, \mathfrak{h})$ such that $\alpha(h) \in \{0, 1, 2\}$ for all $\alpha \in \mathrm{B}$ (cf. Prop. 5). The element x belongs to the subalgebra of $\mathfrak{g}$ generated by the $\mathfrak{g}^\alpha$, $\alpha \in \mathrm{B}$.

c) Deduce from *a*) and *b*), and the Jacobson-Morozov theorem, a new proof of the fact that every nilpotent element of $\mathfrak{g}$ is contained in a Borel subalgebra (cf. §10, Cor. 2 of Th. 1).

¶ 11) Let (x, h, y) be a principal $\mathfrak{sl}_2$-triplet in the semi-simple Lie algebra $\mathfrak{g}$. Give $\mathfrak{g}$ the $\mathfrak{sl}(2, k)$-module structure defined by this triplet. Show that the module thus defined is isomorphic to $\bigoplus_{i=1}^{l} \mathrm{V}(2k_i - 2)$, where the k_i are the characteristic degrees of the algebra of invariant polynomial functions on $\mathfrak{g}$. (Reduce to the case in which $\mathfrak{g}$ is splittable simple. Use Cor. 1 of Th. 1 of §8, no. 3, and Chap. VI, §4, Exerc. 6 *c*).)[20]

¶ 12) Assume that $\mathfrak{g}$ is semi-simple. Let $x \in \mathfrak{g}$ and let s (resp. n) be the semi-simple (resp. nilpotent) component of x. Let $\mathfrak{a}_x$ (resp. $\mathfrak{a}_s$) be the commutant of x (resp. s) in $\mathfrak{g}$.

a) Show that n is a nilpotent element of the semi-simple algebra $\mathscr{D}(\mathfrak{a}_s)$, and that the commutant of n in $\mathfrak{a}_s$ is equal to $\mathfrak{a}_x$. Deduce that $\dim \mathfrak{a}_x < \dim \mathfrak{a}_s$ if $n \neq 0$, i.e. if x is not semi-simple.

b) Show that $\dim \mathfrak{a}_x = \mathrm{rk}(\mathfrak{g})$ if and only if n is a principal nilpotent element of $\mathscr{D}(\mathfrak{a}_s)$.

c) Put $\mathrm{G} = \mathrm{Aut}_e(\mathfrak{g})$. Show that, for all $\lambda \in k$, there exists $\sigma_\lambda \in \mathrm{G}$ such that $\sigma_\lambda x = s + \lambda^2 n$. (Show that, if $n \neq 0$, there exists an $\mathfrak{sl}_2$-triplet in $\mathfrak{a}_s$ of which the first component is n, and deduce a homomorphism $\varphi : \mathbf{SL}(2, k) \to \mathrm{G}$; take σ_λ to be the image under φ of a suitable diagonal element of $\mathbf{SL}(2, k)$.) Deduce that s belongs to the closure of $\mathrm{G}.x$ in the Zariski topology.

d) Show that, if x is not semi-simple, x does not belong to the closure of $\mathrm{G}.s$ in the Zariski topology (use the inequality $\dim \mathfrak{a}_x < \dim \mathfrak{a}_s$, cf. *a*)).

---

[20]For more details on Exercises 6 to 11, see: B. KOSTANT, The principal three-dimensional subgroup and the Betti numbers of a complex simple Lie group, *Amer. J. Math.*, Vol. LXXXI (1959), pp. 973-1032.

e) Assume that k is algebraically closed. Prove the equivalence of the following properties:
(i) x is semi-simple;
(ii) $G.x$ is closed in $\mathfrak{g}$ in the Zariski topology.
(The implication (ii) $\Longrightarrow$ (i) follows from *c*). If (i) is satisfied, and if x' is in the closure of $G.x$, Exerc. 15 of §8 shows that the semi-simple component s' of x' belongs to $G.x$, so x' belongs to the closure of $G.s'$; conclude by applying *d*) to x' and s'.)

f) Assume that k is algebraically closed. Let F_x be the set of elements $y \in \mathfrak{g}$ such that $f(x) = f(y)$ for every invariant polynomial function f on $\mathfrak{g}$; we have $y \in F_x$ if and only if the semi-simple component of y is G-conjugate to s (§8, Exerc. 17). Show that F_x is the union of a finite number of orbits of G, and that this number is $\leq 3^{l(x)}$, where $l(x)$ is the rank of $\mathscr{D}(\mathfrak{a}_s)$. Only one of these orbits is closed: that of s; only one is open in F_x: that consisting of the elements $y \in F_x$ such that $\dim \mathfrak{a}_y = \mathrm{rk}(\mathfrak{g})$.

13) Assume that $\mathfrak{g}$ is semi-simple.

a) Let (x, h, y) be a principal $\mathfrak{sl}_2$-triplet in $\mathfrak{g}$, and let $\mathfrak{b}$ be the Borel subalgebra containing x (Exerc. 7). Show that $\mathfrak{b}$ is contained in $\mathrm{Im}\,\mathrm{ad}\, x$.

b) Assume that k is algebraically closed. Show that, for any element z in $\mathfrak{g}$, there exist $x, t \in \mathfrak{g}$, with x principal nilpotent, such that $z = [x, t]$ (apply *a*) to a Borel subalgebra containing z).

14) Assume that $\mathfrak{g}$ is semi-simple. Let $\mathfrak{p}$ be a parabolic subalgebra of $\mathfrak{g}$, and let f_1 and f_2 be two homomorphisms from $\mathfrak{g}$ to a finite dimensional Lie algebra. Show that $f_1|\mathfrak{p} = f_2|\mathfrak{p}$ implies that $f_1 = f_2$. (Reduce to the case in which $\mathfrak{g}$ is split, then to the case in which $\mathfrak{g} = \mathfrak{sl}(2, k)$, and use Lemma 1 of no. 1.)

¶ 15) Assume that $\mathfrak{g}$ is semi-simple.

a) Let x be a nilpotent element of $\mathfrak{g}$. Show that x is contained in $\mathrm{Im}(\mathrm{ad}\, x)^2$ (use Prop. 2). Deduce that $(\mathrm{ad}\, x)^2 = 0$ implies $x = 0$.

b) Let (x, h, y) be an $\mathfrak{sl}_2$-triplet in $\mathfrak{g}$, and let $\mathfrak{s} = kx \oplus kh \oplus ky$. Prove the equivalence of the following conditions:
(i) $\mathrm{Im}(\mathrm{ad}\, x)^2 = k.x$;
(ii) the $\mathfrak{s}$-module $\mathfrak{g}/\mathfrak{s}$ is a sum of simple modules of dimension 1 or 2;
(iii) the only eigenvalues of $\mathrm{ad}_{\mathfrak{g}}\, h$ distinct from 0, 1 and -1 are 2 and -2, and their multiplicity is equal to 1.

c) Assume that $\mathfrak{g}$ is splittable simple. Let $\mathfrak{h}$ be a splitting Cartan subalgebra of $\mathfrak{g}$, B a basis of $R(\mathfrak{g}, \mathfrak{h})$, and γ the highest root of $R(\mathfrak{g}, \mathfrak{h})$ relative to B. Let (x, h, y) be an $\mathfrak{sl}_2$-triplet such that $h \in \mathfrak{h}_{\mathbf{Q}}$ and $\alpha(h) \geq 0$ for all $\alpha \in B$ (cf. Prop. 5). Show that conditions (i), (ii), (iii) of *b*) are satisfied if and only if $h = H_\gamma$, in which case $x \in \mathfrak{g}^\gamma$ and $y \in \mathfrak{g}^{-\gamma}$.

d) Retain the hypotheses of *c*), and put $G = \mathrm{Aut}_0(\mathfrak{g})$. Show that the $\mathfrak{sl}_2$-triplets satisfying (i), (ii) and (iii) are G-conjugate (use Exerc. 10). Show that, if x is a non-zero nilpotent element of $\mathfrak{g}$ satisfying (i), the closure of G.x in the Zariski topology is equal to $\{0\} \cup \mathrm{G}.x$.

16) Let (x, h, y) be an $\mathfrak{sl}_2$-triplet in $\mathfrak{g}$. Show that, if -2 is a square in k, the elements $x - y$ and h are conjugate by an element of $\mathrm{Aut}_e(\mathfrak{g})$ (reduce to the case in which $\mathfrak{g} = \mathfrak{sl}(2, k)$). Deduce that, if $\mathfrak{g}$ is semi-simple, $x - y$ is semi-simple, and that it is regular if and only if h is regular.

¶ 17) Let (x, h, y) be a principal $\mathfrak{sl}_2$-triplet in the semi-simple algebra $\mathfrak{g}$. For all $i \in \mathbf{Z}$, denote by $\mathfrak{g}_i$ the eigenspace of ad h relative to the eigenvalue i; we have $\mathfrak{g} = \bigoplus_{i \in \mathbf{Z}} \mathfrak{g}_i$, and $\mathfrak{g}_i = 0$ if i is odd. The direct sum $\mathfrak{b}$ of the $\mathfrak{g}_i$, $i \geq 0$, is a Borel subalgebra of $\mathfrak{g}$, and $\mathfrak{g}_0$ is a Cartan subalgebra.

a) Show that, for all $z \in \mathfrak{b}$, the commutant of $y + z$ in $\mathfrak{g}$ is of dimension $l = \mathrm{rk}(\mathfrak{g})$.

b) Let I be the algebra of invariant polynomial functions on $\mathfrak{g}$, and $\mathrm{P}_1, \ldots, \mathrm{P}_l$ homogeneous elements of I generating I; put $\deg(\mathrm{P}_i) = k_i = m_i + 1$. Show that the commutant $\mathfrak{c}$ of x in $\mathfrak{g}$ has a basis $x_1, \ldots, x_l$ with $x_i \in \mathfrak{g}_{2m_i}$ (cf. Exerc. 11).

c) Let $i \in [1, l]$, and let J_i (resp. K_i) be the set of $j \in (1, l)$ such that $m_j = m_i$ (resp. $m_j < m_i$). Let $f_i \in k[\mathrm{X}_1, \ldots, \mathrm{X}_l]$ be the polynomial such that

$$f_i(a_1, \ldots, a_l) = \mathrm{P}_i(y + \sum_{j=1}^{j=l} a_j x_j) \quad \text{for } (a_j) \in k^l.$$

Show that f_i is the sum of a linear form L_i in the X_j, $j \in \mathrm{J}_i$, and a polynomial in the X_j, $j \in \mathrm{K}_i$. (If $t \in k^*$, the automorphism of $\mathfrak{g}$ equal to t^i on $\mathfrak{g}_i$ belongs to $\mathrm{Aut}_0(\mathfrak{g})$ and takes y to $t^{-2}y$ and x_j to $t^{2m_j}x_j$. Use the invariance of P_i under this automorphism.)

d) Let P be the map from $\mathfrak{g}$ to k^l defined by the P_i. If $z \in \mathfrak{g}$, denote by $\mathrm{D}_z\mathrm{P} : \mathfrak{g} \to k^l$ the tangent linear map of P at z (Chap. VII, App. I, no. 2).

Show that $\mathfrak{c} \cap \mathrm{Im}\,\mathrm{ad}(y - x) = 0$ (decompose $\mathfrak{g}$ into a direct sum of simple submodules relative to the subalgebra generated by the given $\mathfrak{sl}_2$-triplet). Deduce that the restriction of $\mathrm{D}_{y-x}\mathrm{P}$ to $\mathfrak{c}$ is an isomorphism from $\mathfrak{c}$ to k^l (use the preceding exercise and Exerc. 18 *a*) of §8). Show, by using this result, that the determinant of the linear forms L_i defined in *c*) is $\neq 0$, and deduce the following results:

d_1) the polynomials $f_1, \ldots, f_l$ are algebraically independent and generate $k[\mathrm{X}_1, \ldots, \mathrm{X}_l]$;

d_2) the map $z \mapsto \mathrm{P}(y + z)$ from $\mathfrak{c}$ to k^l is bijective polynomial, and the inverse map is polynomial;

d_3) for all $z \in y + \mathfrak{c}$, the linear map $\mathrm{D}_z\mathrm{P}|\mathfrak{c}$ is of rank l.

In particular, the map $\mathrm{P} : \mathfrak{g} \to k^l$ is surjective.

e) If k is algebraically closed, every element of $\mathfrak{g}$ whose commutant is of dimension l is conjugate under $\mathrm{Aut}_0(\mathfrak{g})$ to a unique element of $y + \mathfrak{c}$.

f) Give an example of a simple Lie algebra with no principal $\mathfrak{sl}_2$-triplet, for which the map P is not surjective (take $k = \mathbf{R}$ and $l = 1$).

§13

1) The dimensions ≤ 80 of splittable simple Lie algebras are:

3 ($A_1 = B_1 = C_1$), 8 (A_2), 10 ($B_2 = C_2$), 14 (G_2), 15 ($A_3 = D_3$), 21 (B_3 and C_3), 24 (A_4), 28 (D_4), 35 (A_5), 36 (B_4 and C_4), 45 (D_5), 48 (A_6), 52 (F_4), 55 (B_5 and C_5), 63 (A_7), 66 (D_6), 78 (B_6, C_6 and E_6), 80 (A_8).

2) Let $(\mathfrak{g}, \mathfrak{h})$ be a split simple Lie algebra, and B a basis of $R(\mathfrak{g}, \mathfrak{h})$. The simple $\mathfrak{g}$-modules $E(\lambda)$, $\lambda \in P_{++} - \{0\}$, of minimum dimension are those for which λ is one of the following weights:

ϖ_1 (A_1); ϖ_1 and ϖ_l (A_l, $l \geq 2$); ϖ_1 (B_l and C_l, $l \geq 2$); ϖ_1, ϖ_3 and ϖ_4 (D_4); ϖ_1 (D_l, $l \geq 5$); ϖ_1 and ϖ_6 (E_6); ϖ_7 (E_7); ϖ_8 (E_8); ϖ_4 (F_4); ϖ_1 (G_2).

Any two such modules can be transformed into each other by an automorphism of $\mathfrak{g}$.

Type E_8 is the only one for which the adjoint representation is of minimum dimension.

3) *a*) Define an isomorphism from $\mathfrak{sl}(4, k)$ to the orthogonal algebra $\mathfrak{o}_S(6, k)$. (Use the fact that the representation $\bigwedge^2 \sigma$ of no. 1.V is orthogonal and of dimension 6.) The two types of irreducible representation of $\mathfrak{sl}(4, k)$ of degree 4 correspond to the two semi-spinor representations of $\mathfrak{o}_S(6, k)$.

b) Define an isomorphism from $\mathfrak{sp}(4, k)$ to the orthogonal algebra $\mathfrak{o}_S(5, k)$. (Use the fact that the representation σ_2 of no. 3.V is orthogonal and of dimension 5.) The irreducible representation of $\mathfrak{sp}(4, k)$ of degree 4 corresponds to the spinor representation of $\mathfrak{o}_S(5, k)$.

c) Define an isomorphism $\mathfrak{sl}(2, k) \times \mathfrak{sl}(2, k) \rightarrow \mathfrak{o}_S(4, k)$ by using the tensor product of the identity representations of the two $\mathfrak{sl}(2, k)$ factors. Recover this result by means of Chap. I, §6, Exerc. 26.

4) Let S be the square matrix of order n

$$(\delta_{i,n+1-i}) = \begin{pmatrix} 0 & 0 & \cdots & 0 & 1 \\ 0 & 0 & \cdots & 1 & 0 \\ \vdots & \vdots & \ddots & \vdots & \vdots \\ 0 & 1 & \cdots & 0 & 0 \\ 1 & 0 & \cdots & 0 & 0 \end{pmatrix}.$$

The elements of $\mathfrak{o}_S(n, k)$ are the matrices (a_{ij}) that are anti-symmetric with respect to the second diagonal:

$$a_{i,j} = -a_{n+1-j,n+1-i} \quad \text{for every pair } (i, j).$$

The algebra $\mathfrak{o}_S(n, k)$ is splittable simple of type $D_{n/2}$ if n is even and ≥ 6, and of type $B_{(n-1)/2}$ if n is odd and ≥ 5. The diagonal (resp. upper triangular) elements of $\mathfrak{o}_S(n, k)$ form a splitting Cartan (resp. Borel) subalgebra.

5) The notations are those of no. 1.IV, type A_l. Show that, if n is ≥ 0, $\mathbf{S}^n\sigma$ is an irreducible representation of $\mathfrak{sl}(l+1,k)$ of highest weight $n\varpi_1$. (Realise $\mathbf{S}^n\sigma$ on the space of homogeneous polynomials of degree n in $X_0,\ldots,X_l$, and observe that, up to homothety, the only polynomial f such that $\partial f/\partial X_i = 0$ for $i \geq 1$ is X_0^n.) Show that all the weights of this representation are of multiplicity 1.

6) The notations are those of no. 2.IV, type B_l. Show that, if $1 \leq r \leq l-1$, the dimension of $E(\varpi_r)$ is $\binom{2l+1}{r}$, and deduce another proof of the fact that $\bigwedge^r \sigma$ is a fundamental representation of highest weight ϖ_r.

7) The notations are those of no. 2.VII, type B_l. Show that $\mathbf{O}_0^+(\Psi)$ is the commutator subgroup of $\mathbf{SO}(\Psi)$. (Remark that $\mathbf{O}(\Psi)$ is equal to $\{\pm 1\} \times \mathbf{SO}(\Psi)$, hence has the same commutator subgroup as $\mathbf{SO}(\Psi)$, and apply *Algebra*, Chap. IX, §9, Exerc. 11 *b*).) Deduce that $\mathrm{Aut}_e(\mathfrak{g}) = \mathbf{O}_0^+(\Psi)$.

8) The notations are those of no. 3.IV, type C_l $(l \geq 1)$. Show that $\mathbf{S}^2\sigma$ is equivalent to the adjoint representation of $\mathfrak{g}$.

9) The notations are those of no. 3.VII, type C_l $(l \geq 1)$. In particular, we identify $\mathrm{Aut}_e(\mathfrak{g})$ with a subgroup of $\mathbf{Sp}(\Psi)/\{\pm 1\}$. Show that the image in $\mathbf{Sp}(\Psi)/\{\pm 1\}$ of a symplectic transvection (*Algebra*, Chap. IX, §4, Exerc. 6) belongs to $\mathrm{Aut}_e(\mathfrak{g})$. Deduce that $\mathrm{Aut}_e(\mathfrak{g}) = \mathbf{Sp}(\Psi)/\{\pm 1\}$ (*Algebra*, Chap. IX, §5, Exerc. 11), and that $\mathrm{Aut}(\mathfrak{g})/\mathrm{Aut}_e(\mathfrak{g})$ can be identified with k^*/k^{*2}.

¶ 10) The notations are those of no. 4.IV, type D_l $(l \geq 2)$.

a) Let x and y be the elements of $\bigwedge^l V$ defined by

$$x = e_1 \wedge \cdots \wedge e_{l-1} \wedge e_l \quad \text{and} \quad y = e_1 \wedge \cdots \wedge e_{l-1} \wedge e_{-l}.$$

The element x is primitive of weight $2\varpi_l$ and y is primitive of weight $2\varpi_{l-1}$. The submodule X (resp. Y) of $\bigwedge^l V$ generated by x (resp. y) is isomorphic to $E(2\varpi_l)$ (resp. $E(2\varpi_{l-1})$). Show, by calculating dimensions, that $\bigwedge^l V = X \oplus Y$; in particular, $\bigwedge^l V$ is the sum of two non-isomorphic simple modules.

b) Let $e = e_1 \wedge \cdots \wedge e_l \wedge e_{-1} \wedge \cdots \wedge e_{-l} \in \bigwedge^{2l} V$, and Ψ_l the extension of Ψ to $\bigwedge^l V$. Let $z \in \bigwedge^l V$. Prove the equivalences:

$$z \in X \iff z \wedge t = \Psi_l(z,t)e \quad \text{for all } t \in \textstyle\bigwedge^l V$$

$$z \in Y \iff z \wedge t = -\Psi_l(z,t)e \quad \text{for all } t \in \textstyle\bigwedge^l V.$$

(If X′ and Y′ denote the subspaces defined by the right-hand sides, prove first that X′ and Y′ are stable under $\mathfrak{g}$ and contain x and y, respectively.)

c) Assume that z is *pure* (*Algebra*, Chap. III, §11, no. 13), and denote by M_z the l-dimensional subspace of V associated to it. Show that M_z is totally isotropic if and only if z belongs to X or to Y. (Use the fact that, when M_z is totally isotropic, there exists an orthogonal transformation that transforms it into M_x; when M_z is not totally isotropic, construct an l-vector t such that $z \wedge t = 0, \Psi_l(z,t) = 1$, and apply $b)$ above.) When $z \in X$ (resp. $z \in Y$), the dimension of $M_x/(M_x \cap M_z)$ is an even (resp. odd) integer, cf. *Algebra*, Chap. IX, §6, Exerc. 18 d).

d) Let s be a direct (resp. inverse) similarity of V. Show that $\bigwedge^l s$ leaves X and Y stable (resp. interchanges X and Y).

11) The notations are those of no. 4.VII, type D_l ($l \geq 3$). Show that $\mathbf{O}_0^+(\Psi)$ is the commutator subgroup of $\mathbf{SO}(\Psi)$. (Apply *Algebra*, Chap. IX, §6, Exerc. 17 b) and *Algebra*, Chap. IX, §9, Exerc. 11 b).) Deduce that $\mathrm{Aut}_e(\mathfrak{g})$ is equal to the image of $\mathbf{O}_0^+(\Psi)$ in $\mathbf{SO}(\Psi)/\{\pm 1\}$. Moreover, -1 belongs to $\mathbf{O}_0^+(\Psi)$ if and only if l is even or -1 is a square in k (*Algebra*, Chap. IX, §9, Exerc. 11 c)).

12) The Killing form of $\mathfrak{sl}(n,k)$ is $(X,Y) \mapsto 2n\,\mathrm{Tr}(XY)$. That of $\mathfrak{sp}(n,k)$, n even, is $(X,Y) \mapsto (n+2)\mathrm{Tr}(XY)$. That of $\mathfrak{o}_S(n,k)$, where S is non-degenerate symmetric of rank n, is $(X,Y) \mapsto (n-2)\mathrm{Tr}(XY)$.

13) The algebra of invariant polynomial functions of $\mathfrak{g}$ is generated:

a) in case A_l, by the functions $X \mapsto \mathrm{Tr}(X^i)$, $2 \leq i \leq l+1$;

b) in case B_l, by the functions $X \mapsto \mathrm{Tr}(X^{2i})$, $1 \leq i \leq l$;

c) in case C_l, by the functions $X \mapsto \mathrm{Tr}(X^{2i})$, $1 \leq i \leq l$;

d) in case D_l, by the functions $X \mapsto \mathrm{Tr}(X^{2i})$, $1 \leq i \leq l-1$, and by one of the two polynomial functions $\tilde{f}$ such that $\tilde{f}(X)^2 = (-1)^l \det(X)$.

14) a) Let G be the group associated to $\mathfrak{g} = \mathfrak{sl}(n,k)$ by the procedure of §7, Exerc. 26. The natural $\mathfrak{g}$-module structure on k^n gives rise to a homomorphism $\varphi : G \to \mathbf{GL}(n,k)$. Use *loc. cit.* h) to prove that φ is injective, and *loc. cit.* f) to prove that

$$\mathrm{Im}(\varphi) = \mathbf{SL}(n,k).$$

b) Let E be a finite dimensional $\mathfrak{sl}(n,k)$-module, and ρ the corresponding representation of $\mathfrak{sl}(n,k)$. Show that there exists a unique representation $\pi : \mathbf{SL}(n,k) \to \mathbf{GL}(E)$ such that $\pi(e^x) = e^{\rho(x)}$ for every nilpotent element x of $\mathfrak{sl}(n,k)$ (use a)). We say that ρ and π are *compatible*. Generalize the results proved for $n = 2$ in §1, no. 4.

c) Assume that k is $\mathbf{R}, \mathbf{C}$ or a complete field for a discrete valuation with residue field of characteristic $\neq 0$. Show that ρ and π are compatible if and only if π is a homomorphism of Lie groups such that $L(\pi) = \rho$ (same method as in Exerc. 18 b) of §1).

d) Prove analogous results for $\mathfrak{sp}(2n,k)$ and $\mathbf{Sp}(2n,k)$.

¶ 15) Let V be a vector space of finite dimension ≥ 2, $\mathfrak{g}$ a Lie subalgebra of End(V) and θ an element of $\mathfrak{g}$. We make the following assumptions:
(i) V is a semi-simple $\mathfrak{g}$-module;

(ii) θ is of rank 1 (i.e. $\dim \operatorname{Im}(\theta) = 1$);
(iii) the line $\operatorname{Im}(\theta)$ generates the $\mathrm{U}(\mathfrak{g})$-module V.

a) Show that these assumptions are satisfied (for a suitable choice of θ) when $\mathfrak{g} = \mathfrak{sl}(\mathrm{V})$, when $\mathfrak{g} = \mathfrak{gl}(\mathrm{V})$, or when there exists a non-degenerate alternating bilinear form Ψ on V such that $\mathfrak{g} = \mathfrak{sp}(\Psi)$ or $\mathfrak{g} = k.1 \oplus \mathfrak{sp}(\Psi)$. In each of these cases, θ can be taken to be a nilpotent element; in the second case (and only in that case) θ can be taken to be a semi-simple element.

b) We shall prove that the four cases above are the only ones possible. We are reduced immediately to the case in which k is algebraically closed. Show that V is then a simple $\mathfrak{g}$-module, and that $\mathfrak{g} = \mathfrak{c} \oplus \mathfrak{s}$, where $\mathfrak{s}$ is semi-simple, and $\mathfrak{c} = 0$ or $\mathfrak{c} = k.1$; show that V is not isomorphic to a tensor product of $\mathfrak{g}$-modules of dimension ≥ 2; deduce that $\mathfrak{s}$ is simple.

c) Choose a Cartan subalgebra $\mathfrak{h}$ of $\mathfrak{s}$, and a basis B of $\mathrm{R}(\mathfrak{s}, \mathfrak{h})$. Let λ be the highest weight (with respect to B) of the $\mathfrak{s}$-module V, and let e be a non-zero element of V of weight λ. The highest weight of the dual module V^* is $\lambda^* = -w_0\lambda$ (§7, no. 5); let e^* be a non-zero element of V^* of weight λ^*. Identify $\mathrm{V} \otimes \mathrm{V}^*$ with $\operatorname{End}(\mathrm{V})$ as usual. Show that there exist $x \in \mathrm{V}, y \in \mathrm{V}^*$ such that $x \otimes y \in \mathfrak{g}$ and $\langle x, e^* \rangle \neq 0, \langle e, y \rangle \neq 0$ (take the conjugate of θ by e^n, where n is a suitable nilpotent element of $\mathfrak{g}$). Use the fact that $\mathfrak{g}$ is an $\mathfrak{h}$-submodule of $\mathrm{V} \otimes \mathrm{V}^*$ to conclude that $\mathfrak{g}$ contains $e \otimes e^*$. Deduce that $\lambda + \lambda^* = \tilde{\alpha}$, where $\tilde{\alpha}$ is the highest root of $\mathfrak{s}$.

d) Show that $\mathfrak{s}$ is not of type B_l $(l \geq 2)$, D_l $(l \geq 4)$, $\mathrm{E}_6, \mathrm{E}_7, \mathrm{E}_8, \mathrm{F}_4, \mathrm{G}_2$ (by Chap. VI, *Tables*, $\tilde{\alpha}$ would be a fundamental weight, and hence could not be of the form $\lambda + \lambda^*$ above). Deduce that $\mathfrak{s}$ is either of type A_l or of type C_l, and that in the first case $\lambda = \varpi_1$ or $\varpi_1 = \varpi_1^*$, and in the second case $\tilde{\alpha} = 2\varpi_1 = \varpi_1 + \varpi_1^*$; since $\mathfrak{c} = 0$ or $k.1$, this indeed gives the four possibilities in *a*)[21].

¶ 16) Let $\mathfrak{g}$ be an absolutely simple Lie algebra of type A_l $(l \geq 2)$, $\bar{k}$ an algebraic closure of k, and $\pi : \operatorname{Gal}(\bar{k}/k) \to \operatorname{Aut}(\bar{\mathrm{R}}, \bar{\mathrm{B}})$ the homomorphism defined in Exerc. 8 of §5.

a) Assume that π is trivial. Show that there exist exactly two two-sided ideals $\mathfrak{m}$ and $\mathfrak{m}'$ of $\mathrm{U}(\mathfrak{g})$ such that $\mathrm{D} = \mathrm{U}(\mathfrak{g})/\mathfrak{m}$ and $\mathrm{D}' = \mathrm{U}(\mathfrak{g})/\mathfrak{m}'$ are central simple algebras of dimension $(l+1)^2$ (use Exerc. 8 of §7). The principal anti-automorphism of $\mathrm{U}(\mathfrak{g})$ interchanges $\mathfrak{m}$ and $\mathfrak{m}'$; in particular, D' is isomorphic to the opposite of D. The composite $\mathfrak{g} \to \mathrm{U}(\mathfrak{g}) \to \mathrm{D}$ identifies $\mathfrak{g}$ with the Lie subalgebra $\mathfrak{sl}_{\mathrm{D}}$ of D consisting of the elements of trace zero. Moreover, $\mathfrak{g}$ is splittable (and hence isomorphic to $\mathfrak{sl}(l+1, k)$) if and only if D is isomorphic to $\mathbf{M}_{l+1}(k)$.

Conversely, if Δ is a central simple algebra of dimension $(l+1)^2$, the Lie algebra $\mathfrak{sl}_{\Delta}$ is absolutely simple of type A_l, and the corresponding homomor-

[21] For more details on this exercise, see: V. W. GUILLEMIN, D. QUILLEN and S. STERNBERG, The classification of the irreducible complex algebras of infinite type, *J. Analyse Math.*, Vol. XVII (1967), pp. 107-112.

phism π is trivial. Two such algebras $\mathfrak{sl}_\Delta$ and $\mathfrak{sl}_{\Delta'}$ are isomorphic if and only if Δ and Δ' are isomorphic or anti-isomorphic.

b) Assume that π is non-trivial. Since $\mathrm{Aut}(\bar{\mathrm{R}}, \bar{\mathrm{B}})$ has two elements, the kernel of π is an open subgroup of $\mathrm{Gal}(\bar{k}/k)$ of index 2, which corresponds by Galois theory to a quadratic extension k_1 of k. Denote the non-trivial involution of k_1 by $x \mapsto \bar{x}$.

Show that there exists a unique two-sided ideal $\mathfrak{m}$ of $\mathrm{U}(\mathfrak{g})$ such that $\mathrm{D} = \mathrm{U}(\mathfrak{g})/\mathfrak{m}$ is a simple algebra of dimension $2(l+1)^2$ whose centre is a quadratic extension of k (same method). The centre of D can be identified with k_1. The ideal $\mathfrak{m}$ is stable under the principal anti-automorphism of $\mathrm{U}(\mathfrak{g})$; that defines by passage to the quotient an involutive anti-automorphism σ of D such that $\sigma(x) = \bar{x}$ for all $x \in k_1$. The composite map $\mathfrak{g} \to \mathrm{U}(\mathfrak{g}) \to \mathrm{D}$ identifies $\mathfrak{g}$ with the Lie subalgebra $\mathfrak{su}_{\mathrm{D},\sigma}$ of D consisting of the elements x such that $\sigma(x) = -x$ and $\mathrm{Tr}_{\mathrm{D}/k_1}(x) = 0$. Conversely, if Δ is a central simple k_1-algebra of dimension $(l+1)^2$, equipped with an involutive anti-automorphism σ whose restriction to k_1 is $x \mapsto \bar{x}$, the Lie algebra $\mathfrak{su}_{\Delta,\sigma}$ is absolutely simple of type A_l, and the corresponding homomorphism π is that associated to k_1. Two such algebras $\mathfrak{su}_{\Delta,\sigma}$ and $\mathfrak{su}_{\Delta',\sigma'}$ are isomorphic if and only if there exists a k-isomorphism $f : \Delta \to \Delta'$ such that $\sigma' \circ f = f \circ \sigma$.

Show that, when $\mathrm{D} = \mathbf{M}_{l+1}(k_1)$, there exists an invertible hermitian matrix H of degree $l+1$, unique up to multiplication by an element of k^*, such that

$$\sigma(x) = H.{}^t\bar{x}.H^{-1}$$

for all $x \in \mathbf{M}_{l+1}(k_1)$; the algebra $\mathfrak{g}$ can then be identified with the algebra $\mathfrak{su}(l+1, H)$ consisting of the matrices x such that $x.H + H.{}^t\bar{x} = 0$ and $\mathrm{Tr}(x) = 0$.

¶ 17) Let $\mathfrak{g}$ be an absolutely simple Lie algebra of type B_l (resp. $\mathrm{C}_l, \mathrm{D}_l$), with $l \geq 2$ (resp. $l \geq 3, l \geq 4$). Assume further that, when $\mathfrak{g}$ is of type D_4, the image of the homomorphism $\pi : \mathrm{Gal}(\bar{k}/k) \to \mathrm{Aut}(\bar{\mathrm{R}}, \bar{\mathrm{B}})$ defined in Exerc. 8 of §5 is of order ≤ 2. Show that there exists a central simple algebra D of dimension $(2l+1)^2$ (resp. $4l^2, 4l^2$) and an involutive anti-automorphism σ of D such that $\mathfrak{g}$ is isomorphic to the Lie subalgebra of D consisting of the elements x such that $\sigma(x) = -x$ and $\mathrm{Tr}_{\mathrm{D}}(x) = 0$ (same method as in Exerc. 16 *a*)).[22]

18) Let V be a finite dimensional vector space, Q a non-degenerate quadratic form on V, and Ψ the symmetric bilinear form associated to Q. Denote the Lie algebra $\mathfrak{o}(\Psi)$ by $\mathfrak{g}$ and the extension of Ψ to $\bigwedge^2(\mathrm{V})$ by Ψ_2.

a) Show that there exists an isomorphism of vector spaces $\theta : \bigwedge^2(\mathrm{V}) \to \mathfrak{g}$ characterized by the following equivalent properties:

[22]For more details on Exercises 16 and 17, see: N. JACOBSON, *Lie Algebras*, Interscience Publ. (1962), Chap. X and G. B. SELIGMAN, *Modular Lie Algebras*, Springer-Verlag (1967), Chap. IV.

(i) For a, b and x in V, we have $\theta(a \wedge b).x = a.\Psi(x,b) - b.\Psi(x,a)$.

(ii) For x, y in V and u in $\bigwedge^2(\mathrm{V})$ we have $\Psi_2(x \wedge y, u) = \Psi(x, \theta(u).y)$.

Let σ be the identity representation of $\mathfrak{g}$ on V; then θ is an isomorphism of $\bigwedge^2(\mathrm{V})$ with the adjoint representation of $\mathfrak{g}$.

b) Define the linear map $f : \mathfrak{g} \to \mathrm{C}^+(\mathrm{Q})$ as in Lemma 1 of no. 2. Show that $f\theta(a \wedge b) = \frac{1}{2}(ab - ba)$ for a, b in V, and deduce a new proof of assertions (iii), (iv) and (v) of Lemma 1 of no. 2.

c) The notations $l, \mathrm{F}, \mathrm{F}', e_0, \mathrm{N}, \lambda$ and ρ are those of no. 2. Choose $e \neq 0$ in $\bigwedge(\mathrm{F}')$ and define the bilinear form Φ on N by

$$x \wedge y = (-1)^{p(p+1)/2}\Phi(x,y).e \quad (x \in \textstyle\bigwedge^p(\mathrm{F}'), y \in \mathrm{N}).$$

Show that $\Phi(\lambda(a).x, y) + \Phi(x, \lambda(a).y) = 0$ for x, y in N and a in $\mathrm{F} \oplus \mathrm{F}'$. Deduce that Φ is invariant under the representation ρ of $\mathfrak{g}$ (remark that the Lie algebra $\rho(\mathfrak{g})$ is generated by $\lambda(\mathrm{F} \oplus \mathrm{F}')$ by *a*) and *b*) above).

19) Let V be a finite dimensional vector space and $\mathrm{E} = \mathrm{V} \oplus \mathrm{V}^*$. Define a non-degenerate bilinear form Φ on E by

$$\Phi((x,x^*),(y,y^*)) = \langle x, y^* \rangle + \langle y, x^* \rangle.$$

Put $\mathrm{N} = \bigwedge(\mathrm{V})$ and $\mathrm{Q}(x) = \frac{1}{2}\Phi(x,x)$ for $x \in \mathrm{E}$. Denote the spinor representation by $\lambda : \mathrm{C}(\mathrm{Q}) \to \mathrm{End}(\mathrm{N})$ as in no. 2.IV, define $f : \mathfrak{o}(\Phi) \to \mathrm{C}^+(\mathrm{Q})$ as in Lemma 1 of no. 2, and denote by ρ the linear representation $\lambda \circ f$ of the algebra $\mathfrak{o}(\Phi)$ on N.

a) Associate to any endomorphism u of V the endomorphism $\tilde{u}$ of E by the formula $\tilde{u}(x,x^*) = (u(x), -{}^tu(x^*))$. Show that $u \mapsto \tilde{u}$ is a homomorphism of Lie algebras from $\mathfrak{gl}(\mathrm{V})$ to $\mathfrak{o}(\Phi)$; moreover, for u in $\mathfrak{gl}(\mathrm{V})$, $\rho(\tilde{u})$ is the unique derivation of the algebra $\bigwedge(\mathrm{V}) = \mathrm{N}$ that coincides with u on V.

b) Let Ψ be a non-degenerate alternating bilinear form on V and $\gamma : \mathrm{V} \to \mathrm{V}^*$ the isomorphism defined by $\Psi(x,y) = \langle x, \gamma(y) \rangle$ for x, y in V. Show that the endomorphisms $\tilde{X}_+$ and $\tilde{X}_-$ of E defined by

$$\tilde{X}_+(x,x^*) = (\gamma^{-1}(x^*), 0), \qquad \tilde{X}_-(x,x^*) = (0, -\gamma(x))$$

belongs to $\mathfrak{o}(\Phi)$. Put $\tilde{H} = (-1)\tilde{}$. Show that $(\tilde{H}, \tilde{X}_+, \tilde{X}_-)$ is an $\mathfrak{sl}_2$-triplet in the Lie algebra $\mathfrak{o}(\Phi)$.

c) Show that ρ takes the endomorphisms $\tilde{H}, \tilde{X}_+, \tilde{X}_-$ of $\mathfrak{o}(\Phi)$ to the endomorphisms of N denoted by H, X_+ and X_-, respectively, in no. 3.IV. Deduce that (H, X_+, X_-) is an $\mathfrak{sl}_2$-triplet in the Lie algebra $\mathfrak{gl}(\mathrm{N})$.

SUMMARY OF SOME IMPORTANT PROPERTIES OF SEMI-SIMPLE LIE ALGEBRAS

In this summary, $\mathfrak{g}$ denotes a semi-simple Lie algebra over k.

CARTAN SUBALGEBRAS

1) Let E be the set of commutative subalgebras of $\mathfrak{g}$ that are reductive in $\mathfrak{g}$; this is also the set of commutative subalgebras of $\mathfrak{g}$ all of whose elements are semi-simple. The Cartan subalgebras of $\mathfrak{g}$ are the maximal elements of E.

2) Let x be a regular element of $\mathfrak{g}$. Then x is semi-simple. There exists a unique Cartan subalgebra of $\mathfrak{g}$ containing x; it is the commutant of x in $\mathfrak{g}$.

3) Let x be a semi-simple element of $\mathfrak{g}$. Then x belongs to a Cartan subalgebra of $\mathfrak{g}$. The element x is regular if and only if the dimension of the commutant of x is equal to the rank of $\mathfrak{g}$.

4) Let $\mathfrak{h}$ be a Cartan subalgebra of $\mathfrak{g}$. Then $\mathfrak{h}$ is said to be splitting if $\mathrm{ad}_{\mathfrak{g}}x$ is triangularizable for all $x \in \mathfrak{h}$. Moreover, $\mathfrak{g}$ is said to be splittable if $\mathfrak{g}$ has a splitting Cartan subalgebra (this is the case if k is algebraically closed). A split semi-simple Lie algebra is a pair $(\mathfrak{g}, \mathfrak{h})$ where $\mathfrak{g}$ is a semi-simple Lie algebra and $\mathfrak{h}$ is a splitting Cartan subalgebra of $\mathfrak{g}$.

In the remainder of this summary, $(\mathfrak{g}, \mathfrak{h})$ denotes a split semi-simple Lie algebra.

ROOT SYSTEMS

5) For any element α of the dual $\mathfrak{h}^*$ of $\mathfrak{h}$, let $\mathfrak{g}^\alpha$ be the set of $x \in \mathfrak{g}$ such that $[h, x] = \alpha(h)x$ for all $h \in \mathfrak{h}$. If $\alpha = 0$, then $\mathfrak{g}^\alpha = \mathfrak{h}$. Any $\alpha \in \mathfrak{h}^* - \{0\}$ such that $\mathfrak{g}^\alpha \neq 0$ is called a root of $(\mathfrak{g}, \mathfrak{h})$. Denote by $\mathrm{R}(\mathfrak{g}, \mathfrak{h})$ (or simply by R) the set of roots of $(\mathfrak{g}, \mathfrak{h})$. This is a reduced root system in $\mathfrak{h}^*$ in the sense of Chap. VI, §1, no. 4. The algebra $\mathfrak{g}$ is simple if and only if R is irreducible.

6) For all $\alpha \in \mathrm{R}$, $\mathfrak{g}^\alpha$ is of dimension 1. The vector space $[\mathfrak{g}^\alpha, \mathfrak{g}^{-\alpha}]$ is contained in $\mathfrak{h}$, is of dimension 1, and contains a unique element H_α such that $\alpha(H_\alpha) = 2$; we have $H_\alpha = \alpha^\vee$ (Chap. VI, §1, no. 1); the set of H_α, for $\alpha \in \mathrm{R}$, is the inverse root system $\mathrm{R}^\vee$ of R.

7) We have $\mathfrak{g} = \mathfrak{h} \oplus \bigoplus_{\alpha \in \mathrm{R}} \mathfrak{g}^\alpha$. There exists a family $(X_\alpha)_{\alpha \in \mathrm{R}}$ such that, for all $\alpha \in \mathrm{R}$, we have $X_\alpha \in \mathfrak{g}^\alpha$ and $[X_\alpha, X_{-\alpha}] = -H_\alpha$. Every $x \in \mathfrak{g}$ can be written uniquely in the form

$$x = h + \sum_{\alpha \in \mathrm{R}} \lambda_\alpha X_\alpha, \qquad \text{where } h \in \mathfrak{h}, \lambda_\alpha \in k.$$

The bracket of two elements can be calculated by means of the formulas

$$\begin{aligned} [h, X_\alpha] &= \alpha(h) X_\alpha \\ [X_\alpha, X_\beta] &= 0 \quad \text{if } \alpha + \beta \notin \mathrm{R} \cup \{0\} \\ [X_\alpha, X_{-\alpha}] &= -H_\alpha \\ [X_\alpha, X_\beta] &= \mathrm{N}_{\alpha\beta} X_{\alpha+\beta} \quad \text{if } \alpha + \beta \in \mathrm{R}, \end{aligned}$$

the $\mathrm{N}_{\alpha\beta}$ being non-zero elements of k.

8) Let B be a basis of R. The algebra $\mathfrak{g}$ is generated by the X_α and the $X_{-\alpha}$ for $\alpha \in \mathrm{B}$. We have $[X_\alpha, X_{-\beta}] = 0$ if $\alpha, \beta \in \mathrm{B}$ and $\alpha \neq \beta$. Let $(n(\alpha, \beta))_{\alpha, \beta \in \mathrm{B}}$ be the Cartan matrix of R (relative to B). We have $n(\alpha, \beta) = \alpha(H_\beta)$. If $\alpha, \beta \in \mathrm{B}$ and $\alpha \neq \beta$, $n(\alpha, \beta)$ is a negative integer and

$$(\operatorname{ad} X_\beta)^{1-n(\alpha,\beta)} X_\alpha = 0 \quad \text{and} \quad (\operatorname{ad} X_{-\beta})^{1-n(\alpha,\beta)} X_{-\alpha} = 0.$$

9) If $\alpha, \beta, \alpha+\beta \in \mathrm{R}$, let $q_{\alpha\beta}$ be the largest integer j such that $\beta - j\alpha \in \mathrm{R}$. The family $(X_\alpha)_{\alpha \in \mathrm{R}}$ in 7) can be chosen so that $\mathrm{N}_{\alpha,\beta} = \mathrm{N}_{-\alpha,-\beta}$ if $\alpha, \beta, \alpha+\beta \in \mathrm{R}$. Then $\mathrm{N}_{\alpha\beta} = \pm(q_{\alpha\beta} + 1)$. There exists an involutive automorphism θ of $\mathfrak{g}$ that takes X_α to $X_{-\alpha}$ for all $\alpha \in \mathrm{R}$; we have $\theta(h) = -h$ for all $h \in \mathfrak{h}$. The $\mathbf{Z}$-submodule $\mathfrak{g}_{\mathbf{Z}}$ of $\mathfrak{g}$ generated by the H_α and the X_α is a $\mathbf{Z}$-Lie subalgebra of $\mathfrak{g}$, and the canonical map $\mathfrak{g}_{\mathbf{Z}} \otimes_{\mathbf{Z}} k \to \mathfrak{g}$ is an isomorphism.

The pair $(\mathfrak{g}, \mathfrak{h})$ can be obtained by extension of scalars from a split semi-simple $\mathbf{Q}$-Lie algebra.

10) The Weyl group, group of weights, ... of R is called the Weyl group, group of weights, ... of $(\mathfrak{g}, \mathfrak{h})$. The Weyl group will be denoted by W in what follows. We consider its operation, not only on $\mathfrak{h}^*$, but also on $\mathfrak{h}$ (by transport of structure). If $\mathfrak{h}_{\mathbf{Q}}$ (resp. $\mathfrak{h}^*_{\mathbf{Q}}$) denotes the $\mathbf{Q}$-vector subspace of $\mathfrak{h}$ (resp. $\mathfrak{h}^*$) generated by the H_α (resp. the α), then $\mathfrak{h}$ (resp. $\mathfrak{h}^*$) can be canonically identified with $\mathfrak{h}_{\mathbf{Q}} \otimes_{\mathbf{Q}} k$ (resp. $\mathfrak{h}^*_{\mathbf{Q}} \otimes_{\mathbf{Q}} k$), and $\mathfrak{h}^*_{\mathbf{Q}}$ can be identified with the dual of $\mathfrak{h}_{\mathbf{Q}}$. When we speak of the Weyl chambers of R, these are understood to be in $\mathfrak{h}_{\mathbf{R}} = \mathfrak{h}_{\mathbf{Q}} \otimes_{\mathbf{Q}} \mathbf{R}$ or $\mathfrak{h}^*_{\mathbf{R}} = \mathfrak{h}^*_{\mathbf{Q}} \otimes_{\mathbf{Q}} \mathbf{R}$.

11) Let Φ be the Killing form of $\mathfrak{g}$. If $\alpha + \beta \neq 0$, $\mathfrak{g}^\alpha$ and $\mathfrak{g}^\beta$ are orthogonal with respect to Φ. The restriction of Φ to $\mathfrak{g}^\alpha \times \mathfrak{g}^{-\alpha}$ is non-degenerate. If $x, y \in \mathfrak{h}$, then $\Phi(x, y) = \sum_{\alpha \in \mathrm{R}} \alpha(x)\alpha(y)$. We have $\Phi(H_\alpha, H_\beta) \in \mathbf{Z}$. The restriction of Φ to $\mathfrak{h}$ is non-degenerate and invariant under W; its restriction to $\mathfrak{h}_{\mathbf{Q}}$ is positive.

12) The root system of $(\mathfrak{g}, \mathfrak{h})$ depends, up to isomorphism, only on $\mathfrak{g}$ and not on $\mathfrak{h}$. By abuse of language, the Weyl group, group of weights, ... of $(\mathfrak{g}, \mathfrak{h})$ are also called the Weyl group, group of weights, ... of $\mathfrak{g}$.

If R_1 is a reduced root system, there exists a split semi-simple Lie algebra $(\mathfrak{g}_1, \mathfrak{h}_1)$ such that $R(\mathfrak{g}_1, \mathfrak{h}_1)$ is isomorphic to R_1; it is unique, up to isomorphism.

The classification of splittable semi-simple Lie algebras is thus reduced to that of root systems.

SUBALGEBRAS

13) If $P \subset R$, put $\mathfrak{g}^P = \bigoplus_{\alpha \in P} \mathfrak{g}^\alpha$ and $\mathfrak{h}_P = \sum_{\alpha \in P} kH_\alpha$. Let $P \subset R$, $\mathfrak{h}'$ a vector subspace of $\mathfrak{h}$, and $\mathfrak{a} = \mathfrak{h}' \oplus \mathfrak{g}^P$. Then $\mathfrak{a}$ is a subalgebra of $\mathfrak{g}$ if and only if P is a closed subset of R and $\mathfrak{h}'$ contains $\mathfrak{h}_{P\cap(-P)}$; $\mathfrak{a}$ is reductive in $\mathfrak{g}$ if and only if $P = -P$; and $\mathfrak{a}$ is solvable if and only if $P \cap (-P) = \varnothing$.

14) Let P be a closed subset of R, and $\mathfrak{b} = \mathfrak{h} \oplus \mathfrak{g}^P$. The following conditions are equivalent:

(i) $\mathfrak{b}$ is a maximal solvable subalgebra of $\mathfrak{g}$;

(ii) $P \cap (-P) = \varnothing$ and $P \cup (-P) = R$;

(iii) there exists a chamber C of R such that $P = R_+(C)$ (cf. Chap. VI, §1, no. 6).

A Borel subalgebra of $(\mathfrak{g}, \mathfrak{h})$ is a subalgebra of $\mathfrak{g}$ containing $\mathfrak{h}$ and satisfying the above conditions. A subalgebra $\mathfrak{b}$ of $\mathfrak{g}$ is called a Borel subalgebra of $\mathfrak{g}$ if there exists a splitting Cartan subalgebra $\mathfrak{h}'$ of $\mathfrak{g}$ such that $\mathfrak{b}$ is a Borel subalgebra of $(\mathfrak{g}, \mathfrak{h}')$; if k is algebraically closed, this is equivalent to saying that $\mathfrak{b}$ is a maximal solvable subalgebra of $\mathfrak{g}$.

Let $\mathfrak{b} = \mathfrak{h} \oplus \mathfrak{g}^{R_+(C)}$ be a Borel subalgebra of $(\mathfrak{g}, \mathfrak{h})$. The largest nilpotent ideal of $\mathfrak{b}$ is $[\mathfrak{b}, \mathfrak{b}] = \mathfrak{g}^{R_+(C)}$. Let B be the basis of R associated to C; the algebra $[\mathfrak{b}, \mathfrak{b}]$ is generated by the $\mathfrak{g}^\alpha$ for $\alpha \in B$.

If $\mathfrak{b}, \mathfrak{b}'$ are Borel subalgebras of $\mathfrak{g}$, there exists a Cartan subalgebra of $\mathfrak{g}$ contained in $\mathfrak{b} \cap \mathfrak{b}'$; such a subalgebra is splitting.

15) Let P be a closed subset of R, and $\mathfrak{p} = \mathfrak{h} \oplus \mathfrak{g}^P$. The following conditions are equivalent:

(i) $\mathfrak{p}$ contains a Borel subalgebra of $(\mathfrak{g}, \mathfrak{h})$;

(ii) $P \cup (-P) = R$;

(iii) there exists a chamber C of R such that $P \supset R_+(C)$.

A parabolic subalgebra of $(\mathfrak{g}, \mathfrak{h})$ is a subalgebra of $\mathfrak{g}$ containing $\mathfrak{h}$ and satisfying the above conditions. A subalgebra $\mathfrak{p}$ of $\mathfrak{g}$ is called parabolic if there exists a splitting Cartan subalgebra $\mathfrak{h}'$ of $\mathfrak{g}$ such that $\mathfrak{p}$ is a parabolic subalgebra of $(\mathfrak{g}, \mathfrak{h}')$.

Let $\mathfrak{p} = \mathfrak{h} \oplus \mathfrak{g}^P$ be a parabolic subalgebra of $(\mathfrak{g}, \mathfrak{h})$, Q the set of $\alpha \in P$ such that $-\alpha \notin P$, and $\mathfrak{s} = \mathfrak{h} \oplus \mathfrak{g}^{P\cap(-P)}$. Then $\mathfrak{p} = \mathfrak{s} \oplus \mathfrak{g}^Q$, $\mathfrak{s}$ is reductive in $\mathfrak{g}$, and $\mathfrak{g}^{\mathrm{Q}}$ is the largest nilpotent ideal of $\mathfrak{p}$ and the nilpotent radical of $\mathfrak{p}$. The centre of $\mathfrak{p}$ is 0.

AUTOMORPHISMS

16) The subgroup of $\mathrm{Aut}(\mathfrak{g})$ generated by the $e^{\mathrm{ad}\,x}$, with x nilpotent, is the group $\mathrm{Aut}_e(\mathfrak{g})$ of elementary automorphisms of $\mathfrak{g}$; it is a normal subgroup of $\mathrm{Aut}(\mathfrak{g})$; it is equal to its derived group.

If $\bar{k}$ is an algebraic closure of k, the group $\mathrm{Aut}(\mathfrak{g})$ embeds naturally in $\mathrm{Aut}(\mathfrak{g} \otimes_k \bar{k})$. Put

$$\mathrm{Aut}_0(\mathfrak{g}) = \mathrm{Aut}(\mathfrak{g}) \cap \mathrm{Aut}_e(\mathfrak{g} \otimes_k \bar{k});$$

this is a normal subgroup of $\mathrm{Aut}(\mathfrak{g})$, independent of the choice of $\bar{k}$. We have

$$\mathrm{Aut}_e(\mathfrak{g}) \subset \mathrm{Aut}_0(\mathfrak{g}) \subset \mathrm{Aut}(\mathfrak{g}).$$

The derived group of $\mathrm{Aut}_0(\mathfrak{g})$ is $\mathrm{Aut}_e(\mathfrak{g})$. In the Zariski topology, $\mathrm{Aut}(\mathfrak{g})$ and $\mathrm{Aut}_0(\mathfrak{g})$ are closed in $\mathrm{End}_k(\mathfrak{g})$, $\mathrm{Aut}_0(\mathfrak{g})$ is the connected component of the identity in $\mathrm{Aut}(\mathfrak{g})$, and $\mathrm{Aut}_e(\mathfrak{g})$ is dense in $\mathrm{Aut}_0(\mathfrak{g})$.

Let B be a basis of R, and Aut(R, B) the group of automorphisms of R that leave B stable. Then $\mathrm{Aut}(\mathfrak{g})$ is the semi-direct product of a subgroup isomorphic to Aut(R, B) and $\mathrm{Aut}_0(\mathfrak{g})$; in particular, $\mathrm{Aut}(\mathfrak{g})/\mathrm{Aut}_0(\mathfrak{g})$ is isomorphic to Aut(R, B), which is itself isomorphic to a group of automorphisms of the Dynkin graph of $\mathfrak{g}$.

17) A framing of $\mathfrak{g}$ is a triplet $(\mathfrak{h}', \mathrm{B}, (X_\alpha)_{\alpha\in\mathrm{B}})$, where $\mathfrak{h}'$ is a splitting Cartan subalgebra of $\mathfrak{g}$, B is a basis of $\mathrm{R}(\mathfrak{g}, \mathfrak{h}')$, and where, for all $\alpha \in \mathrm{B}$, X_α is a non-zero element of $\mathfrak{g}^\alpha$. The group $\mathrm{Aut}_0(\mathfrak{g})$ operates simply-transitively on the set of framings of $\mathfrak{g}$.

The group $\mathrm{Aut}_e(\mathfrak{g})$ operates transitively on the set of pairs $(\mathfrak{k}, \mathfrak{b})$, where $\mathfrak{k}$ is a splitting Cartan subalgebra of $\mathfrak{g}$ and $\mathfrak{b}$ is a Borel subalgebra of $(\mathfrak{g}, \mathfrak{k})$.

18) Denote by $\mathrm{Aut}(\mathfrak{g}, \mathfrak{h})$ the set of $s \in \mathrm{Aut}(\mathfrak{g})$ such that $s(\mathfrak{h}) = \mathfrak{h}$. Put

$$\mathrm{Aut}(\mathfrak{g}, \mathfrak{h}) = \mathrm{Aut}_e(\mathfrak{g}) \cap \mathrm{Aut}(\mathfrak{g}, \mathfrak{h}), \quad \mathrm{Aut}_0(\mathfrak{g}, \mathfrak{h}) = \mathrm{Aut}_0(\mathfrak{g}) \cap \mathrm{Aut}(\mathfrak{g}, \mathfrak{h}).$$

If $s \in \mathrm{Aut}(\mathfrak{g}, \mathfrak{h})$, the contragredient map of $s|\mathfrak{h}$ is an element of the group A(R) of automorphisms of R; denote this element by $\varepsilon(s)$; the map ε is a homomorphism from $\mathrm{Aut}(\mathfrak{g}, \mathfrak{h})$ to A(R). We have $\mathrm{Aut}_0(\mathfrak{g}) = \mathrm{Aut}_e(\mathfrak{g}).\mathrm{Ker}\,\varepsilon$, and

$$\varepsilon(\mathrm{Aut}_0(\mathfrak{g}, \mathfrak{h})) = \varepsilon(\mathrm{Aut}_e(\mathfrak{g}, \mathfrak{h})) = \mathrm{W}.$$

Let $\mathrm{T_P} = \mathrm{Hom}(\mathrm{P(R)}, k^*)$, $\mathrm{T_Q} = \mathrm{Hom}(\mathrm{Q(R)}, k^*)$. The injection of Q(R) into P(R) defines a homomorphism from $\mathrm{T_P}$ to $\mathrm{T_Q}$; let $\mathrm{Im}(\mathrm{T_P})$ be its image. If $t \in \mathrm{T_Q}$, let $f(t)$ be the endomorphism of $\mathfrak{g}$ such that, for all $\alpha \in \mathrm{R} \cup \{0\}$,

$f(t)|\mathfrak{g}^\alpha$ is the homothety with ratio $t(\alpha)$; we have $f(t) \in \mathrm{Aut}_0(\mathfrak{g},\mathfrak{h})$ and f is an injective homomorphism from $\mathrm{T_Q}$ to $\mathrm{Aut}_0(\mathfrak{g},\mathfrak{h})$. The sequences

$$\{1\} \longrightarrow \mathrm{T_Q} \xrightarrow{f} \mathrm{Aut}(\mathfrak{g},\mathfrak{h}) \xrightarrow{\varepsilon} \mathrm{A(R)} \longrightarrow \{1\}$$

and

$$\{1\} \longrightarrow \mathrm{T_Q} \xrightarrow{f} \mathrm{Aut}_0(\mathfrak{g},\mathfrak{h}) \xrightarrow{\varepsilon} \mathrm{W} \longrightarrow \{1\}$$

are exact. We have $f(\mathrm{Im}(\mathrm{T_P})) \subset \mathrm{Aut}_e(\mathfrak{g},\mathfrak{h})$; f defines, by passage to the quotient, a surjective[23] homomorphism $\mathrm{T_Q}/\mathrm{Im}(\mathrm{T_P}) \to \mathrm{Aut}_0(\mathfrak{g})/\mathrm{Aut}_e(\mathfrak{g})$. In the Zariski topology, $f(\mathrm{T_Q})$ is closed in $\mathrm{Aut}(\mathfrak{g})$, and $f(\mathrm{Im}(\mathrm{T_P}))$ is dense in $f(\mathrm{T_Q})$.

FINITE DIMENSIONAL MODULES

19) Let V be a finite dimensional $\mathfrak{g}$-module. For all $\mu \in \mathfrak{h}^*$, let V^μ be the set of $v \in \mathrm{V}$ such that $h.v = \mu(h)v$ for all $h \in \mathfrak{h}$. The dimension of V^μ is called the multiplicity of μ in V; if it is ≥ 1, i.e. if $\mathrm{V}^\mu \neq 0$, μ is said to be a weight of V. We have $\mathrm{V} = \bigoplus_{\mu\in\mathfrak{h}^*} \mathrm{V}^\mu$. Every weight of V belongs to P(R). If μ is a weight of V, and if $w \in \mathrm{W}$, $w\mu$ is a weight of V of the same multiplicity as μ. If $v \in \mathrm{V}^\mu$ and $x \in \mathfrak{g}^\alpha$, then $x.v \in \mathrm{V}^{\mu+\alpha}$.

20) Let B be a basis of R. Giving B determines an order relation on $\mathfrak{h}^*_{\mathbf{Q}}$: the elements of $\mathfrak{h}^*_{\mathbf{Q}}$ that are ≥ 0 are the linear combinations of elements of B with rational coefficients ≥ 0. Denote by $\mathrm{Q_+(R)}$ (resp. $\mathrm{R_+}$) the set of positive elements of Q(R) (resp. of R).

Let V be a finite dimensional simple $\mathfrak{g}$-module. Then V has a highest weight λ. This weight is of multiplicity 1, and it is a dominant weight: if $\alpha \in \mathrm{R_+}$, $\lambda(H_\alpha)$ is an integer ≥ 0. We have $\mathfrak{g}^\alpha\mathrm{V}^\lambda = 0$ if $\alpha \in \mathrm{R_+}$. Every weight of V is of the form $\lambda - \nu$ with $\nu \in \mathrm{Q_+(R)}$; conversely, if a weight is dominant and is of the form $\lambda - \nu$ with $\nu \in \mathrm{Q_+(R)}$, then it is a weight of V.

21) Two finite dimensional simple $\mathfrak{g}$-modules with the same highest weight are isomorphic. Every dominant weight is the highest weight of a finite dimensional simple $\mathfrak{g}$-module.

Every finite dimensional simple $\mathfrak{g}$-module is absolutely simple.

22) Let Φ be the Killing form of $\mathfrak{g}$, $\mathrm{C} \in \mathrm{U}(\mathfrak{g})$ the corresponding Casimir element, $\langle\cdot,\cdot\rangle$ the inverse form on $\mathfrak{h}^*$ of $\Phi|\mathfrak{h}\times\mathfrak{h}$, and $\rho = \frac{1}{2}\sum_{\alpha\in\mathrm{R_+}}\alpha$. Let V be a finite dimensional simple $\mathfrak{g}$-module, of highest weight λ. Then $\mathrm{C_V}$ is the homothety with ratio $\langle\lambda,\lambda+2\rho\rangle$.

23) Let V be a finite dimensional $\mathfrak{g}$-module, and V^* its dual. Then $\mu \in \mathfrak{h}^*$ is a weight of V^* if and only if $-\mu$ is a weight of V, and the multiplicity of $\mu \in \mathrm{V}^*$

[23]This homomorphism is, in fact, bijective (§7, Exerc. 26 *d*)).

is equal to the multiplicity of $-\mu$ in V. If V is simple of highest weight λ, V^* is simple of highest weight $-w_0\lambda$, where w_0 is the element of W that takes B to $-$B.

24) Let V be a finite dimensional simple $\mathfrak{g}$-module of highest weight λ, and $\mathscr{B}$ the vector space of $\mathfrak{g}$-invariant bilinear forms on V. Let m be the integer $\sum_{\alpha \in R_+} \lambda(H_\alpha)$, and let $w_0 \in W$ be as in 23). If $w_0\lambda \neq -\lambda$, V and V^* are not isomorphic, and $\mathscr{B} = 0$. If $w_0\lambda = -\lambda$, then $\dim \mathscr{B} = 1$ and every non-zero element of $\mathscr{B}$ is non-degenerate; if m is even (resp. odd), every element of $\mathscr{B}$ is symmetric (resp. alternating).

25) Let $\mathbf{Z}[P]$ be the algebra of the group $P = P(R)$ with coefficients in $\mathbf{Z}$. If $\lambda \in P$, denote by e^λ the corresponding element of $\mathbf{Z}[P]$; the e^λ, $\lambda \in P$, form a $\mathbf{Z}$-basis of $\mathbf{Z}[P]$, and $e^\lambda e^\mu = e^{\lambda+\mu}$ for $\lambda, \mu \in P$.

Let V be a finite dimensional $\mathfrak{g}$-module. The character of V, denoted by $\operatorname{ch} V$, is the element $\sum_{\mu \in P} (\dim V^\mu) e^\mu$ of $\mathbf{Z}[P]$; this element belongs to the subalgebra $\mathbf{Z}[P]^W$ of $\mathbf{Z}[P]$ consisting of the elements invariant under W. We have

$$\operatorname{ch}(V \oplus V') = \operatorname{ch} V + \operatorname{ch} V' \quad \text{and} \quad \operatorname{ch}(V \otimes V') = (\operatorname{ch} V).(\operatorname{ch} V').$$

Two finite dimensional $\mathfrak{g}$-modules with the same character are isomorphic.

For all $\alpha \in B$, let V_α be a simple $\mathfrak{g}$-module with highest weight the fundamental weight ϖ_α corresponding to α. The elements $\operatorname{ch} V_\alpha$, $\alpha \in B$, are algebraically independent and generate the $\mathbf{Z}$-algebra $\mathbf{Z}[P]^W$.

26) Let ρ be half the sum of the roots ≥ 0. For all $w \in W$, let $\varepsilon(w)$ be the determinant of w, equal to ± 1. If V is a finite dimensional simple $\mathfrak{g}$-module of highest weight λ, then

$$\left(\sum_{w \in W} \varepsilon(w) e^{w\rho}\right).\operatorname{ch} V = \sum_{w \in W} \varepsilon(w) e^{w(\lambda+\rho)},$$

and

$$\dim V = \prod_{\alpha \in R_+} \frac{\langle \lambda + \rho, H_\alpha \rangle}{\langle \rho, H_\alpha \rangle}.$$

27) For all $\nu \in P$, let $\mathfrak{P}(\nu)$ be the number of families $(n_\alpha)_{\alpha \in R_+}$, where the n_α are integers ≥ 0 such that $\nu = \sum_{\alpha \in R_+} n_\alpha \alpha$. Let V be a finite dimensional simple $\mathfrak{g}$-module of highest weight λ. If $\mu \in P$, the multiplicity of μ in V is

$$\sum_{w \in W} \varepsilon(w) \mathfrak{P}(w(\lambda + \rho) - (\mu + \rho)).$$

28) Let V, V', V'' be finite dimensional simple $\mathfrak{g}$-modules, λ, μ, ν their highest weights. In $V \otimes V'$, the isotypical component of type V'' has length

$$\sum_{w,w' \in W} \varepsilon(ww')\mathfrak{P}(w(\lambda+\rho)+w'(\mu+\rho)-(\nu+2\rho)).$$

In particular, if $\nu = \lambda + \mu$, the isotypical component in question is simple, and generated by $(V \otimes V')^{\lambda+\mu} = V^{\lambda} \otimes V'^{\mu}$.

INVARIANT POLYNOMIAL FUNCTIONS

29) The algebra of polynomial functions on $\mathfrak{g}$ can be identified with the symmetric algebra $\mathbf{S}(\mathfrak{g}^*)$ of $\mathfrak{g}^*$, and hence is a $\mathfrak{g}$-module in a canonical way; hence the notion of an invariant polynomial function on $\mathfrak{g}$. Let $f \in \mathbf{S}(\mathfrak{g}^*)$. Then f is invariant if and only if $f \circ s = f$ for all $s \in \mathrm{Aut}_0(\mathfrak{g})$, or that $f \circ s = f$ for all $s \in \mathrm{Aut}_e(\mathfrak{g})$.

30) Let $I(\mathfrak{g}^*)$ be the algebra of invariant polynomial functions on $\mathfrak{g}$, and $\mathbf{S}(\mathfrak{h}^*)^W$ the algebra of W-invariant polynomial functions on $\mathfrak{h}$. Let

$$i : \mathbf{S}(\mathfrak{g}^*) \to \mathbf{S}(\mathfrak{h}^*)$$

be the restriction homomorphism. The map $i|I(\mathfrak{g}^*)$ is an isomorphism from $I(\mathfrak{g}^*)$ to $\mathbf{S}(\mathfrak{h}^*)^W$. If l is the rank of $\mathfrak{g}$, there exist l homogeneous elements of $I(\mathfrak{g}^*)$ that are algebraically independent and generate the algebra $I(\mathfrak{g}^*)$.

31) An element a of $\mathfrak{g}$ is nilpotent if and only if $f(a) = 0$ for every homogeneous element f of $I(\mathfrak{g}^*)$ of degree > 0.

32) Let $s \in \mathrm{Aut}(\mathfrak{g})$. Then s belongs to $\mathrm{Aut}_0(\mathfrak{g})$ if and only if $f \circ s = f$ for all $f \in I(\mathfrak{g}^*)$.

$\mathfrak{sl}_2$-TRIPLETS

33) An $\mathfrak{sl}_2$-triplet in $\mathfrak{g}$ is a sequence (x, h, y) of elements of $\mathfrak{g}$ distinct from $(0,0,0)$ and such that $[h,x] = 2x, [h,y] = -2y, [x,y] = -h$. Then x, y are nilpotent in $\mathfrak{g}$, and h is semi-simple in $\mathfrak{g}$.

34) Let x be a non-zero nilpotent element of $\mathfrak{g}$. There exist $h, y \in \mathfrak{g}$ such that (x, h, y) is an $\mathfrak{sl}_2$-triplet.

35) Let (x, h, y) and (x', h', y') be $\mathfrak{sl}_2$-triplets in $\mathfrak{g}$. The following conditions are equivalent:
a) there exists $s \in \mathrm{Aut}_e(\mathfrak{g})$ such that $sx = x'$;
b) there exists $s \in \mathrm{Aut}_e(\mathfrak{g})$ such that $sx = x', sh = h', sy = y'$.

36) If k is algebraically closed, conditions *a*) and *b*) of 35) are equivalent to:
c) there exists $s \in \mathrm{Aut}_e(\mathfrak{g})$ such that $sh = h'$.
Moreover, the number of conjugacy classes, relative to $\mathrm{Aut}_e(\mathfrak{g})$, of non-zero nilpotent elements of $\mathfrak{g}$ is at most equal to 3^l, where l is the rank of $\mathfrak{g}$.

37) A nilpotent element x of $\mathfrak{g}$ is called principal if the dimension of the centralizer of x is equal to the rank of $\mathfrak{g}$. There exist principal nilpotent elements in $\mathfrak{g}$. If k is algebraically closed, the principal nilpotent elements of $\mathfrak{g}$ are conjugate under $\mathrm{Aut}_e(\mathfrak{g})$.

CHAPTER IX
Compact Real Lie Groups

In this chapter[1], *the expression "Lie group" means "finite dimensional Lie group over the field of real numbers", the expression "Lie algebra" means, unless stated otherwise, "finite dimensional Lie algebra over the field of real numbers", the expression "real Lie algebra" (resp. "complex Lie algebra") means "finite dimensional Lie algebra over the field of real numbers (resp. "finite dimensional Lie algebra over the field of complex numbers").*

We denote by G_0 *the identity component of a topological group* G. *We denote by* C(G) *the centre of a group* G, *by* D(G) *its derived group, and by* $N_G(H)$ *or* N(H) *(resp.* $Z_G(H)$ *or* Z(H)) *the normalizer (resp. centralizer) of a subset* H *of a group* G.

§ 1. COMPACT LIE ALGEBRAS

1. INVARIANT HERMITIAN FORMS

In this number, the letter k denotes the field **R** or **C**. Let V be a finite dimensional k-vector space, Φ a separating[2] positive hermitian form on V, G a group, $\mathfrak{g}$ an **R**-Lie algebra, $\rho : G \to \mathbf{GL}(V)$ a group homomorphism, $\varphi : \mathfrak{g} \to \mathfrak{gl}(V)$ a homomorphism of **R**-Lie algebras.

a) The form Φ is invariant under G (resp. $\mathfrak{g}$) if and only if $\rho(g)$ is *unitary* with respect to Φ for all $g \in G$ (resp. $\varphi(x)$ is *anti-hermitian*[3] with respect to Φ for all $x \in \mathfrak{g}$). Indeed, denote by a^* the adjoint of an endomorphism a of V with respect to Φ; for g in G, x in $\mathfrak{g}$, u and v in V, we have

$$\Phi(\rho(g)u, \rho(g)v) = \Phi(\rho(g)^*\rho(g)u, v),$$
$$\Phi(\varphi(x)u, v) + \Phi(u, \varphi(x)v) = \Phi((\varphi(x) + \varphi(x)^*).u, v);$$

[1] Throughout this chapter, references to *Algebra*, Chap. VIII, are to the new edition (in preparation)

[2] Recall (*Algebra*, Chap. IX, in preparation) that a hermitian form H on V is said to be *separating* (or non-degenerate) if, for every non-zero element u of V, there exists $v \in V$ such that $H(u, v) \neq 0$.

[3] An element $a \in \mathrm{End}(V)$ is said to be *anti-hermitian* with respect to Φ if the adjoint a^* of a with respect to Φ is equal to $-a$. When $k = \mathbf{C}$ (resp. $k = \mathbf{R}$) this also means that the endomorphism ia of V (resp. of $\mathbf{C} \otimes_{\mathbf{R}} V$) is hermitian.

thus, $\Phi(\rho(g)u, \rho(g)v) = \Phi(u,v)$ for all u, v in V if and only if $\rho(g)^*\rho(g) = \mathrm{Id}_V$; similarly, $\Phi(\varphi(x)u, v) + \Phi(u, \varphi(x)v) = 0$ for all u, v in V if and only if $\varphi(x) + \varphi(x)^* = 0$, hence the stated assertion.

b) If the form Φ is invariant under G (resp. $\mathfrak{g}$), the orthogonal complement of a stable subspace of V is stable; in particular, the representation ρ (resp. φ) is then semi-simple (cf. *Algebra*, Chap. IX); moreover, for all $g \in \mathrm{G}$ (resp. $x \in \mathfrak{g}$), the endomorphism $\rho(g)$ (resp. $\varphi(x)$) of V is then semi-simple, with eigenvalues of absolute value 1 (resp. with purely imaginary eigenvalues); indeed $\rho(g)$ is unitary (resp. $i\varphi(x)$ is hermitian, cf. *Algebra*, Chap. IX).

c) Assume that $k = \mathbf{R}$. If G is a connected Lie group, ρ a morphism of Lie groups, $\mathfrak{g}$ the Lie algebra of G and φ the homomorphism induced by ρ, then Φ is invariant under G if and only if it is invariant under $\mathfrak{g}$ (Chap. III, §6, no. 5, Cor. 3).

d) There exists a separating positive hermitian form on V invariant under G if and only if the subgroup $\rho(\mathrm{G})$ of $\mathbf{GL}(\mathrm{V})$ is relatively compact (*Integration*, Chap. VII, §3, no. 1, Prop. 1).

2. CONNECTED COMMUTATIVE REAL LIE GROUPS

Let G be a connected commutative (real) Lie group. The exponential map

$$\exp_{\mathrm{G}} : \mathrm{L}(\mathrm{G}) \to \mathrm{G}$$

is a morphism of Lie groups, surjective with discrete kernel (Chap. III, §6, no. 4, Prop. 11), hence the fact that L(G) is a connected covering of G.

a) The following conditions are equivalent: G is simply-connected, $\exp_{\mathrm{G}}$ is an isomorphism, G is isomorphic to $\mathbf{R}^n$ ($n = \dim \mathrm{G}$). In this case, transporting the vector space structure of L(G) to G by the isomorphism $\exp_{\mathrm{G}}$ gives a vector space structure on G, which is the only one compatible with the topological group structure of G. Simply-connected commutative Lie groups are called *vector* (Lie) groups; unless stated otherwise, they are always given the $\mathbf{R}$-vector space structure defined above.

b) Denote by $\Gamma(\mathrm{G})$ the kernel of $\exp_{\mathrm{G}}$. By *General Topology*, Chap. VII, §1, no. 1, Th. 1, the group G is compact if and only if $\Gamma(\mathrm{G})$ is a *lattice* in L(G), in other words (*loc. cit.*) if the rank of the free $\mathbf{Z}$-module $\Gamma(\mathrm{G})$ is equal to the dimension of G. Conversely, if L is a finite dimensional $\mathbf{R}$-vector space and Γ a lattice in L, the quotient topological group L/Γ is a compact connected commutative Lie group.

The compact connected commutative Lie groups are called *real tori*, or (in this chapter) *tori*.

c) In the general case, let E be the vector subspace of L(G) generated by $\Gamma(\mathrm{G})$, and let V be a complementary subspace. Then G is the direct product of its Lie subgroups exp(E) and exp(V); the first is a torus, the second is vector. Finally, every compact subgroup of G is contained in exp(E) (since its projection onto exp(V) is necessarily reduced to the identity element); thus, the subgroup exp(E) is the unique *maximal* compact subgroup of G.

For example, take $G = \mathbf{C}^*$; identify L(G) with $\mathbf{C}$ so that the exponential map of G is $x \mapsto e^x$. Then $\Gamma(G) = 2\pi i\mathbf{Z}, E = i\mathbf{R}$, and so $\exp(E) = \mathbf{U}$; if we take $V = \mathbf{R}$, then $\exp(V) = \mathbf{R}_+^*$ and we recover the isomorphism $\mathbf{C}^* \to \mathbf{U} \times \mathbf{R}_+^*$ constructed in *General Topology*, Chap. VIII, §1, no. 3, Prop. 1.

d) Note finally that $\exp_G : L(G) \to G$ is a universal covering of G, hence $\Gamma(G)$ can be identified naturally with the fundamental group of G.

3. COMPACT LIE ALGEBRAS

PROPOSITION 1. *Let* $\mathfrak{g}$ *be a (real) Lie algebra. The following conditions are equivalent:*

(i) $\mathfrak{g}$ *is isomorphic to the Lie algebra of a compact Lie group.*

(ii) *The group* $\mathrm{Int}(\mathfrak{g})$ (Chap. III, §6, no. 2, Def. 2) *is compact.*

(iii) $\mathfrak{g}$ *has an invariant bilinear form* (Chap. I, §3, no. 6) *that is symmetric, positive and separating.*

(iv) $\mathfrak{g}$ *is reductive* (Chap. I, §6, no. 4, Def. 4); *for all* $x \in \mathfrak{g}$, *the endomorphism* $\mathrm{ad}\, x$ *is semi-simple, with purely imaginary eigenvalues.*

(v) $\mathfrak{g}$ *is reductive and its Killing form* B *is negative.*

(i) $\Longrightarrow$ (ii): if $\mathfrak{g}$ is the Lie algebra of a compact Lie group G, the group $\mathrm{Int}(\mathfrak{g})$ is separating and isomorphic to a quotient of the compact group G_0 (Chap. III, §6, no. 4, Cor. 4), hence is compact.

(ii) $\Longrightarrow$ (iii): if the group $\mathrm{Int}(\mathfrak{g})$ is compact, there exists a symmetric bilinear form on $\mathfrak{g}$ that is positive, separating and invariant under $\mathrm{Int}(\mathfrak{g})$ (no. 1), hence also invariant under the adjoint representation of $\mathfrak{g}$.

(iii) $\Longrightarrow$ (iv): if (iii) is satisfied, the adjoint representation of $\mathfrak{g}$ is semi-simple (no. 1), hence $\mathfrak{g}$ is reductive; moreover, the endomorphisms $\mathrm{ad}\, x$, for $x \in \mathfrak{g}$, have the indicated properties (no. 1).

(iv) $\Longrightarrow$ (v): for all $x \in \mathfrak{g}$, $B(x, x) = \mathrm{Tr}((\mathrm{ad}\, x)^2)$; consequently, $B(x, x)$ is the sum of the squares of the eigenvalues of $\mathrm{ad}\, x$, and hence is negative if these are purely imaginary.

(v) $\Longrightarrow$ (i): assume that $\mathfrak{g}$ is reductive, hence the product of a commutative subalgebra $\mathfrak{c}$ and a semi-simple subalgebra $\mathfrak{s}$ (Chap. I, §6, no. 4, Prop. 5). The Killing form of $\mathfrak{s}$ is the restriction of the form B to $\mathfrak{s}$, hence is negative and separating if B is negative. The subgroup $\mathrm{Int}(\mathfrak{s})$ of $\mathbf{GL}(\mathfrak{s})$ is closed (it is the identity component of $\mathrm{Aut}(\mathfrak{s})$, Chap. III, §10, no. 2, Cor. 2) and leaves the separating positive form $-B$ invariant; thus, it is compact, and $\mathfrak{s}$ is isomorphic to the Lie algebra of the compact Lie group $\mathrm{Int}(\mathfrak{s})$. Further, since $\mathfrak{c}$ is commutative, it is isomorphic to the Lie algebra of a torus T. Thus $\mathfrak{g}$ is isomorphic to the Lie algebra of the compact Lie group $\mathrm{Int}(\mathfrak{s}) \times T$.

DEFINITION 1. *A* compact *Lie algebra*[4] *is a Lie algebra that has properties* (i) *to* (v) *of Proposition 1.*

Thus, the compact Lie algebras are the products of a commutative algebra with a compact semi-simple algebra. In other words, a Lie algebra is compact if and only if it is reductive and its derived Lie algebra is compact.

The Lie algebra of a compact Lie group is compact.

PROPOSITION 2. *a) The product of a finite number of Lie algebras is a compact Lie algebra if and only if each factor is compact.*

b) A subalgebra of a compact Lie algebra is compact.

c) Let $\mathfrak{h}$ *be an ideal of a compact Lie algebra* $\mathfrak{g}$. *Then the algebra* $\mathfrak{g}/\mathfrak{h}$ *is compact and the extension* $\mathfrak{h} \to \mathfrak{g} \to \mathfrak{g}/\mathfrak{h}$ *is trivial.*

Assertions *a)* and *b)* follow from the characterization (iii) of Prop. 1. Part *c)* follows from *a)* and the fact that, in a reductive Lie algebra, every ideal is a direct factor (Chap. I, §6, no. 4, Cor. of Prop. 5).

PROPOSITION 3. *Let* G *be a Lie group of which the group of connected components is finite. The following conditions are equivalent:*

(i) *The Lie algebra* L(G) *is compact.*

(ii) *The group* Ad(G) *is compact.*

(iii) *There exists a separating positive symmetric bilinear form on* L(G) *invariant under the adjoint representation of* G.

*(iv) G *has a riemannian metric invariant under left and right translations*.*

(i) $\Longrightarrow$ (ii): if L(G) is compact, the group $\mathrm{Ad}(G_0) = \mathrm{Int}(L(G))$ is compact; since it has finite index in Ad(G), this latter group is also compact.

(ii) $\Longrightarrow$ (iii): this follows from no. 1.

(iii) $\Longrightarrow$ (i): since $\mathrm{Int}(L(G)) \subset \mathrm{Ad}(G)$, this follows from the characterization (iii) of Prop. 1.

(iii) $\Longleftrightarrow$ (iv): this follows from Chap. III, §3, no. 13.

4. GROUPS WHOSE LIE ALGEBRA IS COMPACT

THEOREM 1. (H. Weyl) *Let* G *be a connected Lie group whose Lie algebra is compact semi-simple. Then* G *is compact and its centre is finite.*

Since G is semi-simple, its centre D is discrete. Moreover, the quotient group G/D is isomorphic to Ad(G) (Chap. III, §6, no. 4, Cor. 4), hence compact (Prop. 3). Finally, the group G/D is equal to its derived group (Chap. III, §9, no. 2, Cor. of Prop. 4). The theorem now follows from *Integration*, Chap. VII, §3, no. 2, Prop. 5.

PROPOSITION 4. *Let* G *be a connected Lie group whose Lie algebra is compact. There exist a torus* T, *a simply-connected compact semi-simple Lie*

[4] Note that a real topological vector space cannot be a compact topological space unless it is reduced to 0.

group S, *a vector group* V *and a surjective morphism* $f : \mathrm{V} \times \mathrm{T} \times \mathrm{S} \to \mathrm{G}$ *with finite kernel. If* G *is compact, the group* V *is reduced to the identity element.*

Let C (resp. S) be a simply-connected Lie group whose Lie algebra is isomorphic to the centre (resp. the derived algebra) of L(G). Then C is a vector group, S is a compact group with finite centre (Th. 1) and G can be identified with the quotient of $\mathrm{C} \times \mathrm{S}$ by a discrete subgroup D, which is central (*Integration*, Chap. VII, §3, no. 2, Lemma 4). Since the image of the projection of D onto S is central, hence finite, $\mathrm{D} \cap \mathrm{C}$ is of finite index in D. Let C' be the vector subspace of C generated by $\mathrm{D} \cap \mathrm{C}$, and V a complementary subspace. Then the group $\mathrm{T} = \mathrm{C}'/(\mathrm{D} \cap \mathrm{C})$ is a torus, and G is isomorphic to the quotient of the product group $\mathrm{V} \times \mathrm{T} \times \mathrm{S}$ by a finite group.

If G is compact, so is $\mathrm{V} \times \mathrm{T} \times \mathrm{S}$ (*General Topology*, Chap. III, §4, no. 1, Cor. 2 of Prop. 2), hence so is V, which implies that $\mathrm{V} = \{e\}$.

COROLLARY 1. *Let* G *be a connected compact Lie group. Then* $\mathrm{C(G)}_0$ *is a torus,* D(G) *is a connected compact semi-simple Lie group and the morphism* $(x, y) \mapsto xy$ *from* $\mathrm{C(G)}_0 \times \mathrm{D(G)}$ *to* G *is a finite covering.*

With the notation in Prop. 4, we have $\mathrm{V} = \{e\}$ and the subgroups $f(\mathrm{T})$ and $f(\mathrm{S})$ of G are compact, hence closed. Thus it suffices to show that $f(\mathrm{T}) = \mathrm{C(G)}_0$, $f(\mathrm{S}) = \mathrm{D(G)}$. Now, $\mathrm{L(G)} = \mathrm{L}(f(\mathrm{T})) \times \mathrm{L}(f(\mathrm{S}))$; since S is semi-simple and T is commutative, this implies that $\mathrm{L}(f(\mathrm{T})) = \mathscr{C}(\mathrm{L(G)}) = \mathrm{L}(\mathrm{C(G)}_0)$ (Chap. III, §9, no. 3, Prop. 8) and $\mathrm{L}(f(\mathrm{S})) = \mathscr{D}\mathrm{L(G)} = \mathrm{L(D(G))}$ (Chap. III, §9, no. 2, Cor. of Prop. 4), hence the stated assertion.

COROLLARY 2. *The centre and the fundamental group of a connected compact semi-simple Lie group are finite. Its universal covering is compact.*

With the notation in Prop. 4, the groups V and T are reduced to the identity element; thus S is a universal covering of G, and the fundamental group of G is isomorphic to $\mathrm{Ker}\, f$, hence finite. The centre D of G is discrete since G is semi-simple, so D is finite.

COROLLARY 3. *The fundamental group of a connected compact Lie group* G *is a* **Z**-*module of finite type, of rank equal to the dimension of* C(G).

Indeed, with the notations in Cor. 1, the fundamental group of $\mathrm{C(G)}_0$ is isomorphic to $\mathbf{Z}^n$, with $n = \dim \mathrm{C(G)}_0$, and the fundamental group of D(G) is finite (Cor. 2).

COROLLARY 4. *Let* G *be a connected compact Lie group. The following conditions are equivalent:*

(i) G *is semi-simple;*

(ii) C(G) *is finite;*

(i) $\pi_1(\mathrm{G})$ *is finite.*

If G *is simply-connected, it is semi-simple.*

This follows from Cor. 1 to 3.

COROLLARY 5. *Let* G *be a connected compact Lie group. Then* Int(G) *is the identity component of the Lie group* Aut(G) (Chap. III, §10, no. 2).

Let $f \in \mathrm{Aut}(G)_0$. Then f induces an automorphism f_1 of $C(G)_0$ and an automorphism f_2 of D(G), and we have $f_1 \in \mathrm{Aut}(C(G)_0)_0$, $f_2 \in \mathrm{Aut}(D(G))_0$. Since $\mathrm{Aut}(C(G)_0)$ is discrete (*General Topology*, Chap. VII, §2, no. 4, Prop. 5), we have $f_1 = \mathrm{Id}$; since D(G) is semi-simple, by Chap. III, §10, no. 2, Cor. 2 of Th. 1 there exists an element g of D(G) such that $f_2(x) = gxg^{-1}$ for all $x \in D(G)$. For all $x \in C(G)_0$, we have $gxg^{-1} = x = f_1(x)$; since $G = C(G)_0.D(G)$, it follows that $gxg^{-1} = f(x)$ for all $x \in G$, so $f = \mathrm{Int}\, g$.

PROPOSITION 5. *Let* G *be a Lie group whose Lie algebra is compact.*

a) Assume that G *is connected. Then* G *has a largest compact subgroup* K; *it is connected. There exists a closed central vector subgroup* (no. 2) N *of* G *such that* G *is the direct product* N × K.

b) Assume that the group of connected components of G *is finite. Then:*

(i) *Every compact subgroup of* G *is contained in a maximal compact subgroup.*

(ii) *If* K_1 *and* K_2 *are two maximal compact subgroups of* G, *there exists* $g \in G$ *such that* $K_2 = gK_1g^{-1}$.

(iii) *Let* K *be a maximal compact subgroup of* G. *Then* $K \cap G_0$ *is equal to* K_0; *it is the largest compact subgroup of* G_0.

(iv) *There exists a closed central vector subgroup* N *of* G_0, *normal in* G, *such that, for any maximal compact subgroup* K *of* G, G_0 *is the direct product of* K_0 *by* N, *and* G *is the semi-direct product of* K *by* N.

a) We retain the notations of Prop. 4. The projection of Ker f onto V is a finite subgroup of the vector group V, hence is reduced to the identity element. It follows that Ker f is contained in T × S, hence that G is the direct product of the vector group $N = f(V)$ and the compact group $K = f(T \times S)$. Every compact subgroup of G has a projection onto N that is reduced to the identity element, hence is contained in K. This proves *a*).

b) Assume now that G/G_0 is finite. By *a*), G_0 is the direct product of its largest compact subgroup M and a vector subgroup P; the subgroup M of G is clearly normal. Let $\mathfrak{n}$ be a vector subspace complement of L(M) in L(G), stable under the adjoint representation of G (no. 1 and no. 3, Prop. 3); this is an ideal of L(G) and we have $L(G) = L(M) \times \mathfrak{n}$. Let N be the integral subgroup of G with Lie algebra $\mathfrak{n}$; by Chap. III, §6, no. 6, Prop. 14, it is normal in G. The projection of L(G) onto L(P) with kernel L(M) induces an isomorphism from $\mathfrak{n}$ to L(P); it follows that the projection of G_0 onto P induces an étale morphism from N to P; since P is simply-connected, this is an isomorphism, and N is a vector group. The morphism $(x, y) \mapsto xy$ from M × N to G_0 is an injective étale morphism (since $M \cap N$ is reduced to the identity element), hence an isomorphism. It follows that N is a *closed* subgroup of G and that the quotient G/N is compact, since G_0/N is compact and G/G_0 is finite (*General Topology*, Chap. III, §4, no. 1, Cor. 2 of Prop. 2).

By *Integration*, Chap. VII, §3, no. 2, Prop. 3, every compact subgroup of G is contained in a maximal compact subgroup, these are conjugate, and for any maximal compact subgroup K of G, G is the semi-direct product of K by N. Since G_0 contains N, it is the semi-direct product of N by $G_0 \cap K$; it follows that $G_0 \cap K$ is connected, hence equal to K_0, since $K/(G_0 \cap K)$ is isomorphic to G/G_0, hence finite; finally, K_0 is clearly the largest compact subgroup of G_0 by *a*).

COROLLARY. *If* N *satisfies the conditions of b)* (iv), *and if* K_1 *and* K_2 *are two maximal compact subgroups of* G, *there exists* $n \in N$ *such that* $nK_1n^{-1} = K_2$.

Indeed, by (ii) there exists an element $g \in G$ such that $gK_1g^{-1} = K_2$; by (iv), there exists $n \in N$ and $k \in K_1$ such that $g = nk$. The element n then has the required properties.

§2. MAXIMAL TORI OF COMPACT LIE GROUPS

1. CARTAN SUBALGEBRAS OF COMPACT ALGEBRAS

Lemma 1. Let G *be a Lie group,* K *a compact subgroup of* G, *and* F *an invariant bilinear form on* L(G). *Let* $x, y \in L(G)$. *There exists an element* k *of* K *such that* $F(u, [(\mathrm{Ad}\,k)(x), y]) = 0$ *for all* $u \in L(K)$.

The function $v \mapsto F((\mathrm{Ad}\,v)(x), y)$ from K to $\mathbf{R}$ is continuous, and hence has a minimum at some point $k \in K$. Let $u \in L(K)$ and put

$$h(t) = F((\mathrm{Ad}\exp(tu).k)(x), y), \qquad t \in \mathbf{R}.$$

We have $h(t) \geq h(0)$ for all t; moreover, by Chap. III, §3, no. 12, Prop. 44,

$$\frac{dh}{dt}(0) = F([u, (\mathrm{Ad}\,k)(x)], y) = F(u, [(\mathrm{Ad}\,k)(x), y]),$$

hence the lemma (*Functions of a Real Variable*, Chap. I, §1, no. 7, Prop. 7).

THEOREM 1. *Let* $\mathfrak{g}$ *be a compact Lie algebra. The Cartan subalgebras of* $\mathfrak{g}$ (Chap. VII, §2, no. 1, Def. 1) *are its maximal commutative subalgebras; in particular,* $\mathfrak{g}$ *is the union of its Cartan subalgebras. The group* $\mathrm{Int}(\mathfrak{g})$ *operates transitively on the set of Cartan subalgebras of* $\mathfrak{g}$.

Since $\mathfrak{g}$ is reductive, its Cartan subalgebras are commutative (Chap. VII, §2, no. 4, Cor. 3 of Th. 2). Conversely, let $\mathfrak{t}$ be a commutative subalgebra of $\mathfrak{g}$. By §1, no. 3, Prop. 1, $\mathrm{ad}\,x$ is semi-simple for all $x \in \mathfrak{t}$; by Chap. VII, §2, no. 3, Prop. 10, there exists a Cartan subalgebra of $\mathfrak{g}$ containing $\mathfrak{t}$. This proves the first assertion of the theorem.

Now let $\mathfrak{t}$ and $\mathfrak{t}'$ be two Cartan subalgebras of $\mathfrak{g}$. We prove that there exists $u \in \mathrm{Int}(\mathfrak{g})$ such that $u(\mathfrak{t}) = \mathfrak{t}'$. By Prop. 1 of §1, no. 3, we can assume that $\mathfrak{g}$ is of the form L(G), where G is a connected compact Lie group, and can choose a separating invariant symmetric bilinear form F on $\mathfrak{g}$. Let x (resp. x') be a regular element of $\mathfrak{g}$ such that $\mathfrak{t} = \mathfrak{g}^0(x)$ (resp. $\mathfrak{t}' = \mathfrak{g}^0(x')$) (Chap. VII, §3, no. 3, Th. 2). Applying Lemma 1 with K = G, we see that there exists $k \in \mathrm{G}$ such that $[(\mathrm{Ad}\, k)(x), x']$ is orthogonal to $\mathfrak{g}$ with respect to F, and hence is zero; then $(\mathrm{Ad}\, k)(x) \in \mathfrak{g}^0(x') = \mathfrak{t}'$, so $\mathfrak{g}^0((\mathrm{Ad}\, k)(x)) = \mathfrak{t}'$ since $(\mathrm{Ad}\, k)(x)$ is regular. We conclude that $(\mathrm{Ad}\, k)(\mathfrak{t}) = \mathfrak{t}'$, hence the theorem.

COROLLARY. *Let* $\mathfrak{t}$ *and* $\mathfrak{t}'$ *be Cartan subalgebras of* $\mathfrak{g}$, $\mathfrak{a}$ *a subset of* $\mathfrak{t}$, *and* u *an automorphism of* $\mathfrak{g}$ *that takes* $\mathfrak{a}$ *into* $\mathfrak{t}'$. *There exists an element* v *of* $\mathrm{Int}(\mathfrak{g})$ *such that* $u \circ v$ *takes* $\mathfrak{t}$ *to* $\mathfrak{t}'$, *and coincides with* u *on* $\mathfrak{a}$.

Put $\mathrm{G} = \mathrm{Int}(\mathfrak{g})$, and consider the fixer $\mathrm{Z_G}(\mathfrak{a})$ of $\mathfrak{a}$ in G; this is a Lie subgroup of G, whose Lie algebra $\mathfrak{z}_{\mathfrak{g}}(\mathfrak{a})$ consists of the elements of $\mathfrak{g}$ that commute with every element of $\mathfrak{a}$ (Chap. III, §9, no. 3, Prop. 7). Then $\mathfrak{t}$ and $u^{-1}(\mathfrak{t}')$ are two Cartan subalgebras of the compact Lie algebra $\mathfrak{z}_{\mathfrak{g}}(\mathfrak{a})$. By Th. 1, there exists an element v of $\mathrm{Z_G}(\mathfrak{a})$ such that $v(\mathfrak{t}) = u^{-1}(\mathfrak{t}')$; any such element has the desired properties.

2. MAXIMAL TORI

Let G be a Lie group. A *torus* of G is a closed subgroup that is a torus (§1, no. 2), in other words any commutative connected compact subgroup. The maximal elements of the set of tori of G, ordered by inclusion, are called the *maximal tori* of G.

THEOREM 2. *Let* G *be a connected compact Lie group.*

a) The Lie algebras of the maximal tori of G *are the Cartan subalgebras of* L(G).

b) Let T_1 *and* T_2 *be two maximal tori of* G. *There exists* $g \in \mathrm{G}$ *such that* $\mathrm{T}_2 = g\mathrm{T}_1 g^{-1}$.

c) G *is the union of its maximal tori.*

Let $\mathfrak{t}$ be a Cartan subalgebra of L(G); the integral subgroup of G whose Lie algebra is $\mathfrak{t}$ is closed (Chap. VII, §2, no. 1, Cor. 4 of Prop. 4) and commutative (Th. 1), and hence is a torus of G. If T is a maximal torus of G, its Lie algebra is commutative, hence is contained in a Cartan subalgebra of L(G) (Th. 1). It follows that the maximal tori of G are exactly the integral subgroups of G associated to the Cartan subalgebras of L(G), hence *a*). Assertion *b*) follows from Th. 1, since the canonical homomorphism $\mathrm{G} \to \mathrm{Int}(\mathrm{L(G)})$ is surjective (Chap. III, §6, no. 4, Cor. 4 of Prop. 10).

Denote by X the union of the maximal tori of G, and let T be a maximal torus of G. The continuous map $(g, t) \mapsto gtg^{-1}$ from $\mathrm{G} \times \mathrm{T}$ to G has image X, which is therefore *closed* in G; thus, to prove *c*), it suffices to prove that X is open in G; since X is invariant under inner automorphisms, it suffices to show that, for all $a \in \mathrm{T}$, X is a neighbourhood of a. We argue by induction on the dimension of G and distinguish two cases:

1) *a is not central in* G. Let H be the identity component of the centralizer of a in G; this is a connected compact subgroup of G distinct from G, which contains T, and hence a. Since $\mathrm{Ad}\, a$ is semi-simple (§1, no. 1), the Lie algebra of H is the nilspace of $\mathrm{Ad}\, a - 1$; it now follows from Chap. VII, §4, no. 2, Prop. 4, that the union Y of the conjugates of H is a neighbourhood of a. By the induction hypothesis, $\mathrm{H} \subset \mathrm{X}$, and hence $\mathrm{Y} \subset \mathrm{X}$; thus, X is a neighbourhood of a.

2) *a is central in* G. It suffices to prove that $a \exp x$ belongs to X for all x in L(G). Now every element x of L(G) belongs to a Cartan subalgebra of G (Th. 1); the corresponding integral subgroup T′ contains $\exp x$; since it is conjugate to T, it contains a and hence $a \exp x$, as required.

COROLLARY 1. *a) The exponential map of* G *is surjective.*

b) For all $n \geq 1$, the map $g \mapsto g^n$ from G *to itself is surjective.*

Indeed, exp(L(G)) contains all the maximal tori of G, hence *a*). Assertion *b*) follows from the formula $(\exp x)^n = \exp nx$ for x in L(G).

Remark 1. There exists a *compact* subset K of L(G) such that $\exp_{\mathrm{G}}(\mathrm{K}) = \mathrm{G}$. Indeed, if T is a maximal torus of G, there exists a compact subset $\mathrm{C} \subset \mathrm{L(T)}$ such that $\exp_{\mathrm{T}}(\mathrm{C}) = \mathrm{T}$; it suffices to take $\mathrm{K} = \bigcup_{g \in \mathrm{G}} (\mathrm{Ad}\, g)(\mathrm{C})$.

COROLLARY 2. *The intersection of the maximal tori of* G *is the centre of* G.

Let x be an element of the centre of G; by Th. 2 *c*), there exists a maximal torus T of G containing x; then x belongs to all the conjugates of T, hence to all the maximal tori of G. Conversely, if x belongs to all the maximal tori of G, it commutes with every element of G by Th. 2 *c*).

COROLLARY 3. *Let $g \in \mathrm{G}$, and let* C *be its centralizer. Then g belongs to C_0; the group C_0 is the union of the maximal tori of* G *containing g.*

There exists a maximal torus T of G containing g (Th. 2 *c*)), and hence contained in C_0. Moreover, the group C_0 is a connected compact Lie group, and hence the union of its maximal tori (Th. 2 *c*)); these all contain g (Cor. 2), hence are exactly the maximal tori of G containing g.

COROLLARY 4. *Let $g \in \mathrm{G}$. If g is regular* (Chap. VII, §4, no. 2, Def. 2), *it belongs to a unique maximal torus, which is the identity component of its centralizer. Otherwise, it belongs to infinitely-many maximal tori.*

Since $\mathrm{Ad}\, g$ is semi-simple, the dimension of the nilspace of $\mathrm{Ad}\, g - 1$ is also that of the centralizer C of g. By *loc. cit.*, Prop. 8, and Th. 1, g is regular if and only if C_0 is a maximal torus of G. The conclusion now follows from Cor. 3.

COROLLARY 5. *a) Let* S *be a torus of* G. *The centralizer of* S *is connected; it is the union of the maximal tori of* G *containing* S.

b) Let $\mathfrak{s}$ *be a commutative subalgebra of* L(G). *The fixer of* $\mathfrak{s}$ *in* G *is connected; it is the union of the maximal tori of* G *whose Lie algebras contain* $\mathfrak{s}$.

To prove *a*), it suffices to prove that if an element g of G centralizes S, there exists a maximal torus of G containing S and g. Now, if C is the centralizer of g, we have $g \in C_0$ (Cor. 3) and $S \subset C_0$; if T is a maximal torus of the connected compact Lie group C_0 containing S, we have $g \in T$ (Cor. 2), hence *a*). Assertion *b*) follows from *a*) applied to the closure of the integral subgroup with Lie algebra $\mathfrak{s}$, in view of Chap. III, §9, no. 3, Prop. 9.

Remark 2. It follows from Cor. 5 that a maximal torus of G is a maximal commutative subgroup. The converse is not true: for example, in the group $\mathbf{SO}(3, \mathbf{R})$, the maximal tori are of dimension 1, and thus cannot contain the subgroup of diagonal matrices, which is isomorphic to $(\mathbf{Z}/2\mathbf{Z})^2$. Moreover, if $g \in \mathbf{SO}(3, \mathbf{R})$ is a non-scalar diagonal matrix, g is a regular element of $\mathbf{SO}(3, \mathbf{R})$ whose centralizer is not connected (cf. Cor. 4).

COROLLARY 6. *The maximal tori of* G *are their own centralizers, and are the fixers of their Lie algebras.*

Let T be a maximal torus of G and C its centralizer; since L(T) is a Cartan subalgebra of L(G), we have L(T) = L(C), hence C = T since C is connected (Cor. 5).

COROLLARY 7. *Let* T *and* T′ *be two maximal tori of* G, A *a subset of* T *and* s *an automorphism of* G *that takes* A *into* T′. *There exists* $g \in G$ *such that* $s \circ (\operatorname{Int} g)$ *takes* T *to* T′ *and coincides with* s *on* A.

Let C be the centralizer of A. Then T and $s^{-1}(T')$ are two maximal tori of C_0; every element g of C_0 such that $(\operatorname{Int} g)(T) = s^{-1}(T')$ has the desired properties.

COROLLARY 8. *Let* H *be a compact Lie group,* T *a maximal torus of* H. *Then* $H = N_H(T).H_0$, *and the injection of* $N_H(T)$ *into* H *induces an isomorphism from* $N_H(T)/N_{H_0}(T)$ *to* H/H_0.

Let $h \in H$. Then $h^{-1}Th$ is a maximal torus of H_0, hence (Th. 2) there exists $g \in H_0$ such that $hg \in N_H(T)$; thus h belongs to $N_H(T).H_0$, hence the first assertion. The second follows immediately.

Remarks. 3) Let G be a connected Lie group whose Lie algebra is compact. The *Cartan subgroups* of G are the integral subgroups whose Lie algebras are the Cartan subalgebras of L(G) (the Cartan subgroups of a connected compact group are thus its maximal tori). Theorem 2 and its corollaries remain valid for G, if we replace everywhere the expression "maximal torus" by "Cartan subgroup". This follows immediately from the fact that, in view of Prop. 5 of §1, no. 4, G is the direct product of a vector group V and a connected compact group K and that the Cartan subgroups of G are the products of V with the maximal tori of K. Moreover, note that it follows from Cor. 6 above that the Cartan subgroups of G can also be defined as the fixers of the Cartan subalgebras of L(G).

* 4) Part *c*) of Theorem 2 can also be proved in the following way. Give G an invariant riemannian metric (§1, no. 3, Prop. 3). Then, for any element g of G, there exists a maximal geodesic passing through g and the identity element of G (Hopf-Rinow theorem), and it can be verified that the closure of such a geodesic is a subtorus of G. *

3. MAXIMAL TORI OF SUBGROUPS AND QUOTIENT GROUPS

PROPOSITION 1. *Let* G *and* G′ *be two connected compact Lie groups.*

a) *Let* $f : G \to G'$ *be a surjective morphism of Lie groups. The maximal tori of* G′ *are the images under* f *of the maximal tori of* G. *If the kernel of* f *is central in* G (for example discrete), *the maximal tori of* G *are the inverse images under* f *of the maximal tori of* G′.

b) *Let* H *be a connected closed subgroup of* G. *Every maximal torus of* H *is the intersection with* H *of a maximal torus of* G.

c) *Let* H *be a connected closed normal subgroup of* G. *The maximal tori of* H *are the intersections with* H *of the maximal tori of* G.

a) Let T be a maximal torus of G; then L(T) is a Cartan subalgebra of L(G) (no. 2, Th. 2 *a*)), so $L(f(T))$ is a Cartan subalgebra of L(G′) (Chap. VII, §2, no. 1, Cor. 2 of Prop. 4); it follows that $f(T)$ is a maximal torus of G′ (no. 2, Th. 2 *a*)). If $\operatorname{Ker} f$ is central in G, it is contained in T (Cor. 2 of Th. 2), so $T = f^{-1}(f(T))$.

Conversely, let T′ be a maximal torus of G′; we show that there exists a maximal torus T of G such that $f(T) = T'$. Let T_1 be a maximal torus of G; then $f(T_1)$ is a maximal torus of G′ and there exists $g' \in G'$ such that $T' = g'f(T_1)g'^{-1}$ (Th. 2 *b*)); if $g \in G$ is such that $f(g) = g'$, we have $T' = f(T)$ with $T = gT_1g^{-1}$.

b) Let S be a maximal torus of H; this is a torus of G so there exists a maximal torus T of G containing S. Then $T \cap H$ is a commutative subgroup of H containing S, hence is equal to S (no. 2, Remark 2).

c) By §1, no. 3, Prop. 2 *c*), L(G) is the direct product of L(H) with an ideal; thus, the Cartan subalgebras of L(H) are the intersections with L(H) of the Cartan subalgebras of L(G). Thus, for any maximal torus T of G, $T \cap H$ contains a maximal torus S of H and $S = T \cap H$ (no. 2, Remark 2).

Remarks. 1) Proposition 1 generalizes immediately to connected groups with compact Lie algebras. In particular, if G is a connected Lie group whose Lie algebra is compact, the Cartan subgroups of G (cf. Remark 3, no. 2) are exactly the inverse images of the maximal tori of the connected compact Lie group Ad(G) (under the canonical homomorphism from G to Ad(G)).

2) Let G be a connected compact Lie group, $\tilde{D}(G)$ the universal covering of the group D(G) and $f : \tilde{D}(G) \to G$ the composite of the canonical morphisms

from $\tilde{D}(G)$ to $D(G)$ and from $D(G)$ to G. Then the map $T \mapsto f^{-1}(T)$ is a bijection from the set of maximal tori of G to the set of maximal tori of $\tilde{D}(G)$; the inverse bijection associates to a maximal torus $\tilde{T}$ of $\tilde{D}(G)$ the maximal torus $C(G)_0.f(\tilde{T})$ of G.

4. SUBGROUPS OF MAXIMAL RANK

We shall call the rank of a connected Lie group G the rank of its Lie algebra, and we shall denote it by $\mathrm{rk}\,G$. By Th. 2 a), the rank of a connected compact Lie group is the common dimension of its maximal tori.

Let G be a connected compact Lie group and H a closed subgroup of G. If H is connected, then $\mathrm{rk}\,H \leq \mathrm{rk}\,G$ (since the maximal tori of H are tori in G). By Th. 2 c), to say that H is *connected and of maximal rank* (that is, of rank $\mathrm{rk}\,G$) means that H is a *union of maximal tori* of G. We deduce immediately from Proposition 1:

PROPOSITION 2. *Let $f : G \to G'$ be a surjective morphism of connected compact Lie groups whose kernel is central. The maps $H \mapsto f(H)$ and $H' \mapsto f^{-1}(H')$ are inverse bijections between the set of connected closed subgroups of* G *of maximal rank and the analogous set for* G'.

PROPOSITION 3. *Let* G *be a connected compact Lie group, and* H *a connected closed subgroup of maximal rank.*

a) The compact manifold G/H *is simply-connected.*

b) The homomorphism $\pi_1(H) \to \pi_1(G)$, induced by the canonical injection of H *into* G, *is surjective.*

Since H is connected, we have an exact sequence (*General Topology*, Chap. XI, in preparation)

$$\pi_1(H) \to \pi_1(G) \to \pi_1(G/H, \bar{e}) \to 0$$

where $\bar{e}$ is the image in G/H of the identity element of G. Since G/H is connected, this immediately implies the equivalence of assertions *a*) and *b*). Moreover, if $f : G' \to G$ is a surjective morphism of connected compact Lie groups whose kernel is central, proving the proposition (in the form *a*)) for G is the same as proving it for G' (Prop. 2). Thus, we can first of all replace G by Ad(G), then assume that G is semi-simple, and then by replacing G by a universal covering (§1, no. 4, Cor. 2), assume that G is simply-connected. But then assertion *b*) is trivial.

PROPOSITION 4. *Let* G *be a compact Lie group,* H *a connected closed subgroup of* G *of maximal rank and* N *the normalizer of* H *in* G. *Then* H *is of finite index in* N *and is the identity component of* N.

Indeed, the Lie algebra of H contains a Cartan subalgebra of L(G). Thus, by Chap. VII, §2, no. 1, Cor. 4 of Prop. 4, H is the identity component of N. Since N is compact, H is of finite index in N.

Remarks. 1) Every integral subgroup H of G such that $\operatorname{rk} H = \operatorname{rk} G$ is *closed*: indeed, the preceding proof shows that H is the identity component of its normalizer, which is a closed subgroup of G.

2) With the notations of Prop. 4, every closed subgroup H′ of G containing H and such that (H′ : H) is finite normalizes H, and hence is contained in N; similarly, the normalizer of H′ is contained in N. In particular, N is its own normalizer.

5. WEYL GROUP

Let G be a connected compact Lie group and T a maximal torus of G. Denote by $N_G(T)$ the normalizer of T in G; by Prop. 4 (no. 4), the quotient group $N_G(T)/T$ is finite. We denote it by $W_G(T)$, or by W(T), and call it the *Weyl group* of the maximal torus T of G, or the Weyl group of G relative to T. Since T is commutative, the operation of $N_G(T)$ on T by inner automorphisms of G induces by passage to the quotient an operation, called the *canonical* operation, of the group $W_G(T)$ on the Lie group T. By Cor. 6 of Th. 2 of no. 2, this operation is *faithful*: the associated homomorphism $W_G(T) \to \operatorname{Aut} T$ is *injective*.

If T′ is another maximal torus of G and if $g \in G$ is such that $\operatorname{Int} g$ maps T to T′ (no. 2, Th. 2 *b*)), then $\operatorname{Int} g$ induces an isomorphism a_g from $W_G(T)$ to $W_G(T')$ and $a_g(s)(gtg^{-1}) = gs(t)g^{-1}$ for all $s \in W_G(T)$ and all $t \in T$.

PROPOSITION 5. *a) Every conjugacy class of* G *meets* T.

b) The intersections with T *of the conjugacy classes of* G *are the orbits of the Weyl group.*

Let $g \in G$; by Th. 2 of no. 2, there exists $h \in G$ such that $g \in hTh^{-1}$, hence *a*). By definition of the Weyl group, any two elements in the same orbit of $W_G(T)$ on T are conjugate in G; conversely, let a, b be two elements of T conjugate under G. There exists $h \in G$ such that $b = hah^{-1}$; applying Cor. 7 of Th. 2 (no. 2) with $A = \{a\}$, $s = \operatorname{Int} h$, $T' = T$, we see that there exists $g \in G$ such that $\operatorname{Int} hg$ maps T to T and a to b. The class of hg in $W_G(T)$ then maps a to b, hence the proposition.

COROLLARY 1. *The canonical injection of* T *into* G *defines by passage to the quotient a homeomorphism from* $T/W_G(T)$ *to the space* $G/\operatorname{Int}(G)$ *of conjugacy classes of* G.

Indeed, this is a bijective continuous map between two compact spaces (cf. *General Topology*, Chap. III, p. 29, Cor. 1).

COROLLARY 2. *Let* E *be a subset of* G *stable under inner automorphisms. Then* E *is open* (resp. *closed*, resp. *dense*) *in* G *if and only if* $E \cap T$ *is open* (resp. *closed*, resp. *dense*) *in* T.

This follows from Cor. 1 and the fact that the canonical maps $T \to T/W_G(T)$ and $G \to G/\mathrm{Int}(G)$ are open (*General Topology*, Chap. III, p. 10, Lemma 2).

Denote the Lie algebra of G by $\mathfrak{g}$, and that of T by $\mathfrak{t}$. The operation of $W_G(T)$ on T induces a representation, called the *canonical* representation, of the group $W_G(T)$ on the **R**-vector space $\mathfrak{t}$.

PROPOSITION 6. *a) Every orbit of* G *on* $\mathfrak{g}$ *(for the adjoint representation) meets* $\mathfrak{t}$.

b) The intersections with $\mathfrak{t}$ *of the orbits of* G *are the orbits of* $W_G(T)$ *on* $\mathfrak{t}$.

Assertion *a*) follows from Th. 1 (no. 1). Let x, y be two elements of $\mathfrak{t}$ conjugate under Ad(G), and let $h \in G$ be such that $(\mathrm{Ad}\, h)(x) = y$. Applying the corollary of Th. 1 (no. 1) with $\mathfrak{a} = \{x\}$, $u = \mathrm{Ad}\, h, \mathfrak{t}' = \mathfrak{t}$, we see that there exists $g \in G$ such that $\mathrm{Ad}\, hg$ maps $\mathfrak{t}$ to $\mathfrak{t}$ and x to y. Then $hg \in N_G(T)$ (Chap. III, §9, no. 4, Prop. 11), and the class of hg in $W_G(T)$ maps x to y, hence the proposition.

COROLLARY. *The canonical injection of* $\mathfrak{t}$ *into* $\mathfrak{g}$ *defines by passage to the quotient a homeomorphism from* $\mathfrak{t}/W_G(T)$ *to* $\mathfrak{g}/\mathrm{Ad}(G)$.

Denote this map by j; it is bijective and continuous (Prop. 6). We have a commutative diagram

$$\begin{array}{ccc} \mathfrak{t} & \xrightarrow{i} & \mathfrak{g} \\ {\scriptstyle p}\downarrow & & \downarrow{\scriptstyle q} \\ \mathfrak{t}/W_G(T) & \xrightarrow{j} & \mathfrak{g}/\mathrm{Ad}(G) \end{array}$$

where p and q are quotient maps, and i is the canonical injection. Since i and q are proper (*General Topology*, Chap. I, §10, no. 1, Prop. 2 and *General Topology*, Chap. III, §4, no. 1, Prop. 2 c)) and since p is surjective, it follows that j is proper (*General Topology*, Chap. I, §10, no. 1, Prop. 5), and hence is a homeomorphism.

PROPOSITION 7. *Let* H *be a closed subgroup of* G *containing* T.

a) Denote by $W_H(T)$ *the subgroup* $N_H(T)/T$ *of* $W_G(T)$*; the group* H/H_0 *is isomorphic to the quotient group* $W_H(T)/W_{H_0}(T)$.

b) H *is connected if and only if every element of* $W_G(T)$ *that has a representative in* H *belongs to* $W_{H_0}(T)$.

Assertion *a*) follows from Cor. 8 of Th. 2 (no. 2), and assertion *b*) is a particular case of *a*).

6. MAXIMAL TORI AND COVERING OF HOMOMORPHISMS

Let G be a connected compact Lie group, T a maximal torus of G. Consider the derived group D(G) of G and its universal covering $\tilde{D}(G)$; let $p : \tilde{D}(G) \to G$ be the composite of the canonical morphisms $\tilde{D}(G) \to D(G)$ and $D(G) \to G$. Then $\tilde{D}(G)$ is a connected compact Lie group (§1, no. 4, Cor. 2 of Prop. 4); moreover, the inverse image $\tilde{T}$ of T under p is a maximal torus of $\tilde{D}(G)$ (no. 3, Prop. 1).

Lemma 2. Let H *be a Lie group,* $f_T : T \to H$ *and* $\tilde{f} : \tilde{D}(G) \to H$ *morphisms of Lie groups such that* $f_T(p(t)) = \tilde{f}(t)$ *for all* $t \in \tilde{T}$. *There exists a unique morphism of Lie groups* $f : G \to H$ *such that* $f \circ p = \tilde{f}$ *and such that the restriction of* f *to* T *is* f_T.

Put $Z = C(G)_0$; by §1, no. 4, Cor. 1 of Prop. 4, the morphism of Lie groups $g : Z \times \tilde{D}(G) \to G$ such that $g(z, x) = z^{-1}p(x)$ is a covering; its kernel consists of the pairs (z, x) such that $p(x) = z$, for which $x \in p^{-1}(Z) \subset \tilde{T}$. Since the morphism $(z, x) \mapsto f_T(z^{-1})\tilde{f}(x)$ from $Z \times \tilde{D}(G)$ to H maps $\operatorname{Ker} g$ to $\{e\}$, there exists a morphism f from G to H such that $f \circ p = \tilde{f}$ and $f(z) = f_T(z)$ for $z \in Z$. But we also have $f(t) = f_T(t)$ for $t \in p(\tilde{T})$; since $T = Z.p(\tilde{T})$, the restriction of f to T is indeed f_T.

PROPOSITION 8. *Let* G *be a connected compact Lie group,* T *a maximal torus of* G, H *a Lie group and* $\varphi : L(G) \to L(H)$ *a homomorphism of Lie algebras. There exists a morphism of Lie groups* $f : G \to H$ *such that* $L(f) = \varphi$ *if and only if there exists a morphism of Lie groups* $f_T : T \to H$ *such that* $L(f_T) = \varphi|L(T)$; *then* $f_T = f|T$.

If $f : G \to H$ is a morphism of Lie groups such that $L(f) = \varphi$, the restriction f_T of f to T is the unique morphism from T to H such that $L(f_T) = \varphi|L(T)$. Conversely, let $f_T : T \to H$ be a morphism of Lie groups such that $L(f_T) = \varphi|L(T)$. Let $\tilde{D}(G)$ and p be as above; the map $L(p)$ induces an isomorphism from $L(\tilde{D}(G))$ to the derived algebra $\mathfrak{b}$ of L(G). There exists a morphism of Lie groups $\tilde{f} : \tilde{D}(G) \to H$ such that $L(\tilde{f}) = (\varphi|\mathfrak{b}) \circ L(p)$ (Chap. III, §6, no. 1, Th. 1). The morphisms $t \mapsto \tilde{f}(t)$ and $t \mapsto f_T(p(t))$ from $\tilde{T}$ to H induce the same homomorphism of Lie algebras, and hence coincide. Applying Lemma 2, we deduce the existence of a morphism $f : G \to H$ such that $L(f)$ and φ coincide on L(T) and $\mathfrak{b}$. Since $L(G) = \mathfrak{b} + L(T)$, we have $L(f) = \varphi$.

PROPOSITION 9. *Let* G *be a connected compact Lie group,* T *a maximal torus of* G, H *a Lie group and* $f : G \to H$ *a morphism. Then* f *is injective if and only if its restriction to* T *is injective.*

Indeed, by Th. 2 (no. 2) the normal subgroup $\operatorname{Ker} f$ of G reduces to the identity element if and only if its intersection with T reduces to the identity element.

§3. COMPACT FORMS OF COMPLEX SEMI-SIMPLE LIE ALGEBRAS

1. REAL FORMS

If $\mathfrak{a}$ is a complex Lie algebra, we denote by $\mathfrak{a}_{[\mathbf{R}]}$ (or sometimes by $\mathfrak{a}$) the real Lie algebra obtained by restriction of scalars. If $\mathfrak{g}$ is a real Lie algebra, we denote by $\mathfrak{g}_{(\mathbf{C})}$ (or sometimes by $\mathfrak{g}_{\mathbf{C}}$) the complex Lie algebra $\mathbf{C} \otimes_{\mathbf{R}} \mathfrak{g}$ obtained by extension of scalars. The homomorphisms of real Lie algebras $\mathfrak{g} \to \mathfrak{a}_{[\mathbf{R}]}$ correspond bijectively to the homomorphisms of complex Lie algebras $\mathfrak{g}_{(\mathbf{C})} \to \mathfrak{a}$: if $f : \mathfrak{g} \to \mathfrak{a}_{[\mathbf{R}]}$ and $g : \mathfrak{g}_{(\mathbf{C})} \to \mathfrak{a}$ correspond, we have $f(x) = g(1 \otimes x)$ and $g(\lambda \otimes x) = \lambda f(x)$ for $x \in \mathfrak{g}, \lambda \in \mathbf{C}$.

DEFINITION 1. *Let $\mathfrak{a}$ be a complex Lie algebra. A real form of $\mathfrak{a}$ is a real subalgebra $\mathfrak{g}$ of $\mathfrak{a}$ that is an* **R***-structure on the* **C***-vector space $\mathfrak{a}$ (Algebra,* Chap. II, §8, no. 1, Def. 1).

This means that the homomorphism of complex Lie algebras $\mathfrak{g}_{(\mathbf{C})} \to \mathfrak{a}$ associated to the canonical injection $\mathfrak{g} \to \mathfrak{a}_{[\mathbf{R}]}$ is bijective. Thus, a real subalgebra $\mathfrak{g}$ of $\mathfrak{a}$ is a real form of $\mathfrak{a}$ if and only if the subspaces $\mathfrak{g}$ and $i\mathfrak{g}$ of the real vector space $\mathfrak{a}$ are complementary. The *conjugation* of $\mathfrak{a}$ relative to the real form $\mathfrak{g}$ is the map $\sigma : \mathfrak{a} \to \mathfrak{a}$ such that

$$\sigma(x + iy) = x - iy, \quad x, y \in \mathfrak{g}. \tag{1}$$

PROPOSITION 1. *a) Let $\mathfrak{g}$ be a real form of $\mathfrak{a}$ and σ the conjugation of $\mathfrak{a}$ relative to $\mathfrak{g}$. Then:*

$$\sigma^2 = \mathrm{Id}_{\mathfrak{a}}, \quad \sigma(\lambda x + \mu y) = \bar{\lambda}\sigma(x) + \bar{\mu}\sigma(y), \quad [\sigma(x), \sigma(y)] = \sigma[x, y] \tag{2}$$

for $\lambda, \mu \in \mathbf{C}$, $x, y \in \mathfrak{a}$. An element x of $\mathfrak{a}$ belongs to $\mathfrak{g}$ if and only if $\sigma(x) = x$.

b) Let $\sigma : \mathfrak{a} \to \mathfrak{a}$ be a map satisfying (2). *Then the set $\mathfrak{g}$ of fixed points of σ is a real form of $\mathfrak{a}$, and σ is the conjugation of $\mathfrak{a}$ relative to $\mathfrak{g}$.*

The proof is immediate.

Note that if B denotes the Killing form of $\mathfrak{a}$, and if $\mathfrak{g}$ is a real form of $\mathfrak{a}$, the restriction of B to $\mathfrak{g}$ is the Killing form of $\mathfrak{g}$; in particular, B is real-valued on $\mathfrak{g} \times \mathfrak{g}$. Assume that $\mathfrak{a}$ is *reductive*; then the real Lie algebra $\mathfrak{g}$ is compact if and only if the restriction of B to $\mathfrak{g}$ is negative (§1, no. 3). In that case we say that $\mathfrak{g}$ is a *compact real form* of $\mathfrak{a}$.

2. REAL FORMS ASSOCIATED TO A CHEVALLEY SYSTEM

In this number, we consider a split semi-simple Lie algebra $(\mathfrak{a}, \mathfrak{h})$ over the field $\mathbf{C}$ (Chap. VIII, §2, no. 1), with root system $\mathrm{R}(\mathfrak{a}, \mathfrak{h}) = \mathrm{R}$, and a Chevalley system $(X_\alpha)_{\alpha\in\mathrm{R}}$ of $(\mathfrak{a}, \mathfrak{h})$ (Chap. VIII, §2, no. 4, Def. 3).

Recall (*loc. cit.*) that the linear map $\theta : \mathfrak{a} \to \mathfrak{a}$ that coincides with $-\mathrm{Id}_\mathfrak{h}$ on $\mathfrak{h}$ and maps X_α to $X_{-\alpha}$ for all $\alpha \in \mathrm{R}$ is an automorphism of $\mathfrak{a}$. Moreover (*loc. cit.*, Prop. 7), if $\alpha, \beta, \alpha + \beta$ are roots, then

$$[X_\alpha, X_\beta] = \mathrm{N}_{\alpha,\beta} X_{\alpha+\beta} \tag{3}$$

with $\mathrm{N}_{\alpha,\beta} \in \mathbf{R}^*$ and

$$\mathrm{N}_{-\alpha,-\beta} = \mathrm{N}_{\alpha,\beta}. \tag{4}$$

Denote by $\mathfrak{h}_0$ the real vector subspace of $\mathfrak{h}$ consisting of the $H \in \mathfrak{h}$ such that $\alpha(H) \in \mathbf{R}$ for all $\alpha \in \mathrm{R}$. Then $\mathfrak{h}_0$ is an $\mathbf{R}$-structure on the complex vector space $\mathfrak{h}$, we have $[X_\alpha, X_{-\alpha}] \in \mathfrak{h}_0$ for all $\alpha \in \mathrm{R}$, and the restriction of the Killing form B of $\mathfrak{a}$ to $\mathfrak{h}_0$ is separating positive (Chap. VIII, §2, no. 2, Remark 2). Moreover,

$$\mathrm{B}(H, X_\alpha) = 0, \quad \mathrm{B}(X_\alpha, X_\beta) = 0 \text{ if } \alpha + \beta \neq 0, \quad \mathrm{B}(X_\alpha, X_{-\alpha}) < 0 \tag{5}$$

(Chap. VIII, §2, no. 2, Prop. 1 and no. 4, Lemma 3).

PROPOSITION 2. *a) The real vector subspace* $\mathfrak{a}_0 = \mathfrak{h}_0 + \sum_{\alpha\in\mathrm{R}} \mathbf{R}X_\alpha$ *of* $\mathfrak{a}$ *is a real form of* $\mathfrak{a}$, *of which* $\mathfrak{h}_0$ *is a Cartan subalgebra. The pair* $(\mathfrak{a}_0, \mathfrak{h}_0)$ *is a split semi-simple real Lie algebra, of which* (X_α) *is a Chevalley system.*

b) Let σ *be the conjugation of* $\mathfrak{a}$ *relative to* $\mathfrak{a}_0$. *Then* $\sigma \circ \theta = \theta \circ \sigma$. *The set of fixed points of* $\sigma \circ \theta$ *is a compact real form* $\mathfrak{a}_u$ *of* $\mathfrak{a}$, *of which* $i\mathfrak{h}_0$ *is a Cartan subalgebra.*

Part *a*) follows immediately from the preceding. We prove *b*). Since $\sigma \circ \theta$ and $\theta \circ \sigma$ are two semi-linear maps from $\mathfrak{a}$ to $\mathfrak{a}$ that coincide on $\mathfrak{a}_0$, they coincide. Now $\sigma \circ \theta$ satisfies conditions (2) of no. 1, hence is the conjugation of $\mathfrak{a}$ relative to the real form $\mathfrak{a}_u$ consisting of the $x \in \mathfrak{a}$ such that $\sigma \circ \theta(x) = x$ (Prop. 1). For all $\alpha \in \mathrm{R}$ put

$$u_\alpha = X_\alpha + X_{-\alpha}, \quad v_\alpha = i(X_\alpha - X_{-\alpha}). \tag{6}$$

Then the $\mathbf{R}$-vector space $\mathfrak{a}_u$ is generated by $i\mathfrak{h}_0$, the u_α and the v_α. More precisely, if we choose a chamber C of R, then

$$\mathfrak{a}_u = i\mathfrak{h}_0 \oplus \bigoplus_{\alpha\in\mathrm{R}_+(\mathrm{C})} (\mathbf{R}u_\alpha + \mathbf{R}v_\alpha). \tag{7}$$

It is clear that $i\mathfrak{h}_0$ is a Cartan subalgebra of $\mathfrak{a}_u$, and it remains to prove that the restriction of B to $\mathfrak{a}_u$ is negative. Now $i\mathfrak{h}_0$ and the different subspaces of the form $\mathbf{R}u_\alpha \oplus \mathbf{R}v_\alpha$ are orthogonal with respect to B, by (5); the restriction of B to $i\mathfrak{h}_0$ is negative and

$$B(u_\alpha, u_\alpha) = B(v_\alpha, v_\alpha) = 2B(X_\alpha, X_{-\alpha}) < 0, \quad B(u_\alpha, v_\alpha) = 0, \tag{8}$$

hence the conclusion.

Remark. With the preceding notations, we have the following formulas:

$$[h, u_\alpha] = -i\alpha(h)v_\alpha, \quad [h, v_\alpha] = i\alpha(h)u_\alpha, \quad [u_\alpha, v_\alpha] = 2iH_\alpha, \quad (h \in \mathfrak{h}) \tag{9}$$

$$[u_\alpha, u_\beta] = N_{\alpha,\beta}u_{\alpha+\beta} + N_{\alpha,-\beta}u_{\alpha-\beta}, \qquad \alpha \neq \pm\beta, \tag{10}$$

$$[v_\alpha, v_\beta] = -N_{\alpha,\beta}u_{\alpha+\beta} + N_{\alpha,-\beta}u_{\alpha-\beta}, \qquad \alpha \neq \pm\beta, \tag{11}$$

$$[u_\alpha, v_\beta] = N_{\alpha,\beta}v_{\alpha+\beta} - N_{\alpha,-\beta}v_{\alpha-\beta}, \qquad \alpha \neq \pm\beta, \tag{12}$$

(in the last three formulas, it is understood, as usual, that $N_{\gamma,\delta} = 0$ if $\gamma + \delta$ is not a root).

Note that $\sum \mathbf{R}u_\alpha$ is a real subalgebra of $\mathfrak{a}$, namely $\mathfrak{a}_0 \cap \mathfrak{a}_u$.

Let Q(R) be the group of radical weights of R (Chap. VI, §1, no. 9). Recall that to any homomorphism $\gamma : Q(R) \to \mathbf{C}^*$ is associated an elementary automorphism $f(\gamma)$ of $\mathfrak{a}$ such that $f(\gamma)(h) = h$ for all $h \in \mathfrak{h}$ and $f(\gamma)X_\alpha = \gamma(\alpha)X_\alpha$ (Chap. VIII, §5, no. 2).

PROPOSITION 3. *Let $\mathfrak{g}$ be a compact real form of $\mathfrak{a}$ such that $\mathfrak{g} \cap \mathfrak{h} = i\mathfrak{h}_0$. There exists a homomorphism $\gamma : Q(R) \to \mathbf{R}_+^*$ such that $\mathfrak{g} = f(\gamma)(\mathfrak{a}_u)$.*

Let τ be the conjugation of $\mathfrak{a}$ relative to $\mathfrak{g}$. By hypothesis $\tau(x) = x$ for $x \in i\mathfrak{h}_0$, so $\tau(x) = -x$ for $x \in \mathfrak{h}_0$. Thus, for all $\alpha \in R$ and all $h \in \mathfrak{h}_0$,

$$[h, \tau(X_\alpha)] = [-\tau(h), \tau(X_\alpha)] = -\tau([h, X_\alpha]) = -\tau(\alpha(h)X_\alpha);$$

it follows that $[h, \tau(X_\alpha)] = -\alpha(h)\tau(X_\alpha)$ for all $h \in \mathfrak{h}_0$, hence also for all $h \in \mathfrak{h}$. Hence there exists $c_\alpha \in \mathbf{C}^*$ such that $\tau(X_\alpha) = c_\alpha X_{-\alpha}$. Since $[X_\alpha, X_{-\alpha}] \in \mathfrak{h}_0$, we have $[\tau(X_\alpha), \tau(X_{-\alpha})] = -[X_\alpha, X_{-\alpha}]$, so $c_\alpha c_{-\alpha} = 1$; similarly, formulas (3) and (4) give $c_{\alpha+\beta} = c_\alpha c_\beta$ if $\alpha, \beta, \alpha+\beta$ are roots. By Chap. VI, §1, no. 6, Cor. 2 of Prop. 19, there exists a homomorphism $\delta : Q(R) \to \mathbf{C}^*$ such that $\delta(\alpha) = c_\alpha$ for all $\alpha \in R$.

We now show that each c_α is strictly positive. Indeed, $c_\alpha B(X_\alpha, X_{-\alpha}) = B(X_\alpha, \tau(X_\alpha))$, and since $B(X_\alpha, X_{-\alpha})$ is negative, it suffices to show that $B(z, \tau(z)) < 0$ for every non-zero element z of $\mathfrak{a}$; but every element of $\mathfrak{a}$ can be written as $x + iy$, with x and y in $\mathfrak{g}$, and

$$B(x + iy, \tau(x + iy)) = B(x + iy, x - iy) = B(x, x) + B(y, y),$$

hence the stated assertion, the restriction of B to $\mathfrak{g}$ being negative and separating by hypothesis.

It follows that the homomorphism δ takes values in $\mathbf{R}_+^*$; hence there exists a homomorphism $\gamma : Q(R) \to \mathbf{R}_+^*$ such that $\delta = \gamma^{-2}$. Then $f(\gamma)^{-1}(\mathfrak{g})$ is a real form of $\mathfrak{a}$; the corresponding conjugation is $\tau' = f(\gamma)^{-1} \circ \tau \circ f(\gamma)$. For all $\alpha \in \mathrm{R}$, we have

$$\tau'(X_\alpha) = f(\gamma)^{-1}(\tau(c_\alpha^{-1/2} X_\alpha)) = f(\gamma)^{-1}(c_\alpha^{1/2} X_{-\alpha}) = X_{-\alpha},$$

and $\tau'(h) = \tau(h) = h$ for $h \in i\mathfrak{h}_0$; it follows that τ' is the conjugation with respect to $\mathfrak{a}_u$, and hence that $f(\gamma)^{-1}(\mathfrak{g}) = \mathfrak{a}_u$.

3. CONJUGACY OF COMPACT FORMS

THEOREM 1. *Let $\mathfrak{a}$ be a complex semi-simple Lie algebra.*

a) $\mathfrak{a}$ has compact (resp. *splittable*) *real forms.*

b) The group $\mathrm{Int}(\mathfrak{a})$ *operates transitively on the set of compact* (resp. *splittable*) *real forms of $\mathfrak{a}$.*

Let $\mathfrak{h}$ be a Cartan subalgebra of $\mathfrak{a}$. Then $(\mathfrak{a}, \mathfrak{h})$ is split (Chap. VIII, §2, no. 1, Remark 2), and has a Chevalley system (X_α) (Chap. VIII, §4, no. 4, Cor. of Prop. 5). Part *a)* now follows from Prop. 2. Let $\mathfrak{g}$ be a compact real form of $\mathfrak{a}$; we show that there exists $v \in \mathrm{Int}(\mathfrak{a})$ such that $v(\mathfrak{a}_u) = \mathfrak{g}$. Let $\mathfrak{t}$ be a Cartan subalgebra of $\mathfrak{g}$; then $\mathfrak{t}_{(\mathbf{C})}$ is a Cartan subalgebra of $\mathfrak{a}$; since $\mathrm{Int}(\mathfrak{a})$ operates transitively on the set of Cartan subalgebras of $\mathfrak{a}$ (Chap. VII, §3, no. 2, Th. 1), we are reduced to the case in which $\mathfrak{t}_{(\mathbf{C})} = \mathfrak{h}$. Since $\mathfrak{g}$ is a compact form, the eigenvalues of the endomorphisms ad h, for $h \in \mathfrak{t}$, are purely imaginary (§1, no. 3, Prop. 1), so the roots $\alpha \in \mathrm{R}$ map $\mathfrak{t}$ to $i\mathbf{R}$; this implies that $\mathfrak{t} = i\mathfrak{h}_0$. Then, by Prop. 3 (no. 2), there exists $v \in \mathrm{Int}(\mathfrak{a})$ such that $v(\mathfrak{a}_u) = \mathfrak{g}$, hence *b)* in the case of compact forms. Finally, let $\mathfrak{m}_1$ and $\mathfrak{m}_2$ be two splittable real forms of $\mathfrak{a}$. There exist framings $(\mathfrak{m}_1, \mathfrak{h}_1, \mathrm{B}_1, (X_\alpha^1))$ and $(\mathfrak{m}_2, \mathfrak{h}_2, \mathrm{B}_2, (X_\alpha^2))$ (Chap. VIII, §4, no. 1). These extend in an obvious way to bases e_1 and e_2 of $\mathfrak{a}$. An automorphism of $\mathfrak{a}$ that maps e_1 to e_2 maps $\mathfrak{m}_1$ to $\mathfrak{m}_2$; thus, it suffices to apply Prop. 5 of Chap. VIII, §5, no. 3, to obtain the existence of an element u of $\mathrm{Aut}_0(\mathfrak{a}) = \mathrm{Int}(\mathfrak{a})$ such that $u(\mathfrak{m}_1) = \mathfrak{m}_2$.

Remark. We shall see much later a general classification of real forms of a complex semi-simple Lie algebra.

COROLLARY 1. *Let $\mathfrak{g}$ and $\mathfrak{g}'$ be two compact real Lie algebras. Then $\mathfrak{g}$ and $\mathfrak{g}'$ are isomorphic if and only if the complex Lie algebras $\mathfrak{g}_{(\mathbf{C})}$ and $\mathfrak{g}'_{(\mathbf{C})}$ are isomorphic.*

The condition is clearly necessary. Conversely, assume that $\mathfrak{g}_{(\mathbf{C})}$ and $\mathfrak{g}'_{(\mathbf{C})}$ are isomorphic. Let $\mathfrak{c}$ (resp. $\mathfrak{c}'$) be the centre of $\mathfrak{g}$ (resp. $\mathfrak{g}'$) and $\mathfrak{s}$ (resp. $\mathfrak{s}'$) the derived algebra of $\mathfrak{g}$ (resp. $\mathfrak{g}'$). Then $\mathfrak{c}_{(\mathbf{C})}$ and $\mathfrak{c}'_{(\mathbf{C})}$ are the centres of $\mathfrak{g}_{(\mathbf{C})}$ and $\mathfrak{g}'_{(\mathbf{C})}$, respectively, and hence are isomorphic; it follows that the commutative algebras $\mathfrak{c}$ and $\mathfrak{c}'$ are isomorphic. Similarly, $\mathfrak{s}_{(\mathbf{C})}$ and $\mathfrak{s}'_{(\mathbf{C})}$ are isomorphic, hence $\mathfrak{s}$ and $\mathfrak{s}'$, which are compact real forms of two isomorphic complex semi-simple Lie algebras, are isomorphic by Th. 1 *b)*.

COROLLARY 2. *Let* $\mathfrak{a}$ *be a complex Lie algebra. The following conditions are equivalent:*

(i) $\mathfrak{a}$ *is reductive.*

(ii) *There exists a compact real Lie algebra* $\mathfrak{g}$ *such that* $\mathfrak{a}$ *is isomorphic to* $\mathfrak{g}_{(\mathbf{C})}$.

(iii) *There exists a compact Lie group* G *such that* $\mathfrak{a}$ *is isomorphic to* $\mathrm{L}(\mathrm{G})_{(\mathbf{C})}$.

By Def. 1 of §1, no. 3, conditions (ii) and (iii) are equivalent and imply (i). If $\mathfrak{a}$ is reductive, it is the direct product of a commutative algebra, which clearly has a compact real form, and a semi-simple algebra which has one by Th. 1 *a*), hence (i) implies (ii).

COROLLARY 3. *Let* $\mathfrak{a}_1$ *and* $\mathfrak{a}_2$ *be two complex semi-simple Lie algebras. The compact real forms of* $\mathfrak{a}_1 \times \mathfrak{a}_2$ *are the products* $\mathfrak{g}_1 \times \mathfrak{g}_2$*, where, for* $i = 1, 2$*,* $\mathfrak{g}_i$ *is a compact real form of* $\mathfrak{a}_i$.

Indeed, there exists a compact real form $\mathfrak{g}_1$ (resp. $\mathfrak{g}_2$) of $\mathfrak{a}_1$ (resp. $\mathfrak{a}_2$); then $\mathfrak{g}_1 \times \mathfrak{g}_2$ is a compact real form of $\mathfrak{a}_1 \times \mathfrak{a}_2$. The corollary now follows from Th. 1 *b*), applied to $\mathfrak{a}_1, \mathfrak{a}_2$ and $\mathfrak{a}_1 \times \mathfrak{a}_2$.

Note that it follows from Cor. 3 above that a compact real Lie algebra $\mathfrak{g}$ is simple if and only if the complex Lie algebra $\mathfrak{g}_{(\mathbf{C})}$ is simple. We say that $\mathfrak{g}$ is of *type* A_n, or B_n, ..., if $\mathfrak{g}_{(\mathbf{C})}$ is of type A_n, or B_n, ... (Chap. VIII, §2, no. 2). By Cor. 1 above, *two compact simple real Lie algebras are isomorphic if and only if they are of the same type.*

Let G be an almost simple connected compact Lie group (Chap. III, §9, no. 8, Def. 3). We say that G is of type A_n, or B_n, ..., if its Lie algebra is of type A_n, or B_n, ... Two simply-connected almost simple compact Lie groups are isomorphic if and only if they are of the same type.

4. EXAMPLE I: COMPACT ALGEBRAS OF TYPE A_n

Let V be a finite dimensional complex vector space and Φ a separating positive hermitian form on V. The *unitary group* associated to Φ (cf. *Algebra*, Chap. IX) is the subgroup $\mathbf{U}(\Phi)$ of $\mathbf{GL}(\mathrm{V})$ consisting of the automorphisms of the complex Hilbert space (V, Φ); this is a (real) Lie subgroup of the group $\mathbf{GL}(\mathrm{V})$, whose Lie algebra is the subalgebra $\mathfrak{u}(\Phi)$ of the real Lie algebra $\mathfrak{gl}(\mathrm{V})$ consisting of the endomorphisms x of V such that $x^* = -x$ (Chap. III, §3, no. 10, Cor. 2 of Prop. 37), where x^* denotes the adjoint of x relative to Φ. Since the group $\mathbf{U}(\Phi)$ is compact (§1, no. 1), $\mathfrak{u}(\Phi)$ is a *compact* real Lie algebra. Similarly, the *special unitary* group $\mathbf{SU}(\Phi) = \mathbf{U}(\Phi) \cap \mathbf{SL}(\mathrm{V})$ is a compact Lie subgroup of $\mathbf{SL}(\mathrm{V})$, whose Lie algebra is $\mathfrak{su}(\Phi) = \mathfrak{u}(\Phi) \cap \mathfrak{sl}(\mathrm{V})$.

When $\mathrm{V} = \mathbf{C}^n$ and Φ is the usual hermitian form (for which the canonical basis of $\mathbf{C}^n$ is orthonormal), we write $\mathbf{U}(n, \mathbf{C})$, $\mathbf{SU}(n, \mathbf{C})$, $\mathfrak{u}(n, \mathbf{C})$, $\mathfrak{su}(n, \mathbf{C})$ instead of $\mathbf{U}(\Phi)$, $\mathbf{SU}(\Phi)$, $\mathfrak{u}(\Phi)$, $\mathfrak{su}(\Phi)$. The elements of $\mathbf{U}(n, \mathbf{C})$ (resp. $\mathfrak{u}(n, \mathbf{C})$)

are the matrices $A \in \mathrm{M}_n(\mathbf{C})$ such that $A.{}^t\bar{A} = I_n$ (resp. $A = -{}^t\bar{A}$), which are said to be unitary (resp. anti-hermitian).

PROPOSITION 4. *a) The compact real forms of the complex Lie algebra* $\mathfrak{sl}(\mathrm{V})$ *are the algebras* $\mathfrak{su}(\Phi)$, *where* Φ *belongs to the set of separating positive hermitian forms on the complex vector space* V.

b) The algebras $\mathfrak{u}(\Phi)$ *are the compact real forms of* $\mathfrak{gl}(\mathrm{V})$.

Let Φ be a separating positive hermitian form on V. For all $x \in \mathfrak{gl}(\mathrm{V})$, put $\sigma(x) = -x^*$ (where x^* is the adjoint of x relative to Φ). Then σ satisfies conditions (2) of Prop. 1 of no. 1, so the set $\mathfrak{u}(\Phi)$ (resp. $\mathfrak{su}(\Phi)$) of fixed points of σ on $\mathfrak{gl}(\mathrm{V})$ (resp. $\mathfrak{sl}(\mathrm{V})$) is a compact real form of $\mathfrak{gl}(\mathrm{V})$ (resp. $\mathfrak{sl}(\mathrm{V})$). Since $\mathbf{GL}(\mathrm{V})$ operates transitively on the set of separating positive hermitian forms on V (*Algebra*, Chap. IX) and on the set of compact real forms of $\mathfrak{sl}(\mathrm{V})$ (no. 3, Th. 1 and Chap. VIII, §13, no. 1 (VII)), Prop. 4 is proved.

COROLLARY. *Every compact simple real Lie algebra of type* A_n $(n \geq 1)$ *is isomorphic to* $\mathfrak{su}(n+1, \mathbf{C})$.

Indeed, every complex Lie algebra of type A_n is isomorphic to $\mathfrak{sl}(n+1, \mathbf{C})$ (Chap. VIII, §13, no. 1).

Remarks. 1) We have $\mathfrak{gl}(\mathrm{V}) = \mathfrak{sl}(\mathrm{V}) \times \mathbf{C}.1_{\mathrm{V}}$, $\mathfrak{u}(\Phi) = \mathfrak{su}(\Phi) \times \mathbf{R}.i1_{\mathrm{V}}$; the compact real forms of $\mathfrak{gl}(\mathrm{V})$ are the $\mathfrak{su}(\Phi) \times \mathbf{R}.\alpha 1_{\mathrm{V}}$, $\alpha \in \mathbf{C}^*$.

2) If the complex Lie algebra $\mathfrak{a} = \mathfrak{sl}(n, \mathbf{C})$ is equipped with the splitting and Chevalley system introduced in Chap. VIII, §13, no. 1 (IX), then, with the notations in no. 2,

$$\mathfrak{a}_u = \mathfrak{su}(n, \mathbf{C}), \quad \mathfrak{a}_0 = \mathfrak{sl}(n, \mathbf{R}), \quad \mathfrak{a}_u \cap \mathfrak{a}_0 = \mathfrak{o}(n, \mathbf{R}).$$

5. EXAMPLE II: COMPACT ALGEBRAS OF TYPE B_n AND D_n

Let V be a finite dimensional real vector space and Q a separating positive quadratic form on V. The *orthogonal group* associated to Q (*Algebra*, Chap. IX) is the subgroup $\mathbf{O}(\mathrm{Q})$ of $\mathbf{GL}(\mathrm{V})$ consisting of the automorphisms of the real Hilbert space (V, Q); this is a Lie subgroup of $\mathbf{GL}(\mathrm{V})$, whose Lie algebra is the subalgebra $\mathfrak{o}(\mathrm{Q})$ of $\mathfrak{gl}(\mathrm{V})$ consisting of the endomorphisms x of V such that $x^* = -x$ (Chap. III, §3, no. 10, Cor. 2 of Prop. 37), x^* denoting the adjoint of x relative to Q. Since the group $\mathbf{O}(\mathrm{Q})$ is compact, $\mathfrak{o}(\mathrm{Q})$ is thus a *compact* real Lie algebra. Put $\mathbf{SO}(\mathrm{Q}) = \mathbf{O}(\mathrm{Q}) \cap \mathbf{SL}(\mathrm{V})$; this is a closed subgroup of finite index of $\mathbf{O}(\mathrm{Q})$ (of index 2 if $\dim \mathrm{V} \neq 0$), hence also with Lie algebra $\mathfrak{o}(\mathrm{Q})$.

When $\mathrm{V} = \mathbf{R}^n$ and Q is the usual quadratic form (for which the canonical basis of $\mathbf{R}^n$ is orthonormal), we write $\mathbf{O}(n, \mathbf{R})$, $\mathbf{SO}(n, \mathbf{R})$, $\mathfrak{o}(n, \mathbf{R})$ instead of $\mathbf{O}(\mathrm{Q}), \mathbf{SO}(\mathrm{Q}), \mathfrak{o}(\mathrm{Q})$. The elements of $\mathbf{O}(n, \mathbf{R})$ (resp. $\mathfrak{o}(n, \mathbf{R})$) are the matrices

$A \in \mathrm{M}_n(\mathbf{R})$ such that $A.{}^tA = I_n$ (resp. $A = -{}^tA$), which are said to be orthogonal (resp. anti-symmetric).

Let $\mathrm{V}_{(\mathbf{C})}$ be the complex vector space associated to V and let $\mathrm{Q}_{(\mathbf{C})}$ be the quadratic form on $\mathrm{V}_{(\mathbf{C})}$ associated to Q. Identify $\mathfrak{gl}(\mathrm{V})_{(\mathbf{C})}$ with $\mathfrak{gl}(\mathrm{V}_{(\mathbf{C})})$; then $\mathfrak{o}(\mathrm{Q})_{(\mathbf{C})}$ is identified with $\mathfrak{o}(\mathrm{Q}_{(\mathbf{C})})$: this is clear since the map $x \mapsto x^* + x$ from $\mathfrak{gl}(\mathrm{V}_{(\mathbf{C})})$ to itself is **C**-linear. Since $\mathfrak{o}(\mathrm{Q}_{(\mathbf{C})})$ is of type B_n if $\dim \mathrm{V} = 2n+1$, $n \geq 1$, and of type D_n if $\dim \mathrm{V} = 2n$, $n \geq 3$ (Chap. VIII, §13, nos. 2 and 4), we deduce:

PROPOSITION 5. *Every compact simple real Lie algebra of type* B_n, $n \geq 1$ (resp. *of type* D_n, $n \geq 3$) *is isomorphic to* $\mathfrak{o}(2n+1, \mathbf{R})$ (resp. $\mathfrak{o}(2n, \mathbf{R})$).

6. COMPACT GROUPS OF RANK 1

By *General Topology*, Chap. VIII, §1, no. 4, Prop. 3, Prop. 4 and Remark 4, the topological group $\mathbf{SU}(2, \mathbf{C})$ is isomorphic to the topological group $\mathbf{S}_3$ of quaternions of norm 1, and the quotient of $\mathbf{SU}(2, \mathbf{C})$ by the subgroup Z consisting of the matrices I_2 and $-I_2$ is isomorphic to the topological group $\mathbf{SO}(3, \mathbf{R})$. Note that Z is the centre of $\mathbf{SU}(2, \mathbf{C})$: indeed, since $\mathbf{H} = \mathbf{R}.\mathbf{S}_3$, every element of the centre of the group $\mathbf{S}_3$ is in the centre $\mathbf{R}$ of the algebra $\mathbf{H}$ and hence belongs to the group with two elements $\mathbf{S}_3 \cap \mathbf{R} = \{-1, 1\}$.

PROPOSITION 6. *Every compact semi-simple real Lie algebra of rank* 1 *is isomorphic to* $\mathfrak{su}(2, \mathbf{C})$ *and to* $\mathfrak{o}(3, \mathbf{R})$. *Every connected semi-simple compact Lie group of rank* 1 *is isomorphic to* $\mathbf{SU}(2, \mathbf{C})$ *if it is simply-connected, and to* $\mathbf{SO}(3, \mathbf{R})$ *if not.*

The first assertion follows from the Cor. of Prop. 4 and Prop. 5. Since $\mathbf{SU}(2, \mathbf{C})$ is homeomorphic to $\mathbf{S}_3$ (*General Topology*, Chap. VIII, §1, no. 4, Remark 4), hence is simply-connected (*General Topology*, Chap. XI, in preparation), every simply-connected compact semi-simple Lie group of rank 1 is isomorphic to $\mathbf{SU}(2, \mathbf{C})$; every connected compact semi-simple Lie group of rank 1 that is not simply-connected is isomorphic to a quotient of $\mathbf{SU}(2, \mathbf{C})$ by a subgroup of Z that does not reduce to the identity element, hence to $\mathbf{SO}(3, \mathbf{R})$.

Remark. We have seen above that $\mathbf{SU}(2, \mathbf{C})$ is simply-connected and that $\pi_1(\mathbf{SO}(3, \mathbf{R}))$ is of order 2. We shall see much later that these results generalize to $\mathbf{SU}(n, \mathbf{C})$ $(n \geq 1)$ and $\mathbf{SO}(n, \mathbf{R})$ $(n \geq 3)$, respectively (cf. also §3, Exerc. 4 and 5).

Recall (Chap. VIII, §1, no. 1) that the canonical basis of $\mathfrak{sl}(2, \mathbf{C})$ is the basis (X_+, X_-, H), where

$$X_+ = \begin{pmatrix} 0 & 1 \\ 0 & 0 \end{pmatrix}, \quad X_- = \begin{pmatrix} 0 & 0 \\ -1 & 0 \end{pmatrix}, \quad H = \begin{pmatrix} 1 & 0 \\ 0 & -1 \end{pmatrix}.$$

We thus obtain a basis (U, V, iH) of $\mathfrak{su}(2, \mathbf{C})$, also called canonical, by putting

$$U = X_+ + X_- = \begin{pmatrix} 0 & 1 \\ -1 & 0 \end{pmatrix}, \quad V = i(X_+ - X_-) = \begin{pmatrix} 0 & i \\ i & 0 \end{pmatrix},$$
$$iH = \begin{pmatrix} i & 0 \\ 0 & -i \end{pmatrix}.$$

We have

$$[iH, U] = 2V, \quad [iH, V] = -2U, \quad [U, V] = 2iH. \tag{13}$$

If B denotes the Killing form of $\mathfrak{su}(2, \mathbf{C})$ an immediate calculation gives

$$\mathrm{B}(aU + bV + ciH, a'U + b'V + c'iH) = -8(aa' + bb' + cc'), \tag{14}$$

so that, if we identify $\mathfrak{su}(2, \mathbf{C})$ with $\mathbf{R}^3$ by means of the canonical basis, the adjoint representation of $\mathbf{SU}(2, \mathbf{C})$ defines a homomorphism $\mathbf{SU}(2, \mathbf{C}) \to \mathbf{SO}(3, \mathbf{R})$ (cf. above).

Further, note that $\mathbf{R}iH$ is a Cartan subalgebra of $\mathfrak{su}(2, \mathbf{C})$, that the maximal torus T of $\mathbf{SU}(2, \mathbf{C})$ that corresponds to it consists of the diagonal matrices $\begin{pmatrix} a & 0 \\ 0 & \bar{a} \end{pmatrix}$, where $a\bar{a} = 1$, and that the exponential map

$$\exp : \mathbf{R}iH \to \mathrm{T}$$

maps xH, for $x \in \mathbf{R}i$, to the matrix $\begin{pmatrix} \exp(x) & 0 \\ 0 & \exp(-x) \end{pmatrix}$, and thus has kernel $\mathbf{Z}.K$, where K is the element of $\mathfrak{su}(2, \mathbf{C})$ defined by

$$K = 2\pi iH = \begin{pmatrix} 2\pi i & 0 \\ 0 & -2\pi i \end{pmatrix}. \tag{15}$$

Further, the centre of $\mathbf{SU}(2, \mathbf{C})$ consists of the identity and $\exp(K/2)$.

Put

$$\theta = \begin{pmatrix} 0 & -1 \\ 1 & 0 \end{pmatrix} \in \mathbf{SU}(2, \mathbf{C}). \tag{16}$$

By Chap. VIII, §1, no. 5,

$$\theta^2 = \begin{pmatrix} -1 & 0 \\ 0 & -1 \end{pmatrix}, \quad (\operatorname{Int} \theta)t = t^{-1}, \quad t \in \mathrm{T}, \tag{17}$$

$$(\operatorname{Ad} \theta)X_+ = X_-, \quad (\operatorname{Ad} \theta)X_- = X_+, \quad (\operatorname{Ad} \theta)U = U, \quad (\operatorname{Ad} \theta)V = -V. \tag{18}$$

Finally, for $t = \begin{pmatrix} a & 0 \\ 0 & \bar{a} \end{pmatrix} \in \mathrm{T}$, we have

$$(\operatorname{Ad} t)X_+ = a^2 X_+, \quad (\operatorname{Ad} t)X_- = a^{-2} X_-, \quad (\operatorname{Ad} t)H = H, \tag{19}$$
$$(\operatorname{Ad} t)U = \mathscr{R}(a^2)U + \mathscr{I}(a^2)V, \quad (\operatorname{Ad} t)V = -\mathscr{I}(a^2)U + \mathscr{R}(a^2)V. \tag{20}$$

§4. ROOT SYSTEM ASSOCIATED TO A COMPACT GROUP

In paragraphs 4 *to* 8, *we denote by* G *a connected compact Lie group and by* T *a maximal torus of* G. *We denote by* $\mathfrak{g}$ (resp. $\mathfrak{t}$) *the Lie algebra of* G (resp. T), *by* $\mathfrak{g}_{\mathbf{C}}$ (resp. $\mathfrak{t}_{\mathbf{C}}$) *the complexified Lie algebra of* $\mathfrak{g}$ (resp. $\mathfrak{t}$), *and by* W *the Weyl group of* G *relative to* T (§2, no. 5).

1. THE GROUP X(H)

Let H be a compact Lie group. Denote by X(H) the (commutative) group of continuous homomorphisms from H to the topological group $\mathbf{C}^*$. By Chap. III, §8, no. 1, Th. 1, the elements of X(H) are morphisms of Lie groups; for all $a \in \mathrm{X(H)}$, the differential of a is an $\mathbf{R}$-linear map $\mathrm{L}(a) : \mathrm{L(H)} \to \mathrm{L}(\mathbf{C}^*)$. From now on we identify the Lie algebra of $\mathbf{C}^*$ with $\mathbf{C}$ in such a way that the exponential map of $\mathbf{C}^*$ coincides with the map $z \mapsto e^z$ from $\mathbf{C}$ to $\mathbf{C}^*$. Then, to any element $a \in \mathrm{X(H)}$ is associated an element $\mathrm{L}(a) \in \mathrm{Hom}_{\mathbf{R}}(\mathrm{L(H)}, \mathbf{C})$; we denote by $\delta(a)$ the element of $\mathrm{Hom}_{\mathbf{C}}(\mathrm{L(H)}_{(\mathbf{C})}, \mathbf{C})$ that corresponds to it (that is, whose restriction to $\mathrm{L(H)} \subset \mathrm{L(H)}_{(\mathbf{C})}$ is equal to $\mathrm{L}(a)$).

For all $x \in \mathrm{L(H)}$ and all $a \in \mathrm{X(H)}$, we have

$$a(\exp_{\mathrm{H}} x) = e^{\delta(a)(x)},$$

by functoriality of the exponential map (Chap. III, §6, no. 4, Prop. 10).

We shall often denote the group X(H) additively; in that case, we denote the element $a(g)$ of $\mathbf{C}^*$ by g^a. With this notation, we have the formulas

$$g^{a+b} = g^a g^b, \quad g \in \mathrm{H}, \quad a, b \in \mathrm{X(H)},$$

and

$$(\exp_{\mathrm{H}} x)^a = e^{\delta(a)(x)}, \quad x \in \mathrm{L(H)}, \ a \in \mathrm{X(H)}.$$

Since H is compact, the elements of X(H) take values in the subgroup $\mathbf{U} = \mathbf{U}(1, \mathbf{C})$ of complex numbers of absolute value 1, so that X(H) can be identified with the group of continuous (or analytic) homomorphisms from H to $\mathbf{U}$. It follows that, for all $a \in \mathrm{L(H)}$, the map $\mathrm{L}(a)$ takes values in the subspace $\mathbf{R}i$ of $\mathbf{C}$, so $\delta(a)$ maps L(H) to $\mathbf{R}i$.

If H is commutative, X(H) is simply the (discrete) dual group of H (*Spectral Theories*, Chap. II, §1, no. 1). If H is commutative and finite, X(H) can be identified with the dual finite group $\mathrm{D(H)} = \mathrm{Hom}_{\mathbf{Z}}(\mathrm{H}, \mathbf{Q}/\mathbf{Z})$ (where, as in *Algebra*, Chap. VII, §4, no. 9, Example 1, we identify $\mathbf{Q}/\mathbf{Z}$ with a subgroup of $\mathbf{C}^*$ by the homomorphism $r \mapsto \exp(2\pi i r)$).

For any morphism $f : \mathrm{H} \to \mathrm{H}'$ of compact Lie groups, we denote by $\mathrm{X}(f)$ the homomorphism $a \mapsto a \circ f$ from $\mathrm{X(H')}$ to X(H). If K is a closed normal subgroup of the compact Lie group H, we have an exact sequence of $\mathbf{Z}$-modules $0 \to X(H/K) \to X(H) \to X(K)$.

PROPOSITION 1. *For any compact Lie group* H, *the* $\mathbf{Z}$*-module* X(H) *is of finite type. It is free if* H *is connected.*

Assume first that H is connected; every element of X(H) vanishes on the derived group D(H) of H, hence we have a homomorphism $X(H/D(H)) \to X(H)$. But H/D(H) is connected and commutative, hence is a torus, and X(H/D(H)) is a free $\mathbf{Z}$-module of finite type (*Spectral Theories*, Chap. II, §2, no. 1, Cor. 2 of Prop. 1). In the general case, it follows from the exactness of the sequence

$$0 \to X(H/H_0) \to X(H) \to X(H_0),$$

where $X(H_0)$ is free of finite type and $X(H/H_0)$ is finite, that X(H) is of finite type.

PROPOSITION 2. *Let* H *be a commutative compact Lie group, and* $(a_i)_{i \in I}$ *a family of elements of* X(H); *the* a_i *generate* X(H) *if and only if the intersection of the* $\operatorname{Ker} a_i$ *reduces to the identity element.*

By *Spectral Theories*, Chap. II, §1, no. 7, Th. 4, the orthogonal complement of the kernel of a_i is the subgroup A_i of X(H) generated by a_i; by *loc. cit.*, Cor. 2 of Th. 4, the orthogonal complement of $\bigcap \operatorname{Ker} a_i$ is the subgroup of X(H) generated by the A_i, hence the proposition.

2. NODAL GROUP OF A TORUS

The *nodal group* of a torus S, denoted by $\Gamma(S)$, is the kernel of the exponential map $L(S) \to S$. This is a discrete subgroup of L(S), whose rank is equal to the dimension of S, and the $\mathbf{R}$-linear map $\mathbf{R} \otimes_{\mathbf{Z}} \Gamma(S) \to L(S)$ that extends the canonical injection of $\Gamma(S)$ into L(S) is bijective. It induces by passage to the quotient an isomorphism $\mathbf{R}/\mathbf{Z} \otimes_{\mathbf{Z}} \Gamma(S) \to S$.

For example, the nodal group $\Gamma(\mathbf{U})$ of $\mathbf{U}$ is the subgroup $2\pi i\mathbf{Z}$ of $L(\mathbf{U}) = i\mathbf{R}$.

For any morphism of tori $f : S \to S'$, denote by $\Gamma(f)$ the homomorphism $\Gamma(S) \to \Gamma(S')$ induced by $L(f)$. We have a commutative diagram

$$\begin{array}{ccccccccc} 0 & \longrightarrow & \Gamma(S) & \longrightarrow & L(S) & \xrightarrow{\exp_S} & S & \longrightarrow & 0 \\ & & \big\downarrow{\scriptstyle \Gamma(f)} & & \big\downarrow{\scriptstyle L(f)} & & \big\downarrow{\scriptstyle f} & & \\ 0 & \longrightarrow & \Gamma(S') & \longrightarrow & L(S') & \xrightarrow{\exp_{S'}} & S' & \longrightarrow & 0. \end{array} \tag{1}$$

Let $a \in X(S)$; applying the preceding to the morphism from S to $\mathbf{U}$ defined by a, we see that the $\mathbf{C}$-linear map $\delta(a) : L(S)_{(\mathbf{C})} \to \mathbf{C}$ of no. 1 maps $\Gamma(S)$ to $2\pi i\mathbf{Z}$. Thus, we can define a $\mathbf{Z}$-*bilinear* form on $X(S) \times \Gamma(S)$ by putting

$$\langle a, X \rangle = \frac{1}{2\pi i}\delta(a)(X), \quad a \in X(S),\ X \in \Gamma(S). \tag{2}$$

PROPOSITION 3. *The bilinear form* $(a, X) \mapsto \langle a, X \rangle$ *on* $\mathrm{X(S)} \times \Gamma(\mathrm{S})$ *is invertible.*

Recall (*Algebra*, Chap. IX) that, by definition, this means that the linear maps $\mathrm{X(S)} \to \mathrm{Hom}_{\mathbf{Z}}(\Gamma(\mathrm{S}), \mathbf{Z})$ and $\Gamma(\mathrm{S}) \to \mathrm{Hom}_{\mathbf{Z}}(\mathrm{X(S)}, \mathbf{Z})$ associated to this bilinear form are bijective.

It is immediate that if the conclusion of the proposition is true for two tori, it is also true for their product. Thus, since every torus of dimension n is isomorphic to $\mathbf{U}^n$, we are reduced to the case in which $\mathrm{S} = \mathbf{U}$. In this particular case, the assertion is immediate.

Let $f : \mathrm{S} \to \mathrm{S}'$ be a morphism of tori. Then, each of the linear maps $\mathrm{X}(f) : \mathrm{X(S')} \to \mathrm{X(S)}$ and $\Gamma(f) : \Gamma(\mathrm{S}) \to \Gamma(\mathrm{S}')$ is the transpose of the other: for all $a' \in \mathrm{X(S')}$ and all $X \in \Gamma(\mathrm{S})$,

$$\langle X(f)(a'), X \rangle = \langle a', \Gamma(f)(X) \rangle. \tag{3}$$

PROPOSITION 4. *Let* S *and* S′ *be tori. Denote by* $\mathrm{M(S, S')}$ *the group of morphisms of Lie groups from* S *to* S′. *The maps* $f \mapsto \mathrm{X}(f)$ *and* $f \mapsto \Gamma(f)$ *are isomorphisms of groups from* $\mathrm{M(S, S')}$ *to* $\mathrm{Hom}_{\mathbf{Z}}(\mathrm{X(S')}, \mathrm{X(S)})$ *and to* $\mathrm{Hom}_{\mathbf{Z}}(\Gamma(\mathrm{S}), \Gamma(\mathrm{S'}))$, *respectively.*

If f is a morphism of Lie groups from S to S′, the homomorphism $\mathrm{X}(f)$ is simply the *dual* of f in the sense of *Spectral Theories*, Chap. II, §1, no. 7. The map $\varphi \mapsto \hat{\varphi}$ from $\mathrm{Hom}_{\mathbf{Z}}(\mathrm{X(S')}, \mathrm{X(S)})$ to $\mathrm{M(S, S')}$ defined in *loc. cit.* is the inverse of the map $f \mapsto \mathrm{X}(f)$ from $\mathrm{M(S, S')}$ to $\mathrm{Hom}_{\mathbf{Z}}(\mathrm{X(S')}, \mathrm{X(S)})$; the latter is thus bijective. If we identify $\Gamma(\mathrm{S})$ (resp. $\Gamma(\mathrm{S'})$) with the dual **Z**-module of X(S) (resp. X(S′)) (Prop. 3), $\Gamma(f)$ coincides with the transpose of the homomorphism $\mathrm{X}(f)$, hence the proposition.

Remarks. 1) Let $f : \mathrm{S} \to \mathrm{S}'$ be a morphism of tori. The snake diagram (*Algebra*, Chap. X, §1, no. 2) associated to (1) gives an exact sequence

$$0 \longrightarrow \mathrm{Ker}\,\Gamma(f) \longrightarrow \mathrm{Ker}\,\mathrm{L}(f) \longrightarrow \mathrm{Ker}\,f \xrightarrow{d} \tag{4}$$

$$\xrightarrow{d} \mathrm{Coker}\,\Gamma(f) \longrightarrow \mathrm{Coker}\,\mathrm{L}(f) \longrightarrow \mathrm{Coker}\,f \longrightarrow 0.$$

In particular, assume that f is surjective, with finite kernel N, so that we have an exact sequence

$$0 \longrightarrow \mathrm{N} \xrightarrow{i} \mathrm{S} \xrightarrow{f} \mathrm{S}' \longrightarrow 0,$$

where i is the canonical injection. Then, $\mathrm{L}(f)$ is bijective, and (4) gives an isomorphism $\mathrm{N} \to \mathrm{Coker}\,\Gamma(f)$, hence an exact sequence

$$0 \longrightarrow \Gamma(\mathrm{S}) \xrightarrow{\Gamma(f)} \Gamma(\mathrm{S}') \longrightarrow \mathrm{N} \longrightarrow 0. \tag{5}$$

Moreover, by *Spectral Theories*, Chap. II, §1, no. 7, Th. 4, the sequence

$$0 \longrightarrow \mathrm{X(S')} \xrightarrow{\mathrm{X}(f)} \mathrm{X(S)} \xrightarrow{\mathrm{X}(i)} \mathrm{X(N)} \longrightarrow 0 \tag{6}$$

is exact.

2) By Prop. 4, the map $f \mapsto \Gamma(f)(2\pi i)$ from $\mathrm{M}(\mathbf{U}, \mathrm{S})$ to $\Gamma(\mathrm{S})$ is an isomorphism; if $a \in \mathrm{X}(\mathrm{S}) = \mathrm{M}(\mathrm{S}, \mathbf{U})$ and $f \in \mathrm{M}(\mathbf{U}, \mathrm{S})$, then the composite $a \circ f \in \mathrm{M}(\mathbf{U}, \mathbf{U})$ is the endomorphism $u \mapsto u^r$, where $r = \langle a, \Gamma(f)(2\pi i)\rangle$. We shall identify $\mathrm{M}(\mathbf{U}, \mathbf{U}) = \mathrm{X}(\mathbf{U})$ with $\mathbf{Z}$ from now on, the element r of $\mathbf{Z}$ being associated to the endomorphism $u \mapsto u^r$; thus, with the notations above,

$$a \circ f = \langle a, \Gamma(f)(2\pi i)\rangle.$$

3) To the exact sequence $0 \to \Gamma(\mathrm{S}) \to \mathrm{L}(\mathrm{S}) \xrightarrow{\exp_\mathrm{S}} \mathrm{S} \to 0$ is associated an isomorphism from $\Gamma(\mathrm{S})$ to the fundamental group of S, called *canonical* in the sequel. For any morphism of tori $f : \mathrm{S} \to \mathrm{S}'$, $\Gamma(f)$ can then be identified via the canonical isomorphisms $\Gamma(\mathrm{S}) \to \pi_1(\mathrm{S})$ and $\Gamma(\mathrm{S}') \to \pi_1(\mathrm{S}')$ with the homomorphism $\pi_1(f) : \pi_1(\mathrm{S}) \to \pi_1(\mathrm{S}')$ induced by f. Note that this gives another interpretation of the exact sequence (5) (cf. *General Topology*, Chap. XI, in preparation).

4) The homomorphisms of $\mathbf{Z}$-modules $\delta : \mathrm{X}(\mathrm{S}) \to \mathrm{Hom}_\mathbf{C}(\mathrm{L}(\mathrm{S})_{(\mathbf{C})}, \mathbf{C})$ and $\iota : \Gamma(\mathrm{S}) \to \mathrm{L}(\mathrm{S})_{(\mathbf{C})}$ (ι is induced by the canonical injection of $\Gamma(\mathrm{S})$ into $\mathrm{L}(\mathrm{S})$) extend to isomorphisms of $\mathbf{C}$-vector spaces

$$u : \mathbf{C} \otimes \mathrm{X}(\mathrm{S}) \to \mathrm{Hom}_\mathbf{C}(\mathrm{L}(\mathrm{S})_{(\mathbf{C})}, \mathbf{C})$$

$$v : \mathbf{C} \otimes \Gamma(\mathrm{S}) \to \mathrm{L}(\mathrm{S})_{(\mathbf{C})}$$

which we shall call *canonical* in the sequel. Note that, if we extend the pairing between $\mathrm{X}(\mathrm{S})$ and $\Gamma(\mathrm{S})$ by $\mathbf{C}$-linearity to a bilinear form $\ll\ ,\ \gg$ on $(\mathbf{C} \otimes \mathrm{X}(\mathrm{S})) \times (\mathbf{C} \otimes \Gamma(\mathrm{S}))$, then

$$\langle u(a), v(b)\rangle = 2\pi i \ll a, b \gg .$$

3. WEIGHTS OF A LINEAR REPRESENTATION

In this number k denotes one of the fields $\mathbf{R}$ or $\mathbf{C}$.

Let V be a finite dimensional vector space over k, and $\rho : \mathrm{G} \to \mathbf{GL}(\mathrm{V})$ a continuous (hence real-analytic, Chap. III, §8, no. 1, Th. 1) representation of the connected compact Lie group G on V. Define a complex vector space $\tilde{\mathrm{V}}$ and a continuous representation $\tilde{\rho} : \mathrm{G} \to \mathbf{GL}(\tilde{\mathrm{V}})$ as follows: if $k = \mathbf{C}$, set $\tilde{\mathrm{V}} = \mathrm{V}$, $\tilde{\rho} = \rho$; if $k = \mathbf{R}$, set $\tilde{\mathrm{V}} = \mathrm{V}_{(\mathbf{C})}$ and $\tilde{\rho}$ to be the composite of ρ with the canonical homomorphism $\mathbf{GL}(\mathrm{V}) \to \mathbf{GL}(\tilde{\mathrm{V}})$.

For all $\lambda \in \mathrm{X}(\mathrm{G})$, denote by $\tilde{\mathrm{V}}_\lambda(\mathrm{G})$ the vector subspace of $\tilde{\mathrm{V}}$ consisting of the $v \in \tilde{\mathrm{V}}$ such that $\tilde{\rho}(g)v = g^\lambda v$ for all $g \in \mathrm{G}$ (cf. Chap. VII, §1, no. 1). By *loc. cit.*, Prop. 3, the sum of the $\tilde{\mathrm{V}}_\lambda(\mathrm{G})$ (for λ belonging to $\mathrm{X}(\mathrm{G})$) is direct. Moreover:

Lemma 1. If G *is commutative,* $\tilde{\mathrm{V}}$ *is the direct sum of the* $\tilde{\mathrm{V}}_\lambda(\mathrm{G})$ *for* $\lambda \in \mathrm{X}(\mathrm{G})$.

Since ρ is semi-simple (§1, no. 1), it suffices to prove the lemma in the case in which ρ is simple. In that case, the commutant Z of $\rho(G)$ in $End(\tilde{V})$ reduces to homotheties (*Algebra*, Chap. VIII, §3, no. 2, Th. 1); thus, the image of the homomorphism $\tilde{\rho}$ is contained in the subgroup $\mathbf{C}^*.1_V$ of $GL(\tilde{V})$, and there exists $\lambda \in X(G)$ such that $\tilde{V} = \tilde{V}_\lambda(G)$.

DEFINITION 1. *The weights of the representation* ρ *of* G, *relative to a maximal torus* T *of* G, *are the elements* λ *of* X(T) *such that* $\tilde{V}_\lambda(T) \neq 0$.

Denote by $P(\rho, T)$, or by $P(\rho)$ if there is no possibility of confusion over the choice of T, the set of weights of ρ relative to T. By Lemma 1,

$$\tilde{V} = \bigoplus_{\lambda \in P(\rho,T)} \tilde{V}_\lambda(T). \tag{7}$$

Let T′ be another maximal torus of G and g an element of G such that $(\operatorname{Int} g)T = T'$ (§2, no. 2, Th. 2). For all $\lambda \in X(T)$,

$$\tilde{\rho}(g)(\tilde{V}_\lambda(T)) = \tilde{V}_{\lambda'}(T'), \quad \text{where } \lambda' = X(\operatorname{Int} g^{-1})(\lambda). \tag{8}$$

Consequently,

$$X(\operatorname{Int} g)(P(\rho, T')) = P(\rho, T). \tag{9}$$

The Weyl group $W = W_G(T)$ operates on the left on the **Z**-module X(T) by $w \mapsto X(w^{-1})$; thus, for $t \in T, \lambda \in X(T), w \in W$, we have $t^{w\lambda} = (w^{-1}(t))^\lambda$.

PROPOSITION 5. *The set* $P(\rho, T)$ *is stable under the operation of the Weyl group* W. *Let* $n \in N_G(T)$, *and let* w *be its class in* W; *for* $\lambda \in X(T)$, *we have* $\rho(n)(\tilde{V}_\lambda(T)) = \tilde{V}_{w\lambda}(T)$ *and* $\dim \tilde{V}_{w\lambda}(T) = \dim \tilde{V}_\lambda(T)$.

Formula (9), with $T' = T, g = n$, implies that $P(\rho, T)$ is stable under w; further, $\tilde{\rho}(n)$ induces an isomorphism from $\tilde{V}_\lambda(T)$ to $\tilde{V}_{w\lambda}(T)$ (formula (8)), hence the proposition.

PROPOSITION 6. *The homomorphism* $\rho : G \to \mathbf{GL}(V)$ *is injective if and only if* $P(\rho, T)$ *generates the* **Z**-*module* X(T).

The homomorphism ρ is injective if and only if its restriction to T is injective (§2, no. 6, Prop. 9). Further, since the canonical homomorphism $\mathbf{GL}(V) \to \mathbf{GL}(\tilde{V})$ is injective, we can replace ρ by $\tilde{\rho}$. It then follows from (7) that the kernel of the restriction of ρ to T is the intersection of the kernels of the elements of $P(\rho, T)$. Thus, the conclusion follows from Prop. 2 of no. 1.

The linear representation $L(\rho)$ of $\mathfrak{t}$ in $\mathfrak{gl}(\tilde{V})$ extends to a homomorphism of **C**-Lie algebras

$$\tilde{L}(\rho) : \mathfrak{t}_\mathbf{C} \to \mathfrak{gl}(\tilde{V}).$$

Moreover, recall (no. 1) that we have associated to every element λ of X(T) a linear form $\delta(\lambda)$ on $\mathfrak{t}_{\mathbf{C}}$ such that

$$(\exp_{\mathrm{T}} x)^{\lambda} = e^{\delta(\lambda)(x)}, \qquad x \in \mathfrak{t}. \tag{10}$$

Recall finally (Chap. VII, §1, no. 1) that, for any map $\mu : \mathfrak{t}_{\mathbf{C}} \to \mathbf{C}$, we denote by $\tilde{\mathrm{V}}_{\mu}(\mathfrak{t}_{\mathbf{C}})$ the vector subspace of $\tilde{\mathrm{V}}$ consisting of the v such that $(\tilde{\mathrm{L}}(\rho)(u))(v) = \mu(u).v$ for all $u \in \mathfrak{t}_{\mathbf{C}}$.

We now deduce from (7) and *loc. cit.*, Prop. 3:

PROPOSITION 7. *a) For all* $\lambda \in \mathrm{X(T)}$, *we have* $\tilde{\mathrm{V}}_{\lambda}(\mathrm{T}) = \tilde{\mathrm{V}}_{\delta(\lambda)}(\mathfrak{t}_{\mathbf{C}})$.

b) The map $\delta : \mathrm{X(T)} \to \mathrm{Hom}_{\mathbf{C}}(\mathfrak{t}_{\mathbf{C}}, \mathbf{C})$ *induces a bijection from* $\mathrm{P}(\rho, \mathrm{T})$ *to the set of weights of* $\mathfrak{t}_{\mathbf{C}}$ *on* $\tilde{\mathrm{V}}$.

Note first that, if W operates on $\mathfrak{t}_{\mathbf{C}}$ by associating to any element w of W the endomorphism $\mathrm{L}(w)_{(\mathbf{C})}$ of $\mathfrak{t}_{\mathbf{C}}$, the map δ is compatible with the operation of W on X(T) and $\mathrm{Hom}_{\mathbf{C}}(\mathfrak{t}_{\mathbf{C}}, \mathbf{C})$.

Assume now that $k = \mathbf{R}$. Denote by σ the conjugation of $\tilde{\mathrm{V}}$ relative to V, defined by $\sigma(x + iy) = x - iy$ for x, y in V; for every complex vector subspace E of $\tilde{\mathrm{V}}$, the smallest subspace of $\tilde{\mathrm{V}}$ rational over $\mathbf{R}$ and containing E is $\mathrm{E} + \sigma(\mathrm{E})$. In particular, for all $\lambda \in \mathrm{X(T)}$, there exists a real vector subspace $\mathrm{V}(\lambda)$ of V such that the subspace $\mathrm{V}(\lambda)_{(\mathbf{C})}$ of $\tilde{\mathrm{V}}$ is $\tilde{\mathrm{V}}_{\lambda}(\mathrm{T}) + \tilde{\mathrm{V}}_{-\lambda}(\mathrm{T})$ (note that $\sigma(\tilde{\mathrm{V}}_{\lambda}(\mathrm{T})) = \tilde{\mathrm{V}}_{-\lambda}(\mathrm{T})$). We have $\mathrm{V}(\lambda) = \mathrm{V}(-\lambda)$, and the $\mathrm{V}(\lambda)$ are the isotypical components of the representation of T on V induced by ρ.

4. ROOTS

The *roots* of G relative to T are the non-zero weights of the adjoint representation of G. The set of roots of G relative to T is denoted by R(G, T), or simply by R if there is no risk of confusion. By Prop. 6, the map

$$\delta : \mathrm{X(T)} \to \mathfrak{t}_{\mathbf{C}}^{*}$$

($\mathfrak{t}_{\mathbf{C}}^{*}$ denotes the dual of the complex vector space $\mathfrak{t}_{\mathbf{C}}$) maps R(G, T) bijectively onto the set $\mathrm{R}(\mathfrak{g}_{\mathbf{C}}, \mathfrak{t}_{\mathbf{C}})$ of roots of the split reductive algebra $(\mathfrak{g}_{\mathbf{C}}, \mathfrak{t}_{\mathbf{C}})$ (Chap. VIII, §2, no. 2, Remark 4). If we put

$$\mathfrak{g}^{\alpha} = (\mathfrak{g}_{\mathbf{C}})_{\alpha}(\mathrm{T}) = (\mathfrak{g}_{\mathbf{C}})_{\delta(\alpha)}(\mathfrak{t}_{\mathbf{C}}), \tag{11}$$

for all $\alpha \in \mathrm{R}$, then each $\mathfrak{g}^{\alpha}$ is of dimension 1 over $\mathbf{C}$ (*loc. cit.*, Th. 1) and

$$\mathfrak{g}_{\mathbf{C}} = \mathfrak{t}_{\mathbf{C}} \oplus \bigoplus_{\alpha \in \mathrm{R}} \mathfrak{g}^{\alpha}. \tag{12}$$

For each $\alpha \in \mathrm{R}$, denote by $\mathrm{V}(\alpha)$ the 2-dimensional subspace of $\mathfrak{g}$ such that $\mathrm{V}(\alpha)_{(\mathbf{C})} = \mathfrak{g}^{\alpha} + \mathfrak{g}^{-\alpha}$; the non-zero isotypical components of $\mathfrak{g}$ for the adjoint representation of T are $\mathfrak{t}$ and the $\mathrm{V}(\alpha)$. Further, let K be the quadratic form associated to the Killing form of $\mathfrak{g}$; it is negative (§1, no. 3, Prop. 1) and its restriction $K(\alpha)$ to $V(\alpha)$ is negative and separating. For each element t of T, $\mathrm{Ad}\,t$ leaves $K(\alpha)$ stable, and hence gives a morphism of Lie groups

$$\iota_\alpha : \mathrm{T} \to \mathbf{SO}(\mathrm{K}(\alpha)).$$

There exists a *unique isomorphism* $\rho_\alpha : \mathbf{U} \to \mathbf{SO}(\mathrm{K}(\alpha))$ such that $\iota_\alpha = \rho_\alpha \circ \alpha$. Indeed, let X be a non-zero element of $\mathfrak{g}^\alpha$, and let Y be the image of X under the conjugation of $\mathfrak{g}_\mathbf{C}$ relative to $\mathfrak{g}$; then $Y \in \mathfrak{g}^{-\alpha}$, and we obtain a basis (U, V) of $V(\alpha)$ by putting $U = X + Y, V = i(X - Y)$; the matrix of the endomorphism of $V(\alpha)$ induced by $\mathrm{Ad}\,t$, $t \in \mathrm{T}$, with respect to the basis (U, V) is

$$\begin{pmatrix} \mathscr{R}(t^\alpha) & -\mathscr{I}(t^\alpha) \\ \mathscr{I}(t^\alpha) & \mathscr{R}(t^\alpha) \end{pmatrix},$$

hence the assertion.

PROPOSITION 8. *Let* $\mathrm{Q(R)}$ *be the subgroup of* $\mathrm{X(T)}$ *generated by the roots of* G.

a) The centre $\mathrm{C(G)}$ *of* G *is a closed subgroup of* T, *equal to the intersection of the kernels of the roots. The canonical map* $\mathrm{X(T/C(G))} \to \mathrm{X(T)}$ *is injective with image* $\mathrm{Q(R)}$.

b) The compact group $\mathrm{C(G)}$ *is isomorphic to the dual of the discrete group* $\mathrm{X(T)/Q(R)}$ (*Spectral Theories*, Chap. II, §1, no. 1, Def. 2).

c) $\mathrm{C(G)}$ *reduces to the identity element if and only if* $\mathrm{Q(R)}$ *is equal to* $\mathrm{X(T)}$.

By §2, no. 2, Cor. 2 of Th. 2, C(G) is contained in T. Since this is the kernel of the adjoint representation, it is the intersection of the kernels of the roots, in other words the orthogonal complement of the subgroup Q(R) of X(T). Thus, the proposition follows from *Spectral Theories*, Chap. II, §1, no. 7, Th. 4 and no. 5, Th. 2.

PROPOSITION 9. *Every automorphism of the Lie group* G *that induces the identity on* T *is of the form* $\mathrm{Int}\,t$, *with* $t \in \mathrm{T}$.

Assume first of all that C(G) reduces to the identity element, in other words that $\mathrm{X(T)} = \mathrm{Q(R)}$ (Prop. 8). Let f be an automorphism of G inducing the identity on T, and $\varphi = \mathrm{L}(f)_{(\mathbf{C})}$; then φ is an automorphism of $\mathfrak{g}_\mathbf{C}$ inducing the identity on $\mathfrak{t}_\mathbf{C}$. By Chap. VIII, §5, no. 2, Prop. 2, there exists a unique homomorphism $\theta : \mathrm{Q(R)} \to \mathbf{C}^*$ such that φ induces on each $\mathfrak{g}^\alpha$ the homothety with ratio $\theta(\alpha)$. Since φ leaves stable the real form $\mathfrak{g}$ of $\mathfrak{g}_\mathbf{C}$, it commutes with the conjugation σ of $\mathfrak{g}_\mathbf{C}$ with respect to $\mathfrak{g}$; but $\sigma(\mathfrak{g}^\alpha) = \mathfrak{g}^{-\alpha}$, so $\theta(-\alpha) = \overline{\theta(\alpha)}$ for all $\alpha \in \mathrm{R}$. This implies that $\theta(\alpha)\overline{\theta(\alpha)} = \theta(\alpha)\theta(-\alpha) = 1$. It follows that θ takes values in $\mathbf{U}$, and hence corresponds by duality to an element t of T such that $(\mathrm{Ad}\,t)_{(\mathbf{C})} = \varphi$, so $\mathrm{Int}\,t = f$.

In the general case, the preceding applies to the group G/C(G), whose centre reduces to the identity element, and to its maximal torus T/C(G).

It follows that, if f is an automorphism of G inducing the identity on T, there exists an element t of T such that f and $\operatorname{Int} t$ induce by passage to the quotient the same automorphism of G/C(G). But, since the canonical morphism $D(G) \to G/C(G)$ is a finite covering (§1, no. 4, Cor. 1 of Prop. 4), f and $\operatorname{Int} t$ induce the same automorphism of D(G), hence of $D(G) \times C(G)$, and hence also of G (*loc. cit.*).

COROLLARY. *Let u be an automorphism of* G *and* H *the closed subgroup of* G *consisting of the fixed points of u. Then, the automorphism u is inner if and only if* H_0 *is of maximal rank.*

If u is equal to $\operatorname{Int} g$, with $g \in G$, the subgroup $H_0 = Z(g)_0$ is of maximal rank (§2, no. 2, Cor. 3). Conversely, if H contains a maximal torus S, the automorphism u is of the form $\operatorname{Int} s$ with $s \in S$ (Prop. 9).

5. NODAL VECTORS AND INVERSE ROOTS

Lemma 2. Let S *be a closed subgroup of* T *and* Z(S) *its normalizer in* G.

(i) $R(Z(S)_0, T)$ *is the set of* $\alpha \in R(G, T)$ *such that* $\alpha(S) = \{1\}$.

(ii) *The centre of* $Z(S)_0$ *is the intersection of the* $\operatorname{Ker} \alpha$ *for* $\alpha \in R(Z(S)_0, T)$.

(iii) *If* S *is connected,* Z(S) *is connected.*

The Lie algebra $L(Z(S))_{(\mathbf{C})}$ consists of the invariants of S on $\mathfrak{g}_{\mathbf{C}}$ (Chap. III, §9, no. 3, Prop. 8), and hence is the direct sum of $\mathfrak{t}_{\mathbf{C}}$ and the $\mathfrak{g}^{\alpha}$ for which $\alpha(S) = \{1\}$, hence (i). Assertion (ii) follows from Prop. 8 (no. 4), and assertion (iii) has already been proved (§2, no. 2, Cor. 5 of Th. 2).

THEOREM 1. *Let* $\alpha \in R(G, T)$. *The centralizer* Z_α *of the kernel of* α *is a* connected *closed subgroup of* G; *its centre is* $\operatorname{Ker} \alpha$; *its derived group* $D(Z_\alpha) = S_\alpha$ *is a connected closed semi-simple subgroup of* G *of rank* 1. *We have* $R(Z_\alpha, T) = \{\alpha, -\alpha\}$ *and* $\dim Z_\alpha = \dim T + 2$.

Let Z'_α be the centralizer of $(\operatorname{Ker} \alpha)_0$. By Lemma 2, this is a connected closed subgroup of G, and $R(Z'_\alpha, T)$ is the set of $\beta \in R(G, T)$ such that $\beta((\operatorname{Ker} \alpha)_0) = \{1\}$. Clearly, $\{\alpha, -\alpha\} \subset R(Z'_\alpha, T)$. Conversely, let $\beta \in R(Z'_\alpha, T)$; since $(\operatorname{Ker} \alpha)_0$ is of finite index in $\operatorname{Ker} \alpha$, there exists an integer $r \neq 0$ such that $t^{r\beta} = 1$ for $t \in \operatorname{Ker} \alpha$. From the exactness of the sequence

$$0 \longrightarrow \mathbf{Z} \longrightarrow X(T) \longrightarrow X(\operatorname{Ker} \alpha) \longrightarrow 0$$

corresponding by duality to the exact sequence

$$0 \longrightarrow \operatorname{Ker} \alpha \longrightarrow T \xrightarrow{\alpha} \mathbf{U} \longrightarrow 0,$$

it follows that $r\beta$ is a multiple of α; by Chap. VIII, §2, no. 2, Th. 2 (i), this implies that $\beta \in \{\alpha, -\alpha\}$. Thus, $R(Z'_\alpha, T) = \{\alpha, -\alpha\}$. It follows (Lemma 2) that the centre of Z'_α is $\operatorname{Ker} \alpha$, so $Z'_\alpha = Z_\alpha$. Finally, by Cor. 1 of Prop. 4 (§1, no. 4), $\mathrm{D}(\mathrm{Z}_\alpha)$ is a connected closed semi-simple subgroup of G; it is of rank 1 because $\mathscr{D}\mathrm{L}(\mathrm{Z}_\alpha)_{(\mathbf{C})} = \mathfrak{g}^\alpha + \mathfrak{g}^{-\alpha} + [\mathfrak{g}^\alpha, \mathfrak{g}^{-\alpha}]$.

COROLLARY. *There exists a morphism of Lie groups $\nu : \mathbf{SU}(2, \mathbf{C}) \to \mathrm{G}$ with the following properties:*

a) The image of ν commutes with the kernel of α.

b) For all $a \in \mathbf{U}$, we have $\nu\begin{pmatrix} a & 0 \\ 0 & \bar{a} \end{pmatrix} \in \mathrm{T}$ and $\alpha \circ \nu\begin{pmatrix} a & 0 \\ 0 & \bar{a} \end{pmatrix} = a^2$.

If ν_1 and ν_2 are two morphisms from $\mathbf{SU}(2, \mathbf{C})$ to G with the preceding properties, there exists $a \in \mathbf{U}$ such that $\nu_2 = \nu_1 \circ \mathrm{Int}\begin{pmatrix} a & 0 \\ 0 & \bar{a} \end{pmatrix}$.

By Th. 1 and Prop. 6 of §3, no. 6, there exists a morphism of Lie groups $\nu : \mathbf{SU}(2, \mathbf{C}) \to \mathrm{S}_\alpha$ that is surjective with discrete kernel. Then $\nu^{-1}(\mathrm{T} \cap \mathrm{S}_\alpha)$ is a maximal torus of $\mathbf{SU}(2, \mathbf{C})$ (§2, no. 3, Prop. 1). Since the maximal tori of $\mathbf{SU}(2, \mathbf{C})$ are conjugate (§2, no. 2, Th. 2), we can assume, replacing ν by $\nu \circ \mathrm{Int}\, s$ (with $s \in \mathbf{SU}(2, \mathbf{C})$) if necessary, that $\nu^{-1}(\mathrm{T} \cap \mathrm{S}_\alpha)$ is the group of diagonal matrices in $\mathbf{SU}(2, \mathbf{C})$. Then $\nu\begin{pmatrix} a & 0 \\ 0 & \bar{a} \end{pmatrix} \in \mathrm{T}$ for all $a \in \mathbf{U}$, and the map

$$\begin{pmatrix} a & 0 \\ 0 & \bar{a} \end{pmatrix} \mapsto \alpha \circ \nu\begin{pmatrix} a & 0 \\ 0 & \bar{a} \end{pmatrix}$$

is a root of $\mathbf{SU}(2, \mathbf{C})$, and hence is equal to one of the two maps $\begin{pmatrix} a & 0 \\ 0 & \bar{a} \end{pmatrix} \mapsto a^2$ or $\begin{pmatrix} a & 0 \\ 0 & \bar{a} \end{pmatrix} \mapsto a^{-2}$ (§3, no. 6, formulas (19)). In the first case, the homomorphism ν has the required properties; in the second case, the homomorphism $\nu \circ \mathrm{Int}\,\theta$ has them (*loc. cit.*, formulas (18)).

If ν_1 and ν_2 are two morphisms from $\mathbf{SU}(2, \mathbf{C})$ to G satisfying the stated conditions, they both map $\mathbf{SU}(2, \mathbf{C})$ into S_α (condition *a*)), hence are both universal coverings of S_α. Hence, there exists an automorphism φ of $\mathbf{SU}(2, \mathbf{C})$ such that $\nu_2 = \nu_1 \circ \varphi$, and we conclude by using Prop. 9 of no. 4.

It follows from the preceding corollary that the homomorphism ν_{T} from U to T, defined by $\nu_{\mathrm{T}}(a) = \nu\begin{pmatrix} a & 0 \\ 0 & \bar{a} \end{pmatrix}$ for $a \in \mathbf{U}$, is independent of the choice of ν. Denote by $K_\alpha \in \Gamma(\mathrm{T})$ the image under $\Gamma(\nu_{\mathrm{T}})$ of the element $2\pi i$ of $\Gamma(\mathbf{U}) = 2\pi i\mathbf{Z}$; it is called the *nodal vector associated to the root* α. We have $\langle \alpha, K_\alpha \rangle = 2$, in other words (no. 2, formula (2)) $\delta(\alpha)(K_\alpha) = 4\pi i$; since K_α belongs to the intersection of $\mathfrak{t}$ and the $\mathrm{L}(\mathrm{S}_\alpha)_{(\mathbf{C})}$, we have

$$K_\alpha = 2\pi i H_{\delta(\alpha)}, \tag{13}$$

where $H_{\delta(\alpha)}$ is the *inverse root* associated to the root $\delta(\alpha)$ of $(\mathfrak{g}_{\mathbf{C}}, \mathfrak{t}_{\mathbf{C}})$ (Chap. VIII, §2, no. 2). In other words, when $\Gamma(\mathrm{T}) \otimes \mathbf{R}$ is identified with the dual of $\mathrm{X(T)} \otimes \mathbf{R}$ *via* the pairing $\langle\ ,\ \rangle$, K_α is identified with the *inverse root* $\alpha^\vee \in (\mathrm{X(T)} \otimes \mathbf{R})^*$.

Remark. For all $x \in \mathbf{R}$, we have

$$\nu\begin{pmatrix} \exp(2\pi i x) & 0 \\ 0 & \exp(-2\pi i x) \end{pmatrix} = \nu_{\mathrm{T}}(e^{2\pi i x}) = \exp(xK_\alpha). \tag{14}$$

In particular:

$$\nu\begin{pmatrix} -1 & 0 \\ 0 & -1 \end{pmatrix} = \nu_{\mathrm{T}}(-1) = \exp\Big(\frac{1}{2}K_\alpha\Big). \tag{15}$$

It follows that ν is *injective* if and only if $K_\alpha \notin 2\Gamma(\mathrm{T})$, in other words if there exists $\lambda \in \mathrm{X(T)}$ such that $\langle \lambda, K_\alpha \rangle \notin 2\mathbf{Z}$. When $\mathfrak{g}_{\mathbf{C}}$ is simple, ν is injective unless $\mathfrak{g}_{\mathbf{C}}$ is of type B_n, $\mathrm{C(G)} = \{1\}$ and α is a short root (cf. Chap. VI, Plates).

In the remainder of this paragraph we denote by $\mathrm{R}^\vee(\mathrm{G},\mathrm{T})$ the set of nodal vectors K_α for $\alpha \in \mathrm{R}(\mathrm{G},\mathrm{T})$. This is a subset of $\Gamma(\mathrm{T})$ that the canonical injection of $\Gamma(\mathrm{T})$ into $\mathfrak{t}_{\mathbf{C}}$ identifies with the homothety with ratio $2\pi i$ of the inverse root system $\mathrm{R}^\vee(\mathfrak{g}_{\mathbf{C}}, \mathfrak{t}_{\mathbf{C}}) = \{H_{\delta(\alpha)}\}$ of $\delta(\mathrm{R})$. It follows that $\mathrm{R}^\vee(\mathrm{G},\mathrm{T})$ generates the $\mathbf{R}$-vector space $\mathrm{L}(\mathrm{T} \cap \mathrm{D(G)})$, and hence that its orthogonal complement in $\mathrm{X(T)}$ is $\mathrm{X}(\mathrm{T}/(\mathrm{T} \cap \mathrm{D(G)}))$.

Denote by $\mathrm{Aut(T)}$ the group of automorphisms of the Lie group T; the Weyl group $\mathrm{W} = \mathrm{W_G(T)}$ (§2, no. 5) can be identified with a subgroup of $\mathrm{Aut(T)}$. On the other hand, recall (Chap. VIII, §2, no. 2, Remark 4) that the Weyl group $\mathrm{W}(\mathfrak{g}_{\mathbf{C}}, \mathfrak{t}_{\mathbf{C}})$ of the split reductive algebra $(\mathfrak{g}_{\mathbf{C}}, \mathfrak{t}_{\mathbf{C}})$ operates on $\mathfrak{t}_{\mathbf{C}}$, and thus is canonically identified with a subgroup of $\mathbf{GL}(\mathfrak{t}_{\mathbf{C}})$.

PROPOSITION 10. *The map* $u \mapsto \mathrm{L}(u)_{(\mathbf{C})}$ *from* $\mathrm{Aut(T)}$ *to* $\mathbf{GL}(\mathfrak{t}_{\mathbf{C}})$ *induces an isomorphism from* W *to the Weyl group of the split reductive Lie algebra* $(\mathfrak{g}_{\mathbf{C}}, \mathfrak{t}_{\mathbf{C}})$. *For all* $\alpha \in \mathrm{R}$, $\mathrm{W}_{\mathrm{Z}_\alpha}(\mathrm{T})$ *is of order* 2, *and the image under the preceding isomorphism of the non-identity element of* $\mathrm{W}_{\mathrm{Z}_\alpha}(\mathrm{T})$ *is the reflection* $s_{H_{\delta(\alpha)}}$.

The map under consideration is injective. It remains to show that its image is equal to $\mathrm{W}(\mathfrak{g}_{\mathbf{C}}, \mathfrak{t}_{\mathbf{C}})$.

Let $g \in \mathrm{N_G(T)}$. With the notations in Chap. VIII, §5, no. 2, we have $\mathrm{Ad}\, g \in \mathrm{Aut}(\mathfrak{g}_{\mathbf{C}}, \mathfrak{t}_{\mathbf{C}}) \cap \mathrm{Int}(\mathfrak{g}_{\mathbf{C}})$, so $\mathrm{Ad}\, g \in \mathrm{Aut}_0(\mathfrak{g}_{\mathbf{C}}, \mathfrak{t}_{\mathbf{C}})$ (*loc. cit.*, no. 5, Prop. 11). By *loc. cit.*, no. 2, Prop. 4, the automorphism of $\mathfrak{t}_{\mathbf{C}}$ induced by $\mathrm{Ad}\, g$ belongs to $\mathrm{W}(\mathfrak{g}_{\mathbf{C}}, \mathfrak{t}_{\mathbf{C}})$. Thus, the image of W in $\mathbf{GL}(\mathfrak{t}_{\mathbf{C}})$ is contained in $\mathrm{W}(\mathfrak{g}_{\mathbf{C}}, \mathfrak{t}_{\mathbf{C}})$.

Let $\alpha \in \mathrm{R}(\mathrm{G},\mathrm{T})$, and let $\nu : \mathbf{SU}(2, \mathbf{C}) \to \mathrm{G}$ be a morphism of Lie groups having the properties in the Cor. of Th. 1. The image under ν of the element θ of $\mathbf{SU}(2, \mathbf{C})$ has the following properties (§3, no. 6, formulas (17)):

a) $(\mathrm{Int}\, \nu(\theta))(t) = t$ if $t \in \mathrm{Ker}\, \alpha$,
b) $(\mathrm{Int}\, \nu(\theta))(t) = t^{-1}$ if $t \in \mathrm{T} \cap \mathrm{S}_\alpha$.

It follows that $\mathrm{Ad}\, \nu(\theta)$ induces the identity on $\mathrm{Ker}\, \delta(\alpha) \subset \mathfrak{t}_{\mathbf{C}}$, and induces the map $x \mapsto -x$ on $[\mathfrak{g}^\alpha, \mathfrak{g}^{-\alpha}]$, hence coincides with the reflection $s_{H_{\delta(\alpha)}}$. Thus, the image of W contains all the $s_{H_{\delta(\alpha)}}$, and hence is equal to $W(\mathfrak{g}_C, \mathfrak{t}_C)$. In particular $W_{Z_\alpha}(T)$ is of order 2, and hence consists of the identity and $\operatorname{Int} \nu(\theta)$. This completes the proof of the proposition.

COROLLARY. *Assume that* G *is semi-simple. Then every element of* G *is the commutator of two elements of* G.

Let c be a Coxeter transformation of the Weyl group $W(\mathfrak{g}_C, \mathfrak{t}_C)$ (Chap. V, §6, no. 1), and let n be an element of $N_G(T)$ whose class in W is identified with c by the isomorphism defined in the proposition. Denote by f_c the morphism $t \mapsto (n, t)$ from T to T; for $x \in \mathfrak{t}_C$, we have $L(f_c)_{(C)}(x) = (\operatorname{Ad} n)(x) - x = c(x) - x$.

By Th. 1 of Chap. V, §6, no. 2, the endomorphism c of $\mathfrak{t}_C$ has no eigenvalue equal to 1. Consequently, $L(f_c)$ is surjective, and hence so is f_c. It follows that every element of T is the commutator of two elements of G, which implies the corollary in view of Th. 2, §2, no. 2.

6. FUNDAMENTAL GROUP

In the following proposition, $f(G, T)$ denotes the homomorphism from $\Gamma(T)$ to $\pi_1(G)$ that is the composite of the canonical isomorphism from $\Gamma(T)$ to $\pi_1(T)$ (no. 2, Remark 3) and the homomorphism $\pi_1(\iota)$, where ι is the canonical injection $T \to G$.

PROPOSITION 11. *The homomorphism* $f(G, T) : \Gamma(T) \to \pi_1(G)$ *is surjective. Its kernel is the subgroup* $N(G, T)$ *of* $\Gamma(T)$ *generated by the family of nodal vectors* $(K_\alpha)_{\alpha \in R(G,T)}$.

The homomorphism $f(G, T)$ is surjective by Prop. 3 (§2, no. 4). We denote by $A(G, T)$ the assertion: "the kernel of $f(G, T)$ is generated by the K_α" which it remains to prove, and distinguish several cases:

a) G *is simply-connected.* Let $\rho : \mathfrak{g}_C \to \mathfrak{gl}(V)$ be a linear representation of $\mathfrak{g}_C$ on a finite dimensional complex vector space V. Restricting to $\mathfrak{g}$, we obtain a representation of $\mathfrak{g}$ on the real vector space $V_{(\mathbf{R})}$; since G is simply-connected, there exists an analytic linear representation π of G on $V_{(\mathbf{R})}$ such that $\rho = L(\pi)$. It follows from Prop. 7 of no. 3 that the image $\delta(X(T))$ of $X(T)$ in $\mathfrak{t}_C^*$ contains all the weights of ρ on V. This being true for every representation ρ of $\mathfrak{g}_C$, it follows from Chap. VIII, §7, no. 2, Th. 1 that $\delta(X(T))$ contains the group of weights of $\delta(R)$, which is by definition the set of $\lambda \in \mathfrak{t}_C^*$ such that $\lambda(H_{\delta(\alpha)}) \in \mathbf{Z}$ for all $\alpha \in R$, in other words, $\lambda(K_\alpha) \in 2\pi i\mathbf{Z}$ for all $\alpha \in R$. Thus, the group $X(T)$ contains all the elements λ of $X(T) \otimes \mathbf{Q}$ such that $\langle \lambda, K_\alpha \rangle \in \mathbf{Z}$ for all $\alpha \in R$, which implies by duality that $\Gamma(T)$ is generated by the K_α, hence the assertion $A(G, T)$.

b) G *is the direct product of a simply-connected group* G′ *and a torus* S. Then T is the direct product of a maximal torus T′ of G′ with S, $\Gamma(T)$ can be identified with $\Gamma(T') \times \Gamma(S)$, $\pi_1(G)$ with $\pi_1(G') \times \pi_1(S)$, and $f(G, T)$ with the homomorphisms with components $f(G', T')$ and $f(S, S)$. Since $f(S, S)$ is bijective, the canonical map $\Gamma(T') \to \Gamma(T)$ maps $\operatorname{Ker} f(G', T')$ bijectively onto $\operatorname{Ker} f(G, T)$. Moreover, the K_α belong to the Lie algebra of the derived group G′ of G, hence to the image of $\Gamma(T')$, so it is immediate that $A(G', T')$ implies $A(G, T)$, hence assertion $A(G, T)$, in view of *a*).

c) *General case.* There exists a surjective morphism $p : G' \to G$ with finite kernel, where G′ is the direct product of a simply-connected group by a torus (§1, no. 4, Prop. 4). If T′ is the inverse image of T in G′ (this is a maximal torus of G′ by §2, no. 3, Prop. 1), and N the kernel of p, we have exact sequences $0 \to N \to G' \to G \to 0$ and $0 \to N \to T' \to T \to 0$, hence a commutative diagram with exact rows (no. 2, Remark 1 and *General Topology*, Chap. XI, in preparation)

$$\begin{array}{ccccccccc} 0 & \longrightarrow & \Gamma(T') & \longrightarrow & \Gamma(T) & \longrightarrow & N & \longrightarrow & 0 \\ & & \big\downarrow{\scriptstyle f(G',T')} & & \big\downarrow{\scriptstyle f(G,T)} & & \big\downarrow{\scriptstyle \mathrm{Id}_N} & & \\ 0 & \longrightarrow & \pi_1(G') & \longrightarrow & \pi_1(G) & \longrightarrow & N & \longrightarrow & 0. \end{array}$$

It follows immediately from the snake diagram (*Algebra*, Chap. X, p. 4, Prop. 2) that $A(G', T')$ implies $A(G, T)$, hence the proposition, in view of *b*).

COROLLARY 1. G *is simply-connected if and only if the family* $(K_\alpha)_{\alpha \in R(G,T)}$ *generates* $\Gamma(T)$.

COROLLARY 2. *Let* H *be a connected closed subgroup of* G *containing* T; *there is an exact sequence*

$$0 \longrightarrow N(H, T) \longrightarrow N(G, T) \longrightarrow \pi_1(H) \longrightarrow \pi_1(G) \longrightarrow 0.$$

This follows from *Algebra*, Chap. X, p. 4, Prop. 2 (snake diagram), applied to the commutative diagram

$$\begin{array}{ccccccccc} 0 & \longrightarrow & N(H,T) & \longrightarrow & \Gamma(T) & \longrightarrow & \pi_1(H) & \longrightarrow & 0 \\ & & \big\downarrow & & \big\downarrow & & \big\downarrow & & \\ 0 & \longrightarrow & N(G,T) & \longrightarrow & \Gamma(T) & \longrightarrow & \pi_1(G) & \longrightarrow & 0. \end{array}$$

Remark. It can be shown (cf. Exercise 2 of §5) that $\pi_2(G/H)$ is zero. The exactness of the preceding sequence then gives an isomorphism from $\pi_2(G/H)$ to $N(G, T)/N(H, T)$.

COROLLARY 3. *The homomorphism* $\pi_1(D(G)) \to \pi_1(G)$ *corresponding to the inclusion of* D(G) *into* G *induces an isomorphism from* $\pi_1(D(G))$ *to the torsion subgroup of* $\pi_1(G)$.

Indeed, $T \cap D(G)$ is a maximal torus of D(G) (§2, no. 3, Prop. 1 *c*)); from the exact sequence

$$0 \longrightarrow \Gamma(\mathrm{T} \cap \mathrm{D(G)}) \longrightarrow \Gamma(\mathrm{T}) \longrightarrow \Gamma(\mathrm{T}/(\mathrm{T} \cap \mathrm{D(G)})) \longrightarrow 0$$

and Proposition 11, we obtain an exact sequence

$$0 \longrightarrow \pi_1(\mathrm{D(G)}) \longrightarrow \pi_1(\mathrm{G}) \longrightarrow \Gamma(\mathrm{T}/(\mathrm{T} \cap \mathrm{D(G)})) \longrightarrow 0,$$

hence the corollary, since $\pi_1(\mathrm{D(G)})$ is finite and $\Gamma(\mathrm{T}/(\mathrm{T} \cap \mathrm{D(G)}))$ is free.

7. SUBGROUPS OF MAXIMAL RANK

Recall (Chap. VI, §1, no. 7) that a subset P of $\mathrm{R} = \mathrm{R(G,T)}$ is said to be closed if $(\mathrm{P}+\mathrm{P}) \cap \mathrm{R} \subset \mathrm{P}$, and symmetric if $\mathrm{P} = -\mathrm{P}$.

PROPOSITION 12. *Let $\mathscr{H}$ be the set of connected closed subgroups of* G *containing* T, *ordered by inclusion. The map* $\mathrm{H} \mapsto \mathrm{R(H,T)}$ *is an increasing bijection from $\mathscr{H}$ to the set of symmetric closed subsets of* R(G, T), *ordered by inclusion.*

If $H \in \mathscr{H}$, then $\mathrm{L(H)}_{(\mathbf{C})}$ is the direct sum of $\mathfrak{t}_{\mathbf{C}}$ and the $\mathfrak{g}^\alpha$ for $\alpha \in \mathrm{R(H,T)}$; since this is a reductive subalgebra in $\mathfrak{g}_{\mathbf{C}}$, the subset R(H, T) of R satisfies the stated conditions (Chap. VIII, §3, no. 1, Lemma 2 and Prop. 2). Conversely, if P is a subset of R satisfying these conditions, then $\mathfrak{t}_{\mathbf{C}} \oplus \sum_{\alpha \in \mathrm{P}} \mathfrak{g}^\alpha$ is a subalgebra of $\mathfrak{g}_{\mathbf{C}}$ (*loc. cit.*) which is rational over **R** (no. 3), and hence of the form $\mathfrak{h}_{(\mathbf{C})}$, where $\mathfrak{h}$ is a subalgebra of $\mathfrak{g}$. Let H(P) be the integral subgroup of G defined by $\mathfrak{h}$; it is closed (§2, no. 4, Remark 1). We verify immediately that the maps $\mathrm{H} \mapsto \mathrm{R(H,T)}$ and $\mathrm{P} \mapsto \mathrm{H(P)}$ are increasing and inverses of each other.

COROLLARY 1. *There are only finitely-many closed subgroups of* G *containing* T.

Let H be such a subgroup; then $\mathrm{H}_0 \in \mathscr{H}$, and $\mathscr{H}$ is finite. Moreover, H is a subgroup of $\mathrm{N_G(H_0)}$ containing H_0, and $\mathrm{N_G(H_0)/H_0}$ is finite (§2, no. 4, Prop. 4 and Remark 2).

COROLLARY 2. *Let* H *be a connected closed subgroup of* G *containing* T, *and let* $\mathrm{W_G^H(T)}$ *be the stabilizer in* $\mathrm{W_G(T)}$ *of the subset* R(H, T) *of* R. *The group* $\mathrm{N_G(H)/H}$ *is isomorphic to the quotient group* $\mathrm{W_G^H(T)/W_H(T)}$.

Indeed, it follows from Prop. 7 of §2, no. 5, applied to $\mathrm{N_G(H)}$, that $\mathrm{N_G(H)/H}$ is isomorphic to $\mathrm{W_{N(H)}(T)/W_H(T)}$, where $\mathrm{W_{N(H)}(T)}$ is the set of elements of $\mathrm{W_G(T)}$ whose representatives in $\mathrm{N_G(T)}$ normalize H. Let $n \in \mathrm{N_G(T)}$, and let w be its class in $\mathrm{W_G(T)}$. By Chap. III, §9, no. 4, Prop. 11, n normalizes H if and only if $(\mathrm{Ad}\, n)(\mathrm{L(H)}) = \mathrm{L(H)}$; in view of Prop. 5 of no. 3, this also means that the subset R(H, T) of R is stable under w, hence the corollary.

Remark 1. The group $\mathrm{W_G^H(T)}$ is also the stabilizer in $\mathrm{W_G(T)}$ of the subgroup C(H) of T: this follows from Prop. 8 of no. 4.

PROPOSITION 13. *Let* H *be a connected closed subgroup of* G *of maximal rank, and* C *its centre. Then* C *contains the centre of* G, *and* H *is the identity component of the centralizer of* C.

Let S be a maximal torus of H. Since the centre of G is contained in S, it is contained in C. Put $L = Z(C)_0$; this is a connected closed subgroup of G containing H, hence is of maximal rank, and its centre is equal to C. Denote by R_H and R_L the root systems of H and L, respectively, relative to S; then $R_H \subset R_L \subset R(G, S)$. Since $C(H) = C(L)$, Prop. 8 (no. 4) implies the equality $Q(R_H) = Q(R_L)$; but $Q(R_H) \cap R_L = R_H$ (Chap. VI, §1, no. 7, Prop. 23), so $R_H = R_L$ and $H = L$ (Prop. 12).

Remark 2. Say that a subgroup C of G is *radical* if there exists a maximal torus S of G and a subset P of R(G, S) such that $C = \bigcap_{\alpha \in P} \operatorname{Ker} \alpha$. It follows from Prop. 13 and Lemma 2 of no. 5 that *the map* $H \mapsto C(H)$ *induces a bijection from the set of connected closed subgroups of maximal rank to the set of radical subgroups of* G. The inverse bijection is the map $C \mapsto Z(C)_0$.

COROLLARY. *The set of* $g \in G$ *such that* $T \cap gTg^{-1} \neq C(G)$ *is the union of a finite number of closed analytic submanifolds of* G *distinct from* G.

Indeed, put $A_g = T \cap gTg^{-1}$; we have $T \subset Z(A_g)$ and $gTg^{-1} \subset Z(A_g)$. Hence, there exists $x \in Z(A_g)$ such that $xTx^{-1} = gTg^{-1}$ (§2, no. 2, Th. 2), which implies that $g \in Z(A_g).N_G(T)$. Denote by $\mathscr{A}$ the finite (Cor. 1) set of closed subgroups of G containing T and distinct from G, and put $X = \bigcup_{H \in \mathscr{A}} H.N_G(T)$; this is a finite union of closed submanifolds of G, distinct from G. If $A_g \neq C(G)$, then $Z(A_g) \in \mathscr{A}$, and g belongs to X. Conversely, if $g \in H.N_G(T)$, with $H \in \mathscr{A}$, then A_g contains C(H), so $A_g \neq C(G)$ (Prop. 13).

PROPOSITION 14. *Let* X *be a subset of* T, *and let* R_X *be the set of roots* $\alpha \in R(G, T)$ *such that* $\alpha(X) = \{1\}$. *The group* $Z_G(X)/Z_G(X)_0$ *is isomorphic to the quotient of the subgroup of* $W_G(T)$ *fixing* X *by the subgroup generated by the reflections* s_α *for* $\alpha \in R_X$.

Put $H = Z_G(X)$; since $L(H)_{(\mathbf{C})}$ is the set of points of $\mathfrak{g}_{\mathbf{C}}$ fixed by Ad(X), it is the sum of $\mathfrak{t}_{\mathbf{C}}$ and the $\mathfrak{g}^\alpha$ for which $\alpha(X) = \{1\}$. Consequently, $R(H_0, T) = R_X$, so $W_{H_0}(T)$ is generated by the reflections s_α for $\alpha \in R_X$. It now suffices to apply Prop. 7 of §2, no. 5.

We shall see below (§5, no. 3, Th. 1) that if G is simply-connected and X reduces to a point, the centralizer Z(X) is connected.

8. ROOT DIAGRAMS

DEFINITION 2. *A root diagram* (or simply a diagram, if there is no risk of confusion) *is a triple* $D = (M, M_0, R)$ *where:*

(RD_0) M *is a free* **Z**-*module of finite type and the submodule* M_0 *is a direct factor of* M;

(RD_{I}) R *is a finite subset of* M; $R \cup M_0$ *generates the* **Q**-*vector space* $\mathbf{Q} \otimes M$;

(RD_{II}) *for all* $\alpha \in R$, *there exists an element* $\alpha^\vee$ *of* $M^* = \mathrm{Hom}_{\mathbf{Z}}(M, \mathbf{Z})$ *such that* $\alpha^\vee(M_0) = 0$, $\alpha^\vee(\alpha) = 2$ *and the endomorphism* $x \mapsto x - \alpha^\vee(x)\alpha$ *of* M *leaves* R *stable.*

By Chap. VI, §1, no. 1, for all $\alpha \in R$ the element $\alpha^\vee$ of M^* is uniquely determined by α; we denote by s_α the endomorphism $x \mapsto x - \alpha^\vee(x)\alpha$ of M. Moreover (*loc. cit.*), the **Q**-vector space $\mathbf{Q} \otimes M$ is the direct sum of $\mathbf{Q} \otimes M_0$ and the vector subspace V(R) generated by R, and R is a root system in V(R) (*loc. cit.*, Def. 1).

The elements of R are called the *roots* of the root diagram D, and the elements $\alpha^\vee$ of M^* the *inverse roots.* The group generated by the automorphisms s_α of M is called the *Weyl group* of D and is denoted by W(D); the elements of W(D) induce the identity on M_0, and induce on V(R) the transformations of the Weyl group of the root system R.

Examples. 1) For every free **Z**-module of finite type M, the triple $(M, M, \varnothing)$ is a root diagram.

2) If $D = (M, M_0, R)$ is a root diagram, let M_0^* be the orthogonal complement of V(R) in M^*, and let $R^\vee$ be the set of inverse roots of D. Then $D^\vee = (M^*, M_0^*, R^\vee)$ is a root diagram, called the *inverse* of D. For all $\alpha \in R$, the symmetry $s_{\alpha^\vee}$ of M^* is the contragredient automorphism of the symmetry s_α of M; the map $w \mapsto {}^t w^{-1}$ is an isomorphism from W(D) to $W(D^\vee)$. Moreover, $V(R^\vee)$ can be naturally identified with the dual of the **Q**-vector space V(R), $R^\vee$ then being identified with the inverse root system of R.

If the dual of M^* is identified with M, the inverse diagram of $D^\vee$ is identified with D.

3) Let $(\mathfrak{g}, \mathfrak{h})$ be a split reductive **Q**-Lie algebra, and $M \subset \mathfrak{h}$ a *permissible lattice* (Chap. VIII, §12, no. 6, Def. 1). Let M_0 be the subgroup of M orthogonal to the roots of $(\mathfrak{g}, \mathfrak{h})$ and $R^\vee$ the set of the H_α, $\alpha \in R(\mathfrak{g}, \mathfrak{h})$. Then $(M, M_0, R^\vee)$ is a root diagram, and $(M^*, M_0^*, R(\mathfrak{g}, \mathfrak{h}))$ is the inverse diagram.

4) Let V be a vector space over **Q** and R a root system in V; denote by P(R) the group of weights of R and by Q(R) the group of radical weights of R (Chap. VI, §1, no. 9). Then (Q(R), 0, R) and (P(R), 0, R) are root diagrams. A root diagram (M, M_0, S) is isomorphic to a diagram of the form (Q(R), 0, R) (resp. (P(R), 0, R)) if and only if M is generated by S (resp. M^* is generated by $S^\vee$).

For every subgroup X of P(R) containing Q(R), (X, 0, R) is a root diagram and, up to isomorphism, every diagram (M, M_0, S) such that $M_0 = 0$, in other words such that S generates a subgroup of M of finite index, arises in this way.

The root diagram (M, M_0, R) is said to be *reduced* if the root system R is reduced (in other words (Chap. VI, §1, no. 4) if the relations $\alpha, \beta \in R$, $\lambda \in \mathbf{Z}$,

$\beta = \lambda\alpha$ imply that $\lambda = 1$ or $\lambda = -1$). The diagrams in Examples 1) and 3) are reduced.

9. COMPACT LIE GROUPS AND ROOT SYSTEMS

With the terminology introduced in the preceding number, an important part of the results of numbers 4 and 5 can be summarized in the following theorem:

THEOREM 2. *a)* $(\mathrm{X(T)}, \mathrm{X(T/(T \cap D(G))}, \mathrm{R(G,T)}))$ *is a reduced root diagram; its Weyl group consists of the* $\mathrm{X}(w)$ *for* $w \in \mathrm{W}$; *the group* $\mathrm{X(C(G))}$ *is isomorphic to the quotient of* $\mathrm{X(T)}$ *by the subgroup generated by* $\mathrm{R(G,T)}$.

b) $(\Gamma(\mathrm{T}), \Gamma(\mathrm{C(G)}_0), \mathrm{R}^\vee(\mathrm{G,T}))$ *is a reduced root diagram; its Weyl group consists of the* $\Gamma(w)$, *for* $w \in \mathrm{W}$; *the group* $\pi_1(\mathrm{G})$ *is isomorphic to the quotient of* $\Gamma(\mathrm{T})$ *by the subgroup generated by* $\mathrm{R}^\vee(\mathrm{G,T})$.

c) If each of the **Z**-*modules* $\mathrm{X(T)}$ *and* $\Gamma(\mathrm{T})$ *is identified with the dual of the other* (no. 2, Prop. 3), *each of the preceding root diagrams is identified with the inverse of the other.*

Denote by $\mathrm{D^*(G,T)}$ the diagram $(\mathrm{X(T)}, \mathrm{X(T/(T \cap D(G))}, \mathrm{R(G,T)}))$ and by $\mathrm{D_*(G,T)}$ the diagram $(\Gamma(\mathrm{T}), \Gamma(\mathrm{C(G)}_0), \mathrm{R}^\vee(\mathrm{G,T}))$; these are called the *contravariant diagram* and the *covariant diagram* of G (relative to T), respectively.

Examples. 1) If G is semi-simple of rank 1, then $\mathrm{D^*(G,T)}$ and $\mathrm{D_*(G,T)}$ are necessarily isomorphic to one of the two diagrams $\Delta_2 = (\mathbf{Z}, 0, \{2, -2\})$, $\Delta_1 = (\mathbf{Z}, 0, \{1, -1\})$. If G is isomorphic to $\mathbf{SU}(2, \mathbf{C})$, $\mathrm{D_*(G,T)}$ is isomorphic to Δ_1 (since G is simply-connected) so $\mathrm{D^*(G,T)}$ is isomorphic to Δ_2. If G is isomorphic to $\mathbf{SO}(3, \mathbf{R})$, $\mathrm{D^*(G,T)}$ is isomorphic to Δ_1 (since $\mathrm{C(G)} = \{1\}$), so $\mathrm{D_*(G,T)}$ is isomorphic to Δ_2.

2) If G and G′ are two connected compact Lie groups, with maximal tori T and T′, respectively, and if $\mathrm{D^*(G,T)} = (\mathrm{M}, \mathrm{M}_0, \mathrm{R})$ and $\mathrm{D^*(G',T')} = (\mathrm{M}', \mathrm{M}'_0, \mathrm{R}')$, then $\mathrm{D^*(G \times G', T \times T')}$ can be identified with $(\mathrm{M} \oplus \mathrm{M}', \mathrm{M}_0 \oplus \mathrm{M}'_0, \mathrm{R} \cup \mathrm{R}')$. Similarly for the covariant diagrams.

3) Let N be a closed subgroup of T, central in G, and let $(\mathrm{M}, \mathrm{M}_0, \mathrm{R})$ be the contravariant diagram of G relative to T. Then the contravariant diagram of G/N relative to T/N can be identified with $(\mathrm{M}', \mathrm{M}'_0, \mathrm{R})$, where M′ is the subgroup of M consisting of the λ such that $\lambda(\mathrm{N}) = \{1\}$ and $\mathrm{M}'_0 = \mathrm{M}' \cap \mathrm{M}_0$.

4) Similarly, let N be a finite abelian group, and $\varphi : \pi_1(\mathrm{G}) \to \mathrm{N}$ a surjective homomorphism. Let G′ be the covering of G associated to this homomorphism; this is a connected compact Lie group, of which N is a central subgroup (*General Topology*, Chap. XI, in preparation), and G can be naturally identified with G′/N. Let T′ be the maximal torus of G′ that is the inverse image of T. If $(\mathrm{P}, \mathrm{P}_0, \mathrm{S})$ is the covariant diagram of G relative to T, the covariant diagram of G′ relative to T′ can be identified with $(\mathrm{P}', \mathrm{P}'_0, \mathrm{S})$, where P′ is the kernel of the composite homomorphism $\varphi \circ f(\mathrm{G,T}) : \mathrm{P} \to \mathrm{N}$ (cf. no. 6, Prop. 11), and $\mathrm{P}'_0 = \mathrm{P}_0 \cap \mathrm{P}'$.

Remarks. 1) Let $\mathfrak{c}$ be the centre of $\mathfrak{g}_{\mathbf{C}}$; then $\mathfrak{c} = \mathrm{L}(\mathrm{C}(\mathrm{G}))_{(\mathbf{C})}$. We have the following relations between the diagrams of G relative to T and the direct and inverse root systems of the split reductive algebra $(\mathfrak{g}_{\mathbf{C}}, \mathfrak{t}_{\mathbf{C}})$:

a) The canonical isomorphism from $\mathbf{C} \otimes \Gamma(\mathrm{T})$ to $\mathfrak{t}_{\mathbf{C}}$ induces a bijection from $\mathbf{C} \otimes \Gamma(\mathrm{C}(\mathrm{G})_0)$ to $\mathfrak{c}$ and a bijection from $1 \otimes \mathrm{R}^\vee(\mathrm{G}, \mathrm{T})$ to $2\pi i.\mathrm{R}^\vee(\mathfrak{g}_{\mathbf{C}}, \mathfrak{t}_{\mathbf{C}})$.

b) The canonical isomorphism from $\mathbf{C} \otimes \mathrm{X}(\mathrm{T})$ to the dual $\mathfrak{t}^*_{\mathbf{C}}$ of $\mathfrak{t}_{\mathbf{C}}$ induces a bijection from $\mathbf{C} \otimes \mathrm{X}(\mathrm{T}/(\mathrm{T} \cap \mathrm{D}(\mathrm{G})))$ to the orthogonal complement of $\mathfrak{t}_{\mathbf{C}} \cap \mathscr{D}(\mathfrak{g})_{\mathbf{C}}$, and a bijection from $1 \otimes \mathrm{R}(\mathrm{G}, \mathrm{T})$ to $\mathrm{R}(\mathfrak{g}_{\mathbf{C}}, \mathfrak{t}_{\mathbf{C}})$.

2) Assume that the group G is semi-simple; denote by R (resp. $\mathrm{R}^\vee$) the root system $\mathrm{R}(\mathrm{G}, \mathrm{T})$ (resp. $\mathrm{R}^\vee(\mathrm{G}, \mathrm{T})$), so that we have inclusions

$$\mathrm{Q}(\mathrm{R}) \subset \mathrm{X}(\mathrm{T}) \subset \mathrm{P}(\mathrm{R}) \qquad \mathrm{Q}(\mathrm{R}^\vee) \subset \Gamma(\mathrm{T}) \subset \mathrm{P}(\mathrm{R}^\vee).$$

The finite abelian groups $\mathrm{P}(\mathrm{R})/\mathrm{Q}(\mathrm{R})$ and $\mathrm{P}(\mathrm{R}^\vee)/\mathrm{Q}(\mathrm{R}^\vee)$ are in duality (Chap. VI, §1, no. 9); if $\mathrm{M}^\wedge$ denotes the dual group of the finite abelian group M, we deduce from the preceding *canonical isomorphisms*

$$\Gamma(\mathrm{T})/\mathrm{Q}(\mathrm{R}^\vee) \to \pi_1(\mathrm{G}) \qquad \mathrm{P}(\mathrm{R}^\vee)/\Gamma(\mathrm{T}) \to \mathrm{C}(\mathrm{G})$$
$$\mathrm{P}(\mathrm{R})/\mathrm{X}(\mathrm{T}) \to (\pi_1(\mathrm{G}))^\wedge \qquad \mathrm{X}(\mathrm{T})/\mathrm{Q}(\mathrm{R}) \to (\mathrm{C}(\mathrm{G}))^\wedge.$$

In particular, *the product of the orders of* $\pi_1(\mathrm{G})$ *and* $\mathrm{C}(\mathrm{G})$ *is equal to the connection index f of* $\mathrm{R}(\mathrm{G}, \mathrm{T})$ (*loc. cit.*).

Now let G' be another connected compact Lie group, T' a maximal torus of G'. Let $f : \mathrm{G} \to \mathrm{G}'$ be an isomorphism of Lie groups such that $f(\mathrm{T}) = \mathrm{T}'$; denote by f_T the isomorphism from T to T' that it defines. Then $\mathrm{X}(f_\mathrm{T})$ is an isomorphism from $\mathrm{D}^*(\mathrm{G}', \mathrm{T}')$ to $\mathrm{D}^*(\mathrm{G}, \mathrm{T})$, denoted by $\mathrm{D}^*(f)$, and $\Gamma(f_\mathrm{T})$ is an isomorphism from $\mathrm{D}_*(\mathrm{G}, \mathrm{T})$ to $\mathrm{D}_*(\mathrm{G}', \mathrm{T}')$, denoted by $\mathrm{D}_*(f)$. If $t \in \mathrm{T}$, and if we put $g = f \circ \mathrm{Int}\, t = (\mathrm{Int}\, f(t)) \circ f$, then $\mathrm{D}^*(g) = \mathrm{D}^*(f)$, $\mathrm{D}_*(g) = \mathrm{D}_*(f)$.

PROPOSITION 15. *Let* φ *be an isomorphism from* $\mathrm{D}^*(\mathrm{G}', \mathrm{T}')$ *to* $\mathrm{D}^*(\mathrm{G}, \mathrm{T})$ (resp. *from* $\mathrm{D}_*(\mathrm{G}, \mathrm{T})$ *to* $\mathrm{D}_*(\mathrm{G}', \mathrm{T}')$). *There exists an isomorphism* $f : \mathrm{G} \to \mathrm{G}'$ *such that* $f(\mathrm{T}) = \mathrm{T}'$ *and* $\varphi = \mathrm{D}^*(f)$ (resp. $\varphi = \mathrm{D}_*(f)$); *if* f_1 *and* f_2 *are two such isomorphisms, there exists an element t of* T *such that* $f_2 = f_1 \circ \mathrm{Int}\, t$.

The second assertion follows immediately from Prop. 9 (no. 4); we prove the first for the covariant diagrams, for example. Denote by $\mathfrak{g}'$ (resp. $\mathfrak{t}'$) the Lie algebra of G' (resp. T'), and by $\mathfrak{g}'_{\mathbf{C}}$ (resp. $\mathfrak{t}'_{\mathbf{C}}$) its complexified Lie algebra. By Chap. VIII, §4, no. 4, Th. 2 (i), there exists an isomorphism $\psi : \mathfrak{g}_{\mathbf{C}} \to \mathfrak{g}'_{\mathbf{C}}$ that maps $\mathfrak{t}_{\mathbf{C}}$ to $\mathfrak{t}'_{\mathbf{C}}$ and induces on $\Gamma(\mathrm{T}) \subset \mathfrak{t}_{\mathbf{C}}$ the given isomorphism $\varphi : \Gamma(\mathrm{T}) \to \Gamma(\mathrm{T}')$. Then $\mathfrak{g}$ and $\psi^{-1}(\mathfrak{g}')$ are two compact forms of $\mathfrak{g}_{\mathbf{C}}$ that have the same intersection $\mathfrak{t}$ with $\mathfrak{t}_{\mathbf{C}}$; by §3, no. 2, Prop. 3, there exists an inner automorphism θ of $\mathfrak{g}_{\mathbf{C}}$ inducing the identity on $\mathfrak{t}_{\mathbf{C}}$ and such that $\theta(\mathfrak{g}) = \psi^{-1}(\mathfrak{g}')$. By replacing ψ by $\psi \circ \theta$, we can assume that ψ maps $\mathfrak{g}$ to $\mathfrak{g}'$. Further, by Prop. 4 of no. 2, there exists a unique morphism $f_\mathrm{T} : \mathrm{T} \to \mathrm{T}'$ such that $\Gamma(f_\mathrm{T}) = \varphi$. Then the restriction of ψ to $\mathfrak{t}$ is $\mathrm{L}(f_\mathrm{T})$, and by §2, no. 6, Prop. 8, there exists a unique morphism $f : \mathrm{G} \to \mathrm{G}'$ that induces f_T on T and ψ on $\mathfrak{g}_{\mathbf{C}}$. Applying the preceding to φ^{-1} and ψ^{-1} we obtain an inverse morphism to f, which is therefore an isomorphism. Then $D_*(f) = \Gamma(f_T) = \varphi$, hence the proposition.

Note that, if T and T′ are two maximal tori of G, the diagrams $D^*(G, T)$ and $D^*(G, T')$ are isomorphic (if $g \in G$ is such that $gTg^{-1} = T'$, then $\operatorname{Int} g$ is an isomorphism from G to G that maps T to T′). Denote by $D^*(G)$ the isomorphism class of $D^*(G, T)$ (cf. *Theory of Sets*, Chap. II, §6, no. 2); this is a root diagram that depends only on G and is called the contravariant diagram of G. The covariant diagram $D_*(G)$ of G is defined similarly, and we obtain:

COROLLARY. *Two connected compact Lie groups* G *and* G′ *are isomorphic if and only if the diagrams* $D^*(G)$ *and* $D^*(G')$ (resp. $D_*(G)$ *and* $D_*(G')$) *are equal.*

PROPOSITION 16. *For every reduced root diagram* D, *there exists a connected compact Lie group* G *such that* $D^*(G)$ (resp. $D_*(G)$) *is isomorphic to* D.

a) By replacing D, if necessary, by its inverse diagram, we are reduced to constructing G such that $D^*(G)$ is isomorphic to D. Put $D = (M, M_0, R)$; then $\mathbf{Q} \otimes M$ is the direct sum of $\mathbf{Q} \otimes M_0$ and the vector subspace $V(R)$ generated by R. Moreover, since the inverse roots take integer values on M, the projection of M on $V(R)$ parallel to $\mathbf{Q} \otimes M_0$ is contained in the group of weights $P(R)$ of R, so that M is a subgroup of $M_0 \oplus P(R)$ of finite index. Denote by D′ the diagram $(M_0 \oplus P(R), M_0, R)$.

b) Let $\mathfrak{a}$ be a complex semi-simple Lie algebra whose canonical root system is isomorphic to $R \subset \mathbf{C} \otimes V(R)$ (Chap. VIII, §4, no. 3), and let $\mathfrak{g}_1$ be a compact real form of $\mathfrak{a}$ (§3, no. 2, Th. 1). Let G_1 be a simply-connected real Lie group whose Lie algebra is isomorphic to $\mathfrak{g}_1$; then G_1 is compact (§1, no. 4, Th. 1). Let T_1 be a maximal torus of G_1. By Th. 1, the diagram $D^*(G_1, T_1)$ is isomorphic to $(P(R), 0, R)$.

c) Let T_0 be a torus of dimension equal to the rank of M_0; then $D^*(T_0, T_0)$ is isomorphic to $(M_0, M_0, \varnothing)$, so $D^*(G_1 \times T_0, T_1 \times T_0)$ is isomorphic to D′ (Example 2).

d) Finally, let N be the finite subgroup of $T_1 \times T_0$ orthogonal to M. Put $G = (G_1 \times T_0)/N$, $T = (T_1 \times T_0)/N$. Then G is a connected compact Lie group, T a maximal torus of G, and $D(G, T)$ is isomorphic to D (Example 3).

Scholium. The classification of connected compact Lie groups up to isomorphism is thus reduced to that of reduced root diagrams. The connected compact semi-simple Lie groups correspond to the reduced root diagrams (M, M_0, R) *such that* $M_0 = 0$; *giving such a diagram is equivalent to giving a reduced root system* R *in a vector space* V *over* $\mathbf{Q}$ *and a subgroup* M *of* V *such that* $Q(R) \subset M \subset P(R)$.

Remark 3. Let T′ be another maximal torus of G, B (resp. B′) a basis of the root system R(G, T) (resp. R(G′, T′)) (Chap. VI, §1, no. 5, Def. 2). There exist elements $g \in$ G such that Int g maps T onto T′ and B onto B′, and these elements form a unique coset modulo Int(T) (since T and T′ are conjugate, we can assume that T = T′, and it suffices to apply Chap. VI, §1, no. 5, Remark 4 and Prop. 9 of no. 4). It follows that the isomorphism from T to T′ induced by Int g is independent of the choice of g; consequently the same is true of $D_*(\mathrm{Int}\, g)$ and $D^*(\mathrm{Int}\, g)$. Paraphrasing Chap. VIII, §5, no. 3, Remark 2, *mutatis mutandis*, we can now define the canonical maximal torus of G, the canonical covariant and contravariant root diagrams of G, ...

10. AUTOMORPHISMS OF A CONNECTED COMPACT LIE GROUP

Denote by Aut(G) the Lie group of automorphisms of G (Chap. III, §10, no. 2), and by Aut(G, T) the closed subgroup of Aut(G) consisting of the elements u such that $u(\mathrm{T}) = \mathrm{T}$. We have seen (§1, no. 4, Cor. 5 of Prop. 4) that the identity component of Aut(G) is the subgroup Int(G) of inner automorphisms; denote by $\mathrm{Int}_{\mathrm{G}}(\mathrm{H})$ the image in Int(G) of a subgroup H of G.

Let D be the covariant diagram of G relative to T; denote by Aut(D) the group of its automorphisms, and by W(D) its Weyl group. The map $u \mapsto D_*(u)$ is a homomorphism from Aut(G, T) to Aut(D). Prop. 15 of no. 9 immediately gives:

PROPOSITION 17. *The homomorphism* $\mathrm{Aut(G,T)} \to \mathrm{Aut(D)}$ *is surjective, with kernel* $\mathrm{Int}_{\mathrm{G}}(\mathrm{T})$.

Note that $\mathrm{Aut(G,T)} \cap \mathrm{Int(G)} = \mathrm{Int}_{\mathrm{G}}(\mathrm{N}_{\mathrm{G}}(\mathrm{T}))$ and that the image of $\mathrm{Int}_{\mathrm{G}}(\mathrm{N}_{\mathrm{G}}(\mathrm{T}))$ in Aut(D) is W(D) (no. 5, Prop. 10). Thus, Proposition 17 gives an isomorphism

$$\mathrm{Aut(G,T)}/(\mathrm{Aut(G,T)} \cap \mathrm{Int(G)}) \to \mathrm{Aut(D)}/\mathrm{W(D)}.$$

Further, Aut(G) = Int(G).Aut(G, T). Indeed, if u belongs to Aut(G), $u(\mathrm{T})$ is a maximal torus of T, hence is conjugate to T, and there exists an inner automorphism v of G such that $u(\mathrm{T}) = v(\mathrm{T})$, in other words $v^{-1}u \in \mathrm{Aut(G,T)}$. It follows that Aut(G)/Int(G) can be identified with $\mathrm{Aut(G,T)}/(\mathrm{Aut(G,T)} \cap \mathrm{Int(G)})$, so in view of the preceding we have an exact sequence

$$1 \to \mathrm{Int(G)} \to \mathrm{Aut(G)} \to \mathrm{Aut(D)}/\mathrm{W(D)} \to 1. \qquad (16)$$

Consequently:

PROPOSITION 18. *The group* Aut(G)/Int(G) *is isomorphic to* Aut(D)/W(D).

In particular, assume that G is semi-simple; the group Aut(D) can then be identified with the subgroup of A(R(G, T)) (Chap. VI, §1, no. 1) consisting of the elements u such that $u(X(T)) \subset X(T)$, and the subgroup W(D) can be identified with W(R(G, T)).

COROLLARY. *If* G *is simply-connected, or if* C(G) *reduces to the identity element, the group* Aut(G)/Int(G) *is isomorphic to the group of automorphisms of the Dynkin graph of* R(G, T).

This follows from the preceding and Chap. VI, §4, no. 2, Cor. of Prop. 1.

We are now going to show that the extension (16) admits *sections*.

For all $\alpha \in R(G,T)$, denote by $V(\alpha)$ the 2-dimensional vector subspace of $\mathfrak{g}$ such that $V(\alpha)_{(\mathbf{C})} = \mathfrak{g}^{\alpha} + \mathfrak{g}^{-\alpha}$; denote by K the quadratic form associated to the Killing form of $\mathfrak{g}$.

DEFINITION 3. *A framing of* (G, T) *is a pair* $(B, (U_\alpha)_{\alpha \in B})$, *where* B *is a basis of* R(G, T) (Chap. VI, § 1, no. 5, Def. 2) *and where, for all* $\alpha \in B$, U_α *is an element of* $V(\alpha)$ *such that* $K(U_\alpha) = -1$.

A framing of G is a maximal torus T of G together with a framing of (G, T).

Lemma 3. Let B_0 *be a basis of* R(G, T). *The group* $\mathrm{Int}_G(T)$ *operates simply-transitively on the set of framings of* (G, T) *of the form* $(B_0, (U_\alpha)_{\alpha \in B_0})$.

For all $\alpha \in B_0$, denote by $K(\alpha)$ the restriction of the quadratic form K to $V(\alpha)$; the operation of T on $V(\alpha)$ defines a morphism $\iota_\alpha : T \to \mathbf{SO}(K(\alpha))$. We have seen in no. 4 that $\mathbf{SO}(K(\alpha))$ can be identified with $\mathbf{U}$ in such a way that ι_α is identified with the root α. Since B_0 is a basis of R, it is a basis of the $\mathbf{Z}$-module Q(R) generated by the roots, hence a basis of the submodule X(T/C(G)) of X(T). It follows that the product of the morphisms ι_α induces an isomorphism from T/C(G) to the product of the groups $\mathbf{SO}(K(\alpha))$. But the latter group operates simply-transitively on the set of framings of (G, T) whose first component is B_0.

PROPOSITION 19. *The group* Int(G) *operates simply-transitively on the set of framings of* G.

Let $e = (T, B, (U_\alpha))$ and $e' = (T', B', (U'_\alpha))$ be two framings of G. There exist elements g in G such that $(\mathrm{Int}\, g)(T) = T'$, and these elements form a single coset modulo $N_G(T)$. Thus, we can assume that $T = T'$, and we must prove that there exists a unique element of $\mathrm{Int}_G(N_G(T))$ that transforms e to e'. By Chap. VI, §1, no. 5, Remark 4, there exists a unique element w of W(R) such that $w(B) = B'$. Since W(R) can be identified with $N_G(T)/T$, there exists $n \in N_G(T)$ such that $w = \mathrm{Int}\, n$, and n is uniquely determined modulo T. Thus, we can assume that $B = B'$, and we must prove that there exists a unique element of $\mathrm{Int_G(T)}$ that transforms e to e', which is simply Lemma 3.

COROLLARY. *Let* e *be a framing of* (G,T) *and let* E *be the group of automorphisms of* G *that leave* e *stable. Then* Aut(G) *is the semi-direct product of* E *by* Int(G), *and* Aut(G, T) *is the semi-direct product of* E *by* $\mathrm{Int(G)} \cap \mathrm{Aut(G,T)} = \mathrm{Int_G(N_G(T))}$.

Indeed, every element of Aut(G) transforms e into a framing of G. By Prop. 19, every coset of Aut(G) modulo Int(G) meets E in a single point, hence the first assertion. The second is proved in the same way.

Remark. Let G and G′ be two connected compact Lie groups, and let $e = (\mathrm{T},\mathrm{B},(U_\alpha))$ and $e' = (\mathrm{T}',\mathrm{B}',(U'_\alpha))$ be framings of G and G′, respectively. Let X be the set of isomorphisms from G to G′ that take e to e'. The map $f \mapsto \mathrm{D}^*(f)$ (resp. $\mathrm{D}_*(f)$) is a bijection from X to the set of isomorphisms from $\mathrm{D}^*(\mathrm{G}',\mathrm{T}')$ to $\mathrm{D}^*(\mathrm{G},\mathrm{T})$ (resp. $\mathrm{D}_*(\mathrm{G},\mathrm{T})$ to $\mathrm{D}_*(\mathrm{G}',\mathrm{T}')$) that map B′ to B (resp. B to B′). Indeed, this follows immediately from Prop. 15 and Lemma 3.

§5. CONJUGACY CLASSES

We retain the notations of §4.

1. REGULAR ELEMENTS

By Cor. 4 of Th. 2 of §2, no. 2, the *regular* elements g of G can be characterized by either of the following properties:

a) The subalgebra of $\mathfrak{g}$ fixed by $\mathrm{Ad}\, g$ is a Cartan subalgebra.

b) $\mathrm{Z}(g)_0$ is a maximal torus of G.

The set of regular elements of G is open and dense in G.

In the remainder of this paragraph, we denote by G_r (resp. T_r) the set of points of G (resp. T) that are regular in G. An element g of G belongs to T_r if and only if $\mathrm{Z}(g)_0$ is equal to T; every element of G_r is conjugate to an element of T_r (§2, no. 2, Th. 2).

An element t of T belongs to T_r if and only if $t^\alpha \neq 1$ for every root $\alpha \in \mathrm{R(G,T)}$; consequently, $\mathrm{T}-\mathrm{T}_r$ is the union of the subtori $\mathrm{Ker}\,\alpha$ for α in R(G, T).

PROPOSITION 1. *Put* $n = \dim \mathrm{G}$. *There exists a compact real-analytic manifold* V *of dimension* $n-3$ *and an analytic map* $\varphi : \mathrm{V} \to \mathrm{G}$ *whose image is* $\mathrm{G}-\mathrm{G}_r$.

Let $\alpha \in \mathrm{R(G,T)}$; put $\mathrm{V}_\alpha = (\mathrm{G}/\mathrm{Z}(\mathrm{Ker}\,\alpha)) \times (\mathrm{Ker}\,\alpha)$, and let φ_α be the morphism from V_α to G such that, for all $g \in \mathrm{G}$ and all $t \in \mathrm{Ker}\,\alpha$, we have

$\varphi_\alpha(\bar{g}, t) = gtg^{-1}$ (we denote by $\bar{g}$ the coset of g modulo $\mathrm{Z}(\mathrm{Ker}\,\alpha)$). Then V_α is a compact real-analytic manifold of dimension

$$\begin{aligned}\dim \mathrm{V}_\alpha &= \dim \mathrm{G} - \dim \mathrm{Z}(\mathrm{Ker}\,\alpha) + \dim \mathrm{Ker}\,\alpha \\ &= n - (\dim \mathrm{T} + 2) + (\dim \mathrm{T} - 1) = n - 3\end{aligned}$$

(§4, no. 5, Th. 1); φ_α is a morphism of real-analytic manifolds, and the image of φ_α consists of the elements of G conjugate to an element of $\mathrm{Ker}\,\alpha$. It now suffices to take V to be the sum of the manifolds V_α, and φ to be the morphism inducing φ_α on each V_α.

Remark. Call an element g of G *very regular* if $\mathrm{Z}(g)$ is a maximal torus of G. If $g \in \mathrm{T}$, g is very regular if and only if $w(g) \neq g$ for every non-identity element w of $\mathrm{W}_\mathrm{G}(\mathrm{T})$ (§4, no. 7, Prop. 14). Thus, the set of very regular elements is a dense open subset of G (§2, no. 5, Cor. 2 of Prop. 5).

2. CHAMBERS AND ALCOVES

Denote by $\mathfrak{t}_r$ the set of elements $x \in \mathfrak{t}$ such that $\exp x$ is regular, in other words belongs to T_r. An element x of $\mathfrak{t}$ belongs to $\mathfrak{t} - \mathfrak{t}_r$ if and only if there exists a root $\alpha \in \mathrm{R}(\mathrm{G}, \mathrm{T})$ such that $\delta(\alpha)(x) \in 2\pi i\mathbf{Z}$. For each root $\alpha \in \mathrm{R}(\mathrm{G}, \mathrm{T})$ and each integer n, denote by $\mathrm{H}_{\alpha,n}$ the set of $x \in \mathfrak{t}$ such that $\delta(\alpha)(x) = 2\pi i n$. The $\mathrm{H}_{\alpha,n}$ are called the *singular hyperplanes* of $\mathfrak{t}$, and $\mathfrak{t} - \mathfrak{t}_r$ is the union of the singular hyperplanes. The *alcoves* of $\mathfrak{t}$ are the connected components of $\mathfrak{t}_r$, and the *chambers* are the connected components of the complement in $\mathfrak{t}$ of the union of those singular hyperplanes that pass through the origin (that is, the $\mathrm{H}_{\alpha,0} = \mathrm{Ker}\,\delta(\alpha)$, $\alpha \in \mathrm{R}(\mathrm{G}, \mathrm{T})$).

We have $\Gamma(\mathrm{T}) \subset \mathfrak{t} - \mathfrak{t}_r$; denote by $\mathrm{N}(\mathrm{G}, \mathrm{T})$ the subgroup of $\Gamma(\mathrm{T})$ generated by the nodal vectors (§4, no. 5); by Prop. 11 of §4, no. 6, the quotient $\Gamma(\mathrm{T})/\mathrm{N}(\mathrm{G}, \mathrm{T})$ can be identified with the fundamental group of G.

Finally, denote by W the Weyl group of G relative to T, considered as a group of automorphisms of T and of $\mathfrak{t}$, and denote by W_a (resp. W'_a) the group of automorphisms of the affine space $\mathfrak{t}$ generated by W and the translations $t_\gamma : x \mapsto x + \gamma$ for $\gamma \in \mathrm{N}(\mathrm{G}, \mathrm{T})$ (resp. for $\gamma \in \Gamma(\mathrm{T})$).

Let $w \in \mathrm{W}$, $\gamma \in \Gamma(\mathrm{T})$, $\alpha \in \mathrm{R}(\mathrm{G}, \mathrm{T})$ and $n \in \mathbf{Z}$. We have:

$$w(\mathrm{H}_{\alpha,n}) = \mathrm{H}_{w\alpha,n}, \quad t_\gamma(\mathrm{H}_{\alpha,n}) = \mathrm{H}_{\alpha,n+\langle\gamma,\alpha\rangle}.$$

It follows that, for all chambers C and all $w \in \mathrm{W}$, $w(\mathrm{C})$ is a chamber and that for all alcoves A and all $w \in \mathrm{W}'_a$, $w(\mathrm{A})$ is an alcove. Note that, when $\mathrm{X}(\mathrm{T}) \otimes \mathbf{R}$ is identified with $\mathfrak{t}^*$ *via* the isomorphism $(2\pi i)^{-1}\delta$, the alcoves of $\mathfrak{t}$ and the group W_a are the alcoves and the affine Weyl group associated to the root system $\mathrm{R}(\mathrm{G}, \mathrm{T})$ (Chap. VI, §2, no. 1).

PROPOSITION 2. *a) The group* W_a (resp. W'_a) *is the semi-direct product of* W *by* $\mathrm{N}(\mathrm{G}, \mathrm{T})$ (resp. $\Gamma(\mathrm{T})$); *the subgroup* W_a *of* W'_a *is normal.*

b) The group W (resp. W_a) *operates simply-transitively on the set of chambers* (resp. *alcoves*).

c) Let C *be a chamber and* A *an alcove. Then* $\overline{C}$ (resp. $\overline{A}$, resp. A) *is a fundamental domain for the operation of* W *on* $\mathfrak{t}$ (resp. *of* W_a *on* $\mathfrak{t}$, resp. *of* W_a *on* $\mathfrak{t} - \mathfrak{t}_r$). *If* $x \in \mathfrak{t}_r$ *and* $w \in W_a$ *are such that* $w(x) = x$, *then* $w = \mathrm{Id}$.

d) For every chamber C, *there exists a unique alcove* A *such that* $A \subset C$ *and* $0 \in \overline{A}$. *For every alcove* A, *there exists a unique* $\gamma \in N(G, T)$ *such that* $\gamma \in \overline{A}$.

If $w \in W$ and $\gamma \in \Gamma(T)$, then $wt_\gamma w^{-1} = t_{w(\gamma)}$ and $wt_\gamma w^{-1} t_\gamma^{-1} = t_{w(\gamma)-\gamma}$, with $w(\gamma) - \gamma \in N(G, T)$; this immediately implies *a*). The rest of the proposition follows from Chap. VI, §1, no. 5 and §2, nos. 1 and 2.

COROLLARY 1. *Let* A *be an alcove of* $\mathfrak{t}$, $\overline{A}$ *its closure, and* H_A *the stabilizer of* A *in* W'_a.

a) The group W'_a *is the semi-direct product of* H_A *by* W_a.

b) The exponential map $\overline{A} \to T$ *and the canonical injection* $T \to G$ *induce by passage to the quotients and to subsets homeomorphisms*

$$\overline{A}/H_A \to T/W \to G/\mathrm{Int}(G)$$
$$A/H_A \to T_r/W \to G_r/\mathrm{Int}(G).$$

Let $w' \in W'_a$; then $w'(A)$ is an alcove of $\mathfrak{t}$, and there exists (Prop. 2 *b*)) a unique element w of W_a such that $w(A) = w'(A)$, that is $w^{-1}w' \in H_A$. Since W_a is normal in W'_a, this proves *a*).

The canonical injection of $\overline{A}$ into $\mathfrak{t}$ induces a continuous bijection $\theta : \overline{A} \to \mathfrak{t}/W_a$ (Prop. 2 *c*)), and this is a homeomorphism since $\overline{A}$ is compact. Since W_a is normal in W'_a, the group H_A operates canonically on $\mathfrak{t}/W_a$ (*Algebra*, Chap. I, §5, no. 4) and $\mathfrak{t}/W'_a$ can be identified with the quotient $(\mathfrak{t}/W_a)/H_A$; the map θ is compatible with the operations of H_A, hence induces by passage to the quotient a homeomorphism $\overline{A}/H_A \to \mathfrak{t}/W'_a$. Further, $\exp_\Gamma$ induces a homeomorphism from $\mathfrak{t}/\Gamma(T)$ to T, hence also a homeomorphism from $\mathfrak{t}/W'_a$ to T/W. Assertion *b*) follows from that and Cor. 1 of Prop. 5 of §2, no. 4.

Remarks. 1) The group H_A can be identified naturally with $\Gamma(T)/N(G,T)$, hence also with $\pi_1(G)$. Thus, it reduces to the identity element when G is simply-connected.

2) Let $x \in A$; then $\exp x \in T_r$, so $Z(\exp x)_0 = T$. Further, $\exp x$ is *very regular* (no. 1, Remark) if and only if $w(x) \neq x$ for all $w \in W'_a$ distinct from the identity. By Cor. 1, this also means that $h(x) \neq x$ for all $h \in H_A$ distinct from the identity. In particular, if G is simply-connected, then $Z_G(t) = T$ for all $t \in T_r$, and every regular element of G is very regular.

3) The special points of W_a (Chap. VI, §2, no. 2) are the elements x of $\mathfrak{t}$ such that $\delta(\alpha)(x) \in 2\pi i\mathbf{Z}$ for all $\alpha \in R(G,T)$ (*loc. cit.*, Prop. 3), that is such that $\exp x \in C(G)$ (§4, no. 4, Prop. 8). For such an element x we have $wx - x \in N(G,T)$ for all $w \in W$ (Chap. VI, §1, no. 9, Prop. 27), so the

stabilizers of x in W_a and W'_a coincide. Let S be the set of special points of $\overline{A}$; it follows from the preceding and Cor. 1 that the group H_A operates freely on S, and that the exponential map induces a bijection from S/H_A to C(G).

COROLLARY 2. *Let* C *be a chamber of* $\mathfrak{t}$ *and* $\overline{C}$ *its closure. The canonical injections* $\overline{C} \to \mathfrak{t} \to \mathfrak{g}$ *induce by passage to the quotient homeomorphisms*

$$\overline{C} \to \mathfrak{t}/W \to \mathfrak{g}/\mathrm{Ad}(G).$$

The canonical maps $\overline{C} \to \mathfrak{t}$ and $\mathfrak{t} \to \mathfrak{t}/W$ are proper (*General Topology*, Chap. III, §4, no. 1, Prop. 2 *c*)). The map $\overline{C} \to \mathfrak{t}/W$ is continuous, proper and bijective (Prop. 2 *c*)); thus, it is a homeomorphism, hence the corollary in view of the Cor. of Prop. 6 of §2, no. 5.

Remark 4. Denote by $\mathfrak{g}_{\mathrm{reg}}$ the set of regular elements of $\mathfrak{g}$ (Chap. VII, §2, no. 2, Def. 2) and put $\mathfrak{t}_{\mathrm{reg}} = \mathfrak{t} \cap \mathfrak{g}_{\mathrm{reg}}$. For $x \in \mathfrak{t}$, we have

$$\det(X - \mathrm{ad}_{\mathfrak{g}} x) = X^{\dim \mathfrak{t}} \prod_{\alpha \in R(G,T)} (X - \delta(\alpha)(x)),$$

and hence $\mathfrak{t}_{\mathrm{reg}}$ is the set of elements x of $\mathfrak{t}$ such that $\delta(\alpha)(x) \neq 0$ for all $\alpha \in R(G,T)$, that is the union of the chambers of $\mathfrak{t}$ (so $\mathfrak{t}_r \subset \mathfrak{t}_{\mathrm{reg}}$). Consequently $\overline{C} \cap \mathfrak{t}_{\mathrm{reg}} = C$, so we have homeomorphisms

$$C \to \mathfrak{t}_{\mathrm{reg}}/W \to \mathfrak{g}_{\mathrm{reg}}/\mathrm{Ad}(G).$$

COROLLARY 3. *Assume that* G *is simply-connected; let* g *be a regular element of* G. *There exist a maximal torus* S *of* G, *and an alcove* A *of* L(S), *both uniquely determined, such that* $g \in \exp(A)$ *and* $0 \in \overline{A}$.

We can assume that g belongs to T_r (§2, no. 2, Th. 2). Let x be an element of $\mathfrak{t}_r$ such that $\exp x = g$, and let A′ be the alcove of $\mathfrak{t}$ containing x. The alcoves A of $\mathfrak{t}$ such that $g \in \exp(A)$ are the alcoves $A' - \gamma$ for $\gamma \in \Gamma(T)$; thus, the assertion follows from Prop. 2 *d*).

3. AUTOMORPHISMS AND REGULAR ELEMENTS

Lemma 1. Let u *be an automorphism of* G, *and* H *the set of its fixed points.*

a) H *is a closed subgroup of* G.

b) *If* H_0 *is central in* G, *then* G *is commutative* (*so* G = T).

Assertion *a*) is clear. To prove *b*), we can replace G by D(G) (§1, Cor. 1 of Prop. 4), and hence can assume that G is semi-simple. Then, if H_0 is central in G, we have $L(H) = \{0\}$, so the endomorphism $L(u) - \mathrm{Id}$ of $\mathfrak{g}$ is bijective. Let f be the endomorphism of the manifold G defined by $f(g) = u(g)^{-1}g$ for $g \in G$; it is étale, for if $g \in G$ and $x \in \mathfrak{g}$, we have $T(f)(xg) = u(g)^{-1}(x - L(u)(x))g$, so the tangent map of f at g is bijective. It follows that the image of f is open and compact, hence coincides with G since G is connected. Now let E be a framing of G (§4, no. 10, Def. 3) and $u(\mathrm{E})$ its image under u. By Prop. 19 of *loc. cit.*, there exists an element h of G such that $(\operatorname{Int} h)(\mathrm{E}) = u(\mathrm{E})$. Let $g \in \mathrm{G}$ be such that $h = f(g) = u(g)^{-1}g$; then

$$u \circ \operatorname{Int} g = (\operatorname{Int} u(g)) \circ u = \operatorname{Int} g \circ (\operatorname{Int} h)^{-1} \circ u,$$

so the framing $(\operatorname{Int} g)(\mathrm{E})$ is stable under u. If $(\operatorname{Int} g)(\mathrm{E}) = (\mathrm{T}_1, \mathrm{B}, (U_\alpha)_{\alpha \in \mathrm{B}})$, then $\sum U_\alpha \in \mathrm{L(H)}$; since $\mathrm{L(H)} = \{0\}$, this implies that $\mathrm{B} = \varnothing$, so $\mathrm{G} = \mathrm{T}_1$, and G is commutative.

Lemma 2. Let x be an element of T *and* S *a subtorus of* T. *If the identity component of* $\mathrm{Z}(x) \cap \mathrm{Z(S)}$ *reduces to* T, *there exists an element s of* S *such that xs is regular.*

For all $\alpha \in \mathrm{R(G, T)}$, let S_α be the submanifold of S consisting of the elements s of S such that $(xs)^\alpha = 1$. If there is no element s of S such that xs is regular, S is the union of the submanifolds S_α, hence is equal to one of them. Hence there exists α in R(G, T) such that $(xs)^\alpha = 1$ for all $s \in \mathrm{S}$; but this implies that $x^\alpha = 1$ and $\alpha|\mathrm{S} = 1$, so $\mathrm{Z}(x) \cap \mathrm{Z(S)} \supset \mathrm{Z}(\operatorname{Ker} \alpha)$, hence the lemma.

Lemma 3. Assume that G *is simply-connected. Let* C *be a chamber of* $\mathfrak{t}$, *and u an automorphism of* G *such that* T *and* C *are stable under u. Then the set of points of* T *fixed by u is connected.*

Since G is simply-connected, $\Gamma(\mathrm{T})$ is generated by the nodal vectors K_α ($\alpha \in \mathrm{R(G, T)}$), and hence has a basis consisting of the family of the K_α for which α belongs to the basis B(C) defined by C (Chap. VI, §1, no. 10). Thus, it suffices to prove that, if φ is an automorphism of the torus T leaving stable a basis of $\Gamma(\mathrm{T})$, the set of fixed points of φ is connected. Decomposing this basis into the disjoint union of the orbits of the group generated by φ, we are reduced to the case in which $\mathrm{T} = \mathbf{U}^n$ and φ is the automorphism $(z_1, \ldots, z_n) \mapsto (z_2, \ldots, z_n, z_1)$; in this case the fixed points of φ are the points $(z, z, \ldots, z)$ for $z \in \mathbf{U}$, which form a connected subgroup of T.

PROPOSITION 3. *Let u be an automorphism of* G, *and let x be a point of* G *fixed by u.*

a) There exists an element a of $\mathfrak{g}$, *fixed by* $\mathrm{L}(u)$ *and by* $\operatorname{Ad} x$, *such that $x \exp a$ is regular.*

b) There exists a regular element g of G *fixed by u and commuting with x.*

Let H be the group of fixed points of u, S a maximal torus of $\mathrm{Z}(x) \cap \mathrm{H}$, and K the identity component of $\mathrm{Z(S)} \cap \mathrm{Z}(x)$. This is a connected closed subgroup of G; further, by Cor. 5 of Th. 2 of §2, no. 2, there exist maximal tori of G containing S and x, so K is of maximal rank and contains S and x. On the other hand, K is stable under u since S and x are; denote by V the set of fixed points of u in K. Then

$$S \subset V_0 \subset K \cap H \subset Z(S) \cap Z(x) \cap H,$$

so V_0 is contained in the centralizer of S in $(Z(x) \cap H)_0$; but the latter reduces to S (*loc. cit.*, Cor. 6), hence finally $V_0 = S$. Lemma 1 now implies that K is commutative, hence is a maximal torus of G (since it is connected and of maximal rank). It contains S and x, and is equal to the identity component of $Z(S) \cap Z(x)$; assertion *a)* now follows from Lemma 2. We deduce *b)* by taking $g = x \exp a$.

COROLLARY. *Let* $\mathfrak{s}$ *be a compact Lie algebra, and let* φ *be an automorphism of* $\mathfrak{s}$. *There exists a regular element of* $\mathfrak{s}$ *fixed by* φ.

Replacing $\mathfrak{s}$ by $\mathscr{D}\mathfrak{s}$, we can assume that $\mathfrak{s}$ is semi-simple. Let S be a compact simply-connected Lie group with Lie algebra $\mathfrak{s}$, and let u be the automorphism of S such that $L(u) = \varphi$. Prop. 3 implies the existence of an element a of $\mathfrak{s}$, fixed by φ, such that $\exp a$ is regular in S; in particular, a is regular in $\mathfrak{s}$ (no. 2, remark 4).

THEOREM 1. *Let* u *be an automorphism of a connected compact Lie group* G.

a) The identity component of the group of fixed points of u *contains a regular element of* G.

b) There exists a maximal torus K *of* G *and a chamber of* L(K) *that are stable under* u.

c) If G *is simply-connected, the set of fixed points of* u *is connected.*

Assertion *a)* is the particular case $x = e$ of Prop. 3. We assume now that G is *simply-connected* and prove *b)* and *c)*. Let x be an element of G fixed by u, and let g be a regular element of G, fixed by u and commuting with x (Prop. 3). The centralizer K of g is a maximal torus of G (no. 2, Remark 2), stable under u, and containing x and g. By Cor. 3 of Prop. 2 of no. 2, there exists a unique alcove A of L(K) such that $g \in \exp A$ and $0 \in \overline{A}$; since g is fixed by u, $L(u)$ leaves A, and hence also the chamber of L(K) containing A, stable. This proves *b)*; further, the set of points of K fixed by u is connected (Lemma 3) and contains x and e, hence *c)* (*General Topology*, Chap. I, §11, no. 1, Prop. 2).

It remains to prove *b)* in the general case. Now, if $\tilde{D}(G)$ is the universal covering of D(G), and if $f : \tilde{D}(G) \to G$ is the canonical morphism, there exists an automorphism $\tilde{u}$ of $\tilde{D}(G)$ such that $f \circ \tilde{u} = u \circ f$. If $\tilde{K}$ is a maximal torus of $\tilde{D}(G)$ and $\tilde{C}$ a chamber of $L(\tilde{K})$, stable under $\tilde{u}$ (this exists by what has already been proved), there exists (§2, no. 3, Remark 2) a unique maximal torus K of G and a unique chamber C of L(K) such that $\tilde{K} = f^{-1}(K)$ and $\tilde{C} = L(f)^{-1}(C)$, and we see immediately that K and C are stable under u, hence assertion *b)* in the general case.

COROLLARY 1. *Assume that the* $\mathbf{Z}$*-module* $\pi_1(G)$ *is torsion-free.*

a) The centralizer of every element of G *is connected.*

b) Any two commuting elements of G *belong to the same maximal torus.*

By Cor. 3 of Prop. 11 of §4, no. 6, D(G) is simply-connected. We have $G = C(G)_0.D(G)$; let $x \in G$; write $x = uv$, with $u \in C(G)_0$ and $v \in D(G)$. Then $Z(x) = C(G)_0.Z_{D(G)}(v)$. By Th. 1 *c*), $Z_{D(G)}(v)$ is connected, so $Z(x)$ is connected, hence *a*). Thus, by Cor. 3 of Th. 2 of §2, no. 2, $Z(x)$ is the union of the maximal tori of G containing x, hence *b*).

COROLLARY 2. *Let Γ be a compact subgroup of* Aut(G) *with the following property:*

(*) *There exist elements $u_1, \dots, u_n$ of Γ such that, for all i, the closure Γ_i of the subgroup of Γ generated by $u_1, \dots, u_i$ is a normal subgroup of Γ, and $\Gamma_n = \Gamma$.*

Then, there exists a maximal torus of G *stable under the operation of Γ.*

We argue by induction on the dimension of G. Clearly, we can assume that $u_1 \neq \mathrm{Id}$; then the subgroup H of fixed points of u_1 is distinct from G, and is stable under the operation of Γ. Moreover, since Γ is compact, the image of Γ in $\mathrm{Aut}(H_0)$ is a quotient of Γ, hence it also satisfies condition (*). By the induction hypothesis, there exists a maximal torus S of H stable under Γ. The centralizer K of S in G is connected (§2, no. 2, Cor. 5) and stable under Γ; this is a maximal torus of G, since H_0 contains a regular element of G (Th. 1 *a*)) which is conjugate to an element of S (*loc. cit.*, Cor. 4).

COROLLARY 3. *Let* H *be a Lie group and Γ a compact subgroup of* H. *Assume that* H_0 *is compact and that Γ satisfies condition* (*) *of Cor. 2. Then there exists a maximal torus* T *of* H_0 *such that* $\Gamma \subset N_H(T)$.

COROLLARY 4. *Every nilpotent subgroup of a compact Lie group is contained in the normalizer of a maximal torus.*

Let H be a compact Lie group, N a nilpotent subgroup of H. Then the closure Γ of N is also a nilpotent group (Chap. III, §9, no. 1, Cor. 2 of Prop. 1), and it suffices, in view of Cor. 3, to prove that Γ satisfies condition (*). Now Γ_0 is a connected compact nilpotent Lie group, hence is a torus (§1, no. 4, Cor. 1 of Prop. 4), and there exists an element u_1 of Γ generating a dense subgroup of Γ_0 (*General Topology*, Chap. VII, §1, no. 3, text preceding Prop. 8). The finite group Γ/Γ_0 is nilpotent and there exist $\tilde{u}_2, \dots, \tilde{u}_n \in \Gamma/\Gamma_0$ generating Γ/Γ_0 and such that the subgroup of Γ/Γ_0 generated by $(\tilde{u}_2, \dots, \tilde{u}_r)$ is normal for $r = 2, \dots, n$ (*Algebra*, Chap. I, §6, no. 5, Th. 1 and no. 7, Th. 4). Then, if $u_2, \dots, u_n$ are representatives of $\tilde{u}_2, \dots, \tilde{u}_n$ in Γ, the sequence $(u_1, \dots, u_n)$ has the required properties.

Example. Take $G = \mathbf{U}(n, \mathbf{C})$. We shall see later that the subgroup of diagonal matrices in G is a maximal torus of G and that its normalizer is the set of *monomial* matrices (*Algebra*, Chap. II, §10, no. 7, Example II) in G.

We conclude that, if Φ is a separating positive hermitian form on a finite dimensional complex vector space V and Γ is a nilpotent subgroup of $\mathbf{U}(\Phi)$, there exists a basis of V for which the matrices of the elements of Γ are monomial (*"Blichtfeldt's theorem"*).

4. THE MAPS $(\mathrm{G}/\mathrm{T}) \times \mathrm{T} \to \mathrm{G}$ AND $(\mathrm{G}/\mathrm{T}) \times \mathrm{A} \to \mathrm{G}_r$

The map $(g, t) \mapsto gtg^{-1}$ from $\mathrm{G} \times \mathrm{T}$ to G induces by passage to the quotient a morphism of analytic manifolds

$$f : (\mathrm{G}/\mathrm{T}) \times \mathrm{T} \to \mathrm{G},$$

which is surjective (§2, no. 2, Th. 2). By restriction, f induces a surjective morphism

$$f_r : (\mathrm{G}/\mathrm{T}) \times \mathrm{T}_r \to \mathrm{G}_r.$$

By composition with $\mathrm{Id}_{\mathrm{G}/\mathrm{T}} \times \exp_{\mathrm{T}}$, we obtain morphisms, also surjective,

$$\varphi : (\mathrm{G}/\mathrm{T}) \times \mathfrak{t} \to \mathrm{G},$$
$$\varphi_r : (\mathrm{G}/\mathrm{T}) \times \mathfrak{t}_r \to \mathrm{G}_r;$$

finally, if A is an alcove of $\mathfrak{t}$, φ_r induces a surjective morphism

$$\varphi_{\mathrm{A}} : (\mathrm{G}/\mathrm{T}) \times \mathrm{A} \to \mathrm{G}_r.$$

We define a *right operation of* W *on* G/T as follows: let $w \in \mathrm{W}$ and $u \in \mathrm{G}/\mathrm{T}$; lift w to an element n of $\mathrm{N}_{\mathrm{G}}(\mathrm{T})$ and u to an element g of G. Then the image of gn in G/T does not depend on the choice of n and g; we denote it by $u.w$.

For this operation, W operates *freely* on G/T: indeed, with the preceding notations, assume that $u.w = u$; then $gn \in g\mathrm{T}$, so $n \in \mathrm{T}$ and $w = 1$.

We define a right operation of W on $(\mathrm{G}/\mathrm{T}) \times \mathrm{T}$ by

$$(u, t).w = (u.w, w^{-1}(t)), \quad u \in \mathrm{G}/\mathrm{T},\ t \in \mathrm{T},\ w \in \mathrm{W}$$

and a right operation of W'_a on $(\mathrm{G}/\mathrm{T}) \times \mathfrak{t}$ by

$$(u, x).\omega = (u.\overline{\omega}, \omega^{-1}(x)), \qquad u \in \mathrm{G}/\mathrm{T},\ x \in \mathfrak{t},\ \omega \in \mathrm{W}'_a,$$

where $\overline{\omega}$ is the image of ω in the quotient $\mathrm{W}'_a/\Gamma(\mathrm{T}) = \mathrm{W}$.

If A is an alcove of $\mathfrak{t}$, and if H_{A} is the subgroup of W'_a that stabilizes A, we obtain by restriction an operation of H_{A} on $(\mathrm{G}/\mathrm{T}) \times \mathrm{A}$.

These different operations are compatible with the morphisms f, φ and φ_{A}: for $u \in \mathrm{G}/\mathrm{T}, t \in \mathrm{T}, x \in \mathfrak{t}, y \in \mathrm{A}, w \in \mathrm{W}, \omega \in \mathrm{W}'_a, h \in \mathrm{H}_{\mathrm{A}}$, we have

$$f((u, t).w) = f(u, t), \quad \varphi((u, x).\omega) = \varphi(u, x), \quad \varphi_{\mathrm{A}}((u, y).h) = \varphi_{\mathrm{A}}(u, y).$$

Lemma 4. Let $g \in \mathrm{G}$, $t \in \mathrm{T}$, *and let* $\bar{g}$ *be the image of* g *in* G/T. *Identify the tangent space of* G/T (resp. T, resp. G) *at* $\bar{g}$ (resp. t, resp. gtg^{-1}) *with*

$\mathfrak{g}/\mathfrak{t}$ (resp. $\mathfrak{t}$, resp. $\mathfrak{g}$) *by means of the left translation* $\gamma(g)$ *by* g (resp. t, resp. gtg^{-1}). *The tangent linear map of* f *at* $(\bar{g}, t)$ *is then identified with the linear map* $f' : (\mathfrak{g}/\mathfrak{t}) \times \mathfrak{t} \to \mathfrak{g}$ *defined as follows: if* $z \in \mathfrak{g}, x \in \mathfrak{t}$, *and if* $\bar{z}$ *denotes the image of* z *in* $\mathfrak{g}/\mathfrak{t}$, *then*

$$f'(\bar{z}, x) = (\mathrm{Ad}\, gt^{-1})(z - (\mathrm{Ad}\, t)z + x).$$

Let F be the map from $G \times T$ to T such that $F(g, t) = gtg^{-1}$. Since $F \circ (\gamma(g), \mathrm{Id}_T) = \mathrm{Int}\, g \circ F$, we have $T_{(g,t)}(F)(gz, tx) = T_t(\mathrm{Int}\, g) \circ T_{(e,t)}(F)(z, tx)$; by Chap. III, §3, no. 12, Prop. 46,

$$T_{(e,t)}(F)(z, tx) = t((\mathrm{Ad}\, t^{-1})z - z) + tx = t((\mathrm{Ad}\, t^{-1})(z - (\mathrm{Ad}\, t)z + x))$$

and consequently

$$T_{(g,t)}(F)(gz, tx) = gtg^{-1}((\mathrm{Ad}\, gt^{-1})(z - (\mathrm{Ad}\, t)z + x)).$$

The lemma follows immediately from this formula by passage to the quotient.

PROPOSITION 4. *a) Let* $g \in G, t \in T, x \in \mathfrak{t}$, *and let* $\bar{g}$ *be the image of* g *in* G/T. *The following conditions are equivalent:*

(i) $t \in T_r$ (resp. $x \in \mathfrak{t}_r$).

(i *bis*) *The element* $f(\bar{g}, t)$ (resp. $\varphi(\bar{g}, x)$) *is regular in* G.

(ii) *The map* f (resp. φ) *is a submersion at the point* $(\bar{g}, t)$ (resp. $(\bar{g}, x)$).

(ii *bis*) *The map* f (resp. φ) *is étale at the point* $(\bar{g}, t)$ (resp. $(\bar{g}, x)$).

b) The map f_r (resp. φ_r, resp. φ_A) *makes* $(G/T) \times T_r$ (resp. $(G/T) \times \mathfrak{t}_r$, resp. $(G/T) \times A$) *a principal covering of* G_r *with group* W (resp. W'_a, resp. H_A).

a) The equivalence of (i) and (i *bis*) is clear; that of (ii) and (ii *bis*) follows from the relations $\dim((G/T) \times T) = \dim((G/T) \times \mathfrak{t}) = \dim(G)$. By Lemma 4, f is a submersion at the point $(\bar{g}, t)$ if and only if $\mathfrak{g} = \mathfrak{t} + \mathrm{Im}(\mathrm{Ad}\, t - \mathrm{Id})$, which means that t is regular. Finally, since $\varphi = f \circ (\mathrm{Id}_{G/T} \times \exp_T)$, φ is étale at the point $(\bar{g}, x)$ if and only if f is étale at the point $(\bar{g}, \exp x)$, which by the preceding means that x belongs to $\mathfrak{t}_r$.

b) The morphisms $f_r, \varphi_r, \varphi_A$ are thus étale. On the other hand, W operates freely on G/T, and *a fortiori* on $(G/T) \times T$. Let g, g' in G and t, t' in T_r be such that $f(\bar{g}, t) = f(\bar{g}', t')$; then $\mathrm{Int}\, g^{-1}g'$ maps t' to t, and hence normalizes T, since $T = Z(t)_0 = Z(t')_0$, and the class w of $g^{-1}g'$ in W maps $(\bar{g}, t)$ to $(\bar{g}', t')$. It follows that f_r is a principal covering with group W; this immediately implies that φ_r is a principal covering with group W'_a, and hence by restriction to the connected component $(G/T) \times A$ of $(G/T) \times \mathfrak{t}_r$, that φ_A is a principal covering with group H_A.

Remarks. 1) By Prop. 3 of §2, no. 4, the manifold $(G/T) \times A$ is simply-connected. It follows that φ_A is a universal covering of G_r; since $\pi_1(G_r)$ is canonically isomorphic to $\pi_1(G)$ (no. 1, Prop. 1 and *General Topology*,

Chap. XI, in preparation), we recover the fact that $\pi_1(G)$ can be identified with H_A (that is with $\Gamma(T)/N(G,T)$).

2) The restriction of φ_A to $W \times A \subset (G/T) \times A$ makes $W \times A$ a principal covering of T_r with group H_A. We thus recover Cor. 1 of Prop. 2 of no. 2.

3) Denote by $\mathfrak{g}_r$ the inverse image of G_r under the exponential map and by $\varepsilon : \mathfrak{g}_r \to G_r$ the map induced by $\exp_G$. The map $(g, x) \mapsto (\mathrm{Ad}\, g)(x)$ from $G \times \mathfrak{t}_r$ to $\mathfrak{g}_r$ defines by passage to the quotient a map $\psi_r : (G/T) \times \mathfrak{t}_r \to \mathfrak{g}_r$. We have $\varepsilon \circ \psi_r = \varphi_r$. Let $w \in W, \gamma \in \Gamma(T)$ and $\omega \in W'_a$ be such that $\omega(z) = w(z) + \gamma$ for all $z \in \mathfrak{t}$; then $\psi_r((\bar{g}, x)\omega) = \psi_r(\bar{g}, x) - (\mathrm{Ad}\, g)(\gamma)$ for $g \in G, x \in \mathfrak{t}_r$, so $\psi_r((\bar{g}, x)\omega) = \psi_r(\bar{g}, x)$ if and only if $\gamma = 0$. It follows (cf. *General Topology*, Chap. XI, in preparation) that ψ_r is a principal covering of $\mathfrak{g}_r$ with group W, and that $\varepsilon : \mathfrak{g}_r \to G_r$ is a covering associated to the principal covering φ_r, with fibre isomorphic to the W'_a-set W'_a/W.

§6. INTEGRATION ON COMPACT LIE GROUPS

We retain the notations of §4; *put* $w(G) = \mathrm{Card}(W_G(T))$. *Denote by* dg (resp. dt) *the Haar measure on* G (resp. T) *with total mass* 1, *and by* n (resp. r) *the dimension of* G (resp. T).

1. PRODUCT OF ALTERNATING MULTILINEAR FORMS

Let A be a commutative ring and M an A-module. For each integer $r \geq 0$, denote by $\mathrm{Alt}^r(M)$ the A-module of alternating r-linear forms on M; it can be identified with the dual of the A-module $\bigwedge^r(M)$ (*Algebra*, Chap. III, §7, no. 4, Prop. 7). Let $u \in \mathrm{Alt}^s(M)$ and $v \in \mathrm{Alt}^r(M)$; recall (*Algebra*, Chap. III, §11, no. 2, Example 3) that the alternating product of u and v is the element $u \wedge v \in \mathrm{Alt}^{s+r}(M)$ defined by

$$(u \wedge v)(x_1, \ldots, x_{s+r}) = \sum_{\sigma \in \mathfrak{S}_{s,r}} \varepsilon_\sigma u(x_{\sigma(1)}, \ldots, x_{\sigma(s)}) v(x_{\sigma(s+1)}, \ldots, x_{\sigma(s+r)}),$$

where $\mathfrak{S}_{s,r}$ is the subset of the symmetric group $\mathfrak{S}_{s+r}$ consisting of the permutations whose restrictions to $[1, s]$ and $[s+1, s+r]$ are increasing.

Now let

$$0 \longrightarrow M' \xrightarrow{i} M \xrightarrow{p} M'' \longrightarrow 0$$

be an exact sequence of free A-modules, of ranks $r, r+s$ and s, respectively.

Lemma 1. There exists an A*-bilinear map from* $\mathrm{Alt}^s(M'') \times \mathrm{Alt}^r(M')$ *to* $\mathrm{Alt}^{s+r}(M)$, *denoted by* $(u, v) \mapsto u \cap v$, *and characterized by either of the following two properties:*

a) Denote by $u_1 \in \mathrm{Alt}^s(\mathrm{M})$ *the form* $(x_1, \ldots, x_s) \mapsto u(p(x_1), \ldots, p(x_s))$, *and let* $v_1 \in \mathrm{Alt}^r(\mathrm{M})$ *be a form such that* $v_1(i(x'_1), \ldots, i(x'_r)) = v(x'_1, \ldots, x'_r)$ *for* $x'_1, \ldots, x'_r$ *in* M'; *then* $u \cap v = u_1 \wedge v_1$.

b) For all $x_1, \ldots, x_s$ *in* M *and* $x'_1, \ldots, x'_r$ *in* M',

$$(u \cap v)(x_1, \ldots, x_s, i(x'_1), \ldots, i(x'_r)) = u(p(x_1), \ldots, p(x_s))v(x'_1, \ldots, x'_r). \quad (1)$$

The map $\varphi : \mathrm{Alt}^s(\mathrm{M}'') \otimes_{\mathrm{A}} \mathrm{Alt}^r(\mathrm{M}') \to \mathrm{Alt}^{s+r}(\mathrm{M})$ *such that* $\varphi(u \otimes v) = u \cap v$ *is an isomorphism of free* A*-modules of rank one.*

The existence of a form v_1 satisfying condition *a)* follows from the fact that $\wedge^r(i)$ induces an isomorphism from $\wedge^r(\mathrm{M}')$ to a direct factor submodule of $\wedge^r(\mathrm{M})$ (*Algebra*, Chap. III, §7, no. 2). Let v_1 be such a form; put $u \cap v = u_1 \wedge v_1$. Formula (1) is then satisfied, since if we put $i(x'_k) = x_{s+k}$ for $1 \leq k \leq r$, the only element σ of $\mathfrak{S}_{s,r}$ such that $p(x_{\sigma(i)}) \neq 0$ for $1 \leq i \leq s$ is the identity permutation. On the other hand, formula (1) determines $u \cap v$ uniquely: indeed, let $(e'_1, \ldots, e'_r)$ be a basis of M', $(f''_1, \ldots, f''_s)$ a basis of M'', and $f_1, \ldots, f_s$ elements of M such that $p(f_i) = f''_i$ for $1 \leq i \leq s$. Then $(f_1, \ldots, f_s, i(e'_1), \ldots, i(e'_r))$ is a basis of M (*Algebra*, Chap. II, §1, no. 11, Prop. 21), and formula (1) can be written

$$(u \cap v)(f_1, \ldots, f_s, i(e'_1), \ldots, i(e'_r)) = u(f''_1, \ldots, f''_s)v(e'_1, \ldots, e'_r); \quad (2)$$

but an element of $\mathrm{Alt}^{s+r}(\mathrm{M})$ is determined by its value on a basis.

It follows from the preceding that each of the conditions *a)* and *b)* determines the product $u \cap v$ uniquely; it is clear that this product is bilinear. Finally, the last assertion of the lemma follows from formula (2).

2. INTEGRATION FORMULA OF H. WEYL

Let e be the identity element of G and $\bar{e}$ its class in G/T. Identify the tangent space of G at e with $\mathfrak{g}$, the tangent space of T at e with $\mathfrak{t}$ and the tangent space of G/T at $\bar{e}$ with $\mathfrak{g}/\mathfrak{t}$. Denote by $(u, v) \mapsto u \cap v$ the **R**-bilinear map

$$\mathrm{Alt}^{n-r}(\mathfrak{g}/\mathfrak{t}) \times \mathrm{Alt}^r(\mathfrak{t}) \to \mathrm{Alt}^n(\mathfrak{g})$$

defined in number 1.

Recall (Chap. III, §3, no. 13, Prop. 50) that the map $\omega \mapsto \omega(e)$ is an isomorphism from the vector space of left-invariant differential forms of degree n (resp. r) on G (resp. T) to the space $\mathrm{Alt}^n(\mathfrak{g})$ (resp. $\mathrm{Alt}^r(\mathfrak{t})$). Further, observe that, since every connected compact subgroup of $\mathbf{R}^*$ reduces to the identity element, $\det \mathrm{Ad}\, g = 1$ for all $g \in \mathrm{G}$, so that the left-invariant differential forms of degree n on G are also right-invariant and invariant under inner automorphisms (Chap. III, §3, no. 16, Cor. of Prop. 54): we shall speak simply of G-invariant differential forms from now on.

Similarly, it follows from Chap. III, §3, no. 16, Prop. 56 and the preceding that the map $\omega \mapsto \omega(\bar{e})$ is an isomorphism from the space of G-invariant differential forms of degree $n - r$ on G/T to the space $\mathrm{Alt}^{n-r}(\mathfrak{g}/\mathfrak{t})$.

If $\omega_{G/T}$ is a G-invariant differential form of degree $n-r$ on G/T, and ω_T an invariant differential form of degree r on T, denote by $\omega_{G/T} \cap \omega_T$ the unique invariant differential form of degree n on G such that

$$(\omega_{G/T} \cap \omega_T)(e) = \omega_{G/T}(\bar{e}) \cap \omega_T(e).$$

Recall finally that $f : (G/T) \times T \to G$ denotes the morphism of manifolds induced by the map $(g,t) \mapsto gtg^{-1}$ from $G \times T$ to G by passage to the quotient (§5, no. 4). If α and β are differential forms on G/T and T, respectively, denote simply by $\alpha \wedge \beta$ the form $\mathrm{pr}_1^*\alpha \wedge \mathrm{pr}_2^*\beta$ on $(G/T) \times T$.

For $t \in T$, denote by $\mathrm{Ad}_{\mathfrak{g}/\mathfrak{t}}(t)$ the endomorphism of $\mathfrak{g}/\mathfrak{t}$ induced by $\mathrm{Ad}\, t$ by passage to the quotient. Put

$$\delta_G(t) = \det(\mathrm{Ad}_{\mathfrak{g}/\mathfrak{t}}(t) - 1) = \prod_{\alpha \in R(G,T)} (t^\alpha - 1). \tag{3}$$

Let $x \in \mathfrak{t}$ and $\alpha \in R(G,T)$; denote by $\hat{\alpha}$ the element $(2\pi i)^{-1}\delta(\alpha)$ of $\mathfrak{t}^*$, so that

$$((\exp x)^\alpha - 1)((\exp x)^{-\alpha} - 1) = (e^{2\pi i \hat{\alpha}(x)} - 1)(e^{-2\pi i \hat{\alpha}(x)} - 1) = 4\sin^2 \pi\hat{\alpha}(x).$$

If $R_+(G,T)$ denotes the set of positive roots of R(G, T) relative to a basis B, we have

$$\delta_G(\exp x) = \prod_{\alpha \in R_+(G,T)} 4\sin^2 \pi\hat{\alpha}(x),$$

so, in particular, $\delta_G(t) > 0$ for all $t \in T_r$. We remark also that $\delta_G(t) = \delta_G(t^{-1})$ for $t \in T$.

PROPOSITION 1. *Let $\omega_G, \omega_{G/T}$ and ω_T be invariant differential forms on G, G/T and T, respectively, of respective degrees $n, n-r$ and r. If $\omega_G = \omega_{G/T} \cap \omega_T$, then*

$$f^*(\omega_G) = \omega_{G/T} \wedge \delta_G\omega_T.$$

Clearly we can assume that $\omega_{G/T}$ and ω_T are non-zero; then the differential form $(u,t) \mapsto \omega_{G/T}(u) \wedge \omega_T(t)$ on $(G/T) \times T$ is of degree n and everywhere non-zero; hence there exists a numerical function δ on $(G/T) \times T$ such that

$$f^*(\omega_G)(u,t) = \delta(u,t)\omega_{G/T}(u) \wedge \omega_T(t).$$

Observe now that, for $h \in G$, $u \in G/T$, $t \in T$, we have $f(h.u,t) = (\mathrm{Int}\, h)f(u,t)$; since ω_G is invariant under inner automorphisms, it follows immediately that $\delta(h.u,t) = \delta(u,t)$, so $\delta(u,t) = \delta(\bar{e},t)$.

Denote by $p : \mathfrak{g} \to \mathfrak{g}/\mathfrak{t}$ the quotient map and by $\varphi : \mathfrak{g}/\mathfrak{t} \to \mathfrak{g}$ the map defined by

$$\varphi(p(X)) = (\mathrm{Ad}\, t^{-1})X - X \quad \text{for } X \in \mathfrak{g};$$

recall (§5, no. 4, Lemma 4) that the tangent map

$$T_{(e,t)}(f) : T_e(G/T) \times T_t(T) \to T_t(G)$$

takes (z, tH) to $t(\varphi(z) + H)$ for $z \in \mathfrak{g}/\mathfrak{t}, H \in \mathfrak{t}$.

Let $z_1, \ldots, z_{n-r}$ be elements of $\mathfrak{g}/\mathfrak{t}$, $H_1, \ldots, H_r$ elements of $\mathfrak{t}$. Then

$$\begin{aligned}
&f^*\omega_G(\bar{e}, t)(z_1, \ldots, z_{n-r}, tH_1, \ldots, tH_r)\\
&\quad = \omega_G(t)(t\varphi(z_1), \ldots, t\varphi(z_{n-r}), tH_1, \ldots, tH_r) \text{ by the calculation of } T_{(\bar{e},t)}(f)\\
&\quad = \omega_G(e)(\varphi(z_1), \ldots, \varphi(z_{n-r}), H_1, \ldots, H_r) \quad \text{since } \omega_G \text{ is invariant}\\
&\quad = \omega_{G/T}(\bar{e})(p\varphi(z_1), \ldots, p\varphi(z_{n-r})).\omega_T(e)(H_1, \ldots, H_r) \quad \text{(no. 1, Lemma 1)}\\
&\quad = \det(p\varphi)\omega_{G/T}(\bar{e})(z_1, \ldots, z_{n-r}).\omega_T(e)(H_1, \ldots, H_r)\\
&\quad = \delta_G(t)\omega_{G/T}(\bar{e})(z_1, \ldots, z_{n-r}).\omega_T(t)(tH_1, \ldots, tH_r)\\
&\qquad\qquad\qquad\qquad \text{since } \omega_T \text{ is invariant}\\
&\quad = \delta_G(t)(\omega_{G/T} \wedge \omega_T)(\bar{e}, t)(z_1, \ldots, z_{n-r}, tH_1, \ldots, tH_r),
\end{aligned}$$

so $f^*\omega_G(\bar{e}, t) = \delta_G(t)(\omega_{G/T} \wedge \omega_T)(\bar{e}, t)$; thus, $\delta(\bar{e}, t) = \delta_G(t)$, hence the proposition.

Give the manifolds G, T and G/T the orientations defined by the forms ω_G, ω_T and $\omega_{G/T}$, respectively. These forms define invariant measures on G, T and G/T (Chap. III, §3, no. 16, Props. 55 and 56), also denoted by ω_G, ω_T and $\omega_{G/T}$.

Lemma 2. If $\omega_G = \omega_{G/T} \cap \omega_T$, *then*

$$\int_G \omega_G = \int_{G/T} \omega_{G/T} \cdot \int_T \omega_T.$$

Denote by π the canonical morphism from G to G/T. Let $g \in G$, and let $t_1, \ldots, t_{n-r}$ be elements of $T_{\pi(g)}(G/T)$. Identify the fibre $\pi^{-1}(\pi(g)) = gT$ with T by the translation $\gamma(g)$. The relation $\omega_G = \omega_{G/T} \cap \omega_T$ now implies the equality (*Differentiable and Analytic Manifolds, Results*, 11.4.5):

$$\omega_G \llcorner (t_1, \ldots, t_{n-r}) = (\omega_{G/T}(t_1, \ldots, t_{n-r}))\omega_T.$$

Thus $\int_\pi \omega_G = \left(\int_T \omega_T\right) \omega_{G/T}$ (*Differentiable and Analytic Manifolds, Results*, 11.4.6), and

$$\int_G \omega_G = \int_{G/T} \int_\pi \omega_G = \int_T \omega_T \cdot \int_{G/T} \omega_{G/T}$$

(*Differentiable and Analytic Manifolds, Results*, 11.4.8).

Lemma 3. The inverse image on $(G/T) \times T_r$ *of the measure* dg *on* G_r *under the local homeomorphism* f_r (*Integration*, Chap. V, § 6, no. 6) *is the measure* $\mu \otimes \delta_G\, dt$, *where* μ *is the unique* G*-invariant measure on* G/T *of total mass* 1.

Choose an invariant differential form ω_{T} (resp. $\omega_{\mathrm{G/T}}$) on T (resp. G/T) of maximal degree, such that the measure defined by ω_{T} (resp. $\omega_{\mathrm{G/T}}$) is equal to dt (resp. μ). Put $\omega_{\mathrm{G}} = \omega_{\mathrm{G/T}} \cap \omega_{\mathrm{T}}$. Lemma 2 implies that the measure defined by ω_{G} is equal to dg. Let U be an open subset of $(\mathrm{G/T}) \times \mathrm{T}_r$ such that f_r induces an isomorphism from U to an open subset V of G_r. Let φ be a continuous function with compact support in V; denote also by φ the extension of φ to G_r which vanishes outside V. We have

$$\begin{aligned}\int_{\mathrm{V}} \varphi\, dg = \int_{\mathrm{V}} \varphi\, \omega_{\mathrm{G}} &= \int_{\mathrm{U}} (\varphi \circ f_r) f_r^*(\omega_{\mathrm{G}}) \\ &= \int_{\mathrm{U}} (\varphi \circ f_r) \omega_{\mathrm{G/T}} \wedge \delta_{\mathrm{G}} \omega_{\mathrm{T}} \qquad \text{(Prop. 1)} \\ &= \int_{\mathrm{U}} (\varphi \circ f_r) d\mu . \delta_{\mathrm{G}}\, dt,\end{aligned}$$

hence the lemma.

THEOREM 1 (H. Weyl). *The measure dg on* G *is the image under the map* $(g,t) \mapsto gtg^{-1}$ *from* $\mathrm{G} \times \mathrm{T}$ *to* G *of the measure* $dg \otimes \frac{1}{w(\mathrm{G})} \delta_{\mathrm{G}}\, dt$, *where*

$$\delta_{\mathrm{G}}(t) = \det(\mathrm{Ad}_{\mathfrak{g}/\mathfrak{t}}(t) - 1) = \prod_{\alpha \in \mathrm{R(G,T)}} (t^{\alpha} - 1).$$

Equivalently (*Integration*, Chap. V, §6, no. 3, Prop. 4), dg is the image under the map $f : (\mathrm{G/T}) \times \mathrm{T} \to \mathrm{G}$ of the measure $\mu \otimes \frac{1}{w(\mathrm{G})} \delta_{\mathrm{G}}\, dt$.

We prove the last assertion. It follows from §5, no. 1 and *Differentiable and Analytic Manifolds, Results*, 10.1.3 c) that $\mathrm{G} - \mathrm{G}_r$ is negligible in G and $\mathrm{T} - \mathrm{T}_r$ is negligible in T. Further, the map f_r makes $(\mathrm{G/T}) \times \mathrm{T}_r$ a principal covering of G_r, with group W (§5, no. 4, Prop. 4 *b*)). The theorem now follows from Lemma 3 and *Integration*, Chap. V, §6, no. 6, Prop. 11.

COROLLARY 1. (i) *Let φ be an integrable function on* G *with values in a Banach space or in* $\overline{\mathbf{R}}$. *For almost all $t \in \mathrm{T}$, the function $g \mapsto \varphi(gtg^{-1})$ on* G *is integrable for dg. The function $t \mapsto \delta_{\mathrm{G}}(t) \int_{\mathrm{G}} \varphi(gtg^{-1})\, dg$ is integrable on* T, *and we have*

$$\int_{\mathrm{G}} \varphi(g)\, dg = \frac{1}{w(\mathrm{G})} \int_{\mathrm{T}} \left(\int_{\mathrm{G}} \varphi(gtg^{-1}) dg \right) \delta_{\mathrm{G}}(t) dt \tag{4}$$

("Hermann Weyl's integration formula").

(ii) *Let φ be a positive measurable function on* G. *For almost all $t \in \mathrm{T}$, the function $g \mapsto \varphi(gtg^{-1})$ on* G *is measurable. The function $t \mapsto \int_{\mathrm{G}}^* \varphi(gtg^{-1}) dg$ on* T *is measurable, and we have*

$$\int_{\mathrm{G}}^* \varphi(g) dg = \frac{1}{w(\mathrm{G})} \int_{\mathrm{T}}^* \left(\int_{\mathrm{G}}^* \varphi(gtg^{-1}) dg \right) \delta_{\mathrm{G}}(t) dt. \tag{5}$$

Since the map f is induced by passage to the quotient from the map $(g,t) \mapsto gtg^{-1}$ from $\mathrm{G} \times \mathrm{T} \to \mathrm{G}$, it suffices to apply *Integration*, Chap. V, §5, 6, 8 and *Integration*, Chap. VII, §2.

COROLLARY 2. *Let* φ *be a central function on* G (that is, such that $\varphi(gh) = \varphi(hg)$ for all g and h in G) *with values in a Banach space or in* $\overline{\mathbf{R}}$.

a) φ *is measurable if and only if its restriction to* T *is measurable.*

b) φ *is integrable if and only if the function* $(\varphi|\mathrm{T})\delta_{\mathrm{G}}$ *is integrable on* T, *and in that case we have*

$$\int_{\mathrm{G}} \varphi(g)dg = \frac{1}{w(\mathrm{G})} \int_{\mathrm{T}} \varphi(t)\delta_{\mathrm{G}}(t)dt. \tag{6}$$

Denote by $p : \mathrm{G}/\mathrm{T} \times \mathrm{T} \to \mathrm{T}$ the second projection. We have $\varphi \circ f = (\varphi|\mathrm{T}) \circ p$; further, the image under p of the measure $\mu \otimes \frac{1}{w(\mathrm{G})}\delta_{\mathrm{G}}\, dt$ is $\frac{1}{w(\mathrm{G})}\delta_{\mathrm{G}}\, dt$. The corollary now follows from Th. 1 above and Th. 1 of *Integration*, Chap. V, §6, no. 2, applied to the two proper maps f and p.

COROLLARY 3. *Let* H *be a connected closed subgroup of* G *containing* T, $\mathfrak{h}$ *its Lie algebra, and* dh *the Haar measure on* H *of total mass* 1. *Let* φ *be an integrable central function on* G, *with values in a Banach space or in* $\overline{\mathbf{R}}$. *Then the function* $h \mapsto \varphi(h)\det(\mathrm{Ad}_{\mathfrak{g}/\mathfrak{h}}(h) - 1)$ *is integrable and central on* H *and we have*

$$\int_{\mathrm{G}} \varphi(g)dg = \frac{w(\mathrm{H})}{w(\mathrm{G})} \int_{\mathrm{H}} \varphi(h)\det(\mathrm{Ad}_{\mathfrak{g}/\mathfrak{h}}(h) - 1)dh. \tag{7}$$

Indeed, the function $h \mapsto \varphi(h)\det(\mathrm{Ad}_{\mathfrak{g}/\mathfrak{h}}(h) - 1)$ is a central function on H whose restriction to T is the function $t \mapsto \varphi(t)\delta_{\mathrm{G}}(t)\delta_{\mathrm{H}}(t)^{-1}$. Thus, the corollary follows from Cor. 2 applied to G and to H.

Remarks. 1) If we take $\varphi = 1$ in Cor. 3, we obtain

$$\int_{\mathrm{H}} \det(\mathrm{Ad}_{\mathfrak{g}/\mathfrak{h}}(h) - 1)dh = w(\mathrm{G})/w(\mathrm{H}) \tag{8}$$

and in particular

$$\int_{\mathrm{T}} \delta_{\mathrm{G}}(t)dt = w(\mathrm{G}). \tag{9}$$

2) Let ν be the measure on the quotient T/W defined by

$$\int_{\mathrm{T}/\mathrm{W}} \psi(\tau)d\nu(\tau) = \frac{1}{w(\mathrm{G})} \int_{\mathrm{T}} \psi(\pi(t))\delta_{\mathrm{G}}(t)dt,$$

where π denotes the canonical projection of T onto T/W. Cor. 2 means that the homeomorphism $\mathrm{T}/\mathrm{W} \to \mathrm{G}/\mathrm{Int}(\mathrm{G})$ (§2, no. 5, Cor. 1 of Prop. 5) transports the measure ν to the image of the measure dg under the canonical projection $G \to G/\mathrm{Int}(G)$.

3) Assume that G is simply-connected. Let A be an alcove of $\mathfrak{t}$, and dx the Haar measure on $\mathfrak{t}$ such that $\int_A dx = 1$. Then the measure ν can also be obtained by transporting the measure $\frac{1}{w(G)} \prod_{\alpha \in R_+(G,T)} 4\sin^2 \pi\hat{\alpha}(x)dx$ on $\overline{A}$ by the homeomorphism $\overline{A} \to T/W$ (§5, no. 2, Cor. 1 of Prop. 2).

Example. Take G to be the group $\mathbf{SU}(2, \mathbf{C})$ and T to be the subgroup of diagonal matrices (§3, no. 6); identify $\mathfrak{t}$ with $\mathbf{R}$ by the choice of basis $\{iH\}$ of $\mathfrak{t}$ (*loc. cit.*). Put $A =]0, \pi[$; this is an alcove of $\mathfrak{t}$. The interval $\overline{A} = [0, \pi]$ can be identified with the space of conjugacy classes of G, the element θ of $\overline{A}$ corresponding to the conjugacy class of $\begin{pmatrix} e^{i\theta} & 0 \\ 0 & e^{-i\theta} \end{pmatrix}$. Let $d\theta$ be Lebesgue measure on $[0, \pi]$; it follows from the preceding that the image on $\overline{A}$ of the Haar measure on G is the measure $\frac{2}{\pi}\sin^2\theta\, d\theta$.

3. INTEGRATION ON LIE ALGEBRAS

PROPOSITION 2. *Let* H *be a* (real) *Lie group of dimension* m, $\mathfrak{h}$ *its Lie algebra. Let* ω_H *be a right-invariant differential form of degree* m *on* H, *and let* $\omega_{\mathfrak{h}}$ *be the translation-invariant differential form on* $\mathfrak{h}$, *of degree* m, *that coincides with* $\omega_H(e)$ *at the origin. We have*

$$(\exp_H)^*\omega_H = \lambda_{\mathfrak{h}}\omega_{\mathfrak{h}} \tag{10}$$

where $\lambda_{\mathfrak{h}}$ *is the* $\mathrm{Ad}(H)$-*invariant function on* $\mathfrak{h}$ *such that*

$$\lambda_{\mathfrak{h}}(x) = \det\left(\sum_{p \geq 0} \frac{1}{(p+1)!}(\mathrm{ad}\ x)^p\right) \quad \text{for } x \in \mathfrak{h}.$$

Let $x, x_1, \ldots, x_m$ be elements of $\mathfrak{h}$. We have

$$(\exp^*\omega_H)_x(x_1, \ldots, x_m) = (\omega_H(\exp x))(T_x(\exp)(x_1), \ldots, T_x(\exp)(x_m)).$$

Denote by $\varpi(x) : \mathfrak{h} \to \mathfrak{h}$ the right differential of the exponential at x (Chap. III, §3, no. 17, Def. 8); by definition,

$$T_x(\exp)(y).(\exp x)^{-1} = \varpi(x).y \quad \text{for all } y \in \mathfrak{h}.$$

The form ω_H being right invariant, we obtain

$$\begin{aligned}(\omega_H(\exp x))&(T_x(\exp)(x_1), \ldots, T_x(\exp)(x_m)) \\ &= \omega_H(e)(\varpi(x).x_1, \ldots, \varpi(x).x_m) = (\det \varpi(x))\omega_{\mathfrak{h}}(x_1, \ldots, x_m);\end{aligned}$$

thus, $\exp^*\omega_H = \lambda_{\mathfrak{h}}\omega_{\mathfrak{h}}$, with $\lambda_{\mathfrak{h}}(x) = \det \varpi(x) = \det\frac{\exp \mathrm{ad}\ x - 1}{\mathrm{ad}\ x}$ (Chap. III, §6, no. 4, Prop. 12).

Let $h \in \mathrm{H}$; since $\mathrm{Ad}\, h$ is an automorphism of $\mathfrak{h}$, we have

$$\mathrm{ad}\,((\mathrm{Ad}\, h)(x)) = \mathrm{Ad}\, h \circ \mathrm{Ad}\, x \circ (\mathrm{Ad}\, h)^{-1},$$

so $\lambda_{\mathfrak{h}}((\mathrm{Ad}\, h)(x)) = \lambda_{\mathfrak{h}}(x)$. Thus, the function $\lambda_{\mathfrak{h}}$ is invariant under $\mathrm{Ad(H)}$; this completes the proof of the proposition.

Remark. Consider the function $\lambda_{\mathfrak{g}}$ associated to a compact Lie group G; in view of §2, no. 1, Th. 1, to calculate $\lambda_{\mathfrak{g}}$ it suffices to know its values on $\mathfrak{t}$. But, for $x \in \mathfrak{t}$, the endomorphism $\mathrm{ad}\, x$ of $\mathfrak{g}$ is semi-simple, and has eigenvalues 0 (with multiplicity r) and, for all $\alpha \in \mathrm{R(G,T)}$, $\delta(\alpha)(x)$ (with multiplicity 1). It follows immediately that

$$\lambda_{\mathfrak{g}}(x) = \prod_{\alpha \in \mathrm{R(G,T)}} \frac{e^{\delta(\alpha)(x)} - 1}{\delta(\alpha)(x)} = \frac{\delta_{\mathfrak{g}}(x)}{\pi_{\mathfrak{g}}(x)} \tag{11}$$

with $\delta_{\mathfrak{g}}(x) = \delta_{\mathrm{G}}(\exp x)$ and $\pi_{\mathfrak{g}}(x) = \prod_{\alpha \in \mathrm{R(G,T)}} \delta(\alpha)(x) = \det \mathrm{ad}_{\,\mathfrak{g}/\mathfrak{t}}(x)$.

Let $\omega_{\mathrm{G/T}}$ be an invariant differential form of degree $n - r$ on G/T and $\omega_{\mathfrak{t}}$ a translation-invariant differential form of degree r on $\mathfrak{t}$. With the notation of no. 1, denote by $\omega_{\mathrm{G/T}} \cap \omega_{\mathfrak{t}}$ the unique translation-invariant differential form $\omega_{\mathfrak{g}}$ of degree n on $\mathfrak{g}$ such that $\omega_{\mathfrak{g}}(0) = \omega_{\mathrm{G/T}}(\bar{e}) \cap \omega_{\mathfrak{t}}(0)$.

Finally, denote by $\psi : (\mathrm{G/T}) \times \mathfrak{t} \to \mathfrak{g}$ the morphism of manifolds induced by the map $(g, x) \mapsto (\mathrm{Ad}\, g)(x)$ from $\mathrm{G} \times \mathfrak{t}$ to $\mathfrak{g}$ by passage to the quotient.

PROPOSITION 3. *Let $\omega_{\mathfrak{g}}$, $\omega_{\mathfrak{t}}$, $\omega_{\mathrm{G/T}}$ be invariant differential forms on $\mathfrak{g}$, $\mathfrak{t}$, G/T, respectively, of respective degrees $n, r, n-r$. If $\omega_{\mathfrak{g}} = \omega_{\mathrm{G/T}} \cap \omega_{\mathfrak{t}}$, we have*

$$\psi^* \omega_{\mathfrak{g}} = \omega_{\mathrm{G/T}} \wedge \pi_{\mathfrak{g}} \omega_{\mathfrak{t}}$$

where $\pi_{\mathfrak{g}}$ is the function on $\mathfrak{t}$ defined by $\pi_{\mathfrak{g}}(x) = \prod_{\alpha \in \mathrm{R(G,T)}} \delta(\alpha)(x)$.

Denote by ω_{G} (resp. ω_{T}) the invariant differential form of maximum degree on G (resp. T) that coincides with $\omega_{\mathfrak{g}}$ (resp. $\omega_{\mathfrak{t}}$) at the origin. Consider the commutative diagram

$$\begin{array}{ccc} (\mathrm{G/T}) \times \mathfrak{t} & \xrightarrow{\psi} & \mathfrak{g} \\ \Big\downarrow {\scriptstyle (\mathrm{Id}, \exp_{\mathrm{T}})} & & \Big\downarrow {\scriptstyle \exp_{\mathrm{G}}} \\ (\mathrm{G/T}) \times \mathrm{T} & \xrightarrow{f} & \mathrm{G} \end{array} .$$

In view of Prop. 1 of no. 2 and the relation $\exp_{\mathrm{T}}^* \omega_{\mathrm{T}} = \omega_{\mathfrak{t}}$, we deduce the equality

$$\psi^* \exp_{\mathrm{G}}^* \omega_{\mathrm{G}} = \omega_{\mathrm{G/T}} \wedge \delta_{\mathfrak{g}} \omega_{\mathfrak{t}}.$$

By Prop. 2, $\psi^* \exp_{\mathrm{G}}^* \omega_{\mathrm{G}} = (\psi^* \lambda_{\mathfrak{g}}) \psi^* \omega_{\mathfrak{g}}$. Since the function $\lambda_{\mathfrak{g}}$ is invariant under $\mathrm{Ad(G)}$, we have

$$(\psi^*\lambda_{\mathfrak{g}})(\bar{g},x) = \lambda_{\mathfrak{g}}(x) = \frac{\delta_{\mathfrak{g}}(x)}{\pi_{\mathfrak{g}}(x)} \quad \text{for } \bar{g} \in \mathrm{G/T},\ x \in \mathfrak{t}.$$

It follows that the forms $\psi^*\omega_{\mathrm{G}}(\bar{g},x)$ and $\omega_{\mathrm{G/T}}(\bar{g}) \wedge \pi_{\mathfrak{g}}(x)\omega_{\mathfrak{t}}(x)$ coincide where $\delta_{\mathfrak{g}}(x)$ is non-zero, that is on the dense open subset $(\mathrm{G/T}) \times \mathfrak{t}_r$; thus, they are equal, hence the proposition.

Choose invariant differential forms ω_{G} on G and ω_{T} on T, of maximum degree, such that $|\omega_{\mathrm{G}}| = dg$ and $|\omega_{\mathrm{T}}| = dt$; denote by $\omega_{\mathfrak{g}}$ (resp. $\omega_{\mathfrak{t}}$) the translation-invariant differential form on $\mathfrak{g}$ (resp. $\mathfrak{t}$) that coincides with $\omega_{\mathrm{G}}(e)$ (resp. $\omega_{\mathrm{T}}(e)$) at the origin, and dz (resp. dx) the Haar measure $|\omega_{\mathfrak{g}}|$ (resp. $|\omega_{\mathfrak{t}}|$). Reasoning as in no. 2, *mutatis mutandis*, gives the following proposition:

PROPOSITION 4. *The measure dz on $\mathfrak{g}$ is the image under the proper map $(g,x) \mapsto (\mathrm{Ad}\,g)(x)$ from $\mathrm{G} \times \mathfrak{t}$ to $\mathfrak{g}$ of the measure $dg \otimes \frac{1}{w(\mathrm{G})}\pi_{\mathfrak{g}}\,dx$.*

We leave to the reader the statement and proof of the analogues of Cor. 1 to 3 and Remarks 1 to 3 of no. 2. For example, let φ be an integrable function on $\mathfrak{g}$ (with values in a Banach space or $\overline{\mathbf{R}}$); then

$$\int_{\mathfrak{g}} \varphi(z)dz = \frac{1}{w(\mathrm{G})}\int_{\mathfrak{t}}\left(\int_{\mathrm{G}} \varphi((\mathrm{Ad}\,g)x)dg\right)\pi_{\mathfrak{g}}(x)dx, \tag{12}$$

and, in particular, if φ is invariant under Ad(G),

$$\int_{\mathfrak{g}} \varphi(z)dz = \frac{1}{w(\mathrm{G})}\int_{\mathfrak{t}} \varphi(x)\pi_{\mathfrak{g}}(x)dx. \tag{13}$$

4. INTEGRATION OF SECTIONS OF A VECTOR BUNDLE

In this number and the next, we denote by X a real manifold of class C^r $(1 \leq r \leq \infty)$, locally of finite dimension.

Let Y be a manifold of class C^r. If $r < \infty$, consider the map $f \mapsto j^r(f)$ from $\mathscr{C}^r(\mathrm{X};\mathrm{Y})$ to $\mathscr{C}(\mathrm{X};\mathrm{J}^r(\mathrm{X},\mathrm{Y}))$ (*Differentiable and Analytic Manifolds, Results*, 12.3.7). The inverse image under this map of the topology of compact convergence on $\mathscr{C}(\mathrm{X};\mathrm{J}^r(\mathrm{X},\mathrm{Y}))$ is called the *topology of compact C^r-convergence* on $\mathscr{C}^r(\mathrm{X};\mathrm{Y})$; it is the upper bound of the topologies of uniform C^r-convergence on K (*Differentiable and Analytic Manifolds, Results*, 12.3.10), where K runs through the set of compact subsets of X.

When $r = \infty$, we call the *topology of compact C^∞-convergence* on $\mathscr{C}^\infty(\mathrm{X};\mathrm{Y})$ the upper bound of the topologies of compact C^k-convergence, in other words the coarsest topology for which the canonical injections $\mathscr{C}^\infty(\mathrm{X};\mathrm{Y}) \to \mathscr{C}^k(\mathrm{X};\mathrm{Y})$ are continuous for $0 \leq k < \infty$.

Let E be a real vector bundle with base X, of class C^r, and let $\mathscr{S}^r(\mathrm{X};\mathrm{E})$ be the vector space of sections of E of class C^r. In this number we give $\mathscr{S}^r(\mathrm{X};\mathrm{E})$ the topology induced by the topology of compact C^r-convergence on $\mathscr{C}^r(\mathrm{X};\mathrm{E})$, also called the topology of compact C^r-convergence; it makes

$\mathscr{S}^r(X;E)$ into a *complete* separated locally convex topological vector space (cf. *Differentiable and Analytic Manifolds, Results*, 15.3.1 and *Spectral Theories*, in preparation).

Now let H be a Lie group, $m : H \times X \to X$ a law of left operation of class C^r; put $hx = m(h,x)$ for $h \in H, x \in X$. Let E be a vector H-bundle with base X, of class C^r (Chap. III, §1, no. 8, Def. 4). For $s \in \mathscr{S}^r(X;E)$ and $h \in H$, denote by ${}^h s$ the section $x \mapsto h.s(h^{-1}x)$ of E; the map $(h,s) \mapsto {}^h s$ is a law of operation of H on the space $\mathscr{S}^r(X;E)$.

Lemma 4. The law of operation $H \times \mathscr{S}^r(X;E) \to \mathscr{S}^r(X;E)$ *is continuous.*

In view of the definition of the topology of $\mathscr{S}^r(X;E)$ and *General Topology*, Chap. X, §3, no. 4, Th. 3, it suffices to prove that for any integers $k \leq r$, the map $f : H \times X \times \mathscr{S}^k(X;E) \to J^k(X;E)$ such that $f(h,x,s) = j^k_x({}^h s)$ is continuous. For $h \in H$, denote by τ_h (resp. θ_h) the automorphism $x \mapsto hx$ of X (resp. of E). Define maps

$$\begin{aligned} f_1 &: H \times X \to J^k(X,X) \\ f_2 &: H \times E \to J^k(E,E) \\ g &: H \times X \times \mathscr{S}^k(X;E) \to J^k(X,E) \end{aligned}$$

by $f_1(h,x) = j^k_x(\tau_h), f_2(h,v) = j^k_v(\theta_h), g(h,x,s) = j^k_{hx}(s)$. We have

$$f(h,x,s) = f_2(h, s(h^{-1}x)) \circ g(h^{-1},x,s) \circ f_1(h^{-1},x),$$

and consequently it suffices, by *Differentiable and Analytic Manifolds, Results*, 12.3.6, to prove that f_1, f_2 and g are continuous.

Now g is the composite map

$$H \times X \times \mathscr{S}^k(X;E) \xrightarrow{(m,\mathrm{Id})} X \times \mathscr{S}^k(X;E) \xrightarrow{(\mathrm{Id},j^k)} X \times \mathscr{C}(X; J^k(X,E)) \xrightarrow{\varepsilon} J^k(X,E)$$

with $\varepsilon(x,u) = u(x)$; the map ε being continuous (*General Topology*, Chap. X, §3, no. 4, Cor. 1 of Th. 3), g is continuous.

Let $(h_0,x_0) \in H \times X$; we shall prove that f_1 is continuous at (h_0,x_0). There exist charts (U,ψ,F) and (V,χ,F') of X and an open subset Ω of H such that $x_0 \in U, h_0 \in \Omega$ and $m(\Omega \times U) \subset V$. By using the expression for $J^k(X,X)$ in these charts, we are reduced to proving, for $1 \leq l \leq k$, the continuity at (h_0,x_0) of the map $(h,x) \mapsto \Delta^l_x(\tau_h)$ from $\Omega \times U$ to $P_l(F;F')$, with $\Delta^l_x(\tau_h)(v) = \frac{1}{l!}D^l\tau_h(x).v$ for $v \in F$ (*Differentiable and Analytic Manifolds, Results*, 12.2). But $D^l\tau_h(x)$ is simply the lth partial derivative of $m(h,x)$ with respect to x, which is continuous by hypothesis; consequently, f_1 is continuous. The proof that f_2 is continuous is similar, hence the lemma.

PROPOSITION 5. *Assume that the group* H *is compact and denote by* dh *the Haar measure on* H *of total mass* 1. *Let* s *be a section of* E *of class* C^r.

Denote by $s^\sharp$ *the vector integral* $\int_{\mathrm{H}} {}^h s\, dh$. *Then* $s^\sharp$ *is a section of* E *of class* C^r, *invariant under* H; *for* $x \in \mathrm{X}$, *we have* $s^\sharp(x) = \int_{\mathrm{H}} hs(h^{-1}x)\, dh \in \mathrm{E}_x$. *The endomorphism* $s \mapsto s^\sharp$ *of* $\mathscr{S}^r(\mathrm{X};\mathrm{E})$ *is a projection onto the subspace of* H-*invariant sections.*

Consider the map $h \mapsto {}^h s$ from H to $\mathscr{S}^r(\mathrm{X};\mathrm{E})$; it is continuous by Lemma 4. Since the space $\mathscr{S}^r(\mathrm{X};\mathrm{E})$ is separated and complete, the integral $s^\sharp = \int_{\mathrm{H}} {}^h s\, dh$ belongs to $\mathscr{S}^r(\mathrm{X};\mathrm{E})$ (*Integration*, Chap. III, §3, no. 3, Cor. 2). The linear map $s \mapsto s(x)$ from $\mathscr{S}^r(\mathrm{X};\mathrm{E})$ to E_x being continuous, we have $s^\sharp(x) = \int_{\mathrm{H}} {}^h s(x)\, dh$ for all $x \in \mathrm{X}$. It is clear that $s^\sharp$ is invariant under H; if s is an H-invariant section, we have $s^\sharp = s$, hence the last assertion.

COROLLARY 1. *Let* F *be a Banach space,* $\rho : \mathrm{H} \to \mathbf{GL}(\mathrm{F})$ *an analytic linear representation,* $f \in \mathscr{C}^r(\mathrm{X};\mathrm{F})$. *For* $x \in \mathrm{X}$, *put*

$$f^\sharp(x) = \int_{\mathrm{H}} \rho(h).f(h^{-1}x)\, dh.$$

Then $f^\sharp$ *is a morphism of class* C^r *from* X *to* F, *compatible with the operations of* H; *for* $x \in \mathrm{X}$, *we have (with* τ_h *denoting the automorphism* $x \mapsto hx$ *of* X*)*

$$d_x f^\sharp = \int_{\mathrm{H}} (\rho(h) \circ d_{h^{-1}x} f \circ \mathrm{T}_x(\tau_{h^{-1}}))\, dh \in \mathscr{L}(\mathrm{T}_x(\mathrm{X});\mathrm{F}). \tag{14}$$

The first assertion follows from the proposition applied to the bundle $\mathrm{X} \times \mathrm{F}$, equipped with the law of operation $(h;(x,f)) \mapsto (hx, \rho(h).f)$. The second follows from *Integration*, Chap. III, §3, no. 2, Prop. 2, by applying to the vector integral $f^\sharp$ the homomorphism $d_x : \mathscr{C}^r(\mathrm{X};\mathrm{F}) \to \mathscr{L}(\mathrm{T}_x(\mathrm{X});\mathrm{F})$ which is continuous by the definition of the topology of compact C^r-convergence.

COROLLARY 2. *Let* F *be a Banach space,* $f \in \mathscr{C}^r(\mathrm{X};\mathrm{F})$; *put*

$$f^\sharp(x) = \int_{\mathrm{H}} f(hx)\, dh$$

for $x \in \mathrm{X}$. *The function* $f^\sharp$ *is of class* C^r, *and* $f^\sharp(hx) = f^\sharp(x)$ *for* $x \in \mathrm{X}$, $h \in \mathrm{H}$.

COROLLARY 3. *Let* F *be a Banach space,* p *an integer* ≥ 0, ${}^k\Omega^p(\mathrm{X};\mathrm{F})$ *the space of differential forms of degree* p *on* X, *with values in* F, *and of class* C^k $(2 \leq k+1 \leq r)$. *For* $\omega \in {}^k\Omega^p(\mathrm{X};\mathrm{F})$, *put* $\omega^\sharp = \int_{\mathrm{H}} \tau_h^* \omega\, dh$. *Then the map* $\omega \mapsto \omega^\sharp$ *is a projection on* ${}^k\Omega^p(\mathrm{X};\mathrm{F})$ *whose image is the subspace of* H-*invariant forms. We have* $d(\omega^\sharp) = (d\omega)^\sharp$ *for all* $\omega \in {}^k\Omega^p(\mathrm{X};\mathrm{F})$.

The first assertion follows from the proposition applied to the vector H-bundle $\mathrm{Alt}^p(\mathrm{T}(\mathrm{X});\mathrm{F})$ (Chap. III, §1, no. 8, Examples). To prove the second assertion, it suffices, in view of *Integration*, Chap. III, §3, no. 2, Prop. 2, to prove that the map $d : {}^k\Omega^p(\mathrm{X};\mathrm{F}) \to {}^{k-1}\Omega^{p+1}(\mathrm{X};\mathrm{F})$ is continuous when the first (resp. second) space is given the topology of compact C^k-convergence (resp. C^{k-1}-convergence). But this follows immediately from the definition of these topologies by means of semi-norms (*Spectral Theories*, in preparation) and the fact that d is a differential operator of order ≤ 1 (*Differentiable and Analytic Manifolds, Results*, 14.4.2).

5. INVARIANT DIFFERENTIAL FORMS

Let X be a locally finite dimensional real manifold of class C^∞, and let $(g, x) \mapsto gx$ be a law of left operation of class C^∞ of a connected compact Lie group G on X. For $g \in G$, denote by τ_g the automorphism $x \mapsto gx$ of X. Denote by $\Omega(X)$ the algebra of real differential forms of class C^∞ on X (*Differentiable and Analytic Manifolds, Results*, 8.3.1).

For any element ξ of $\mathfrak{g}$, denote by D_ξ the corresponding vector field on X (Chap. III, §3, no. 5) and by $\theta(\xi), i(\xi)$ the corresponding operators on $\Omega(X)$, so that we have the formulas (*Differentiable and Analytic Manifolds, Results*, 8.4.5 and 8.4.7)

$$\theta(\xi)\omega = d(i(\xi)\omega) + i(\xi)d\omega \tag{15}$$

$$\frac{d}{dt}(\tau^*_{\exp t\xi}\omega) = \tau^*_{\exp t\xi}(\theta(\xi)\omega). \tag{16}$$

A differential form $\omega \in \Omega(X)$ is invariant if $\tau_g^*\omega = \omega$ for all $g \in G$; by formula (16), this is equivalent to $\theta(\xi)\omega = 0$ for all $\xi \in \mathfrak{g}$. Denote by $\Omega(X)^G$ the space of invariant differential forms on X; if $\omega \in \Omega(X)^G$, we have $d\omega \in \Omega(X)^G$, so $\Omega(X)^G$ is a *subcomplex* of the *complex* $(\Omega(X), d)$.

THEOREM 2. *The canonical injection* $\iota : \Omega(X)^G \to \Omega(X)$ *is a homotopism of complexes* (*Algèbre*, Chap. X, p. 33, déf. 5); *the map* $\omega \mapsto \omega^\natural = \int_G \tau_g^*\omega\, dg$ *is a homotopism, inverse to it up to homotopy. In particular, the map* $H(\iota) : H(\Omega(X)^G) \to H(\Omega(X))$ *is bijective.*

By Cor. 3 of no. 4, the map $\omega \mapsto \omega^\natural$ is a morphism of complexes from $\Omega(X)$ to $\Omega(X)^G$ that induces the identity on the subcomplex $\Omega(X)^G$; thus, to prove the theorem it suffices to construct a homomorphism $s : \Omega(X) \to \Omega(X)$, graded of degree -1, such that

$$\omega^\natural - \omega = (d \circ s + s \circ d)(\omega) \quad \text{for all } \omega \in \Omega(X). \tag{17}$$

By Lemma 1 of *Integration*, Chapter IX, §2, no. 4 and Remark 1 of §2, no. 2, there exists a positive measure $d\xi$ on $\mathfrak{g}$ of compact support whose image under the exponential map is equal to dg. For $\omega \in \Omega(X)$, put

$$s(\omega) = \int_0^1 \left\{ \int_{\mathfrak{g}} \tau^*_{\exp t\xi}(i(\xi)\omega).d\xi \right\} dt;$$

we have to show that formula (17) is satisfied. As in the proof of Cor. 1 (no. 4), we verify the formula

$$ds(\omega) = \int_0^1 \left\{ \int_{\mathfrak{g}} \tau^*_{\exp t\xi} d(i(\xi)\omega).d\xi \right\} dt.$$

We now deduce from formulas (15) and (16) the equalities

$$\begin{aligned}
ds(\omega) + s(d\omega) &= \int_0^1 \left\{ \int_{\mathfrak{g}} \tau^*_{\exp t\xi}(d(i(\xi)\omega) + i(\xi)d\omega).d\xi \right\} dt \\
&= \int_0^1 \left\{ \int_{\mathfrak{g}} \tau^*_{\exp t\xi}(\theta(\xi)\omega).d\xi \right\} dt \\
&= \int_{\mathfrak{g}} \left\{ \int_0^1 \frac{d}{dt}(\tau^*_{\exp t\xi}\omega)dt \right\} d\xi \\
&= \int_{\mathfrak{g}} (\tau^*_{\exp \xi}\omega - \omega)\, d\xi \\
&= \omega^\sharp - \omega,
\end{aligned}$$

hence Th. 2.

We apply the theorem in the case X = G, for the action of G by left translations. Recall (Chap. III, §3, no. 13, Prop. 50) that associating to a differential form on G its value at the identity element gives an isomorphism from $\Omega(\mathrm{G})^{\mathrm{G}}$ to the graded algebra $\mathrm{Alt}(\mathfrak{g})$ of alternating multilinear forms on $\mathfrak{g}$. Identify $\Omega(\mathrm{G})^{\mathrm{G}}$ with $\mathrm{Alt}(\mathfrak{g})$ by means of this isomorphism. The operator d is then given by the formula (Chap. III, §3, no. 14, Prop. 51)

$$\begin{aligned}
&d\omega(a_1, \dots, a_{p+1}) \\
&\quad = \sum_{i<j} (-1)^{i+j} \omega([a_i, a_j], a_1, \dots, a_{i-1}, a_{i+1}, \dots, a_{j-1}, a_{j+1}, \dots, a_{p+1})
\end{aligned}$$

for ω in $\mathrm{Alt}^p(\mathfrak{g})$ and $a_1, \dots, a_{p+1}$ in $\mathfrak{g}$.

For $\xi \in \mathfrak{g}$, let L_ξ be the corresponding left-invariant vector field (defined by means of the action of G on itself by *right* translations, cf. Chap. III, §3, no. 6). The operators $\theta(\mathrm{L}_\xi), i(\mathrm{L}_\xi)$ commute with the action of G on $\Omega(\mathrm{G})$ defined by left translation, and hence induce operators $\theta(\xi), i(\xi)$ on $\Omega(\mathrm{G})^{\mathrm{G}}$; with the preceding identifications, these are expressed by the formulas (*Differentiable and Analytic Manifolds, Results*, 8.3.2 and 8.4.2)

$$(\theta(\xi)\omega)(a_1, \dots, a_p) = -\sum_i \omega(a_1, \dots, a_{i-1}, [\xi, a_i], a_{i+1}, \dots, a_p)$$

$$(i(\xi)\omega)(a_1, \dots, a_{p-1}) = \omega(\xi, a_1, \dots, a_{p-1})$$

for ω in $\mathrm{Alt}^p(\mathfrak{g})$ and $a_1, \dots, a_p$ in $\mathfrak{g}$.

The subcomplex ${}^{\mathrm{G}}\Omega(\mathrm{G})^{\mathrm{G}}$ of *biinvariant* forms (Chap. III, §3, no. 13) can be identified with the subcomplex $\mathrm{Alt}(\mathfrak{g})^{\mathrm{G}}$ of alternating multilinear forms on $\mathfrak{g}$ invariant under the adjoint representation (that is, such that $\theta(\xi)\omega = 0$ for all $\xi \in \mathfrak{g}$). Thus, we have a commutative diagram of complexes

$$\begin{array}{ccccc} {}^{\mathrm{G}}\Omega(\mathrm{G})^{\mathrm{G}} & \longrightarrow & \Omega(\mathrm{G})^{\mathrm{G}} & \longrightarrow & \Omega(\mathrm{G}) \\ \downarrow & & \downarrow & & \\ \mathrm{Alt}(\mathfrak{g})^{\mathrm{G}} & \longrightarrow & \mathrm{Alt}(\mathfrak{g}) & & \end{array} \tag{18}$$

where the horizontal arrows are the canonical injections, and the vertical arrows are the isomorphisms induced by the map $\omega \mapsto \omega(e)$.

COROLLARY 1. *a) In the diagram* (18), *all the morphisms are homotopisms.*

b) Let $\omega \in \mathrm{Alt}(\mathfrak{g})$. *Then* ω *belongs to* $\mathrm{Alt}(\mathfrak{g})^{\mathrm{G}}$ *if and only if* $d\omega = 0$ *and* $d(i(\xi)\omega) = 0$ *for all* $\xi \in \mathfrak{g}$. *The differential of the complex* $\mathrm{Alt}(\mathfrak{g})^{\mathrm{G}}$ *is zero.*

c) The graded vector space $\mathrm{H}(\Omega(\mathrm{G}))$ *is isomorphic to* $\mathrm{Alt}(\mathfrak{g})^{\mathrm{G}}$.

The theorem, applied to the action of G on G by left translations (resp. to the action $((g,h);x) \mapsto gxh^{-1}$ of $\mathrm{G}\times\mathrm{G}$ on G) implies that the canonical injection $\Omega(\mathrm{G})^{\mathrm{G}} \to \Omega(\mathrm{G})$ (resp. ${}^{\mathrm{G}}\Omega(\mathrm{G})^{\mathrm{G}} \to \Omega(\mathrm{G})$) is a homotopism; in view of *Algèbre*, Chap. X, p. 34, Cor., assertion *a*) follows.

We prove *b*). By Prop. 51 of Chap. III, §3, no. 14, we have $d\alpha = -d\alpha$, that is $d\alpha = 0$, for every differential form α on G that is left and right invariant. Thus, if $\omega \in \mathrm{Alt}(\mathfrak{g})^{\mathrm{G}}$, then $d\omega = 0$, and consequently $d(i(\xi)\omega) = \theta(\xi)\omega - i(\xi)d\omega = 0$. Conversely, if $d\omega = 0$ and $d(i(\xi)d\omega) = 0$, then $\theta(\xi)\omega = 0$.

Assertion *c*) follows from *a*) and *b*).

Remark. Consider the subcomplexes $\mathrm{Z}(\Omega(\mathrm{G}))$ and $\mathrm{B}(\Omega(\mathrm{G}))$ of $\Omega(\mathrm{G})$ (*Algèbre*, Chap. X, p. 25). It follows from the formula giving the differential of the product of two forms (*Differentiable and Analytic Manifolds, Results*, 8.3.5) that $\mathrm{Z}(\Omega(\mathrm{G}))$ is a subalgebra of $\Omega(\mathrm{G})$ and that $\mathrm{B}(\Omega(\mathrm{G}))$ is an ideal of $\mathrm{Z}(\Omega(\mathrm{G}))$; consequently, the exterior product induces a graded algebra structure on $\mathrm{H}(\Omega(\mathrm{G}))$. The preceding now gives an isomorphism of graded *algebras* $\mathrm{H}(\Omega(\mathrm{G})) \to \mathrm{Alt}(\mathfrak{g})^{\mathrm{G}}$.

Let H be a closed subgroup of G; we apply Th. 2 to $\mathrm{X} = \mathrm{G}/\mathrm{H}$. By Chap. III, §1, no. 8, Cor. 1 of Prop. 17, the G-invariant differential forms on G/H can be identified with the H-invariant elements of $\mathrm{Alt}(\mathrm{T}_e(\mathrm{G}/\mathrm{H}))$, that is with the elements of $\mathrm{Alt}(\mathfrak{g})$ that are H-invariant and annihilated by the operators $i(\xi)$ for all $\xi \in \mathrm{L}(\mathrm{H})$. Consequently:

COROLLARY 2. *Let* H *be a closed subgroup of* G.

a) The canonical injection $\Omega(\mathrm{G}/\mathrm{H})^{\mathrm{G}} \to \Omega(\mathrm{G}/\mathrm{H})$ *is a homotopism.*

b) The complex $\Omega(\mathrm{G}/\mathrm{H})^{\mathrm{G}}$ *can be identified with the subcomplex of* $\mathrm{Alt}(\mathfrak{g})$ *consisting of the elements* ω *of* $\mathrm{Alt}(\mathfrak{g})$ *that are invariant under the adjoint representation of* H *and are such that* $i(\xi)\omega = 0$ *for all* $\xi \in \mathrm{L}(\mathrm{H})$. *If, in addition,* H *is connected, this subcomplex consists of the* $\omega \in \mathrm{Alt}(\mathfrak{g})$ *such that* $\theta(\xi)\omega = 0$ *and* $i(\xi)\omega = 0$ *for all* $\xi \in \mathrm{L}(\mathrm{H})$.

§7. IRREDUCIBLE REPRESENTATIONS OF CONNECTED COMPACT LIE GROUPS

We retain the notations of §6. *A representation of* G *is a continuous (hence analytic) homomorphism from* G *to a group* **GL**(V), *where* V *is a* finite dimensional complex *vector space. Every representation of* G *is semi-simple* (§1, no. 1).

Choose a chamber C *of* $\mathfrak{t}$ (§5, no. 2), *and put* $\Gamma(\mathrm{T})_{++} = \overline{\mathrm{C}} \cap \Gamma(\mathrm{T})$.

1. DOMINANT CHARACTERS

Denote by X_+ the set of elements λ of X(T) such that $\langle \lambda, x \rangle \geq 0$ for all $x \in \Gamma(\mathrm{T})_{++}$, that is, such that the linear form $\delta(\lambda) : \mathfrak{t}_{\mathbf{C}} \to \mathbf{C}$ maps the chamber C of $\mathfrak{t}$ to $i\mathbf{R}_+$.

Give X(T) the ordered group structure for which the positive elements are those of X_+; put $\mathrm{R}_+ = \mathrm{R}(\mathrm{G},\mathrm{T}) \cap \mathrm{X}_+$ and $\mathrm{R}_- = -\mathrm{R}_+$. The elements of R_+ are called *positive roots*, those of R_- negative roots; every root is either positive or negative (Chap. VI, §1, no. 6, Th. 3). A positive root that is not the sum of two positive roots is said to be *simple*; every positive root is a sum of simple roots (*loc. cit.*); the simple roots form a basis of the subgroup of X(T) generated by the roots, a subgroup that can be identified with X(T/C(G)) (§4, no. 4); the reflections with respect to the simple roots generate the Weyl group $\mathrm{W} = \mathrm{W}_{\mathrm{G}}(\mathrm{T})$ (Chap. VI, §1, no. 5, Th. 2).

Lemma 1. Let λ *be an element of* X(T). *The following conditions are equivalent:*

(i) $\lambda - w(\lambda) \geq 0$ (resp. > 0) *for all* $w \in \mathrm{W}$ *such that* $w \neq 1$;

(ii) *for all* $w \in \mathrm{W}$ *such that* $w \neq 1$, $\lambda - w(\lambda)$ *is a linear combination with positive coefficients* (resp. *positive coefficients not all zero*) *of simple roots*;

(iii) $\langle \lambda, K_\alpha \rangle \geq 0$ (resp. > 0) *for every positive root* α;

(iv) $\langle \lambda, K_\alpha \rangle \geq 0$ (resp. > 0) *for every simple root* α.

The equivalence if (iii) and (iv) is immediate. Since the set of the K_α can be identified with the inverse root system of R(G, T) (§4, no. 5), the equivalence of (i) and (iii) follows from Chap. VI, §1, no. 6, Prop. 18 and Cor. The implication (ii) $\Rightarrow$ (i) is trivial, and the opposite implication follows from *loc. cit.*

Denote by X_{++} the set of elements of X(T) such that $\langle \lambda, K_\alpha \rangle \geq 0$ for every positive root α. The elements of X_{++} are said to be *dominant*. They form a fundamental domain for the operation of W on X(T) (Chap. VI, §1, no. 10). We have $\mathrm{X}_{++} \subset \mathrm{X}_+$.

If G is simply-connected, for each simple root α there exists an element ϖ_α of X(T) such that $\langle \varpi_\beta, K_\alpha \rangle = \delta_{\alpha\beta}$ for every simple root β, that is, $s_\alpha(\varpi_\alpha) = \varpi_\alpha - \alpha$, $s_\beta(\varpi_\alpha) = \varpi_\alpha$ for every simple root $\beta \neq \alpha$; the ϖ_α are called the *fundamental* dominant weights; they form a basis of the commutative group X(T) and of the commutative monoid X_{++}; more precisely, every element λ of X(T) can be written in the form $\lambda = \sum_{\alpha} \langle \lambda, K_\alpha \rangle \varpi_\alpha$.

Denote by ρ the element of $X(T) \otimes \mathbf{Q}$ such that

$$2\rho = \sum_{\alpha \in R_+} \alpha.$$

We have $\langle \rho, K_\alpha \rangle = 1$ for every simple root α (Chap. VI, §1, no. 10, Prop. 29). If G is simply-connected, ρ is the sum of the fundamental dominant weights.

2. HIGHEST WEIGHT OF AN IRREDUCIBLE REPRESENTATION

Associate to any representation $\tau : G \to \mathbf{GL}(V)$ the homomorphism $L(\tau)_{(\mathbf{C})}$ from the $\mathbf{C}$-Lie algebra $\mathfrak{g}_{\mathbf{C}}$ to End(V) extending the linear representation $L(\tau)$ of $\mathfrak{g}$ on the real vector space underlying V (Chap. III, §3, no. 11). By Prop. 7 of §4, no. 3, the map δ from X(T) to $\mathrm{Hom}_{\mathbf{C}}(\mathfrak{t}_{\mathbf{C}}, \mathbf{C}) = \mathfrak{t}_{\mathbf{C}}^*$ induces a bijection from the set of weights of τ relative to T to the set of weights of $L(\tau)_{(\mathbf{C})}$ relative to the Cartan subalgebra $\mathfrak{t}_{\mathbf{C}}$ of $\mathfrak{g}_{\mathbf{C}}$.

Lemma 2. Let φ be a linear representation of the complex Lie algebra $\mathfrak{g}_{\mathbf{C}}$ on a finite dimensional complex vector space V. *There exists a representation τ of* G *on* V *such that $L(\tau)_{(\mathbf{C})} = \varphi$ if and only if φ is semi-simple and the weights of $\mathfrak{t}_{\mathbf{C}}$ on* V *belong to* $\delta(X(T))$.

If there exists a representation τ of G such that $L(\tau)_{(\mathbf{C})} = \varphi$, then φ is semi-simple because G is connected and τ is semi-simple (Chap. III, §6, no. 5, Cor. 2 of Prop. 13), and the weights of $\mathfrak{t}_{\mathbf{C}}$ on V belong to the image of δ. Thus, the condition is necessary; we shall prove that it is sufficient. If φ is semi-simple, V is the direct sum of the $V_\mu(\mathfrak{t}_{\mathbf{C}})$, where μ belongs to the set of weights of $\mathfrak{t}_{\mathbf{C}}$ on V (Chap. VII, §2, no. 4, Cor. 3 of Th. 2); if all the weights belong to the image of δ, there exists a representation τ_T of T on V such that $L(\tau_T)_{(\mathbf{C})} = \varphi|\mathfrak{t}_{\mathbf{C}}$: indeed, it suffices to put $\tau_T(t)v = t^\lambda v$ for $t \in T$ and $v \in V_{\delta(\lambda)}(\mathfrak{t}_{\mathbf{C}})$. The lemma now follows from Prop. 8 of §2, no. 6.

THEOREM 1. *a) Let $\tau : G \to \mathbf{GL}(V)$ be an irreducible representation of* G. *Then the set of weights of τ* (relative to T) *has a largest element λ, which is dominant, and the space $V_\lambda(T)$ is of dimension* 1.

b) Two irreducible representations of G *are equivalent if and only if their highest weights are equal.*

c) For every dominant element λ of X(T), *there exists an irreducible representation of* G *of highest weight λ.*

By Lemma 2, the equivalence classes of irreducible representations of G correspond bijectively to the classes of finite dimensional irreducible representations of $\mathfrak{g}$ whose weights belong to $\delta(X(T))$.

Denote by $\mathscr{C}\mathfrak{g}_{\mathbf{C}}$ the centre and by $\mathscr{D}\mathfrak{g}_{\mathbf{C}}$ the derived Lie algebra of $\mathfrak{g}_{\mathbf{C}}$, so that $\mathfrak{g}_{\mathbf{C}} = \mathscr{C}\mathfrak{g}_{\mathbf{C}} \oplus \mathscr{D}\mathfrak{g}_{\mathbf{C}}$. For every bilinear form μ on $\mathfrak{t}_{\mathbf{C}} \cap \mathscr{D}\mathfrak{g}_{\mathbf{C}}$, denote by $\mathrm{E}(\mu)$ the simple $\mathscr{D}\mathfrak{g}_{\mathbf{C}}$-module introduced in Chap. VIII, §6, no. 3; for every linear form ν on $\mathscr{C}\mathfrak{g}_{\mathbf{C}}$, denote by $\mathbf{C}(\nu)$ the 1-dimensional $\mathscr{C}\mathfrak{g}_{\mathbf{C}}$-module over $\mathbf{C}$ associated to it. Then the $\mathfrak{g}_{\mathbf{C}}$-modules $\mathbf{C}(\nu) \otimes \mathrm{E}(\mu)$ are simple, and by Chap. VIII, §7, no. 2, Cor. 2 of Th. 1 and *Algebra*, Chap. VIII, §11, no. 1, Th. 1, every finite dimensional simple $\mathfrak{g}_{\mathbf{C}}$-module is isomorphic to one of the modules $\mathbf{C}(\nu) \otimes \mathrm{E}(\mu)$; moreover (*loc. cit.*) $\mathbf{C}(\nu) \otimes \mathrm{E}(\mu)$ is finite dimensional if and only if $\mu(H_\alpha)$ is a positive integer for every simple root α. If we denote by $\nu + \mu$ the linear form on $\mathfrak{t}_{\mathbf{C}}$ that induces ν on $\mathscr{C}\mathfrak{g}_{\mathbf{C}}$ and μ on $\mathfrak{t}_{\mathbf{C}} \cap \mathscr{D}\mathfrak{g}_{\mathbf{C}}$, then $(\nu + \mu)(H_\alpha) = \mu(H_\alpha)$; moreover, the weights of $\mathbf{C}(\nu) \otimes \mathrm{E}(\mu)$ are the $\nu + \lambda$, where λ is any weight of $\mathrm{E}(\mu)$, and are thus of the form $\nu + \mu - \theta$, with $\theta \in \delta(\mathrm{X}_+)$ (Chap. VIII, §6, no. 2, Lemma 2).

We conclude that the $\mathfrak{g}$-module $\mathbf{C}(\nu) \otimes \mathrm{E}(\mu)$ is finite dimensional if and only if $(\nu + \mu)(H_\alpha)$ is a positive integer for every simple root α, and that its weights belong to $\delta(\mathrm{X}(\mathrm{T}))$ if and only if $\nu + \mu$ belongs to $\delta(\mathrm{X}(\mathrm{T}))$. The conjunction of these two conditions means that $\nu + \mu$ belongs to $\delta(\mathrm{X}_{++})$; in that case, $\nu+\mu$ is the highest weight of $\mathbf{C}(\nu) \otimes \mathrm{E}(\mu)$. Thus, we have constructed for every dominant weight λ of $\mathrm{X}(\mathrm{T})$ an irreducible representation of G of highest weight λ, and have thus obtained, up to equivalence, all the irreducible representations of G. Since the vectors of weight $\nu + \mu$ in $\mathbf{C}(\nu) \otimes \mathrm{E}(\mu)$ form a subspace of dimension 1, the proof is complete.

COROLLARY. *The group* G *has a (finite dimensional)* faithful *linear representation.*

Observe first that every element of $\mathrm{X}(\mathrm{T})$ is equal to the difference of two dominant weights: more precisely, let ϖ be an element of X_{++} such that $\langle \varpi, K_\alpha \rangle > 0$ for every simple root α; for all $\lambda \in \mathrm{X}(\mathrm{T})$ there exists a positive integer n such that $\langle \lambda + n\varpi, K_\alpha \rangle \geq 0$ for every simple root α, that is (no. 1, Lemma 1) $\lambda + n\varpi \in \mathrm{X}_{++}$.

Consequently, there exists a finite family $(\lambda_i)_{i \in I}$ of elements of X_{++} generating the $\mathbf{Z}$-module $\mathrm{X}(\mathrm{T})$. For $i \in I$, let τ_i be an irreducible representation of G of highest weight λ_i (Th. 1); let the representation τ be the direct sum of the τ_i. By construction the set $\mathrm{P}(\tau, \mathrm{T})$ of weights of τ (relative to T) generates the $\mathbf{Z}$-module $\mathrm{X}(\mathrm{T})$. It now follows from Prop. 6 of §4, no. 3 that the homomorphism τ is injective, hence the corollary.

Remarks. 1) Let $\mathfrak{n}_+$ be the subalgebra of $\mathfrak{g}_{\mathbf{C}}$ that is the sum of the $\mathfrak{g}^\alpha$ for $\alpha > 0$. Let $\tau : \mathrm{G} \to \mathbf{GL}(\mathrm{V})$ be an irreducible representation, $\lambda \in \mathrm{X}_{++}$ its highest weight and $\tau' : \mathfrak{g}_{\mathbf{C}} \to \mathfrak{gl}(\mathrm{V})$ the representation induced by τ. Then $\mathrm{V}_\lambda(\mathrm{T})$ is the subspace consisting of the vectors v in V such that $\tau'(x)v = 0$ for all $x \in \mathfrak{n}_+$.

Indeed, this follows from the corresponding statement for the $\mathfrak{g}_{\mathbf{C}}$-modules $\mathbf{C}(\nu) \otimes \mathrm{E}(\mu)$ (Chap. VIII, §6, no. 2, Prop. 3).

2) Let $\Theta(G)$ be the algebra of continuous representative functions on G with values in **C** (*Algebra*, Chap. VIII). Let G operate on $\Theta(G)$ by left and right translations. For each $\lambda \in X_{++}$, let $(V_\lambda, \tau_\lambda)$ be an irreducible representation of G of highest weight λ (Th. 1), and $(V_\lambda^*, \check{\tau}_\lambda)$ the contragredient representation (Chap. III, §3, no. 11); by *Spectral Theories*, the representation of $G \times G$ on $\Theta(G)$ is isomorphic to the direct sum of the representations $(V_\lambda \otimes V_\lambda^*, \tau_\lambda \otimes \check{\tau}_\lambda)$ for all λ in X_{++}. Remark 1 now implies the following statement: *Let* $\lambda \in X_{++}$, *and let* E_λ *be the vector subspace of* $\Theta(G)$ *consisting of the continuous representative functions* f *on* G *such that* $f(gt) = \lambda(t)^{-1} f(g)$ *for all* $g \in G$ *and all* $t \in T$, *and such that* $f * x = 0$ *for all* $x \in \mathfrak{n}_- = \bigoplus_{\alpha<0} \mathfrak{g}^\alpha$. *Then* E_λ *is stable under left translations, and the representation of* G *on* E_λ *by left translations is irreducible of highest weight* λ.

3) Let $\tau : G \to \mathbf{GL}(V)$ be an irreducible representation. There exists an element ν of $X(C(G))$ such that $\tau(s)v = \nu(s)v$ for all $s \in C(G), v \in V$: indeed, $\tau(C(G))$ is contained in the commutant of $\tau(G)$, which is equal to $\mathbf{C}^*.1_V$ (*Algebra*, Chap. VIII, §3, no. 2, Th. 1). For every weight λ of τ, the restriction of λ to $C(G)$ is equal to ν.

4) The definitions and statements of Chap. VIII, §7, nos. 2 to 5 can be generalized without difficulty to the present situation; we leave the details to the reader.

PROPOSITION 1. *Let* $\tau : G \to \mathbf{GL}(V)$ *be an irreducible representation of* G *of highest weight* $\lambda \in X_{++}$. *Let* m *be the integer* $\sum_{\alpha \in R_+} \langle \lambda, K_\alpha \rangle$, *and let* w_0 *be the element of the Weyl group such that* $w_0(R_+) = R_-$ (Chap. VI, § 6, Cor. 3 of Prop. 17). *There are three possible cases*:

a) $w_0(\lambda) = -\lambda$ *and* m *is even. Then there exists a* G*-invariant separating symmetric bilinear form on* V; *the representation* τ *is of real type* (Appendix II).

b) $w_0(\lambda) \neq -\lambda$. *Every* G*-invariant bilinear form on* V *is zero*; *the representation* τ *is of complex type* (loc. cit.).

c) $w_0(\lambda) = -\lambda$ *and* m *is odd. Then there exists a* G*-invariant separating alternating bilinear form on* V; *the representation* τ *is of quaternionic type* (loc. cit.).

If the restriction of τ *to* $C(G)_0$ *is not trivial, we are in case b)*.

A bilinear form B on V is invariant under G if and only if it is invariant under $\mathfrak{g}_\mathbf{C}$ (Chap. III, §6, no. 5, Cor. 3). Thus, if G is semi-simple, the proposition follows from Chap. VIII, §7, no. 5, Prop. 12 and Prop. 3 of Appendix II.

In the general case, put $C(G)_0 = S$, and identify $X(T/S)$ with a subgroup of $X(T)$ (stable under W). If $\tau(S) = \{1_V\}$, τ induces by passage to the quotient a representation $\tau' : G/S \to \mathbf{GL}(V)$ of highest weight λ; in this case the proposition follows from the preceding, applied to G/S.

Assume that $\tau(S) \neq \{1_V\}$. There exists a non-zero element ν of $X(S)$ such that $\tau(s) = \nu(s)_V$ for all $s \in S$ (Remark 3). Then ν is the image of λ under the restriction homomorphism $X(T) \to X(S)$; since W operates trivially on $X(S)$, the equality $w_0(\lambda) = -\lambda$ implies that $\nu = -\nu$, which is impossible: hence $w_0(\lambda) \neq -\lambda$. On the other hand, if B is a G-invariant bilinear form on V, we have, for all x, y in V and s in S,

$$B(\nu(s)x, \nu(s)y) = B(x, y) = \nu(s)^2 B(x, y)$$

which implies that $B = 0$, hence the proposition.

Let $\mathfrak{S}_{\mathbf{R}}(G)$ be the set of classes of irreducible continuous representations of G on finite dimensional real vector spaces. Prop. 1 and the results of Appendix II give a bijection $\Phi : X_{++}/\Sigma \to \mathfrak{S}_{\mathbf{R}}(G)$, where Σ denotes the subgroup $\{1, -w_0\}$ of $\mathrm{Aut}(X(T))$. More precisely, let $\lambda \in X_{++}$, and let E_λ be a representation of G of highest weight λ; then

$$\Phi(\{\lambda, -w_0(\lambda)\}) = E_{\lambda[\mathbf{R}]} \text{ if } \lambda \neq -w_0(\lambda) \text{ or if } \sum_{\alpha \in R_+} \langle \lambda, K_\alpha \rangle \notin 2\mathbf{Z}$$

$$\Phi(\{\lambda, -w_0(\lambda)\}) = E'_\lambda \text{ if } \lambda = -w_0(\lambda) \text{ and } \sum_{\alpha \in R_+} \langle \lambda, K_\alpha \rangle \in 2\mathbf{Z}$$

where E'_λ is an **R**-structure on E_λ invariant under G.

3. THE RING R(G)

Let R(G) be the ring of representations (continuous, on finite dimensional complex vector spaces) of G (*Algebra*, Chap. VIII, §10, no. 6). If τ is a representation of G, denote by $[\tau]$ its class in R(G); if τ and τ' are two representations of G, we have, by definition,

$$[\tau] + [\tau'] = [\tau \oplus \tau']$$
$$[\tau][\tau'] = [\tau \otimes \tau'].$$

Since every representation of G is semi-simple, the **Z**-module R(G) is free and admits as a basis the set of classes of irreducible representations of G, a set that can be identified with X_{++} by Th. 1. The map $\tau \mapsto L(\tau)_{(\mathbf{C})}$ induces a homomorphism l from the ring R(G) to the ring $\mathscr{R}(\mathfrak{g}_{\mathbf{C}})$ of representations of $\mathfrak{g}_{\mathbf{C}}$ (Chap. VIII, §7, no. 6).

Let $\tau : G \to \mathbf{GL}(V)$ be a representation of G; we consider the gradation $(V_\lambda(T))_{\lambda \in X(T)}$ of the **C**-vector space V. Denote by Ch(V), or by Ch(τ), the *character* of the graded vector space V (Chap. VIII, §7, no. 7); if $(e^\lambda)_{\lambda \in X(T)}$ denotes the canonical basis of the algebra $\mathbf{Z}[X(T)] = \mathbf{Z}^{(X(T))}$, we have, by definition,

$$\mathrm{Ch}(\tau) = \sum_{\lambda \in X(T)} (\dim V_\lambda(T))\, e^\lambda.$$

We define in the same way (*loc. cit.*) a homomorphism of rings, also denoted by Ch, from R(G) to $\mathbf{Z}[X(T)]$. If G is semi-simple, we have a commutative diagram

$$\begin{array}{ccc} R(G) & \xrightarrow{\text{Ch}} & \mathbf{Z}[X(T)] \\ \downarrow{\iota} & & \downarrow{\tilde{\delta}} \\ \mathscr{R}(\mathfrak{g}_{\mathbf{C}}) & \xrightarrow{\text{ch}} & \mathbf{Z}[P] \end{array} \tag{1}$$

where P denotes the group of weights of $R(\mathfrak{g}_{\mathbf{C}}, \mathfrak{t}_{\mathbf{C}})$ and $\tilde{\delta}$ the homomorphism induced by δ.

The Weyl group W operates by automorphisms on the group X(T), and hence on the ring $\mathbf{Z}[X(T)]$. By Prop. 5 of §4, no. 3, the image of Ch is contained in the subring $\mathbf{Z}[X(T)]^W$ consisting of the elements invariant under W.

PROPOSITION 2. *The homomorphism* Ch *induces an isomorphism from* R(G) *to* $Z[X(T)]^W$.

For $\lambda \in X_{++}$, denote by $[\lambda]$ the class in R(G) of the irreducible representation of highest weight λ. Since the family $([\lambda])_{\lambda \in X_{++}}$ is a basis of the $\mathbf{Z}$-module R(G), it suffices to prove the following assertion:

The family $(\text{Ch}[\lambda])_{\lambda \in X_{++}}$ *is a basis of the* $\mathbf{Z}$*-module* $\mathbf{Z}[X(T)]^W$.

For every element $u = \sum_{\lambda} a_\lambda e^\lambda$ of $\mathbf{Z}[X(T)]$, a term $a_\lambda e^\lambda$ in u is called maximal if λ is a maximal element of the set of $\mu \in X(T)$ such that $a_\mu \neq 0$. Th. 1 implies that $\text{Ch}[\lambda]$ has a unique maximal term, namely e^λ. Thus, the proposition follows from the following lemma.

Lemma 3. For each $\lambda \in X_{++}$, *let* C_λ *be an element of* $\mathbf{Z}[X(T)]^W$ *having unique maximal term* e^λ. *Then the family* $(C_\lambda)_{\lambda \in X_{++}}$ *is a basis of* $\mathbf{Z}[X(T)]^W$.

The proof is identical to that of Prop. 3 of Chap. VI, §3, no. 4, replacing A by $\mathbf{Z}$, P by X(T) and $P \cap \overline{C}$ by X_{++}.

Let $\Theta(G)$ (resp. $\Theta(T)$) be the $\mathbf{C}$-algebra of continuous representative functions on G (resp. T), and let $Z\Theta(G)$ (resp. $\Theta(T)^W$) be the subalgebra consisting of the central (resp. W-invariant) functions. The restriction map $\Theta(G) \to \Theta(T)$ induces a homomorphism of rings $r : Z\Theta(G) \to \Theta(T)^W$. On the other hand, the map that associates to a representation τ its character (that is, the function $g \mapsto \text{Tr}\,\tau(g)$) extends to a homomorphism of $\mathbf{C}$-algebras $\text{Tr} : \mathbf{C} \otimes_{\mathbf{Z}} R(G) \to Z\Theta(G)$ which, by *Spectral Theories*, is an *isomorphism*. Similarly, the canonical injection $X(T) \to \Theta(T)$ induces an isomorphism of $\mathbf{C}$-algebras $\iota : \mathbf{C}[X(T)] \to \Theta(T)$, which induces an isomorphism $\iota : \mathbf{C}[X(T)]^W \to \Theta(T)^W$. The diagram

$$\begin{array}{ccc} \mathbf{C} \otimes_{\mathbf{Z}} R(G) & \xrightarrow{1 \otimes \text{Ch}} & \mathbf{C}[X(T)]^W \\ \downarrow{\text{Tr}} & & \downarrow{\iota} \\ Z\Theta(G) & \xrightarrow{r} & \Theta(T)^W \end{array} \tag{2}$$

is *commutative*: indeed, for every representation $\tau : \mathrm{G} \to \mathbf{GL}(\mathrm{V})$ and all $t \in \mathrm{T}$,

$$\operatorname{Tr}\tau(t) = \sum_{\lambda \in \mathrm{X}(\mathrm{T})} (\dim \mathrm{V}_\lambda(\mathrm{T}))\, \lambda(t) = \iota(\operatorname{Ch}\tau)(t),$$

that is, $(r \circ \operatorname{Tr})[\tau] = (\iota \circ \operatorname{Ch})[\tau]$.

We can now deduce from the proposition the following result.

COROLLARY. *The restriction map* $r : \mathrm{Z}\Theta(\mathrm{G}) \to \Theta(\mathrm{T})^{\mathrm{W}}$ *is bijective.*

4. CHARACTER FORMULA

In this number, we write the group $\mathrm{X}(\mathrm{T})$ multiplicatively, and regard its elements as complex-valued functions on T. *We assume that the element* ρ *of* $\mathrm{X}(\mathrm{T}) \otimes \mathbf{Q}$ *belongs to* $\mathrm{X}(\mathrm{T})$.

Denote by $\mathrm{L}^2(\mathrm{T})$ the Hilbert space of classes of square-integrable complex functions on T, and by $\Theta(\mathrm{T})$ the subspace consisting of the continuous representative functions. By *Spectral Theories*, $\mathrm{X}(\mathrm{T})$ is an orthonormal basis of $\mathrm{L}^2(\mathrm{T})$ and an algebraic basis of $\Theta(\mathrm{T})$.

For $f \in \mathrm{L}^2(\mathrm{T})$ and $w \in \mathrm{W}$, denote by ${}^w f$ the element of $\mathrm{L}^2(\mathrm{T})$ defined by ${}^w f(t) = f(w^{-1}(t))$; thus, for $\lambda \in \mathrm{X}(\mathrm{T})$ we have ${}^w\lambda = w(\lambda)$. Denote by $\varepsilon : \mathrm{W} \to \{1, -1\}$ the *signature* (the unique homomorphism such that $\varepsilon(s) = -1$ for every reflection s); for $f \in \mathrm{L}^2(\mathrm{T})$, put

$$\mathrm{J}(f) = \sum_{w \in \mathrm{W}} \varepsilon(w)\, {}^w f.$$

If $\lambda \in \mathrm{X}_{++}$, the characters ${}^w(\lambda\rho)$ are distinct; indeed, it suffices to prove that ${}^w(\lambda\rho) \neq \lambda\rho$ for all $w \neq 1$; but this follows from Lemma 1 (no. 1) and the fact that $\langle \lambda\rho, K_\alpha \rangle = \langle \lambda, K_\alpha \rangle + 1 > 0$ for every positive root α. Consequently,

$$\| \mathrm{J}(\lambda\rho) \|^2 = \operatorname{Card}(\mathrm{W}) = w(\mathrm{G}).$$

An element $f \in \mathrm{L}^2(\mathrm{T})$ is said to be *anti-invariant if* ${}^w f = \varepsilon(w) f$ for all $w \in \mathrm{W}$ (that is, if ${}^s f = -f$ for every reflection s). We shall show that $\frac{1}{w(\mathrm{G})}\mathrm{J}$ is the *orthogonal projection* of $\mathrm{L}^2(\mathrm{T})$ onto the subspace of anti-invariant elements. Indeed, let f, f' be in $\mathrm{L}^2(\mathrm{T})$, with f' anti-invariant; then $\mathrm{J}(f)$ is anti-invariant and

$$\begin{aligned}\langle f', \mathrm{J}(f) \rangle &= \sum_{w \in \mathrm{W}} \varepsilon(w) \langle f', {}^w f \rangle = \sum_{w \in \mathrm{W}} \langle {}^w f', {}^w f \rangle \\ &= \sum_{w \in \mathrm{W}} \langle f', f \rangle = w(\mathrm{G}) \langle f', f \rangle.\end{aligned}$$

PROPOSITION 3. *The elements* $J(\lambda\rho)/\sqrt{w(G)}$, *for* $\lambda \in X_{++}$, *form an orthonormal basis of the subspace of anti-invariant elements of* $L^2(T)$, *and an algebraic basis of the subspace of anti-invariant elements of* $\Theta(T)$.

The proof is identical to that of Chap. VI, §3, no. 3, Prop. 1.

By Chap. VI, *loc. cit.*, Prop. 2,

$$J(\rho) = \rho \prod_{\alpha>0} (1 - \alpha^{-1}) = \rho^{-1} \prod_{\alpha>0} (\alpha - 1), \tag{3}$$

so

$$J(\rho)\overline{J(\rho)} = \prod_{\alpha} (\alpha - 1). \tag{4}$$

By Cor. 2 of Th. 1 (§6, no. 2), we deduce:

Lemma 4. If φ *and* ψ *are two continuous central functions on* G,

$$\int_G \overline{\varphi(g)}\psi(g)\,dg = \frac{1}{w(G)} \int_T \overline{(\varphi(t)J(\rho)(t))}.(\psi(t)J(\rho)(t))\,dt.$$

For all $\lambda \in X_{++}$, denote by χ_λ the character of an irreducible representation of G of highest weight λ.

THEOREM 2 (H. Weyl). *For all* $\lambda \in X_{++}$, *we have* $J(\rho).\chi_\lambda|T = J(\lambda\rho)$.

The function $J(\rho).\chi_\lambda|T$ is anti-invariant under W, and is a linear combination with integer coefficients of elements of X(T). Thus, by Chap. VI, §3, no. 3, Prop. 1, it can be written as $\sum_\mu a_\mu J(\mu\rho)$, where μ belongs to X_{++}, and the a_μ are integers all but finitely-many of which are zero; since $\int_G |\chi_\lambda(g)|^2\,dg = 1$
(*Spectral Theories*), it follows from Prop. 3 and Lemma 4 that $\sum_\mu (a_\mu)^2 = 1$; thus, the a_μ are all zero, except one which must be equal to 1 or -1. But the coefficient of λ in $\chi_\alpha|T$ is equal to 1 (Th. 1), hence the coefficient of $\lambda\rho$ in $J(\rho).\chi_\lambda|T$ is equal to 1 (Chap. VI, §3, no. 3, Remark 2), which implies that $a_{\lambda\rho} = 1$, hence the theorem.

COROLLARY 1. *With the notations of* no. 3, *we have in* $\mathbf{Z}[X(T)]$,

$$\left(\sum_{w\in W} \varepsilon(w)e^{w\rho}\right) \mathrm{Ch}[\lambda] = \sum_{w\in W} \varepsilon(w)e^{w\lambda}e^{w\rho} \quad \text{for all } \lambda \in X_{++}.$$

This follows from the theorem and commutative diagram (2) (no. 3).

COROLLARY 2. *For all* $\lambda \in X_{++}$ *and every regular element* t *of* T,

$$\chi_\lambda(t) = \frac{\sum_w \varepsilon(w)\lambda(wt)\rho(wt)}{\sum_w \varepsilon(w)\rho(wt)} \tag{5}$$

where the two sums are over the elements w of W.

Indeed, $\mathrm{J}(\rho)(t)$ is non-zero, since t is regular (formula (4)).

If φ is a central function on G, the restriction of φ to T is invariant under W, so $\mathrm{J}(\rho).\varphi|\mathrm{T}$ is anti-invariant under W. Further, by *Spectral Theories* and Th. 1, the family $(\chi_\lambda)_{\lambda\in \mathrm{X}_{++}}$ is an algebraic basis of the space of central representative functions on G and an orthonormal basis of the space $\mathrm{ZL}^2(\mathrm{G})$ of classes of square-integrable central functions on G.

Thus, from Prop. 3 and Th. 2 we deduce:

COROLLARY 3. *The map which associates to any continuous central function φ on* G *the function $w(\mathrm{G})^{-1/2}\mathrm{J}(\rho)(\varphi|\mathrm{T})$ induces an isomorphism from the space of central representative functions on* G *to the space of anti-invariant elements of $\Theta(\mathrm{T})$; it extends by continuity to an isomorphism of Hilbert spaces from $\mathrm{ZL}^2(\mathrm{G})$ to the subspace of anti-invariant elements of $\mathrm{L}^2(\mathrm{T})$.*

COROLLARY 4. *Let φ be a continuous central function on* G. *Then,*

$$\int_\mathrm{G} \overline{\chi_\lambda(g)}\varphi(g)\,dg = \int_\mathrm{T} \overline{\lambda(t)} \prod_{\alpha>0}(1-\alpha(t)^{-1})\varphi(t)\,dt = \int_\mathrm{T} \overline{\lambda\rho(t)}.\varphi(t)\mathrm{J}(\rho)(t)\,dt.$$

Indeed, by Lemma 4 and Th. 2,

$$\begin{aligned}\int_\mathrm{G} \overline{\chi_\lambda(g)}\varphi(g)\,dg &= \frac{1}{w(\mathrm{G})}\int_\mathrm{T} \overline{\chi_\lambda(t)\mathrm{J}(\rho)(t)}(\varphi(t)\mathrm{J}(\rho)(t))\,dt \\ &= \frac{1}{w(\mathrm{G})}\int_\mathrm{T} \overline{\mathrm{J}(\lambda\rho)(t)}\varphi(t)\mathrm{J}(\rho)(t)\,dt.\end{aligned}$$

But the function $t \mapsto \varphi(t)\mathrm{J}(\rho)(t)$ is anti-invariant and $\frac{1}{w(\mathrm{G})}\mathrm{J}(\lambda\rho)$ is the orthogonal projection of $\lambda\rho$ onto the subspace of anti-invariant elements of $\mathrm{L}^2(\mathrm{T})$, so

$$\frac{1}{w(\mathrm{G})}\int_\mathrm{T} \overline{\mathrm{J}(\lambda\rho)(t)}\varphi(t)\mathrm{J}(\rho)(t)\,dt = \int_\mathrm{T} \overline{\lambda\rho(t)}\varphi(t)\mathrm{J}(\rho)(t)\,dt;$$

finally, by formula (3), we have $\overline{\rho(t)}\mathrm{J}(\rho)(t) = \prod\limits_{\alpha>0}(1-\alpha(t)^{-1})$, hence the corollary.

Remarks. 1) For all $w \in \mathrm{W}$, put $\rho_w = {}^w\rho/\rho$; then

$$\sum_w \varepsilon(w)\rho_w = \prod_{\alpha>0}(1-\alpha^{-1}) = \rho^{-2}\prod_{\alpha>0}(\alpha-1). \tag{6}$$

If t is a regular element of T, we deduce from (5) that

$$\chi_\lambda(t) = \frac{\sum_w \varepsilon(w)\, {}^w\lambda(t)\rho_w(t)}{\sum_w \varepsilon(w)\rho_w(t)} = \frac{\sum_w \varepsilon(w)\, {}^w\lambda(t)\rho_w(t)}{\prod_{\alpha>0}(1-\alpha(t)^{-1})}. \tag{7}$$

Note that ρ_w is a linear combination of roots with integer coefficients, and hence belongs to X(T) even if we do not assume that $\rho \in \mathrm{X(T)}$. It follows that formula (7) is valid without the assumption that $\rho \in \mathrm{X(T)}$: indeed, to prove this we replace G by a suitable connected covering, and are then reduced to Cor. 2.

2) Similarly, the first equality of Cor. 4 remains valid without the assumption that $\rho \in \mathrm{X(T)}$.

3) We can deduce Th. 2 from the analogous infinitesimal statement (Chap. VIII, §9, no. 1, Th. 1); the same is true for Th. 3 of the next number (which is the analogue of Th. 2 of *loc. cit.*, no. 2).

5. DEGREE OF IRREDUCIBLE REPRESENTATIONS

We now return to the additive notation for the group X(T) and we no longer assume that ρ belongs to X(T).

THEOREM 3. *The dimension of the space of an irreducible representation of* G *of highest weight* λ *is given by*

$$\chi_\lambda(e) = \prod_{\alpha\in \mathrm{R}_+} \frac{\langle \lambda+\rho, K_\alpha\rangle}{\langle \rho, K_\alpha\rangle}.$$

Put $\gamma = \frac{1}{2}\sum\limits_{\alpha>0} K_\alpha$, so $\delta(\alpha)(\gamma) = 2\pi i$ for every simple root α (Chap. VI, §1, no. 10, Prop. 29). The line $\mathbf{R}\gamma$ is not contained in any of the hyperplanes $\operatorname{Ker}\delta(\alpha)$, so $\exp(z\gamma)$ is a regular element of G for all sufficiently small $z \in \mathbf{R}^*$; for all $\mu \in \mathrm{X(T)}$ and all $z \in \mathbf{R}$, we have

$$\mathrm{J}(\mu)(\exp(z\gamma)) = \sum_{w\in \mathrm{W}} \varepsilon(w)\, e^{z\delta(\mu)(w^{-1}\gamma)}.$$

Let us accept provisionally the following lemma:

Lemma 5. We have

$$\mathrm{J}(\mu)(\exp(z\gamma)) = e^{z\delta(\mu)(\gamma)} \prod_{\alpha>0}(1-e^{-z\delta(\mu)(K_\alpha)}).$$

Thus, the function $\mathrm{J}(\mu)(\exp(z\gamma))$ is the product of a function that tends to 1 when z tends to 0 and of

$$z^{\mathrm{N}} \prod_{\alpha>0} \delta(\mu)(K_\alpha) = (2\pi i z)^{\mathrm{N}} \prod_{\alpha>0} \langle \mu, K_\alpha\rangle$$

where $\mathrm{N} = \operatorname{Card} \mathrm{R}_+$.

Assume first of all that $\rho \in \mathrm{X(T)}$; applying Cor. 2 of Th. 2 we see that, as z tends to 0, $\chi_\lambda(z\gamma)$ tends to

$$\prod_{\alpha>0}\langle\lambda+\rho, K_\alpha\rangle / \prod_{\alpha>0}\langle\rho, K_\alpha\rangle,$$

hence the theorem in this case.

In the general case, it suffices to remark that, in proving Th. 3, we can always replace G by a suitable connected covering and thus reduce to the preceding case.

We now prove Lemma 5. Let $z \in \mathbf{C}$; denote by φ_z the map from $\mathfrak{t}$ to the $\mathbf{C}$-algebra $\mathrm{Map}(\mathrm{X(T)}, \mathbf{C})$ of maps from $\mathrm{X(T)}$ to $\mathbf{C}$ that associates to $H \in \mathfrak{t}$ the map

$$\varphi_z(H) : \mu \mapsto \mu(\exp zH) = e^{z\delta(\mu)(H)}.$$

We have $\varphi_z(H+H') = \varphi_z(H)\varphi_z(H')$, so there exists a homomorphism of rings

$$\psi_z : \mathbf{Z}[\mathfrak{t}] \to \mathrm{Map}(\mathrm{X(T)}, \mathbf{C})$$

such that $\psi_z(e^H)(\mu) = e^{z\delta(\mu)(H)}$. On the other hand, by Chap. VI, §3, no. 3, Prop. 2, we have the following relation in $\mathbf{Z}[\mathfrak{t}]$:

$$\sum_{w\in \mathrm{W}} \varepsilon(w)\, e^{w\gamma} = e^\gamma \prod_{\alpha>0}(1-e^{-K_\alpha}).$$

Applying the homomorphism ψ_z, and using the equality

$$\psi_z(e^{w\gamma})(\mu) = e^{z\delta(\mu)(w\gamma)} = e^{z\delta(w^{-1}\mu)(\gamma)},$$

we deduce the stated formula.

COROLLARY 1. *Let* $\|\ \|$ *be a norm on* $\mathrm{X(T)}\otimes\mathbf{R}$. *For all* $\lambda \in \mathrm{X}_{++}$, *let* $d(\lambda)$ *be the dimension of the space of an irreducible representation of* G *of highest weight* λ.

a) $\sup_{\lambda\in\mathrm{X}_{++}} d(\lambda)/\|\lambda+\rho\|^{\mathrm{N}} < \infty$, *where* $\mathrm{N} = 1/2(\dim \mathrm{G} - \dim \mathrm{T})$.

b) If G *is semi-simple,* $\inf_{\lambda\in\mathrm{X}_{++}} d(\lambda)/\|\lambda+\rho\| > 0$.

a) For all $\alpha \in \mathrm{R}_+$, there exists $\mathrm{A}_\alpha > 0$ with $|\langle\lambda+\rho, K_\alpha\rangle| \leq \mathrm{A}_\alpha \|\lambda+\rho\|$, hence $d(\lambda)/\|\lambda+\rho\|^{\mathrm{N}} \leq \prod_{\alpha>0} \mathrm{A}_\alpha/\langle\rho, K_\alpha\rangle$.

b) Assume that G is semi-simple, denote by $\beta_1, \dots, \beta_r$ the simple roots and put $N_i = K_{\beta_i}$. Then

$$d(\lambda) \geq \prod_{i=1}^r \frac{\langle\lambda+\rho, N_i\rangle}{\langle\rho, N_i\rangle} = \prod_{i=1}^r \langle\lambda+\rho, N_i\rangle;$$

since $\langle \lambda + \rho, N_i \rangle \geq \langle \rho, N_i \rangle = 1$, this implies that

$$d(\lambda) \geq \sup_i |\langle \lambda + \rho, N_i \rangle|.$$

If G is semi-simple, $x \mapsto \sup|\langle x, N_i \rangle|$ is a norm on $X(T) \otimes \mathbf{R}$, necessarily equivalent to the given norm, hence *b*).

COROLLARY 2. *Assume that* G *is semi-simple; let d be an integer. The set of classes of representations of* G *of dimension* $\leq d$ *is finite.*

Cor. 1 *b*) implies that the set X_d of elements λ of X_{++} such that $d(\lambda) \leq d$ is finite. For all λ in X_d, let V_λ be an irreducible representation of highest weight λ; every representation of dimension $\leq d$ is isomorphic to a direct sum $\bigoplus_{\lambda \in X_d} V_\lambda^{n_\lambda}$, with $n_\lambda \leq d$, hence the corollary.

6. CASIMIR ELEMENTS

By Prop. 3 of §1, no. 3, there exist *negative* symmetric bilinear forms on $\mathfrak{g}$, separating and invariant under Ad(G) (if G is semi-simple, we can take the Killing form of $\mathfrak{g}$, for example). Let F be such a form. Recall (Chap. I, §3, no. 7) that the Casimir element associated to F is the element Γ of the centre of the enveloping algebra $U(\mathfrak{g})$ such that, for any basis (e_i) of $\mathfrak{g}$ satisfying $F(e_i, e_j) = -\delta_{ij}$, we have $\Gamma = -\sum e_i^2$.

In the remainder of this chapter we shall call the *Casimir elements* of G the elements of $U(\mathfrak{g})$ obtained in this way from *negative* separating invariant symmetric bilinear forms on $\mathfrak{g}$. If Γ is a Casimir element of G and if $\tau : G \to \mathbf{GL}(V)$ is an irreducible representation of G, the endomorphism Γ_V of V is a homothety (*Algebra*, Chap. VIII, §3, no. 2, Th. 1), whose ratio we shall denote by $\tilde{\Gamma}(\tau)$.

PROPOSITION 4. *Let* Γ *be the Casimir element of* G.

a) *If* τ *is an irreducible representation of* G, $\tilde{\Gamma}(\tau)$ *is real and positive. If* τ *is not the trivial representation,* $\tilde{\Gamma}(\tau) > 0$.

b) *There exists a unique quadratic form* Q_Γ *on* $X(T) \otimes \mathbf{R}$ *such that, for every irreducible representation* τ *of* G,

$$\tilde{\Gamma}(\tau) = Q_\Gamma(\lambda + \rho) - Q_\Gamma(\rho),$$

where λ *is the highest weight of* τ. *The form* Q_Γ *is positive, separating and invariant under* W.

Let F be a separating negative symmetric bilinear form on $\mathfrak{g}$ defining Γ. Let $\tau : G \to \mathbf{GL}(V)$ be an irreducible representation of G; let $\langle\ ,\ \rangle$ be a Hilbert scalar product on V invariant under G (§1, no. 1), and (e_i) a basis of $\mathfrak{g}$ such that $F(e_i, e_j) = -\delta_{ij}$. Then, for every element v of V not invariant under G we have

$$\tilde{\Gamma}(\tau)\langle v, v\rangle = \langle v, \Gamma_{\mathrm{V}}(v)\rangle = -\sum_i \langle v, (e_i)^2_{\mathrm{V}} v\rangle = \sum_i \langle v, (e_i)^*_{\mathrm{V}}(e_i)_{\mathrm{V}} v\rangle$$
$$= \sum_i \langle (e_i)_{\mathrm{V}} v, (e_i)_{\mathrm{V}} v\rangle > 0,$$

hence *a*).

Let B be the inverse form on $\mathfrak{t}^*_{\mathbf{C}}$ of the restriction to $\mathfrak{t}_{\mathbf{C}}$ of the bilinear form on $\mathfrak{g}_{\mathbf{C}}$ induced by F by extension of scalars. By the Cor. of Prop. 7 of Chap. VIII, §6, no. 4, we have[5] $\tilde{\Gamma}(\tau) = \mathrm{B}(\delta(\lambda), \delta(\lambda+2\rho))$. Extend $\delta : \mathrm{X}(\mathrm{T}) \to \mathfrak{t}^*_{\mathbf{C}}$ to an $\mathbf{R}$-linear map from $\mathrm{X}(\mathrm{T}) \otimes \mathbf{R}$ to $\mathfrak{t}^*_{\mathbf{C}}$ and let Q_Γ be the quadratic form $x \mapsto \mathrm{B}(\delta(x), \delta(x))$ on $\mathrm{X}(\mathrm{T}) \otimes \mathbf{R}$; it is separating and invariant under W, and

$$\tilde{\Gamma}(\tau) = \mathrm{B}(\delta(\lambda+\rho), \delta(\lambda+\rho)) - \mathrm{B}(\delta(\rho), \delta(\rho)) = \mathrm{Q}_\Gamma(\lambda+\rho) - \mathrm{Q}_\Gamma(\rho).$$

We show that the form Q_Γ is *positive*. Indeed, if $x \in \mathrm{X}(\mathrm{T}) \otimes \mathbf{Q}$, the element $\delta(x)$ of $\mathfrak{t}^*_{\mathbf{C}}$ takes purely imaginary values on $\mathfrak{t}$, hence real values on $i\mathfrak{t}$; we conclude by remarking that, for $y \in i\mathfrak{t}$, we have $\mathrm{F}(y, y) \geq 0$.

It remains to prove the uniqueness assertion in *b*). Let Q be a quadratic form on $\mathrm{X}(\mathrm{T}) \otimes \mathbf{R}$ satisfying the required condition, and let Φ (resp. Φ_Γ) be the bilinear form associated to Q (resp. Q_Γ). For $\lambda, \mu \in \mathrm{X}_{++}$, we have

$$\Phi(\lambda, \mu) = (\mathrm{Q}(\lambda+\mu+\rho) - \mathrm{Q}(\rho)) - (\mathrm{Q}(\lambda+\rho) - \mathrm{Q}(\rho)) - (\mathrm{Q}(\mu+\rho) - \mathrm{Q}(\rho))$$
$$= \Phi_\Gamma(\lambda, \mu).$$

Since $\mathrm{X}(\mathrm{T})_{++}$ generates the $\mathbf{R}$-vector space $\mathrm{X}(\mathrm{T}) \otimes \mathbf{R}$, we have $\Phi = \Phi_\Gamma$, so $\mathrm{Q} = \mathrm{Q}_\Gamma$.

Remark. Let $x \in \mathfrak{g}$. There exists a strictly positive real number A such that, for every irreducible representation $\tau : \mathrm{G} \to \mathbf{GL}(\mathrm{V})$ and every Hilbert structure on V invariant under G,

$$\| \mathrm{L}(\tau)(x) \|^2 \leq \mathrm{A}.\tilde{\Gamma}(\tau).$$

Indeed, with the notations in the preceding proof, we can choose the basis (e_i) of $\mathfrak{g}$ so that $x = ae_1$, $a \in \mathbf{R}$. Then, for $v \in \mathrm{V}$, we have

$$\langle x_{\mathrm{V}} v, x_{\mathrm{V}} v\rangle = |a|^2 \langle e_1 v, e_1 v\rangle \leq |a|^2 \tilde{\Gamma}(\tau)\langle v, v\rangle.$$

§8. FOURIER TRANSFORM

We retain the notations and conventions of the preceding paragraph.

[5] The proof of *loc. cit.*, which has been stated only for split semi-simple Lie algebras, is valid without change in the case of split reductive Lie algebras.

1. FOURIER TRANSFORMS OF INTEGRABLE FUNCTIONS

In this number, we recall some definitions and results from *Spectral Theories*[6].

Denote by $\hat{\mathrm{G}}$ the set of classes of irreducible representations of G (on finite dimensional complex vector spaces). For all $u \in \hat{\mathrm{G}}$, denote by E_u the space of u and $d(u)$ its dimension. There exist separating positive hermitian forms on E_u invariant under u, and any two such forms are proportional. Denote by A^* (resp. $\| \mathrm{A} \|_\infty$) the adjoint (resp. the norm) of an element A of $\mathrm{End}(\mathrm{E}_u)$ relative to one of these forms; for all $g \in \mathrm{G}$, we have $u(g)^* = u(g)^{-1} = u(g^{-1})$ and $\| u(g)\|_\infty = 1$; for all $x \in \mathfrak{g}$, we have $u(x)^* = -u(x) = u(-x)$.

Give $\mathrm{End}(\mathrm{E}_u)$ the Hilbert space structure for which the scalar product is

$$\langle \mathrm{A}|\mathrm{B}\rangle = d(u)\mathrm{Tr}(\mathrm{A}^*\mathrm{B}) = d(u)\mathrm{Tr}(\mathrm{B}\mathrm{A}^*), \tag{1}$$

and put

$$\|\mathrm{A}\|_2 = \langle \mathrm{A}|\mathrm{A}\rangle^{1/2} = (d(u)\mathrm{Tr}(\mathrm{A}^*\mathrm{A}))^{1/2}. \tag{2}$$

We have

$$\sqrt{d(u)}\|\mathrm{A}\|_\infty \leq \|\mathrm{A}\|_2 \leq d(u)\|\mathrm{A}\|_\infty, \tag{3}$$

so

$$|\langle \mathrm{A}|\mathrm{B}\rangle| \leq d(u)^2\|\mathrm{A}\|_\infty\|\mathrm{B}\|_\infty. \tag{4}$$

For all $g \in \mathrm{G}$, we have $\|u(g)\|_2 = d(u)$.

Denote by $\mathrm{F}(\hat{\mathrm{G}})$ the algebra $\prod_{u\in\hat{\mathrm{G}}} \mathrm{End}(\mathrm{E}_u)$. Denote by $\mathrm{L}^2(\hat{\mathrm{G}})$ the Hilbert sum of the Hilbert spaces $\mathrm{End}(\mathrm{E}_u)$; this is the space of families $\mathrm{A}= (\mathrm{A}_u)\in\mathrm{F}(\hat{\mathrm{G}})$ such that $\sum_u \|\mathrm{A}_u\|_2^2 < \infty$, with the scalar product

$$\langle \mathrm{A}|\mathrm{B}\rangle = \sum_{u\in\hat{\mathrm{G}}} \langle \mathrm{A}_u|\mathrm{B}_u\rangle = \sum_{u\in\hat{\mathrm{G}}} d(u)\mathrm{Tr}(\mathrm{A}_u^*\mathrm{B}_u). \tag{5}$$

Denote the Hilbert norm on $\mathrm{L}^2(\hat{\mathrm{G}})$ also by $\| \ \|_2$, so that $\|\mathrm{A}\|_2^2 = \sum_{u\in\hat{\mathrm{G}}} \|\mathrm{A}_u\|_2^2$ for $\mathrm{A} \in \mathrm{L}^2(\hat{\mathrm{G}})$.

If f is an integrable complex function on G, put

$$u(f) = \int_\mathrm{G} f(g)u(g)\, dg \in \mathrm{End}(\mathrm{E}_u) \tag{6}$$

for all $u \in \hat{\mathrm{G}}$. We have $\|u(f)\|_\infty \leq \int_\mathrm{G} |f(g)|\, dg = \|f\|_1$. The *Fourier cotransform* of f, denoted by $\overline{\mathscr{F}}(f)$, is the family $(u(f))_{u\in\hat{\mathrm{G}}} \in \mathrm{F}(\hat{\mathrm{G}})$. If $f \in \mathrm{L}^2(\hat{\mathrm{G}})$,

[6] See note [1], §7, p. 66.

$$\|f\|_2^2 = \sum_{u\in\hat{G}} \langle u(f)|u(f)\rangle = \|\overline{\mathscr{F}}(f)\|_2^2,$$

so $\overline{\mathscr{F}}$ induces an isometric linear map from the Hilbert space $L^2(G)$ to the Hilbert space $L^2(\hat{G})$: in other words, for f and f' in $L^2(G)$, we have

$$\int_G \overline{f(g)} f'(g)\, dg = \langle \overline{\mathscr{F}}(f)|\overline{\mathscr{F}}(f')\rangle = \sum_{u\in\hat{G}} d(u)\mathrm{Tr}(u(f)^* u(f')). \tag{7}$$

For f and f' in $L^1(G)$, the convolution product $f * f'$ of f and f' is defined by

$$(f * f')(h) = \int_G f(hg^{-1}) f'(g)\, dg = \int_G f(g) f'(g^{-1}h)\, dg$$

(the integral makes sense for almost all $h \in G$).

We have $f * f' \in L^1(G)$ and, for all $u \in \hat{G}$, $u(f * f') = u(f)u(f')$, so

$$\overline{\mathscr{F}}(f * f') = \overline{\mathscr{F}}(f).\overline{\mathscr{F}}(f'). \tag{8}$$

Conversely, let $A = (A_u)_{u\in\hat{G}}$ be an element of $F(\hat{G})$; for all $u \in \hat{G}$, let $\mathscr{F}_u A$ be the (analytic) function on G defined by

$$(\mathscr{F}_u A)(g) = \langle u(g)|A_u\rangle = d(u)\mathrm{Tr}(A_u u(g)^{-1}). \tag{9}$$

If $A \in L^2(\hat{G})$, the family $(\mathscr{F}_u A)_{u\in\hat{G}}$ is summable in $L^2(G)$; the *Fourier transform* of A, denoted by $\mathscr{F}(A)$, is the sum of this family. The maps $\overline{\mathscr{F}}$ and $\mathscr{F}$ are inverse isomorphisms between the Hilbert spaces $L^2(G)$ and $L^2(\hat{G})$.

In other words:

PROPOSITION 1. *Every square integrable complex function f on G is the sum in the Hilbert space $L^2(G)$ of the family $(f_u)_{u\in\hat{G}}$ where, for all $h \in G$ and all $u \in \hat{G}$,*

$$f_u(h) = \langle u(h)|u(f)\rangle$$
$$= d(u)\int_G f(g)\mathrm{Tr}(u(gh^{-1}))\, dg = d(u)\int_G f(gh)\mathrm{Tr}(u(g))\, dg. \tag{10}$$

For all $u \in \hat{G}$ choose an orthonormal basis B_u of E_u, and denote by $(u_{ij}(g))$ the matrix of $u(g)$ in this basis. Prop. 1 also means that the family of functions $\sqrt{d(u)}u_{ij}$, for u in $\hat{G}$ and i, j in B_u, is an *orthonormal basis* of the space $L^2(G)$.

If f is an integrable function on G such that the family (f_u) is uniformly summable, then the sum of this family is a continuous function which coincides almost everywhere with f; in other words, if we assume in addition that f is continuous, then for all $h \in G$,

$$f(h) = \sum_{u \in \hat{\mathrm{G}}} d(u) \int_{\mathrm{G}} f(gh) \mathrm{Tr}(u(g))\, dg. \tag{11}$$

Conversely, let $\mathrm{A} \in \mathrm{F}(\hat{\mathrm{G}})$; if the family $(\mathscr{F}_u \mathrm{A})_{u \in \hat{\mathrm{G}}}$ is uniformly summable, the function

$$g \mapsto \sum_{u \in \hat{\mathrm{G}}} (\mathscr{F}_u \mathrm{A})(g) = \sum_{u \in \hat{\mathrm{G}}} d(u) \mathrm{Tr}(\mathrm{A}_u u(g)^{-1})$$

is a continuous function on G whose Fourier cotransform is A.

Let f be an integrable function on G, and let $s \in \mathrm{G}$. Denote by $\gamma(s)f$ and $\delta(s)f$ the functions on G defined by $\gamma(s)f = \varepsilon_s * f$, $\delta(s)f = f * \varepsilon_{s^{-1}}$, that is,

$$(\gamma(s)f)(g) = f(s^{-1}g), \quad (\delta(s)f)(g) = f(gs) \ \text{ for } g \in \mathrm{G},$$

(Chap. III, §3, no. 4 and *Integration*, Chap. VII, §1, no. 1). We have

$$u(\gamma(s)f) = \int_{\mathrm{G}} f(s^{-1}g) u(g)\, dg = \int_{\mathrm{G}} f(g) u(sg)\, dg,$$

so

$$u(\gamma(s)f) = u(s)u(f), \tag{12}$$

and similarly

$$u(\delta(s^{-1})f) = u(f)u(s). \tag{13}$$

When G is commutative, $\hat{\mathrm{G}}$ is the underlying set of the dual group of G (*Spectral Theories*, Chap. II, §1, no. 1), $d(u) = 1$ for all $u \in \hat{\mathrm{G}}$, and we recover the definitions of the Fourier transform given in *Spectral Theories*, Chap. II.

2. FOURIER TRANSFORMS OF INFINITELY-DIFFERENTIABLE FUNCTIONS

Recall (Chap. III, §3, no. 1, Def. 2) that U(G) denotes the algebra of distributions on G with support contained in $\{e\}$. The canonical injection of $\mathfrak{g}$ into U(G) extends to an isomorphism from the enveloping algebra of the Lie algebra $\mathfrak{g}$ to U(G) (*loc. cit.*, no. 7, Prop. 25); from now on we identify these two algebras by this isomorphism. If f is an infinitely-differentiable complex function on G and if $t \in \mathrm{U(G)}$, we denote by $\mathrm{L}_t f$ and $\mathrm{R}_t f$ the functions on G defined by

$$\mathrm{L}_t f(g) = \langle \varepsilon_g * t, f \rangle, \quad \mathrm{R}_t f(g) = \langle t * \varepsilon_g, f \rangle$$

(cf. *loc. cit.*, no. 6). For all $g \in \mathrm{G}$,

$$\mathrm{L}_t \circ \gamma(g) = \gamma(g) \circ \mathrm{L}_t, \quad \mathrm{R}_t \circ \delta(g) = \delta(g) \circ \mathrm{R}_t.$$

Let $u \in \hat{G}$; denote by E_u the space of u. The morphism of Lie groups $u : G \to \mathbf{GL}(E_u)$ gives by differentiation a homomorphism of (real) Lie algebras $\mathfrak{g} \to \mathrm{End}(E_u)$, hence a homomorphism of algebras, also denoted by u, from $U(G)$ to $\mathrm{End}(E_u)$. If $t \in U(G)$ and if f is an infinitely-differentiable function on G, then

$$u(L_t f) = u(f)u(t^\vee), \quad u(R_t f) = u(t^\vee)u(f), \tag{14}$$

where $t^\vee$ denotes the image of t under the principal anti-automorphism of U(G) (Chap. I, §2, no. 4); indeed, it suffices to verify this for $t \in \mathfrak{g}$, in which case it follows by differentiation from formulas (12) and (13) (cf. Chap. III, §3, no. 7, Prop. 27).

For all $u \in \hat{G}$, denote by $\lambda(u)$ the highest weight of u (§7, no. 2, Th. 1), so $u \mapsto \lambda(u)$ is a bijective map from $\hat{G}$ to the set X_{++} of dominant elements of X(T).

Let $\Gamma \in U(G)$ be a Casimir element of G (§7, no. 6); for all $u \in \hat{G}$, the endomorphism $u(\Gamma)$ of E_u is a homothety, whose ratio we denote by $\tilde{\Gamma}(u)$, so we have a map $u \mapsto \tilde{\Gamma}(u)$ from $\hat{G}$ to $\mathbf{C}$.

If φ and ψ are two functions on $\hat{G}$ with positive real values, denote by "$\varphi \preccurlyeq \psi$" or "$\varphi(u) \preccurlyeq \psi(u)$" the relation "there exists $M > 0$ such that $\varphi(u) \leq M\psi(u)$ for all $u \in \hat{G}$"; this is a pre-order relation on the set of functions on $\hat{G}$ with positive real values.

PROPOSITION 2. *Let $m \mapsto \|m\|$ be a norm on the* $\mathbf{R}$*-vector space* $\mathbf{R}\otimes X(T)$ *and Γ a Casimir element of* G. *Let φ be a function on $\hat{G}$ with positive real values.*

a) The following conditions are equivalent:

(i) *There exists an integer $n > 0$ such that $\varphi(u) \preccurlyeq (\|\lambda(u)\| + 1)^n$* (resp. *for every integer $n > 0$, we have $\varphi(u) \preccurlyeq (\|\lambda(u)\| + 1)^{-n}$*).

(ii) *There exists an integer $n > 0$ such that $\varphi(u) \preccurlyeq (\tilde{\Gamma}(u) + 1)^n$* (resp. *for every integer $n > 0$, we have $\varphi(u) \preccurlyeq (\tilde{\Gamma}(u) + 1)^{-n}$*).

b) If G *is semi-simple, conditions* (i) *and* (ii) *are also equivalent to:*

(iii) *There exists an integer $n > 0$ such that $\varphi(u) \preccurlyeq d(u)^n$* (resp. *for every integer $n > 0$, we have $\varphi(u) \preccurlyeq d(u)^{-n}$*).

Note first of all that condition (i) is clearly independent of the choice of norm. Thus we can use the norm defined by the quadratic form Q_Γ associated to Γ (§7, no. 6, Prop. 4). Then

$$0 \leq \tilde{\Gamma}(u) = \|\lambda(u) + \rho\|^2 - \|\rho\|^2,$$

so $\tilde{\Gamma}(u) + 1 \preccurlyeq (\|\lambda(u)\| + 1)^2 \preccurlyeq \tilde{\Gamma}(u) + 1$, hence *a*).

Further, if G is semi-simple,

$$\|\lambda(u) + \rho\| \preccurlyeq d(u) \preccurlyeq \|\lambda(u) + \rho\|^N, \quad \text{where } N = 1/2(\dim G - \dim T)$$

(§7, no. 5, Cor. 1 of Th. 3), so $\|\lambda(u)\| + 1 \preccurlyeq d(u) \preccurlyeq (\|\lambda(u)\| + 1)^N$, hence *b*).

It follows from Prop. 2 that condition (i) is independent of the choice of maximal torus, chamber, and norm, and that condition (ii) is independent of the choice of Casimir element. A function φ satisfying conditions (i) and (ii) is said to be *moderately increasing* (resp. *rapidly decreasing*). The product of two moderately increasing functions is moderately increasing; the product of a moderately increasing function and a rapidly decreasing function is rapidly decreasing. If φ is rapidly decreasing, the family $(\varphi(u))_{u\in\hat{\mathrm{G}}}$ is summable.

Examples. The function $u \mapsto d(u)$ is moderately increasing (§7, no. 5, Cor. 1 of Th. 3); for any norm $\| \ \|$ on $\mathbf{R}\otimes \mathrm{X}(\mathrm{T})$, the function $u \mapsto \|\lambda(u)\|$ is moderately increasing. For any Casimir element $\varGamma$, the function $u \mapsto \tilde{\varGamma}(u)$ is moderately increasing; more generally:

PROPOSITION 3. *For all $t \in \mathrm{U}(\mathrm{G})$, the functions $u \mapsto \|u(t)\|_\infty$ and $u \mapsto \|u(t)\|_2$ on $\hat{\mathrm{G}}$ are moderately increasing.*

Since the product of two moderately increasing functions is moderately increasing, it suffices to prove this when $t \in \mathfrak{g}$: in that case the assertion follows from the Remark in §7, no. 6 and the inequality

$$\|u(t)\|_2 \leq d(u)\|u(t)\|_\infty.$$

THEOREM 1. *a) Let f be an infinitely-differentiable complex function on G. Then the family $(f_u)_{u\in\hat{\mathrm{G}}}$, where $f_u(g) = \langle u(g)|u(f)\rangle$, is uniformly summable on G and, for all $h \in \mathrm{G}$,*

$$f(h) = \sum_{u\in\hat{\mathrm{G}}} \langle u(h)|u(f)\rangle = \sum_{u\in\hat{\mathrm{G}}} d(u) \int_{\mathrm{G}} f(g)\mathrm{Tr}(u(gh^{-1}))\, dg.$$

b) Let f be an integrable function on G; then f is equal almost everywhere to an infinitely-differentiable function if and only if the function $u \mapsto \|u(f)\|_\infty$ is rapidly decreasing on $\hat{\mathrm{G}}$.

Let f be an infinitely-differentiable function on G, and let $\varGamma$ be a Casimir element for G; by formula (14),

$$\tilde{\varGamma}(u)^n u(f) = u(f)u(\varGamma)^n = u((\mathrm{L}_\varGamma)^n f)$$

for all $n \geq 0$, and consequently

$$\tilde{\varGamma}(u)^n \|u(f)\|_\infty \leq \|(\mathrm{L}_\varGamma)^n f\|_1 \leq \sup_{g\in\mathrm{G}} |((\mathrm{L}_\varGamma)^n f)(g)|; \tag{15}$$

thus, the function $u \mapsto \|u(f)\|_\infty$ is indeed rapidly decreasing.

Conversely, let $\mathrm{A} = (\mathrm{A}_u)_{u\in\hat{\mathrm{G}}}$ be an element of $\mathrm{F}(\hat{\mathrm{G}})$ such that the function $u \mapsto \|\mathrm{A}_u\|_\infty$ is rapidly decreasing. Put $f_u(g) = \langle u(g)|\mathrm{A}_u\rangle$; the function $g \mapsto f_u(g)$ is analytic, hence infinitely-differentiable. By Chap. III, §3, no. 7, Prop. 27,

$$(\mathrm{L}_x f_u)(g) = \langle u(g)u(x)|\mathrm{A}_u\rangle$$

for all $x \in \mathfrak{g}$. Let $t \in \mathrm{U(G)}$; by the preceding formula,

$$(\mathrm{L}_t f_u)(g) = \langle u(g)u(t)|\mathrm{A}_u\rangle$$

and consequently

$$|(\mathrm{L}_t f_u)(g)| = |\langle u(g)u(t)|\mathrm{A}_u\rangle \leq d(u)^2\|u(t)\|_\infty\|u(g)\|_\infty\|\mathrm{A}_u\|_\infty = d(u)^2\|u(t)\|_\infty\|\mathrm{A}_u\|_\infty.$$

Since $d(u)$ and $\|u(t)\|_\infty$ are moderately increasing (Prop. 3) and $\|\mathrm{A}_u\|_\infty$ is rapidly decreasing, the function $u \mapsto \sup_g |(\mathrm{L}_t f_u)(g)|$ is rapidly decreasing; thus, the family $(\mathrm{L}_t f_u)_{u\in\hat{\mathrm{G}}}$ is uniformly summable. It follows[7] that the sum of the family (f_u) is an infinitely-differentiable function on G, whose Fourier cotransform is (A_u), hence the theorem.

Denote by $\mathscr{S}(\hat{\mathrm{G}})$ the vector subspace of $\mathrm{L}^2(\hat{\mathrm{G}})$ consisting of the families $\mathrm{A} = (\mathrm{A}_u)_{u\in\hat{\mathrm{G}}}$ such that the function $u \mapsto \|\mathrm{A}_u\|_\infty$ is rapidly decreasing on $\hat{\mathrm{G}}$. It follows from the theorem that the maps $\overline{\mathscr{F}} : f \mapsto (u(f))_{u\in\hat{\mathrm{G}}}$ and $\mathscr{F} : \mathrm{A} \mapsto \sum_{u\in\hat{\mathrm{G}}} \langle u(g)|\mathrm{A}_u\rangle$ induce inverse isomorphisms between the complex vector spaces $\mathscr{C}^\infty(\mathrm{G};\mathbf{C})$ and $\mathscr{S}(\hat{\mathrm{G}})$. Give the space $\mathscr{C}^\infty(\mathrm{G};\mathbf{C})$ the topology of uniform C^∞-convergence (§6, no. 4) which can be defined by the family of semi-norms $f \mapsto \sup_{g\in\mathrm{G}} |\mathrm{L}_t f(g)|$ for $t \in \mathrm{U(G)}$, and the space $\mathscr{S}(\hat{\mathrm{G}})$ the topology defined by the sequence of semi-norms $p_n : \mathrm{A} \mapsto \sup_{u\in\hat{\mathrm{G}}}(\tilde{\Gamma}(u)+1)^n\|\mathrm{A}_u\|_\infty$. Formula (15) of the preceding proof shows that $\overline{\mathscr{F}}$ is continuous. Let $t \in \mathrm{U(G)}$, and let $\mathrm{A} = (\mathrm{A}_u)_{u\in\hat{\mathrm{G}}}$ be an element of $\mathscr{S}(\hat{\mathrm{G}})$; put $f_n(g) = \langle u(g)|\mathrm{A}_u\rangle$. Let p be an integer such that $\sum_{u\in\hat{\mathrm{G}}} \tilde{\Gamma}(u)^{-p} = \mathrm{M} < \infty$. By the preceding proof, there exists a positive integer m such that, for all $g \in \mathrm{G}$,

$$|(\mathrm{L}_t f_u)(g)| \leq d(u)^2\|u(t)\|_\infty\|\mathrm{A}_u\|_\infty \leq m.(1+\tilde{\Gamma}(u))^m\tilde{\Gamma}(u)^{-p}\|\mathrm{A}_u\|_\infty$$

so $|(\mathrm{L}_t\mathscr{F}(\mathrm{A}))(g)| \leq m\mathrm{M}p_m(\mathrm{A})$; this proves that $\mathscr{F}$ is continuous. Consequently:

COROLLARY. *The maps* $\overline{\mathscr{F}} : f \mapsto (u(f))_{u\in\hat{\mathrm{G}}}$ *and* $\mathscr{F} : \mathrm{A} \mapsto \sum_{u\in\hat{\mathrm{G}}} \langle u(g)|\mathrm{A}_u\rangle$ *induce inverse isomorphisms between the topological vector spaces* $\mathscr{C}^\infty(\mathrm{G};\mathbf{C})$ *and* $\mathscr{S}(\hat{\mathrm{G}})$.

[7] This follows from the fact that the space $\mathscr{C}^\infty(\mathrm{G};\mathbf{C})$, with the topology of uniform C^∞-convergence (§6, no. 4), is *complete.*

3. FOURIER TRANSFORMS OF CENTRAL FUNCTIONS

For all $u \in \hat{\mathrm{G}}$, denote by χ_u the *character* of u; thus,

$$\chi_u(g) = \mathrm{Tr}(u(g)), \quad (g \in \mathrm{G}). \tag{16}$$

Recall from *Spectral Theory* the formulas

$$\chi_u * \chi_v = 0 \quad (u, v \in \hat{\mathrm{G}}, u \neq v) \tag{17}$$

$$\chi_u * \chi_u = \frac{1}{d(u)}\chi_u \quad (u \in \hat{\mathrm{G}}). \tag{18}$$

For all $u \in \hat{\mathrm{G}}$, denote by ε_u the identity map of E_u. Recall (§7, no. 4) that $\mathrm{ZL}^2(\mathrm{G})$ denotes the subspace of $\mathrm{L}^2(\mathrm{G})$ consisting of the classes of the functions f that are central, that is, such that $f \circ \mathrm{Int}\, s = f$ for all $s \in \mathrm{G}$, or equivalently that $\gamma(s)f = \delta(s^{-1})f$ for all $s \in \mathrm{G}$.

PROPOSITION 4. *Let $f \in \mathrm{L}^2(\mathrm{G})$. Then f is central if and only if $u(f)$ is a homothety for all $u \in \hat{\mathrm{G}}$. In that case*

$$u(f) = \frac{\varepsilon_u}{d(u)} \int_{\mathrm{G}} f(g)\chi_u(g)\, dg. \tag{19}$$

By Prop. 1 (no. 1), to say that f is central means that $u(\gamma(s)f) = u(\delta(s^{-1})f)$ for all $s \in \mathrm{G}$ and all $u \in \hat{\mathrm{G}}$; but this can also be written as $u(s)u(f) = u(f)u(s)$ for all $s \in \mathrm{G}$ and all $u \in \hat{\mathrm{G}}$ (formulas (12) and (13)), hence the first assertion of Prop. 4 (Schur's lemma). If $u(f)$ is a homothety, then $u(f) = \lambda_u \varepsilon_u$ with

$$\lambda_u = \frac{1}{d(u)}\mathrm{Tr}(u(f)) = \frac{1}{d(u)} \int_{\mathrm{G}} f(g)\mathrm{Tr}(u(g))\, dg = \frac{1}{d(u)} \int_{\mathrm{G}} f(g)\chi_u(g)\, dg.$$

Consequently, for $f \in \mathrm{ZL}^2(\mathrm{G})$ we have

$$u(f) = \langle \overline{\chi}_u | f \rangle \frac{\varepsilon_u}{d(u)}, \tag{20}$$

so

$$\overline{\mathscr{F}}(f) = \left(\langle \overline{\chi}_u | f \rangle \frac{\varepsilon_u}{d(u)} \right)_{u \in \hat{\mathrm{G}}} \tag{21}$$

with

$$\|\overline{\mathscr{F}}(f)\|_2^2 = \sum_u \|\langle \overline{\chi}_u | f \rangle \frac{\varepsilon_u}{d(u)}\|_2^2 = \sum_u |\langle \overline{\chi}_u | f \rangle|^2.$$

Conversely, if φ is a square-integrable complex function on $\hat{\mathrm{G}}$, the element $\left(\frac{\varphi(u)}{d(u)}\varepsilon_u\right)_{u \in \hat{\mathrm{G}}}$ of $\mathrm{F}(\hat{\mathrm{G}})$ belongs to $\mathrm{L}^2(\hat{\mathrm{G}})$, and we have (formula (9))

$$\left(\mathscr{F}_u\left(\frac{\varphi(u)}{d(u)}\varepsilon_u\right)\right)(g) = d(u)\mathrm{Tr}\left(\frac{\varphi(u)}{d(u)}\varepsilon_u u(g)^{-1}\right) = \varphi(u)\overline{\chi}_u(g),$$

so

$$\mathscr{F}\left(\left(\frac{\varphi(u)}{d(u)}\varepsilon_u\right)\right) = \sum_{u\in\hat{\mathrm{G}}} \varphi(u)\overline{\chi}_u. \tag{22}$$

Note, in particular, that formulas (20) and (21) give, for u, v in $\hat{\mathrm{G}}$,[8]

$$u(\overline{\chi}_v) = 0 \quad \text{if } u \neq v, \tag{23}$$

$$u(\overline{\chi}_u) = \frac{\varepsilon_u}{d(u)} \in \mathrm{End}(\mathrm{E}_u), \tag{24}$$

$$\overline{\mathscr{F}}(\chi_u) = \frac{\varepsilon_u}{d(u)} \in \mathrm{End}(\mathrm{E}_u) \subset \mathrm{F}(\hat{\mathrm{G}}). \tag{25}$$

PROPOSITION 5. *Let f be a continuous central function on* G. *Then f is infinitely-differentiable if and only if the function $u \mapsto |\langle\chi_u|f\rangle|$ is rapidly decreasing on $\hat{\mathrm{G}}$; in that case,*

$$f(g) = \sum_{u\in\hat{\mathrm{G}}} \langle\chi_u|f\rangle\chi_u(g)$$

for all $g \in$ G.

By Th. 1 *b*), the function $\overline{f}$ is infinitely-differentiable if and only if the function $u \mapsto \|u(\overline{f})\|_\infty$ is rapidly decreasing; but, by (20),

$$\|u(\overline{f})\|_\infty = \frac{1}{d(u)}|\langle\chi_u|f\rangle|,$$

hence the first assertion, since the functions $d(u)$ and $\frac{1}{d(u)}$ are moderately increasing.

Assume that f is infinitely-differentiable; by Th. 1 *a*), $f(g) = \sum_{u\in\hat{\mathrm{G}}} f_u(g)$ for all $g \in$ G, so

$$f_u(g) = \langle u(g)|u(f)\rangle = d(u)\mathrm{Tr}(u(g)^{-1}.u(f)) = d(u)\mathrm{Tr}\left(u(g)^{-1}\langle\overline{\chi}_u|f\rangle\frac{\varepsilon_u}{d(u)}\right)$$
$$= \langle\overline{\chi}_u|f\rangle\mathrm{Tr}(u(g)^{-1}) = \langle\overline{\chi}_u|f\rangle\overline{\chi}_u(g).$$

Hence, $f(g) = \sum_{u\in\hat{\mathrm{G}}} \langle\overline{\chi}_u|f\rangle\overline{\chi}_u(g)$; but, for all $u \in \hat{\mathrm{G}}$, the contragredient representation u' of u satisfies $\overline{\chi}_u = \chi_{u'}$ and the map $u \mapsto u'$ is a permutation of $\hat{\mathrm{G}}$; so we also have $f(g) = \sum_{u\in\hat{\mathrm{G}}} \langle\chi_u|f\rangle\chi_u(g)$, hence the proposition.

---

[8] We embed $\mathrm{End}(\mathrm{E}_u)$ in the product $\mathrm{F}(\hat{\mathrm{G}}) = \prod_{v\in\hat{\mathrm{G}}} \mathrm{End}(\mathrm{E}_v)$ by associating to any $\mathrm{A} \in \mathrm{End}(\mathrm{E}_u)$ the family $(\mathrm{A}_v)_{v\in\hat{\mathrm{G}}}$ such that $\mathrm{A}_u = \mathrm{A}$ and $\mathrm{A}_v = 0$ for $v \neq u$.

COROLLARY. *Let f be a continuous central function on G. Then f is infinitely-differentiable if and only if the restriction of f to T is infinitely-differentiable.*

Indeed, by Cor. 4 of §7, no. 4,

$$\langle \chi_u | f \rangle = \int_{\mathrm{G}} \overline{\lambda(u)(t)} \varphi(t)\, dt, \quad \text{where} \quad \varphi(t) = \prod_{\alpha>0} (1 - \alpha(t)^{-1}) f(t).$$

If $f|\mathrm{T}$ is infinitely-differentiable, so is φ; by Prop. 5, applied to the group T, the function $\mu \mapsto \int_{\mathrm{T}} \overline{\mu(t)} \varphi(t)\, dt$ on $\hat{\mathrm{T}} = \mathrm{X}(\mathrm{T})$ is then rapidly decreasing, and so is the function $u \mapsto \langle \chi_u | f \rangle$; hence the function f is infinitely-differentiable (Prop. 5). The converse is clear.

4. CENTRAL FUNCTIONS ON G AND FUNCTIONS ON T

Denote by $\mathscr{C}(\mathrm{G})$ the space of continuous complex-valued functions on G and by $\mathscr{C}^\infty(\mathrm{G})$ the subspace of infinitely-differentiable functions. Then we have a sequence of inclusions

$$\Theta(\mathrm{G}) \subset \mathscr{C}^\infty(\mathrm{G}) \subset \mathscr{C}(\mathrm{G}) \subset \mathrm{L}^2(\mathrm{G}).$$

Denote by $\mathrm{Z}\Theta(\mathrm{G}), \mathrm{Z}\mathscr{C}^\infty(\mathrm{G}), \mathrm{Z}\mathscr{C}(\mathrm{G}), \mathrm{ZL}^2(\mathrm{G})$, respectively, the subspaces consisting of the central functions in these various spaces. Introduce similarly the spaces $\Theta(\mathrm{T}), \mathscr{C}^\infty(\mathrm{T}), \mathscr{C}(\mathrm{T}), \mathrm{L}^2(\mathrm{T})$; for any space E in this list, denote by E^{W} (resp. $\mathrm{E}^{-\mathrm{W}}$) the subspace consisting of the invariant (resp. anti-invariant) elements for the operation of the Weyl group W. We have a commutative diagram

$$\begin{array}{ccc}
\mathrm{Z}\mathscr{C}(\mathrm{G}) & \xrightarrow{a_c} & \mathscr{C}(\mathrm{T})^{\mathrm{W}} \\
\uparrow & & \uparrow \\
\mathrm{Z}\mathscr{C}^\infty(\mathrm{G}) & \xrightarrow{a_\infty} & \mathscr{C}^\infty(\mathrm{T})^{\mathrm{W}} \\
\uparrow & & \uparrow \\
\mathrm{Z}\Theta(\mathrm{G}) & \xrightarrow{a_\Theta} & \Theta(\mathrm{T})^{\mathrm{W}}
\end{array}$$

where the vertical arrows represent the canonical injections, and the maps a_c, a_∞, a_Θ are induced by the restriction map from $\mathscr{C}(\mathrm{G})$ to $\mathscr{C}(\mathrm{T})$.

The maps a_c, a_∞, a_Θ are *bijective* (§2, no. 5, Cor. 1 of Prop. 5, §8, no. 3, Cor. of Prop. 5, and §7, no. 3, Cor. of Prop. 2).

Assume now that the semi-sum ρ of the positive roots belongs to X(T) and consider the map b which to every continuous function φ on T associates $\varphi.\mathrm{J}(\rho)$. We have a commutative diagram

$$\begin{array}{ccccc}
ZL^2(G) & & \xrightarrow{\quad\quad u \quad\quad} & & L^2(T)^{-W} \\
\uparrow & & & & \uparrow \\
Z\mathscr{C}(G) & \xrightarrow{a_c} & \mathscr{C}(T)^W & \xrightarrow{b_c} & \mathscr{C}(T)^{-W} \\
\uparrow & & \uparrow & & \uparrow \\
Z\mathscr{C}^\infty(G) & \xrightarrow{a_\infty} & \mathscr{C}^\infty(T)^W & \xrightarrow{b_\infty} & \mathscr{C}^\infty(T)^{-W} \\
\uparrow & & \uparrow & & \uparrow \\
Z\Theta(G) & \xrightarrow{a_\Theta} & \Theta(T)^W & \xrightarrow{b_\Theta} & \Theta(T)^{-W}
\end{array}$$

where the vertical arrows are the canonical inclusions, the maps b_c, b_∞, b_Θ are induced by b, and u extends $b_c \circ a_c$ by continuity (§7, no. 4, Cor. 3 of Th. 2). The maps u and b_Θ are *bijective* (*loc. cit.*); so is b_∞ (Exerc. 5); on the other hand, b_c is not surjective in general (Exerc. 6).

§9. COMPACT LIE GROUPS OPERATING ON MANIFOLDS

In this paragraph, X *denotes a separated, locally finite dimensional, real manifold of class* C^r $(1 \leq r \leq \omega)$.

1. EMBEDDING OF A MANIFOLD IN THE NEIGHBOURHOOD OF A COMPACT SET

Lemma 1. Let T *and* T′ *be two topological spaces,* A *and* A′ *compact subsets of* T *and* T′*, respectively,* W *a neighbourhood of* A × A′ *in* T × T′*. There exists an open neighbourhood* U *of* A *in* T *and an open neighbourhood* U′ *of* A′ *in* T′ *such that* U × U′ ⊂ W.

Let $x \in A$; there exist open subsets U_x of T and U'_x of T′ such that $\{x\} \times A' \subset U_x \times U'_x \subset W$: indeed, the compact subset $\{x\} \times A'$ of T × T′ can be covered by a finite number of open sets contained in W, of the form $U_i \times U'_i$, with $x \in U_i$; it suffices to take $U_x = \bigcap_i U_i$ and $U'_x = \bigcup_i U'_i$.

Since A is compact, there exist points $x_1, \ldots, x_m$ of A such that $A \subset \bigcup_i U_{x_i}$; put $U = \bigcup_i U_{x_i}$ and $U' = \bigcap_i U'_{x_i}$. Then $A \times A' \subset U \times U' \subset W$, hence the lemma.

In the remainder of this number, Y denotes a separated manifold of class C^r.

PROPOSITION 1. *Let* $\varphi : X \to Y$ *be a morphism of class* C^r*,* A *a compact subset of* X*. The following conditions are equivalent:*

(i) *The restriction of* φ *to* A *is injective, and* φ *is an immersion at every point of* A;

(ii) *there exists an open neighbourhood* U *of* A *such that* φ *induces an embedding of* U *into* Y.

When these conditions are satisfied, φ is said to be an *embedding in the neighbourhood of* A.

We prove that (i) implies (ii), the other implication being clear. Assuming (i), there exists an open neighbourhood V of X containing A such that the restriction of φ to V is an immersion (*Differentiable and Analytic Manifolds, Results*, 5.7.1). Denote by Γ the set of points (x, y) in $\mathrm{V} \times \mathrm{V}$ such that $\varphi(x) = \varphi(y)$, and by Δ the diagonal in $\mathrm{V} \times \mathrm{V}$. Then Δ is an open subset of Γ: indeed, for all $x \in \mathrm{V}$, there exists an open neighbourhood U_x of x such that the restriction of φ to U_x is injective, that is, such that $\Gamma \cap (\mathrm{U}_x \times \mathrm{U}_x) = \Delta \cap (\mathrm{U}_x \times \mathrm{U}_x)$.

Since Y is separated, Γ is closed in $\mathrm{V} \times \mathrm{V}$; consequently the complement W of $\Gamma - \Delta$ in $\mathrm{V} \times \mathrm{V}$ is open. By assumption, W contains $\mathrm{A} \times \mathrm{A}$; Lemma 1 implies that there exists an open subset U′ of V containing A such that $\mathrm{U}' \times \mathrm{U}' \subset \mathrm{W}$, that is, such that the restriction of φ to U′ is injective. Moreover, there exists an open neighbourhood U of A whose closure is compact and contained in U′ (*General Topology*, Chap. I, §9, no. 7, Prop. 10). Then φ induces a homeomorphism from $\overline{\mathrm{U}}$ to $\varphi(\overline{\mathrm{U}})$, and consequently from U to $\varphi(\mathrm{U})$; it follows that the restriction of φ to U is an embedding (*Differentiable and Analytic Manifolds, Results*, 5.8.3).

PROPOSITION 2. *Assume that the manifold* Y *is paracompact; let* A *be a subset of* X, *and let* $\varphi : X \to \mathrm{Y}$ *be a morphism of class* C^r *that induces a homeomorphism from* A *to* $\varphi(\mathrm{A})$, *and that is étale at every point of* A. *Then there exists an open neighbourhood* U *of* A *such that* φ *induces an isomorphism from* U *to an open submanifold of* Y.

Restricting X and Y if necessary, we can assume that φ is étale and surjective. Denote by $\sigma : \varphi(\mathrm{A}) \to \mathrm{A}$ the inverse homeomorphism of $\varphi|\mathrm{A}$. Since Y is metrizable (*Differentiable and Analytic Manifolds, Results*, 5.1.6), $\varphi(\mathrm{A})$ admits a fundamental system of paracompact neighbourhoods; hence, by *General Topology*, Chap. XI, there exist an open neighbourhood V of $\varphi(\mathrm{A})$ in Y and a continuous map $s : \mathrm{V} \to \mathrm{X}$, that coincides with σ on $\varphi(\mathrm{A})$ and is such that $\varphi(s(y)) = y$ for all $y \in \mathrm{V}$. Moreover, s is topologically étale, so $s(\mathrm{V})$ is an open set U containing A. Then φ induces a homeomorphism φ' from U to V; by *Differentiable and Analytic Manifolds, Results*, 5.7.8, φ' is an isomorphism.

In the remainder of this number, we assume that $r \neq \omega$.

PROPOSITION 3. *Let* A *be a compact subset of* X. *The set* $\mathscr{P}$ *of morphisms* $\varphi \in \mathscr{C}^r(\mathrm{X};\mathrm{Y})$ *that are embeddings in the neighbourhood of* A *is open in* $\mathscr{C}^r(\mathrm{X};\mathrm{Y})$ *in the topology of compact* C^r*-convergence* (§ 6, no. 4).

Clearly, it suffices to prove the proposition for $r = 1$.

a) We show first that the subset J of $\mathscr{C}^1(\mathrm{X};\mathrm{Y})$ consisting of the morphisms that are immersions at every point of A is open. Consider the map $j_{\mathrm{A}} : \mathscr{C}^1(\mathrm{X};\mathrm{Y}) \times \mathrm{A} \to \mathrm{J}^1(\mathrm{X};\mathrm{Y})$ such that $j_{\mathrm{A}}(\varphi, x) = j^1_x(\varphi)$ (*Differentiable and Analytic Manifolds, Results*, 12.1).

By definition of the topology on $\mathscr{C}^1(\mathrm{X};\mathrm{Y})$, the map $\tilde{j}_{\mathrm{A}} : \varphi \to j_{\mathrm{A}}(\varphi, .)$ from $\mathscr{C}^1(\mathrm{X};\mathrm{Y})$ to $\mathscr{C}(\mathrm{A};\mathrm{J}^1(\mathrm{X};\mathrm{Y}))$ is continuous; it now follows from *General Topology*, Chap. X, §3, no. 4, Th. 3, that j_{A} is continuous.

On the other hand, let M be the set of jets j in $\mathrm{J}^1(\mathrm{X};\mathrm{Y})$ whose tangent map $\mathrm{T}(j) : \mathrm{T}_{\mathbf{s}(j)}(\mathrm{X}) \to \mathrm{T}_{\mathbf{b}(j)}(\mathrm{Y})$ (*Differentiable and Analytic Manifolds, Results*, 12.3.4) is injective. The set M is open in $\mathrm{J}^1(\mathrm{X};\mathrm{Y})$; indeed, it suffices to verify this assertion when X is an open subset of a finite dimensional vector space E, and Y is an open subset of a Banach space F; we are then reduced (*Differentiable and Analytic Manifolds, Results*, 12.3.1) to proving that the set of injective continuous linear maps is open in $\mathscr{L}(\mathrm{E};\mathrm{F})$, which follows from *Spectral Theory*, Chap. III, §2, no. 7, Prop. 16.

We conclude from the preceding that the set $j_{\mathrm{A}}^{-1}(\mathrm{M})$ is open in $\mathscr{C}^1(\mathrm{X};\mathrm{Y}) \times \mathrm{A}$, hence that its complement $\mathscr{F}$ is closed. Since A is compact, the projection $\mathrm{pr}_1 : \mathscr{C}^1(\mathrm{X};\mathrm{Y}) \times \mathrm{A} \to \mathscr{C}^1(\mathrm{X};\mathrm{Y})$ is a proper morphism, hence closed; consequently, the set J, which is equal to $\mathscr{C}^1(\mathrm{X};\mathrm{Y}) - \mathrm{pr}_1(\mathscr{F})$, is open in $\mathscr{C}^1(\mathrm{X};\mathrm{Y})$.

b) Let H be the subset of $\mathrm{J} \times \mathrm{A} \times \mathrm{A}$ consisting of the elements (f, x, y) such that $f(x) = f(y)$. It is clear that H contains $\mathrm{J} \times \Delta$, where Δ denotes the diagonal in the product $\mathrm{A} \times \mathrm{A}$; we show that $\mathrm{H}' = \mathrm{H} - (\mathrm{J} \times \Delta)$ is *closed* in $\mathrm{J} \times \mathrm{A} \times \mathrm{A}$. Since $\mathscr{P}$ is the complement in J of the image of H' under the proper projection $\mathrm{pr}_1 : \mathrm{J} \times \mathrm{A} \times \mathrm{A} \to \mathrm{J}$, this will imply the proposition.

The topology of $\mathscr{C}^1(\mathrm{X};\mathrm{Y})$ being finer than the topology of compact convergence, the map $(\varphi, x) \mapsto \varphi(x)$ from $\mathscr{C}^1(\mathrm{X};\mathrm{Y}) \times \mathrm{A}$ to Y is continuous (*General Topology*, Chap. X, §3, no. 4, Cor. 1 of Th. 3); it follows that H is closed in $\mathrm{J} \times \mathrm{A} \times \mathrm{A}$. Hence, it suffices to show that $\mathrm{J} \times \Delta$ is open in H, in other words, that for all $\varphi \in \mathrm{J}$ and all $x \in \mathrm{A}$ there exists a neighbourhood Ω of φ in J and a neighbourhood B of x in X such that, for any morphism ψ in Ω, the restriction of ψ to $\mathrm{A} \cap \mathrm{B}$ is *injective*.

Thus, the proposition follows from the following lemma:

Lemma 2. Let x be a point of X, $\varphi : \mathrm{X} \to \mathrm{Y}$ *a morphism of class* C^1 *that is an immersion at x. There exists a neighbourhood Ω of φ in* $\mathscr{C}^1(\mathrm{X};\mathrm{Y})$ *and a neighbourhood* B *of x in* X *such that, for all $\psi \in \Omega$, the restriction of ψ to* B *is injective.*

Let U be a relatively compact open neighbourhood of x isomorphic to a finite dimensional vector space, and such that $\varphi(\overline{\mathrm{U}})$ is contained in the domain V of a chart. The set Ω_0 of $\psi \in \mathscr{C}^1(\mathrm{X};\mathrm{Y})$ such that $\psi(\overline{\mathrm{U}}) \subset \mathrm{V}$ is open in $\mathscr{C}^1(\mathrm{X};\mathrm{Y})$, and the restriction map $\Omega_0 \to \mathscr{C}^1(\mathrm{U};\mathrm{V})$ is continuous; we are thus reduced to proving the lemma when $\mathrm{X} = \mathrm{U}$ and $\mathrm{Y} = \mathrm{V}$, in other words, we can assume that X is a finite dimensional vector space and Y is a Banach space. Choose norms on X and Y.

The linear map $D\varphi(x) : X \to Y$ is injective; denote its conorm by q (*Spectral Theory*, Chap. III, §2, no. 6), so that, by definition, we have $\|D\varphi(x).t\| \geq q\|t\|$ for all $t \in X$. Let $\varepsilon \in \mathbf{R}$ be such that $0 < \varepsilon < q/2$, and let B be a closed ball with centre x such that $\|D\varphi(u) - D\varphi(x)\| \leq \varepsilon$ for all $u \in B$. Denote by Ω the subset of $\mathscr{C}^1(X;Y)$ consisting of the morphisms ψ such that $\|D\psi(u) - D\varphi(u)\| \leq \varepsilon$ for all $u \in B$; it is open by definition of the topology of $\mathscr{C}^1(X;Y)$. For $\psi \in \Omega$, put $\psi_0 = \psi - D\varphi(x)$. We have $\|D\psi_0(u)\| \leq 2\varepsilon$ for all $u \in B$, and consequently $\|\psi_0(u) - \psi_0(v)\| \leq 2\varepsilon\|u - v\|$ for all u and v in B (*Differentiable and Analytic Manifolds, Results*, 2.2.3). It follows that

$$\|\psi(u) - \psi(v)\| \geq \|D\varphi(x).(u - v)\| - \|\psi_0(u) - \psi_0(v)\| \geq (q - 2\varepsilon)\|u - v\|.$$

Consequently, the restriction of ψ to B is injective, hence the lemma.

PROPOSITION 4. *Let* A *be a compact subset of* X. *There exist a finite dimensional vector space* E *and a morphism* $\varphi \in \mathscr{C}^r(X;E)$ $(r \neq \omega)$ *that is an embedding in the neighbourhood of* A.

Let $(U_i, \varphi_i, E_i)_{i \in I}$ be a finite family of charts of X whose domains cover A. We extend φ_i to a map from X to E_i (also denoted by φ_i) by putting $\varphi_i(x) = 0$ for $x \notin U_i$. Let $(V_i)_{i \in I}$ be a covering of A by open subsets of X such that $\overline{V}_i \subset U_i$ for all $i \in I$ (the existence of such a covering follows from *General Topology*, Chap. IX, §4, no. 3, Cor. 1 of Th. 3, applied to the compact space X′ obtained by adjoining to X a point at infinity and the covering of X′ consisting of the open sets U_i $(i \in I)$ and $X' - A$). For all $i \in I$, let α_i be a numerical function of class C^r on X, equal to 1 at every point of V_i, and with support contained in U_i (*Differentiable and Analytic Manifolds, Results*, 5.3.6).

Consider the map $\varphi : X \to \bigoplus_{i \in I}(E_i \oplus \mathbf{R})$ defined by

$$\varphi(x) = (\alpha_i(x)\varphi_i(x), \alpha_i(x))_{i \in I}.$$

For all $i \in I$, the map $\alpha_i\varphi_i$ is of class C^r (since its restrictions to U_i and to the complement of the support of α_i are), and its restriction to V_i is an embedding; it follows that φ is a morphism of class C^r and is an immersion at every point of A. We show that the restriction of φ to A is injective. Let x, y be two points of A such that $\varphi(x) = \varphi(y)$, and let $i \in I$ be such that $x \in V_i$. Then $\alpha_i(x) = 1$, so $\alpha_i(y) = 1$, which implies that $y \in U_i$; but we also have $\varphi_i(x) = \varphi_i(y)$, so $x = y$ since φ_i induces an embedding of U_i into E_i.

It can be shown[9] that every separated manifold, countable at infinity and of pure dimension n, embeds in $\mathbf{R}^{2n}$; for a weaker result, cf. Exercise 2.

[9] See H. WHITNEY, The self-intersections of a smooth n-manifold in $2n$-space, *Ann. of Math.*, 45 (1944), pp. 220-246.

2. EQUIVARIANT EMBEDDING THEOREM

In this number, we assume that $r \neq \omega$.

Lemma 3. Let G *be a compact topological group operating continuously on a topological space* X; *let* A *be a subset of* X, *stable under* G, *and* W *a neighbourhood of* A. *Then, there exists an open neighbourhood* V *of* A *stable under* G *and contained in* W.

Put $\mathrm{F} = \mathrm{X} - \mathring{\mathrm{W}}$ and $\mathrm{V} = \mathrm{X} - \mathrm{GF}$. Then V is open (*General Topology*, Chap. III, §4, no. 1, Cor. 1 of Prop. 1), stable under G, and $\mathrm{A} \subset \mathrm{V} \subset \mathrm{W}$.

THEOREM 1. *Let* G *be a compact Lie group,* $(g, x) \mapsto gx$ *a law of left operation of class* C^r *of* G *on* X, *and* A *a compact subset of* X. *There exists an analytic linear representation* ρ *of* G *on a finite dimensional vector space* E, *a morphism* $\varphi : \mathrm{X} \to \mathrm{E}$ *of class* C^r, *compatible with the operations of* G, *and an open neighbourhood* U *of* A, *stable under* G, *such that the restriction of* φ *to* U *is an embedding.*

Replacing A by the compact set GA, we are reduced to the case in which A is stable under G.

Let E_0 be a finite dimensional vector space such that there exists an element of $\mathscr{C}^r(\mathrm{X};\mathrm{E}_0)$ that is an embedding in the neighbourhood of A (no. 1, Prop. 4); the set $\mathscr{P}$ of morphisms having this property is a non-empty open subset of $\mathscr{C}^r(\mathrm{X};\mathrm{E}_0)$ (no. 1, Prop. 3). Consider the continuous linear representation of the compact group G on the space $\mathscr{C}^r(\mathrm{X};\mathrm{E}_0)$ (§6, no. 4, Lemma 4). By the Peter-Weyl theorem (*Spectral Theory*, in preparation), the union of the finite dimensional subspaces stable under G is dense in $\mathscr{C}^r(\mathrm{X};\mathrm{E}_0)$; hence, there exists an element φ_0 of $\mathscr{P}$ such that the maps $x \mapsto \varphi_0(gx)$, for $g \in \mathrm{G}$, generate a *finite dimensional* vector subspace E_1 of $\mathscr{C}^r(\mathrm{X};\mathrm{E}_0)$, which is clearly stable under the operation of G.

Take E to be the space $\mathrm{Hom}_{\mathbf{R}}(\mathrm{E}_1, \mathrm{E}_0)$, ρ to be the representation of G on E induced by the action on E_1, and $\varphi : \mathrm{X} \to \mathrm{E}$ to be the map that associates to $x \in \mathrm{X}$ the linear map $\psi \mapsto \psi(x)$ from E_1 to E_0. This is a morphism of class C^r; for $x \in \mathrm{X}$, $g \in \mathrm{G}$, $\psi \in \mathrm{E}_1$, we have (denoting by $\tau(g)$ the automorphism $x \mapsto gx$ of X):

$$\varphi(gx)(\psi) - \psi(gx) = \varphi(x)(\psi \circ \tau(g)) = (\rho(g)\varphi(x))(\psi).$$

Let $\alpha : \mathrm{Hom}_{\mathbf{R}}(\mathrm{E}_1, \mathrm{E}_0) \to \mathrm{E}_0$ be the linear map $u \mapsto u(\varphi_0)$; we have $\alpha \circ \varphi = \varphi_0$, so φ is an embedding in the neighbourhood of A because φ_0 is one. Hence, there exists an open neighbourhood U of A such that the restriction of φ to U is an embedding; we can choose U stable under G by Lemma 3, hence the theorem.

COROLLARY 1. *Assume that* X *is compact. There exists an analytic linear representation* ρ *of* G *on a finite dimensional vector space* E *and an embedding* $\varphi : \mathrm{X} \to \mathrm{E}$ *such that* $\varphi(gx) = \rho(g)\varphi(x)$ *for* $g \in \mathrm{G}, x \in \mathrm{X}$.

COROLLARY 2. *Let* H *be a closed subgroup of* G. *There exists an analytic linear representation of* G *on a finite dimensional vector space* E *and a point* $v \in \mathrm{E}$ *with fixer* H.

Apply Cor. 1 to the canonical operation of G on the compact manifold G/H. This gives an analytic linear representation $\rho : \mathrm{G} \to \mathbf{GL}(\mathrm{E})$ and an embedding $\varphi : \mathrm{G/H} \to \mathrm{E}$ such that $\varphi(gx) = \rho(g)\varphi(x)$, $g \in \mathrm{G}, x \in \mathrm{G/H}$. Let $\bar{e} \in \mathrm{G/H}$ be the class of $e \in \mathrm{G}$, and $v = \varphi(\bar{e})$ its image. For all $g \in \mathrm{G}$, we have $\rho(g)v = v \iff \varphi(g\bar{e}) = \varphi(\bar{e}) \iff g\bar{e} = \bar{e} \iff g \in \mathrm{H}$.

COROLLARY 3. *Assume that* X *is paracompact. There exists a real Hilbert space* E, *a continuous unitary representation*[10] ρ *of* G *on* E *and an embedding* $\varphi : \mathrm{X} \to \mathrm{E}$ *of class* C^r *such that* $\varphi(gx) = \rho(g)\varphi(x)$ *for all* $g \in \mathrm{G}$ *and all* $x \in \mathrm{X}$.

The space X/G is locally compact (*General Topology*, Chap. III, §4, no. 5, Prop. 11). Its connected components are the images of the connected components of X, which are countable at infinity (*General Topology*, Chap. I, §9, no. 10, Th. 5); thus, they are themselves countable at infinity, which implies that X/G is paracompact (*loc. cit.*). Hence, there exists a locally finite covering $(\mathrm{U}'_\alpha)_{\alpha \in \mathrm{I}}$ of X/G by relatively compact open sets, and a covering $(\mathrm{V}'_\alpha)_{\alpha \in \mathrm{I}}$ such that $\overline{\mathrm{V}}'_\alpha \subset \mathrm{U}'_\alpha$ for all $\alpha \in \mathrm{I}$ (*General Topology*, Chap. IX, §4, no. 3, Cor. 1 of Th. 3); taking the inverse image, we obtain two locally finite coverings $(\mathrm{U}_\alpha)_{\alpha \in \mathrm{I}}$ and $(\mathrm{V}_\alpha)_{\alpha \in \mathrm{I}}$ of X by relatively compact open sets stable under G, such that $\overline{\mathrm{V}}_\alpha \subset \mathrm{U}_\alpha$ for all $\alpha \in \mathrm{I}$.

For all $\alpha \in \mathrm{I}$, there exists a representation ρ_α of G on a finite dimensional real vector space E_α and a morphism $\varphi_\alpha \in \mathscr{C}^r(\mathrm{X};\mathrm{E}_\alpha)$, compatible with the operations of G, whose restriction to U_α is an embedding (Th. 1). For $\alpha \in \mathrm{I}$, let a_α be a numerical function of class C^r on X, equal to 1 on V_α and to 0 outside U_α (*Differentiable and Analytic Manifolds, Results*, 5.3.6). Put $b_\alpha(x) = \int_{\mathrm{G}} a_\alpha(gx)\,dg$ for $x \in \mathrm{X}$. The function b_α is of class C^r, invariant under G (§6, no. 4, Cor. 2), equal to 1 on V_α and to 0 outside U_α. Give each E_α a Hilbert scalar product invariant under G (§1, no. 1), and $\mathbf{R}$ its canonical Hilbert structure; let E be the Hilbert sum of the family $(\mathrm{E}_\alpha \oplus \mathbf{R})_{\alpha \in \mathrm{I}}$, and let ρ be the representation of G on E induced by the ρ_α and the trivial action of G on $\mathbf{R}$. For $x \in \mathrm{X}$, put $\varphi(x) = (b_\alpha(x)\varphi_\alpha(x), b_\alpha(x))_{\alpha \in \mathrm{I}}$. Then φ is a morphism of class C^r from X to E, compatible with the operations of G; we verify as in the proof of Prop. 4 (no. 1) that φ is an embedding, which implies the corollary.

[10]That is (*Spectral Theory*, in preparation) a continuous linear representation (*Integration*, Chap. VIII, §2, no. 1) such that the operator $\rho(g)$ is unitary for all $g \in \mathrm{G}$.

3. TUBES AND TRANSVERSALS

Lemma 4. Let H *be a compact Lie group,* $\rho : \mathrm{H} \to \mathbf{GL}(\mathrm{V})$ *a continuous (hence analytic) representation of* H *on a finite dimensional real vector space, and* W *a neighbourhood of the origin in* V*. There exists an open neighbourhood* B *of the origin, contained in* W *and stable under* H*, and an analytic isomorphism* $u : \mathrm{V} \to \mathrm{B}$*, compatible with the operations of* H*, such that* $u(0) = 0$ *and* $\mathrm{D}u(0) = \mathrm{Id}_{\mathrm{V}}$.

Choose a scalar product on V invariant under H (§1, no. 1). There exists a real number $r > 0$ such that the open ball B of radius r is contained in W; it is clearly stable under H. Put $u(v) = r(r^2 + \|v\|^2)^{-1/2}v$ for all $v \in \mathrm{V}$; then u is a bijective analytic map from V to B, compatible with the operations of H, and its inverse map $w \mapsto r(r^2 - \|w\|^2)^{-1/2}w$ is analytic. Moreover, $u(0) = 0$ and $\mathrm{D}u(0) = \mathrm{Id}_{\mathrm{V}}$.

PROPOSITION 5. *Let* H *be a compact Lie group,* $(h, x) \mapsto hx$ *a law of left operation of class* C^r *of* H *on* X*, and* x *a point of* X *fixed under the operation of* H*. Then the group* H *operates linearly on the vector space* $\mathrm{T} = \mathrm{T}_x(\mathrm{X})$*; there exists an open embedding* $\varphi : \mathrm{T} \to \mathrm{X}$ *of class* C^r*, compatible with the operations of* H*, such that* $\varphi(0) = x$ *and* $\mathrm{T}_0(\varphi)$ *is the identity map of* T.

Let $(\mathrm{U}, \psi, \mathrm{E})$ be a chart of X at x, such that U is stable under H (no. 2, Lemma 3) and such that $\psi(x) = 0$. Identify E with T by means of $\mathrm{T}_x(\psi)$, and put

$$\psi^\sharp(y) = \int_{\mathrm{H}} h.\psi(h^{-1}y)\, dh \qquad \text{for } y \in \mathrm{U},$$

where dh is the Haar measure on H of total mass 1.

Then (§6, no. 4, Cor. 1) $\psi^\sharp$ is a morphism of class C^r from U to T, compatible with the operations of H, such that $\psi^\sharp(x) = 0$ and $d_x\psi^\sharp = \mathrm{Id}_{\mathrm{T}}$. Hence, there exists an open set $\mathrm{U}' \subset \mathrm{U}$ containing x, and an open neighbourhood V of 0 in T, such that $\psi^\sharp$ induces an isomorphism $\theta : \mathrm{U}' \to \mathrm{V}$. Restricting U′ and V if necessary, we can assume that they are stable under H and that there exists an isomorphism $u : \mathrm{T} \to \mathrm{V}$ compatible with the operations of H (Lemma 4). It now suffices to take $\varphi = \theta^{-1} \circ u$.

Recall (*Differentiable and Analytic Manifolds, Results*, 6.5.1) that if G is a Lie group, H a Lie subgroup of G and Y a manifold on which H operates on the left, we denote by $\mathrm{G} \times {}^{\mathrm{H}}\mathrm{Y}$ the quotient of the product manifold $\mathrm{G} \times \mathrm{Y}$ by the right operation $((g, y), h) \mapsto (gh, h^{-1}y)$ of H; this is a manifold on which the Lie group G operates naturally on the left; the projection $\mathrm{G} \times {}^{\mathrm{H}}\mathrm{Y} \to \mathrm{G}/\mathrm{H}$ is a bundle with fibre Y. Further, if Y is a finite dimensional vector space on which H operates linearly, $\mathrm{G} \times {}^{\mathrm{H}}\mathrm{Y}$ has a natural structure of vector G-bundle with base G/H (*Differentiable and Analytic Manifolds, Results*, 7.10.2).

Let G be a Lie group operating *properly* on the manifold X (*General Topology*, Chap. III, §4, no. 1, Def. 1) such that the law of operation $(g, x) \mapsto gx$

is of class C^r. Then, for every point x of X, the orbit Gx of x is a closed submanifold of X, isomorphic to the Lie homogeneous space G/G_x, where G_x is the fixer of x in G (cf. Chap. III, §1, no. 7, Prop. 14 (ii), and *General Topology*, Chap. III, §4, no. 2, Prop. 4); this is a compact Lie group (*loc. cit.*).

PROPOSITION 6. *Assume that the manifold* X *is paracompact; let* x *be a point of* X, G_x *its fixer. There exists a finite dimensional analytic linear representation* $\tau : G_x \rightarrow \mathbf{GL}(W)$, *and an open embedding* $\alpha : G \times^{G_x} W \rightarrow X$ *of class* C^r, *compatible with the operations of* G, *that maps the class of* $(e, 0) \in G \times W$ *to* x.

Put $T = T_x(X)$. Let W be a complementary subspace of $T_x(Gx)$ in T, stable under G_x (for example, the orthogonal complement of $T_x(Gx)$ with respect to a scalar product on T invariant under G_x). On the other hand, let $\varphi : T \rightarrow X$ be a morphism with the properties stated in Prop. 5 (relative to $H = G_x$). Consider the morphism $\lambda : G \times W \rightarrow X$ defined by $\lambda(g, w) = g\varphi(w)$. It induces by passage to the quotient a morphism $\mu : G \times^{G_x} W \rightarrow X$ of class C^r, compatible with the operations of G, that maps the class z of $(e, 0)$ to x.

We show that μ is étale at the point z. We have

$$\begin{aligned} \dim(G \times^{G_x} W) &= \dim(G) + \dim(W) - \dim(G_x) \\ &= \dim(Gx) + \dim(W) = \dim(T), \end{aligned}$$

so it suffices to show that μ is submersive at z, or equivalently that λ is submersive at $(e, 0)$. But, the tangent map $T_{(e,0)}(\lambda) : T_e(G) \oplus W \rightarrow T$ is equal to $T_e(\rho(x)) + i$, where $\rho(x)$ is the orbital map $g \mapsto gx$ and i the canonical injection from W to T; since $\operatorname{Im} T_e(\rho(x)) = T_x(Gx)$, the map $T_{(e,0)}(\lambda)$ is surjective, and μ is étale at z.

We are going to show that there exists an open neighbourhood Ω of Gz in $G \times^{G_x} W$, *stable under* G, such that μ induces an isomorphism from Ω onto an open subset of X. This will imply the proposition: indeed, the inverse image of Ω in $G \times W$ is stable under G, and hence is of the form $G \times B$, where B is an open subset of W containing the origin and stable under G_x; restricting Ω if necessary, we can assume that there exists an isomorphism $u : W \rightarrow B$, compatible with the operations of G_x (Lemma 4). It is clear that the composite morphism $\alpha : G \times^{G_x} W \xrightarrow{(\mathrm{Id}, u)} G \times^{G_x} B \xrightarrow{\mu} X$ satisfies the conditions in the statement of the proposition.

Thus, the proposition is a consequence of the following lemma:

Lemma 5. *Let* Z *be a separated manifold of class* C^r, *equipped with a law of left operation* $m : G \times Z \rightarrow Z$ *of class* C^r, *and* $\mu : Z \rightarrow X$ *a morphism (of class* C^r*) compatible with the operations of* G. *Let* z *be a point of* Z, *and* $x = \mu(z)$. *Assume that* μ *is étale at* z, *and that the fixer of* z *in* G *is equal to the fixer* G_x *of* x. *Then, there exists an open neighbourhood* Ω *of the orbit* Gz, *stable under* G, *such that* μ *induces an isomorphism from* Ω *onto an open subset of* X.

Since μ is compatible with the operations of G, it is étale at every point of Gz; since the canonical map $G/G_x \to Gx$ is a homeomorphism, so is the map from Gz to Gx induced by μ. Hence, it follows from Prop. 2 of no. 1 that there exists an open neighbourhood U of Gz in Z such that μ induces an open embedding of U into X.

Since G operates properly on X, there exists an open neighbourhood V of x and a compact subset K of G such that $gV \cap V = \varnothing$ for $g \notin K$ (*General Topology*, Chap. III, §4, no. 4, Prop. 7); in particular, $e \in K$. The set W_1 of points $y \in Z$ such that $Ky \subset U$ is open in Z: indeed, $Z - W_1$ is the image of the closed set $(K \times Z) - m^{-1}(U)$ under the proper projection $pr_2 : K \times Z \to Z$. Put $W = W_1 \cap \mu^{-1}(V)$; this is an open subset of Z, containing z, and satisfying the following conditions:

(i) $KW \subset U$, and in particular $W \subset U$;

(ii) $\mu(W) \subset V$.

Put $\Omega = GW$ and consider the restriction of μ to Ω. This is an étale morphism, since every point of Ω is conjugate under G to a point of U. We show that it is injective: let g, h in G and u, v in W be such that $\mu(gu) = \mu(hv)$. Put $k = g^{-1}h$; then $\mu(u) = k\mu(v)$, so $k \in K$ by (ii). But kv and u belong to U by (i); thus, $u = kv$ because the restriction of μ to V is injective, so $gu = hv$. Hence, the restriction of μ to Ω is injective, and consequently (*Differentiable and Analytic Manifolds, Results*, 5.7.8) is an isomorphism onto an open submanifold of X, which completes the proof.

Under the conditions of Prop. 6, the image of α is an open neighbourhood T of the orbit A of x, equipped with the structure of vector bundle with base A, for which the zero section is the orbit A itself. Such a neighbourhood is called a *linear tube* (around the orbit in question). For each point $a \in A$, the fibre Y_a of this vector bundle is a submanifold of X, stable under the fixer G_a of a, and such that the morphism from $G \times^{G_a} Y_a$ to X that maps the class of $(g, y) \in G \times Y_a$ to $gy \in X$ induces a morphism of class C^r from $G \times^{G_a} Y_a$ to T. Then Y_a is said to be the *transversal* at a of the tube T. We remark that the tangent space at a of Y_a is canonically isomorphic to Y_a and that it is a complement of $T_a(A)$ in $T_a(X)$; thus, the vector bundle T with base A is canonically isomorphic to the normal bundle of A in X (*Differentiable and Analytic Manifolds, Results*, 8.1.3).

4. ORBIT TYPES

Let G be a topological group operating continuously on a separated topological space E. For every point x of E, denote by G_x the fixer of x in G, and assume that the canonical map $G/G_x \to Gx$ is a homeomorphism; this is notably the case in the following two situations:

a) the topologies of G and E are discrete;

b) G operates properly on E (*General Topology*, Chap. III, §4, no. 2, Prop. 4), for example, G is compact (*General Topology*, Chap. III, §4, no. 1, Prop. 2).

Denote by $\mathscr{T}$ the set of conjugacy classes of closed subgroups of G. For every $x \in \mathrm{E}$, we call the *orbit type* of x, or sometimes the type of x, the class of G_x in $\mathscr{T}$; two points of the same orbit are of the same orbit type (*Algebra*, Chap. I, §5, no. 2, Prop. 2); two orbits are of the same type if and only if they are isomorphic as G-sets (*Algebra*, Chap. I, §5, no. 5, Th. 1). For every $t \in \mathscr{T}$, denote by $\mathrm{E}_{(t)}$ the set of points of E of type t, that is, the union of the orbits of type t; this is a stable subset of E. For $\mathrm{H} \in t$, we also write $\mathrm{E}_{(\mathrm{H})}$ for $\mathrm{E}_{(t)}$; for example, $\mathrm{E}_{(\mathrm{G})}$ is the closed subspace of E consisting of the points fixed by G.

Give $\mathscr{T}$ the following preorder relation:

$$t \leq t' \iff \text{there exists } \mathrm{H} \in t \text{ and } \mathrm{H}' \in t' \text{ such that } \mathrm{H} \supset \mathrm{H}'.$$

Let Ω and Ω' be two orbits of G on E, t and t' their types; then $t \leq t'$ if and only if there exists a G-morphism (necessarily surjective and continuous) from Ω' to Ω.

Let x, x' be in E, and t, t' their types; then $t \leq t'$ if and only if there exists $a \in \mathrm{G}$ such that $a\mathrm{G}_{x'}a^{-1} \subset \mathrm{G}_x$.

Lemma 6. Let G *be a Lie group.*

a) *Every decreasing sequence of compact subgroups of* G *is stationary.*

b) *Let* H *and* H′ *be two compact subgroups of* G *such that* $\mathrm{H} \subset \mathrm{H}'$ *and such that there exists an isomorphism* (*of topological groups*) *from* H′ *to* H. *Then* $\mathrm{H} = \mathrm{H}'$.

c) *With the relation* $t \leq t'$, *the set* $\mathscr{T}$ *is a noetherian ordered set* (*Theory of Sets*, Chap. III, §6, no. 5, text preceding Prop. 7).

a) Let $(\mathrm{H}_i)_{i \geq 1}$ be a decreasing sequence of compact subgroups of G; these are Lie subgroups of G (Chap. III, §8, no. 2, Th. 2). The sequence of integers $(\dim \mathrm{H}_i)_{i \geq 1}$ is decreasing, hence stationary, so there exists an integer N such that the subgroups H_i have the same identity component for $i \geq \mathrm{N}$. Then the decreasing sequence of positive integers $(\mathrm{H}_i : (\mathrm{H}_i)_0)_{i \geq \mathrm{N}}$ is stationary, so $\mathrm{H}_i = \mathrm{H}_{i+1}$ for i sufficiently large.

b) Let f be an isomorphism from H′ to H. The sequence $(f^n(\mathrm{H}))_{n \geq 0}$ is a decreasing sequence of compact subgroups of G, so $f^n(\mathrm{H}) = f^{n+1}(\mathrm{H})$ for n sufficiently large, by *a*). Since f is an isomorphism, this implies that $f(\mathrm{H}) = \mathrm{H} = f(\mathrm{H}')$, so $\mathrm{H} = \mathrm{H}'$.

c) Let $t, t' \in \mathscr{T}$ be such that $t \leq t'$ and $t' \leq t$. Then, there exist $\mathrm{H}, \mathrm{H}_1 \in t$ and $\mathrm{H}', \mathrm{H}'_1 \in t'$ such that $\mathrm{H} \supset \mathrm{H}'$ and $\mathrm{H}_1 \subset \mathrm{H}'_1$. Let g and g' be two elements of G such that $\mathrm{H}_1 = g\mathrm{H}g^{-1}$ and $\mathrm{H}'_1 = g'\mathrm{H}'g'^{-1}$; put $u = g'^{-1}g$. Then

$$u\mathrm{H}u^{-1} \subset \mathrm{H}' \subset \mathrm{H};$$

by b), this implies that $u\mathrm{H}u^{-1} = \mathrm{H}$, so $\mathrm{H}' = \mathrm{H}$ and $t' = t$. Thus, the set $\mathscr{T}$ is ordered, and it is noetherian by a).

THEOREM 2. *Let* G *be a Lie group operating properly on* X*, such that the law of operation* $(g, x) \mapsto gx$ *is of class* C^r*. Assume that* X *is paracompact.*

a) The map which associates to any point of X *its orbit type has the following semi-continuity property: let* $x \in \mathrm{X}$ *and let* $t \in \mathscr{T}$ *be its orbit type; there exists a stable open neighbourhood* U *of* x *such that, for any* $u \in \mathrm{U}$*, the type of* u *is* $\geq t$*.*

b) For all $t \in \mathscr{T}$*,* $\mathrm{X}_{(t)}$ *is a submanifold of* X*, the equivalence relation on* $\mathrm{X}_{(t)}$ *induced by the operation of* G *is regular (Differentiable and Analytic Manifolds, Results,* 5.9.5)*, and the morphism* $\mathrm{X}_{(t)} \to \mathrm{X}_{(t)}/\mathrm{G}$ *is a bundle.*

c) Assume that X/G *is connected. Then the set of orbit types of elements of* X *has a largest element* τ*; moreover,* $\mathrm{X}_{(\tau)}$ *is a dense open subset of* X *and* $\mathrm{X}_{(\tau)}/\mathrm{G}$ *is connected.*

Let x be a point of X and $t \in \mathscr{T}$ its type. To prove a) and b), we can replace X by a stable open set containing x, and hence (Prop. 6) can assume that X is of the form $\mathrm{G} \times^{\mathrm{H}} \mathrm{W}$, where W is the space of a finite dimensional analytic linear representation of a compact subgroup H of G, the point x being the image $p(e, 0)$ of $(e, 0) \in \mathrm{G} \times \mathrm{W}$ under the canonical projection $p : \mathrm{G} \times \mathrm{W} \to \mathrm{G} \times^{\mathrm{H}} \mathrm{W}$. If $u = p(g, y) \in \mathrm{G} \times^{\mathrm{H}} \mathrm{W}$ and $a \in \mathrm{G}$, then $au = u$ if and only if there exists $h \in \mathrm{H}$ with $(ag, y) = (gh^{-1}, hy)$, that is, if $a \in g\mathrm{H}_y g^{-1}$. Thus, $\mathrm{G}_u = g\mathrm{H}_y g^{-1}$; in particular, $\mathrm{G}_x = \mathrm{H}$, so G_u is conjugate to a subgroup of G_x, which proves that the type of u is $\geq t$, hence a).

Moreover, u is of type t if and only if G_u is conjugate to H in G, or equivalently that H_y is conjugate to H in G; by Lemma 6 b), this means that $\mathrm{H}_y = \mathrm{H}$, and hence that y is fixed by H. If W′ is the vector subspace of W consisting of the elements fixed by H, it follows that $\mathrm{X}_{(t)}$ can be identified with $\mathrm{G} \times^{\mathrm{H}} \mathrm{W}'$, and hence also with $\mathrm{G}/\mathrm{H} \times \mathrm{W}'$, hence b).

To prove c), observe that the assumption that X/G is connected implies that X is pure of finite dimension: indeed, for all $k \geq 0$, denote by X_k the set of points $x \in \mathrm{X}$ such that $\dim_x \mathrm{X} = k$; then X_k is open and closed in X, and stable under G, so X is equal to one of the X_k.

We now prove c) by induction on the dimension of X, the assertion being clear for $\dim \mathrm{X} = 0$. Let τ be a maximal element among the orbit types of the points of X (such an element exists by Lemma 6 c)). We shall prove the following:

c′) For every subset A *of* $\mathrm{X}_{(t)}$*, open and closed in* $\mathrm{X}_{(\tau)}$ *and stable under* G*, the closure* $\overline{\mathrm{A}}$ *of* A *in* X *is open.*

This assertion implies c). Indeed, note first that $\mathrm{X}_{(\tau)}$ is open in X, by a); assertion c') implies that $\overline{\mathrm{X}}_{(\tau)}$ is open and closed in X, hence equal to X since it is stable under G and X/G is connected. Let A be a non-empty open and closed subset of $\mathrm{X}_{(\tau)}$ stable under G; by c'), $\overline{\mathrm{A}}$ is open and closed in X and stable under G, hence equal to X; this implies that A is dense in

$X_{(\tau)}$, hence equal to $X_{(\tau)}$. Consequently, every non-empty open and closed subset of $X_{(\tau)}/G$ is equal to $X_{(\tau)}/G$, which proves that $X_{(\tau)}/G$ is connected. Finally, since $X_{(\tau)}$ is dense in X, it follows from *a*) that every point of X is of type $\leq \tau$; in other words, τ is the largest element among the orbit types of the points of X.

We now prove *c'*). We can assume that A is non-empty; let $x \in \overline{A}$. It suffices to prove that $\overline{A}$ is a neighbourhood of x. For this we can, as above, assume that $X = G \times {}^H W$ with H compact, x being the canonical image of $(e, 0)$. Assume first that H operates trivially on W: then X can be identified with $(G/H) \times W$, and $X_{(\tau)}/G = X/G$ is homeomorphic to W, hence connected; thus, $A/G = X/G$, so $A = X$. Assume from now on that H does not operate trivially on W. Choose a scalar product on W invariant under the compact group H; let S be the unit sphere in W (the set of vectors of norm 1). Note that S/H is connected: indeed, if $\dim(W) \geq 2$, S is connected, and if $\dim(W) = 1$, S is a space of two points on which H operates non-trivially. Put $Y = G \times {}^H S$; this is a closed submanifold of X, stable under G, of codimension 1, and Y/G, which is homeomorphic to S/H, is connected. Thus, by the induction hypothesis, there exists a maximal orbit type θ for Y, the set $Y_{(\theta)}$ is open and dense in Y, and $Y_{(\theta)}/G$ is connected.

Consider the operation of $\mathbf{R}_+^*$ on X induced by passage to the quotient by the law of operation $(\lambda, (g, w)) \mapsto (g, \lambda w)$ of $\mathbf{R}_+^*$ on $G \times W$. Two points of X conjugate under this operation are of the same orbit type; consequently, $X_{(\theta)}$ contains $\mathbf{R}_+^* Y_{(\theta)}$, which is a dense open subset of X. But $X_{(\tau)}$ is open by *a*), and hence meets $X_{(\theta)}$, so $\theta = \tau$.

On the other hand, the homeomorphism $(\lambda, w) \mapsto \lambda w$ from $\mathbf{R}_+^* \times S$ to $W - \{0\}$ (*General Topology*, Chap. VI, §2, no. 3, Prop. 3) induces a homeomorphism from $\mathbf{R}_+^* \times (S/H)$ to $(\mathbf{R}_+^* S)/H$, hence also from $\mathbf{R}_+^* \times (Y/G)$ to $(\mathbf{R}_+^* Y)/G$, and from $\mathbf{R}_+^* \times (Y_{(\theta)}/G)$ to $(\mathbf{R}_+^* Y_{(\theta)})/G$. Thus, $(\mathbf{R}_+^* Y_{(\theta)})/G$ is connected, and $X_{(\tau)}/G$, which contains a connected dense subset, is itself connected (*General Topology*, Chap. I, §11, no. 1, Prop. 1). Consequently, A is equal to $X_{(\tau)}$, hence is dense in X, and $\overline{A}$ is a neighbourhood of x. This completes the proof of the theorem.

With the notations in Th. 2 *c*), the points of $X_{(\tau)}$ are said to be *principal* and their orbits are called *principal orbits*. If x is a point of X, and if $G \times {}^{G_x} W$ is a linear tube in X around the orbit of x, the point x is principal if and only if G_x operates trivially on W, that is, if the tube is of the form $(G/G_x) \times W$.

Examples. 1) Let G be a connected compact Lie group, operating on itself by inner automorphisms. The fixer of an element x of G is simply the centralizer $Z(x)$ of x in G; it contains every maximal torus containing x. It follows that the largest orbit type τ is the conjugacy class of the maximal tori of G. The open set $G_{(\tau)}$ is the set of *very regular* elements of G (§5, no. 1, Remark). Assume that G is simply-connected. Then $G_{(\tau)}$ is equal to the set G_r of regular elements of G (§5, no. 2, Remark 2); if A is an alcove of a Cartan subalgebra $\mathfrak{t}$ of $\mathfrak{g} = \mathrm{L(G)}$, *the composite map* $\pi : \mathrm{A} \xrightarrow{\exp} \mathrm{G}_r \longrightarrow \mathrm{G}_r/\mathrm{Int(G)}$ *is an isomorphism of analytic manifolds.* Indeed, this is a homeomorphism (§5, no. 2, Cor. of Prop. 2); let $a \in \mathrm{A}$, put $t = \exp a$ and identify $\mathrm{T}_t(\mathrm{G})$ with $\mathfrak{g}$ by means of the translation $\gamma(t)$. The tangent map $\mathrm{T}_a(\pi)$ can then be identified with the composite of the canonical injection $\mathfrak{t} \to \mathfrak{g}$ and the quotient map $\mathfrak{g} \to \mathfrak{g}/\mathrm{Im}(\mathrm{Ad}\, t^{-1} - 1)$. Since t is regular, $\mathrm{T}_a(\pi)$ is an isomorphism, hence the stated result (*Differentiable and Analytic Manifolds, Results*, 5.7.8).

2) Let E be a real euclidean affine space, $\mathfrak{H}$ a set of hyperplanes of E, W the group of displacements of E generated by the orthogonal reflections with respect to the hyperplanes of $\mathfrak{H}$. Assume that $\mathfrak{H}$ is stable under W and that the group W, with the discrete topology, operates properly on E.

The preceding can be applied to the operation of W on E. The fixer of a point x of E is the subgroup of W generated by the reflections with respect to the hyperplanes of $\mathfrak{H}$ containing x (Chap. V, §3, no. 3, Prop. 2). Consequently, the largest orbit type τ is the class of the subgroup $\{\mathrm{Id_E}\}$, and $\mathrm{E}_{(\tau)}$ is the union of the chambers of E. Note that in this case the covering $\mathrm{E}_{(\tau)} \to \mathrm{E}_{(\tau)}/\mathrm{W}$ is *trivial*, and in particular $\mathrm{E}_{(\tau)}$ is not connected if $\mathfrak{H}$ is non-empty.

APPENDIX I

STRUCTURE OF COMPACT GROUPS

1. EMBEDDING A COMPACT GROUP IN A PRODUCT OF LIE GROUPS

PROPOSITION 1. *Every compact topological group* G *is isomorphic to a closed subgroup of a product of compact Lie groups.*

Denote by $\hat{\mathrm{G}}$ the set of classes of irreducible continuous unitary representations of G on finite dimensional complex Hilbert spaces (*Spectral Theory*, in preparation). For all $u \in \hat{\mathrm{G}}$, let H_u be the space of u and $\rho_u : \mathrm{G} \to \mathbf{U}(\mathrm{H}_u)$ the homomorphism associated to u. By the Peter-Weyl theorem (*Spectral Theory*, in preparation), the continuous homomorphism $\rho = (\rho_u)_{u \in \hat{\mathrm{G}}}$ from G to $\prod_{u \in \hat{\mathrm{G}}} \mathbf{U}(\mathrm{H}_u)$ is injective; since G is compact, ρ induces an isomorphism from G onto a closed subgroup of the group $\prod_{u \in \hat{\mathrm{G}}} \mathbf{U}(\mathrm{H}_u)$.

COROLLARY 1. *Let* V *be a neighbourhood of the identity element of* G. *Then* V *contains a closed normal subgroup* H *of* G *such that the quotient* G/H *is a Lie group.*

Let $(\mathrm{K}_\lambda)_{\lambda \in \mathrm{L}}$ be a family of compact Lie groups such that G can be identified with a closed subgroup of $\prod_{\lambda \in \mathrm{L}} \mathrm{K}_\lambda$; for $\lambda \in \mathrm{L}$, demote by $p_\lambda : \mathrm{G} \to \mathrm{K}_\lambda$

the restriction to G of the canonical projection. There exists a finite subset $J \subset L$, and for each $\lambda \in J$ a neighbourhood V_λ of the origin in K_λ, such that V contains $\bigcap_{\lambda \in J} p_\lambda^{-1}(V_\lambda)$. It now suffices to put $H = \bigcap_{\lambda \in J} \operatorname{Ker}(p_\lambda)$.

Denote by $(H_\alpha)_{\alpha \in I}$ the decreasing filtered family of closed normal subgroups of G, such that the quotient G/H_α is a Lie group. Consider the projective system of compact Lie groups G/H_α (cf. *General Topology*, Chap. III, §7, no. 2, Prop. 2).

COROLLARY 2. *The canonical map* $G \to \varprojlim_\alpha G/H_\alpha$ *is an isomorphism of topological groups.*

Indeed, Cor. 1 implies that condition (PA) of *General Topology*, Chap. III, §7, no. 2, is satisfied; the assertion now follows from Prop. 2 of *loc. cit.*

COROLLARY 3. G *is a Lie group if and only if there exists a neighbourhood of the identity element* e *of* G *that contains no normal subgroup distinct from* $\{e\}$.

The necessity of this condition has already been proved (Chap. III, §4, no. 2, Cor. 1 of Th. 2), and the sufficiency is an immediate consequence of Cor. 1.

2. PROJECTIVE LIMITS OF LIE GROUPS

Lemma 1. Let $(G_\alpha, f_{\alpha\beta})$ *be a projective system of topological groups relative to a filtered index set* I, *and* G *its limit. Assume that the canonical maps* $f_\alpha : G \to G_\alpha$ *are surjective.*

a) The subgroups $\overline{D(G_\alpha)}$ (resp. $C(G_\alpha)$, resp. $C(G_\alpha)_0$) *form a projective system of subsets of* G_α.

b) We have $\overline{D(G)} = \varprojlim_\alpha \overline{D(G_\alpha)}$ *and* $C(G) = \varprojlim_\alpha C(G_\alpha)$.

c) If G_α *is compact for all* $\alpha \in I$, *then* $C(G)_0 = \varprojlim_\alpha C(G_\alpha)_0$.

Let α, β be two elements of I, with $\alpha \leq \beta$. Then $f_{\alpha\beta}(D(G_\beta)) \subset D(G_\alpha)$, and $f_{\alpha\beta}(C(G_\beta)) \subset C(G_\alpha)$ since $f_{\alpha\beta}$ is surjective; since $f_{\alpha\beta}$ is continuous, it follows that $f_{\alpha\beta}(\overline{D(G_\beta)}) \subset \overline{D(G_\alpha)}$ and $f_{\alpha\beta}(C(G_\beta)_0) \subset C(G_\alpha)_0$, hence *a*). Since f_α is surjective, $f_\alpha(D(G)) = D(G_\alpha)$ (*Algebra*, Chap. I, §6, no. 2, Prop. 6), so $\overline{D(G)} = \varprojlim \overline{D(G_\alpha)}$ (*General Topology*, Chap. I, §4, no. 4, Cor. of Prop. 9). The surjectivity of f_α also implies the inclusion $f_\alpha(C(G)) \subset C(G_\alpha)$ and hence $C(G) \subset \varprojlim C(G_\alpha)$; the opposite inclusion is immediate. Finally, assertion *c*) follows from *b*) and *General Topology*, Chap. III, §7, no. 2, Prop. 4).

Lemma 2. Let $(S_a)_{a \in A}, (T_b)_{b \in B}$ *be two finite families of almost simple, simply-connected Lie groups* (Chap. III, § 9, no. 8, Def. 3), $u : \prod_{a \in A} S_a \to \prod_{b \in B} T_b$

a surjective morphism. Then there exist an injective map $l : \mathrm{B} \to \mathrm{A}$ and isomorphisms $u_b : \mathrm{S}_{l(b)} \to \mathrm{T}_b$ $(b \in \mathrm{B})$ such that $u((s_a)_{a\in\mathrm{A}}) = (u_b(s_{l(b)}))_{b\in\mathrm{B}}$ for every element $(s_a)_{a\in\mathrm{A}}$ of $\prod_{a\in\mathrm{A}} \mathrm{S}_a$.

Denote by $\mathfrak{s}_a$ (resp. $\mathfrak{t}_b$) the Lie algebra of S_a (resp. T_b) for $a \in \mathrm{A}$ (resp. $b \in \mathrm{B}$), and consider the homomorphism $\mathrm{L}(u) : \prod_{a\in\mathrm{A}} \mathfrak{s}_a \to \prod_{b\in\mathrm{B}} \mathfrak{t}_b$. Its kernel is an ideal of the semi-simple Lie algebra $\prod_{a\in\mathrm{A}} \mathfrak{s}_a$, and hence is of the form $\prod_{a\in\mathrm{A}''} \mathfrak{s}_a$, with $\mathrm{A}'' \subset \mathrm{A}$ (Chap. I, §6, no. 2, Cor. 1). Put $\mathrm{A}' = \mathrm{A} - \mathrm{A}''$. By restriction, $\mathrm{L}(u)$ induces an isomorphism $f : \prod_{a\in\mathrm{A}'} \mathfrak{s}_a \to \prod_{b\in\mathrm{B}} \mathfrak{t}_b$. By *loc. cit.*, for all $a \in \mathrm{A}'$ the ideal $f(\mathfrak{s}_a)$ is equal to one of the $\mathfrak{t}_b$; hence, there exists a bijection $l : \mathrm{B} \to \mathrm{A}'$ such that $f(\mathfrak{s}_{l(b)}) = \mathfrak{t}_b$ for $b \in \mathrm{B}$, and f induces an isomorphism $f_b : \mathfrak{s}_{l(b)} \to \mathfrak{t}_b$. Since the groups S_a and T_b are simply-connected, there exist isomorphisms $u_b : \mathrm{S}_{l(b)} \to \mathrm{T}_b$ such that $\mathrm{L}(u_b) = f_b$ for $b \in \mathrm{B}$ (Chap. III, §6, no. 3, Th. 3).

Denote by $\tilde{u} : \prod_{a\in\mathrm{A}} \mathrm{S}_a \to \prod_{b\in\mathrm{B}} \mathrm{T}_b$ the morphism defined by $\tilde{u}((s_a)_{a\in\mathrm{A}}) = (u_b(s_{l(b)}))_{b\in\mathrm{B}}$. By construction, $\mathrm{L}(\tilde{u}) = f = \mathrm{L}(u)$, so $\tilde{u} = u$, which proves the lemma.

Lemma 3. Under the hypotheses of Lemma 1, assume that the G_α are simply-connected compact Lie groups. Then, the topological group G is isomorphic to the product of a family of almost simple, simply-connected compact Lie groups.

For all $\alpha \in \mathrm{I}$, the group G_α is the direct product of a finite family of almost simple, simply-connected subgroups $(\mathrm{S}_\alpha^\lambda)_{\lambda\in\mathrm{L}_\alpha}$ (Chap. III, §9, no. 8, Prop. 28). Let $\beta \in \mathrm{I}$, $\beta \geq \alpha$. By Lemma 2, there exists a map $l_{\beta\alpha} : \mathrm{L}_\alpha \to \mathrm{L}_\beta$ such that $f_{\alpha\beta}(\mathrm{S}_\beta^{l_{\beta\alpha}(\lambda)}) = \mathrm{S}_\alpha^\lambda$ for $\lambda \in \mathrm{L}_\alpha$. We have $l_{\gamma\beta} \circ l_{\beta\alpha} = l_{\gamma\alpha}$ for $\alpha \leq \beta \leq \gamma$, so $(\mathrm{L}_\alpha, l_{\beta\alpha})$ is an inductive system of sets relative to I. Let L be its limit; the maps $l_{\beta\alpha}$ being injective, L_α can be identified with a subset of L, so that $\mathrm{L} = \bigcup_{\alpha\in\mathrm{I}} \mathrm{L}_\alpha$.

Let $\lambda \in \mathrm{L}$. Put $\mathrm{S}_\alpha^\lambda = \{1\}$ when $\lambda \notin \mathrm{L}_\alpha$, and denote by $\varphi_{\alpha\beta}^\lambda : \mathrm{S}_\beta^\lambda \to \mathrm{S}_\alpha^\lambda$ the morphism induced by $f_{\alpha\beta}$; this gives a projective system of topological groups $(\mathrm{S}_\alpha^\lambda, \varphi_{\alpha\beta}^\lambda)$, whose limit is isomorphic to S_λ. The canonical homomorphism of topological groups

$$\varprojlim_{\alpha\in\mathrm{I}} \left(\prod_{\lambda\in\mathrm{L}} \mathrm{S}_\alpha^\lambda\right) \to \prod_{\lambda\in\mathrm{L}} \left(\varprojlim_{\alpha\in\mathrm{I}} \mathrm{S}_\alpha^\lambda\right)$$

is bijective (*Theory of Sets*, Chap. III, §7, no. 3, Cor. 2); it is thus an isomorphism since the groups in question are compact. But the first of these groups can be identified with G and the second with the product of the S_λ, hence the lemma.

3. STRUCTURE OF CONNECTED COMPACT GROUPS

Let G be a commutative compact group. Recall (*Spectral Theory*, Chap. II, §1, no. 9, Prop. 11) that G is then isomorphic to the dual topological group of a *discrete* commutative group $\hat{G}$. The group G is connected if and only if $\hat{G}$ is torsion-free (*Spectral Theory*, Chap. II, §2, no. 2, Cor. 1 of Prop. 4).

The following properties are equivalent (*Spectral Theory*, Chap. II, §2, no. 2, Cor. 2 of Prop. 4 and §1, no. 9, Cor. 2 of Prop. 11):

(i) G is totally discontinuous;

(ii) $\hat{G}$ is a torsion group;

(iii) the topological group G is isomorphic to the limit of a projective system of finite (commutative) groups, each having the discrete topology.

The proposition below generalizes Cor. 1 of Prop. 4 of §1, no. 4.

PROPOSITION 2. *Let* G *be a connected compact group.*

a) $C(G)_0$ *is a commutative connected compact group*; D(G) *is a connected compact group, equal to its derived group.*

b) The continuous homomorphism $(x, y) \mapsto xy$ *from* $C(G)_0 \times D(G)$ *to* G *is surjective and its kernel is a central subgroup of* $C(G)_0 \times D(G)$ *that is compact and totally discontinuous.*

c) There exists a family $(S_\lambda)_{\lambda \in L}$ *of almost simple compact Lie groups and a surjective continuous homomorphism* $\prod_{\lambda \in L} S_\lambda \to D(G)$, *whose kernel is a totally discontinuous, compact, central subgroup.*

Let $(G_\alpha, f_{\alpha\beta})$ be a projective system of compact Lie groups, relative to a filtered set I, such that G is isomorphic to $\varprojlim G_\alpha$ and such that the canonical maps $f_\alpha : G \to G_\alpha$ are surjective (Cor. 2 of Prop. 1). For $\alpha \in I$, let $\pi_\alpha : \tilde{D}(G_\alpha) \to D(G_\alpha)$ be a universal covering of the group $D(G_\alpha)$. The $f_{\alpha\beta}$ induce morphisms $\tilde{f}_{\alpha\beta} : \tilde{D}(G_\beta) \to \tilde{D}(G_\alpha)$, and $(\tilde{D}(G_\alpha), \tilde{f}_{\alpha\beta})$ is a projective system of topological groups satisfying the hypotheses of Lemma 3.

It follows from this lemma that the topological group $\varprojlim \tilde{D}(G_\alpha)$ is isomorphic to the product of a family $(S_\lambda)_{\lambda \in L}$ of almost simple compact Lie groups. By Lemma 1, the limit of the projective system of homomorphisms (π_α) can be identified with a continuous homomorphism $\pi : \prod_{\lambda \in L} S_\lambda \to \overline{D(G)}$, which is surjective (*General Topology*, Chap. I, §9, no. 6, Cor. 2 of Prop. 8).

Now observe that the group $\prod_{\lambda \in L} S_\lambda$ is equal to its derived group: this follows from §4, no. 5, Cor. of Prop. 10. The same is true for $\overline{D(G)}$, since π is surjective. Consequently, $D(G) \supset D(\overline{D(G)}) = \overline{D(G)}$. Thus, the group D(G) is compact and equal to its derived group; this proves *a*), since the assertions concerning $C(G)_0$ are trivial.

On the other hand, the kernel of $\pi : \prod_{\lambda \in L} S_\lambda \to D(G)$ can be identified with $\varprojlim \operatorname{Ker}(\pi_\alpha)$ (*Algebra*, Chap. II, §6, no. 1, Remark 1), and thus with a compact, totally discontinuous, central subgroup, hence *c*).

We prove *b*). For all α in I, the morphism $s_\alpha : C(G_\alpha)_0 \times D(G_\alpha) \to G_\alpha$ such that $s_\alpha(x,y) = xy$ for $x \in C(G_\alpha)_0, y \in D(G_\alpha)$, is surjective and its kernel is a finite central subgroup (§1, no. 4, Cor. 1 of Prop. 4). The s_α form a projective system of maps, whose limit can, by the preceding, be identified with the homomorphism $(x,y) \mapsto xy$ from $C(G)_0 \times D(G)$ to G. We now see as before that this map is surjective and that its kernel is central and totally discontinuous, hence *b*).

COROLLARY. *Every solvable connected compact group is commutative.*

Indeed, the derived group is then solvable and equal to its derived group (Prop. 2 *a*)), hence reduced to the identity element.

APPENDIX II

REPRESENTATIONS OF REAL, COMPLEX OR QUATERNIONIC TYPE

1. REPRESENTATIONS OF REAL ALGEBRAS

Denote by σ the automorphism $\alpha \mapsto \bar{\alpha}$ of **C**; if W is a complex vector space, denote by $\overline{W}$ the **C**-vector space $\sigma_*(W)$ (that is, the group W with the law of operation $(\alpha, w) \mapsto \bar{\alpha}w$ for $\alpha \in \mathbf{C}, w \in W$).

PROPOSITION 1. *Let* A *be an* **R**-*algebra* (*associative and unital*) *and* V *a finite dimensional simple* A-*module over* **R**. *Then, we must be in one of the following three situations:*

α) *The commutant of* V (*Algebra*, Chap. VIII, § 5, no. 1) *is isomorphic to* **R**, *and the* $A_{(\mathbf{C})}$-*module* $V_{(\mathbf{C})}$ *is simple*;

β) *the commutant of* V *is isomorphic to* **C**; *the* $A_{(\mathbf{C})}$-*module* $V_{(\mathbf{C})}$ *is the direct sum of two non-isomorphic simple* $A_{(\mathbf{C})}$-*submodules which are interchanged by* $\sigma \otimes 1_V$;

γ) *the commutant of* V *is isomorphic to* **H**; *the* $A_{(\mathbf{C})}$-*module* $V_{(\mathbf{C})}$ *is the direct sum of two isomorphic simple* $A_{(\mathbf{C})}$-*submodules that are interchanged by* $\sigma \otimes 1_V$.

The commutant E of V is a field, a finite extension of **R** (*Algebra*, Chap. VIII, §3, no. 2, Prop. 2), hence isomorphic to **R**, **C** or **H** (*Algebra*, Chap. VIII, §15). The $A_{(\mathbf{C})}$-module $V_{(\mathbf{C})}$ is semi-simple (*Algebra*, Chap. VIII, §11, no. 4), and its commutant can be identified with $\mathbf{C} \otimes_{\mathbf{R}} E$ (*Algebra*, Chap. VIII, §11, no. 2, Lemma 1).

If E is isomorphic to **R**, the commutant of $V_{(\mathbf{C})}$ is isomorphic to **C**, and $V_{(\mathbf{C})}$ is a simple $A_{(\mathbf{C})}$-module (*Algebra*, Chap. VIII, §11, no. 4).

If E is not isomorphic to **R**, it contains a field isomorphic to **C**; it follows that V has an $A_{(\mathbf{C})}$-module structure, denoted by V^c. Then V^c is a simple

$A_{(\mathbf{C})}$-module, and the **C**-linear map $\psi : V_{(\mathbf{C})} \to V^c \oplus \overline{V}^c$ such that $\psi(\alpha \otimes v) = (\alpha v, \bar{\alpha} v)$ for $\alpha \in \mathbf{C}$, $v \in V$, is an isomorphism (*Algebra*, Chap. V, §10, no. 4, Prop. 8). Moreover, $\sigma \otimes 1_V$ corresponds under this isomorphism to the **R**-automorphism $(v, v') \mapsto (v', v)$ of $V^c \oplus \overline{V}^c$, and hence interchanges the two $A_{(\mathbf{C})}$-submodules $\psi^{-1}(V^c)$ and $\psi^{-1}(\overline{V}^c)$.

The commutant $E_{(\mathbf{C})}$ of $V_{(\mathbf{C})}$ thus contains $\mathbf{C} \times \mathbf{C}$, operating by homotheties on $V^c \oplus \overline{V}^c$. There is no isomorphism of $A_{(\mathbf{C})}$-modules from V^c to $\overline{V}^c$ if and only if $E_{(\mathbf{C})}$ reduces to $\mathbf{C} \times \mathbf{C}$, that is, if E is isomorphic to **C**. This completes the proof.

PROPOSITION 2. *Let* A *be an* **R***-algebra (associative and unital), and* W *a finite dimensional simple* $A_{(\mathbf{C})}$*-module over* **C**. *Then, we must be in one of the following three situations:*

a) There exists an $A_{(\mathbf{C})}$*-isomorphism* θ *from* W *to* $\overline{W}$ *with* $\theta \circ \theta = 1_W$. *Then the set* V *of fixed points of* θ *is an* **R***-structure on* W, *and a simple* A*-module with commutant* $\mathbf{R}.1_V$. *Moreover,* $W_{[\mathbf{R}]}$ *is the direct sum of two isomorphic simple* A*-modules.*

b) The $A_{(\mathbf{C})}$*-modules* W *and* $\overline{W}$ *are not isomorphic; then* $W_{[\mathbf{R}]}$ *is a simple* A*-module with commutant* $\mathbf{C}.1_W$.

c) There exists an $A_{(\mathbf{C})}$*-isomorphism* θ *from* W *to* $\overline{W}$ *with* $\theta \circ \theta = -1_W$. *Then the* A*-module* $W_{[\mathbf{R}]}$ *is simple, and its commutant is the field* $\mathbf{C}.1_W \oplus \mathbf{C}.\theta$, *which is isomorphic to* **H**.

The complex vector space $\mathrm{Hom}_{A_{(\mathbf{C})}}(W, \overline{W})$ is of dimension ≤ 1 (*Algebra*, Chap. VIII, §3, no. 2); if $\theta \in \mathrm{Hom}_{A_{(\mathbf{C})}}(W, \overline{W})$, the endomorphism $\theta \circ \theta$ of W is a homothety, with ratio $\alpha \in \mathbf{C}$. For all $w \in W$, we have $\alpha\theta(w) = \theta \circ \theta \circ \theta(w) = \theta(\alpha w) = \bar{\alpha}\theta(w)$, so α is real. If $\theta' = \lambda\theta$, with $\lambda \in \mathbf{C}$, then $\theta' \circ \theta' = |\lambda|^2 \theta \circ \theta$; thus, exactly one of the following three possibilities is realised:

a) There exists $\theta \in \mathrm{Hom}_{A_{(\mathbf{C})}}(W, \overline{W})$ with $\theta \circ \theta = 1_W$;

b) $\mathrm{Hom}_{A_{(\mathbf{C})}}(W, \overline{W}) = \{0\}$;

c) There exists $\theta \in \mathrm{Hom}_{A_{(\mathbf{C})}}(W, \overline{W})$ with $\theta \circ \theta = -1_W$.

In case *a*), the set V of fixed points of θ is an **R**-structure on W (*Algebra*, Chap. V, p. 61, Prop. 7); since $V_{(\mathbf{C})}$ is isomorphic to W, the A-module V is simple with commutant $\mathbf{R}.1_V$ (Prop. 1), and $W_{[\mathbf{R}]}$ is not simple.

Conversely, if $W_{[\mathbf{R}]}$ is not simple, let V be a simple A-submodule of $W_{[\mathbf{R}]}$; since the $A_{(\mathbf{C})}$-module W is simple, $V + iV = W$ and $V \cap iV = \{0\}$, that is, $W = V \oplus iV$. Thus, V is an **R**-structure on W, and the isomorphism θ from W to $\overline{W}$ such that $\theta(v + iv') = v - iv'$ for v and v' in V satisfies $\theta \circ \theta = 1_W$.

Consequently, in cases *b*) and *c*), the A-module $W_{[\mathbf{R}]}$ is simple; by Prop. 1, its commutant E is isomorphic to **C** in case *b*), and to **H** in case *c*). Moreover, it is clear that E contains $\mathbf{C}.1_W$, and $\mathbf{C}.\theta$ in case *c*), hence the proposition.

With the assumptions in the proposition, the $A_{(\mathbf{C})}$-module W is said to be of *real, complex or quaternionic type* (relative to A) in case *a*), *b*) or *c*), respectively.

For K = $\mathbf{R}$ or $\mathbf{C}$, denote by $\mathfrak{S}_{\mathrm{K}}(\mathrm{A})$ the set of classes of finite dimensional simple $\mathrm{A}_{(\mathrm{K})}$-modules over K. The group $\Gamma = \mathrm{Gal}(\mathbf{C}/\mathbf{R})$ operates on $\mathfrak{S}_{\mathbf{C}}(\mathrm{A})$; the two preceding propositions establish a *bijective correspondence* between $\mathfrak{S}_{\mathbf{R}}(\mathrm{A})$ and the quotient set $\mathfrak{S}_{\mathbf{C}}(\mathrm{A})/\Gamma$.

2. REPRESENTATIONS OF COMPACT GROUPS

Let G be a compact topological group, and let $\rho : \mathrm{G} \to \mathbf{GL}(\mathrm{W})$ be a continuous representation of G on a finite dimensional complex vector space. We shall say that ρ is irreducible of real, complex or quaternionic type if this is the case for the $\mathbf{C}^{(\mathrm{G})}$-module W (relative to the algebra $\mathrm{A} = \mathbf{R}^{(\mathrm{G})}$). Let H be a separating positive hermitian form on W, invariant under G.

PROPOSITION 3. *Assume that ρ is irreducible.*

a) The representation ρ is of real type if and only if there exists a non-zero symmetric bilinear form B *on* W, *invariant under* G. *In this case the form* B *is separating; the set* V *of $w \in \mathrm{W}$ such that $\mathrm{H}(w, x) = \mathrm{B}(w, x)$ for all $x \in \mathrm{W}$ is an* $\mathbf{R}$*-structure on* W *invariant under* G.

b) The representation ρ is of complex type if and only if there exists no non-zero bilinear form on W *invariant under* G.

c) The representation ρ is of quaternionic type if and only if there exists a non-zero alternating bilinear form on W, *invariant under* G; *such a form is necessarily separating.*

For $\theta \in \mathrm{Hom}_{\mathbf{C}^{(\mathrm{G})}}(\mathrm{W}, \overline{\mathrm{W}})$ and $x, y \in \mathrm{W}$, put $\mathrm{B}_\theta(x, y) = \mathrm{H}(\theta x, y)$. Then B_θ is a bilinear form on W, invariant under G, and separating if θ is non-zero. Denote by $\mathscr{B}(\mathrm{W})^{\mathrm{G}}$ the space of bilinear forms on W invariant under G; the map $\theta \mapsto \mathrm{B}_\theta$ from $\mathrm{Hom}_{\mathbf{C}^{(\mathrm{G})}}(\mathrm{W}, \overline{\mathrm{W}})$ to $\mathscr{B}(\mathrm{W})^{\mathrm{G}}$ is an isomorphism of $\mathbf{C}$-vector spaces. This implies, in particular, assertion *b*).

Let θ be a $\mathbf{C}^{(\mathrm{G})}$-isomorphism from W to $\overline{\mathrm{W}}$ such that $\theta \circ \theta = \alpha_{\mathrm{W}}$, with $\alpha \in \{-1, +1\}$ (Prop. 2); since $\mathscr{B}(\mathrm{W})^{\mathrm{G}}$ is of dimension 1, there exists $\varepsilon \in \mathbf{C}$ such that

$$\mathrm{B}_\theta(y, x) = \varepsilon \mathrm{B}_\theta(x, y) \quad \text{for all } x, y \text{ in W}.$$

Iterating, we obtain $\mathrm{B}_\theta(y, x) = \varepsilon \mathrm{B}_\theta(x, y) = \varepsilon^2 \mathrm{B}_\theta(y, x)$, so $\varepsilon^2 = 1$ and $\varepsilon \in \{-1, +1\}$. Moreover, for x in W,

$$\mathrm{H}(\theta x, \theta x) = \mathrm{B}_\theta(x, \theta x) = \varepsilon \mathrm{B}_\theta(\theta x, x) = \varepsilon \mathrm{H}(\theta \circ \theta(x), x) = \varepsilon\alpha \mathrm{H}(x, x)$$

so $\varepsilon\alpha > 0$ since H is positive, that is, $\varepsilon = \alpha$. Assertions *a*) and *c*) now follow from Prop. 2.

Denote by dg the Haar measure of total mass 1 on G.

Lemma 1. Let W^{G} *be the subspace of* W *consisting of the elements invariant under* G. *The endomorphism* $\int_{\mathrm{G}} \rho(g)\, dg$ *of* W *is a projection with image* W^{G}, *compatible with the operations of* G. *In particular,*

$$\dim W^G = \int_G \operatorname{Tr} \rho(g)\, dg.$$

Put $p = \int_G \rho(g)\, dg$; for $h \in G$,

$$\rho(h) \circ p = \int_G \rho(hg)\, dg = \int_G \rho(g)\, dg = p$$

and similarly $p \circ \rho(h) = p$. Thus, p is compatible with the operations of G, and its image is contained in W^G. If $w \in W^G$, we have $p(w) = \int_G \rho(g) w\, dg = w$, hence the lemma.

Lemma 2. Let u *be an endomorphism of a finite dimensional vector space* E *over a field* K. *Then*

$$\operatorname{Tr} u^2 = \operatorname{Tr} \mathbf{S}^2(u) - \operatorname{Tr} \mathsf{\Lambda}^2(u).$$

Let $\chi_u(X) = \prod_{i=1}^{n} (X - \alpha_i)$ be a decomposition of the characteristic polynomial of u into linear factors in a suitable extension of K. We have $\operatorname{Tr} u^2 = \sum_i \alpha_i^2$, $\operatorname{Tr} \Lambda^2(u) = \sum_{i<j} \alpha_i\alpha_j$, $\operatorname{Tr} \mathbf{S}^2(u) = \sum_{i \leq j} \alpha_i\alpha_j$ (cf. *Algebra*, Chap. VII, §5, no. 5, Cor. 3), hence the result.

PROPOSITION 4. *Assume that* ρ *is irreducible. Then,* ρ *is of real* (resp. *complex,* resp. *quaternionic*) *type if and only if the integral* $\int_G \operatorname{Tr} \rho(g^2)\, dg$ *is equal to* 1 (resp. 0, resp. -1).

Denote by $\check{\rho}$ the contragredient representation of ρ on W^* (defined by $\check{\rho}(g) = {}^t\rho(g^{-1})$). Applying Lemma 2 to $\check{\rho}(g)$ and integrating over G gives

$$\int_G \operatorname{Tr} \rho(g^2)\, dg = \int_G \operatorname{Tr} {}^t\rho(g^{-2})\, dg = \int_G \operatorname{Tr} \mathbf{S}^2(\check{\rho}(g)) dg - \int_G \operatorname{Tr} \mathsf{\Lambda}^2(\check{\rho}(g)) dg$$

hence, by Lemma 1,

$$\int_G \operatorname{Tr} \rho(g^2) dg = \dim(\mathbf{S}^2 W^*)^G - \dim(\mathsf{\Lambda}^2 W^*)^G.$$

But $\mathbf{S}^2 W^*$ (resp. $\mathsf{\Lambda}^2 W^*$) can be identified with the space of symmetric (resp. alternating) bilinear forms on W. Thus, the proposition follows immediately from Prop. 3.

EXERCISES

§1

1) Let G be a finite dimensional, connected, commutative *complex* Lie group, and let V be its Lie algebra.

a) The map $\exp_G : V \to G$ is a surjective homomorphism; its kernel Γ is a discrete subgroup of V.

b) G is compact if and only if Γ is a lattice in V; then G is said to be a *complex torus*.

c) Let Γ be the discrete subgroup of $\mathbf{C}^2$ generated by the elements $e_1 = (1,0), e_2 = (0,1), e_3 = (\sqrt{2}, i)$; put $G = \mathbf{C}^2/\Gamma$, $H = (\Gamma + \mathbf{C}e_1)/\Gamma$. Show that H is isomorphic to $\mathbf{C}^*$, that G/H is a complex torus of dimension 1, but that G contains no non-zero complex torus.

d) Every finite dimensional connected compact complex Lie group is a complex torus (cf. Chap. III, §6, no. 3, Prop. 6).

2) Let H be the set of complex numbers τ such that $\mathscr{I}(\tau) > 0$.

a) Show that an analytic law of left operation of the discrete group $\mathbf{SL}(2, \mathbf{Z})$ on H can be defined by putting $\gamma\tau = \frac{a\tau+b}{c\tau+d}$ for all $\gamma = \begin{pmatrix} a & b \\ c & d \end{pmatrix} \in \mathbf{SL}(2, \mathbf{Z})$ and $\tau \in H$.

b) For $\tau \in H$, denote by T_τ the complex torus $\mathbf{C}/(\mathbf{Z} + \mathbf{Z}\tau)$. Show that the map $\tau \mapsto T_\tau$ induces by passage to the quotient a bijective map from the set $H/\mathbf{SL}(2, \mathbf{Z})$ to the set of isomorphism classes of complex tori of dimension one.

3) Let G be an integral subgroup of $\mathbf{O}(n)$ such that the identity representation of G on $\mathbf{R}^n$ is irreducible. Show that G is closed (write G in the form $K \times N$, where K is compact and N is commutative, and prove that $\dim \overline{N} \leq 1$).

4) Show that the conditions in Prop. 3 (no. 3) are equivalent to each of the following conditions:
(ii′) The group Ad(G) is relatively compact in Aut(L(G)).
(ii″) The group Ad(G) is relatively compact in End(L(G)).

(v) Every neighbourhood of the identity element e of G contains a neighbourhood of e stable under inner automorphisms (cf. *Integration*, §3, no. 1, Prop. 1).

5) Let A be the closed subgroup of $\mathbf{GL}(3, \mathbf{R})$ consisting of the matrices (a_{ij}) such that $a_{ij} = 0$ for $i > j$, $a_{ii} = 1$ for $1 \leq i \leq 3$, $a_{12} \in \mathbf{Z}$, $a_{23} \in \mathbf{Z}$; let $\theta \in \mathbf{R} - \mathbf{Q}$ and let B be the subgroup of A consisting of the matrices (a_{ij}) such that $a_{12} = a_{23} = 0$ and $a_{13} \in \theta\mathbf{Z}$, and let $G = A/B$.

a) Show that G is a Lie group of dimension one, whose Lie algebra is compact.

b) Show that $C(G) = \overline{D(G)} = G_0$, and that G_0 is the largest compact subgroup of G.

c) Show that G is not the semi-direct product of G_0 with any subgroup.

6) Show that a (real) Lie algebra is compact if and only if it admits a basis $(e_\lambda)_{\lambda \in L}$ for which the structure constants $\gamma_{\lambda\mu\nu}$ are anti-symmetric in λ, μ, ν (that is, they satisfy $\gamma_{\lambda\mu\nu} = -\gamma_{\mu\lambda\nu} = -\gamma_{\lambda\nu\mu}$).

7) An *involutive Lie algebra* is a (real) Lie algebra $\mathfrak{g}$ equipped with an automorphism s such that $s \circ s = 1_{\mathfrak{g}}$. Denote by $\mathfrak{g}^+$ (resp. $\mathfrak{g}^-$) the eigenspace of $\mathfrak{g}$ relative to the eigenvalue $+1$ (resp. -1) of s.

a) Show that $\mathfrak{g}^+$ is a subalgebra of $\mathfrak{g}$ and that $\mathfrak{g}^-$ is a $\mathfrak{g}^+$-module; we have $[\mathfrak{g}^-, \mathfrak{g}^-] \subset \mathfrak{g}^+$, and $\mathfrak{g}^+$ and $\mathfrak{g}^-$ are orthogonal with respect to the Killing form.

b) Show that the following conditions are equivalent:
(i) The $\mathfrak{g}^+$-module $\mathfrak{g}^-$ is simple.
(ii) The algebra $\mathfrak{g}^+$ is maximal among the subalgebras of $\mathfrak{g}$ distinct from $\mathfrak{g}$.
If these conditions are satisfied, and if $\mathfrak{g}^+$ contains no non-zero ideal of $\mathfrak{g}$, the involutive Lie algebra $(\mathfrak{g}, s)$ is said to be *irreducible*.

c) Assume that $\mathfrak{g}$ is semi-simple. Show that the following conditions are equivalent:
(i) The only ideals of $\mathfrak{g}$ stable under s are $\{0\}$ and $\mathfrak{g}$;
(ii) $\mathfrak{g}$ is either simple or the sum of two simple ideals interchanged by s.
Show that these conditions are satisfied when $(\mathfrak{g}, s)$ is irreducible (observe that $\mathfrak{g}$ is the direct sum of ideals stable under s and satisfying (ii)).

d) Assume from now on that $\mathfrak{g}$ is compact semi-simple. Show that the involutive Lie algebra $(\mathfrak{g}, s)$ is irreducible if and only if s is different from the identity and $(\mathfrak{g}, s)$ satisfies the equivalent conditions in *c*) (let $\mathfrak{p}$ be a $\mathfrak{g}^+$-submodule of $\mathfrak{g}^-$ and $\mathfrak{q}$ its orthogonal complement with respect to the Killing form; observe that $[\mathfrak{p}, \mathfrak{q}] = 0$ and deduce that $\mathfrak{p} + [\mathfrak{p}, \mathfrak{p}]$ is an ideal of $\mathfrak{g}$).

e) Prove that $\mathfrak{g}$ is the direct sum of a family $(\mathfrak{g}_i)_{0 \leq i \leq n}$ of ideals stable under s, such that $\mathfrak{g}_0$ is fixed by s and such that the involutive algebras $(\mathfrak{g}_i, s|\mathfrak{g}_i)$ are irreducible for $1 \leq i \leq n$.

8) Let G be a compact semi-simple Lie group, u an automorphism of G of order two, K the identity component of the set of fixed points of u, and X the manifold G/K (the homogeneous space X is then said to be *symmetric*).

a) Show that, if G is almost simple, then K is maximal among the connected closed subgroups of G distinct from G; in other words (Chap. III, §3, Exerc. 8), the operation of G on X is primitive. The symmetric space X is then said to be *irreducible*.

b) Assume that the manifold X is simply-connected; show that it is then isomorphic to a product of manifolds each of which is isomorphic either to a Lie group or to an irreducible symmetric space.

9) Let $\mathfrak{a}$ be a real or complex Lie algebra, and let G be a compact subgroup of $\mathrm{Aut}(\mathfrak{a})$. Prove that $\mathfrak{a}$ has a Levi subalgebra (Chap. I, §6, no. 8, Def. 7) stable under G (reduce to the case in which the radical of $\mathfrak{a}$ is abelian, and use *Integration*, Chap. VII, §3, no. 2, Lemma 2).

§2

1) Let G be a connected compact Lie group, g an element of G.

a) Show that there exists an integer $n \geq 1$ such that the centralizer $\mathrm{Z}(g^n)$ is connected (prove that, for suitable n, the closed subgroup generated by g^n is a torus).

b) Assume that the dimension of $\mathrm{Ker}(\mathrm{Ad}\, g^n - 1)$ is independent of n $(n \geq 1)$. Prove that $\mathrm{Z}(g)$ is connected.

c) If g^n is regular for all $n \geq 1$, $\mathrm{Z}(g)$ is connected.

2) Show that every connected compact Lie group G is the semi-direct product of its derived group by a torus (observe that, if T is a maximal torus of G, then $\mathrm{D(G)} \cap \mathrm{T}$ is a torus).

3) Let G be a compact Lie group, $\mathfrak{g}$ its Lie algebra, $\mathfrak{s}$ a vector subspace of $\mathfrak{g}$ such that, for x, y, z in $\mathfrak{s}$, we have $[x, [y, z]] \in \mathfrak{s}$. Let $\mathscr{L}$ be the set of commutative subalgebras of $\mathfrak{g}$ contained in $\mathfrak{s}$. Show that the identity component of the stabilizer of $\mathfrak{s}$ in G operates transitively on the set of maximal elements of $\mathscr{L}$ (argue as in the proof of Th. 1).

4) Let G be a connected compact Lie group, T a maximal torus of G and S a sub-torus of T. Denote by Σ (resp. F) the stabilizer (resp. the fixer) of S in $\mathrm{W_G(T)}$.

a) Prove that the group $\mathrm{N_G(S)/Z_G(S)}$ is isomorphic to the quotient Σ/F.

b) Let H be a connected closed subgroup of G, such that S is a maximal torus of H. Show that every element of $\mathrm{W_H(S)}$, considered as an automorphism of S, is the restriction to S of an element of Σ.

5) Let G be a connected compact Lie group and S a torus of G. Show that the following conditions are equivalent:

(i) S is contained in a unique maximal torus;

(ii) $\mathrm{Z_G(S)}$ is a maximal torus;

(iii) S contains a regular element.

6) Let G be a connected compact Lie group, T a maximal torus of G, $\mathfrak{g}$ (resp. $\mathfrak{t}$) the Lie algebra of G (resp. T), and $i : \mathfrak{t} \to \mathfrak{g}$ the canonical injection. Prove that the map ${}^t i : \mathfrak{g}^* \to \mathfrak{t}^*$ induces by passage to the quotient a homeomorphism from $\mathfrak{g}^*/\mathrm{G}$ to $\mathfrak{t}^*/\mathrm{W_G(T)}$.

¶ 7) Let X and Y be two real manifolds of class C^r $(1 \leq r \leq \omega)$, separated, connected and of dimension n; let $f : \mathrm{X} \to \mathrm{Y}$ be a proper morphism of class C^r, equipped with an orientation F (*Differentiable and Analytic Manifolds, Results*, 10.2.5).

a) Show that there exists a unique real number d such that $\int_{\mathrm{X}} f^*\alpha = d \int_{\mathrm{Y}} \alpha$ for every twisted differential form α of degree n on Y of class C^1 and of compact support (use *Differentiable and Analytic Manifolds, Results*, 11.2.4).

b) Let $y \in \mathrm{Y}$ be such that f is étale at every point of $f^{-1}(y)$. For $x \in f^{-1}(y)$, put $\nu_x(f) = 1$ (resp. $\nu_x(f) = -1$) if the maps F_x and $\tilde{f}_x$ (*Differentiable and Analytic Manifolds, Results*, 10.2.5, Example *b*)) from $\mathrm{Or}(\mathrm{T}_x(\mathrm{X}))$ to $\mathrm{Or}(\mathrm{T}_y(\mathrm{Y}))$ coincide (resp. are opposite). Prove that $d = \sum_{x \in f^{-1}(y)} \nu_x(f)$ and, in particular, that $d \in \mathbf{Z}$.

We call d the *degree* of f and denote it by $\deg f$. If $\mathrm{X} = \mathrm{Y}$ and X is orientable, we can take for F the orientation that preserves the orientation of X.

c) Show that, if $\deg f \neq 0$ then f is surjective.

d) If there exists $x \in \mathrm{X}$ such that $f^{-1}(f(x)) = \{x\}$ and such that f is étale at x, then f is surjective.

8) Let G be a connected compact Lie group, dg the normalized Haar measure on G.

a) Let $f : \mathrm{G} \to \mathrm{G}$ be a morphism of manifolds; for $g \in \mathrm{G}$, the differential $\mathrm{T}_g(f)$ can be identified by left translations with a linear map from L(G) to L(G). Prove the formulas

$$\deg f . \int_{\mathrm{G}} \varphi(g)\, dg = \int_{\mathrm{G}} \varphi(f(g)) \det \mathrm{T}_g(f)\, dg$$

for every integrable (complex-valued) function φ on G, and

$$\deg f = \int_{\mathrm{G}} \det \mathrm{T}_g(f)\, dg.$$

b) Let $\psi_k : \mathrm{G} \to \mathrm{G}$ be the map $g \mapsto g^k$. Prove the formula

$$\deg \psi_k = \int_{\mathrm{G}} \det(1 + \mathrm{Ad}\, g + \cdots + (\mathrm{Ad}\, g)^{k-1}) dg.$$

c) Let T be a maximal torus of G, and T_r the set of regular elements of T. Show that $\exp_{\mathrm{G}}^{-1}(\mathrm{T}_r) \subset \mathrm{L(T)}$ and $\psi_k^{-1}(\mathrm{T}_r) \subset \mathrm{T}_r$ for all $k \geq 1$.

d) Show that $\deg \psi_k = k^{\dim(\mathrm{T})}$; deduce the equality

$$\int_G \det(1 + \mathrm{Ad}\, g + \cdots + (\mathrm{Ad}\, g)^{k-1})dg = k^{\dim(\mathrm{T})}.$$

9) This exercise is devoted to another proof of Th. 2. Let G be a connected compact Lie group, T a maximal torus of G, N its normalizer. Denote by $\mathrm{G} \times^{\mathrm{N}} \mathrm{T}$ the quotient manifold of $\mathrm{G} \times \mathrm{T}$ by N for the operation defined by $(g, t).n = (gn, n^{-1}tn)$ (*Differentiable and Analytic Manifolds, Results*, 6.5.1).

a) Show that the morphism $(g, t) \mapsto gtg^{-1}$ from $\mathrm{G}\times\mathrm{T}$ to G defines by passage to the quotient an analytic morphism $f : \mathrm{G} \times^{\mathrm{N}} \mathrm{T} \to \mathrm{G}$.

b) Let θ be an element of T whose set of powers is dense in T (*General Topology*, Chap. VII, §1, no. 3, Cor. 2 of Prop. 7). Show that $f^{-1}(\theta) = \{x\}$, where x is the class of (e, θ) in $\mathrm{G} \times^{\mathrm{N}} \mathrm{T}$, and that f is étale at x.

c) Conclude from Exerc. 7 *d*) that f is surjective, and deduce another proof of Th. 2.

10) * In this exercise, we use the following result from algebraic topology (*Lefschetz's formula*[11]): let X be a finite dimensional compact manifold, and f a morphism from X to itself. Assume that the set F of points x of X such that $f(x) = x$ is finite, and that for all $x \in \mathrm{F}$ the number $\delta(x) = \det(1 - \mathrm{T}_x(f))$ is non-zero. If $\mathrm{H}^i(f)$ denotes the endomorphism of the vector space $\mathrm{H}^i(\mathrm{X}, \mathbf{R})$ induced by f (for $i \geq 0$), then

$$\sum_{x\in\mathrm{F}} \delta(x)/|\delta(x)| = \sum_{i\geq 0}(-1)^i \mathrm{Tr}\,\mathrm{H}^i(f).$$

Let G be a connected compact Lie group, T a maximal torus of G. For $g \in \mathrm{G}$, denote by $\tau(g)$ the automorphism of the manifold G/T induced by left multiplication by g.

a) Let t be an element of T such that the subgroup generated by t is dense in T. Show that the fixed points of $\tau(t)$ are the classes $n\mathrm{T}$ for $n \in \mathrm{N_G(T)}$. Deduce that $(\mathrm{N_G(T)} : \mathrm{T}) = \sum_i (-1)^i \dim_{\mathbf{R}} \mathrm{H}^i(\mathrm{G/T}, \mathbf{R})$.

b) Let g be an arbitrary element of G; show that the set of fixed points of $\tau(g)$ is non-empty, and deduce another proof of Th. 2.*

11) Let G be a compact Lie group. A subgroup S of G is said to be *of type* (C) if it is equal to the closure of a cyclic subgroup that is of finite index in its normalizer.

a) Let S be a subgroup of type (C); show that $\mathrm{S_0}$ is a maximal torus of $\mathrm{G_0}$, and that S is the direct product of $\mathrm{S_0}$ and a finite cyclic subgroup.

b) Prove that every element g of G is contained in a subgroup of type (C) (consider the group generated by g and a maximal torus of $\mathrm{Z}(g)_0$).

[11] For a proof of this theorem, see for example S. LEFSCHETZ, Intersections and transformations of complexes and manifolds, *Trans. Amer. Math. Soc.* 28 (1926), pp. 1-49.

c) Let S be a subgroup of type (C), and s an element of S whose set of powers is dense in S. Show that every element of $s\mathrm{G}_0$ is conjugate under $\mathrm{Int}(\mathrm{G}_0)$ to an element of $s\mathrm{S}_0$ (use the method of Exerc. 10).

d) Denote by $p : \mathrm{G} \to \mathrm{G}/\mathrm{G}_0$ the quotient map. Show that the map $\mathrm{S} \mapsto p(\mathrm{S})$ induces a bijection from the set of conjugacy classes of subgroups of G of type (C) to the set of conjugacy classes of cyclic subgroups of G/G_0.

e) Let S be a subgroup of G of type (C); denote by S_ρ the set of elements of S whose class in S/S_0 generates S/S_0. Show that two elements of S_ρ that are conjugate in G are conjugate in $\mathrm{N}_\mathrm{G}(\mathrm{S})$.

§3

1) Let G be a compact Lie group of dimension > 0. Then every finite commutative subgroup of G is cyclic if and only if G is isomorphic to $\mathbf{U}$, $\mathbf{SU}(2, \mathbf{C})$ or the normalizer of a maximal torus in $\mathbf{SU}(2, \mathbf{C})$.

2) Denote by K one of the fields $\mathbf{R}$, $\mathbf{C}$ or $\mathbf{H}$, and by n an integer ≥ 1. Give the space K_s^n the usual hermitian form.

a) Show that $\mathbf{U}(n, \mathrm{K})$ is a compact real Lie group.

b) Show that the sphere of radius 1 in K_s^n is a (real) Lie homogeneous space for $\mathbf{U}(n, \mathrm{K})$; the fixer of a point is isomorphic to $\mathbf{U}(n-1, \mathrm{K})$.

c) Deduce that the groups $\mathbf{U}(n, \mathbf{C})$ and $\mathbf{U}(n, \mathbf{H})$ are connected, and that $\mathbf{O}(n, \mathbf{R})$ has two connected components.

d) Show that the groups $\mathbf{SO}(n, \mathbf{R})$ and $\mathbf{SU}(n, \mathbf{C})$ are connected.

e) Show that the group $\mathbf{O}(n, \mathbf{R})$ (resp. $\mathbf{U}(n, \mathbf{C})$) is the semi-direct product of $\mathbf{Z}/2\mathbf{Z}$ (resp. $\mathbf{T}$) by $\mathbf{SO}(n, \mathbf{R})$ (resp. $\mathbf{SU}(n, \mathbf{C})$).

3) *a*) Show that the Lie algebra of the real Lie group $\mathbf{U}(n, \mathbf{H})$ is the set of matrices $x \in \mathrm{M}_n(\mathbf{H})$ such that ${}^t\bar{x} = -x$, with the bracket $[x, y] = xy - yx$. Denote it by $\mathfrak{u}(n, \mathbf{H})$.

b) Identify $\mathbf{C}$ with the subfield $\mathbf{R}(i)$ of $\mathbf{H}$, and $\mathbf{C}^{2n}$ with $\mathbf{H}^n$ by the isomorphism $(z_1, \dots, z_{2n}) \mapsto (z_1 + jz_{n+1}, \dots, z_n + jz_{2n})$. Prove the equality $\mathbf{U}(n, \mathbf{H}) = \mathbf{U}(2n, \mathbf{C}) \cap \mathbf{Sp}(2n, \mathbf{C})$.

c) Deduce from *b*) that every compact simple (real) Lie algebra of type C_n is isomorphic to $\mathfrak{u}(n, \mathbf{H})$.

4) Let n be an integer ≥ 1.

a) Show that the group $\mathbf{SU}(n, \mathbf{C})$ is simply-connected (use Exerc. 2).

b) Show that the centre of $\mathbf{SU}(n, \mathbf{C})$ consists of the matrices $\lambda.I_n$ for $\lambda \in \mathbf{C}$, $\lambda^n = 1$.

c) Every almost simple compact Lie group of type A_n is isomorphic to the quotient of $\mathbf{SU}(n+1, \mathbf{C})$ by the cyclic subgroup consisting of the matrices

$\zeta^k.I_{n+1}$ $(0 \leq k < d)$, where d divides $n+1$ and ζ is a primitive dth root of unity.

d) Prove that the group $\mathbf{SL}(n, \mathbf{C})$ is simply-connected (use Chap. III, §6, no. 9, Th. 6).

5) For n an integer ≥ 1, denote by $\mathbf{Spin}(n, \mathbf{R})$ the reduced Clifford group associated to the usual quadratic form on $\mathbf{R}^n$ (*Algebra*, Chap. IX, §9, no. 5).

a) Show that $\mathbf{Spin}(n, \mathbf{R})$ is a compact Lie group and that the surjective homomorphism $\varphi : \mathbf{Spin}(n, \mathbf{R}) \to \mathbf{SO}(n, \mathbf{R})$ (*loc. cit.*) is analytic, with kernel $\{+1, -1\}$.

b) Show that, for $n \geq 2$, $\mathbf{Spin}(n, \mathbf{R})$ is connected and simply-connected (use Exerc. 2). The group $\pi_1(\mathbf{SO}(n, \mathbf{R}))$ is cyclic of order 2.

c) Let Z_n be the centre of $\mathbf{Spin}(n, \mathbf{R})$. Show that $Z_n = \{+1, -1\}$ if n is odd, and $Z_n = \{+1, -1, \varepsilon, -\varepsilon\}$ if n is even, where $\varepsilon = e_1 \ldots e_n$ is the product of the elements of the canonical basis of $\mathbf{R}^n$; the group Z_{2r} is isomorphic to $(\mathbf{Z}/2\mathbf{Z})^2$ (resp. to $\mathbf{Z}/4\mathbf{Z}$) if r is even (resp. odd).

d) Prove that every almost simple connected compact Lie group of type B_n $(n \geq 2)$ is isomorphic to $\mathbf{Spin}(2n+1, \mathbf{R})$ or $\mathbf{SO}(2n+1, \mathbf{R})$.

e) If r is odd (resp. even) and ≥ 2, every almost simple connected compact Lie group of type D_r is isomorphic to $\mathbf{Spin}(2r, \mathbf{R})$, $\mathbf{SO}(2r, \mathbf{R})$ or $\mathbf{SO}(2r, \mathbf{R})/\{\pm I_{2r}\}$ (resp. to one of the preceding groups or $\mathbf{Spin}(2r, \mathbf{R})/\{1, \varepsilon\}$).

6) *a*) Show that the compact Lie group $\mathbf{U}(n, \mathbf{H})$ is connected and simply-connected (use Exerc. 2), and that its centre is $\{\pm I_n\}$.

b) Every almost simple connected compact Lie group of type C_n $(n \geq 3)$ is isomorphic to $\mathbf{U}(n, \mathbf{H})$ or $\mathbf{U}(n, \mathbf{H})/\{\pm I_n\}$.

7) Let A be the algebra of Cayley octonions (*Algebra*, Chap. III, App., no. 3), with the basis $(e_i)_{0 \leq i \leq 7}$ of *loc. cit.* Denote by V the subspace of pure octonions generated by $e_1, \ldots, e_7$ and by E the subspace of V generated by $e_1, e_2, e_3, e_5, e_6, e_7$. Identify the subalgebra of A generated by e_0, e_4 with the field $\mathbf{C}$ of complex numbers, and denote by G the topological group of automorphisms of the unital algebra A.

a) Denote by Q the quadratic form on V induced by the Cayley norm, so that $(e_i)_{1 \leq i \leq 7}$ is an orthonormal basis of V. Prove that the map $\sigma \mapsto \sigma|V$ is an injective homomorphism from G to the group $\mathbf{SO}(Q)$, which is isomorphic to $\mathbf{SO}(7, \mathbf{R})$.

b) Show that the multiplication of A gives E the structure of a $\mathbf{C}$-vector space of dimension 3, of which $\{e_1, e_2, e_3\}$ is a basis. Denote by Φ the hermitian form on E for which this basis is orthonormal. Let H be the fixer of e_4 in G. Show that the map $\sigma \mapsto \sigma|E$ is an isomorphism from H to the group $\mathbf{SU}(\Phi)$, which is isomorphic to $\mathbf{SU}(3, \mathbf{C})$. The map $\sigma \mapsto \sigma(e_4)$ induces an embedding of G/H in the sphere of V, which is isomorphic to $\mathbf{S}_6$.

c) Let T be the torus in H consisting of the automorphisms σ such that $\sigma(e_0) = e_0$, $\sigma(e_1) = \alpha e_1$, $\sigma(e_2) = \beta e_2$, $\sigma(e_3) = \gamma e_3$, $\sigma(e_4) = e_4$, $\sigma(e_5) = \bar{\alpha} e_5$, $\sigma(e_6) = \bar{\beta} e_6$, $\sigma(e_7) = \bar{\gamma} e_7$, where α, β, γ are three complex numbers of modulus 1 such that $\alpha\beta\gamma = 1$. Let N be the normalizer of T in G; show that N/T is of order 12 (note that each element of N must stabilize the set of the $\pm e_i$, $i \neq 0$); deduce that G is of rank 2.

d) Show that G is connected semi-simple of type G_2 and that G/H can be identified with $\mathbf{S}_6$ (show that $G_0 \neq H$; deduce that G_0 is of type G_2, then use a dimension argument). Every compact group of type G_2 is isomorphic to G.

8) Let G be an almost simple, connected, compact Lie group of type A_n, B_n, C_n, D_n or G_2. Prove that $\pi_2(G) = 0$ and $\pi_3(G) = \mathbf{Z}$ (use Exerc. 2 to 7 and the fact that $\pi_i(\mathbf{S}_n)$ is zero for $i < n$ and cyclic for $i = n$ (cf. *General Topology*, Chap. XI).

9) Let $\mathfrak{g}$ be a compact Lie algebra, $\mathfrak{t}$ a Cartan subalgebra of $\mathfrak{g}$; let $(X_\alpha)_{\alpha\in R}$ be a Chevalley system in the split reductive algebra $(\mathfrak{g}_\mathbf{C}, \mathfrak{t}_\mathbf{C})$ such that X_α and $X_{-\alpha}$ are conjugate (with respect to $\mathfrak{g}$) for all $\alpha \in R$.

a) Let $\mathscr{T}$ be a $\mathbf{Z}$-submodule of $\mathfrak{t}_\mathbf{C}$ containing the iH_α ($\alpha \in R$) and such that $\alpha(\mathscr{T}) \subset \mathbf{Z}i$ for all $\alpha \in R$. Show that the $\mathbf{Z}$-submodule $\mathscr{G}$ of $\mathfrak{g}_\mathbf{C}$ generated by $\mathscr{T}$ and the elements u_α and v_α ($\alpha \in R$) is a $\mathbf{Z}$-Lie subalgebra of $\mathfrak{g}$.

b) Assume that $\mathfrak{g}$ (resp. $\mathfrak{t}$) is the Lie algebra of a compact group G (resp. of a maximal torus T of G). Let $\Gamma(T)$ be the kernel of the homomorphism $\exp_T : \mathfrak{t} \to T$. Show that the $\mathbf{Z}$-module $(2\pi)^{-1}\Gamma(T)$ satisfies the hypotheses of *a*).

c) Let $\langle\,,\rangle$ be an invariant scalar product on $\mathfrak{g}$; let μ (resp. τ) be the Haar measure on $\mathfrak{g}$ (resp. $\mathfrak{t}$) corresponding to the Lebesgue measure when $\mathfrak{g}$ (resp. $\mathfrak{t}$) is identified with a space $\mathbf{R}^n$ by means of an orthonormal basis. Denote also by μ (resp. τ) the measure on $\mathfrak{g}/\mathscr{G}$ (resp. $\mathfrak{t}/\mathscr{T}$) that is the quotient of μ (resp. τ) by the normalized Haar measure on $\mathscr{G}$ (resp. $\mathscr{T}$). Prove the formula $\mu(\mathfrak{g}/\mathscr{G}) = \tau(\mathfrak{t}/\mathscr{T}). \prod_{\alpha\in R_+} \langle iH_\alpha, iH_\alpha\rangle$.

§4

1) Take G to be one of the groups $\mathbf{SU}(n, \mathbf{C})$ or $\mathbf{U}(n, \mathbf{H})$; identify $\mathfrak{g}_\mathbf{C}$ with $\mathfrak{sl}(n, \mathbf{C})$ or $\mathfrak{sp}(2n, \mathbf{C})$, respectively, cf. §3, no. 4 and Exerc. 3. We use the notations of Chap. VIII, §13, nos. 1 and 3, with $k = \mathbf{C}$.

a) Show that the subgroup T of G consisting of the diagonal matrices with complex coefficients is a maximal torus, and that $L(T)_{(\mathbf{C})} = \mathfrak{h}$.

b) Identify X(T) with a subgroup of $\mathfrak{h}^*$ by means of the homomorphism δ. Show that the linear forms ε_i and the dominant weights ϖ_j belong to X(T). If $t = \mathrm{diag}(t_1, \ldots, t_n) \in T$, then $\varepsilon_i(t) = t_i$ and $\varpi_j(t) = t_1 \ldots t_j$ for $1 \leq i, j \leq n$.

c) Deduce from *b*) another proof of the fact that the groups $\mathbf{SU}(n, \mathbf{C})$ and $\mathbf{U}(n, \mathbf{H})$ are simply-connected (cf. §3, Exerc. 4 and 6).

2) Take $\mathrm{G} = \mathbf{SO}(n, \mathbf{R})$, with $n \geq 3$; put $n = 2l + 1$ if n is odd, and $n = 2l$ if n is even. The algebra $\mathfrak{g}_{\mathbf{C}}$ can be identified with $\mathfrak{o}(n, \mathbf{C})$; we use the notations of Chap. VIII, §13, nos. 2 and 4. Denote by $(f_i)_{1 \leq i \leq n}$ the canonical basis of $\mathbf{R}^n$. Put $e_j = \frac{1}{\sqrt{2}}(f_{2j-1} + if_{2j})$ and $e_{-j} = \frac{1}{\sqrt{2}}(f_{2j-1} - if_{2j})$ for $1 \leq j \leq l$, and $e_0 = i\sqrt{2} f_{2l+1}$ if n is odd; choose the Witt basis of $\mathbf{C}^n$ given by $e_1, \ldots, e_l, e_{-l}, \ldots, e_{-1}$ if n is even (resp. $e_1, \ldots, e_l, e_0, e_{-l}, \ldots, e_{-1}$ if n is odd).

a) Let H_i be the subspace of $\mathbf{R}^n$ generated by f_{2i-1} and f_{2i} $(1 \leq i \leq l)$; show that the subgroup of G consisting of the elements g such that $g(\mathrm{H}_i) \subset \mathrm{H}_i$ and $\det(g|\mathrm{H}_i) = 1$ for $1 \leq i \leq l$ is a maximal torus T of G, and that $\mathrm{L(T)}_{(\mathbf{C})} = \mathfrak{h}$.

b) Identify X(T) with a subgroup of $\mathfrak{h}^*$ by means of δ. Show that the linear forms ε_i belong to X(T); if $t \in \mathrm{T}$ and if the restriction of t to H_j is a rotation through an angle θ_j, then $\varepsilon_j(t) = e^{i\theta_j}$. The weights $\varpi_1, \ldots, \varpi_{l-2}$; $2\varpi_{l-1}, 2\varpi_l, \varpi_{l-1} \pm \varpi_l$ belong to X(T). If n is odd, ϖ_{l-1} belongs to X(T).

c) Let $\tilde{\mathrm{G}} = \mathbf{Spin}(n, \mathbf{R})$ and let $\varphi : \tilde{\mathrm{G}} \to \mathrm{G}$ be the canonical covering. For $\boldsymbol{\theta} = (\theta_1, \ldots, \theta_l) \in \mathbf{R}^l$, put $t(\boldsymbol{\theta}) = \prod_{i=1}^{l} (\cos\theta_i - f_{2i-1} f_{2i} \sin\theta_i) \in \tilde{\mathrm{G}}$. Show that the set of the $t(\boldsymbol{\theta})$ for $\boldsymbol{\theta} \in \mathbf{R}^l$ is a maximal torus $\tilde{\mathrm{T}}$ of $\tilde{\mathrm{G}}$, such that $\varphi(\tilde{\mathrm{T}}) = \mathrm{T}$. If $\mathrm{X}(\tilde{\mathrm{T}})$ is identified with a subgroup of $\mathfrak{h}^*$, we have $\varepsilon_j(t(\boldsymbol{\theta})) = e^{2i\theta_j}$.

d) Show that the weights ϖ_{l-1} and ϖ_l belong to $\mathrm{X}(\tilde{\mathrm{T}})$; deduce that $\mathbf{Spin}(n,\mathbf{R})$ is simply-connected (cf. §3, Exerc. 5).

3) *a*) Show that the automorphism $\sigma : \mathrm{A} \mapsto \overline{\mathrm{A}}$ of $\mathbf{SU}(n, \mathbf{C})$ is not inner for $n \geq 3$. Every non-inner automorphism of $\mathbf{SU}(n, \mathbf{C})$ is of the form $(\mathrm{Int}\, g) \circ \sigma$, $g \in \mathbf{SU}(n, \mathbf{C})$.

b) Show that for every almost simple connected compact group G of type A_n $(n \geq 2)$, the group Aut(G)/Int(G) is cyclic of order 2 (cf. §3, Exerc. 4).

4) Let n be an integer ≥ 2.

a) Let $g \in \mathbf{O}(2n, \mathbf{R})$ with $\det g = -1$; show that the automorphism $\mathrm{Int}\, g$ of $\mathbf{O}(2n, \mathbf{R})$ induces an automorphism of $\mathbf{SO}(2n, \mathbf{R})$ that is not inner.

b) For $n \geq 2$, the group $\mathrm{Aut}(\mathbf{SO}(2n, \mathbf{R}))$ is equal to $\mathrm{Int}(\mathbf{O}(2n, \mathbf{R}))$ (which is isomorphic to $\mathbf{O}(2n, \mathbf{R})/\{\pm I_{2n}\}$).

c) Establish an analogous result for $\mathbf{Spin}(2n, \mathbf{R})$ and $\mathbf{SO}(2n, \mathbf{R})/\{\pm I_{2n}\}$. If n is even and $\neq 2$, every automorphism of the group $\mathbf{Spin}(2n, \mathbf{R})/\{1, \varepsilon\}$ (§3, Exerc. 5) is inner.

5) Let R be a reduced and irreducible root system.

a) Show that the set of roots in R of greatest length is a symmetric closed subset of R, stable under W(R).

b) Show that every non-empty subset P of R that is closed, symmetric and stable under W(R) is equal to R or to the set of roots of greatest length in R.

c) Assume that $P \neq R$. Show that the root system P is of type D_l (resp. $(A_1)^l, D_4, A_2$) if R is of type B_l (resp. C_l, F_4, G_2).

6) Let H be a closed subgroup of G containing $N_G(T)$.

a) Show that H is equal to its own normalizer in G, and that H/H_0 is isomorphic to $W/W_{H_0}(T)$.

b) Show that $R(H_0, T)$ is stable under W. Conversely, if K is a connected closed subgroup containing T such that R(K, T) is stable under W, the normalizer of K contains that of T.

c) Assume that G is almost simple. Prove that H is equal to $N_G(T)$, or to G, or that we are (up to isomorphism) in one of the following situations:
α) (resp. α')) $G = \mathbf{Spin}(2l+1, \mathbf{R})$ (resp. $G = \mathbf{SO}(2l+1, \mathbf{R})$) and H_0 is the fixer of a non-zero vector of $\mathbf{R}^{2l+1}$, isomorphic to $\mathbf{Spin}(2l, \mathbf{R})$ (resp. $\mathbf{SO}(2l, \mathbf{R})$);
β) $G = U(l, \mathbf{H})$ and H_0 is the subgroup D consisting of the diagonal matrices;
β') $G = U(l, \mathbf{H})/\{\pm 1\}$ and $H_0 = D/\{\pm 1\}$;
γ) G is of type F_4 and H_0 is isomorphic to $\mathbf{Spin}(8, \mathbf{R})$;
δ) G is of type G_2 and H_0 is the subgroup (isomorphic to $\mathbf{SU}(3, \mathbf{C})$) defined in Exerc. 7 of §3.

7) Let $\tau : G \to \mathbf{GL}(V)$ be a continuous representation of G on a finite dimensional real vector space. Assume that for all $\lambda \in X(T)$, we have $\dim_{\mathbf{C}} \tilde{V}_\lambda \leq 1$.

a) Show that the representation τ is the direct sum of a finite family $(\tau_i)_{1 \leq i \leq s}$ of irreducible representations, mutually non-isomorphic, whose commutants K_i $(1 \leq i \leq s)$ are isomorphic to $\mathbf{R}$ or $\mathbf{C}$.

b) There exists a $\mathbf{C}$-vector space structure on V for which the operations of $\tau(G)$ are $\mathbf{C}$-linear if and only if $K_1, \ldots, K_s$ are isomorphic to $\mathbf{C}$; there are then 2^s structures of this type.

8) Let H be a connected closed subgroup of G of maximum rank, $\mathfrak{h}$ its Lie algebra, X the manifold G/H and V the tangent space of X at the point corresponding to the class of H; identify V with the quotient $\mathfrak{g}/\mathfrak{h}$.

a) If j is an almost complex structure (*Differentiable and Analytic Manifolds, Results*, 8.8.3) on X, denote by $V''(j)$ the subspace of the complex vector space $\mathbf{C} \otimes V$ consisting of the elements u such that $j(u) = -iu$, and by $\mathfrak{q}(j)$ the subspace of $\mathfrak{g}_{\mathbf{C}}$ that is the inverse image of $V''(j)$, so that the canonical map $V \to \mathfrak{g}_{\mathbf{C}}/\mathfrak{q}(j)$ is a $\mathbf{C}$-linear isomorphism when V is given the $\mathbf{C}$-vector space structure induced by j. Prove that the map $j \mapsto \mathfrak{q}(j)$ is a bijection from the set of almost complex structures on X invariant under G to the set of complex subspaces $\mathfrak{p}$ of $\mathfrak{g}_{\mathbf{C}}$ satisfying the following conditions:

$$\mathfrak{p} + \bar{\mathfrak{p}} = \mathfrak{g}_{\mathbf{C}} \tag{1}$$
$$\mathfrak{p} \cap \bar{\mathfrak{p}} = \mathfrak{h}_{\mathbf{C}} \tag{2}$$
$$[\mathfrak{h}, \mathfrak{p}] \subset \mathfrak{p}. \tag{3}$$

b) There exists such a structure on X if and only if the commuting fields of the irreducible subrepresentations of the adjoint representation of H on V are all isomorphic to **C**; in that case, there are 2^s such structures, where s is the number of irreducible subrepresentations of V (use Exerc. 7).

c) Let j be an almost complex structure on X, invariant under G, and $\mathfrak{p} = \mathfrak{q}(j)$ the associated subspace. Then, j is *integrable* (that is, associated to a complex-analytic manifold structure on X, cf. *Differentiable and Analytic Manifolds, Results*, 8.8.5 to 8.8.8) if and only if $\mathfrak{p}$ satisfies the condition

$$[\mathfrak{p}, \mathfrak{p}] \subset \mathfrak{p}. \tag{4}$$

d) There exists a complex structure (that is, an integrable almost complex structure) on X invariant under G if and only if H is the centralizer of a torus in G (show that conditions (1) to (4) above imply that $\mathfrak{p}$ is a parabolic subalgebra of $\mathfrak{g}_{\mathbf{C}}$ (Chap. VIII, §3, no. 5)); in that case, these complex structures correspond bijectively to the parabolic subalgebras $\mathfrak{p}$ of $\mathfrak{g}_{\mathbf{C}}$ that are the direct sum of $\mathfrak{h}_{\mathbf{C}}$ and their unipotent radical (cf. Chap. VIII, §3, no. 4).

e) Prove that there exist exactly Card(W) complex structures on G/T invariant under G; if σ and σ' are two such structures, there exists a unique element $w \in$ W such that the canonical operation of w on G/T (by inner automorphisms) transforms σ into σ'. If w is a non-identity element of W, the operation of w on G/T is not **C**-analytic for any complex structure on G/T invariant under G.

f) Determine the complex structures on $\mathbf{S}^2$ invariant under $\mathbf{SO}(3)$.

g)* With the notations of Exerc. 8 *d*), let G_c be the complexification of G and P the complex Lie subgroup of G_c with Lie algebra $\mathfrak{p}$. Show that the canonical map $G/H \to G_c/P$ is an isomorphism of complex-analytic manifolds.*

9) Let H be a connected closed subgroup of G, of maximum rank, distinct from G and maximal for these properties. Denote by Z the quotient group C(H)/C(G).

a) We have $\dim Z \leq 1$; if $\dim Z = 0$, then Z is of order 2, 3 or 5 (reduce to the case in which G is almost simple, with trivial centre; apply Chap. VI, §4, Exerc. 4).

b) Assume that G is almost simple and that Z is of order 3 or 5; put $z =$ Card(Z) and $\pi =$ Card$(\pi_1(\mathrm{H}))$/Card$(\pi_1(\mathrm{G}))$. Show that we must be in one of the following seven cases:
(i) G is of type G_2, H is of type A_2, and $z = 3, \pi = 1$;
(ii) G is of type F_4, H is of type $A_2 \times A_2$, and $z = \pi = 3$;
(iii) G is of type E_6, H is of type $A_2 \times A_2 \times A_2$, and $z = \pi = 3$;

(iv) G is of type E_7, H is of type $A_2 \times A_5$, and $z = \pi = 3$;
(v) G is of type E_8, H is of type A_8, and $z = \pi = 3$;
(vi) G is of type E_8, H is of type $A_2 \times E_6$, and $z = \pi = 3$;
(vii) G is of type E_8, H is of type $A_4 \times A_4$, and $z = \pi = 5$.
(Use *loc. cit.* and the plates of Chap. VI; to calculate π, remark that if f and f' denote the connection indices of G and H, respectively, then $z\pi f = f'$.)

c) In each of the preceding cases, determine the group H.

10) We retain the notations of the preceding exercise.

a) Assume that $\dim Z = 1$. Then there are exactly two complex structures on G/H invariant under G; there exists an automorphism of G that leaves H stable and interchanges these two structures (use Exerc. 8).

b) Determine the complex structures on $\mathbf{P}_n(\mathbf{C})$ invariant under $\mathbf{SU}(n+1, \mathbf{C})$.

c) Assume that $\dim Z = 0$ and $\mathrm{Card}(Z) \neq 2$. Show that there exist almost complex structures on G/H invariant under G (if z is an element of C(H) that is not central in G, $\mathrm{Int}\, z$ induces an automorphism of G/H of odd order (Exerc. 9); use Exerc. 8 *b*)). Show that there is no complex structure on G/H invariant under G (use Exerc. 8 *d*)).

d) There is no complex structure on $\mathbf{S}_6$ invariant under $\mathbf{SO}(7, \mathbf{R})$ (use Exerc. 7 of §3).

11) We retain the notations of Exerc. 9 and 10.

a) The homogeneous space G/H is symmetric (§1, Exerc. 8) if and only if Z is of order 2 or of dimension 1. The symmetric space G/H is then irreducible (*loc. cit.*).

b) Assume that $\dim Z = 1$; denote by X the complex-analytic manifold G/H. Show that we are then in one of the following situations:
(i) G is of type A_l and D(H) of type $A_{p-1} \times A_{l-p}$; X is isomorphic to the grassmannian $\mathbf{G}_p(\mathbf{C}^{l+1})$.
(ii) G is of type B_l and D(H) of type B_{l-1}; X is isomorphic to the submanifold of $\mathbf{P}_{2l}(\mathbf{C})$ defined by the vanishing of a separating quadratic form (*smooth projective quadric*).
(ii′) G is of type D_l and D(H) of type D_{l-1}; X is isomorphic to a smooth projective quadric in $\mathbf{P}_{2l-1}(\mathbf{C})$.
(iii) G is of type C_l and D(H) of type A_{l-1}; X is isomorphic to the submanifold of $\mathbf{G}_l(\mathbf{C}^{2l})$ consisting of the maximal isotropic subspaces for the usual alternating bilinear form on $\mathbf{C}^{2l}$.
(iv) G is of type D_l and D(H) of type A_{l-1}; X is isomorphic to the submanifold of $\mathbf{G}_l(\mathbf{C}^{2l})$ consisting of the maximal isotropic subspaces for the usual symmetric bilinear form on $\mathbf{C}^{2l}$.
(v) G is of type E_6 and D(H) of type D_5.
(vi) G is of type E_7 and D(H) of type E_6.

c) Assume that $\mathrm{Card}(Z) = 2$; give the list of possible situations. If G is of type A_l, B_l, C_l or D_l, show that the real manifold G/H is isomorphic to a grassmannian manifold $G_p(K^q)$, with $K = \mathbf{R}, \mathbf{C}$ or $\mathbf{H}$.

¶ 12) Assume that G is simply-connected; for all $\alpha \in R(G,T)$, put $t_\alpha = \exp(\frac{1}{2}K_\alpha)$. Put $N = N_G(T)$, and denote by $\varphi : N \to W$ the canonical map. Let B be a basis of R(G, T); choose an element n_α of $(N \cap S_\alpha) - (T \cap S_\alpha)$ for all $\alpha \in B$.

a) Show that $\varphi(n_\alpha) = s_\alpha$ and $n_\alpha^2 = t_\alpha$, so that $n_\alpha^4 = 1$.

b) Let α and β be two distinct elements of B, and let $m_{\alpha\beta}$ be the order of $s_\alpha s_\beta$ in W. Prove that

$$\begin{aligned} n_\alpha n_\beta &= n_\beta n_\alpha && \text{if } m_{\alpha\beta} = 2 \\ n_\alpha n_\beta n_\alpha &= n_\beta n_\alpha n_\beta && \text{if } m_{\alpha\beta} = 3 \\ (n_\alpha n_\beta)^2 &= (n_\beta n_\alpha)^2 && \text{if } m_{\alpha\beta} = 4 \\ (n_\alpha n_\beta)^3 &= (n_\beta n_\alpha)^3 && \text{if } m_{\alpha\beta} = 6 \end{aligned}$$

(if for example $m_{\alpha\beta} = 3$, then $(s_\alpha s_\beta)s_\alpha(s_\alpha s_\beta)^{-1} = s_\beta$ and $s_\alpha s_\beta(\alpha) = \beta$; show that $n_\alpha n_\beta n_\alpha n_\beta^{-1} n_\alpha^{-1} n_\beta^{-1}$ belongs to S_β, and conclude by remarking that $S_\alpha \cap S_\beta \cap T = \{e\}$).

c) Deduce from b) that there exists a unique section $\nu : W \to N$ of φ such that $\nu(s_\alpha) = n_\alpha$ and $\nu(ww') = \nu(w)\nu(w')$ if $l(ww') = l(w) + l(w')$ (where $l(w)$ denotes the length of w with respect to the generating system $(s_\alpha)_{\alpha \in B}$; cf. Chap. VI, §1, no. 5, Prop. 5). Put $n_w = \nu(w)$.

d) Let W^* be the subgroup of N generated by the n_α; show that $W^* \cap T$ is the subgroup T_2 of T consisting of the elements of order ≤ 2 and that W can be identified with W^*/T_2.

e) Let $w \in W$ be such that $w^2 = 1$; show that $n_w^2 = \prod_{\alpha \in R_w} t_\alpha$, where R_w is the set of positive roots α such that $w(\alpha) < 0$ (write $w = s_{\alpha_1} \ldots s_{\alpha_r}$, with $r = l(w)$ and $\alpha_i \in B$; apply c) and Chap. VI, §1, no. 6, Cor. 2 of Prop. 17).

f) Assume that G is almost simple. Let c be a Coxeter transformation in W and h the Coxeter number of W (Chap. VI, §1, no. 11). Show that $n_c^h = \prod_{\alpha \in R_+} t_\alpha$ (use c), e) and Exerc. 2 of Chap. V, §6).

13) a) Choose a basis B of R(G, T), and denote by R_+ the set of positive roots of R(G, T). Prove that $z_G = \prod_{\alpha \in R_+} \exp(\frac{1}{2}K_\alpha)$ is an element of C(G), independent of the choice of T and of B; we have $z_G^2 = e$.

b) Let H be another connected compact Lie group. Then $z_{G \times H} = (z_G, z_H)$; prove that, if $f : G \to H$ is a surjective morphism of Lie groups, then $f(z_G) = z_H$.

c) Put $R = R(G,T)$; assume that G is simply-connected, so that X(T) can be identified with P(R). Show that the kernel of the homomorphism $\chi \mapsto \chi(z_G)$

from X(T) to $\{1, -1\}$ is the subgroup P′(R) defined in Exerc. 8 of Chap. VI, §1.

d) If $G = \mathbf{SU}(n, \mathbf{C})$, then $z_G = (-1)^{n+1} I_n$;
if $G = \mathbf{SU}(n, \mathbf{H})$, then $z_G = -I_n$;
if $G = \mathbf{Spin}(n, \mathbf{R})$ with $n \equiv 3, 4, 5$ or $6 \pmod 8$, then $z_G = -1$;
if $G = \mathbf{Spin}(n, \mathbf{R})$ with $n \equiv 0, 1, 2, 7 \pmod 8$, then $z_G = 1$;
if G is of type E_6, E_8, F_4 or G_2, then $z_G = e$;
if G is simply-connected of type E_7, then z_G is the unique non-identity element of C(G).
(Use Chap. VI, §4, Exerc. 5.)

14) Assume that the group G is *almost simple*; denote the Coxeter number of R(G, T) by h (Chap. VI, §1, no. 11). An element g of G is said to be a *Coxeter element* if there exists a maximal torus S of G such that g belongs to $N_G(S)$ and its class in $W_G(S)$ is a Coxeter transformation (*loc. cit.*).

a) Show that any two Coxeter elements are conjugate (argue as in the proof of the Cor. of Prop. 10, no. 5).

b) Every Coxeter element g is regular and satisfies $g^h = z_G$, where z_G is the element of C(G) defined in Exerc. 13; in particular, g is of order h or $2h$ according as z_G is or is not equal to e (use Exerc. 12 *f*)).

c) An element $g \in G$ is a Coxeter element if and only if the automorphism $\operatorname{Ad} g \otimes 1_{\mathbf{C}}$ of $\mathfrak{g}_{\mathbf{C}}$ satisfies the equivalent conditions of Chap. VIII, §5, Exerc. 5 *f*).

d) Show that every regular element g of G such that $g^h \in C(G)$ is a Coxeter element; for $p < h$, there is no regular element k such that $k^p \in C(G)$.

15) Let H be a connected closed subgroup of G. Then H is said to be *clean* if it is not contained in any connected closed subgroup of maximal rank distinct from G.

a) Show that H is clean if and only if its centralizer in G is equal to C(G). In particular, if H is clean then $C(H) = C(G) \cap H$.

b) Assume from now on that H is clean. Show that, for every maximal torus S of H, we have $C(H) = S \cap C(G)$.

c) Let H′ be a connected closed subgroup of G containing H. Prove that $\operatorname{rk} H < \operatorname{rk} H'$, and deduce that H is clean in H′.

d) Let K be a connected closed subgroup of H, clean in H and containing a regular element of G. Show that K is clean in G.

16) Let H be a connected closed subgroup of G such that $T \cap H$ is a maximal torus S of H. Let $\lambda \in R(H, S)$; denote by $R(\lambda)$ the set of roots of R(G, T) whose restriction to S is equal to λ.

a) Show that $R(\lambda)$ is not empty.

b) Let $w \in W_H(S)$; show that there exists an element $\bar{w}$ of $W_G(T)$ such that $R(w\lambda) = \bar{w}R(\lambda)$ (use Exerc. 4 of §2).

c) Let $P(\lambda)$ be the intersection of $R(G, T)$ with the subgroup of $X(T)$ generated by $R(\lambda)$. Show that there exists a closed subgroup G_λ of G containing T such that $P(\lambda) = R(G_\lambda, T)$. Deduce that the reflection $s_\lambda \in W_H(S)$ is the restriction to S of a product of reflections s_α with $\alpha \in P(\lambda)$.

d) Show that the nodal vector $K_\lambda \in L(S)$ associated to λ is a linear combination with integer coefficients of the K_α for $\alpha \in R(\lambda)$.

e) Let B_H be a basis of $R(H, S)$. Show that $R(\lambda)$ is contained in the subgroup of $X(T)$ generated by the union of the $R(\mu)$ for $\mu \in B_H$ (prove that the set of $\lambda \in R(H, S)$ having the stated property is stable under s_μ for all $\mu \in B_H$, and use c)).

¶ 17) We retain the notations of the preceding exercise; assume further that the subgroup H is *clean* (Exerc. 15).

a) Show that $R(G, T)$ is contained in the subgroup of $X(T)$ generated by the union of the $R(\lambda)$ for $\lambda \in B_H$.

b) Let Δ be the identity component of the subgroup of S consisting of the $s \in S$ such that $\lambda(s) = \mu(s)$ for all λ, μ in B_H. Show that there exists a basis $\{\alpha_1, \dots, \alpha_l\}$ of $R(G, T)$ and an integer k, with $0 \leq k \leq l - 1$, such that Δ is the identity component of the set of $t \in T$ satisfying

$$\alpha_1(t) = \cdots = \alpha_k(t) = 1, \ \alpha_{k+1}(t) = \cdots = \alpha_l(t).$$

(Let x be the element of $L(S)$ such that $\delta(\lambda)(x) = 2\pi i$ for all $\lambda \in B_H$; deduce from a) that $\exp x \in C(G)$. Choose $\{\alpha_1, \dots, \alpha_l\}$ so that $i\delta(\alpha_j)(x)$ is zero for $1 \leq j \leq k$ and < 0 for $k + 1 \leq j \leq l$; then show that for all $\lambda \in B_H$ every root in $R(\lambda)$ can be written

$$\alpha_j + n_1\alpha_1 + \cdots + n_k\alpha_k,$$

with $j \geq k + 1$ and $n_i \in \mathbf{N}$. Conclude by using a).)

c) Show that the integer k does not depend on the choice of the tori S, T or of the bases B_H, $\{\alpha_1, \dots, \alpha_l\}$; it is equal to the rank of the derived group of $Z_G(\Delta)$.

18) We retain the notations of Exerc. 16 and 17. The subgroup H is said to be *principal* if it is clean and if the sub-torus Δ contains a regular element (in other words, if $k = 0$).

a) Let K be a connected closed subgroup of H. Show that K is a principal subgroup of G if and only if K is principal in H and H is principal in G (use Exerc. 15 c) and d)).

b) Assume from now on that H is principal. Show that the union of the $R(\mu)$, for $\mu \in B_H$, is a basis of $R(G, T)$.

c) Let $\alpha \in R(G, T)$. Prove that the restriction of α to S is a non-zero integer multiple of a root $\lambda \in R(H, S)$; the root α is a sum of elements of $R(\lambda)$ (show that the intersection of $L(S)$ with the hyperplane $\delta(\alpha) = 0$ is a wall of $L(S)$).

d) Let $\lambda \in \mathrm{R(H, S)}$; show that the nodal vector K_λ is a linear combination with non-negative integer coefficients of the K_α for $\alpha \in \mathrm{R}(\lambda)$ (cf. Exerc. 16 *d*) and Chap. V, §3, no. 5, Lemma 6).

19) We retain the notations of Exerc. 16 to 18; assume that the subgroup H is principal. Give $\mathrm{X(T)} \otimes \mathbf{R}$ (resp. $\mathrm{X(S)} \otimes \mathbf{R}$) a scalar product invariant under $\mathrm{W_G(T)}$ (resp. under $\mathrm{W_H(S)}$). Let λ, μ be two roots in $\mathrm{B_H}$.

a) Show that, if λ and μ are orthogonal, the sets $\mathrm{R}(\lambda)$ and $\mathrm{R}(\mu)$ are orthogonal. Deduce that if G is almost simple then so is H.

b) Assume from now on that $n(\lambda, \mu) = -1$. Show that there exists a surjective map $u : \mathrm{R}(\lambda) \to \mathrm{R}(\mu)$ such that, for $\alpha \in \mathrm{R}(\lambda)$, $u(\alpha)$ is the unique root of $\mathrm{R}(\mu)$ linked (that is, not orthogonal) to α; we have $K_\mu = \sum_{\beta \in \mathrm{R}(\mu)} K_\beta$, and the roots in $\mathrm{R}(\mu)$ are pairwise orthogonal (write K_μ and K_λ as linear combinations of the K_α, then identify the coefficients).

c) If $n(\mu, \lambda) = -1$, the map u is bijective, and the only pairs of linked roots in $\mathrm{R}(\lambda) \cup \mathrm{R}(\mu)$ are the pairs $(\alpha, u(\alpha))$ for $\alpha \in \mathrm{R}(\lambda)$.

d) Assume that $n(\mu, \lambda) = -2$; let $\beta \in \mathrm{R}(\mu)$. Show that $u^{-1}(\beta)$ contains either one or two elements; if $u^{-1}(\beta) = \{\alpha_1, \alpha_2\}$, the root α_i $(i = 1, 2)$ is orthogonal to the other roots in $\mathrm{R}(\lambda)$, and has the same length as β; if $u^{-1}(\beta) = \{\alpha\}$ and $\|\alpha\| = \|\beta\|$, α is linked to a unique root $\alpha' \in \mathrm{R}(\lambda)$, of the same length as α, and $\{\alpha, \alpha'\}$ is orthogonal to the remainder of $\mathrm{R}(\lambda)$; if $u^{-1}(\beta) = \{\alpha\}$ and $\|\alpha\| \neq \|\beta\|$, α is orthogonal to the other roots in $\mathrm{R}(\lambda)$.

e) Study similarly the case $n(\mu, \lambda) = -3$.

20) Assume that the group G is almost simple; let H be a principal subgroup of G, of rank ≥ 2.

a) Show that H is semi-simple of type $\mathrm{B}_h, \mathrm{C}_h, \mathrm{F}_4$ or G_2 (use Exerc. 19).

b) Assume that H is of rank ≥ 3. Show that G is of type $\mathrm{A}_l, \mathrm{D}_l, \mathrm{E}_6, \mathrm{E}_7$ or E_8 (consider the terminal vertices of the Dynkin graph of R(G, T), and apply Exerc. 19).

c) Show that, if $\mathrm{rk\,H} \geq 3$, we are in one of the following situations:

G is of type A_{2l} $(l \geq 3)$ and H of type B_l;
G is of type A_{2l-1} $(l \geq 3)$ and H of type C_l;
G is of type D_l $(l \geq 4)$ and H of type B_{l-1};
G is of type E_6 and H of type F_4.

d) Show that, if H is of type B_2, then G is of type A_3 or A_4.

e) Show that, if H is of type G_2, then G is of type $\mathrm{B}_3, \mathrm{D}_4$ or A_6.

(For a more explicit description of these situations, see §5, Exerc. 5.)

f) Let K be a connected closed subgroup of G containing H; show that either K = G or K = H, or H is of type G_2, K of type B_3 and G of type D_4 or A_6.

21) *a*) Let H be a connected closed subgroup of G of rank 1. Show that the following conditions are equivalent:
(i) H is principal;
(ii) H is clean and contains a regular element of G;
(iii) there exists a principal $\mathfrak{sl}_2$-triplet (x, h, y) in $\mathfrak{g}_{\mathbf{C}}$ (Chap. VIII, §11, no. 4) such that

$$\mathrm{L(H)}_{(\mathbf{C})} = \mathbf{C}x + \mathbf{C}h + \mathbf{C}y.$$

b) Show that G contains a principal subgroup of rank 1, and that any two such subgroups are conjugate (with the notations of Exerc. 17, remark that $\mathrm{S} = \varDelta$).

c) Show that a connected closed subgroup of G is principal if and only if it contains a principal subgroup of G of rank 1 (use Exerc. 18 *a*)).

22) Let H be a principal subgroup of G of rank 1; let $\varGamma$ be the subgroup of Aut(G) consisting of the automorphisms u such that $u(\mathrm{H}) = \mathrm{H}$. Show that $\mathrm{Aut(G)} = \varGamma.\mathrm{Int(G)}$.

§5

1) Assume that G is simply-connected; the *alcoves* of G are the subsets of G of the form exp(A), where A is an alcove of a Cartan subalgebra of $\mathfrak{g}$.

a) Show that the alcoves of G form a partition of G_r.

b) Every alcove of G is contained in a unique maximal torus.

c) Show that the alcoves of G contained in T form a partition of T_r, and that the set of such alcoves is a principal homogeneous space under W.

d) Every conjugacy class of regular elements of G has a unique point in each alcove.

e) Let E be an alcove of G. Show that E is a contractible space; if $\mathfrak{g}$ is simple, E is homeomorphic to an open simplex in a euclidean space.

2) *a*) Let E be a simply-connected topological space and $u : \mathrm{E} \to \mathrm{G}$ a continuous map such that $u(\mathrm{E}) \subset \mathrm{G}_r$. Show that u is homotopic (*General Topology*, Chap. XI) to the constant map with value e (consider the covering $\varphi_r : (\mathrm{G/T}) \times \mathfrak{t}_r \to \mathrm{G}_r$; lift u to $\tilde{u} : \mathrm{E} \to (\mathrm{G/T}) \times \mathfrak{t}$, then use the fact that $\mathfrak{t}$ is contractible).

**b*) Prove that the group $\pi_2(\mathrm{G})$ is zero.
(Let $u : \mathbf{S}_2 \to \mathrm{G}$ be a map of class C^∞; by using Prop. 1 and the transversality theorem, show that u is homotopic to a map with image contained in G_r, then apply *a*).)$_*$

3) Let $f : \tilde{\mathrm{G}} \to \mathrm{G}$ be a universal covering of G and π the kernel of f (isomorphic to $\pi_1(\mathrm{G})$); put $\mathrm{C} = \mathrm{C(G)}$ and $\tilde{\mathrm{C}} = \mathrm{C}(\tilde{\mathrm{G}})$. Let σ be an automorphism of

G, $\tilde{\sigma}$ the automorphism of $\tilde{G}$ induced by σ. Denote by $G_\sigma, C_\sigma, \tilde{G}_\sigma, \tilde{C}_\sigma$ the set of points of $G, C, \tilde{G}, \tilde{C}$ fixed by $\sigma, \sigma, \tilde{\sigma}, \tilde{\sigma}$, respectively.

a) Show that the identity component $(G_\sigma)_0$ of G_σ is $f(\tilde{G}_\sigma)$.

b) Denote by s the endomorphism of the **Z**-module π induced by $\tilde{\sigma}$. Show that the quotient group $G_\sigma/(G_\sigma)_0$ is isomorphic to a subgroup of $\mathrm{Coker}(1-s)$, and in particular is commutative. If $1-s$ is surjective, G_σ is connected.

c) Let $n \in \mathbf{N}$ be such that $s^n = \mathrm{Id}_\pi$. Show that the group $G_\sigma/(G_\sigma)_0$ can be identified with a subgroup of the quotient $\mathrm{Ker}(1+s+\cdots+s^{n-1})/\mathrm{Im}(1-s)$; deduce that it is annihilated by n. If n is prime to the order of the torsion subgroup of π, then G_σ is connected.

d) Recover the results of *b*) and *c*) by using Exerc. 23 of *Algèbre*, Chap. X, p. 194.

e) Show that G_σ is connected in each of the following cases:
(i) G is semi-simple of type A_{2n} ($n \geq 1$) and σ is not inner;
(ii) G is semi-simple of type E_6 and σ is not inner;
(iii) G is semi-simple of type D_4 and σ is a *triality* (that is, of order 3 modulo Int(G)).

f) Define an isomorphism from $C_\sigma/(C_\sigma \cap (G_\sigma)_0)$ to $((1-\tilde{\sigma})\tilde{C} \cap \pi)/(1-\tilde{\sigma})\pi$. Deduce that if $1-s$ is not surjective and $\pi \subset (1-\tilde{\sigma})C$, then C_σ is not contained in $(G_\sigma)_0$. For $G = \mathbf{SO}(2n, \mathbf{R})$ and σ not inner, we have $-I_{2n} \notin (G_\sigma)_0$ and G_σ is not connected.

4) Assume that G is semi-simple. Let e be a framing of G and Φ a group of automorphisms of G respecting this framing; denote by H the subgroup of G consisting of the elements fixed by Φ.

a) Show that H_0 is semi-simple; if G is simply-connected, H is connected (reduce to the case in which G is almost simple and Φ is cyclic, and apply Th. 1 (no. 3) and Chap. VIII, §5, Exerc. 13).

b) Assume that G is almost simple. Show that
if G is of type A_{2l} ($l \geq 1$) and Φ is of order 2, H_0 is of type B_l;
if G is of type A_{2l-1} ($l \geq 2$) and Φ is of order 2, H_0 is of type C_l;
if G is of type D_l ($l \geq 4$) and Φ is of order 2, H_0 is of type B_{l-1};
if G is of type D_4 and Φ is of order 3 or 6, H_0 is of type G_2;
if G is of type E_6 and Φ is of order 2, H_0 is of type F_4.
In each case, determine the groups $\pi_i(H)$, $i = 0, 1$ (cf. Exerc. 3).

c) Prove that the subgroup H_0 is principal (§4, Exerc. 18).

d) Prove that a semi-simple group of type B_3 or A_6 contains a principal subgroup of type G_2 (use *b*) and Exerc. 20 of §4).

5) Assume that the group G is almost simple; let H be a principal connected closed subgroup of G (§4, Exerc. 18). Denote by Φ the group of automorphisms u of G that fix H, and by F the subgroup of elements of G fixed by Φ.

a) Show that there exists a framing of G stable under Φ.

b) Show that we must be in one of the following situations:
(i) $H = F_0$;
(ii) G is of type B_3, H of type G_2 and Φ reduces to the identity;
(iii) G is of type A_6, H of type G_2 and Φ of order 2.
(Use Exerc. 4 and Exerc. 20 of §4.)

¶ 6) Assume that G is simply-connected. Let p be a prime number and g an element of C(G) such that $g^p = e$.

a) Show that there exist elements $u \in T$ and $w \in W$ such that
(i) $w(u)u^{-1} = g$;
(ii) $w^p = 1$;
(iii) $u^p = e$ if $p \neq 2$, $u^p = e$ or g if $p = 2$.
(Let A be an alcove in t; for $i = 0, 1, \dots, p-1$, let $x_i \in \overline{A}$ be such that $\exp x_i = g^i$. Take u to be the element $\exp x$, where x is the barycentre of the facet of $\overline{A}$ whose vertices are the x_i, and w to be the element of W such that $w(A) = A - x$.)

b) Prove that there exist $u \in T$ and $v \in N_G(T)$ such that
(i) $vuv^{-1}u^{-1} = g$;
(ii) $v^p = e$ if $p \neq 2$, $v^p = e$ or g if $p = 2$;
(iii) $u^p = e$ if $p \neq 2$, $u^p = e$ or g if $p = 2$.
(Use the construction in *a*), lifting w to n_w as in Exerc. 12 of §4; take $v = n_w^{p+1}$ if $p \neq 2$, $v = n_w$ if $p = 2$.)

7) Let p be a prime number. Show that the following conditions are equivalent:
(i) p does not divide the order of the torsion subgroup of $\pi_1(G)$;
(ii) for every element g of G of order p, the centralizer of g in G is connected;
(iii) every subgroup of G isomorphic to $(\mathbf{Z}/p\mathbf{Z})^2$ is contained in a maximal torus.
(To prove that (i) $\Longrightarrow$ (ii), use Exerc. 3 *c*); to prove that (iii) $\Longrightarrow$ (i), use Exerc. 6.)

8) Take $G = \mathbf{SO}(8, \mathbf{R})/\{\pm I_8\}$; denote by $(\alpha_i)_{1 \leq i \leq 4}$ a basis of R(G, T) such that α_1, α_3 and α_4 are not orthogonal to α_2 (Chap. VI, Plate IV). Let A be the subgroup of T consisting of the $t \in T$ such that

$$\alpha_1(t) = \alpha_3(t) = \alpha_4(t), \ \alpha_1(t)^2 = \alpha_2(t)^2 = 1.$$

Prove that A is isomorphic to $(\mathbf{Z}/2\mathbf{Z})^2$ and that its centralizer in G is a finite non-commutative group.

9) Let R be an irreducible root system, $R^\vee$ the inverse system, B a basis of R and α the highest root of R (relative to B). Put

$$\alpha = \sum_{\beta \in B} n_\beta \beta \quad \text{and} \quad \alpha^\vee = \sum_{\beta \in B} n_\beta^\vee \beta^\vee \ ;$$

let $\nu(R) = \sup_{\beta \in B} n_\beta^\vee$.

a) Show that the interval $(1, \nu(\mathrm{R}))$ in $\mathbf{N}$ is the union of 1 and the $n_\beta^\vee$ for $\beta \in \mathrm{B}$ (put $\alpha_0 = \alpha$; let $(\alpha_1, \ldots, \alpha_q)$ be a sequence of distinct elements of B such that α_i is not orthogonal to α_{i-1} for $i = 1, \ldots, q$, such that $n_{\alpha_q}^\vee = \nu(\mathrm{R})$ and such that q is maximal with these properties; prove that $n_{\alpha_i}^\vee = i + 1$ for $i = 1, \ldots, q$).

b) Let p be a prime number. Show that the following three properties are equivalent:
(i) $p \leq \nu(\mathrm{R})$;
(ii) there exists $\beta \in \mathrm{B}$ such that $p = n_\beta^\vee$;
(iii) there exists $\beta \in \mathrm{B}$ such that p divides $n_\beta^\vee$.
In this case p is called a *torsion prime number* of R.

c) For each type of irreducible root system, give the value of $\nu(\mathrm{R})$ and the torsion prime numbers. Show that the set of the n_β and that of the $n_\beta^\vee$ coincide except for type G_2.

d) Let R′ be a symmetric closed subset of R that is irreducible as a root system. Show that $\nu(\mathrm{R}') \leq \nu(\mathrm{R})$.

10) Let R be a root system and p a prime number. Show that the following properties are equivalent:
(i) p is a torsion prime number of an irreducible component of R (Exerc. 9 *b*));
(ii) there exists a symmetric closed subset R_1 of R, distinct from R and maximal for these properties, such that $(\mathrm{Q}(\mathrm{R}^\vee) : \mathrm{Q}(\mathrm{R}_1^\vee)) = p$ ($\mathrm{Q}(\mathrm{R}^\vee)$ denotes the $\mathbf{Z}$-module generated by the inverse roots of R and $\mathrm{Q}(\mathrm{R}_1^\vee)$ the submodule generated by the $\alpha^\vee$ for $\alpha \in \mathrm{R}_1$);
(iii) there exists a symmetric closed subset R_1 of R such that the p-torsion submodule of $\mathrm{Q}(\mathrm{R}^\vee)/\mathrm{Q}(\mathrm{R}_1^\vee)$ is non-zero.
(To prove that (i) $\Longrightarrow$ (ii), use Exerc. 4 of Chap. VI, §4; to prove that (iii) $\Longrightarrow$ (i), use Exerc. 9 *d*).)
In this case p is said to be a *torsion prime number* of R.

11) Let p be a prime number. Then p is said to be a *torsion prime number* of G if there exists a connected closed subgroup H of G, of maximum rank, such that the p-torsion subgroup of $\pi_1(\mathrm{H})$ is non-zero. Put $\mathrm{R} = \mathrm{R}(\mathrm{G}, \mathrm{T})$.

a) The torsion prime numbers of G are the torsion prime numbers of R (Exerc. 10) and the prime numbers that divide the order of the torsion group of $\pi_1(\mathrm{G})$.

b) Show that every torsion prime number of G divides $w/l!$, where w is the order of W and l the rank of the semi-simple group D(G) (use Chap. VI, §2, no. 4, Prop. 7).

c) Let $t \in \mathrm{T}$ and $n \in \mathbf{N}$ be such that $t^n \in \mathrm{C}(\mathrm{G})$. Let R_1 be the set of $\alpha \in \mathrm{R}$ such that $\alpha(t) = 1$, and m the order of the torsion group of $\mathrm{Q}(\mathrm{R}^\vee)/\mathrm{Q}(\mathrm{R}_1^\vee)$ (cf. Exerc. 10). Show that every prime factor of m divides n (reduce to the case in which G is simply-connected and almost simple).

d) Assume that G is simply-connected. Let $g \in \mathrm{G}$, $n \in \mathbf{N}$ be such that g^n belongs to C(G) and such that no torsion prime number of G divides n. Show that the derived group of the centralizer of g is simply-connected.

e) Let g be an element of G and p a prime number, such that $g^p = e$ and p is not a torsion prime number of G. Show that the centralizer $\mathrm{Z}(g)$ is connected and that p is not a torsion prime number of $\mathrm{Z}(g)$.

f) Assume that G is simply-connected, and let p be a torsion prime number of G. Prove that there exists an element g of G of order p such that $\pi_1(\mathrm{Z}(g))$ is cyclic of order p (use Exerc. 10 and Exerc. 4 of Chap. VI, §4).

12) Let p be a prime number. Show that the following conditions are equivalent:
(i) p is not a torsion prime number of G;
(ii) for every subgroup F of G isomorphic to $(\mathbf{Z}/p\mathbf{Z})^n$ (for $n \in \mathbf{N}$), the centralizer of F in G is connected;
(ii′) for every subgroup F of G isomorphic to $(\mathbf{Z}/p\mathbf{Z})^2$, the centralizer of F in G is connected;
(iii) every subgroup of G isomorphic to $(\mathbf{Z}/p\mathbf{Z})^n$ for some integer n is contained in a maximal torus;
(iii′) every subgroup of G isomorphic to $(\mathbf{Z}/p\mathbf{Z})^3$ is contained in a maximal torus.
(To prove that (i) $\Longrightarrow$ (ii), use Exerc. 11 *e*); to prove that (iii) $\Longrightarrow$ (i), use Exerc. 11 *f*) and Exerc. 7.)

§6

1) Let R be a reduced root system in a real vector space V. Give the space V a scalar product invariant under W(R), and the space $\mathbf{S}(\mathrm{V})$ the corresponding scalar product (*Topological Vector Spaces*, Chap. V, §3, no. 3, formulas (13) and (14)); thus, for $x_1, \dots, x_n, y_1, \dots, y_n$ in V,

$$(x_1 \dots x_n \,|\, y_1 \dots y_n) = \sum_{\sigma \in \mathfrak{S}_n} (x_1|y_{\sigma(1)}) \dots (x_n|y_{\sigma(n)}).$$

Choose a chamber C of R; put $\mathrm{N} = \mathrm{Card}(\mathrm{R}_+)$, $\rho = \frac{1}{2} \sum_{\alpha \in \mathrm{R}_+} \alpha$ and $w(\mathrm{R}) = \mathrm{Card}(\mathrm{W}(\mathrm{R}))$. Let P be the element $\prod_{\alpha \in \mathrm{R}_+} \alpha$ of $\mathbf{S}^{\mathrm{N}}(\mathrm{V})$.

a) Show that $\mathrm{P} = \frac{1}{\mathrm{N}!} \sum_{w \in \mathrm{W}(\mathrm{R})} \varepsilon(w)(w\rho)^{\mathrm{N}}$ (cf. Chap. VI, §3, no. 3, Prop. 2).

b) Deduce the equality $(\mathrm{P}|\mathrm{P}) = w(\mathrm{R}) \prod_{\alpha \in \mathrm{R}_+} (\rho|\alpha)$.

c) Prove that $(\mathrm{P}|\mathrm{P}) = 2^{-\mathrm{N}} w(\mathrm{R}) \prod_{i=1}^{l} m_i! \prod_{\alpha \in \mathrm{R}_+} (\alpha|\alpha) = 2^{-\mathrm{N}} \prod_{i=1}^{l} (m_i+1)! \prod_{\alpha \in \mathrm{R}_+} (\alpha|\alpha)$, where $m_1, \dots, m_l$ are the exponents of W(R) (use Exerc. 3 of Chap. VIII, §9).

¶ 2) Let V be a finite dimensional real Hilbert space over $\mathbf{R}$. Give the space $\mathbf{S}(V^*)$ of polynomials on V the scalar product defined in *Topological Vector Spaces*, Chap. V, §3, no. 3, formulas (13) and (14) (cf. Exerc. 1). Denote by γ the canonical gaussian measure on V (*Integration*, Chap. IX, §6, nos. 4 to 6); thus, if $(x_i)_{1\le i\le n}$ is an orthonormal basis of V^*, we have $d\gamma(x) = (2\pi)^{-n/2}e^{-(x|x)/2}dx_1\dots dx_n$.

a) Let q be the element of $\mathbf{S}^2(V)$ that defines the scalar product on V^*, and let $\Delta : \mathbf{S}(V^*) \to \mathbf{S}(V^*)$ be the operator given by taking the inner product with q (*Algebra*, Chap. III, §11, no. 9). Show that, for every orthonormal basis $(x_i)_{1\le i\le n}$ of V^* and every polynomial $P \in \mathbf{S}(V^*)$, we have $\Delta(P) = \frac{1}{2}\sum\limits_{i=1}^{n}\frac{\partial^2 P}{\partial x_i^2}$.

b) For $P \in \mathbf{S}(V^*)$, put $P^* = P * \gamma$, so that $P^*(x) = \int_V P(x-y)\,d\gamma(y)$ for $x \in V$. Prove the equality $P^* = \sum\limits_{n=1}^{\infty}\frac{\Delta^n(P)}{n!} = e^{\Delta}(P)$.
(Reduce to proving the analogous formula for the function $x \mapsto e^{i(x|u)}$, for $u \in V$.)

c) Prove the formula $\int \overline{P^*(ix)}Q^*(ix)\,d\gamma(x) = (P|Q)$ for P and Q in $\mathbf{S}(V^*)$ (identified with a subspace of $\mathbf{S}_{\mathbf{C}}((V\otimes\mathbf{C})^*)$).

d) If P is a homogeneous polynomial on V such that $\Delta(P) = 0$, then $\int_V P(x)^2\,d\gamma(x) = (P|P)$.

e) Let W be a finite group of automorphisms of V generated by reflections; denote by H the set of reflections in W. For $h \in H$, let $e_h \in V$ and $f_h \in V^*$ be such that $h(x) = x + f_h(x)e_h$. Show that the polynomial $P = \prod\limits_{h\in H} f_h$ satisfies $\Delta(P) = 0$ (use Chap. V, §5, no. 4, Prop. 5).

3) Let μ be a Haar measure on the additive group $\mathfrak{g}$.

a) There exists a unique Haar measure μ_G on G with the following property: if ω_G and $\omega_{\mathfrak{g}}$ are invariant differential forms of degree n on G and $\mathfrak{g}$, respectively, such that $\omega_G(e) = \omega_{\mathfrak{g}}(0)$ and $|\omega_{\mathfrak{g}}| = \mu$, then $|\omega_G| = \mu_G$. The map $\mu \mapsto \mu_G$ is a bijection from the set of Haar measures on $\mathfrak{g}$ to the analogous set for G.

b) Choose a scalar product $(\ |\ )$ invariant under $\mathfrak{g}$, and a Haar measure τ on $\mathfrak{t}$ such that μ (resp. τ) corresponds to the Lebesgue measure when $\mathfrak{g}$ is identified with $\mathbf{R}^n$ (resp. $\mathfrak{t}$ with $\mathbf{R}^r$) by means of an orthonormal basis. Prove the formula

$$\int_{\mathfrak{t}} \pi_{\mathfrak{g}}(x)e^{-(x|x)/2}\,d\tau(x) = (2\pi)^{n/2}\frac{\tau_T(T)}{\mu_G(G)}w(G).$$

c) Identify X(T) with a subset of the Hilbert space $\mathfrak{t}^*$ by means of the map $(2\pi i)^{-1}\delta$ (§4, no. 2), and put $P(x) = \prod\limits_{\alpha\in R_+}\langle\alpha, x\rangle$ for $x \in \mathfrak{t}$. With the notations of Exerc. 1 and 2, show that

$$\frac{\tau_T(T)}{\mu_G(G)} = (2\pi)^N w(G)^{-1}(P|P) = \pi^N \prod_{\alpha\in R_+}(\alpha|\alpha)\prod_{i=1}^{l} m_i!$$

d) With the notations of Exerc. 9 of §3, let $\mathfrak{g}_{\mathbf{Z}}$ be the **Z**-Lie subalgebra of $\mathfrak{g}$ generated by $(2\pi)^{-1}\Gamma(\mathrm{T})$ and the elements u_α, v_α for $\alpha \in \mathrm{R}$. Prove the equality

$$\mu_{\mathrm{G}}(\mathrm{G}) = \mu(\mathfrak{g}/\mathfrak{g}_{\mathbf{Z}})\frac{2^r\pi^{\mathrm{N}+r}}{\prod\limits_i m_i!}.$$

e) Assume that G is simply-connected. Show that $l = r$ and that

$$\mu_{\mathrm{G}}(\mathrm{G}) = 2^{l/2}\pi^{-\mathrm{N}}f^{1/2}\prod_{\alpha\in\mathrm{R}_+}(\alpha|\alpha)^{-1}\prod_{i=1}^{l}(\alpha_i|\alpha_i)^{-1/2}\prod_i(m_i!)^{-1},$$

where $\{\alpha_1, \ldots, \alpha_l\}$ is a basis of R and f is the connection index of R.

f) Assume further that R is irreducible and that all its roots are of the same length; take the scalar product on $\mathfrak{g}$ to be the opposite of the Killing form. Prove that $\mu_{\mathrm{G}}(\mathrm{G}) = (2\pi)^{\mathrm{N}+r}(2h)^{n/2}f^{1/2}\prod\limits_i(m_i!)^{-1}$.

4) Let X be a differentiable manifold of class C^∞. In this and the following exercises, we shall denote simply by H(X) the graded **R**-space $\mathrm{H}(\Omega(\mathrm{X}))$.

a) Show that $\Omega(\mathrm{X})$ is an associative and anti-commutative graded differential algebra (*Algèbre*, Chap. X, p. 183, Exerc. 18). Deduce that H(X) has a natural associative and anti-commutative graded algebra structure.

b) If X is connected and of dimension p, then $\mathrm{H}^i(\mathrm{X}) = 0$ for $i > p$ and $\dim_{\mathbf{R}} \mathrm{H}^0(\mathrm{X}) = 1$. Moreover, if X is compact, the space $\mathrm{H}^p(\mathrm{X})$ is of dimension one (use *Differentiable and Analytic Manifolds, Results*, 11.2.4).

c) Let Y be another manifold of class C^∞, and $f : \mathrm{X} \to \mathrm{Y}$ a map of class C^∞. The map $f^* : \Omega(\mathrm{Y}) \to \Omega(\mathrm{X})$ is a morphism of complexes (*Differentiable and Analytic Manifolds, Results*, 8.3.5); show that $\mathrm{H}(f^*) : \mathrm{H}(\mathrm{Y}) \to \mathrm{H}(\mathrm{X})$ is a morphism of algebras.

d) Assume that we are given a law of operation of G on X of class C^∞. Show that the subcomplex $\Omega(\mathrm{X})^{\mathrm{G}}$ of invariant forms is a subalgebra of $\Omega(\mathrm{X})$. Deduce that the map $\mathrm{H}(i) : \mathrm{H}(\Omega(\mathrm{X})^{\mathrm{G}}) \to \mathrm{H}(\mathrm{X})$ defined in Th. 2 is an isomorphism of graded algebras.

e) Show that $(\mathrm{Alt}(\mathfrak{g}))^{\mathrm{G}}$ is a subalgebra of $\mathrm{Alt}(\mathfrak{g})$, and that the graded algebra H(G) is isomorphic to it.

5) Denote by H(G) the graded **R**-algebra $\mathrm{H}(\Omega(\mathrm{G}))$ (cf. Exerc. 4).

a) The space $\mathrm{H}^p(\mathrm{G})$ is zero for $p > n$ and of dimension one for $p = n$ and $p = 0$. The space $\mathrm{H}^1(\mathrm{G})$ can be identified canonically with $\mathfrak{c}^*$, where $\mathfrak{c} = \mathrm{L}(\mathrm{C}(\mathrm{G}))$.

b) Assume from now on that G is semi-simple. Show that $\mathrm{H}^1(\mathrm{G}) = \mathrm{H}^2(\mathrm{G}) = 0$ (cf. Chap. I, §6, Exerc. 1).

c) Denote by $\mathrm{B}(\mathfrak{g})$ the space of G-invariant symmetric bilinear forms on $\mathfrak{g}$; for $b \in \mathrm{B}(\mathfrak{g})$ and x, y, z in $\mathfrak{g}$, put $\tilde{b}(x, y, z) = b([x, y], z)$. Show that the map $b \mapsto \tilde{b}$

defines an isomorphism from $B(\mathfrak{g})$ to $H^3(G)$ (let $\omega \in H^3(G)$; prove that, for all $x \in \mathfrak{g}$, there exists a unique linear form $f(x)$ on $\mathfrak{g}$ such that $df(x) = i(x)\omega$, and consider the form $(x, y) \mapsto -\langle y, f(x)\rangle$).

d) Show that the dimension of the **R**-vector space $H^3(G)$ is equal to the number of simple ideals of $\mathfrak{g}$.

6) Put $b_i(G) = \dim_{\mathbf{R}} H^i(G)$ for $i \geq 0$ and, if X denotes an indeterminate, $P_G(X) = \sum_{i\geq 0} b_i(G)X^i$.

a) Prove that $b_i(G) = \int_G \operatorname{Tr} \wedge^i(\operatorname{Ad} g)\, dg$ and $P_G(X) = \int_G \det(1 + X.\operatorname{Ad} g)\, dg$ (cf. Appendix II, Lemma 1).

b) Deduce the equalities $\sum_i b_i(G) = 2^r$, and $\sum_i (-1)^i b_i(G) = 0$ if $\dim G > 0$ (use H. Weyl's formula, or Exerc. 8 of §2).

c) Take $G = \mathbf{U}(n, \mathbf{C})$. Show that $P_G(X)$ is the coefficient of $(X_1 \ldots X_n)^{2n-2}$ in the polynomial $\frac{1}{n!}(1+X)^n \prod_{\substack{1\leq i,j\leq n \\ i\neq j}} (XX_i + X_j)(X_i - X_j)$ (with coefficients in $\mathbf{Z}[X]$).

7) Let K be a connected compact Lie group.

a) Let $f : K \to G$ be a surjective homomorphism with finite kernel. Show that the homomorphism $H(f^*) : H(G) \to H(K)$ is an isomorphism.

b) Show that the algebra $H(G \times K)$ can be identified canonically with the skew tensor product $H(G) \,{}^g\!\otimes H(K)$.

c) Deduce from *a*) and *b*) that H(G) is isomorphic to $H(C(G)_0) \,{}^g\!\otimes H(D(G))$ as an algebra. Show that the algebra $H(C(G)_0)$ is isomorphic to $\wedge(\mathfrak{c}^*)$, with $\mathfrak{c} = L(C(G))$.

¶ 8) Let k be a field of characteristic zero and E a graded left bigebra over k (*Algebra*, Chap. III, §11, no. 4, Def. 3). Assume that E is anti-commutative and anti-cocommutative, and that $E_m = 0$ for m sufficiently large. Denote by P the set of primitive elements of E (cf. Chap. II, §1).

a) Show that every homogeneous element of P is of odd degree (expand $c(x^m)$ for m large). Deduce that there is a canonical morphism $\varphi : \wedge(P) \to E$ of graded left bigebras (the graded left bigebra structure of $\wedge(P)$ being that defined in *Algebra*, Chap. III, §11, Exerc. 6).

b) Show that φ is an isomorphism (adapt the proof of Th. 1 of Chap. II, §1, no. 6).

9) Denote by $m : G \times G \to G$ the map such that $m(g, h) = gh$ for g, h in G. Identify $H(G \times G)$ with $H(G) \,{}^g\!\otimes H(G)$ (Exerc. 7 *b*)), so that m^* defines a homomorphism of algebras $c : H(G) \to H(G) \,{}^g\!\otimes H(G)$.

a) Show that $(H(G), c)$ is an anti-commutative and anti-cocommutative graded left bigebra (observe that the map $g \mapsto g^{-1}$ induces on $H^p(G)$ the multiplication by $(-1)^p$).

b) Let P(G) be the graded subspace of H(G) consisting of the primitive elements; deduce from Exerc. 8 that there is an isomorphism of graded bigebras $\mathsf{\Lambda}(\mathrm{P(G)}) \to \mathrm{H(G)}$.

c) Show that $\dim_{\mathbf{R}} \mathrm{P(G)} = r$ (use Exerc. 6 *b*)).

d) Deduce that the polynomial $\sum_{i \geq 0} b_i(\mathrm{G})\mathrm{X}^i$ is of the form

$$(1+\mathrm{X})^c \prod_{i=1}^{l} (1+\mathrm{X}^{2k_i+1}),$$

where c is the dimension of C(G), l the rank of D(G), and the k_i are integers ≥ 1; we have $k_1 + \cdots + k_l = \frac{1}{2}\mathrm{Card}\,\mathrm{R(G,T)}$[12].

10) Let H be a closed subgroup of G; denote by dh the Haar measure on H of total mass 1. Put $\chi(\mathrm{G/H}) = \sum_{i \geq 0} (-1)^i \dim_{\mathbf{R}} \mathrm{H}^i(\mathrm{G/H})$.

a) Prove the equality $\chi(\mathrm{G/H}) = \int_{\mathrm{H}} \det(1 - \mathrm{Ad}_{\mathfrak{g}/\mathfrak{h}} h)\, dh$ (use *Algèbre*, Chap. X, p. 41, Prop. 11 and Lemma 1 of Appendix II, no. 2).

b) Let $\pi_0(\mathrm{H})$ be the number of connected components of H. Show that

$$\chi(\mathrm{G/H}) = 0 \text{ if } \mathrm{H}_0 \text{ is not of maximum rank}$$
$$\chi(\mathrm{G/H}) = w(\mathrm{G})/w(\mathrm{H}_0)\, \pi_0(\mathrm{H}) \text{ if } \mathrm{H}_0 \text{ is of maximum rank.}$$

11) Let u be an automorphism of G of order two; denote by K the identity component of the set of fixed points of u, $\mathfrak{k}$ its Lie algebra, and X the symmetric space G/K (§1, Exerc. 8).

a) Show that every G-invariant differential form ω on X satisfies $d\omega = 0$ (observe that u induces on $\mathrm{Alt}^p(\mathfrak{g}/\mathfrak{k})$ the multiplication by $(-1)^p$).

b) Deduce that there is an isomorphism from the graded algebra H(G/K) to the graded subalgebra of $\mathrm{Alt}(\mathfrak{g}/\mathfrak{k})$ consisting of the K-invariant elements.

c) Put $b_i(\mathrm{G/K}) = \dim_{\mathbf{R}} \mathrm{H}^i(\mathrm{G/K})$ for $i \geq 0$; for $k \in \mathrm{K}$, denote by $\mathrm{Ad}^- k$ the restriction of $\mathrm{Ad}\, k$ to the eigenspace of $\mathrm{L}(u)$ with eigenvalue -1. Let dk be the Haar measure on K of total mass 1. Prove the formulas $b_i(\mathrm{G/K}) = \int_{\mathrm{K}} \mathrm{Tr}\, \mathsf{\Lambda}^i(\mathrm{Ad}^- k)\, dk$ and $\sum_{i \geq 0} b_i(\mathrm{G/K})\mathrm{X}^i = \int_{\mathrm{K}} \det(1 + \mathrm{X}\,\mathrm{Ad}^- k)\, dk$.

d) Prove that if, in addition, K is of maximum rank, the algebra H(G/K) is zero in odd degrees (observe that $u = \mathrm{Int}\, k$, with $k \in \mathrm{K}$).

e) Calculate the graded algebra $\mathrm{H}(\mathbf{S}_n)$. Deduce that $\mathbf{S}_n$ admits a Lie group structure (compatible with its manifold structure) if and only if n is equal to 1 or 3.

12) Let H be a connected closed subgroup of G and $\mathfrak{h}$ its Lie algebra.

[12] The integers k_i are in fact the exponents of R(G, T). Cf. J. LERAY, Sur l'homologie des groupes de Lie, des espaces homogènes et des espaces fibrés principaux, *Colloque de Topologie de Bruxelles* (1950), pp. 101-115.

a) Let α be an element of $\mathfrak{h}^*$ invariant under H; show that there exists an element $\bar{\alpha}$ of $\mathfrak{g}^*$ invariant under H whose restriction to $\mathfrak{h}$ is equal to α. Show that the element $d\bar{\alpha} \in \mathrm{Alt}^2(\mathfrak{g})$ is annihilated by $i(\eta)$ and $\theta(\eta)$ for all $\eta \in \mathfrak{h}$, and that its class in $\mathrm{H}^2(\mathrm{G}/\mathrm{H})$ does not depend on the choice of $\bar{\alpha}$. This defines a homomorphism $\varphi : \mathrm{H}^1(\mathrm{H}) \to \mathrm{H}^2(\mathrm{G}/\mathrm{H})$.

b) Assume from now on that G is semi-simple. Prove that φ is an isomorphism.

c) Prove that H is semi-simple if and only if $\mathrm{H}^2(\mathrm{G}/\mathrm{H}) = 0$.

d) Without assuming that H is connected, define an isomorphism $(\mathfrak{h}^*)^{\mathrm{H}} \to \mathrm{H}^2(\mathrm{G}/\mathrm{H})$ (apply *b*) to H_0).

13) Let H be a closed subgroup of G; put $\mathrm{X} = \mathrm{G}/\mathrm{H}$ and $n = \dim \mathrm{X}$. Show that the following conditions are equivalent:
(i) There exists a 2-form ω on X such that $d\omega = 0$ and such that the alternating form ω_x on $\mathrm{T}_x(\mathrm{X})$ is separating for all $x \in \mathrm{X}$;
(ii) n is even, and there exists an element ω of $\mathrm{H}^2(\mathrm{G}/\mathrm{H})$ such that $\omega^{n/2} \neq 0$;
(iii) H is the centralizer of a torus in G;
(iv) H is of maximum rank, and there exists a G-invariant complex structure on X;
(v) there exists a complex structure j and a 2-form ω on X such that $d\omega = 0$, $\omega_x(ju, jv) = \omega_x(u, v)$ and $\omega_x(u, ju) > 0$ for all x in X and all non-zero u, v in $\mathrm{T}_x(\mathrm{X})$ (* in other words, a *Kähler structure* on X $_*$);
(vi) there exists a complex structure j and a 2-form ω on X satisfying the conditions in (v) and invariant under G (* that is, a G-invariant Kähler structure on X $_*$).
(To prove that (ii) $\Longrightarrow$ (iii), put $\mathrm{S} = \mathrm{C}(\mathrm{H})_0$ and $\mathrm{Z} = \mathrm{Z}_{\mathrm{G}}(\mathrm{S})$; by using Exerc. 12 *d*), show that the canonical map $\mathrm{H}^2(\mathrm{G}/\mathrm{Z}) \to \mathrm{H}^2(\mathrm{G}/\mathrm{H})$ is surjective, and deduce that Z = H. The equivalence of (iii) and (iv) follows from Exerc. 8 of §4. To prove that (iii) $\Longrightarrow$ (vi), construct an H-invariant separating positive hermitian form on $\mathfrak{g}/\mathrm{L}(\mathrm{H})$ and consider its imaginary part; use Exerc. 11 *a*).)

§7

1) Take G to be the group $\mathbf{U}(n, \mathbf{C})$, and T to be the subgroup consisting of the diagonal matrices; for $t = \mathrm{diag}(t_1, \ldots, t_n)$ in T and $1 \leq i \leq n$, put $\varepsilon_i(t) = t_i$. Denote by σ the identity representation of G on $\mathbf{C}^n$.

a) The group X(T) has basis $\varepsilon_1, \ldots, \varepsilon_n$; show that every element of $\mathbf{Z}[\mathrm{X}(\mathrm{T})]^{\mathrm{W}}$ is of the form $e^{k(\varepsilon_1+\cdots+\varepsilon_n)}\mathrm{P}(e^{\varepsilon_1}, \ldots, e^{\varepsilon_n})$, where k is a relative integer and P a symmetric polynomial in n variables with integer coefficients.

b) Show that the representations $\mathsf{\Lambda}^r \sigma$ $(1 \leq r \leq n)$ are irreducible.

c) Show that the homomorphism $u : \mathbf{Z}[\mathrm{X}_1, \mathrm{X}_2, \ldots, \mathrm{X}_n][\mathrm{X}_n^{-1}] \to \mathrm{R}(\mathrm{G})$ such that $u(\mathrm{X}_i) = [\mathsf{\Lambda}^i \sigma]$ is an isomorphism.

2) Let $\mathrm{G} = \mathbf{SO}(2l+1, \mathbf{R})$, with $l \geq 1$; we use the notations of Exerc. 2 of §4. Denote by σ the representation of G on $\mathbf{C}^{2l+1}$ obtained from the identity representation by extension of scalars.

a) The **Z**-module X(T) has basis $\varpi_1, \dots, \varpi_{l-1}, 2\varpi_l$, as well as $\varepsilon_1, \dots, \varepsilon_l$.

b) Denote by η_i the element $e^{\varepsilon_i} + e^{-\varepsilon_i}$ of $\mathbf{Z}[\mathrm{X(T)}]$. Show that every element of $\mathbf{Z}[\mathrm{X(T)}]^{\mathrm{W}}$ can be written as $\mathrm{P}(\eta_1, \dots, \eta_l)$ where P is a symmetric polynomial in l variables with integer coefficients.

c) Show that the representations $\bigwedge^r \sigma$ $(r \leq 2l+1)$ are irreducible. For $r \leq l$, prove the equality

$$\mathrm{Ch}(\textstyle\bigwedge^r \sigma) = s_r(\eta_1, \dots, \eta_l) + (l - r + 2)s_{r-2}(\eta_1, \dots, \eta_l) + \dots + \\ + \binom{l-r+2k}{k} s_{r-2k}(\eta_1, \dots, \eta_l) + \cdots,$$

where the s_k are the elementary symmetric polynomials in l variables.

d) Show that the homomorphism $u : \mathbf{Z}[\mathrm{X}_1, \dots, \mathrm{X}_l] \to \mathrm{R(G)}$ such that $u(\mathrm{X}_i) = [\bigwedge^i \sigma]$ is an isomorphism.

3) Let $\mathrm{G} = \mathbf{SO}(2l, \mathbf{R})$, with $l \geq 2$; we use the notations of Exerc. 2 of §4. Denote by σ the representation of G on $\mathbf{C}^{2l}$ obtained from the identity representation by extension of scalars.

a) The **Z**-module X(T) has basis $\varepsilon_1, \dots, \varepsilon_l$.

b) Denote by η_i the element $e^{\varepsilon_i} + e^{-\varepsilon_i}$ of $\mathbf{Z}[\mathrm{X(T)}]$ and by δ the element $\prod_{i=1}^{l} (e^{\varepsilon_i} - e^{-\varepsilon_i})$. Show that every element of $\mathbf{Z}[\mathrm{X(T)}]^{\mathrm{W}}$ can be written as $\mathrm{P}(\eta_1, \dots, \eta_l) + \mathrm{Q}(\eta_1, \dots, \eta_l)\delta$, where P and Q are symmetric polynomials in l variables with coefficients in $\frac{1}{2}\mathbf{Z}$.

c) Show that the representations $\bigwedge^r \sigma$ are irreducible for $r \neq l$; the representation $\bigwedge^l \sigma$ is the direct sum of two subrepresentations τ^+ and τ^-, of highest weights $2\varpi_l$ and $2\varpi_{l-1}$, respectively (see Chap. VIII, §13, Exerc. 10).

d) Show that, for $r \leq l$, the element $\mathrm{Ch}(\bigwedge^r \sigma)$ is given by the formula in Exerc. 2 *c*) and that

$$\mathrm{Ch}(\tau^+) = \frac{1}{2}(\delta + \mathrm{Ch}(\textstyle\bigwedge^l \sigma)) \qquad \mathrm{Ch}(\tau^-) = \frac{1}{2}(-\delta + \mathrm{Ch}(\textstyle\bigwedge^l \sigma)).$$

e) Show that the homomorphism $u : \mathbf{Z}[\mathrm{X}_1, \dots, \mathrm{X}_l; \mathrm{Y}] \to \mathrm{R(G)}$ such that $u(\mathrm{X}_i) = [\bigwedge^i \sigma]$ and $u(\mathrm{Y}) = [\tau^+]$ is surjective, and that its kernel is generated by $\mathrm{Y}^2 - \mathrm{YX}_l + \mathrm{A}$, with $\mathrm{A} \in \mathbf{Z}[\mathrm{X}_1, \dots, \mathrm{X}_l]$.

4) Let E be a complex vector space, and let $\mathbf{T}^p_q(\mathrm{E})$ be the space of tensors of type (p, q) on E (*Algebra*, Chap. III, §5, no. 6). Denote by $\mathrm{H}^p_q(\mathrm{E})$ the subspace of $\mathbf{T}^p_q(\mathrm{E})$ consisting of the symmetric tensors (that is, those belonging to the image of $\mathbf{TS}^p(\mathrm{E}) \otimes \mathbf{TS}^q(\mathrm{E}^*)$ in $\mathrm{T}^p_q(\mathrm{E})$ and annihilated by the contractions c^i_j for $i \in [1, p]$ and $j \in [p+1, p+q]$ (*loc. cit.*).

a) Show that $\mathrm{H}^p_q(\mathbf{C}^n)$ is stable under $\mathbf{GL}(n, \mathbf{C})$, and consequently under $\mathbf{SU}(n, \mathbf{C})$; denote this representation of $\mathbf{SU}(n, \mathbf{C})$ by τ^p_q.

b) Show that τ_q^p is an irreducible representation whose highest weight (with the notations of Exerc. 1 of §4) is $p\varpi_1 + q\varpi_{n-1}$.

c) Every irreducible representation of $\mathbf{SU}(3, \mathbf{C})$ is isomorphic to one of the representations τ_q^p.

d) Let $(x_i)_{1\le i\le n}$ be the canonical basis of $\mathbf{C}^n$, $(y_j)_{1\le j\le n}$ the dual basis. Show that $\mathrm{H}_q^p(\mathbf{C}^n)$ can be identified with the space of polynomials $\mathrm{P} \in \mathbf{C}[x_1, \dots, x_n, y_1, \dots, y_n]$, homogeneous of degree p in the x_i and of degree q in the y_i, and such that $\sum_{i=1}^{n} \frac{\partial^2 \mathrm{P}}{\partial x_i \partial y_i} = 0$.

5) Let k be a commutative field of characteristic zero, V a vector space over k, and Ψ a separating quadratic form on V. Let $\Gamma \in \mathbf{S}^2(\mathrm{V}^*)$ (resp. $\Gamma^* \in \mathbf{S}^2(\mathrm{V})$) be the element ssociated to Ψ (resp. the inverse form of Ψ). Denote by Q the endomorphism of $\mathbf{S}(\mathrm{V})$ given by taking the product with Γ^*, by Δ the endomorphism of $\mathbf{S}(\mathrm{V})$ given by taking the inner product with Γ, and by h the endomorphism of $\mathbf{S}(\mathrm{V})$ that reduces on $\mathbf{S}^r(\mathrm{V})$ to multiplication by $-\frac{n}{2}-r$.

a) If $(x_i)_{1\le i\le n}$ is an orthonormal basis of V, then $\mathrm{Q(P)} = \frac{1}{2}\left(\sum x_i^2\right).\mathrm{P}$ and $\Delta(\mathrm{P}) = \frac{1}{2}\sum \frac{\partial^2 \mathrm{P}}{\partial x_i^2}$ for $\mathrm{P} \in \mathbf{S}(\mathrm{V})$.

b) Prove the formulas $[\Delta, \mathrm{Q}] = -h$, $[h, \Delta] = 2\Delta$, $[h, \mathrm{Q}] = -2\mathrm{Q}$.

c) Let H_r be the subspace of $\mathbf{S}^r(\mathrm{V})$ annihilated by Δ ("harmonic homogeneous polynomials of degree r"). Deduce from *b*) that there is a direct sum decomposition, stable under $\mathbf{O}(\Psi)$, given by

$$\mathbf{S}^r(\mathrm{V}) = \mathrm{H}_r \oplus \mathrm{QH}_{r-2} \oplus \mathrm{Q}^2\mathrm{H}_{r-4} \oplus \cdots.$$

d) Show that the representation H_r is irreducible (cf. Chap. VIII, §13, no. 3, (IV)).

e) Take $k = \mathbf{C}$, and $\mathrm{V} = \mathbf{C}^n$ equipped with the usual quadratic form ($n \ge 3$). We thus obtain irreducible representations τ_r of $\mathbf{SO}(n, \mathbf{R})$; with the notations of Exerc. 2 of §4, show that the highest weight of τ_r is $r\varpi_1$.

f) Let Γ be the Casimir element of G obtained from the Killing form. Prove the formula $\Gamma_{\mathbf{S}(\mathrm{V})} = \frac{1}{2n-4}\left(-4\mathrm{Q}\Delta + \left(\mathrm{H} + \frac{n}{2}\mathrm{I}\right)\left(\mathrm{H} + \left(2 - \frac{n}{2}\right)\mathrm{I}\right)\right)$. Calculate $\tilde{\Gamma}(\tau_r)$ and deduce the value of the form Q_Γ (Prop. 4).

6) Assume that G is almost simple. Show that G admits a faithful irreducible representation if and only if it is not isomorphic to $\mathbf{Spin}(4k, \mathbf{R})$ for $k \ge 2$.

7) Let $\tau : \mathrm{G} \to \mathbf{SO}(n, \mathbf{R})$ be a *real* unitary representation of G; denote by $\varphi : \mathbf{Spin}(n, \mathbf{R}) \to \mathbf{SO}(n, \mathbf{R})$ the canonical double covering. Then τ is said to be *spinorial* if there exists a morphism $\tilde{\tau} : \mathrm{G} \to \mathbf{Spin}(n, \mathbf{R})$ such that $\varphi \circ \tilde{\tau} = \tau$.

a) Let Σ be a subset of $\mathrm{P}(\mathrm{T}, \tau)$ such that $\Sigma \cup (-\Sigma) = \mathrm{P}(\mathrm{T}, \tau)$ and $\Sigma \cap (-\Sigma) = \varnothing$; denote by ω_Σ the sum of the elements of Σ. The class ϖ of ω_Σ in $\mathrm{X(T)}/2\mathrm{X(T)}$ is independent of the choice of Σ. Prove that τ is spinorial if and only if $\varpi = 0$.

b) Prove that $\rho \in \mathrm{X(T)}$ if and only if the adjoint representation is spinorial.

8) Let G be a Lie group whose Lie algebra is compact, and which has a finite number of connected components. Show that G has a faithful linear representation on a finite dimensional vector space (write G as the semi-direct product of a compact group K by a vector group N; choose a faithful representation of K on a finite dimensional vector space W and represent G as a subgroup of the affine group of $\mathrm{W} \oplus \mathrm{N}$).

§8

1) Let $\mathrm{G} = \mathbf{SU}(2, \mathbf{C})$, and let σ be the identity representation of G on $\mathbf{C}^2$.

a) The irreducible representations of G are the representations $\tau^n = \mathbf{S}^n\sigma$ for $n \geq 0$.

b) Let e_1, e_2 be the canonical basis of $\mathbf{C}^2$; show that the coefficients of τ^n in the basis $(e_1^i e_2^{n-i})_{0\leq i \leq n}$ are the functions τ_{ij}^n such that, for $g = \begin{pmatrix} \alpha & \beta \\ -\bar\beta & \bar\alpha \end{pmatrix} \in \mathrm{G}$, we have $\tau_{ij}^n(g) = \frac{(-1)^i}{j!}\,\alpha^{i+j-n}\beta^{j-i}\mathrm{P}_{ij}^n(|\alpha|^2)$, with $\mathrm{P}_{ij}^n(t) = \frac{d^j}{dt^j}\,[t^{n-i}(1-t)^i]$ ("Jacobi polynomials").

c) Deduce from b) that the functions $(n+1)^{1/2}\left(\frac{j!(n-j)!}{i!(n-i)!}\right)^{1/2}\tau_{ij}^n$, for i, j, n integers with $0 \leq i \leq n$, $0 \leq j \leq n$, form an orthonormal basis of $\mathrm{L}^2(\mathrm{G})$.

d) For $g = \begin{pmatrix} \alpha & \beta \\ -\bar\beta & \bar\alpha \end{pmatrix} \in \mathrm{G}$, put $\alpha = t^{1/2}e^{i(\varphi-\psi)/2}$, $\beta = (1-t)^{1/2}e^{i(\varphi+\psi)/2}$, with $0 \leq t \leq 1$, $0 \leq \varphi < 2\pi$, $-2\pi \leq \psi < 2\pi$. Show that the normalized Haar measure dg on G is equal to $(8\pi^2)^{-1}dt d\varphi d\psi$.

e) Let a, b in $\frac{1}{2}\mathbf{Z}$ be such that $a - b \in \mathbf{Z}$. Deduce from d) that the polynomials $\mathrm{P}^n_{n/2-a,n/2-b}(t)$, for n an integer of the same parity as $2a$, $n \geq \max(2a, 2b)$, form an orthogonal basis of $\mathrm{L}^2([0,1])$ for the measure $t^{-a-b}(1-t)^{a-b}\,dt$.

2) Let f be a complex-valued function on G of class C^∞. Prove that there exist two complex-valued functions g and φ on G, of class C^∞, such that φ is central and $f = g * \varphi$.

3) Let u be an irreducible representation of G, and let λ be its highest weight. For $x \in \mathfrak{g}$, denote by $\bar{x}$ the unique element of $\overline{\mathrm{C}}$ that is conjugate to x under $\mathrm{Ad(G)}$. Prove the equality $\|u(x)\|_\infty = |\delta(\lambda)(\bar{x})|$.

4) Choose a separating positive quadratic form Q on $\mathfrak{g}$ invariant under $\mathrm{Ad(G)}$. For $x \in \mathfrak{t}$, put $\vartheta_0(x) = \sum_{u \in \Gamma(\mathrm{T})} e^{-\mathrm{Q}(x+u)}$.

a) Show that ϑ_0 is a function of class C^∞ on $\mathfrak{t}$, and that there exists a function ϑ_1 of class C^∞ on T such that $\vartheta_1 \circ \exp_\mathrm{T} = \vartheta_0$.

b) Show that there exists a unique central function ϑ on G, of class C^∞, whose restriction to T is equal to ϑ_1. For every maximal torus S of G and all $x \in L(S)$, we have

$$\vartheta(\exp x) = \sum_{u \in \Gamma(S)} e^{-Q(x+u)}.$$

c) Let A be an alcove of $\mathfrak{t}$, dx a Haar measure on $\mathfrak{t}$, and h a locally integrable function on $\mathfrak{t}$ invariant under the affine Weyl group W'_a (§5, no. 2). Prove the equality

$$\int_A h(x)\,dx = \frac{1}{w(G)} \int_{\mathfrak{t}} h(x) e^{-Q(x)} (\vartheta_0(x))^{-1} dx.$$

d) For $x \in \mathfrak{g}$, put $\xi(x) = \lambda_{\mathfrak{g}}(x) e^{-Q(x)} (\vartheta(\exp x))^{-1}$, with $\lambda_{\mathfrak{g}}(x) = \det \frac{e^{\operatorname{ad} x} - 1}{\operatorname{ad} x}$ (§6, no. 3). Show that ξ is a function of class C^∞ on $\mathfrak{g}$ and that if μ is a Haar measure on $\mathfrak{g}$, the image under $\exp_G$ of the measure $\xi\mu$ is a Haar measure on G (use *c*) as well as Cor. 2 of Th. 1 and Prop. 4 (§6, no. 2 and 3)).

e) Prove the formula $\vartheta_0(x) = m \sum\limits_{\lambda \in X(T)} \exp(\delta(\lambda)(x) - \frac{1}{4} Q'(\delta(\lambda)))$ for $x \in \mathfrak{t}$, where Q′ is the quadratic form on $\mathfrak{t}_{\mathbf{C}}^*$ inverse to the quadratic form on $\mathfrak{t}_{\mathbf{C}}$ induced by Q and where m is a constant that should be calculated (use Poisson's formula, cf. *Spectral Theory*, Chap. II, §1, no. 8). Deduce the equality $\vartheta(t) = m \sum\limits_{\lambda \in X(T)} e^{-Q'(\delta(\lambda))/4} t^\lambda$ for $t \in T$.

5) *a*) Let V be a real vector space, f a non-zero linear form on V, H the kernel of f. Then, a function φ on V of class C^∞ vanishes on H if and only if it can be written as $f\varphi'$, where φ' is a function of class C^∞ on V.

b) Let V be a real vector space, $(f_i)_{i \in I}$ a finite family of non-zero linear forms such that the $H_i = \operatorname{Ker} f_i$ are pairwise distinct. Then, a function φ on V of class C^∞ vanishes on the union of the H_i if and only if it can be written as $\psi \prod\limits_{i \in I} f_i$, where ψ is a function on V of class C^∞.

c) Let T be a torus, $(\alpha_i)_{i \in I}$ a finite family of characters of T distinct from 1, such that the $K_i = \operatorname{Ker} \alpha_i$ are pairwise distinct. Then, a function φ on T of class C^∞ vanishes on the union of the K_i if and only if it can be written as $\psi . \prod\limits_{i \in I} (\alpha_i - 1)$, where ψ is a function of class C^∞ on T (argue locally on T and use *b*)).

d) With the notations of no. 4, prove that the map $b_\infty : \mathscr{C}^\infty(T)^W \to \mathscr{C}^\infty(T)^{-W}$ is bijective.

6) Assume that G is not commutative. Show that the continuous function $J(\rho)^{1/3}$ on T is anti-invariant under W, but does not belong to the image of the map $b_c : \mathscr{C}(T)^W \to \mathscr{C}(T)^{-W}$.

§9

1) Let A be the compact subset of $\mathbf{R}$ consisting of 0 and the real numbers $1/n$, with n an integer ≥ 1. Show that when $\mathscr{C}^r(\mathbf{R};\mathbf{R})$ is given the topology of uniform C^r-convergence on A, the set of morphisms that are embeddings in the neighbourhood of A is not open in $\mathscr{C}^r(\mathbf{R};\mathbf{R})$ (consider a sequence of functions $(f_n)_{n\geq 1}$ such that $f_n(x) = x$ for $x \leq \frac{1}{n+1}$, $f_n(x) = x - \frac{1}{n}$ for $x \geq \frac{1}{n}$).

2) Let X be a separated manifold of class C^r $(1 \leq r \leq \infty)$, countable at infinity, and pure of dimension n.

a) Assume that there exists an embedding φ of X into a finite dimensional vector space V. Show that there exists an embedding of X in $\mathbf{R}^{2n+1}$ (if $\dim V > 2n+1$, prove that there exists a point p of V such that, for all $x \in X$, the straight line joining p to $\varphi(x)$ meets $\varphi(V)$ only in the point $\varphi(x)$, and meets it transversally at this point; deduce that there is an embedding of X in a space of dimension equal to $\dim V - 1$).

b) Show that there exists an embedding of X in $\mathbf{R}^{2n+1}$. (Let $\mathscr{O}$ be the set of open subsets of X, $\mathscr{U}$ the subset of $\mathscr{O}$ consisting of the open sets U such that there exists a morphism $\varphi : X \to \mathbf{R}^{2n+1}$ whose restriction to U is an embedding; by using *a*), show that $\mathscr{U}$ is a quasi-full subset (*General Topology*, Chap. IX, §4, Exerc. 27) of $\mathscr{O}$, and hence is equal to $\mathscr{O}$.)

c) Show that there exists a proper embedding of X in $\mathbf{R}^{2n+1}$. (Construct a proper embedding of X in $\mathbf{R}^{2n+2}$ by means of a proper function on X.)

¶ 3) Let G be a Lie group, H a compact subgroup of G. Assume that G admits a *faithful* (finite dimensional) linear representation.

a) Let $\Theta_H(G)$ be the subalgebra of $\mathscr{C}(H;\mathbf{R})$ consisting of the restrictions to H of the (continuous) representative functions on G. Show that $\Theta_H(G)$ is dense in $\mathscr{C}(H;\mathbf{R})$ for the topology of uniform convergence.

b) Let $f \in \Theta_H(G)$. Show that there exists a representation σ of G on a finite dimensional real vector space such that a coefficient of the representation $\sigma|H$ is not orthogonal to f (*Spectral Theory*, in preparation).

c) Let $\rho : H \to \mathbf{GL}(V)$ be a representation of H on a finite dimensional real vector space. Show that there exists a finite dimensional real vector space W, a representation $\sigma : G \to \mathbf{GL}(W)$ and an injective homomorphism $u : V \to W$, such that $u(\rho(h)v) = \sigma(h)u(v)$ for $h \in H$, $v \in V$.
(Reduce to the case in which ρ is irreducible, and use *b*).)

4) Let G be a Lie group, H a compact subgroup of G, ρ a unitary representation of H on a real Hilbert space V. Prove that there exists a real Hilbert space W, a unitary representation σ of G on W and an isometric injection $u : V \to W$ with closed image, such that $u(\rho(h)v) = \sigma(h)u(v)$ for $h \in H$, $v \in V$ (same method as for Exerc. 3).

5) Let G be a Lie group, H a compact subgroup of G. Assume that there exists a faithful linear representation $\rho : G \to \mathbf{GL}(V)$ of G on a finite dimensional real vector space.

a) Show that there exists a representation σ of $\mathbf{GL}(V)$ on a finite dimensional real vector space W, a separating positive quadratic form q on V and a vector w in W such that $\rho(H) = \mathbf{O}(q) \cap F_w$, where F_w is the fixer of w in $\mathbf{GL}(V)$ (choose a quadratic form q invariant under H and a representation of $\mathbf{O}(q)$ such that $\rho(H)$ is the fixer of a point (Cor. 2 of no. 2), then apply Exerc. 3 *c*)).

b) Deduce from *a*) that there exists a finite dimensional real vector space E, a representation of G on E and a vector $e \in E$ with fixer H (take $E = W \oplus Q$, where Q is the space of quadratic forms on V, and $e = (w, q)$).

6) Let G be a Lie group, H a compact subgroup of G. Show that there exists a unitary representation of G on a real Hilbert space E and a vector $e \in E$ with fixer H (take $E = L^2(G)$).

7) Let G be a Lie group, H a compact subgroup of G, and ρ a unitary representation of H on a real Hilbert space W. Denote by X the manifold $G \times^H W$ (no. 3).

a) Show that there exists a Hilbert space V, a unitary representation σ of G on V and an (analytic) embedding $\varphi : X \to V$ such that $\varphi(gx) = \sigma(g)\varphi(x)$ for $g \in G$, $x \in X$ (use Exerc. 4 and Exerc. 6).

b) Prove that, if W is finite dimensional and if G admits a faithful finite dimensional linear representation, then V can be chosen to be finite dimensional (use Exerc. 3 and Exerc. 5).

¶ 8) Let X be a paracompact manifold of class C^r $(1 \leq r \leq \infty)$, and G a Lie group operating properly on X such that the law of operation $(g, x) \mapsto gx$ is of class C^r.

a) Show that there exists a unitary representation ρ of G on a Hilbert space V and an embedding φ (of class C^r) of X in V, such that $\varphi(gx) = \sigma(g)\varphi(x)$ for $g \in G, x \in X$. (Use Prop. 6, Exerc. 7, and argue as in the proof of Prop. 4.)

b) Make the following additional assumptions:
(i) G admits a finite dimensional linear representation;
(ii) X is countable at infinity, and of bounded dimension;
(iii) X has only a finite number of orbit types for the operation of G.
Prove that there exists a finite covering of X by open sets $(U_i)_{i\in I}$ that are stable under G, compact subgroups $(H_i)_{i\in I}$ of G, and for all $i \in I$ a closed submanifold S_i of U_i, stable under H_i, such that the map $(g, s) \mapsto gs$ induces by passage to the quotient an isomorphism from $G \times^{H_i} S_i$ to U_i (show that the open subsets of X/G whose inverse image in X admits such a covering form a quasi-full subset of the set of open subsets of X/G, cf. *General Topology*, Chap. IX, §4, Exerc. 27).

c) With the assumptions in *b*), prove that there exists a representation of G on a finite dimensional real vector space V and an embedding of X in V of class C^r compatible with the operations of G (argue by induction on the set of subgroups of G, using *b*), Exerc. 2 and Exerc. 7).

9) Let G be a Lie group operating properly on a manifold X. Make one of the following two assumptions:
(i) X/G is compact, and X is countable at infinity and of bounded dimension;
(ii) X is a finite dimensional vector space on which G operates linearly.
Prove that the set of orbit types of elements of X is *finite* (treat the two cases simultaneously by induction on dim X, and use Prop. 6).

10) Let G be the Lie subgroup of $\mathbf{GL}(3, \mathbf{R})$ consisting of the matrices $\begin{pmatrix} 1 & a & b \\ 0 & 1 & 0 \\ 0 & 0 & 1 \end{pmatrix}$, $a, b \in \mathbf{R}$. Show that, for the identity representation of G on $\mathbf{R}^3$, the number of orbit types is infinite.

¶ 11) For any integer $n \geq 1$, define a map φ_n from $(0, \frac{1}{2}(\times \mathbf{T}^2$ to $\mathbf{R}^3$ by putting $\varphi_n(r; \alpha, \beta) = ((n + r\cos 2\pi\beta)\cos 2\pi\alpha,\ (n + r\cos 2\pi\beta)\sin 2\pi\alpha,\ r\sin 2\pi\beta)$. Put

$$\mathrm{T}_n = \varphi_n\left(\left[0, \frac{1}{2}\right[\times \mathbf{T}^2\right), \quad \mathrm{S}_n = \varphi_n(\{0\} \times \mathbf{T}^2).$$

a) Show that the restriction of φ_n to $[0, \frac{1}{2}] \times \mathbf{T}^2$ is an isomorphism (of class C^ω) from $]0, \frac{1}{2}[\times \mathbf{T}^2$ to $\mathrm{T}_n - \mathrm{S}_n$.

b) Let $f_n : \mathrm{T}_n \to \mathrm{T}_n$ be the map that coincides with the identity on S_n and is such that

$$f_n(\varphi_n(r; \alpha, \beta)) = \varphi_n(r; \alpha - (n-1)\beta, -\alpha + n\beta)$$

for $r > 0$. Show that f_n induces an automorphism of $\mathrm{T}_n - \mathrm{S}_n$.

c) Show that there exists a real-analytic manifold structure on $\mathbf{R}^3$ such that the maps $f_n : \mathrm{T}_n \to \mathbf{R}^3$ and the canonical injection $\mathbf{R}^3 - \bigcup_n \mathrm{S}_n \to \mathbf{R}^3$ are analytic. Denote by X the real-analytic manifold defined in this way.

d) For $\vartheta \in \mathbf{T}$, denote by R_ϑ the rotation

$$(x, y, z) \mapsto (x\cos 2\pi\vartheta - y\sin 2\pi\vartheta,\ x\sin 2\pi\vartheta + y\cos 2\pi\vartheta,\ z)$$

of $\mathbf{R}^3$. Prove that putting, for $\theta \in \mathbf{T}$ and $u \in \mathrm{X}$,

$$\vartheta.u = \mathrm{R}_\vartheta(u) \quad \text{if} \quad u \in \mathrm{X} - \bigcup_n \mathrm{S}_n;$$
$$\vartheta.u = \mathrm{R}_{n\vartheta}(u) \quad \text{if} \quad u \in \mathrm{S}_n \text{ for some integer } n \geq 1$$

defines an analytic law of operation of $\mathbf{T}$ on X.

e) Show that the set of orbit types of X (for the operation of $\mathbf{T}$ defined in *d*)) is *infinite*.

f) Show that there is no embedding, compatible with the operation of **T**, of X in a finite dimensional vector space on which **T** operates linearly (use Exerc. 9).

12) Let G be a compact Lie group.

a) Prove that the set of conjugacy classes of normalizers of integral subgroups of G is finite (consider the operation of G on the grassmannian of subspaces of L(G), and apply Exerc. 9).

b) The set of conjugacy classes of (compact) semi-simple subgroups of G is finite (observe that a Lie algebra can contain only finitely-many semi-simple ideals (Chap. I, §6, Exerc. 7) and use *a*)).

13) Let G be a Lie group, H and K two compact subgroups of G. Assume that G admits a faithful finite dimensional linear representation. Show that there exists a finite set F of subgroups of H such that for all $g \in \mathrm{G}$ the subgroup $\mathrm{H} \cap g\mathrm{K}g^{-1}$ is conjugate in H to a subgroup in F (use Exerc. 5 and 9).

14) Assume that the manifold X is paracompact and locally finite dimensional. Let G be a Lie group operating properly on X, H a compact subgroup of G, and t the conjugacy class of H.

a) Show that the set $\mathrm{X_H}$ of points of X whose fixer is equal to H is a locally closed submanifold of X.

b) Show that the map $(g, x) \mapsto gx$ from $\mathrm{G} \times \mathrm{X_H}$ to X induces an isomorphism (of class C^r) from $\mathrm{G} \times^{\mathrm{N(H)}} \mathrm{X_H}$ to $\mathrm{X}_{(t)}$.

¶ 15) Let G be a locally compact topological group operating properly on a topological space E; let ρ be a representation of G on a finite dimensional real vector space V. Denote by dg a right Haar measure on G.

a) Let $\mathscr{P}$ be the set of subsets A of E having the following property: there exists an open covering $(\mathrm{U}_\alpha)_{\alpha \in \mathrm{I}}$ of E such that, for all $\alpha \in \mathrm{I}$, the set of $g \in \mathrm{G}$ such that $g\mathrm{A} \cap \mathrm{U}_\alpha \neq \varnothing$ is relatively compact in G.
Prove that, for every continuous function $f : \mathrm{E} \to \mathrm{V}$ whose support belongs to $\mathscr{P}$, the map $x \mapsto \int_{\mathrm{G}} \rho(g)^{-1} f(gx)\, dg$ is a continuous map from E to V, compatible with the operations of G.

b) Make one of the following assumptions:
(i) the space E/G is regular;
(ii) there exists an open subset U of E and a compact subset K of G such that $\mathrm{E} = \mathrm{GU}$ and $g\mathrm{U} \cap \mathrm{U} = \varnothing$ for $g \notin \mathrm{K}$.
Show that every point x in E has a neighbourhood belonging to $\mathscr{P}$ (in case (i), take $\mathrm{A} = \mathrm{V} \cap \mathrm{W}$, where V is a neighbourhood of x such that $g\mathrm{V} \cap \mathrm{V} = \varnothing$ for g outside a compact subset of G, and W to be a closed neighbourhood of $\mathrm{G}x$ stable under G and contained in GV).
Every point in E has an open neighbourhood stable under G and satisfying (ii).

c) Assume in addition that E is completely regular. Let $x \in \mathrm{E}, v \in \mathrm{V}$ be such that the fixer of x is contained in that of v. Prove that there exists a continuous map $\mathrm{F} : \mathrm{E} \to \mathrm{V}$, compatible with the operations of G, such that $\mathrm{F}(x) = v$. (Let $\mathscr{F}$ be the space of continuous numerical functions on E whose support belongs to $\mathscr{P}$, and let $u : \mathscr{F} \to \mathrm{V}$ be the map $\alpha \mapsto \int_{\mathrm{G}} \alpha(gx)\rho(g^{-1}).v\, dg$. Let C be a convex neighbourhood of v in V; construct a neighbourhood A of x belonging to $\mathscr{P}$ such that $gx \in \mathrm{A}$ implies that $\rho(g^{-1}).v \in \mathrm{C}$, and a function α on E, with support in A, such that $\alpha(x) \neq 0$ and $\int_{\mathrm{G}} \alpha(gx)\, dg = 1$. Show that $u(\alpha)$ belongs to C and deduce that $v \in \mathrm{Im}\, u$.)

16) Let G be a topological group operating properly on a separated topological space E. Let x be a point in E, H its fixer in G, and S a subset of E containing x and stable under H. The group H operates on the right on $\mathrm{G} \times \mathrm{S}$ by the formula $(g, s).h = (gh, h^{-1}s)$ for $g \in \mathrm{G}, h \in \mathrm{H}, s \in \mathrm{S}$; the map $(g, s) \mapsto gs$ induces by passage to the quotient a map from $(\mathrm{G} \times \mathrm{S})/\mathrm{H}$ to X. Then S is said to be a *transversal at* x if this map is a homeomorphism onto an open subset of X.

a) Show that, if G is discrete, there exists a transversal at x.

b) Let F be a separated topological space on which G operates properly, and let $f : \mathrm{E} \to \mathrm{F}$ be a continuous map, compatible with the operations of G, such that the fixer of $f(x)$ in G is equal to H. Show that, if S is a transversal at $f(x)$, then $f^{-1}(\mathrm{S})$ is a transversal at x.

c) Let N be a closed normal subgroup of G, $\pi : \mathrm{E} \to \mathrm{E/N}$ the canonical projection, T a transversal at $\pi(x)$ for the operation of G/N on E/N, and $\mathrm{S} \subset \pi^{-1}(\mathrm{T})$ a transversal at x for the operation of HN on $\pi^{-1}(\mathrm{T})$. Show that S is a transversal at x in E (for the operation of G).

17) Let G be a Lie group. We are going to show that G has the following property:

(T) For every completely regular topological space E on which G operates properly, and every point x in E, there exists a transversal at x (Exerc. 16).

a) Show that every Lie group having a faithful finite dimensional linear representation satisfies (T) (use Exerc. 15 and 16 *b*), as well as Prop. 6).

b) Prove that, if G has a closed normal subgroup N such that G/N satisfies (T) and such that KN satisfies (T) for every compact subgroup K of G, then G satisfies (T) (apply Exerc. 16 *c*)).

c) If G_0 is compact, then G satisfies (T).

d) Show that, if G has a discrete normal subgroup N such that G/N satisfies (T), then G satisfies (T).

e) Show that G satisfies (T) (let N be the kernel of the adjoint representation; prove that $\mathrm{G/N}_0$ satisfies (T), then apply *a*), *b*) and Exerc. 9 of §1).

18) Let G be a Lie group operating properly on a completely regular topological space E.

a) Let $x \in \mathrm{E}$, and let t be its orbit type; show that there exists an open neighbourhood U of x, stable under G, such that for all $u \in \mathrm{U}$ the type of u is $\geq t$.

b) Assume that G operates freely on E; let $\pi : \mathrm{E} \to \mathrm{E}/\mathrm{G}$ be the canonical projection. Show that, for every point z of E/G, there exists an open neighbourhood U of z and a continuous map $s : \mathrm{U} \to \mathrm{E}$ such that $\pi \circ s(u) = u$ for all $u \in \mathrm{U}$.

19) Let G be a Lie group, H a compact subgroup of G. Prove that there exists a neighbourhood V of H such that every subgroup of G contained in V is conjugate to a subgroup of H (apply Exerc. 18 *a*) to the space of compact subsets of G, cf. *Integration*, Chap. VIII, §5, no. 6).

¶ 20) Let G be a compact Lie group, m a positive integer.

a) Show that the set of conjugacy classes of subgroups of G of order $\leq m$ is finite (supposing this to be false, construct a finite group F and a sequence of homomorphisms $\varphi_n : \mathrm{F} \to \mathrm{G}$ such that $\varphi_i(\mathrm{F})$ is not conjugate to $\varphi_j(\mathrm{F})$ for $i \neq j$, and such that $\varphi_n(f)$ tends to a limit $\varphi(f)$ for all $f \in \mathrm{F}$; show that φ is a homomorphism, and that this leads to a contradiction with Exerc. 19).

b) Show that the set of conjugacy classes of subgroups F of G all of whose elements are of order $\leq m$ is finite (let μ be a Haar measure on G, and let U be a symmetric neighbourhood of the identity element such that U^2 contains no non-trivial element of order $\leq m$; prove that $\mathrm{Card}(\mathrm{F}) \leq \mu(\mathrm{G})/\mu(\mathrm{U})$).

¶ 21) Let G be a compact Lie group, T a maximal torus of G.

a) Let $\mathscr{S}$ be a set of closed subgroups of G, stable under conjugation, such that the family of subgroups $(\mathrm{S} \cap \mathrm{T})_{\mathrm{S} \in \mathscr{S}}$ is finite. Show that the set of conjugacy classes of the subgroups S_0, for $\mathrm{S} \in \mathscr{S}$, is finite (by using Exerc. 12 *a*)), reduce to the case in which the subgroups S_0 are normal; consider the groups $\mathrm{C}(\mathrm{S}_0)_0$ and $\mathrm{D}(\mathrm{S}_0)$, and apply Exerc. 12 *b*)).

b) Show that $\mathscr{S}$ is the union of a finite number of conjugacy classes of subgroups of G (by using *a*), reduce to the case in which the subgroups $\mathrm{S} \in \mathscr{S}$ all have the same identity component Σ, which is normal in G; then bound the orders of the elements of the groups S/Σ, and apply Exerc. 20 *b*)).

c) Let E be a separated topological space on which G operates continuously. Show that if the elements of E have only a finite number of orbit types for the operation of T, the same is true for the operation of G.

APPENDIX I

1) Let G be a connected compact group. Denote by $d(\mathrm{G})$ the upper bound of the dimensions of the quotients of G that are Lie groups. Assume that $d(\mathrm{G}) < \infty$.

a) Let K be a closed normal subgroup of G; show that $d(\mathrm{G}/\mathrm{K}) \leq d(\mathrm{G})$, and that $d(\mathrm{G}/\mathrm{K}) = d(\mathrm{G})$ if K is totally discontinuous.

b) Show that D(G) is a Lie group, and that the kernel of the homomorphism $(x, y) \mapsto xy$ from $\mathrm{C(G)}_0 \times \mathrm{D(G)}$ to G is finite.

c) Let $p = d(\mathrm{C(G)}_0)$. Then $p < \infty$; prove that there exists a totally discontinuous compact group D and a homomorphism $i : \mathbf{Z}^p \to \mathrm{D}$ with dense image, such that $\mathrm{C(G)}_0$ is isomorphic to $(\mathbf{R}^p \times \mathrm{D})/\Gamma$, where Γ is the image of $\mathbf{Z}^p$ under the homomorphism $x \mapsto (x, i(x))$ (write $\mathrm{C(G)}_0$ as a projective limit of tori of dimension p).

d) Assume that G is locally connected; show that G is then a Lie group.

INDEX OF NOTATION

The reference numbers indicate respectively the chapter, paragraph and number.

INDEX OF TERMINOLOGY

The reference numbers indicate respectively the chapter, paragraph and number.

Printed in the United States
By Bookmasters